Nanoparticles in Pharmacotherapy

Nanoparticles in Pharmacotherapy

Edited by

Alexandru Mihai Grumezescu

Assistant Professor, Department of Science and Engineering of Oxide Materials and Nanomaterials, Faculty of Applied Chemistry and Materials Science, University Politehnica of Bucharest, Bucharest, Romania

ELSEVIER

Elsevier
Radarweg 29, PO Box 211, 1000 AE Amsterdam, Netherlands
The Boulevard, Langford Lane, Kidlington, Oxford OX5 1GB, United Kingdom
50 Hampshire Street, 5th Floor, Cambridge, MA 02139, United States

British Library Cataloguing-in-Publication Data
A catalogue record for this book is available from the British Library

Library of Congress Cataloging-in-Publication Data
A catalog record for this book is available from the Library of Congress

ISBN: 978-0-12-816504-1

For Information on all Elsevier publications
visit our website at https://www.elsevier.com/books-and-journals

Publisher: Mathew Deans
Acquisition Editor: Simon Holt
Editorial Project Manager: Andrae Akeh
Production Project Manager: Bharatwaj Varatharajan
Cover Designer: Greg Harris

Typeset by MPS Limited, Chennai, India

Contents

List of Contributors

A.A. Adegoke
SARChI, Institute for Water and Wastewater Technology, Durban University of Technology, Durban, South Africa; Department of Microbiology, University of Uyo, Uyo, Nigeria

Muhammad Afzal
Department of Pharmacology, College of Pharmacy, Al Jouf University, Sakaka, Saudi Arabia

Farhan J Ahmad
Department of Pharmaceutics, Nanomedicine Research Lab, School of Pharmaceutical Education and Research, Jamia Hamdard (Hamdard University), New Delhi, India

Fasihuddin Badruddin Ahmad
Department of Chemistry, Faculty of Resource Science and Technology, UNIMAS, Kuching, Malaysia

Maria Gabriela José de Almeida-Cincotto
São Paulo State University (UNESP), School of Pharmaceutical Sciences, Araraquara, São Paulo, Brazil

Ecaterina Andronescu
Department of Science and Engineering of Oxide Materials and Nanomaterials, Faculty of Applied Chemistry and Materials Science, Politehnica University of Bucharest, Bucharest, Romania

Firoz Anwar
Department of Biochemistry, Faculty of Science, King Abdulaziz University, Jeddah, Saudi Arabia

Taruna Arora
Department of Biochemistry, Institute of Home Economics, University of Delhi, New Delhi, India

Sarwar Beg
Department of Pharmaceutics, Nanomedicine Research Lab, School of Pharmaceutical Education and Research, Jamia Hamdard (Hamdard University), New Delhi, India

Gautam Behl
Pharmaceutical and Molecular Biotechnology Research Centre, Department of Science, Waterford Institute of Technology, Waterford, Ireland

Rohit Bhatia
Division of Sustainable Chemistry, Rudraksh Proudhyogiki Sangathan, Delhi, India

Showkat Ahmad Bhawani
Department of Chemistry, Faculty of Resource Science and Technology, UNIMAS, Kuching, Malaysia

Gabriela Bielecka
Department of Applied Pharmacy, Medical University of Lublin, Lublin, Poland

Murali Mohan Bommana
Impax Labs, Middlesex, NJ, United States

Pablo Botella
Instituto de Tecnología Química, Universitat Politècnica de València-Consejo Superior de Investigaciones Científicas, Valencia, Spain

Siddhardha Busi
Department of Microbiology, School of Life Sciences, Pondicherry University, Puducherry, India

Bruna Galdorfini Chiari-Andréo
São Paulo State University (UNESP), School of Pharmaceutical Sciences, Araraquara, São Paulo, Brazil; Araraquara University (UNIARA), Araraquara, São Paulo, Brazil

Leila Aparecida Chiavacci
São Paulo State University (UNESP), School of Pharmaceutical Sciences, Araraquara, São Paulo, Brazil

Chris Edin
Pharmaceutical Manufacturing Technology Centre, Bernal Institute, University of Limerick, Limerick, Ireland

Enas Elmowafy
Department of Pharmaceutics and Industrial Pharmacy, Faculty of Pharmacy, Ain Shams University, Cairo, Egypt

Laurence Fitzhenry
Pharmaceutical and Molecular Biotechnology Research Centre, Department of Science, Waterford Institute of Technology, Waterford, Ireland

Adriana Ganem-Rondero
División de Estudios de Posgrado (Tecnología Farmacéutica), Facultad de Estudios Superiores Cuautitlán, Universidad Nacional Autónoma de México, Cuautitlán Izcalli, Mexico

Oana Gherasim
Department of Science and Engineering of Oxide Materials and Nanomaterials, Faculty of Applied Chemistry and Materials Science, Politehnica University of Bucharest, Bucharest, Romania; Lasers Department, National Institute for Lasers, Plasma and Radiation Physics, Magurele, Romania

Tudor George Gherasim
National Institute of Neurology and Neurovascular Diseases, Bucharest, Romania

Kauveri Vengatajalabathy Gobi
Department of Chemistry, National Institute of Technology, Warangal, India

Kotagiri Yugender Goud
BAE Laboratory, Université de Perpignan Via Domitia, Perpignan, France; Department of Chemistry, National Institute of Technology, Warangal, India

Valentina Grumezescu
Department of Science and Engineering of Oxide Materials and Nanomaterials, Faculty of Applied Chemistry and Materials Science, Politehnica University of Bucharest, Bucharest, Romania; Lasers Department, National Institute for Lasers, Plasma and Radiation Physics, Magurele, Romania

Simerdeep Singh Gupta
Product Development, Novel Laboratories, Inc., Somerset, NJ, United States; Teva Pharmaceuticals, Pomona, NY, United States

Abdul Hafeez
Glocal School of Pharmacy, Glocal University, Saharanpur, India

Akhtar Hayat
BAE Laboratory, Université de Perpignan Via Domitia, Perpignan, France; Interdisciplinary Research Centre in Biomedical Materials (IRCBM), COMSATS Institute of Information Technology (CIIT), Lahore, Pakistan

Mohamad Nasir Mohamad Ibrahim
School of Chemical Sciences, Universiti Sains Malaysia, Pulau Pinang, Malaysia

Javed Iqbal
Pharmaceutical Manufacturing Technology Centre, Bernal Institute, University of Limerick, Limerick, Ireland

Vera Lucia Borges Isaac
São Paulo State University (UNESP), School of Pharmaceutical Sciences, Araraquara, São Paulo, Brazil

Barbara Jachimska
Jerzy Haber Institute of Catalysis and Surface Chemistry, Polish Academy of Sciences, Kraków, Poland

Suresh Kumar Kailasa
Department of Applied Chemistry, S.V. National Institute of Technology, Surat, India

Imran Kazmi
Glocal School of Pharmacy, Glocal University, Saharanpur, India

Bharat Kirthivasan
Product Development, Novel Laboratories, Inc., Somerset, NJ, United States

Janardhan Reddy Koduru
Department of Environmental Engineering, Kwangwoon University, Seoul, Korea

Parveen Kumar
Department of Chemistry, Dyal Singh College, University of Delhi, New Delhi, India

Vikas Kumar
Department of Pharmaceutical Sciences, Shalom Institute of Health and Allied Sciences, Sam Higginbottom University of Agriculture, Technology & Sciences (SHUATS), Allahabad, India

Jean Louis Marty
BAE Laboratory, Université de Perpignan Via Domitia, Perpignan, France

Seema Mehdi
Department of Pharmacology, JSS College of Pharmacy, JSS Academy of Higher Education and Research, Mysuru, India

Arunachalam Muthuraman
Department of Pharmacology, JSS College of Pharmacy, JSS Academy of Higher Education and Research, Mysuru, India; Pharmacology Unit, Faculty of Pharmacy, AIMST University, Semeling, Malaysia

Maha Nasr
Department of Pharmaceutics and Industrial Pharmacy, Faculty of Pharmacy, Ain Shams University, Cairo, Egypt

María Guadalupe Nava-Arzaluz
División de Estudios de Posgrado (Tecnología Farmacéutica), Facultad de Estudios Superiores Cuautitlán, Universidad Nacional Autónoma de México, Cuautitlán Izcalli, Mexico

João Augusto Oshiro Jr
São Paulo State University (UNESP), School of Pharmaceutical Sciences, Araraquara, São Paulo, Brazil

Niall O'Reilly
Pharmaceutical and Molecular Biotechnology Research Centre, Department of Science, Waterford Institute of Technology, Waterford, Ireland

Tae-Jung Park
Department of Chemistry, Institute of Interdisciplinary Convergence Research, Research Institute of Halal Industrialization Technology, Chung-Ang University, Seoul, Republic of Korea

Elizabeth Piñón-Segundo
Laboratorio de Sistemas Farmacéuticos de Liberación Modificada, Facultad de Estudios Superiores Cuautitlán, Universidad Nacional Autónoma de México, Cuautitlán Izcalli, Mexico

Roxana Cristina Popescu
Department of Science and Engineering of Oxide Materials and Nanomaterials, Faculty of Applied Chemistry and Materials Science, Politehnica University of Bucharest, Bucharest, Romania; Department of Life and Environmental Physics, "Horia Hulubei" National Institute of Physics and Nuclear Engineering, Magurele, Romania

Mahfoozur Rahman
Department of Pharmaceutical Sciences, Shalom Institute of Health and Allied Sciences, Sam Higginbottom University of Agriculture, Technology & Sciences (SHUATS), Allahabad, India

Jobina Rajkumari
Department of Microbiology, School of Life Sciences, Pondicherry University, Puducherry, India

Narahari Rishitha
Department of Pharmacology, JSS College of Pharmacy, JSS Academy of Higher Education and Research, Mysuru, India

Eva Rivero-Buceta
Instituto de Tecnología Química, Universitat Politècnica de València-Consejo Superior de Investigaciones Científicas, Valencia, Spain

Jigneshkumar V. Rohit
Department of Chemistry, National Institute of Technology Srinagar, Hazratbal Kashmir, India

Moru Satyanarayana
BAE Laboratory, Université de Perpignan Via Domitia, Perpignan, France; Department of Chemistry, National Institute of Technology, Warangal, India

Amit Shah
Impax Labs, Middlesex, NJ, United States

Abhijat Shikhar
Tansukh Herbals Pvt Ltd., Lucknow, India

G. Singh
SARChI, Institute for Water and Wastewater Technology, Durban University of Technology, Durban, South Africa

Jan Sobczyński
Department of Applied Pharmacy, Medical University of Lublin, Lublin, Poland

Mahmoud Eid Soliman
Department of Pharmaceutics and Industrial Pharmacy, Faculty of Pharmacy, Ain Shams University, Cairo, Egypt

T.A. Stenström
SARChI, Institute for Water and Wastewater Technology, Durban University of Technology, Durban, South Africa

Cibele Yumi Yamada Taniguchi
São Paulo State University (UNESP), School of Pharmaceutical Sciences, Araraquara, São Paulo, Brazil; Araraquara University (UNIARA), Araraquara, São Paulo, Brazil

Gavin Walker
Pharmaceutical Manufacturing Technology Centre, Bernal Institute, University of Limerick, Limerick, Ireland

Amr A. Yakout
Faculty of Science, Chemistry Department, Alexandria University, Alexandria, Egypt

Foreword

Pharmacology and drug design represent one of the most time- and resource-consuming industries. The great success of this field relies on innovation, technology, and development of new tools to obtain improved and much more efficient therapeutic solutions.

On the other hand, nanotechnology is one of the most resourceful and dynamic research fields, enabling the design of many biomedical applications. The idea of a miraculous tiny therapeutic drug able to target a particular cell or organ is not new. However, only in recent years has technological progress allowed for the development of personalized therapeutics involving nanoscale materials. Nanostructures are currently found in numerous health-care and cosmetic products, and they are intensively investigated to develop personalized solutions for diagnosis and therapy. Because of their small size and physicochemical properties, nanoparticles may specifically deliver and control the release of any therapeutic drug; they can be utilized not only to obtain modified surfaces and coatings, but also to detect a specific biomedical modification, which may be crucial for diagnosis and therapy. Although they offer potentially unlimited biomedical applications and the pharma domain has greatly benefited from the progress made in this field, the design of a nanostructured drug is a very challenging work. Many times, the long-term impact and unconsidered side effects may alter the idea of their initial biological application and such aspects are revealed usually after their use in vivo. Also, unpredicted behavior due to interindividual diversity may occur. All these safety related issues and unpredictable impact under certain conditions are the main limitations to the practical use of nanoparticles in pharmacotherapy.

This book is a valuable resource highlighting the newest ideas and progress made on the design, applications, and impact of nanostructures on the pharmaceutical domain.

Alina Maria Holban

Faculty of Biology, University of Bucharest, Bucharest, Romania

Preface

This book offers the most recent findings for nanoparticles (dendrimer based nanocarriers, polymeric micelles, lipoplexes, solid lipid nanoparticles, silver nanoparticles, graphene-based nanomaterials, etc.) used in pharmacotherapy. It discusses the synthesis, characterization, and current or future potential of these nanomaterials applied in the biomedical field. The book entitled, *Nanoparticles in Pharmacotherapy*, contains 17 chapters, as follows:

Chapter 1, entitled Pharmacotherapy and nanotechnology, prepared by Oana Gherasim et al., discusses a few of the current possibilities entailed in nanotechnology-related pharmaceuticals, with a particular emphasis on novel biocompatible systems based on natural-derived chitosan and hyaluronic acid polysaccharides.

Chapter 2, entitled Nanostructures in pharma: elixir to oral medicine, prepared by Murali Mohan Bommana et al., gives an up-to-date overview about fabrication characterization of nanostructures and their in vitro and in vivo models with improved predictive abilities translating to better clinical outcomes.

Chapter 3, entitled Polymeric micelles in biomedical science, prepared by Showkat Ahmad Bhwani et al., presents the recent progress related to the preparation, characterization, and biomedical applications of polymeric micelles. The authors highlight polymeric micelles that can self-assemble into well-defined core–shell structures at concentrations above their critical micelle concentration. These well-defined characteristics of polymeric micelles have made them highly suited for enabling the enhanced permeability and retention (EPR)-mediated passive drug delivery of hydrophobic compounds.

Chapter 4, entitled Current trends in site and target specific delivery of nanomedicine for gene therapy, prepared by Arunachalam Muthuraman et al., highlights a possible way to approach site and target specific delivery of genes via nanotechnology with special attention to gene therapy for genetic disorders, such as hemophilia, retinal disease, infectious disease, metabolic disorders, neurodegenerative disease, and cardiovascular disease. Also, the authors present the limitation of gene delivery and some future directions.

Chapter 5, entitled Nanoparticles for cosmetic use and its application, prepared by Bruna Galdorfini Chiari-Andréo et al., gives an up-to-date overview about the main kinds of nanoparticles employed in the cosmetic area, highlighting the advantages obtained with their use. Also, the authors discuss scale-up and safety assessment, mainly in relation to the pathway followed by the nanoparticle when applied on the skin.

Chapter 6, entitled Nanoparticulate drug delivery systems for the oral administration of macromolecular drugs, prepared by Javed Iqbal et al., presents the recent progress related to the barriers to absorption faced by macromolecular compounds and outlines the advantageous properties of nanoparticulate drug delivery systems (DDS), which allow them to overcome such barriers. Inorganic (e.g., titania and alumina, etc.) and organic (lipid and polymer) based nanoparticles are discussed in detail, paying particular attention to the materials and techniques required to successfully prepare these nanoparticles. Stimuli-responsive, biodegradable, and mucoadhesive polymer-based carrier systems are particularly suited for the encapsulation and delivery of macromolecular drugs and as such, this chapter focuses on the most promising advances in these areas and discusses the future trends in this rapidly expanding field.

Chapter 7, entitled Nanomaterial-based electrochemical sensors in pharmaceutical applications, prepared by Jean Louis Marty et al., provides an overview of the role of carbon nanomaterials in the field of pharmaceutical industry to develop biosensors for detection of various drugs using electrochemical methods.

Chapter 8, entitled The evolution of dendrimers to composite dendrimers: a review of the state of the art, prepared by Maha Nasr et al., discusses composite dendrimers, a newly introduced system made of synthetic dendritic polymers in conjunction with either organic or inorganic entities. In this chapter the published work related to "composite dendrimers" is highlighted, while contrasting their applications to mere dendrimers.

Chapter 9, entitled Physicochemical characterization of PAMAM dendrimer as a multifunctional nanocarriers, prepared by Barbara Jachimska et al., presents the recent progress related to the understanding of self-assembling behavior of branched polyelectrolytes, which are exciting candidates for reversibly tunable swelling materials for drug delivery or smart molecules for surface functionalization.

Chapter 10, entitled Nanostructure lipid carriers, prepared by Jan Sobczyński et al., presents an overview on the current state of art of nanostructure lipid carriers (NLC) with a focus on excipient choice and formulation optimization for different administration routes. Moreover, examples of successful NLC application for various active ingredients are presented.

Chapter 11, entitled Lipid nanocarriers as skin drug delivery systems, prepared by María Guadalupe Nava-Arzaluz et al., provides a general overview of the different lipid nanocarriers proposed for the delivery of active compounds into and through the skin, providing a compilation of the studies performed in recent years.

Chapter 12, entitled Stimuli sensitive systems for camptothecin delivery, prepared by Pablo Botella et al., summarizes the most advanced achievements in the design, development, and synthesis of camptothecin stimuli responsive drug delivery systems, including some remarks on the most important formulations currently under clinical trials.

Chapter 13, entitled Functionalized graphene-based nanomaterials for drug delivery and biomedical applications in cancer chemotherapy, prepared by Mahfoozur Rahman et al., provides an exhaustive and updated account on the application of graphenes for diagnosis and treatment of diverse types of cancer along with their toxicity considerations.

Chapter 14, entitled Antimicrobial activity of silver nanoparticles, prepared by Suresh Kumar Kailasa et al., offers a rational vision of the main achievements acquired for the development of silver nanoparticles such as nanocarriers for antimicrobial therapy.

Chapter 15, entitled Microbially synthesized nanoparticles as next generation antimicrobials: scope and applications, prepared by Siddhardha Busi et al., gives an up-to-date overview about microbial mediated production of nanoparticles and their substantial interest as a potential solution to the growing need for the development of eco-friendly ways to fight microbial resistance and control diseases.

Chapter 16, entitled Biosensors for monitoring pharmaceutical nanocontaminants and drug resistant bacteria in surface water, subsurface water and wastewater effluent for reuse, prepared by Adegoke A.A. et al., ventures to critically assess the existing pharmaceutical nanobiosensors for the monitoring of PPCPs and microbial contaminants in water, the potential for improvement, possible challenges and suggested way forward toward a more effective contaminant free water system.

Chapter 17, entitled Metal oxide nanoparticles in removing residual pharmaceutical products and pathogens from water and wastewater, prepared by Adegoke A.A. et al., overviews the different types of metal oxide nanomaterials used for the monitoring/removal of the pharmaceutical compounds and personal care products (PPCPs) prevalent in wastewater. Further, the feasibility, challenges, and future perspectives of the application of these metal oxide nanoparticles (MONs) in the development of efficient wastewater treatment processes are explored.

Alexandru Mihai Grumezescu
University Politehnica of Bucharest, Bucharest, Romania

CHAPTER

Pharmacotherapy and nanotechnology

1

Oana Gherasim[1,2], Roxana Cristina Popescu[1,3], Tudor George Gherasim[4], Valentina Grumezescu[1,2] and Ecaterina Andronescu[1]

[1]*Department of Science and Engineering of Oxide Materials and Nanomaterials, Faculty of Applied Chemistry and Materials Science, Politehnica University of Bucharest, Bucharest, Romania* [2]*Lasers Department, National Institute for Lasers, Plasma and Radiation Physics, Magurele, Romania* [3]*Department of Life and Environmental Physics, "Horia Hulubei" National Institute of Physics and Nuclear Engineering, Magurele, Romania* [4]*National Institute of Neurology and Neurovascular Diseases, Bucharest, Romania*

CHAPTER OUTLINE

1.1 INTRODUCTION

In the present era, although science is rapidly expanding through human ingenuity, transferring knowledge from parameter-controlled research environments to everyday clinical situations remains a constant challenge in humankind's struggle to achieve a disease-free world. Medical efforts throughout history have managed to surpass many types of pathological processes, thus burdening the modern world with the toughest and most complex medical conditions we have yet to face. Borrowing some fundamentals from imposing nanotechnology, where the human mind tries to go deeper and understand the smallest and most complex building blocks of reality that it can, biomedical research has found no other alternative than to try to understand and manipulate the smallest and most complex processes and entities that it can, in a desperate race toward treatment perfection and omnipotence. As it is known that pharmacological and surgical therapies are the main pillars of modern evidence-based

Nanoparticles in Pharmacotherapy. DOI: https://doi.org/10.1016/B978-0-12-816504-1.00002-8

medical treatments, enhancing these two directions is common sense. In the following lines, for the purpose of this chapter, we will focus on the prospects and possibilities of enhancing and evolving the pharmacological side of modern medicine.

Classical pharmacology, as the reader may be familiar with, in the current trend of personalized medicine is rapidly losing ground, mostly due to its partially controllable tissue volume distribution mechanisms combined with partially specific receptor/cell targeting, that in the end generate undesirable cell collateral damage, or in other words, an unwanted range of side effects. Considering that the goal of a personalized medicine approach is to correctly offer a treatment option that is extremely finely tailored for both the patient's organism and the pathological process at hand and also has very few side effects (ideally none), it is obvious that current therapies lack the requirements. The concept of "efficient-disease-targeting-side-effect-free" treatment options is the Promethean fire of medicine, which in theory, in the context of personalized medicine, should be closer to reality. With this in the background, encouraged by current technological breakthroughs, modern science has turned its attention toward the attractive and promising field of nanopharmacology.

For what concerns the designing and development of novel pharmaceutical formulations, the general requirements involved for such particular systems (such as biocompatibility, solubility, compositional and structural adjustment, tunable degradability, and specialized institutional approval) must be accompanied by some specific considerations, including maximization of therapeutic effects, minimization of unwanted and side effects, and accurate pharmacokinetics and pharmacodynamics (Kang et al., 2015; Khadka et al., 2014; Kim et al., 2013; Sharma and Singh, 2012). In particular, the attractive physicochemical properties, the easy and versatile processability, and the tunable functionality related to macromolecules derived from natural and renewable sources (such as polysaccharides) have convinced the worldwide healthcare professionals and scientific community with respect to their unconventional revaluation toward novel pharmacological formulations. In the following sections of the current chapter we will cover some specific applications related to both chitosan (CS) and hyaluronic acid (HA) biopolymers, selected by us as indisputable representatives of polysaccharides-based nanopharmacology.

CS represents the second most abundant natural-derived polysaccharide, being the deacetylated derivate of chitin, which is extracted from cell walls of fungi, but especially from the exoskeletons of crustaceans and insects (Islam et al., 2017). For what concerns its composition, CS is a cationic biopolymer composed of D-glucosamine and *N*-acetyl-D-glucosamine units (corresponding to deacetylated and acetylated units, respectively) that are β-(1−4)-linked and randomly distributed within a linear structure (Islam et al., 2017). The special intrinsic properties related to CS, including biocompatibility (Cheung et al., 2015; Croisier and Jérôme, 2013; Rodrigues et al., 2012), nonimmunogenicity (Ibrahim et al., 2015; Jayakumar et al., 2010; Sapru et al., 2017), biodegradability

(Kean and Thanou, 2010; Makarios-Laham and Lee, 1995; Ratajska et al., 2003), and antibacterial activity (Andres et al., 2010; Benhabiles et al., 2012; Goy et al., 2016), represent attractive features for numerous biomedical applications. The molecular weight of CS, as well as the degree of removed and substituted acetyl functions and amount of moisture content may represent critical aspects for the final behavior of the developed systems (Campana et al., 2017; Islam et al., 2017), due to their implications with respect to the physicochemical behavior of CS-based materials (including solubility, swelling, and degradation mechanism) and to the related biological behavior (since supplementary functionalizing or grafting protocols may enable the development of novel multifunctional materials for particular biomedical applications). However, besides its utilization as an active excipient in pharmaceutical formulations (which enables the enhancement of drug stability and solubility, as well as improved pharmacokinetics) (Bernkop-Schnürch and Dünnhaupt, 2012; Wang, Y. et al., 2017), the use of CS and CS derivatives in pharmaceutical approaches is of great interest, thanks to their particular biological behavior as analgesic, antitumor, hemostatic, hypocholesterolemic, antimicrobial, antioxidant, mucoadhesive, etc. (Islam et al., 2017).

HA (or hyaluronan) is an anionic polysaccharide with high molecular weight, naturally occurring in different organisms (from bacteria to vertebrates, including humans), having important structural and biological functions. It has D-glucuronic acid and *N*-acetyl-D-glucosamine units within its composition, which are linked by β-(1−4), respectively β-(1−3) bonds (Necas et al., 2008; Zhang et al., 2014). Thanks to its frequent presence within the extracellular matrix of human connective tissues and other anatomical structures and fluids, the biocompatibility, nontoxicity, and nonimmunogenicity related to this biopolymer are indisputable features (Collins and Birkinshaw, 2013; Owen et al., 2017; Tripodo et al., 2015). The easy tailored physicochemical and functional properties of HA (mainly due to molecular weight variation) (Komorowicz et al., 2017; Kim et al., 2018) enable remarkable possibilities toward the synthesis of novel materials and devices for specific and selective biomedical applications. Besides its intrinsic bio-related properties, HA exhibits special features, including biodegradability, tissue and wound regeneration potential, selective targeting of cancer-related overexpressed receptors (such as CD44—cluster of differentiation protein and RHAMM—receptor for hyaluronan-mediated motility), improved cellular attachment, etc. (Dosio et al., 2016; Hemshekhar et al., 2016; Oommen et al., 2014; Selyanin et al., 2015).

While shrimp-derived CS was only recently recognized as generally safe by the US Food and Drug Administration, HA received its organizational approval a while ago so it is currently included in various commercially available products. Still, given the complex pathologies reported in modern society, as well as the limitations related to conventional pharmaceutical therapies, both CS and HA are extensively investigated with the aim to develop unconventional and performance-enhanced pharmaceuticals.

1.2 CARRIERS FOR ANTIINFECTIVE THERAPY

The intrinsic antimicrobial activity related to CS-based nanostructured materials was intensively investigated during the last few decades. Besides the nanosize-related effects exhibited by biomaterials containing CS and CS derivatives (O'Callaghan and Kerry, 2016; Peña et al., 2013; Sarwar et al., 2014), some specific mechanisms were proposed for the antipathogenic-related activity, including (1) electrostatic-guided interactions between positively charged CS and negatively charged pathogens' surface (Jeon et al., 2014; Palma-Guerrero et al., 2010; Tayel et al., 2010; Xing et al., 2009a,b); (2) interactions with microbial genetic cargo and functional macromolecules (Chien et al., 2015; Galvan Marquez et al., 2013; Kulikov et al., 2015; Park et al., 2011); and (3) chelation of metal-containing cellular structures and essential nutrients (Kong et al., 2010; Ma et al., 2017; Mansilla et al., 2013; Xia et al., 2011).

An extensive study regarding the optimal conditions (reagents ratio, temperature, and pH values) for the synthesis of CS nanoparticles (ChNPs) by involving sodium tripolyphosphate (TPP) as a cross-linking agent during the ionic gelation was recently performed by MubarakAli and coworkers (2018). Positively charged and spherical nanosized particles were thus obtained, the zeta potential and mean dimension values being experimentally evaluated as 34.53 mV and 200 nm, respectively. Thanks to the cationic nature of ChNPs, variable inhibitory effects against microbial development were evidenced; the most pronounced impairment was reported against *Escherichia coli* strain when using 40 $\mu g\ mL^{-1}$ ChNPs suspension, while for *Staphylococcus aureus* and *Candida albicans* an improved antipathogenic efficiency was reported when using 80 and 400 $\mu g\ mL^{-1}$ of ChNPs, respectively. In addition, the synthesized ChNPs proved to have antioxidant and noncytotoxic effects on human-derived cells (MubarakAli et al., 2018).

In a study performed by Aleanizy et al. (2018), the ionic gelation procedure was used to obtain spherical nanosized CS particles (with mean size around 105 nm) and negatively charged surface (the zeta potential being evaluated as -32.2 ± 0.92 mV). Very low concentration of the as-synthesized CSNPs (1000 ppm in 5, 10, and 15 μL suspension volumes) resulted in similar antibacterial effects against *Pseudomonas aeruginosa* when compared with commercial azithromycin antibiotic (10 mg). An enhanced antibacterial activity was assessed both under slightly acidic and neutral culture medium conditions. The CSNPs also exhibited antivirulence activity against the pathogenic strain, the measured levels of bacterial toxin pyocyanin being significantly inhibited after the performed assay. Moreover, the CSNPs treatment resulted in an increased bacterial sensitivity for lytic effects caused by anionic molecules (Aleanizy et al., 2018).

The antibacterial efficiency of *N*-derivatives of CS against *E. coli* and *S. aureus* was experimentally proved in a study performed by Bakshi et al. (2018). Chemical-guided methods were applied to pristine CS backbone to modify amine groups, enabling thus the synthesis of methyl CS, trimethyl CS, diethyl methyl CS, and carboxymethyl CS. The alkylation protocol distinctively impacted the crystallinity, morphology, thermal behavior, and surface charge of all CS derivatives, but resulted in significant impairment of bacterial development even at low concentrations (0.1 mg mL^{-1}). Moreover, all specimens were proved to represent highly biocompatible materials, as assessed on human-derived keratinocyte cultures and zebra fish embryos (Bakshi et al., 2018).

Sodium phytate was proposed as an alternative agent for sodium TPP used during the conventional ionic gelation procedure of ChNPs. By varying the molecular weight of CS (low and medium), as well as the incubation time implied during the synthesis procedure, significant modifications regarding the mean particle size and the particle morphology were evidenced. Spherical and positively charged nanoparticles (more than 50 mV zeta potential and less than 100 nm mean particle size) with enhanced pH and thermal stability (compared with conventional nanoparticles) were obtained by using both CS at concentrations below 3 mg mL^{-1}. The optimal CS-based nanosystems proved superior biocompatibility on AML-12 hepatocyte cells at concentrations up to 150 μg mL^{-1} but resulted in significant impairment of bacterial cultures. As it was assessed against distinctive Gram-positive (*S. aureus* and *Bacillus subtilis*) and Gram-negative (*E. coli* and *Salmonella typhi*) pathogens, the CS–sodium phytate nanoparticles exhibited stronger antibacterial activity than conventional CS–TPP particles (Yang et al., 2017).

Solid mixtures based on different molar ratios of CS and coniferous-derived abietic acid (AB) were developed both by physical mixing and solid dispersing procedures in a study performed by Cuzzucoli Crucitti et al. (2018). Complementary compositional and structural data proved superior features when using the second synthesis method, a subtle dependence between the homogenous entrapment of AB within the polysaccharide matrix and the initial CS/AB ratio being evidenced. The systems obtained by using equal molar concentrations from both organic compounds were characterized by nanosized particle dimension (~370 nm) and slightly negative charge (around −5 mV) and resulted in minimum inhibitory concentration against *Staphylococcus epidermidis* below the values corresponding to pristine components. In this particular case, the CS/AB nanosystems proved enhanced antioxidant activity when compared with CS and AB alone, but also good biocompatibility, since the estimated hemolysis activity was much lower than 1% (Cuzzucoli Crucitti et al., 2018).

Natural-derived flavonoids were successfully encapsulated into positively charged CS capsules in a study performed by Omwenga and coworkers (2018). Quercetin and baicalein were significantly entrapped within highly stable spherical nanocapsules (the corresponding association efficiencies were 99% ± 1.2% and 87% ± 5.1%, while the average particle sizes were 194 ± 4 and 187 ± 2 nm, respectively), which resulted in resulting in sustained release of both bioflavonoids from the CS-based matrix. When compared with same concentrations of free flavonoid (which resulted either in lack of bacterial response or in pronounced bactericidal effect), the composite nanosystems proved to significantly affect the molecular signaling pathways of *E. coli* pathogen. Moreover, by applying flavonoid-loaded nanocapsules in concentration of $2.0-2.5 \times 10^{-4}$ mM, significant inhibitory effects against bacterial biofilm development were evidenced, the reported data being comparable with treatment of 1 mM kanamycin. Also, an enhanced biocompatible behavior was assessed for the composite nanoformulations on fibroblast-like cells (Omwenga et al., 2018).

To develop novel therapeutic formulations for ophthalmologic applications, levofloxacin—a representative of fluoroquinolone antibiotics—was efficiently entrapped ($73.63\% \pm 1.54\%$) into CS nanoparticles (161.90 ± 3.32 nm average size) by using the ionotropic gelation method. The synthesized LFX–CH–NPs systems, which possessed exclusive spherical morphology and enhanced stability (thanks to the positive surface charge), successfully formed in situ gels when mixed with alginate solution. The LFX–CH–NPs-containing composite gels proved nonirritating and there were no morphological and functional alterations on chick-derived chorioallantoic membrane and goat-derived eye cornea, respectively. In addition, when compared with levofloxacin solution, a progressive and sustained drug release was noticed for the nanostructured gels, which further enabled enhanced and prolonged antibacterial efficiency against *P. aeruginosa* and *S. aureus* strains (Ameeduzzafar et al., 2018).

Abdelkader and coworkers (2017) used a similar procedure for the facile one-pot synthesis of meropenem-loaded CS nanoparticles. The as-obtained positively charged spherical systems, with average particle size of 261 ± 37.5 nm, resulted in the efficient encapsulation of beta-lactam antibiotic (76.3%) and in its gradual diffusion-guided release profile. Also, the meropenem-loaded CS nanoparticles proved to maintain their physicochemical features after being freeze-dried in the presence of different organic molecules (proposed as cryoprotective compounds). When compared with the free antibiotic, enhanced bacteriostatic and bactericidal effects were proved for the meropenem-loaded CS nanoparticles against various multidrug-resistant strains of *E. coli*, *Klebsiella pneumoniae*, and *S. aureus*. Moreover, the synthesized nanosystems proved a potentiated and sustained efficiency on rat systemic infection caused by meropenem-sensitive *K. pneumoniae* pathogen (Abdelkader et al., 2017).

Composite systems with exclusive spherical morphology, preferential smooth surface, and narrow size distribution consisting of crab-derived CS and alginate were synthesized by Thaya and coworkers (2018) using the ionic complexation procedure. The positively charged CS, responsible for interactions with bacterial cell membranes, enabled the CS/ALG microspheres to exhibit significant antibacterial efficiency against clinically relevant pathogens, such as *S. aureus*, *Enterococcus faecalis*, *P. aeruginosa*, and *Proteus vulgaris*, in a concentration-dependent manner. The composite microspheres also resulted in the substantial impairment of bacterial biofilm starting from initial development stages (24 h), a maximum efficiency being reported after a prolonged period of assessment (Thaya et al., 2018).

A similar synthesis protocol was applied in a recent study to embed vancomycin into composite spherical microspheres based on CS and alginate. When compared with pristine CS samples, increased surface roughness and mean particle size, as well as enhanced swelling behavior, and cumulative glycopeptide antibiotic release under physiologically stimulated environment were reported by increasing the amount of alginate within the composite particles. The intrinsic roughness of the synthesized CS–AL microspheres represented a beneficial feature for cellular interactions, while their composite nature combined the drug release kinetics and degradation profiles of both polysaccharides. Also, in comparison with the air-dried CS–AL systems, an additional lyophilization procedure enabled the obtaining of porous microsystems with superior swelling characteristics and cumulative release of vancomycin of ~78%, as assessed after 21 days (Unagolla and Jayasuriya, 2018).

An intricate experimental protocol was proposed for the synthesis of AGS 7, a gemini surfactant with suitable biocompatibility (as assessed on different human-derived epithelial cultures). The AGS 7 compound was further used for the development of CS-based carriers intended for the encapsulation of vancomycin. The obtained DL_CSSNPs systems exhibited spherical morphology, positive surface charge, as well as nanosized hydrodynamic and physical dimensions (220 ± 5.90 nm, respectively 50–90 nm) and excellent physicochemical stability (for up to 3 months). The enhanced antibiotic encapsulation efficiency (59.89% ± 2.33%) was accompanied by a pH-depended release profile (a cumulative vancomycin release of 94.29% ± 5.30% being reported after 24 hours under slightly acidic conditions). The DL_CSSNPs nanosystems had enhanced and sustained antibacterial activity against *S. aureus* and methicillin-resistant *S. aureus* (MRSA) strains in comparison with bare antibiotic. Still, a moderately improved antibacterial efficiency was observed under slightly acidic conditions. Moreover, the proposed nanocarriers proved their promising inhibitory effects against MRSA development on albino mice (Kalhapure et al., 2017).

Chitosan–dextran composite nanoparticles loaded with broad spectrum ciprofloxacin (Cipro–NPs) were proposed by Chavan and coworkers (2017) as promising carriers for ophthalmic use. Spherical systems with ~350 nm mean particle size were obtained by applying a modified ionic gelation synthesis protocol. The increased encapsulation of ciprofloxacin within composite nanoparticles (estimated up to 83%) encouraged a slight decrease in particle charge, the reported zeta potentials being reduced from +4.06 mV to +3.55 mV (corresponding to drug-free and drug-loaded particles, respectively). In comparison with free ciprofloxacin administration, the sustained release of antibiotic from composite CS–dextran matrix enabled superior antibacterial efficiency against nine pathogens implicated in ophthalmic conditions. Furthermore, the developed nanostructured carriers proved to have nonirritating effects on ocular-like chorioallantoic membrane insulated from chicken egg (Chavan et al., 2017).

1.3 CARRIERS FOR ANTIINFLAMMATORY THERAPY

CS stiff base functionalized with chromone (CSCH) and its corresponding cross-linked derivative obtained by using terephthalaldehyde (CSCH−TP) were recently developed and evaluated as novel antioxidant agents. When compared with pristine polysaccharide, both CS−chromone derivatives exhibited predominant amorphous nature, reduced thermal stability, polymorphic and nanoporous surface, as well as increased scavenging activity with respect to the DPPH (1,1-diphenyl-2-picrylhydrazyl) free radical. Notably, the best functional results (corresponding to the most potent antioxidant activity) were reported for the CSCH derivative, probably thanks to the cumulative intrinsic antioxidant effects of both CS and oxygen-enriched natural-derived chromone (Demetgül and Beyazit, 2018).

The reactions between pristine CS and succinic and glutaric anhydrides enabled Skorik and coworkers (2017) to synthesize CS derivatives with high degree of functional groups substitution (34%, respectively 39%, after 30 minutes of reaction). The resulted *N*-succinyl-and *N*-glutaryl-CS derivatives (SC and GC, respectively) proved similar concentration-dependent antioxidant effects by varying the degree of substitution, the reported values being drastically reduced in comparison with the acknowledged Trolox antioxidant. Both SC and GC (with substitution degrees of 75% and 72%, respectively) proved similar improved inhibitory effects against human-derived platelets, but also enhanced anticoagulant activities (related to standard blood platelet aggregation and coagulation assays). It is worth mentioning that *N*-succinyl-CS derivative resulted in the best anticoagulant properties (Skorik et al., 2017).

The multipotent natural-derived α-mangostin compound was efficiently encapsulated into positively charged quaternized β-ciclodextrin grafted chitosan (QCD-*g*-CS) complexes by using a high shear mixing protocol. Superior entrapment efficiency values were reported by increasing separately the time and speed during the mixing procedure. A biphasic release profile of α-mangostin was evidenced for composite complexes with reduced entrapment efficiency (5% and 15%), while a monotonous and sustained release of the xanthone compound was identified for the QCD-g-CS systems with 75% loading efficiency. As it was assessed on J744A.1 macrophage cell line, the QCD-g-CS complexes with reduced entrapment efficiency of α-mangostin proved the most inhibiting effects against tumor necrosis factor alpha (TNF-α) cytokine, resulting thus in best antiinflammatory activity. In addition, these systems also exhibited the most pronounced antimicrobial effects (Phunpee et al., 2018).

According to Shi and coworkers (2018), composite systems based on CS-modified P(mPEGMA-*co*-MMA), methoxy poly(ethylene glycol) methacrylate-*co*-poly(methylacrylic acid) copolymer, were assessed as suitable carriers for ibuprofen. The obtained CS/NPs particles exhibited exclusive spherical shape and core/shell structure, as well as highly positive surface charge and nanosized dimension (160.7 ± 2.3 nm). The biological evaluation, performed on 293T epithelial cells, evidenced enhanced cellular viability results and improved cellular uptake of drug-free CS/NPs. Also, the ibuprofen-loaded nanoparticles resulted in prolonged release of the nonsteroidal antiinflammatory drug from the composite matrix, the cumulative release being significantly increased under neutral physiological stimulated medium. Moreover, the ibuprofen-loaded CS-modified copolymer systems proved prolonged antifebrile effects in comparison with pristine drug suspension, as it was reported on a rat animal model (Shi et al., 2018).

A multistage Pickering emulsion, solvent evaporation, sonication protocol was applied to encapsulate ibuprofen into CS-coated poly(lactic-*co*-glycolic) acid copolymer. The as-synthesized ibuprofen–PLGA-CS nanosystems, with ~200 nm mean particle size and prolonged stability (as evidenced for up to 90 days), proved distinctive pH-dependent and temperature-dependent release profiles. Still, the proposed carriers enabled a long-term release of ibuprofen (more than 16 hours) under biological simulated condition (Wang, J. et al., 2017).

Composite biocompatible nanoparticles based on β-cyclodextrin grafted with *N*-maleoyl chitosan (CD-*g*-NMCS) were proposed by Hou and coworkers (2017) as suitable carriers for ketoprofen (KTP) nonsteroidal antiinflammatory drug. In comparison with pristine CS and CD-*g*-CS systems, the maleoyl-conjugated samples proved superior water solubility and increased viscosity. Spherical and positively charged particles were obtained, the average size (from 179.2 ± 16.1 to 274.0 ± 9.0 nm), zeta potential (ranging between 36.2 ± 1.9 and 42.4 ± 3.1 mV), and KTP entrapment efficiency ($62.6\% \pm 2.8\%$ and $65.4\% \pm 3.1\%$) being distinctively impacted by varying the anhydride degree of substitution (21.2% and 30.5%, respectively). The improved long-term stability of bare CD-*g*-NMCS nanosystems was evidenced under physiological simulated environment, while for the drug-loaded nanoparticles was assessed the excellent biocompatibility and the sustained release of KTP. The pharmacokinetic parameters obtained for KTP after rat intravenous administration outlined the promising potential of CD-*g*-NMCS nanosystems as carriers for prolonged antiinflammatory use (Hou et al., 2017).

In a study performed by Asasutjarit et al. (2017), diclofenac sodium (DS) was efficiently loaded (up to $92.0 \pm 2.0\%$ w/w) into *N*-trimethyl CS (TMC) spherical nanoparticles (129.3 ± 3.8 nm average size). Different treatments with γ radiation dosages ranging between 5 and 25 kGy did not affect neither the chemical composition of pristine DS and TMC, nor the physicochemical features of the synthesized DC–TMC nanosystems (which size and zeta potential values presented insignificant modifications, while the drug content slightly decreased with ~2% w/w at highest radiation dose). To provide the sterility assurance level of 10^{-6}, the optimal radiation dose of 9.9 kGy was experimentally identified, its effectiveness being demonstrated against various bacterial and fungal pathogens. Comparable drug release profiles were identified both before and after gamma irradiation (10 kGy). The DS–TMC nanoparticles were proposed as sterilizable carriers for ophthalmic applications, since they evaluation both before and after γ irradiation evidenced minimal ocular irritating effects (Asasutjarit et al., 2017).

A double emulsion protocol followed by solvent evaporation was employed in a study performed by Xiao and coworkers (2017) to entrap KPV/BSA (lysine–proline–valine tripeptide/bovine serum albumin) mixture into poly(lactide-*co*-glycolide) matrix (PLGA) modified with CS and polyvinyl alcohol and further coated with HA. The complex HA–KPV–NPs systems presented exclusive spherical morphology and narrow size distribution, electronegative surface, average hydrodynamic diameter of 272.3 nm, and KPV encapsulation efficiency of 39.7% ± 3.6%. The biological assays, performed on Colon-26 epithelial and Raw 264.7 macrophage cell lines, evidenced that the nanostructured HA–KPV–NPs systems did not exhibit cytotoxic effects, but were significantly internalized within the treated cells. The as-treated macrophages resulted in reduced expression of inflammatory markers for prolonged periods, while the healing promotion of adenocarcinoma-derived epithelial layer (Caco-2 cells) was evidenced. The oral administration of HA–KPV–NPs embedded into alginate/CS hydrogel in a mouse animal model resulted in selective and active targeting of inflamed colonic tissue (Xiao et al., 2017).

Promising platforms based on gelatin microcarriers (GMC) grafted with oxidized HA (26.9% ± 0.1%) for the local treatment of ophthalmic conditions were developed by using a combined protocol consisting in emulsion, chemical cross-linking, and lyophilization. When assessed in the presence of corneal-derived epithelial, stromal and endothelial cells for up to 4 days, the GMC–oHA systems proved superior biocompatibility, as evidenced from the preserved morphological features, increased cellular metabolic activity, and absence of inflammatory response. Moreover, the clinical observations reported after the implantation of GMC–oHA materials in the anterior chamber of white rabbits for up to 2 weeks (such as tissue structure, morphology of corneal endothelium, endothelial cells' density, corneal curvature, and thickness) evidenced the potential use of oHA-containing systems as promising noninflammatory materials for various formulations designed for local ophthalmologic treatment (Lai and Ma, 2017).

Dexamethasone (DEX), a highly potent glucocorticoid derivative, was successfully encapsulated into ChNPs coated with HA, by using a modified ionic cross-linking procedure followed by adsorption and ultrafiltration. When compared with pristine DEX-loaded ChNPs (with 305.25 ± 14.29 average size, positive zeta potential, 72.95% ± 3.85% and 14.51% ± 1.02% drug encapsulation and loading efficiency values, respectively), the HA-coated spherical systems presented smooth surface with negative charge and slightly increased nanosized dimension (375.72 ± 13.18 and 400.57 ± 15.23 nm before and after freeze drying by using 5% w/v of mannitol, respectively). The as-obtained CS–DEX–HA nanosystems resulted in a sustained drug release under physiologically simulated conditions, while the corresponding suspensions were evaluated as suitable platforms for ophthalmological applications (in terms of pH, refractive index, surface tension, and viscosity). Also, the cryopreserved CS–DEX–HA nanosystems maintained their physicochemical and functional features for up to 3 months of storage (Kalam, 2016).

1.4 CARRIERS FOR ANTICANCER THERAPY

When referring to organic-based nanoparticles, there are two possible ways to provide the targeting of tumor cells by guided transport of the nanoparticles, namely (1) the passive targeting, and (2) the active targeting. The first strategy considers the enhanced permeability and retention effect (EPR), which is characteristic for cancerous tissue, as it has a higher perfusion of blood and, moreover, because of the enlarged gaps between endothelial cells in the vascular lumen and accompanying reduced lymphatic drainage. In the case of polymeric nanoparticles, the active targeting can be induced by specific functionalization with different molecules that facilitate cellular uptake and internalization within tumor cells, by means of specific recognition pathways. However, in the case of the last-mentioned approach, one can say that it is based on a combination between both methods of targeting, as the particles reach the tumor site by means of the blood flow (where they are able to penetrate the blood vessel walls by the EPR effect), while their cancerous cell internalization is specifically increased (as the functionalizing molecules are recognized as self).

The use of CS in pharmaceutical applications has been exploited thanks to its versatile properties, like biocompatibility and biodegradability, but also thanks to the related immune enhancing effects and certain antitumor activity (Ta et al., 2010). Biocompatibility is an essential requirement for materials and substances intended to be used in biomedical applications, especially for those that directly interact with living tissues and biological fluids. When designing complex nanosystems for therapeutic use, biodegradability not only modulates the delivery of the active substance from the carrier, but also facilitates the elimination of the nanoparticles after they have completed their function.

Moreover, it has been proved that CS has both in vitro and in vivo antitumor activity. Xu et al. (2009) studied the effects of ChNPs on human hepatocellular carcinoma (BEL-7402) tumor xenografts on nude mice. The reported results showed a significant tumor inhibition growth via necrosis, simultaneously with tumor angiogenesis inhibition, due to the lowered levels of vascular endothelial growth factor receptor 2.

Studies involving the potential antitumor effects of CS with different molecular weights have been conducted. Yeh et al. (2013) evaluated the inhibitory potential of different molecular weighted CS particles (from 8,000 to 320,000 kDa) in lung and hepatic metastasis of B16F10 melanoma cells xenograft models in C57BL/6 mice. The highest survival rate was recorded for the groups treated with 86,000 kDa ChNPs (45%), followed by 8,000 kDa groups (40%), after a 5-week treatment. For the treatment scheme employed in this study, the results showed considerably slow growth of the tumor metastasis for the groups exposed to low molecular weight CS.

Maeda and Kimura (2004) showed that the antitumor activity of low molecular weight CS can be linked to its immune modulatory effect in sarcoma 180 xenograft mice. The in vitro evaluation of 21 and 46 kDa CS ($10\ mg\ mL^{-1}$) showed an increased activity of natural killer cells, intestinal intraepithelial lymphocytes, and splenic lymphocytes. However, the intragastric administration of low molecular weight CS ($100\ mg\ kg^{-1}$ dose) showed reduced tumor growth, accompanied by increased activity of natural killer cells, but no activity of intestinal intraepithelial lymphocytes and splenic lymphocytes in vivo.

Contrary to these findings, Cai-qin et al. (2002) previously reported that high molecular weight CS induced a higher inhibition rate in sarcoma 180.

Trying to explain the molecular mechanisms of CS antitumor activity, Zhou et al. (2016) used different CS oligosaccharides (maximum molecular weight of 2,000 kDa) to evaluate the in vitro effect on 10 different tumor cell lines, namely human gastric adenocarcinoma cells (BGC-823 and SGC-7901), human lung adenocarcinoma cells (A549 and NCI-H460), human kidney carcinoma cells (KCC-853 and 786-O), human colon carcinoma cells (HCT-116 and HT-29), and human breast tumor cells (MCF-7 and Bcap-37). Complementary studies were performed in vivo on sarcoma 180 bearing/residual mice. The tumor growth inhibition was associated with an increased expression of the TNF-α.

Gibot et al. (2015) studied the potential anticancer mechanisms of CS against different melanoma cell lines, by coating the culture plates and investigating the adherence, proliferation, and apoptosis of the cells after seeding. The results showed that CS decreases the adherence and proliferation of primary melanoma cells, whereas in the case of metastatic melanoma has a pro-apoptotic effect due to caspase cascade activation. Moreover, the upregulation of other pro-apoptotic proteins (like Bcl-2 and Bcl-XL) has been proved. Also, in the case of the RPMI7951 metastatic melanoma cells, CD95 receptor expression increased following CS exposure, a phenomenon that eventually can lead to FasL-induced apoptosis.

HA has been proved to have implications in cancer development, as some elements from its signaling pathway (HA synthases, HA receptors, HYAL-1 hyaluronidase) proved to promote tumor growth, metastasis, or angiogenesis. Thus these molecules not only became important biomarkers in diagnosis and monitoring of different cancers (especially breast, bladder, endometrial, ovarian, prostate carcinomas), but they also became important targeting elements in therapeutic approaches (Dheer et al., 2016; Lokeshwar et al., 2014). For example, CD44 glycoprotein (cluster of differentiation-44) is overexpressed in pancreatic, breast, and lung cancer cells, and it can be specifically bind to the HA molecule (Safdar et al., 2018).

There are several approaches in the literature exploiting this specificity and inducing targeting abilities by using HA in the design and functionalizing of different nanosystems. Other receptors that can be specifically targeted by HA are LYVE-1 (lymphatic vessel endothelial hyaluronan receptor 1), which is expressed in lymphatic endothelial cells and facilitates hyaluronan transport and might have a role in tumor metastasis, respectively RHAMM (hyaluronan-mediated motility receptor), which is mostly localized inside cell and is transported at its surface in response to stimuli-like cytokines (especially tumor growth factor beta) or wounding signaling (Hardwick et al., 1992; Matthaiolampakis et al., 2015).

Moreover, both HA (anionic) and CS (cationic) are polysaccharides with high affinity for mucosal surfaces, making them promising candidates for the development of drug delivery systems for nasal, pulmonary, or gastrointestinal administration. A combination of the two polymers have been employed not only to induce a specific active targeting to the final nanostructure, but also to modify the degradability rate of CS at pH values above 6 (Salatin and Khosroushahi, 2017).

Wang et al. (2015) exploited cancer-linker properties of both CS and HA, to design nanoparticles for the active transport of 5-fluorouracil chemotherapeutic substance in CD44 over expressing A549 adenocarcinomic human alveolar basal epithelial cells. Facilitated intracellular accumulation of the drug leaded to the apoptosis of the tumor cells by means of mitochondrial damage due to reactive oxygen species (ROS) production.

Similarly, Abruzzo et al. (2016) obtained HA–CS–sulfobutyl–ether–β-cyclodextrin (HA–Chit–SEBC) systems for the transport of the lipophilic drug curcumin. Sulfobutyl–ether–β-cyclodextrin is a vehicle molecule used to obtain conjugates with water insoluble drugs, to increase their water solubility (Kiser et al., 2015). HA–Chit–SEBC water stable nanoparticles showed good encapsulation efficiency for curcumin and led to the change of the lipid organization in the cellular membrane, without any ROS production. The resulted formulations provided selective cytotoxicity against cancer cells, compared with normal cells, by increasing the expression of genes associated to apoptosis process in HT-29 cells.

HA-*g*-PCL (hyaluronic acid grafted polycaprolactone) nanoparticles were obtained for the selective transport of lipophilic drugs paclitaxel and doxorubicin in cancer cells and covered with CS to modulate the delivery of the drugs (Huang et al., 2014). The CS layer started to degrade gradually at physiological pH, while the loaded conjugate was released and able to selectively target cancer cells compared with fibroblast cells. The drug release was accelerated in the tumor area due to the presence of hyaluronidase-1 enzyme, which caused the degradation of HA. In vivo studies showed efficient delivery of the active substances inside the tumor cells with high targeting ability and minimum adverse effects.

Gao et al. (2017) developed layer-by-layer nanoparticles for the targeted delivery of irinotecan and 5-fluorouracil for the in vitro and in vivo treatment of gastric carcinoma. The particles consisted in a core of irinotecan and poly (lactic-*co*-glycolic) acid and a shell of CS and 5-fluorouracil, with different degradability and mechanical properties and HA as targeting element.

Tang et al. (2016) developed CS–HA nanoparticles for the improvement of LY2835219 delivery to colorectal carcinoma cells in vitro. The mechanism of LY2835219 upon cyclin inhibition (CDK4 and CDK6) led to cell cycle arrest in the G1 phase, suppression of the DNA synthesis, and eventually cell growth (Dickson, 2014).

One hot topic in cancer therapy research is gene therapy, which refers to the use of genes to treat or prevent certain diseases. This concept follows three approaches, which consider (1) the replacement of the mutated gene causing the disease, (2) the inactivation of the mutated gene, or (3) the introduction of a new gene with functions that can fight against the disease. The motivation of using nanoparticles in gene therapy approaches for cancer treatment would be to avoid using a viral vector, which is usually necessary in these experimental techniques, to facilitate the introduction of the gene inside the targeted cell. This approach would make the gene therapy treatment method more clinically available (Liu and Zhang, 2011; Soleimani et al., 2016).

CS has been used in such specific applications, thanks to its ability to form polyplexes with nucleic acids (Lallana et al., 2017). However, these structures might be too stable, and have a low degradation rate, enabling thus too slowly the DNA (Oliveira et al., 2014). In this regard, HA has been introduced as a third component, not only to accelerate the release of the nucleic acids, but also to allow a targeting effect (Lu et al., 2011; Yang, 2006).

For example, Lallana and coworkers (2017) obtained nanoparticles based on CS and HA to transfect CD44 positively HCT-166 cells. The influence of RNA/DNA cargo dimension, CS molecular weight, and deacetylation degree, has been studied. The research group of Cosco et al. (2015) has also successfully used CS-based nanoparticles (CS–PLGA for the encapsulation of miR-34a) as transfection agents, both in vitro and in vivo for multiple myeloma cells. The molecular weight of HA proved to influence the transfection efficiency, as complexes containing HA with high molecular mass improved stability and gene intracellular delivery (Sato et al., 2017).

Targeting and stimulation of the immune system, as an approach in cancer treatment, has been intensively exploited lately (Conniot et al., 2014; Evans et al., 2018; Shekarian et al., 2015; Zang et al., 2017). Targeting different immune system cells that are also present in the tumor tissue can facilitate the transport of the nanoparticles at the targeted area and increase their local concentration. On the other hand, signaling different types of cells using antibodies or signaling molecules and stimulating their function may induce or increase their activity against cancer cells. Both CS and HA have been used in the development of such systems (Hanefeld et al., 2014). HA is capable of targeting CD44 receptors on cells, which might have implications in the secretion of different cytokines involved in the recruitment of immune system cells (Maisel et al., 2016) or suppression (Liu et al., 2009; Ruffell and Johnson, 2015).

1.5 CONCLUSIONS

The novel nanotechnology-derived materials have proven to be suitable candidates for unconventional and enhanced pharmacological applications. Whether we are considering to diminish or eliminate the shortcomings related to conventional drug formulations, or to potentiate or induce specific and selective therapeutic effects, tremendous interdisciplinary knowledge, multidisciplinary research studies, and financial efforts have been put toward the design, development, and even clinical implementation of biopolymeric-based pharmaceuticals. In particular, the versatile physicochemical features and tunable biofunctionality related to microstructured and nanostructured systems based on natural-derived CS and HA biopolymers have gained a particular worldwide interest with respect to the emerging strategy of patient-oriented medicine.

REFERENCES

Abdelkader, A., El-Mokhtar, M.A., Abdelkader, O., Hamad, M.A., Elsabahy, M., El-Gazayerly, O.N., 2017. Ultrahigh antibacterial efficacy of meropenem-loaded chitosan nanoparticles in a septic animal model. Carbohydr. Polym. 174, 1041–1050.

Abruzzo, A., Zuccheri, G., Belluti, F., Provenzano, S., Verardi, L., Bigucci, F., et al., 2016. Chitosan nanoparticles for lipophilic anticancer drug delivery: development, characterization and in vitro studies on HT29 cancer cells. Colloids Surf. B Biointerfaces 145, 362–372.

Aleanizy, F.S., Alqahtani, F.Y., Shazly, G., Alfaraj, R., Alsarra, I., Alshamsan, A., et al., 2018. Measurement and evaluation of the effects of pH gradients on the antimicrobial and antivirulence activities of chitosan nanoparticles in *Pseudomonas aeruginosa*. Saudi Pharm. J. 26 (1), 79–83.

Ameeduzzafar, Imam, S.S., Abbas Bukhari, N.S., Ahmad, J., Ali, A., 2018. Formulation and optimization of levofloxacin loaded chitosan nanoparticle for ocular delivery: in-vitro characterization, ocular tolerance and antibacterial activity. Int. J. Biol. Macromol. 108, 650–659.

Andres, Y., Giraud, L., Gerente, C., Le Cloirec, P., 2010. Antibacterial effects of chitosan powder: mechanisms of action. Environ. Technol. 28 (12), 1357–1363.

Asasutjarit, R., Theerachayanan, T., Kewsuwan, P., Veeranondh, S., Fuongfuchat, A., Ritthidej, G.C., 2017. Gamma sterilization of diclofenac sodium loaded *N*-trimethyl chitosan nanoparticles for ophthalmic use. Carbohydr. Polym. 157, 603–612.

Bakshi, P.S., Selvakumar, D., Kadirvelu, K., Kumar, N.S., 2018. Comparative study on antimicrobial activity and biocompatibility of *N*-selective chitosan derivatives. React. Funct. Polym. 124, 149–155.

Benhabiles, M.S., Salah, R., Lounici, H., Drouiche, N., Goosen, M.F.A., Mameri, N., 2012. Antibacterial activity of chitin, chitosan and its oligomers prepared from shrimp shell waste. Food Hydrocoll. 29 (1), 48–56.

Bernkop-Schnürch, A., Dünnhaupt, S., 2012. Chitosan-based drug delivery systems. Eur. J. Pharm. Biopharm. 81 (3), 463–469.

Cai-qin, Q., Yu-min, D., Ling, X., Xiao-hai, G., Ji-lan, Z., Hui-lan, L., 2002. Effect of molecular weight and structure on antitumor activity of oxidized chitosan. Wuhan Univ. J. Nat. Sci. 7 (2), 231–236.

Campana, R., Casettari, L., Ciandrini, E., Illum, L., Baffone, W., 2017. Chitosans inhibit the growth and the adhesion of *Klebsiella pneumoniae* and *Escherichia coli* clinical isolates on urinary catheters. Int. J. Antimicrob. Agents 50, 135–141.

Chavan, C., Bala, P., Pal, K., Kale, S.N., 2017. Cross-linked chitosan–dextran sulphate vehicle system for controlled release of ciprofloxaxin drug: an ophthalmic application. OpenNano 2, 28–36.

Cheung, R.C., Ng, T.B., Wong, J.H., Chan, W.Y., 2015. Chitosan: an update on potential biomedical and pharmaceutical applications. Mar. Drugs 13 (8), 5156–5186.

Chien, R.C., Yen, M.T., Mau, J.L., 2015. Antimicrobial and antitumor activities of chitosan from shiitake stipes, compared with commercial chitosan from crab shells. Carbohydr. Polym. 138, 259–264.

Collins, M.N., Birkinshaw, C., 2013. Hyaluronic acid based scaffolds for tissue engineering—a review. Carbohydr. Polym. 92 (2), 1262–1279.

Conniot, J., Silva, J.M., Fernandes, J.G., Silva, L.C., Gaspar, R., Brocchini, S., et al., 2014. Cancer immunotherapy: nanodelivery approaches for immune cell targeting and tracking. Front. Chem. 2, 105.

Cosco, D., Cilurzo, F., Maiuolo, J., Federico, C., Di Martino, M.T., Cristiano, M.C., et al., 2015. Delivery of miR-34a by chitosan/PLGA nanoplexes for the anticancer treatment of multiple myeloma. Sci. Rep. 5, 17579.

Croisier, F., Jérôme, C., 2013. Chitosan-based biomaterials for tissue engineering. Eur. Polym. J. 49 (4), 780–792.

Cuzzucoli Crucitti, V., Migneco, L.M., Piozzi, A., Taresco, V., Garnett, M., Argent, R.H., et al., 2018. Intermolecular interaction and solid state characterization of abietic acid/chitosan solid dispersions possessing antimicrobial and antioxidant properties. Eur. J. Pharm. Biopharm. 125, 114–123.

Demetgül, C., Beyazit, N., 2018. Synthesis, characterization and antioxidant activity of chitosan–chromone derivatives. Carbohydr. Polym. 181, 812–817.

Dheer, D., Arora, D., Jaglan, S., Rawal, R.K., Shankar, R., 2016. Polysaccharides based nanomaterials for targeted anti-cancer drug delivery. J. Drug Target. 25 (1), 1–16.

Dickson, M.A., 2014. Molecular pathways: CDK4 inhibitors for cancer therapy. Clin. Cancer Res. 20 (13), 3379–3383.

Dosio, F., Arpicco, S., Stella, B., Fattal, E., 2016. Hyaluronic acid for anticancer drug and nucleic acid delivery. Adv. Drug Deliv. Rev. 97, 204–236.

Evans, E.R., Bugga, P., Asthana, V., Drezek, R., 2018. Metallic nanoparticles for cancer immunotherapy. Mater. Today 21 (6), 673–685.

Galvan Marquez, I., Akuaku, J., Cruz, I., Cheetham, J., Golshani, A., Smith, M.L., 2013. Disruption of protein synthesis as antifungal mode of action by chitosan. Int. J. Food Microbiol. 164, 108–112.

Gao, Z., Li, Z., Yan, J., Wang, P., 2017. Irinotecan and 5-fluorouracil-coloaded, hyaluronic acid-modified layer-by-layer nanoparticles for targeted gastric carcinoma therapy. Dovepress Drug Des. Dev. Ther. 11, 2595–2604.

Gibot, L., Chabaud, S., Bouhout, S., Bolduc, S., Auger, F.A., Moulin, V.J., 2015. Anticancer properties of chitosan on human melanoma are cell line dependent. Int. J. Biol. Macromol. 72, 370–379.

Goy, R.C., Morais, S.T.B., Assis, O.B.G., 2016. Evaluation of the antimicrobial activity of chitosan and its quaternized derivative on *E. coli* and *S. aureus* growth. Rev. Bras. Farmacogn. 26 (1), 122–127.

Hanefeld, A., Weigandt, M., Wolf, M., Knolle, P., Schroeder, M., Scherliess, R., et al., 2014. Antigen-loaded chitosan nanoparticles for immunotherapy. WO2015185180A1 patent.

Hardwick, C., Hoare, K., Owens, R., Hohn, H.P., Hook, M., Moore, D., et al., 1992. Molecular cloning of a novel hyaluronan receptor that mediates tumor cell motility. J. Cell Biol. 117 (6), 1343–1350.

Hemshekhar, M., Thushara, R.M., Chandranayaka, S., Sherman, L.S., Kemparaju, K., Girish, K.S., 2016. Emerging roles of hyaluronic acid bioscaffolds in tissue engineering and regenerative medicine. Int. J. Biol. Macromol. 86, 917–928.

Hou, X., Zhang, W., He, M., Lu, Y., Lou, K., Gao, F., 2017. Preparation and characterization of β-cyclodextrin grafted *N*-maleoyl chitosan nanoparticles for drug delivery. Asian J. Pharm. Sci. 12, 558–568.

Huang, P., Yang, C., Liu, J., Wang, W., Guo, S., Li, J., et al., 2014. Improving the oral delivery efficiency of anticancer drugs by chitosan coated polycaprolactone-grafted hyaluronic acid nanoparticles. J. Mater. Chem. B 25 (2), 4021–4033.

Ibrahim, H.M., El-Zairy, E.M.R., Mosaad, R.M.V., 2015. Preparation, characterization and median lethal dose (LD50) of carboxymethyl chitosan as target drug delivery. Int. J. Adv. Res. 3 (1), 865–873.

Islam, S., Rahman Bhuiyan, M.A., Islam, M.N., 2017. Chitin and chitosan: structure, properties and applications in biomedical engineering. J. Polym. Environ. 25 (3), 854–866.

Jayakumar, R., Prabaharan, M., Nair, S.V., Tokura, S., Tamura, H., Selvamurugan, N., 2010. Novel carboxymethyl derivatives of chitin and chitosan materials and their biomedical applications. Progr. Mater. Sci. 55 (7), 675–709.

Jeon, S.J., Oh, M., Yeo, W.S., Galvao, K.N., Jeong, K.C., 2014. Underlying mechanism of antimicrobial activity of chitosan microparticles and implications for the treatment of infectious diseases. PLoS One 9 (3), e92723.

Kalam, M.A., 2016. Development of chitosan nanoparticles coated with hyaluronic acid for topical ocular delivery of dexamethasone. Int. J. Biol. Macromol. 89, 127–136.

Kalhapure, R.S., Jadhav, M., Rambharose, S., Mocktar, C., Singh, S., Renukuntla, J., et al., 2017. pH-responsive chitosan nanoparticles from a novel twin-chain anionic amphiphile for controlled and targeted delivery of vancomycin. Colloids Surf. B: Biointerfaces 158, 650–657.

Kang, L., Gao, Z., Huang, W., Jin, M., Wang, Q., 2015. Nanocarrier-mediated co-delivery of chemotherapeutic drugs and gene agents for cancer treatment. Acta Pharm. Sin. B 5 (3), 169–175.

Kean, T., Thanou, M., 2010. Biodegradation, biodistribution and toxicity of chitosan. Adv. Drug Deliv. Rev. 62 (1), 3–11.

Khadka, P., Ro, J., Kim, H., Kim, I., Kim, J.T., Kim, H., et al., 2014. Pharmaceutical particle technologies: an approach to improve drug solubility, dissolution and bioavailability. Asian J. Pharm. Sci. 9, 304–316.

Kim, K.S., Duncan, B., Creran, B., Rotello, V.M., 2013. Triggered nanoparticles as therapeutics. Nano Today 8, 439–447.

Kim, J., Chang, J.Y., Kim, Y.Y., Kim, M.J., Kho, H.S., 2018. Effects of molecular weight of hyaluronic acid on its viscosity and enzymatic activities of lysozyme and peroxidase. Arch. Oral Biol. 89, 55–64.

Kiser, T.H., Fish, D.N., Aquilante, C.L., Rower, J.E., Wempe, M.F., MacLaren, R., et al., 2015. Evaluation of sulfobutylether-β-cyclodextrin (SBECD) accumulation and voriconazole pharmacokinetics in critically ill patients undergoing continuous renal replacement therapy. Crit. Care 19 (1), 32.

Komorowicz, E., Balázs, N., Varga, Z., Szabó, L., Bóta, A., Kolev, K., 2017. Hyaluronic acid decreases the mechanical stability, but increases the lytic resistance of fibrin matrices. Matrix Biol. 63, 55–68.

Kong, M., Chen, X.G., Xing, K., Park, H.J., 2010. Antimicrobial properties of chitosan and mode of action: a state of the art review. Int. J. Food Microbiol. 144, 51–63.

Kulikov, S.N., Tikhonov, V.E., Bezrodnykh, E.A., Lopatin, S.A., Varlamov, V.P., 2015. Comparative evaluation of antimicrobial activity of oligochitosans against *Klebsiella pneumoniae*. Russian J. Bioorg. Chem. 41, 57–62.

Lai, J.Y., Ma, D.H.K., 2017. Ocular biocompatibility of gelatin microcarriers functionalized with oxidized hyaluronic acid. Mater. Sci. Eng. C 72, 150–159.

Lallana, E., Rios de la Rosa, J.M., Tirella, A., Pelliccia, M., Gennari, A., Stratford, I.J., et al., 2017. Chitosan/hyaluronic acid nanoparticles: rational design revisited for RNA delivery. Mol. Pharm. 14 (7), 2422–2436.

Liu, T., Soong, L., Liu, G., König, R., Chopra, A.K., 2009. CD44 expression positively correlates with Foxp3 expression and suppressive function of CD4 + Treg cells. Biol. Direct 4, 40.

Liu, C., Zhang, N., 2011. Nanoparticles in gene therapy: principles, prospects, and challenges. Prog. Mol. Biol. Trans. Sci. 104, 509–562.

Lokeshwar, V.B., Mirza, S., Jordan, A., 2014. Targeting hyaluronic acid family for cancer chemoprevention and therapy. Adv. Cancer Res. 123, 35–65.

Lu, H.D., Zhao, H.Q., Wang, K., Lv, L.L., 2011. Novel hyaluronic acid–chitosan nanoparticles as non-viral gene delivery vectors targeting osteoarthritis. Int. J. Pharm. 420 (2), 358–365.

Ma, Z., Garrido-Maestu, A., Jeong, K.C., 2017. Application, mode of action, and in vivo activity of chitosan and its micro and nanoparticles as antimicrobial agents: a review. Carbohydr. Polym. 176, 257–265.

Maeda, Y., Kimura, Y., 2004. Antitumor effects of various low-molecular-weight chitosans are due to increased natural killer activity of intestinal intraepithelial lymphocytes in sarcoma 180-bearing mice. J. Nutr. 134 (4), 945–950.

Maisel, D., Birzele, F., Voss, E., Nopora, A., Bader, S., Friess, T., et al., 2016. Targeting tumor cells with anti-CD44 antibody triggers macrophage-mediated immune modulatory effects in a cancer xenograft model. PLoS One 11 (7), e0159716.

Makarios-Laham, I., Lee, T.-C., 1995. Biodegradability of chitin- and chitosan-containing films in soil environment. J. Polym. Environ. 3 (1), 31–36.

Mansilla, A.Y., Albertengo, L., Rodriguez, M.S., Debbaudt, A., Zuniga, A., Casalongue, C. A., 2013. Evidence on antimicrobial properties and mode of action of a chitosan obtained from crustacean exoskeletons on *Pseudomonas syringae* pv. tomato DC3000. Appl. Microbiol. Biotechnol. 97, 6957–6966.

Matthaiolampakis, G., Milane, L., Singh, A., Amiji, M., 2015. Hyaluronic acid targeting of CD44 for cancer therapy: from receptor biology to nanomedicine. J. Drug Target. 23, 605–618.

MubarakAli, D., LewisOscar, F., Gopinath, V., Alharbi, N.S., Alharbi, S.A., Thajuddin, N., 2018. An inhibitory action of chitosan nanoparticles against pathogenic bacteria and fungi and their potential applications as biocompatible antioxidants. Microbial. Pathog. 114, 323–327.

Necas, J., Bartosikova, L., Brauner, P., Kolar, J., 2008. Hyaluronic acid (hyaluronan): a review. Vet. Med. 53 (8), 397–411.

O'Callaghan, K.A.M., Kerry, J.P., 2016. Preparation of low- and medium-molecular weight chitosan nanoparticles and their antimicrobial evaluation against a panel of microorganisms, including cheese-derived cultures. Food Control 69, 256–261.

Oliveira, A.V., Bitoque, D.B., Silva, G.A., 2014. Combining hyaluronic acid with chitosan enhances gene delivery. J. Nanomater. 2014, (Article ID 246347).

Omwenga, E.O., Hensel, A., Shitandi, A., Goycoolea, F.M., 2018. Chitosan nanoencapsulation of flavonoids enhances their *quorum sensing* and biofilm formation inhibitory activities against an *E. coli* Top 10 biosensor. Colloids Surf. B: Biointerfaces 164, 125–133.

Oommen, O.P., Wang, S., Hilborn, J., Varghese, O.P., 2014. Chemical modifications of hyaluronan for therapeutic applications. In: Collins, M.N. (Ed.), Hyaluronic Acid for Biomedical and Pharmaceutical Applications. Smithers Information Ltd, UK, pp. 149–176. , ISBN: 978-190903-077-0.

Owen, S.C., Kuo, J.W., Prestwich, G.D., 2017. Hyaluronic acid. In: Hashmi, S. (Ed.), Reference Module in Materials Science and Materials Engineering—Comprehensive Biomaterials II. Elsevier, Oxford, UK, pp. 306–331. , ISBN: 978-0-12-803581-8.

Palma-Guerrero, J., Lopez-Jimenez, J.A., Perez-Berna, A.J., Huang, I.C., Jansson, H.B., Salinas, J., et al., 2010. Membrane fluidity determines sensitivity of filamentous fungi to chitosan. Mol. Microbiol. 75, 1021–1032.

Park, S.C., Nah, J.W., Park, Y., 2011. pH-dependent mode of antibacterial actions of low molecular weight water-soluble chitosan (LMWSC) against various pathogens. Macromol. Res. 19, 853–860.

Peña, A., Sanchez, N.S., Calahorra, M., 2013. Effects of chitosan on *Candida albicans*: conditions for its antifungal activity. BioMed. Res. Int. 2013, 527549.

Phunpee, S., Suktham, K., Surassmo, S., Jarussophon, S., Rungnim, C., Soottitantawat, A., et al., 2018. Controllable encapsulation of α-mangostin with quaternized β-cyclodextrin grafted chitosan using high shear mixing. Int. J. Pharm. 538, 21–29.

Ratajska, M., Strobin, G., Wiśniewska-Wrona, M., Ciechańska, D., Boryniec, H., Biniaś, D., et al., 2003. Studies on the biodegradation of chitosan in an aqueous medium. Fibres Text. East. Eur. 11 (3), 75–79.

Rodrigues, S., Dionísio, M., Remuñán López, C., Grenha, A., 2012. Biocompatibility of chitosan carriers with application in drug delivery. J. Funct. Biomater. 3 (3), 615–641.

Ruffell, B., Johnson, P., 2015. Hyaluronan induces cell death in activated T cells through CD44. J. Immunol. 181 (10), 7044–7054.

Safdar, M.H., Hussain, Z., Abourehab, M.A.S., Hasan, H., Afzal, S., Thu, H.E., 2018. New developments and clinical transition of hyaluronic acid-based nanotherapeutics for treatment of cancer: reversing multidrug resistance, tumour-specific targetability and improved anticancer efficacy. Artif. Cells Nanomed. Biotechnol. 46 (8), 1967–1980.

Salatin, S., Khosroushahi, A.Y., 2017. Overviews on the cellular uptake mechanism of polysaccharide colloidal nanoparticles. J. Cell. Mol. Med. 21 (9), 1668–1686.

Sapru, S., Ghosh, A.K., Kundu, S.C., 2017. Non-immunogenic, porous and antibacterial chitosan and *Antheraea mylitta* silk sericin hydrogels as potential dermal substitute. Carbohydr. Polym. 167, 196–209.

Sarwar, A., Katas, H., Zin, N.M., 2014. Antibacterial effects of chitosan–tripolyphosphate nanoparticles: impact of particle size molecular weight. J. Nanopart. Res. 16, 1–14.

Sato, T., Nakata, M., Yang, Z., Torizuka, Y., Kishimoto, S., Ishihara, M., 2017. In vitro and in vivo gene delivery using chitosan/hyaluronic acid nanoparticles: influences of molecular mass of hyaluronic acid and lyophilization on transfection efficiency. J. Gene Med. 19, e2968.

Selyanin, M.A., Boykov, P.Y., Khabarov, V.N., 2015. Hyaluronic Acid: Preparation, Properties, Application in Biology and Medicine. Wiley, ISBN: 978-1-118-63379-3.

Sharma, S., Singh, A., 2012. Nanotechnology based targeted drug delivery: current status and future prospects for drug development. In: Kapetanović, I.M. (Ed.), Drug Discovery and Development—Present and Future. InTech, Rijeka (Croatia), pp. 427–462. , ISBN: 978-953-307-615-7.

Shekarian, T., Valsesia-Wittmann, S., Caux, C., Marabelle, A., 2015. Paradigm shift in oncology: targeting the immune system rather than cancer cells. Mutagenesis 30 (2), 205–211.

Shi, Y., Xue, J., Xu, S., You, Y., Yan, X.Q., Zhao, X., et al., 2018. Polyelectrolyte complex nanoparticles based on chitosan and methoxy poly(ethylene glycol) methacrylate-*co*-poly(methylacrylic acid) for oral delivery of ibuprofen. Colloids Surf. B: Biointerfaces 165, 235–242.

Skorik, Y.A., Kritchenkov, A.S., Moskalenko, Y.A., Golyshev, A.A., Raik, S.V., Whaley, A. K., et al., 2017. Synthesis of *N*-succinyl- and *N*-glutaryl-chitosan derivatives and their antioxidant, antiplatelet, and anticoagulant activity. Carbohydr. Polym. 166, 166–172.

Soleimani, M., Majid, A.Z., Aaesha, M.M., Matar, R., 2016. Nanoparticles in gene therapy. Int. J. Integr. Biol. 17, 7–16.

Ta, H.T., Dunstan, D.E., Dass, C.R., 2010. Anticancer activity and therapeutic application of chitosan nanoparticles. In: Kim, S.K. (Ed.), Chitin Chitosan, Oligosaccharides and Their Derivatives, Biological Activities and Applications. CRC Press (Taylor & Francis Group), Boca Raton, FL, pp. 271–284. , ISBN: 9781439816042.

Tang, X., Zeng, B., Gao, J.K., Liu, H.Q., 2016. Molecular mechanism of enhanced anticancer effect of nanoparticle formulated LY2835219 via p16-CDK4/6-pRb pathway in colorectal carcinoma cell line. J. Nanomater. 2016, Article ID 2095878.

Tayel, A.A., Moussa, S., El-Tras, W.F., Knittel, D., Opwis, K., Schollmeyer, E., 2010. Anticandidal action of fungal chitosan against *Candida albicans*. Int. J. Biol. Macromol. 47 (4), 454–457.

Thaya, R., Vaseeharan, B., Sivakamavalli, J., Iswarya, A., Govindarajan, M., Alharbi, N.S., et al., 2018. Synthesis of chitosan–alginate microspheres with high antimicrobial and antibiofilm activity against multi-drug resistant microbial pathogens. Microb. Pathog. 114, 17–24.

Tripodo, G., Trapani, A., Torre, M.L., Giammona, G., Trapani, G., Mandracchia, M., 2015. Hyaluronic acid and its derivatives in drug delivery and imaging: recent advances and challenges. Eur. J. Pharm. Biopharm. 97 (Pt B0), 400–416.

Unagolla, J.M., Jayasuriya, A.C., 2018. Drug transport mechanisms and in vitro release kinetics of vancomycin encapsulated chitosan–alginate polyelectrolyte microparticles as a controlled drug delivery system. Eur. J. Pharm. Sci. 114, 199–209.

Wang, T., Hou, J., Su, C., Zhao, L., Shi, Y., 2015. Hyaluronic acid-coated chitosan nanoparticles induce ROS-mediated tumor cell apoptosis and enhance antitumor efficiency by targeted drug delivery via CD44. J. Nanobiotechnol. 15, 7.

Wang, J., Law, W.C., Chen, L., Chen, D., Tang, C.Y., 2017. Fabrication of monodisperse drug-loaded poly(lactic-*co*-glycolic acid)–chitosan core–shell nanocomposites via pickering emulsion. Composites B 121, 99–107.

Wang, Y., Beck-Broichsitter, M., Yang, M., Rantanen, J., Bohr, A., 2017. Investigation of nanocarriers and excipients for preparation of nanoembedded microparticles. Int. J. Pharm. 526 (1–2), 300–308.

Xia, W., Liu, P., Zhang, J., Chen, J., 2011. Biological activities of chitosan and chitooligosaccharides. Food Hydrocoll. 25, 170–179.

Xiao, B., Xu, Z., Viennois, E., Zhang, Y., Zhang, Z., Zhang, M., et al., 2017. Orally targeted delivery of tripeptide KPV via hyaluronic acid-functionalized nanoparticles efficiently alleviates ulcerative colitis. Mol. Ther. 25 (7), 1628–1640.

Xing, K., Chen, X.G., Kong, M., Liu, C.S., Cha, D.S., Park, H.J., 2009a. Effect of oleoyl-chitosan nanoparticles as a novel antibacterial dispersion system on viability: membrane permeability and cell morphology of *Escherichia coli* and *Staphylococcus aureus*. Carbohydr. Polym. 76, 17–22.

Xing, K., Chen, X.G., Liu, C.S., Cha, D.S., Park, H.J., 2009b. Oleoyl-chitosan nanoparticles inhibits *Escherichia coli* and *Staphylococcus aureus* by damaging the cell membrane and putative binding to extracellular or intracellular targets. Int. J. Food Microbiol. 132, 127–133.

Xu, Y., Wen, Z., Xu, Z., 2009. Chitosan nanoparticles inhibit the growth of human hepatocellular carcinoma xenografts through an angiogenic mechanism. Anticancer Res. 29 (12), 5103–5109.

Yang, Z.H., 2006. Development of chitosan nanoparticles for gene delivery system. Mol. Ther. 13, S76–S77.

Yang, J., Lu, H., Li, M., Liu, J., Zhang, S., Xiong, L., et al., 2017. Development of chitosan–sodium phytate nanoparticles as a potent antibacterial agent. Carbohydr. Polym. 178, 311–321.

Yeh, M.Y., Wu, M.F., Shang, H.S., Shih, Y.L., Chen, Y.L., Hung, H.F., et al., 2013. Effects of chitosan on xenograft models of melanoma in C57BL/6 mice and hepatoma formation in SCID mice. Anticancer Res. 33 (11), 4867–4873.

Zang, X., Zhao, X., Hu, H., Qiao, M., Deng, Y., Chen, D., 2017. Nanoparticles for tumor immunotherapy. Eur. J. Pharm. Biopharm. 115, 243–256.

Zhang, H., Huang, S., Yang, X., Zhai, G., 2014. Current research on hyaluronic acid–drug bioconjugates. Eur. J. Medi. Chem. 86, 310–317.

Zhou, P., Yang, X., Zhang, Y., Du, P., Yuan, S., Yang, D., et al., 2016. Antitumor effects of orally and intraperitoneally administered chitosan oligosaccharides (COSs) on S180-bearing/residual mouse. J. Food Sci. 81 (12), 3035–3042.

CHAPTER

2 Nanostructures in pharma: elixir to oral medicine

Murali Mohan Bommana[1], Bharat Kirthivasan[2], Simerdeep Singh Gupta[2,3], Abhijat Shikhar[4] and Amit Shah[1]

[1]*Impax Labs, Middlesex, NJ, United States* [2]*Product Development, Novel Laboratories, Inc., Somerset, NJ, United States* [3]*Teva Pharmaceuticals, Pomona, NY, United States* [4]*Tansukh Herbals Pvt Ltd., Lucknow, India*

CHAPTER OUTLINE

2.1 NANOCRYSTALS

The pharmaceutical pipeline produces 70% of low soluble molecules with bottleneck procedures for the manufacture of dosage forms to deliver the drug in a reproducible manner with bioavailability in the range to show the efficacy of the actives. The bioavailability is hampered by the low solubility exposure of the active pharmaceutical ingredient (API). Different technologies have surfaced with lot of success in the process of drug comminution to the nanorange, which yields successful dissolution rate and thereby greater exposure. The in vitro and in vivo performance of most drug products is governed by the particle size of their API. The API could be present in their crystalline salt forms, amorphous state, as

Nanoparticles in Pharmacotherapy. DOI: https://doi.org/10.1016/B978-0-12-816504-1.00001-6

different polymorphs, or as a different form such as cocrystals. The form has an impact on their dissolution performance, and thus it affects the bioavailability of the drug product. The rate at which the dissolution occurs may be impacted by various factors such as formulation composition, API particle size, organoleptic properties, etc. This section of the chapter will focus on the effect of particle size on the following three aspects of drug product development:

1. dissolution performance and bioavailability,
2. stability of the product, and
3. manufacturability of the dosage form.

When an API is used for development or commercial manufacturing, the physical properties of that drug dictates the particle size, and particle size range. By definition, when the particle size of the drug crystals is between 1 and 100 nm, the powder is considered to be nanocrystalline (Park, 2007; Muller and Keck, 2004; Rabinow, 2004). Most existing drug products, and the ones being discovered, are poorly water soluble, and a vast majority of them fall in the BCS class II category. This means that those drugs have poor water solubility, but good membrane permeability. A point to be noted in this scenario is that the API should be in solution form to get absorbed. To improve the absorption and thus bioavailability of such an API, the formulation obstacle is to improve their dissolution rate. There have been several attempts to improve the dissolution rate of APIs, out of which particle size reduction is one of the go-to methods because of the following reasons:

- The crystal form of the API does not change in the top-down processing method in most cases. The behavior of dissolution rate is thus as expected.
- The API is not in amorphous state, which makes it physically more stable.
- The particle size remains the same through the shelf life of the product, unless there is unexpected rise in moisture in case of hydrophilic drugs.

2.1.1 MECHANISM OF NANOCRYSTALS IMPROVED PERFORMANCE AND PROPERTIES

2.1.1.1 Surface area

During the micronization process, the particle size of API decreases, and thereby the dissolution rate or velocity of the same. Indeed true, the nanonization increases the surface area tremendously and the process drives the dissolution velocity and therefore increased amount of drug is in solution state and measured performance of the drug in vivo. The whole principle is defined by the modified Noyes–Whitney equation, where the surface area increases are directly proportional to the increased dissolution performance. So, the surface area of the nanocrystals drives the dissolution velocity of the crystals, and hence the concentration of the bulk to the medium is all the time maintained to have a flux to in vivo exposure.

2.1.1.2 Apparent solubility and supersaturation

Saturation solubility defines the intrinsic property of the API. Sometimes the intrinsic solubility could be higher than the thermodynamic solubility based on high energy amorphous or nanocrystalline nature of the API. The elevated solubility known as the "spring effect" could be attributed to the supersaturated state. Polymers like polyvinylpyrrolidine, hypromellose, etc. increase the aqueous solubility and help in maintaining the supersaturation of the drug. Viscosity inducing agents also decrease the crystal growth by inhibition mechanisms, which also maintain the supersaturation of the drug (Peltonen and Strachan, 2015).

2.1.1.3 Amorphous state of the nanocrystals

Amorphous systems have very high energy and show higher supersaturation, which reflects the in vivo exposure of the same, which reflects the bioavailability of the drug. The prerequisite is to maintain the amorphous nature of the API throughout the shelf life of the product, so that it can be FDA approved. Some of the amorphous systems are tough to stabilize and may not surface as a product, because of the stabilization of the system. The excipients should stabilize the amorphous component of the API in the same nature to show higher supersaturation and thereby the in vivo exposure of the drug. The same is the case with the nanocrystals. The API when amorphous or partly amorphous in nature shows higher exposure because of the high supersaturation attained by the nanocrystals in amorphous nature. But even in this case, the general prerequisite is to maintain the amorphous nature throughout the shelf life of the product cycle (Junghanns and Muller, 2008).

2.1.2 MANUFACTURING OF NANOCRYSTALS

Nanocrystals are defined by their size and size distribution. Based on the required size and distribution, they are processed accordingly. The starting size of the powder is always higher, in micron range. This starting material is then processed into particles in nanorange either by bottom-up or top-down technique (Fig. 2.1). In a nutshell, bottom-up technique is where the powder in solution form is dried in a controlled manner to obtain particles in the nanorange. On the other hand, top-down technique is where larger particles are broken down by attrition to nanosize. There is another technique where the API is entrapped as a solution into microemulsion globules, and rapidly dried to obtain nanocrystals.

2.1.3 BOTTOM-UP TECHNIQUES

2.1.3.1 Nanoprecipitation

Nanoprecipitation is a simple technique where API is dissolved in a solvent, and antisolvent is added to it that causes rapid precipitation of particles in nanosize range. The solubility of API in antisolvent is poor. Nanoprecipitation can also be

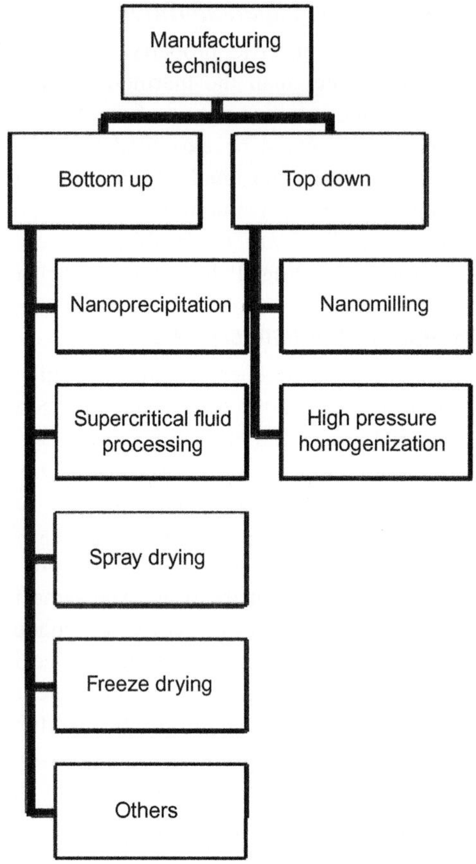

FIGURE 2.1

Schematic of the manufacturing techniques for drug nanocrystals.

combined with sonication to further break the particle size, or prevent agglomeration. Nanoprecipitation is generally used in laboratory to prepare small-scale development batches. It is not a favorable technique for scale up due to excessive use of organic solvents. The prepared nanocrystals have to be stabilized or coated with a crystal inhibitor to keep the nanocrystals in size.

2.1.3.2 Supercritical fluid processing

Supercritical fluid technique employs dissolving the API in supercritical carbon dioxide in presence of small amount of organic solvent. Carbon dioxide behaves as a solvent at supercritical temperature and pressure of—31.1°C and 72.9 atm. When the gas is released in controlled manner, the residue powder can be obtained in nanorange.

2.1.3.3 Spray drying

API is first dissolved in a low boiling point organic solvent, and the solution is sprayed through a fine nozzle to create a mist. Under ambient temperature conditions, the solvent rapidly evaporates, and dry powder in nanosize range is obtained.

2.1.3.4 Freeze drying

The principle of freeze drying to obtain nanocrystals is to freeze the API solution, and lyophilize under vacuum to get rid of the solvent. When this processing is done in controlled manner, the precipitate powder is nanometer size range.

2.1.3.5 Others

There are other bottom-up techniques used such as high gravity controlled precipitation, confined liquid impinging jets, and multiinlet vortex mixing.

2.1.4 TOP-DOWN TECHNIQUES

2.1.4.1 Nanomilling/media milling

Nanomilling utilizes beads to mill the API particles by attrition. The API is suspended in a medium in presence of a stabilizer. The suspension is added to a chamber with beads made of yttrium stabilized glass, zirconium oxide, or cross-linked polystyrene resins. Either the chamber is moved around, or the beads are agitated using a propeller shaft for long time durations. The resultant suspension has nanocrystals of the API, which can be dried using lyophilization, spray drying, etc. Nanomilling is commonly used in R&D, and is scalable. Products that are on the market such as Tricor, Invega, and Emend employ nanomilling.

2.1.4.2 High pressure homogenization

High pressure homogenization is also commonly used to prepare nanoparticles using a top-down method. In principle, drug suspension is passed at high speed onto a homogenizer, in a repetitive manner until the particle size is broken down to nanorange (Keck and Müller, 2006). The technique can be done in both aqueous and nonaqueous media. High pressure homogenization is a robust method that gives narrow particle size distribution.

2.1.5 GENERAL ISSUES: CAUSE, EFFECT, AND SOLUTION

Even though the nanocrystal technology has its merits, there are some issues that should be considered during product development. Following are the challenges commonly faced with nanocrystals during formulation, processing, and use.

2.1.5.1 Crystal growth and agglomeration

Ostwald ripening or coarsening is a process where the total energy of a two-phase system is decreased with an increase in the size scale (Voorhees, 1985). In case of nanocrystals, there is a range of particle size distribution, and thus there are particles of different sizes. With time or in solution, there is diffusion of smaller particles onto larger ones, which eventually leads to increase in particle size.

The increased particle size in nanoparticles could lead to various performance issues. Hecq et al. showed that oral bioavailability of nifedipine is greatly impacted with the particle size of the nanosuspensions (Hecq et al., 2005). With an increase in particle size of the API, its solubility rate greatly drops, which in turn decreases its bioavailability. In their work, Hoet et al. highlighted some risks associated with increase in particle size of nanoparticles (Hoet et al., 2004). Nanoparticles used for parenteral use could block blood vessels and cause embolism if their particle size is increased to 5 μM and above. In case of inhalation of nanoparticles, there could be pulmonary blockage due to crystal growth.

Crystal growth of nanoparticles can be avoided by using stabilizers in the formulation by preventing interfacial interaction between particles. Recently, Serajuddin and Prajapati demonstrated the use of various sugar esters as nanoparticle stabilizers in solid form (Prajapati and Serajuddin, 2016; U.S. Patent Application No. 15/042,282). Surfactants prevent agglomeration and Ostwald ripening in solution form, but may prove ineffective in solid form during storage. Sugar esters on the other hand not only act as stabilizers in solid form, but also provide complete dispersion in aqueous media without agitation. Other factors such as moisture, temperature, and physical agitation should also be considered in case-by-case basis to prevent crystal growth in nanocrystals.

2.1.5.2 Polymorphism

The processing technique to prepare nanocrystals not only affects their size, but also the physical form of the product (Tahara et al., 2016). The top-down method imparts attrition energy onto the particles, which may result in part of the product to convert to its amorphous state. In the bottom-down approach, the chances of producing amorphous state are much higher. The amorphous particles, even though they are in the nanorange, are physically unstable. With time the amorphous particles will eventually convert to a more stable crystalline form. During this conversion, there is lack of control on the particle size. This increased particle size may lead to issues mentioned in the previous section. Frank and Boeck demonstrated that freeze drying of nanosuspensions could also lead to polymorphism, agglomeration, and crystal growth (Frank and Boeck, 2016). Stabilizers such as mannitol and sucrose could be added in small quantities to prevent agglomeration during freeze drying. The processing parameters could be controlled better to avoid polymorphism. Process analytical tools could be used to determine the endpoint of the process so that the material is not pushed to its limits for conversion to take place. If processing cannot be controlled, the nanocrystal product could also be stored at lower temperatures to prevent the conversion to crystal form.

2.1.6 CHARACTERIZATION OF NANOCRYSTALS

Any product or intermediate should be tested to ensure that it meets the desired quality attribute. In case of nanocrystals, the characterization should be to ensure their size, physical, and chemical properties, morphology, and performance. The following tests are widely used in the pharmaceutical industry to characterize nanocrystals.

2.1.6.1 Particle size analysis

Particle size analysis is a key test not only to ensure that the selected formulation and processing parameters are good, but also to evaluate the stability of nanocrystals. In addition to particle size analysis it is also important to look into particle size distribution of the product. The distribution range gives a good guide to formulators about the manufacturing process, and long-term stability of the crystals. Various techniques are used to measure the particle size of nanocrystals, out of which the following are widely used: dynamic light scattering, laser diffraction, and Coulter counter. Dynamic light scattering is a quick method to analyze suspensions of nanocrystals. This technique has its limitations because it can only measure particle size in liquid medium as a suspension. In contrast, laser diffraction can measure particle size in both powder and suspension form. Laser diffraction also has a wider measurement range than dynamic light scattering.

2.1.6.2 Physical testing

Physical testing of the nanocrystals encompasses the analysis of crystalline nature of the particles, their thermal properties, surface charge, and zeta potential. X-ray diffraction (XRD) is used to find out the crystalline nature of particles. It is important for those APIs where there is polymorphism expected as a result of processing. XRD is also a useful tool to differentiate between amorphous and crystalline particles. Differential scanning calorimetry (DSC) is used to analyze the thermal characteristics of solids. The most commonly analyzed tests by DSC are melting point of crystalline solids, and glass transition of amorphous state. Measurement of surface charge and zeta potential is important to understand the behavior of particles during processing and their dispersion performance.

2.1.6.3 Morphology

Scanning electron microscopy, transmission electron microscopy (TEM), and atomic force microscopy (AFM) are used to evaluate particle morphology and size. However, it is difficult to assess particle size distribution, as the sample size is too small.

2.1.6.4 Dispersion testing

In most cases, nanocrystals are formulated for solubility enhancement purposes. When nanoparticle powder is added to aqueous medium, the particles are expected to disperse uniformly in the medium without agglomeration. Dispersion testing is an important tool where particle size of the suspension is measured as a function of time. The research work done by Prajapati et al. describes the dispersion testing method in detail to assess the performance of lipid-based drug delivery systems containing probucol (Prajapati et al., 2013). A similar approach could be used to characterize nanocrystal suspensions for their in vitro and in vivo performance.

2.1.6.5 Chemical analysis

Nanocrystals can be qualitatively analyzed for impurities by HPLC. Other methods for chemical analysis are FTIR, NMR, and mass spectroscopy (Table 2.1).

2.2 POLYMERIC NANOPARTICLES

Polymeric nanoparticles are a relatively new technology with a great targeting potential for personalized treatments in effective medicine. They comprise of particles with size range of 10 nm to 1 μm. The particle size and surface properties of the polymeric nanoparticles dictates the fate of the nanoparticles in vivo. In particular, the field is emerging into all sorts of pharmaceutical, biomedical, and diagnostic applications, based on the merit of the purpose of the targeting potential of the nanoparticles. To name a few, polymeric nanoparticles are synthesized using various methods like emulsification solvent evaporation, emulsification solvent diffusion, nanoprecipitation, salting out method, etc. Different kinds of biodegradable polymers are employed in the manufacture of nanoparticles as specified in the review by Kumar et al. (2013). Further once the nanoparticles are either taken orally or injected they are cleared by the reticulo-endothelial cells by a definite process called opsonization. So to increase the circulation half-life and protect the nanoparticles from macrophages, surface coatings are applied to provide steric hindrance and help in the survival of the nanoparticles and deliver the nanomedicine. Further the surface of the polymer could be embellished with ligands or antibodies to target the nanoparticles to the specific sites of the body.

Different attributes of the nanoparticles like size, surface area and charge, hydrophobicity, or polyethylene glycol (PEG) coating, impart certain properties to the influential uptake or targeting of the nanoparticles. Particle size influences the uptake of the nanoparticles. The bigger particles are cleared by the macrophages, and the smaller particles are cleared by the glomerular filtration. The shape of the nanoparticles affects the procedural endocytosis, with spherical being the best choice. With surface charge positive (zeta potential) have a better chance of internalization because of the interaction with negative cell membranes, and less

Table 2.1 Case Studies of Marketed Nanocrystals/Nanoparticle Technology

Brand	Indication	Technology	Drug Delivery Company
Rapamune	Immunosuppressant	Sirolimus with a molecular weight of 914.2 has a very low aqueous solubility. The dosage form is a tablet with the drug in the form of nanocrystals. The final dosage form tablet release the drug in the form of nanosuspension. The nanocrystals have higher bioavailability than oral suspension.	Elan NanoSystems
Emend	Antiemetic	Aprepitant is the drug of choice with a narrow window of absorption at the upper gastrointestinal tract. Previously high amount of drug was loaded to increase the chances of absorption. This comes with the disadvantage of side effects. So, by using nanonization the dose was reduced, surface area increased to serve the narrow absorption window.	Elan NanoSystems
Tricor	Hypercholesterolemia	Fenofibrate, a lipophilic compound, is practically insoluble in water. The solubility is not influenced by pH, because of the absence of ionizable groups on fenofibrate skeleton. During the fed stage, the drug has 35% more bioavailability than fasting stage. With nanocrystal technology, the food effect was made negligible and bioequivalent in fed and fasting stage.	Abbott Laboratories
Megace ES	Antianorexic	Megestrol acetate is a low water soluble molecule and has a varied absorption. Megace ES is a nanocrystalline product, and has no food effect. The decreased dose volume because of the decrease viscosity effect had better patient compliance of the nanocrystal effect. Thus better bioavailability was achieved compared with the conventional oral suspension of the same (Junghanns and Muller, 2008).	Elan NanoSystems

compared with negative and neutral charged nanoparticles. Further, the surface charge dictates the colloidal stability of the polymeric nanoparticles, with higher surface charges showing a higher stability. The applications of the polymeric nanoparticles are being investigated in the areas of targeting, oral delivery of proteins or peptides, brain targeting, gene delivery, etc.

2.2.1 IN VIVO EVALUATION OF ORAL NANOPARTICLES

2.2.1.1 Types of in vivo evaluation tests

Although in vitro evaluation methods provide valuable insight into drug permeability and toxicity of nanoparticles, in vivo evaluation is crucial for the validation and eventual clinical translation of the nanoparticle formulation into an oral drug delivery system. Noninvasive monitoring options are obviously preferred, but they are limited by the extent of data they provide. The more traditional methods of in vivo testing of nanoparticles include drug release kinetics and efficacy studies (Liu et al., 2016), supplemented by biodistribution, and histopathological studies. It is recommended that, instead of relying on one particular method, researchers use a combination of in vivo methods to bolster the validity of their findings.

2.2.1.2 Pharmacokinetic studies

As a general theme, all pharmacokinetic studies of intestinal drug absorption involve oral administration of the nanoparticle formulation and the appropriate control formulations to animals and analyzing plasma or urine concentrations of the drug using established HPLC methods. In addition to constructing the drug release kinetic curve, the parameters commonly measured are maximum plasma/urine drug concentration (C_{max}), time taken to reach maximum plasma/urine drug concentration (T_{max}), and the area under the drug concentration versus time curve (AUC0−t). Harsha et al. tested a spray-dried formulation of vildagliptin and metformin hydrocholoride loaded nanoparticles for stomach mucoadhesive drug delivery. Male Sprague−Dawley rats (one control group with oral administration of vildagliptin−metformin and five groups with the oral administration of the nanoparticle formulation) were tested for drug retention in the gastrointestinal tract by extracting nanoparticles retained in the stomach and small intestine at different time points. Vildagliptin−metformin was extracted from these samples for quantitative spectroscopic analyses. The drug release studies revealed a sustained action of the drug with the nanoparticle formulation showing significantly higher drug retention compared with the conventional formulation.

Lipid nanoparticles of raloxifene were tested for drug release upon both intravenous and oral administration in female Sprague−Dawley rats (Ravi et al., 2014). The AUC0−t, C_{max}, and the mean residence time observed in the nanoparticle group were significantly higher than the control group (free raloxifene), suggesting an increase in the bioavailability of raloxifene when loaded into lipid nanoparticles. In another study (Venkatesh et al., 2015), bioavailability of drug-loaded poly(lactide-*co*-glycolide) (PLGA) nanoparticles was tested in New Zealand rabbits alongside the standard formulation (suspension) of the drug. The HPLC analysis at different time points revealed the overall high plasma concentration, low clearance, long half-life, and higher relative bioavailability of the drug in the nanoparticle formulation compared with that in the standard suspension.

Sonaje et al. (2009) tested the efficacy of self-assembled, pH-sensitive nanoparticles for oral insulin delivery. Streptozotocin-induced diabetic male Wistar rats were dosed with insulin-loaded nanoparticles; two other cohorts were dosed with oral insulin solution and subcutaneous injection of insulin, respectively. Blood samples from the tail veins were tested for plasma insulin levels using the insulin ELISA kit. Oral administration of insulin-loaded nanoparticles suppressed the initial hyperglycemia in rats, an effect that was significant compared with the oral insulin solution group. The AUC and relative bioavailability calculations revealed that the nanoparticles significantly enhanced the intestinal absorption of insulin, thereby increasing its bioavailability. A similar improvement in the oral bioavailability of insulin was also reported by Cui et al. (2006) when they tested their biodegradable nanoparticles loaded with insulin–phospholipid complex.

Another study evaluated the oral delivery of insulin using insulin–thiomer complex nanoparticles using nondiabetic Sprague–Dawley rats (Deutel et al., 2008). Test rats were orally dosed with insulin nanoparticles formulated as enteric coated tablets as well as suspensions. Serum insulin concentrations and blood glucose levels were detected using the ELISA kit and the blood glucose strips respectively. The oral nanoparticulate suspension showed a significantly higher insulin concentration and blood sugar reduction with a rapid onset of action compared with the enteric coated tablets.

2.2.1.3 Drug efficacy studies

Some studies reported testing methods for the pharmacodynamic efficacy of drugs loaded into nanoparticles. Cetin et al. (2011) tested the hypocalcemic potency of salmon calcitonin (sCT) loaded Eudragit and Eudragit-PLGA nanoparticles upon oral administration in female Sprague–Dawley rats. Plasma samples from rats treated with sCT loaded nanoparticles and an oral solution of sCT were assayed for calcium levels to determine the hypocalcemic effects of the drug. The sCT loaded nanoparticles yielded a significant decrease in the calcium concentrations compared with both the vehicle group (solvent only) and the sCT solution group. Among the nanoparticle formulations tested, the polycationic Eudragit RSPO–sCT combination showed the highest hypocalcemic efficacy.

Padhye and Nagarsenker (2013) evaluated the anticholesterolemic effects of simvastatin loaded solid lipid nanoparticles in a poloxamer induced hyperlipidemia model. An intraperitoneal injection of poloxamer F-127 (1.0 g/kg) was used to induce hyperlipidemia in female Sprague–Dawley rats. Following this, the test rats were dosed with different formulations of drug-loaded lipid nanoparticles.

The positive and negative control rats received the simvastatin drug suspension and plain water respectively. Serum was separated from the blood drawn from the animals at different time points after the induction of hyperlipidemia and analyzed for total cholesterol and high density lipoproteins. A 4- to 4.8-fold increase in the total cholesterol levels was detected upon poloxamer injection, which confirmed the induction of hyperlipidemia. The treatment groups were compared with controls with respect to the increase in total cholesterol (ITC) values, which accounted for differences in basal cholesterol values among the animals. The negative control group had a significantly higher ITC value compared with the drug-loaded nanoparticles and the free drug groups. Moreover, the nanoparticle formulations exhibited a significantly higher cholesterol lowering effect compared with the free drug group, thereby suggesting an increase in drug bioavailability. Sonaje et al. (2009) studied the mucoadhesive properties of their insulin-loaded self-assembled nanoparticles in male Sprague–Dawley rats. They orally administered fluorescently labeled insulin-loaded nanoparticles to the rats, after which they were sacrificed and their intestinal segments examined under confocal laser scanning microscopy. The intensity of the fluorescence observed in the luminal surface of the duodenum was stronger and appeared at a deeper level than those seen at the jejunum and the ileum, thereby indicating that the nanoparticles could effectively adhere to the mucosal surface.

In another drug efficacy study by Chen et al. (2008), *N*-trimethyl chitosan (TMC) nanoparticles loaded with a model vaccine urease were tested for antibody titers in plasma and gastrointestinal secretions. Separate groups of female Kunming mice were immunized with urease loaded nanoparticles, urease + TMC solution, urease solution, and a control vehicle medium. Blood samples were obtained from the orbit venous plexus. The gastric secretions were obtained using cellulose wicks, whereas the intestinal secretions were obtained from scrapings of the intestinal luminal mucosa. The urease-specific IgG in the serum and the urease-specific IgA in the gastric and intestinal mucosa were quantified using ELISA. Both the serum and gastrointestinal antibody titers elicited by the urease loaded nanoparticles were significantly higher than that of the urease solution alone, which indicated that this method could be effective for the oral delivery of vaccines.

Bisht et al. (2008) tested the efficacy of their polymeric nanoparticles loaded with rapamycin in a xenograft model of human pancreatic cancer. Fresh tumor chunks (Panc 198) were subcutaneously implanted bilaterally into the flanks of male CD1 nu/nu athymic mice. Three weeks following the grafting procedure, mice were randomized into the oral nanorapamycin, oral conventional rapamycin (Rapamune) and control groups. The organs and xenograft tumor tissues were harvested upon sacrifice for immunohistochemical and Western-blot studies. Tissue homogenates from the xenograft tumor were used to quantify free rapamycin concentrations using tandem mass spectrometry with electrospray positive ionization. Both oral nanorapamycin and Rapamune showed significant tumor growth retardation compared with the controls. This was also confirmed in the immunohistochemical staining and Western blots, which showed markers of reduced proliferation and mTOR pathway inhibition, respectively. No significant differences were detected in these parameters between the nanorapamycin and Rapamune cohorts.

2.2.1.4 Drug transport and distribution studies

Lipid nanoparticles were tested for their lymphatic transport in female Wistar rats by testing blood levels of the drug in the presence and absence of cycloheximide, which is a known inhibitor of lymphatic transport (Ravi et al., 2014). When the lipid nanoparticles were orally administered to both a control group (no cycloheximide) and the test group (previously dosed with cycloheximide), significantly lower concentration of the drug was detected in the plasma of the test group animals, thereby underscoring the importance of lymphatic transport in the absorption of lipid nanoparticles. In the same study, tissue distribution of the drug when administered in the nanoparticle formulation and in free form was compared. Female Wistar rats were dosed either with the nanoparticle formulation or free drug, and their spleen, liver, kidney, and lung tissues were collected at different time points. Samples extracted after tissue homogenization were tested for drug levels using HPLC. It was found that raloxifene from the nanoparticles accumulated to a lesser extent in the liver compared with free raloxifene, and to a greater extent in the spleen. The authors cited the lymphatic transport and the phagocytic uptake of lipid nanoparticles for the observed effects.

Another study used fluorescently labeled nanoparticles (Semete et al., 2010) for tissue distribution analysis using two different dyes (rhodamine and coumarin) to determine any differential biodistribution based on the dyes. The brain, heart, kidneys, liver, lungs, and spleen were collected, homogenized, and analyzed for fluorescent particles at both excitation and emission wavelengths. Furthermore, confocal microscopy was used to determine whether the observed fluorescence originated from the nanoparticles or from leached dye. Tissues obtained from mice treated with saline only were treated as negative controls and used for data normalization. Most of the particles were detected in the liver, followed by the kidney, heart, and brain. Confocal imaging confirmed that the detected fluorescence originated from the dye in the nanoparticles and not that from free/leached dye. No dye-based differences were reported with respect to biodistribution characteristics.

2.2.1.5 Histopathological examination studies

A histopathological examination was conducted to test for abnormal irritability of the rat mucosa upon administration of the spray-dried formulation of vildagliptin and metformin (Harsha, 2014). Mucosal sections were isolated from male Sprague–Dawley rats treated with the nanoparticle formulation and analyzed under a microscope for visible signs of tissue injury. No abnormal changes were detected in the nanoparticle group.

In another histopathological analysis (Semete et al., 2010), PLGA nanoparticles were compared with silicon dioxide nanoparticles and zinc oxide nanopowder, with the zinc oxide serving as the positive control owing to its known toxic effects. Brain, heart, kidney, liver, lung, spleen, large and small intestines, pancreas, stomach, thymus, and gonadal tissue were sectioned and stained with hematoxylin and eosin for microscopic imaging. The PLGA nanoparticles did not exhibit any signs of lesions at all the concentrations tested. Similar results were noted for the silicon dioxide group and the negative control group (polystyrene beads). The positive control group (zinc oxide), on the other hand, had to be discontinued prior to the final analyses due to widespread animal death.

2.2.2 CHARACTERIZATION OF THE NANOPARTICLES BY VARIOUS TECHNIQUES

There are various techniques used to characterize polymeric nanoparticles for oral delivery. The methods commonly referred to in the literature are mentioned below (Jawahar and Meyyanathan, 2012).

2.2.2.1 Particle size and distribution

2.2.2.1.1 Photon-correlation spectroscopy

The use of photon-correlation spectroscopy to characterize the particle size of nanoparticles has been well discussed in the literature (Pal et al., 2011; Lim et al., 2013).

2.2.2.1.2 Scanning electron microscopy

The literature is replete with the use of scanning electron microscopy to examine the morphology of nanoparticles (Beck-Broichsitter et al., 2012; Puranik et al., 2016; Biswas et al., 2015; Yin and Feng, 2005; Bhavsar et al., 2006).

2.2.2.1.3 Transmission electron microscopy

Although scanning electron microscopy provides valuable information on the size, size distribution, and shape of nanomaterials, TEM provides greater detail at the atomic scale owing to the more powerful electron beams often used with TEM. This difference yields high-resolution imaging, which is useful for the determination of the crystal structure, morphological characteristics (Mohammadpour et al., 2012), and the granularity of the nanoparticulate sample (Hall et al., 2007).

2.2.2.1.4 Cryo-transmission electron microscopy

Cryo-transmission electron microscopy is now becoming a useful tool in the morphological understanding as well as three-dimensional tomography of polymeric nanoparticles particularly because they provide veritable structural information without requiring chemical fixation. Cryo preparation is devised as an accelerated conversion of a thin fluid suspension film into a solid specimen almost devoid of vapor pressure for TEM (Yaghmur et al., 2015; Kuntsche et al., 2011).

Since cryo-TEM can provide high-resolution images without the requirement for deleterious chemical fixation, it is the preferred method to study multiple structures where different morphology and size coexist. It must be remembered that cryo-imaging is not a quantitative methodology and should be a part of a much larger integrated quantitative analysis before any far reaching conclusions are drawn.

2.2.2.1.5 Atomic force microscopy

AFM is used to evaluate the local properties, such as height, friction, and magnetism using a probe (Chogale et al., 2016). AFM helps evaluate the surface roughness nanocomposites while allowing 3D imaging. With AFM, the surface morphology of nanostructures such as polymeric nanoparticles can be quantified (Kumari et al., 2016), which could possibly lead to statistical analysis.

2.2.2.2 Molecular weight

2.2.2.2.1 Gel chromatography

The principle of gel chromatography or size exclusion chromatography is based on separation of components based on their hydrodynamic volume. Although primarily a separation technique, gel chromatography is also used to characterize the molecular weight distribution of nanoparticulate formulations in combination with differential light scattering techniques (Robertson et al., 2016; Zhang et al., 2012).

2.2.2.3 Surface charge

2.2.2.3.1 Zeta potential measurement

The zeta potential is the potential at the electrical double layer surrounding a nanoparticle in solution. High surface charge bearing nanoparticles that also tend to have larger mean particle size are prone to higher risk of phagocytosis. Therefore, along with particle size measurement, the monitoring of mean zeta potential is paramount in the characterization of a nanoparticle batch (Zhu et al., 2016; Sahle et al., 2016; Othman et al., 2016; Honary and Zahir, 2013).

2.2.2.3.2 Laser Doppler anemometry

The concept of the Doppler shift is familiar to most people through the experience of variation in pitch of the sound of a vehicle as it moves close and then away. The shift in the frequency of sound is proportional to the speed of the source of the sound. Light behaves analogously. When light is incident on a moving object that it then reflects, the frequency of the scattered light is shifted by an amount proportional to the speed of the object. Therefore, theoretically, one could estimate the speed of said object by measuring the shift in frequency thus forming the basis of LDA. A flow is seeded with small particles that scatter light. Laser light of known frequency is used to illuminate the neutrally buoyant particles, which will scatter the light on to a photomultiplier tube (PMT). The PMT will be able to amplify the miniscule frequency difference to measurable levels (Greated, 1982).

Laser Doppler anemometry assists in the measurement of velocity of the nanoparticles during electrophoresis and calculates zeta potential employing Smoluchowski approximation (Chandra et al., 2014).

2.2.2.4 Hydrophobicity

The characterization of nanoparticle hydrophobicity involves a combination of techniques that measure surface tension, surface adsorption of probe molecules, and measurement of partitioning characteristics in a reference phase. Contact angle measurements are frequently used to estimate surface tension. Adsorption studies evaluate the relative affinity of the test nanoparticles to a standard hydrophobic material such as Rose Bengal. This technique is considered particularly sensitive as it exploits the high specific surface area of nanoparticles. Finally, the octanol–water partitioning coefficient is the most commonly employed partitioning method for the characterization of nanoparticles (Xiao and Wiesner, 2012).

2.2.2.5 Surface element analysis

2.2.2.5.1 X-ray photon spectroscopy

This method is used to reveal surface properties like elemental composition, chemical status, empirical formula, electronic state or binding modes of surface ligands, and depth analysis or atomic composition with depth. Using this technique, it is possible to estimate the kinetic energy and the number of electrons escaping from the surface layer of the nanoparticles. The limitations of this method include its requirement of an ultrahigh vacuum chamber and the possibility of characterization up to a depth of only 10 nm from the nanoparticle surface (Ghosh and Paria, 2012).

2.2.2.5.2 Nuclear magnetic resonance

Nuclear magnetic resonance (NMR) spectroscopy employs the magnetic properties of certain atomic nuclei to provide information on the physical and chemical properties of the atoms or molecules housing the nuclei. On the one hand, various imaging modalities reveal structural information at the long-range order (e.g., crystal structure). NMR is sensitive to the local environment and is amenable to amorphous, polymeric, and other biomolecules that lack long-range order (Lin et al., 2014). It has been used to determine the structure, purity, functionality, as well as conformational changes seen with interaction of ligands with nanoparticles (Zhang and Yan, 2009). Compared with other imaging techniques, NMR is a noninvasive technique that requires little to no sample preparation, but requires a comparatively large amount of sample to make a reliable measurement.

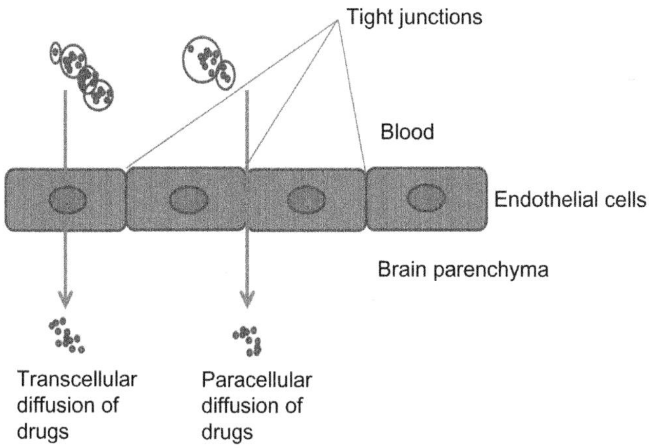

FIGURE 2.2

Cartoon showing the mechanistic transport of nanoparticles across blood–brain barrier (BBB).

2.2.3 NANOPARTICLES–BRAIN DELIVERY

The blood–brain barrier (BBB) acts a protective barrier to the potential entry of the drugs to the brain parenchyma. The endothelial cells with the abundant astrocytes produce tight junctions less permeable to the important drugs to elute into the brain. Potential breakdown strategies were implemented recently to override the tight junctions of the brain endothelial cells and make the protective barrier to open the window for the more restrictive drugs to the less restrictive one to facilitate the entry of the potential medicine for the treatment of the debilitating diseases of the brain (Fig. 2.2).

As seen above in the cartoon, the basic mechanistic evaluation of the transport of drugs across the brain endothelial cells and the blood–brain barrier. Furthermore there are three physical barriers to brain delivery: the blood–brain barrier, blood–tumor–brain barrier, and nose–brain barrier. Blood–brain barrier is ranked the worst of all three based on the passive diffusion or paracellular transport of the actives. Actives greater than 500 Da have restricted access to the entry into the brain and bliss to the drug delivery could be achieved by the interaction of small nanoparticles or targeted moieties on the surface of the nanostructures, which help in the transport of the actives across the brain. In case of oncological issues, the next line of defense is the brain–tumor barrier with lots of efflux transporters and small size pores with restricted access to the drug from the brain parenchyma to the brain tumor medium. With this in action, further bliss in disguise is the nose to brain delivery of the drug active to the brain using the olfactory pathway. The restricted amount of nasal fluid, protective mechanism by the cilia, and the biotransformation of the drug limits the utility of the pathway to potential easy access to the brain.

For the last two decades, much emphasis in research has been on fabrication of different types of nanoparticles. Initially nanoparticles were sterically stabilized by the attachment of polyethylene strands to have a prolonged circulation time and less clearance. This research further progressed by appending ligands and antibodies to the nanoparticles and enhancing the chances of increased payload to the brain cells by attaching to the receptors on the brain endothelial cells. The process of entry of the payload is attainable by designing a delivery system that can deliver the drug in a continuous and extended manner with passive diffusion as the main mechanism of uptake of the payload into the brain. The mechanism of uptake of the nanoparticles was mainly by passive diffusion and the cargo of drugs had to stay near the vicinity of the brain endothelial cells for longer or extended periods of time to enhance the uptake or diffusion of the drugs across the BBB to show their efficacy (Huile, 2016).

Tapan Parikh et al. experimented with the efficacy of the surface charge in delivering the drug sulpiride to the brain. The sulpiride was loaded into the PEGylated nanoparticles and the thiolated cationized bovine serum albumin was surface crowned as a targeting moiety to deliver the drug to the brain. Rhodamine B and Alexa Fluor 488 dye were loaded into nanoparticles and were used as markers for the nanoparticle uptake into the brain parenchyma. Sprague–Dawley rats were injected with the targeted nanoparticles, positive and negative controls, and the targeted nanoparticles showed a better efficacy (fluorescence imaging and concentrations delivered) than the controls. The targeted nanoparticles with surface embellishment with cationized albumin (positive charged zeta potential) could be a promising delivery for the schizophrenic drug, sulpiride (Parikh et al., 2009). Murali Mohan Bommana et al. reported the targeting ability of the immunopegylated nanoparticles to target the fluorescein isothiocyanate (FITC)–dextran to the ventral hippocampal region of the brain. The targeting moiety is OX-26 antibody, which was linked to the surface of the nanoparticles loaded with the marker. The concentrations were analyzed using a new technology married to the characterization of nanoparticles for brain targeting. In vivo brain microdialysis, a tool to characterize the concentration attained at the target site, further corroborated the fluorescence microscopic images of the distribution of the dye (Bommana et al., 2012).

2.3 FUTURE DIRECTIONS AND CONCLUSIONS

The term *nano* has merged with formulation science for pharmaceutical and biomedical applications, introducing the concept of personalized treatments, and care. As discussed in the first part of the chapter, nanocrystals increase the rate of dissolution or solubility and thereby enhance the bioavailability and depletion of food effect of the BCS class II or IV drugs successfully. Apparently, they are scaled up with great process engineering and have been FDA approved and marketed in the United States. On the flip side, the polymeric nanoparticle research is undergoing a brisk pace and a few products have been lined up for approval. In the future, nanomedicine will drive towards more creative, intensive, and personal care, and build up customized treatment with greater patient compliance.

REFERENCES

Beck-Broichsitter, M., Schweiger, C., Schmehl, T., Gessler, T., Seeger, W., Kissel, T., 2012. Characterization of novel spray-dried polymeric particles for controlled pulmonary drug delivery. J. Control. Release 158 (2), 329–335.

Bhavsar, M., Tiwari, S., Amiji, M., 2006. Formulation optimization for the nanoparticles-in-microsphere hybrid oral delivery system using factorial design. J. Control. Release 110, 422–430.

Bisht, S., Feldmann, G., Koorstra, J., Mullendore, M., Alvarez, H., Karikari, C., et al., 2008. In vivo characterization of a polymeric nanoparticle platform with potential oral drug delivery capabilities. Mol. Cancer Ther. 7, 3878–3888.

Biswas, S., Chattopadhyay, M., Sen, K., Saha, M., 2015. Development and characterization of alginate coated low molecular weight chitosan nanoparticles as new carriers for oral vaccine delivery in mice. Carbohydr. Polym. 121, 403–410.

Bommana, M.M., Kirthivasan, B., Squillante, E., 2012. In vivo brain microdialysis to evaluate FITC–dextran encapsulated immunopegylated nanoparticles. Drug Deliv. 19, 298–306.

Cetin, M., Aktas, M., Vural, I., Ozturk, M., 2011. Salmon calcitonin-loaded Eudragit and Eudragit-PLGA nanoparticles: in vitro and in vivo evaluation. J. Microencapsul. 29, 156–166.

Chandra, H.K., Prabha, S., Chandra, R., Ahmed, B., Nimesh, S., 2014. Advances in preparation and characterization of chitosan nanoparticles for therapeutics. Artif. Cell Nanomed. B 44, 305–314.

Chen, F., Zhang, Z., Yuan, F., Qin, X., Wang, M., Huang, Y., 2008. In vitro and in vivo study of N-trimethyl chitosan nanoparticles for oral protein delivery. Int. J. Pharm. 349, 226–233.

Chogale, M., Ghodake, V., Patravale, V., 2016. Performance parameters and characterizations of nanocrystals: a brief review. Pharmaceutics 8, 26.

Cui, F., Shi, K., Zhang, L., Tao, A., Kawashima, Y., 2006. Biodegradable nanoparticles loaded with insulin–phospholipid complex for oral delivery: preparation, in vitro characterization and in vivo evaluation. J. Control. Release 114, 242–250.

Deutel, B., Greindl, M., Thaurer, M., Bernkop-Schnürch, A., 2008. Novel insulin thiomer nanoparticles: in vivo evaluation of an oral drug delivery system. Biomacromolecules 9, 278–285.

Frank, K.J., Boeck, G., 2016. Development of a nanosuspension for iv administration: from miniscale screening to a freeze dried formulation. Eur. J. Pharm. Sci. 87, 112–117.

Ghosh, C.R., Paria, S., 2012. Core/shell nanoparticles: classes, properties, synthesis mechanisms, characterization, and applications. Chem. Rev. 112, 2373–2433.

Greated, C., 1982. Principles and practice of laser-Doppler anemometry: second edition. Opt. Laser Technol. 14, 325.

Hall, J., Dobrovolskaia, M., Patri, A., McNeil, S., 2007. Characterization of nanoparticles for therapeutics. Nanomedicine 2, 789–803.

Harsha, N., 2014. In vitro and in vivo evaluation of nanoparticles prepared by nano spray drying for stomach mucoadhesive drug delivery. Drying Technol. 33, 1199–1209.

Hecq, J., Deleers, M., Fanara, D., Vranckx, H., Amighi, K., 2005. Preparation and characterization of nanocrystals for solubility and dissolution rate enhancement of nifedipine. Int. J. Pharm. 299, 167–177.

Hoet, P.H., Brüske-Hohlfeld, I., Salata, O.V., 2004. Nanoparticles-known and unknown health risks. J. Nanobiotechnol. 2, 12.

Honary, S., Zahir, F., 2013. Effect of zeta potential on the properties of nano-drug delivery systems—a review (part 1). Trop. J. Pharm. Res. 12.

Huile, G., 2016. Progress and perspectives on targeting nanoparticles for brain drug delivery. Acta Pharm. Sinica B 6, 268–286.

Jawahar, N., Meyyanathan, S., 2012. Polymeric nanoparticles for drug delivery and targeting: a comprehensive review. Int. J. Health Allied Sci. 1, 217.

Junghanns, J.A.H., Muller, R.H., 2008. Nanocrystal technology, drug delivery and clinical applications. Int. J. Nanomed. 3, 295–309.

Keck, C.M., Müller, R.H., 2006. Drug nanocrystals of poorly soluble drugs produced by high pressure homogenisation. Eur. J. Pharm. Biopharm. 62, 3–16.

Kumar, G., Dhyana, A., Kothiyal, P., 2013. Review article on targeted polymeric nanoparticles. Am. J. Adv. Drug Deliv. 3, 196–215.

Kumari, C., Bhat, K., Bansal, R., 2016. Evaluation of surface roughness of different restorative composites after polishing using atomic force microscopy. J. Conserv. Dent. 19, 56.

Kuntsche, J., Horst, J., Bunjes, H., 2011. Cryogenic transmission electron microscopy (cryo-TEM) for studying the morphology of colloidal drug delivery systems. Int. J. Pharm. 417, 120–137.

Lim, J., Yeap, S., Che, H., Low, S., 2013. Characterization of magnetic nanoparticle by dynamic light scattering. Nanoscale Res. Lett. 8, 381.

Lin, P., Lin, S., Wang, P., Sridhar, R., 2014. Techniques for physicochemical characterization of nanomaterials. Biotechnol. Adv. 32, 711–726.

Liu, W., Pan, H., Zhang, C., Zhao, L., Zhao, R., Zhu, Y., et al., 2016. Developments in methods for measuring the intestinal absorption of nanoparticle-bound drugs. IJMS 17, 1171.

Mohammadpour, D., Eskandari, R., Avadi, M., Zolfagharian, H., Sadeghi, M., Rezayat, M., 2012. Preparation and in vitro characterization of chitosan nanoparticles containing *Mesobuthus eupeus* scorpion venom as an antigen delivery system. J. Venom. Anim. Toxins Incl. Trop. Dis. 18, 44–52.

Muller, R.H., Keck, C.M., 2004. Challenges and solutions for the delivery of biotech drugs—a review of drug nanocrystal technology and lipid nanoparticles. J. Biotechnol. 113, 151–170.

Othman, R., Vladisavljević, G., Shahmohamadi, H., Nagy, Z., Holdich, R., 2016. Formation of size-tuneable biodegradable polymeric nanoparticles by solvent displacement method using micro-engineered membranes fabricated by laser drilling and electroforming. Chem. Eng. J. 304, 703–713.

Padhye, S., Nagarsenker, M., 2013. Simvastatin solid lipid nanoparticles for oral delivery: formulation development and in vivo evaluation. Indian J. Pharm. Sci. 75, 591–598.

Pal, S., Jana, U., Manna, P., Mohanta, G., Manavalan, R., 2011. Nanoparticle: an overview of preparation and characterization. JAPS 01, 228–234.

Parikh, T., Bommana, M.M., Squillante, E., 2009. Efficacy of surface charge in targeting pegylated nanoparticles of sulpiride to the brain. Eur. J. Pharm. Biopharm. 74, 442–450.

Park, K., 2007. Nanotechnology: what it can do for drug delivery. J. Control. Release 120, 1–2.

Peltonen, L., Strachan, C., 2015. Understanding critical quality attributes for nanocrystals from preparation to delivery. Molecules 20, 22286–22300.

Puranik, A., Pao, L., White, V., Peppas, N., 2016. Synthesis and characterization of pH-responsive nanoscale hydrogels for oral delivery of hydrophobic therapeutics. Eur. J. Pharm. Biopharm. 108, 196–213.

Prajapati, H., Dalrymple, D.M., Serajuddin, A.T., 2013. In vitro dispersion test can serve as a predictive method for assessing performance of lipid-based drug delivery systems. J. Exc. Food Chem. 4, 111–125.

Prajapati, H.N., Serajuddin, A.T., 2016. U.S. Patent Application No. 15/042,282.

Rabinow, B.E., 2004. Nanosuspensions in drug delivery. Nat. Rev. Drug Discov. 3, 785–796.

Ravi, P., Aditya, N., Kathuria, H., Malekar, S., Vats, R., 2014. Lipid nanoparticles for oral delivery of raloxifene: optimization, stability, in vivo evaluation and uptake mechanism. Eur. J. Pharm. Biopharm. 87, 114–124.

Robertson, J., Rizzello, L., Olias, M., 2016. Purification of nanoparticles by size and shape. Sci. Rep. 6, 1–9.

Sahle, F., Balzus, B., Gerecke, C., Kleuser, B., Bodmeier, R., 2016. Formulation and in vitro evaluation of polymeric enteric nanoparticles as dermal carriers with pH-dependent targeting potential. Eur. J. Pharm. Sci. 92, 98–109.

Semete, B., Booysen, L., Lemmer, Y., Kalombo, L., Katata, L., Verschoor, J., et al., 2010. In vivo evaluation of the biodistribution and safety of PLGA nanoparticles as drug delivery systems. Nanomed. Nanotechnol. Biol. Med. 6, 662–671.

Sonaje, K., Lin, Y., Juang, J., Wey, S., Chen, C., Sung, H., 2009. In vivo evaluation of safety and efficacy of self-assembled nanoparticles for oral insulin delivery. Biomaterials 30, 2329–2339.

Tahara, K., Nishikawa, M., Matsui, K., Hisazumi, K., Onodera, R., Tozuka, Y., et al., 2016. In vitro and in vivo characterization of drug nanoparticles prepared using PureNano™ continuous crystallizer to improve the bioavailability of poorly water soluble drugs. Pharm. Res. 33, 2259–2268.

Venkatesh, D., Baskaran, M., Karri, V., Mannemala, S., Radhakrishna, K., Goti, S., 2015. Fabrication and in vivo evaluation of Nelfinavir loaded PLGA nanoparticles for enhancing oral bioavailability and therapeutic effect. Saudi Pharm. J. 23, 667–674.

Voorhees, P.W., 1985. The theory of Ostwald ripening. J. Stat. Phys. 38, 231–252.

Xiao, Y., Wiesner, M., 2012. Characterization of surface hydrophobicity of engineered nanoparticles. J. Hazard. Mater. 215-216, 146–151.

Yaghmur, A., Helvig, S., Azmi, I.D.M., Moghimi, S., 2015. Recent advances in cryo-TEM imaging of soft lipid nanoparticles. AIMS Biophys. 2, 116–130.

Yin, W.,K., Feng, S., 2005. Effects of particle size and surface coating on cellular uptake of polymeric nanoparticles for oral delivery of anticancer drugs. Biomaterials 26, 2713–2722.

Zhang, B., Yan, B., 2009. Analytical strategies for characterizing the surface chemistry of nanoparticles. Anal. Bioanal. Chem. 396, 973–982.

Zhang, J., Haas, R., Leone, A., 2012. Polydispersity characterization of lipid nanoparticles for siRNA delivery using multiple detection size-exclusion chromatography. Anal. Chem. 84, 6088–6096.

Zhu, X., Wu, J., Shan, W., Tao, W., Zhao, L., Lim, J., 2016. Polymeric nanoparticles amenable to simultaneous installation of exterior targeting and interior therapeutic proteins. Angew. Chem. Int. Ed. 55, 3309–3312.

CHAPTER

Polymeric micelles in biomedical science

3

Showkat Ahmad Bhawani[1], Fasihuddin Badruddin Ahmad[1], Mohamad Nasir Mohamad Ibrahim[2] and Amr A. Yakout[3]

[1]Department of Chemistry, Faculty of Resource Science and Technology, UNIMAS, Kuching, Malaysia [2]School of Chemical Sciences, Universiti Sains Malaysia, Pulau Pinang, Malaysia [3]Faculty of Science, Chemistry Department, Alexandria University, Alexandria, Egypt

CHAPTER OUTLINE

Nanoparticles in Pharmacotherapy. DOI: https://doi.org/10.1016/B978-0-12-816504-1.00005-3

3.1 INTRODUCTION

Polymeric micelles (PMs) are nanosized colloidal systems formed by the combination of amphiphilic block or graft copolymers in a solvent (Fig. 3.1). These polymer segments are attached together spontaneously in the form of self-assembly as a result of hydrophobic and ion pair interactions. As compared with their low-molecular weight counterparts (surfactant-based micelles), polymeric micelles can form self-assembly in the form of a well-defined core–shell structures at concentrations above their critical micelle concentration (CMC) and possess a high stability. Due to their unique physicochemical nature, polymeric micelles are extensively used in drug delivery systems. These are used to improve the efficiency of delivery of hydrophobic drugs. Over the last three decades nanomedicine has received a great attention both at the academic as well as industrial level to improve the delivery and stability at the target site. An increasing demand and attention for development of highly efficient as well as site specific drug delivery systems has led to the introduction of polymeric micelles as carriers. The size of polymeric micelles range between 10 and 100 nm smaller than liposomes. The size of the polymeric micelles can be controlled by various factors such as preparation method, molecular weight of block copolymer, aggregation number, and proportion of both hydrophilic and hydrophobic blocks. Polymeric micelles possess some interesting characteristics favorable for efficient delivery of drugs such as in vivo stable and selective delivery to the target site, nontoxicity, nonimmunogenicity, water solubility and lack of long-term target site accumulation.

Polymeric micelles are generally prepared from two types of polymers: block copolymer and graft copolymer. Block copolymers are linear and are made of a

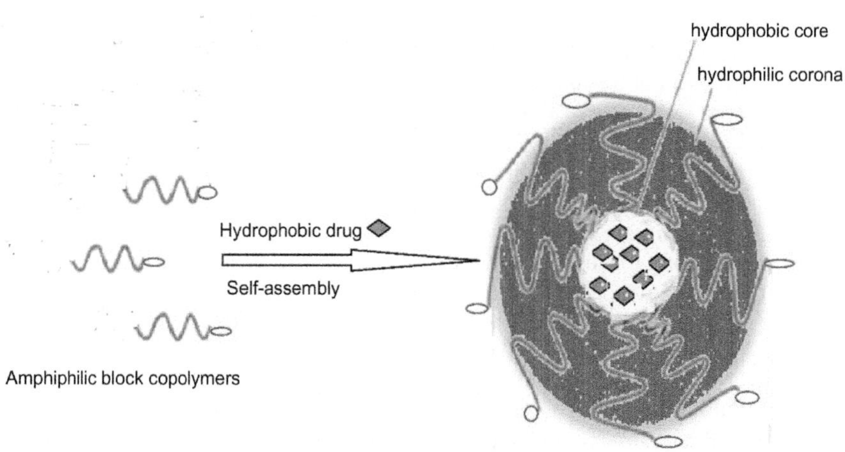

FIGURES 3.1

Polymeric micelle.

sequence of two polymer parts that differ in physicochemical properties such as charge and polarity. The block copolymers are categorized on the basis of their architecture, for example, AB-type diblock, ABA- or BAB triblock, and $(AB)_n$ multiblocks, where A is the hydrophilic block in the solvent and B is the hydrophobic block (Table 3.1). Generally copolymers such as AB or ABA assemble spontaneously into a number of supramolecular structures with different morphology in aqueous solution. The spherical micelles are generally formed from the copolymers of AB type if the length of hydrophilic block exceeds the length of hydrophobic block. The micellar system consists of two parts, that is, core and shell, where they core represents the hydrophobic block and the shell remains in contact with the aqueous system and represents a hydrophilic block. The micelles formed from the block copolymer of AB and ABA type are pharmacologically important. The hydrophilic block commonly used is polyethylene oxide chain because of several advantages such as highly soluble in water, low toxicity, micelles can circulate longer, and delivery of drugs more efficiently at the target sites. On the other hand graft copolymers have side chain blocks grafted with a main polymer chain. The grafted copolymers have some advantages over the block copolymers because they have multigrafted hydrophobic polymer chains along a hydrophilic polymer backbone. The CMC value also plays an important role in drug delivery by polymeric micelles. The lower the CMC values of polymeric micelles the higher will be the pharmacokinetic stability. The kinetically stable polymers dissociate or decay slowly into monomers after reaching the target site and remain for some time in the blood circulation. Conversely, the polymeric micelles with high CMC under extreme dilution in circulating blood can decay into monomers before reaching the target site. The conventional surfactant micelles in delivery of nanomedicines face several problems such as premature disintegration in system circulation (Talelli et al., 2012) due to the dilution in the bloodstream and by the interaction with the blood plasma proteins. This is the main reason that many micellar nanomedicines did not manage to display the improved retention in blood circulation, therapeutic efficacy, and accumulation at target site (Torchilin, 2004a,b; Kwon, 2003; Lu and Park, 2013). The main advantages of polymeric micelles as compared with conventional surfactant micelles are lower CMCs, solubilizing and stabilizing hydrophobic compounds, prolonged circulation times, increased accumulation at target site, and improved therapeutic efficacy (Danson et al., 2004; Hamaguchi et al., 2005; Xu et al., 2013).

3.2 PREPARATION OF POLYMERIC MICELLES

The basic principle of preparation of polymeric micelles is self-assembly of amphiphilic block copolymers in aqueous medium into various shapes such as spherical, cylindrical or worm-like micelles, polymer vesicles, and polymerosomes. The morphology of micelles depends on the hydrophilic volume fraction

Table 3.1 Copolymers for Micelle Formation

Sr. No.	Type of Micelle Forming Copolymers	Example of Polymers	Structure	Representation of Structure
1	Block copolymers	Poly(styrene)-*b*-poly (ethylene oxide)	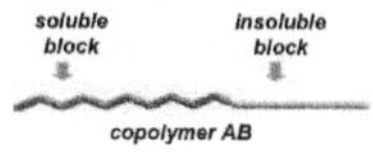	Diblock AAAAAAABBBBBB
		Poly(ethylene oxide)-*b*-poly(propylene oxide)-*b*-poly (ethylene oxide)	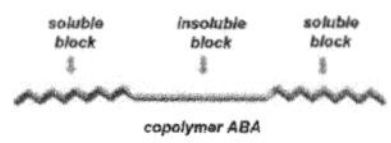	Triblock AAAABBBBBAAAA
2	Graft copolymers	*N*-phthaloylchitosan-*g*-polycaprolactone	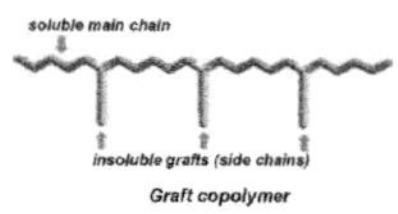	AAAAAAAAAAAAA B B B B B B

(vf) based on the hydrophilic–hydrophobic balance of block copolymers. The value of volume fraction can easily predict the shape of micelles and it has been reported that value of volume fraction more that 45% can generate spherical micelles and value of volume fraction nearly 35% can form polymer vesicles (Erhardt et al., 2001; Riess et al., 1985). Micellization is basically the result of two distinctive forces: one is the attractive force that leads to the assembly of molecules and the other is the repulsive force that prevents unlimited growth of the micelles (Astafieva et al., 1993; Price, 1983). Different amphiphilic copolymers have been investigated for producing micelles, such as *N*-phthaloylcarboxymethylchitosan (Peng and Zhang, 2007), poly(2-ethylhexyl acrylate)-*b*-poly (acrylic acid) (Kriz et al., 1999), poly(tert-butyl acrylate)-*b*-poly(2-vinylpyridine) (Prochazka et al., 1996), poly(ethylene oxide)-*b*-polycaprolactone (Geng and Discher, 2005), poly(ε-caprolactone)-*b*-poly(ethylene glycol)-*b*-poly(ε-caprolactone) (Hu et al., 2004; Xu et al., 2010), poly(ε-caprolactone)-*b*-poly(methacrylic acid) (Lee et al., 2005), poly(ethylene glycol)-*b*-poly(e-caprolactone-*co*-trimethylenecarbonate) (Danhier et al., 2009), poly(aspartic acid)-*b*-polylactide (Ouchi et al., 2004), poly(ethylene glycol)-block-poly(aspartate-hydrazide) (Alani et al., 2010), poly(*N*-isopropyl acrylamide-comethacryl acid)-*g*-poly(D,L-lactide) (Huang et al., 2007), and stearic acid–grafted chitosan oligosaccharide (Hu et al., 2009).

3.2.1 SYNTHETIC STRATEGIES FOR POLYMERIC MICELLES

Polymeric micelles can be generated and classified on the basis of intermolecular forces such as hydrophobic interactions, hydrogen bonding, electrostatic interactions, and van der Waals interactions. On the basis of these interactions polymeric micelles can be formed by conventional approach (hydrophobic interactions), polyion complex (metal complexation), and noncovalent approach (hydrogen bonding).

3.2.1.1 Conventional approach

This approach involves the hydrophobic interactions between the core segment and shell (corona) region in the aqueous medium. The simplest example is the formation of micelles by poly(ethylene oxide)-*b*-poly(propylene oxide)-*b*-poly (ethylene oxide) involving hydrophobic interactions (Bouchemal et al., 2009).

3.2.1.2 Polyion complex micelles

The concept for the formation of polyion complex micelles was first proposed by Harada and Kataoka (Harada and Kataoka, 1997). The electrostatic interactions are the main driving force for the assembly of block copolymers containing one ionic block, one nonionic block, and ionic polymers with opposite charge to that of copolymers. These oppositively charged polymers can penetrate the corona of the micelle and generate a polyionic micelle. The polyion complex micelles have some interesting features favorable for drug delivery such as high drug loading capacity and prolonged circulation in the blood. In addition to this they can be synthesized easily by self-assembly in the aqueous medium and have structural stability. The uniqueness of these micelles is because of the fact that corona of these micelles can entrap a variety of therapeutic agents such as both hydrophobic and hydrophilic compounds, metal complexes, and charged macromolecules. The main advantage of these micelles is in their preparation in aqueous medium, which protects the body from the side effects caused by the residual organic solvents during drug delivery. These micelles have been widely used in the delivery of genes (Nishiyama and Kataoka, 2006), DNA, enzymes, and charged drugs with antisense nucleotides (Ranger et al., 2001; Zhang and Ma, 2009).

3.2.1.3 Noncovalent approach

The main interaction in the formation of noncovalently connected polymeric micelles is due to the hydrogen bonding. These polymers are formed by the self-assembly in which core and shell are noncovalently connected at their homopolymer chain end by hydrogen bonding or metal–ligand interactions (Moughton and O'Reilly, 2008). For example, polymeric micelles prepared in a common solvent, chloroform due to the hydrogen bonding in which poly(4-vinylpyridine) acts as a backbone and carboxyl-terminated polybutadiene as the grafts (Wang et al., 2001).

3.3 PREPARATION OF DRUG-LOADED POLYMERIC MICELLES

The drugs can be loaded or incorporated into the micelles either by chemical conjugation or by physical entrapment techniques (Table 3.2). The chemical

Table 3.2 Incorporation Mode of Different Drugs Into Polymeric Micelles (Jones and Leroux 1999)

Drug	Polymer	Incorporation Mode	Size (nm)	References
Amphotericin B	PEO–PBLA	P	26	Yu et al. (1998)
Antisense oligonucleotide	PEO–P(Lys)	EA	50	Kataoka et al. (1996)
Cisplatin	PEO–P(Asp)	C	16	Yokoyama et al. (1996)
Cyclophosphamide	PEO–P(Lys)		NA	Bader et al. (1984)
Dequalinium	PEO–PE	P	15	Weissig et al. (1998a,b)
Doxorubicin	PEO–P(Asp)	C	50	Yokoyama et al. (1990)
	PEO–P(Asp)		14–31	Yokoyama et al. (1992)
	PEO–P(Asp)		17–24	Yokoyama et al. (1994)
	PEO–PBLA	P	30	Kwon et al. (1995)
	PEO–PDLLA		NA	Piskin et al. (1995)
	PEO–PBLA		37	Kwon et al. (1997)
	PEO–P(Asp)	P + C	NA	Yokoyama et al. (1998a,b)
	PNIPA–PBMA	P	Na	Chung et al. (1998)
	PAA–PMMA		NA	Inoue et al. (1998)
Gd–DTPA–PE ^{111}In–DTPA–SA	PEO–PE		20	Trubetskoy and Torchilin (1996)
Haloperidol	PEO–PPO–PEO		NA	Kabanov et al. (1989)
	PEO–PPO–PEO		15	Kabanov et al. (1992)
Indomethacin	PEO–PBLA		25–29	La et al. (1996)
	PEO–PCL		145–156	Kim et al. (1998)
	PEO–PCL		114–156	Shin et al. (1996)
Iodine derivative of benzoic acid	PEO–P(Lys)	C	80	Trubetskoy et al. (1997)
KRN-5500	PEO–PBLA	P	71	Yokoyama et al. (1998a,b)
	PEO–(C_{16},BLA)			
	PEO–P(Asp,BLA)			
Paclitaxel	PEO–PDLLA	P	NA	Zhang et al. (1996)
	LCC		<100	Miwa et al. (1998)
Plasmid DNA	PEO–P(Lys)	EA	140–150	Katayose and Kataoka (1998)
Soybean trypsin inhibitor	PEO–PE	P	15	Weissig et al. (1998a,b)
Testosterone	PEO–PDLLA		NA	Hagan et al. (1996)
Topoisomerase II inhibitor ellipticine	PEO–PE		NA	Trubetskoy and Torchilin (1995)

NA, *not available;* P, *physical entrapment;* C, *chemical bonding;* EA, *electrostatic association;* DTPA, *diethylenetriaminepenta-acetic acid;* PEO-P(Lys), *poly (ethylene oxide)–poly(L-lysine);* SA, *stearylamine;* LCC, N-*laurel-carboxymethyl-chitosan;* PE, *phosphatidyl ethanolamine;* PAA, *poly(acrylic acid).*

conjugation involves the formation of a covalent bond between the hydrophobic core and specific groups on the drug. These bonds are basically resistant to enzymatic cleavage due to the steric hindrance and need some special space groups to get hydrolyzed (Ulbrich et al., 1987). From this perspective the physical entrapment should be preferred as an incorporation approach. The physical entrapment methods involved in the drug loading are direct dissolution, solvent evaporation, or solution casting and dialysis.

3.3.1 DIRECT DISSOLUTION METHOD

This method is also called the organic solvent free method. In this method the direct dissolution of drug and amphiphilic copolymer takes place in water and then at or above CMC they self-assemble into a drug-loaded polymeric micelle in water (Fig. 3.2). It has been reported that simple incorporation of the drug into micelles in water may not achieve a good level of loading (Kwon et al., 1997, 1995). The high level of loading can be achieved by connecting this with increasing temperature or by preparing an evaporated thin film of drug before the addition of copolymer. For example direct dissolution of poly(3-caprolactone)-*b*-poly (2-aminoethyl methacrylate) [PCL-*b*-PAMA] or poly(3-caprolactone)-*b*-poly[2-(methacryloyl-oxy) ethyl phosphorylcholine] [PCL-*b*-PMC] block copolymers in water.

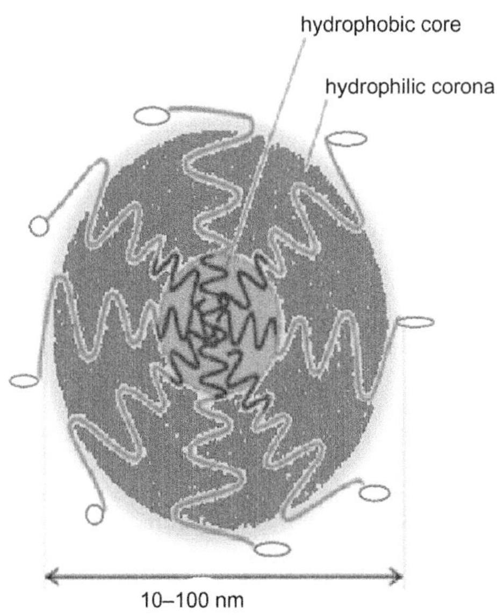

FIGURE 3.2

Drug-loaded polymeric micelle formed from self-assembly of amphiphilic block copolymers.

3.3.2 SOLVENT EVAPORATION

This method is also called the solution casting method. This technique is especially employed for the amphiphilic block copolymers that are insoluble in water. In this method a volatile organic solvent is used for the dissolution of block copolymer and drug. After the evaporation of solvent a thin film of drug and copolymer is obtained. The drug-loaded polymeric micelles are produced by the reconstitution of film with water.

3.3.3 DIALYSIS

This method is commonly used when the core forming blocks are long and hydrophobic. The micelles from such kind of copolymers have high potential to solubilize the large amounts of poorly water-soluble drugs (Mourya et al., 2011). In this technique the solutions of drug and polymer in organic solvent are placed in a dialysis bag and after this the solvent is exchanged with water by immersing the bag into water, inducing assembly (Gohy, 2005; Jette et al., 2004). The main disadvantage of this technique is the use of chlorinated solvents for emulsification and that it is time consuming.

3.3.4 LYOPHILIZATION METHOD

This method is very simple and cost-effective. In this method tert-butanol/water mixture is used to dissolve both the drug as well as polymer and then the solution is lyophilized. The lyophilized product is redispersed in a suitable vehicle to obtain the drug-loaded polymeric micelles (Fournier et al., 2004; La et al., 1996).

3.4 CHARACTERIZATION OF POLYMERIC MICELLES

The important parameters that are used for the characterization of polymeric micelles are CMC, and size, shape, and stability of polymeric micelles.

3.4.1 CRITICAL MICELLE CONCENTRATION

This is the minimum concentration at which polymers align or aggregate in the form of micelle. At this level of concentration the amphiphilic polymers self-assemble in aqueous solution in which hydrophobic chains aggregate in the form of core and hydrophilic chains align towards the aqueous environment. Below the CMC the polymeric chains are not able to self-assemble and may collapse. Therefore, CMC is the key parameter to generate stable micelles and will eventually provide efficient drug carriers that will not disintegrate while interacting with the blood components before reaching the target site. This physical quantity can be determined by various methods (Mittal, 1996; Hiemenz, 1986) such as surface tension measurement (Nakamura et al., 1976; Lieberman et al., 1996), gel permeation chromatography (Weissig et al., 1998a,b), light scattering (Turro and Chung, 1984), and by using fluorescent probes (Astafieva et al., 1993; Turro and Chung, 1984; Wilhelm et al., 1991). Among all the above-mentioned methods the fluorescent probe with pyrene is widely used for the determination of CMC (Kabanov et al., 1989).

3.4.2 SIZE AND SHAPE OF POLYMERIC MICELLES

The size of micelles plays an important role in drug delivery. The smaller size will minimize the risks of embolism in capillaries as compared with large drug carriers. The size of polymeric micelles depends on several factors such as molecular weight of copolymer, relative proportion of hydrophilic and hydrophobic chain, and aggregation number (Trubetskoy and Torchilin, 1996; Shin et al., 1998; Yokoyama et al., 1992). The size of polymeric micelles can be determined by various techniques such as scanning electron microscopy (SEM) (Kim et al., 1998), transmission electron microscopy (TEM) (Yu et al., 1998), and atomic force microscopy (AFM) (Cammas et al., 1997; Kohori et al., 1998). SEM is widely used for the determination of size and shape of colloidal carriers but needs gold coating. While the new imaging technique, atomic force microscopy, enables the determination of size of polymeric micelles without gold coating (Allémann et al., 1998).

The polydispersity index of prepared micelles can be examined by using quasielastic light scattering technique. The blue color observed from the monodisperse micelles on scattering of light indicates good micelle preparation as compared with white colored observed aggregates (Webber, 1998). The sphericity of polymeric micelles (Kataoka et al., 1996; Nagasaki et al., 1998) can be determined by using dynamic light scattering technique.

3.4.3 ELECTROPHORETIC MEASUREMENTS

Isoelectric focusing and two-dimensional polyacrylamide gel electrophoresis have been used for this study (Liu et al., 2008). Polymeric micelles were evaluated on the basis of their nature of surface coverage, which provides information about their clearance and distribution in vivo.

3.4.4 NUCLEAR MAGNETIC RESONANCE

The structure and composition of synthesized polymers used to produce the polymeric micelles can be determined by nuclear magnetic resonance (NMR) technique. The number average molecular weight of the polymers can also be determined by this technique (Zeng and Pitt, 2006).

3.4.5 DIFFERENTIAL SCANNING CALORIMETRY

The glass and melting point temperature and enthalpies of nanomicelles can be determined by differential scanning calorimetry (DSC), which provides information about the nature and speciation of the crystallinity within nanomicelles (Zeng and Pitt, 2006). The polymeric micelles are analyzed by this technique to study the various states of the core and the extent of interactions between drugs and polymers.

3.4.6 FLOW CYTOMETRY ANALYSIS

The binding capacity of the polymeric micelles with the receptors especially in ligand coupled polymeric micelles is determined by flow cytometry analysis (FCA) (Hayama et al., 2007). This technique is based on the counting and examining of microscopic particles when a beam of light (with single wavelength) is directed onto a hydrodynamically focused stream of fluid.

3.4.7 STABILITY OF POLYMERIC MICELLES

The stability of polymeric micelles is also an important component in drug delivery. The polymeric micelles on intravenous injection face different environmental changes such as dilution, exposure to pH and salt changes, and contact with various proteins and cells (Owen et al., 2012). Under these circumstances polymeric micelles should be capable of releasing drugs at the target site and preventing disintegration before reaching the target. The stability of polymeric micelles largely depends on the length of a hydrophobic segment (Domeselaar et al., 2003; Adams and Kwon, 2002) and the tendency to dissociate is related to the composition and cohesion of the hydrophobic core (Gaucher et al., 2005).

3.5 APPLICATION OF POLYMERIC MICELLES IN BIOMEDICAL SCIENCES

Polymeric micelles possess significant characteristics and provide promising applications in the field of biomedical science. This is because of their ability to solubilize hydrophobic drugs and enhance their bioavailability. The polymeric micelles can also reduce harmful side effects and toxicity of various drugs. They are able to enhance the permeability through biological barriers and changes in biodistribution. Polymeric micelles are most promising and effective nanocarrier systems for diagnostic applications and tumor chemotherapy (Alexis et al., 2008). Polymeric micelles have been proposed as delivery vehicles in cancer therapy for drugs such as paclitaxel (Zhang et al., 1996) and doxorubicin (Yokoyama et al., 1990) and for diagnostic agents, for example, complex of gadolinium with a chelating agent (diethylenetriaminepenta-acetic acid, Gd–DTPA) (Trubetskoy and

Torchilin, 1996). The polymeric micelles are also used to aid the diagnostic procedure for lymphatic pathways and for detecting the SLN (Hamoudeh et al., 2008). A few drugs with polymeric micelles as delivery vehicles under clinical trial are listed in Table 3.3.

The most important characteristic of polymeric micelles is their low CMC values that provide them high pharmacokinetic stability. An ideal system (Kabanov et al., 1992) of self-assembling drug delivery system must possess the following properties: (1) the size of aggregate must be around 10 nm, which will enable them to penetrate easily into various biological barriers; (2) should be stable and retain their integrity before reaching the target site; (3) should release the drug at the target site; and (4) the components of the carrier system should be easily flushed out from the body once the therapeutic work has been done. Polymeric micelles possess various advantages that are favorable for their use as drug carriers in pharmaceutical science. The main advantages are (1) high structural stability; (2) size of polymeric micelles in the range of 10–100 nm; (3) can

Table 3.3 Polymeric Micelles in Clinical Trials (Oerlemans et al., 2010)

Drug	Block Copolymer	Target	Diameter	Clinical Phase	References
SN-38	PEG–PGlu (SN-38)	Breast cancer	20 nm	II	Matsumura and Kataoka (2009)
Paclitaxel	PEG–P (aspartate)	Advanced stomach cancer	85 nm	II	Matsumura (2008a,b), Hamaguchi et al. (2007)
Doxorubicin	Pluronic L61 and F127	Adenocarcinoma of esophagus, gastroesophageal junction and stomach	22–27 nm	III	Sutton et al. (2007)
Cisplatin	PEG–PGlu (cisplatin)	Solid tumors	30 nm	III	Matsumura (2008a,b), Wilson et al. (2008)
Paclitaxel	PEG–P(D,L-lactide)	Breast cancer	20–50 nm	IV	Kim et al. (2004), Lee et al. (2008)
		Pancreatic cancer		II	Podoltsev et al. (2008), Saif et al. (2008)

PEG, *poly ethylene glycol;* PGlu, *polyglutamate;* SN-38, *7-ethyl-10-hydroxy-campothecin.*

encapsulate a large number of hydrophobic drugs; and (4) a large number of different chemical species can be incorporated into polymeric micelles.

3.5.1 PASSIVE DRUG TARGETING

The passive targeting of drug carriers is related to achievement of enhanced permeability and retention (EPR) effect and will also amplify the target/nontarget ratio of a delivered drug. This will also diminish the side reactions leading to harmful effects and its access will be limited to action on cancer tissue only. Polymeric micelles are able to achieve passive targeting by modifying them so they are target specific, and will be promising for use as drug carriers and in imaging in the treatment of cancer (Dhembre et al., 2011). Polymeric micelles are able to achieve passive targeting easily because of their small size and they can circulate for longer periods of time in the blood circulatory system upon administration (Vetvicka et al., 2009), which increases the chances that they can pass by the target site. The drug-loaded polymeric micelles exhibit a reduced distribution in nontargeted areas and accumulates to a larger extent at target sites as compared with a free drug (Kwon et al., 1994). The accumulation is due to the increased vascular permeability and impaired lymphatic drainage in malignant or inflamed tissues (Maeda et al., 1992; Jain, 1997). The perivascular accumulation of macromolecules and colloidal drug carriers in the tumor is because of large pores due to the leakier and less permselective tumor vessels (Yuan et al., 1994, 1995).

3.5.2 ACTIVE TARGETING

The aim of active targeting is to augment the drug delivery to the target employing biologically specific interactions such as antigen–antibody binding or locally applied signals such as sonication or heating. The active targeting decreases adverse side effects, because the specific interactions between the targeting component and antigens displayed on target tissue accumulates the drug only in the tumor sites and it allows the drug's cellular uptake through endocytosis (Bae et al., 2005). This can be achieved by modifying the carrier by means of ligand coupling or addition of a pH-sensitive moiety with regard to the biological characteristics of the tumor tissue. The cellular uptake can be increased by the attachment of a ligand to the polymeric micelles' surface via interaction between the ligand and its receptors, which are overexpressed at the cancerous cell surface. The tumor cells upon active targeting show unique characteristics, such as overexpression of cell surface tumor-associated antigens as well as of the tumor-specific antigens and relatively more acidic nature of tumor (pH 7.0) than normal tissue (pH 7.4) (Kedar et al., 2010). A few selected drugs for active targeting in cancer therapy are given in Table 3.4.

Table 3.4 Polymeric Micelles for Active Targeting for Anticancer Therapy (Oerlemans et al., 2010)

Drug	Block Copolymer	Active Ligand	Target	References
Paclitaxel	PEG-*b*-PE	mAb 2C5	Nucleosome-restricted specificity for different cancer cells	Torchilin et al. (2003a,b)
	mPEG-*b*-PCL	Folate	Folate receptor	Park et al. (2005)
	P105 and P105/L101	Folate	Folate receptor	Wang et al. (2007)
Doxorubicin	PEG-*b*-PG	mAb C225	EGF receptor	Vega et al. (2003)
	PEG-*b*-PCL	$\alpha_v\beta_3$ ligand (cRGDfK)	$\alpha_v\beta_3$ integrin	Nasongkla et al. (2004)
	PLGA-*b*-PEG	Folate	Folate receptor	Yoo and Park (2004)
9-Nitrocamptothecin	PEG3350-DSPE: mPEG2300-DSPE (1:100)	Folate	Folate receptor	Han et al. (2009)
17-Allylamino-17-demethoxy geldanamycin	DSPE-PEG3400-SPA	Vasoactive intestinal peptide	$VPAC_1$ receptors	Onyuksel et al. (2009)

PE, *phosphatidyl ethanolamine;* PG, *poly(L-Glu);* PCL, *poly(ε-caprolactone);* cRGDfK, *cyclic(Arg-Gly-Asp-D-Phe-Lys);* [m]PEG-DSPE, *[methoxy-]poly(ethylene glycol)-distearoylphosphatidylethanolamine;* P, *pluronic;* SPA, *succinimidyl propionate.*

3.5.3 POLYMERIC MICELLES FOR DRUG DELIVERY TO THE BRAIN

The drug delivery to the brain and for neurodegenerative diseases such as Parkinson's and Alzheimer's is very complicated due to the presence of the blood–brain barrier (BBB) because it limits the penetration of drugs. This obstacle can be easily penetrated by using polymeric micelles as drug carriers. This can be achieved by employing two different approaches. The first one is based on the modification of polymer micelles with antibodies or ligand molecules capable of transcytosis across brain microvessel endothelial cells, comprising the blood–brain barrier. The second approach is based on the use of Pluronic block copolymers to hinder the drug efflux systems, particularly Pgp, and selectively enhance the permeability of the blood–brain barrier to Pgp substrates (Dhembre et al., 2011). It has been reported that the poloxamer (Pluronic) micelles conjugated with antibodies may improve brain distribution of haloperidol, which is a neuroleptic agent (Kabanov et al., 1992).

3.5.4 POLYMERIC MICELLES FOR NUCLEAR IMAGING

The nuclear imaging modality is the most sensitive technique in which concentration of the isotope should be around 10^{-10} M at the site of interest (Torchilin, 2002). This technique is mainly based on the visualization of tiny amounts of γ-emitting isotopes, such as technetium-99m (^{99m}Tc), indium-111 (^{111}In), and iodine-125 (^{125}I). While fluorine-18 (^{18}F), copper-64 (^{64}Cu) and zirconium-89 (^{89}Zr) are used for single photon emission computed tomography (SPECT) imaging or positron-emitting isotopes for positron emission tomography (PET) imaging. The two main drawbacks for the specificity of nuclear imaging are the limited temporal resolution and the fact that it does not allow visualization of the surrounding anatomy (Nijsen et al., 2007).

The micelles loaded with gamma emitters such as ^{99m}Tc and ^{111}In, have been used and investigated in detail for noninvasive biodistribution studies (Lee et al., 2010; Hoang et al., 2009) because these are easily available and can stay for a longer period in in vivo imaging (Shokeen et al., 2008). The selective *N*-(*N*-(3-diphenylphosphinopropionyl)glycyl) cysteine linker is used for coupling of ^{99m}Tc (Visentin et al., 2004). While molecules like diethylenetriaminepenta-acetic acid (DTPA) or 1,4,7,10-tetraazacyclododecane-1,4,7,10-tetraacetic acid (DOTA), conjugated with polymers are used for the coupling of ^{111}In to micelles. The isotopes used for PET imaging with long half-lives, such as ^{64}Cu (T½ = 12.7 hours), ^{89}Zr (T½ = 78.4 hours), and gallium-67 (^{67}Ga, T½ = 3.2 days), can be directly coupled with chelators like DTPA and DOTA. The biodistribution results of ^{64}Cu-containing micelles in vivo with PET imaging showed enhanced blood circulation time and low accumulation in excretory organs (Pressly et al., 2007; Fukukawa et al., 2008).

3.5.5 POLYMERIC MICELLES FOR MAGNETIC RESONANCE IMAGING

The magnetic resonance imaging (MRI) imaging is used for the detection of changes in magnetization of hydrogen nuclei (^{1}H) in the presence of strong magnetic field under the application of radiofrequency pulses in the body. The paramagnetic elements, such as iron, manganese, gadolinium, and holmium, are usually used as contrast agents for ^{1}H MRI. These elements actually alter the magnetization of hydrogen nuclei, thereby increasing the contrast (Fossheim et al., 1996; Seevinck et al., 2007). The intensity of MRI signals can be increased by using complexing paramagnetic metals with polymeric micelles. Shiraishi et al. reported a twofold relative signal intensity increase in tumor-bearing mice and was maintained for 48 hours when gadolinium was complexed with prepared DOTA-grafted PEG-*b*-poly(L-lysine) micelles (Shiraishi et al., 2009). Encapsulation of hydrophobic iron oxide particles in different micellar systems has been investigated to increase circulation time for MRI (Talelli et al., 2009). Some of the amphiphilic chelating probes such as paramagnetic Gd–DTPA–PE and radioactive ^{111}In–DTPA–SA have been incorporated into PEO (5 kDa)–PE micelles and are used in vivo for MRI and scintigraphy techniques. The size (20 nm) of these micelles played an important role in diagnostic imaging applications (Rangel-Yagui et al., 2005). The mixed micelles produced from monoolein and taurocholate with Mn-mesoproporphyrin seem to have potential as an oral hepatobiliary contrast agent for MRI (Schmeidl et al., 1995).

Fluorine-19 (^{19}F)-containing contrast agents are new promising agents that can replace ^{1}H nuclei contrast agents in MRI. The main advantage of ^{19}F MRI is the absence of a background signal, due to the low concentration of ^{19}F in the body. At present, perfluorooctylbromide and perfluoropolyether are the most commonly used ^{19}F contrast agents (Caruthers et al., 2006; Soman et al., 2008) and are administered as emulsions due to their poor water solubility. The self-assembling fluorinated block copolymers have also been synthesized and showed promising imaging results in vitro (Peng et al., 2009; Du et al., 2008).

3.5.6 POLYMERIC MICELLES FOR COMPUTERIZED TUMOR IMAGING

The computerized tumor (CT) imaging discriminates structures in the body based on the differences in absorption of X-rays by the different tissues. The heavy elements such as iodine, bromine, and barium are used as contrast agents for CT imaging (Zielhuis et al., 2005). The amount of contrast agents required at the site of interest is approximately 10^{-2} M (Torchilin, 2002). The development of carrier devices that enhance the sensitivity of contrast agents is required for CT imaging. However the increasing concentration of contrast agents makes this imaging modality less suitable for molecular imaging. This drawback can be avoided by increasing the circulation time of contrast agents and/or by making them target oriented. The CT iodine containing micelles (mPEG-*b*-iodolysine, iodolysine is iodine-substituted poly-L-lysine) have been developed and were used as so-called blood pool agents (Torchilin et al., 1999). The long circulation intravascular contrast agents could benefit clinical applications like minimally invasive angiography and image guidance of minimally invasive procedures.

3.5.7 SMART POLYMER/STIMULI RESPONSIVE/STIMULI SENSITIVE ASSEMBLIES

The controlled delivery of drugs by polymeric micelles is also an important aspect. The controlled uptake and release of drugs by polymeric micelles must respond to the external stimuli (pH, oxidation/reduction, temperature, light, etc.). This external stimulus allows the polymeric micelles to uptake a drug selectively or release a drug from assembly (Du et al., 2009).

Thermoresponsive assemblies: Polymers with lower critical solution temperature (LCST) are able to form thermoresponsive assemblies. The LCST can be defined as the phase transition temperature above which polymers are hydrophobic and below which they are hydrophilic. The thermoresponsive assemblies undergo a structural change on increasing in temperature by virtue of which the deposition of drug and easier absorption of drug by the cells take place (Zhang et al., 2002). These assemblies can be fabricated by the self-assembly of a hydrophobic block with thermoresponsive block polymers. In such kind of assemblies the thermosensitivity is acquired by the outer shell that leads to hydrophilic character and the inner-core is hydrophobic loaded with water-insoluble drugs. For example, assembly of poly(*N*-isopropylacrylamide-*co*-*N*,*N*-dimethyl-acrylamide) and poly(D,L-lactide) (Akimoto et al., 2009).

pH-responsive/acid sensitive assemblies: These kind of assemblies undergo release or acceptance of protons under the influence of changes in environmental pH. These assemblies contain polymers with acidic or basic groups in their block copolymer at different pH (Du et al., 2009). The most important applications of pH-sensitive assemblies are in drug and gene delivery and as glucose sensors (Aguilar et al., 2007). For example, the assembly of folic acid-functionalized diblock copolymer containing 2-(methacryloy-loxy)-ethyl phosphorylcholine and either 2-(diisopropylamino) ethyl methacrylate (Licciardi et al., 2005).

Redox-responsive assemblies: They are considered as promising vehicles for the targeted intracellular delivery of hydrophobic drugs for cancer therapy (Sun et al., 2010). Therefore, the preparation of redox-responsive assemblies has attracted many researchers in recent years (Du et al., 2009). For example preparation of a disulfide linked dextran-*b*-poly(ε-caprolactone) diblock copolymer by using doxorubicin a category referred to as shell-sheddable micelles (Sun et al., 2010). In the presence of reducing agents the disulfide bond breaks down easily and in this way the shedding process takes place very fast; otherwise they will be stable for a longer period of time.

Light-responsive assemblies: These kind of assemblies respond to the light [ultraviolet (UV) or near infrared (NIR)]. Therefore, no extra chemical additives are used to trigger the release of drugs (Zhanga et al., 2009). The UV-light is used only for the release of drugs from topical preparations of polymeric micelles because it will not penetrate deeply into the body; instead it will get absorbed. While NIR can penetrate deeply into the body since lipids, water, and hemoglobin have low absorption in NIR region (Oerlemans et al., 2010). A chromophore such as liquid crystalline azobenzene is introduced into the hydrophobic block of these assemblies. This chromophore will respond to the light and results in the conformational or structural changes. They have potential application in the intracellular delivery of doxorubicin (Liu et al., 2012). For example, preparation of an assembly from dextran-graft-(2-diazo-1,2-naphthoquin-one).

3.5.8 POLYMERIC IMMUNOMICELLES

The surface of the micelle is attached with the specific antibodies also referred to as immunomicelles, providing many opportunities with respect to diversity of targets and specificity of interactions in drug delivery. Many researchers have grabbed this opportunity to produce covalently linked immunomicelles (Mourya et al., 2011). The antibodies attached to the micelles retain their ability to interact specifically with their antigens. The special character of immunomicelles shows in their behavior in such a way that they can recognize surface of various tumor cells but not the normal cells because these antibodies can bind a broad variety of cancer cells. In this manner they act as specific ligands for delivery and carrying of drugs into tumor cells (Iakaubov and Torchillin, 1997; Torchillin et al., 2003). The efficiency of immunomicelles in targeting of cancer cells was revealed by solubilizing paclitaxel and camptothecin in mixed micelles of polyethylene glycol-phosphatidyl ethanolamine and vitamin K and tailored with antinucleosome monoclonal antibody 2C5 (mAb 2C5) for transport to reach tumor cells in vitro (Torchilin, 2007). This assembly has shown a significantly higher cytotoxicity against various cancer cells in vitro. In in vivo experiments, the intravenous administration of tumor-specific mAb 2C5 immunomicelles loaded with paclitaxel into mice with murine Lewis lung carcinoma has shown increased accumulation of paclitaxel and enhanced inhibition of tumor growth (Torchilin, 2004a,b). Immunomicelles are characterized by dynamic light scattering and free-surface electron microscopy on the basis of their micellar size distribution before and after the attachment of various antibodies (Gao et al., 2002; Hobbs et al., 1998). Fluorescence microscopy is used for the measurement of binding of immunomicelles to the cancerous cells in vitro (Torchilin, 2005).

3.6 CONCLUSION

Over the last few decades a plethora of formulation techniques have been developed for the carrying of drugs to specific sites. The past few decades have also witnessed an explosive development of polymeric micelles for targeted and controlled drug delivery. Polymeric micelles (PM) have shown great potential as carrier of drugs and have enhanced the solubility and the therapeutic potential of hydrophobic drugs. PMs can be used to target tumor sites by passive as well as active mechanisms. Some of the interesting inherent properties of PMs, including size, stability in plasma, longevity in vivo, and pathological characteristics of tumors allow PMs to target the tumor site by a passive mechanism called the EPR effect. The separate functionality at the outer shell is useful for targeting of anticancer drugs to the tumor by active mechanisms. PMs can be conjugated with many ligands such as antibody fragments, epidermal growth factors, α_2-glycoprotein, transferrin, and folate as carrier to target the tumor cells. The advances in polymer chemistry, biomacromolecular engineering, and nanotechnology have developed block copolymers that can produce more stable core–shell structures to further stabilize PMs. The core cross-linked polymeric micelles (CCPMs) are able to overcome several problems associated with conventional micellar formulations. The CCPMs are characterized by improving delivery of drugs by prolonged circulation times, by enhanced accumulation in tumors and at sites of inflammation via the EPR effect, and by increased therapeutic efficacy. The CCPMs have also been developed in such a way that they can entrap drugs covalently within the micellar core, to avoid the premature release in systemic circulation, and will allow sustained and tailorable release kinetics upon accumulation at the pathological site. With advancement in the field of nanotechnology polymeric micelles have emerged as important pharmaceutical carriers as well as a promising class for diagnostic nuclear medicines. The polymeric micelles can easily attach radioisotopes as well as paramagnetic metals to the hydrophilic portions or can noncovalently incorporate them into the hydrophobic core. Some of the approaches involving the application of polymeric micelles as carriers of radiopharmaceuticals, radioisotopes, and contrast agents, for diagnostic/imaging use in nuclear medicine employing PEO and PE polymers labeled with ^{111}In or ^{99m}Tc, have preferably shown promising results especially for SLN imaging, visualization of lymphatic pathways, and diagnosis of some types of tumors, such as breast and prostate cancers.

REFERENCES

Adams, M.L., Kwon, G.S., 2002. The effects of acyl chain length on the micelle properties of poly(ethylene oxide)-block-poly(*N*-hexyl-L-aspartamide)-acyl conjugates. J. Biomater. Sci. Polym. Ed. 13, 991–1006.

Aguilar, M.R., Elvira, C., Gallardo, A., Vázquez, B., Román, J.S., 2007. Smart polymers and their applications as biomaterials. Top. Tissue Eng. 3, 1–27.

Akimoto, J., Nakayama, M., Sakai, K., Okano, T., 2009. Temperature-induced intracellular uptake of thermo responsive polymeric micelles. Biomacromolecules 10, 1331–1336.

Alani, A.G., Bae, Y., Rao, D.A., Kwon, G.S., 2010. Polymeric micelles for the pH-dependent controlled, continuous low dose release of paclitaxel. Biomaterials 31, 1765–1772.

Alexis, F., Rhee, J.W., Richie, J.P., Radovic-Moreno, A.F., Langer, R., Farokhzad, O.C., 2008. New frontiers in nanotechnology for cancer treatment. Urol. Oncol. Sem. Orig. Invest 26, 74–85.

Allémann, E., Leroux, J.C., Gurny, R., 1998. Biodegradable nanoparticles of poly(lactic acid) and poly(lactic-*co*-glycolic acid) for parenteral administration. In: Lieberman, H., Rieger, M., Banker, G. (Eds.), Pharmaceutical Dosage Forms: Disperse Systems, vol. 3. Marcel Dekker, New York, pp. 163–193.

Astafieva, I., Zhong, X., Eisenberg, F.A., 1993. Critical micellization phenomena in block polyelectrolyte solutions. Macromolecules 26, 7339–7352.

Bader, H., Ringsdorf, H., Schmidt, B., 1984. Water soluble polymers in medicine. Angew. Makromol. Chem. 123/124, 457–485.

Bae, Y., Jang, W.D., Nishiyama, N., Fukushima, S., Kataoka, K., 2005. Multifunctional polymeric micelles with folate mediated cancer cell targeting and pH-triggered drug releasing properties for active intracellular drug delivery. Mol. Biosyst. 1, 241–250.

Bouchemal, K., Agnely, F., Koffi, A., Ponchel, G., 2009. A concise analysis of the effect of temperature and propanediol-1,2 on Pluronic F127 micellization using isothermal titration microcalorimetry. J. Colloid Interface Sci. 338, 169–176.

Cammas, S., Suzuki, K., Sone, C., Sakurai, Y., Kataoka, K., Okano, T., 1997. Thermo-responsive polymer nanoparticles with a core–shell micelle structure as site-specic drug carriers. J. Control. Release 48, 157–164.
Caruthers, S.D., Neubauer, A.M., Hockett, F.D., Lamerichs, R., Winter, P.M., Scott, M.J., et al., 2006. In vitro demonstration using 19F magnetic resonance to augment molecular imaging with paramagnetic perfluorocarbon nanoparticles at 1.5 Tesla. Invest Radiol. 41, 305–312.
Chung, J.E., Yamato, M., Yokoyama, M., Aoyagi, T., Sakurai, Y., Okano, T., 1998. Thermo-responsive drug delivery of polymeric micelles incorporating adriamycin. Proc. Int. Symp. Control. Release Bioact. Mater. 25, 380–381.
Danhier, F., Magotteaux, N., Ucakar, B., Lecouturier, N., Brewster, M., Préat, V., 2009. Novel self-assembling PEG-p-(CL-*co*-TMC) polymeric micelles as safe and effective delivery system for paclitaxel. Eur. J. Pharm. Biopharm. 73, 230–238.
Danson, S., Ferry, D., Alakhov, V., Margison, J., Kerr, D., Jowle, D., et al., 2004. Phase I dose escalation and pharmacokinetic study of Pluronic polymer–bound doxorubicin (SP1049C) in patients with advanced cancer. Br. J. Cancer 90, 2085–2091.
Domeselaar, G.H.V., Kwon, G.S., Andrew, L.C., Wishart, D.S., 2003. Application of solid phase peptide synthesis to engineering PEO–peptide block copolymers for drug delivery. Colloids Surf B: Biointerfaces 30, 323–334.
Du, W., Nystrom, A.M., Zhang, L., Powell, K.T., Li, Y., Cheng, C., et al., 2008. Amphiphilic hyperbranched fluoropolymers as nanoscopic 19F magnetic resonance imaging agent assemblies. Biomacromolecules 9, 2826–2833.
Du, J., Rachel, K., Reilly, O., 2009. Advances and challenges in smart and functional polymer vesicles. Soft Matter. 5, 3544–3561.
Dhembre, G.N., Moon, R.S., Kshirsagar, R.V., 2011. A review on polymeric micellar nanocarriers. IJPBS 2, 109–116.
Erhardt, R., Böker, A., Zettl, H., Kaya, H., Pyckhout-Hintzen, W., Krausch, G., et al., 2001. Janus micelles. Macromolecules 34, 1069–1075.
Fossheim, S., Johansson, C., Fahlvik, A.K., Grace, D., Klaveness, J., 1996. Lanthanide-based susceptibility contrast agents: assessment of the magnetic properties. Magn. Reson. Med. 35, 201–206.
Fournier, E., Dufresne, M.H., Smith, D.C., Ranger, M., Leroux, J.C., 2004. A novel one-step drug-loading procedure for watersoluble amphiphilic nanocarriers. Pharm. Res. 21, 962–968.
Fukukawa, K., Rossin, R., Hagooly, A., Pressly, E.D., Hunt, J.N., Messmore, B.W., et al., 2008. Synthesis and characterization of core–shell star copolymers for in vivo PET imaging applications. Biomacromolecules 9, 1329–1339.
Gao, Z., Lukyanov, A.N., Singhal, A., Torchillin, V.P., 2002. Diacyl-polymer micelles as nanocarriers for poorly water soluble anticancer drugs. Nano Lett. 2, 979–982.
Gaucher, G., Dufresne, M.H., Sant, V.P., Kang, N., Maysinger, D., Leroux, J.C., 2005. Block copolymer micelles: preparation, characterization and application in drug delivery. J. Control. Release 109, 169–188.
Geng, Y., Discher, D.E., 2005. Hydrolytic degradation of poly(ethylene oxide)-block-polycaprolactone worm micelles. J. Am. Chem. Soc. 127, 12780–12781.
Gohy, J.F., 2005. Block copolymer micelles. Adv. Polym. Sci. 190, 65–136.
Hagan, S.A., Coombes, G.A., Garnett, M.C., Dunn, S.E., Davies, M.C., Illum, L., et al., 1996. Polylactide–poly(ethylene glycol) copolymers as drug delivery systems, 1.

Characterization of water dispersible micelle-forming systems. Langmuir 12, 2153–2161.

Hamaguchi, T., Matsumura, Y., Suzuki, M., Shimizu, K., Goda, R., Nakamura, I., et al., 2005. NK105, a paclitaxel-incorporating micellar nanoparticle formulation, can extend in vivo antitumour activity and reduce the neurotoxicity of paclitaxel. Br. J. Cancer 92, 1240–1246.

Hamaguchi, T., Kato, K., Yasui, H., Morizane, C., Ikeda, M., Ueno, H., et al., 2007. A phase I and pharmacokinetic study of NK105, a paclitaxel-incorporating micellar nanoparticle formulation. Br. J. Cancer 97, 170–176.

Hamoudeh, M., Kamleh, M.A., Diab, R., Fessi, H., 2008. Radionuclides delivery systems for nuclear imaging and radiotherapy of cancer. Adv. Drug Deliv. Syst. 60, 1329–1346.

Han, X., Liu, J., Liu, M., Xie, C., Zhan, C., Gu, B., et al., 2009. 9-NC-loaded folate-conjugated polymer micelles as tumor targeted drug delivery system: preparation and evaluation in vitro. Int. J. Pharm. 372, 125–131.

Harada, A., Kataoka, K., 1997. Formation of stable and mono dispersive polyion complex micelles in aqueous medium from poly(L-lysine) and poly(ethylene glycol)–poly(aspartic acid) block copolymer. J. Macromol. Sci. Pure Appl. Chem. A 34, 2119–2133.

Hayama, A., Yamamoto, T., Yokoyama, M., Kawano, K., Hattori, Y., Maitani, Y., 2007. Polymeric micelles modified by folate–PEG–lipid for targeted drug delivery to cancer cells in vitro. J. Nanosci. Nanotechnol. 8, 1–6.

Hiemenz, P.C., 1986. Principles of Colloid and Surface Chemistry. Marcel Dekker, New York.

Hoang, B., Lee, H., Reilly, R.M., Allen, C., 2009. Noninvasive monitoring of the fate of ^{111}In-labeled block copolymer micelles by high resolution and high sensitivity microSPECT/CT imaging. Mol. Pharm. 6, 581–592.

Hobbs, S.K., Monsky, W.L., Yuan, F., Roberts, W.G., Griffith, L., Torchillin, V.P., et al., 1998. Regulation of transport pathways in tumor vessels: role of tumor type and microenvironment. Proc. Natl. Acad. Sci. U.S.A. 95, 4607–4612.

Hu, Y., Zhang, L., Cao, Y., Ge, H., Jiang, X., Yang, C., 2004. Degradation behavior of poly(ε-caprolactone)-*b*-poly(ethylene glycol)-*b*-poly(ε-caprolactone) micelles in aqueous solution. Biomacromolecules 5, 1756–1762.

Hu, F.Q., Liu, L.N., Du, Y.Z., Yuan, H., 2009. Synthesis and antitumor activity of doxorubicin conjugated stearic acid-g-chitosan oligosaccharide polymeric micelles. Biomaterials 30, 6955–6963.

Huang, C.K., Lo, C.L., Chen, H.H., Hsiue, G.H., 2007. Multifunctional micelles for cancer cell targeting, distribution imaging, and anticancer drug delivery. Adv. Funct. Mater. 17, 2291–2297.

Iakaubov, L.Z., Torchillin, V.P., 1997. A novel class of a anti-tumor antibodies: nucleosome restricted antinuclear autoantibodies (ANA) from healthy aged nonauto-immune mice. Oncol. Res. 9, 439–446.

Inoue, T., Chen, G., Nakamae, K., Hoffman, A.S., 1998. An AB block copolymer of oligo (methyl methacrylate) and poly(acrylic acid) for micellar delivery of hydrophobic drugs. J. Control. Release 51, 221–229.

Jain, R.K., 1997. Delivery of molecular and cellular medicine to solid tumors. Adv. Drug Deliv. Rev. 26, 71–90.

Jette, K.K., Law, D., Schmitt, E.A., Kwon, G.S., 2004. Preparation and drug loading of poly(ethylene glycol)-block-poly(ε-caprolactone) micelles through the evaporation of a cosolvent azeotrope. Pharm. Res. 21, 1184–1191.

Jones, M.C., Leroux, J.C., 1999. Polymeric micelles a new generation of colloidal drug carriers. Eur. J. Pharm. Biopharm. 48, 101–111.

Kataoka, K., Togawa, H., Harada, A., Yasugi, K., Matsumoto, T., Katayose, S., 1996. Spontaneous formation of polyion complex micelles with narrow distribution from antisense oligonucleotide and cationic block copolymer in physiological saline. Macromolecules 29, 8556–8557.

Kabanov, A.V., Chekhonin, V.P., Alakhov, V.Y., Batrakova, E.V., Lebedev, A.S., Melik-Nubarov, N.S., et al., 1989. The neuroleptic activity of haloperidol increases after its solubilization in surfactant micelles. FEBS Lett. 258, 343–345.

Kabanov, A.V., Batrakova, E.V., Melik-Nubarov, N.S., Fedoseev, N.A., Dorodnich, T.Y. U., Alakhov, V.Y.U., et al., 1992. A new class of drug carriers; micelle poly(oxyethylene)–poly(oxypropylene) block copolymers as microcontainers for drug targeting from blood to brain. J. Control. Release 22, 141–158.

Katayose, S., Kataoka, K., 1998. Remarkable increase in nuclease resistance of plasmid DNA through supramolecular assembly with poly(ethylene glycol)–poly(L-lysine) block copolymer. J. Pharm. Sci. 87, 160–163.

Kedar, U., Phutane, P., Shidhaye, S., Kadam, V., 2010. Advances in polymeric micelles for drug delivery and tumor targeting. Nanomedicine: NBM 6, 714–729.

Kim, S.Y., Shin, I.G., Lee, Y.M., Cho, C.S., Sung, Y.K., 1998. Methoxy poly(ethylene glycol) and e-caprolactone amphiphilic block copolymeric micelle containing indomethacin, II. Micelle formation and drug release behaviors. J. Control. Release 51, 13–22.

Kim, T.Y., Kim, D.W., Chung, J.Y., Shin, S.G., Kim, S.C., Heo, D.S., et al., 2004. Phase I and pharmacokinetic study of Genexol-PM, a cremophor-free, polymeric micelle-formulated paclitaxel, in patients with advanced malignancies. Clin. Cancer Res. 10, 3708–3716.

Kohori, F., Sakai, K., Aoyagi, T., Yokoyama, M., Sakurai, Y., Okano, T., 1998. Preparation and characterization of thermally responsive block copolymer micelles comprising poly(*N*-isopropylacrylamide-*b*-DL-lactide). J. Control. Release 55, 87–98.

Kriz, J., Pletil, J., Tuzar, Z., 1999. Interface affected polymer dynamics: NMR, SANS, and DLS study of the influence of shell-core interactions on the core chain mobility of poly (2-ethylhexyl acrylate)-block-poly(acrylic acid) micelles in water. Macromolecules 32, 397–410.

Kwon, G.S., 2003. Polymeric micelles for delivery of poorly water-soluble compounds. Crit. Rev. Ther. Drug Carrier Syst. 20, 357–403.

Kwon, G.S., Suwa, S., Yokoyama, M., Okano, T., Sakurai, Y., Kataoka, K., 1994. Enhanced tumor accumulation and prolonged circulation times of micelle-forming poly (ethylene oxide-aspartate) block copolymer–adriamycin conjugates. J. Control. Release 29, 17–23.

Kwon, G.S., Naito, M., Yokoyama, M., Okano, T., Sakurai, Y., Kataoka, K., 1995. Physical entrapment of adriamycin in AB block copolymer micelles. Pharm. Res. 12, 192–195.

Kwon, G.S., Naito, M., Yokoyama, M., Okano, T., Sakurai, Y., Kataoka, K., 1997. Block copolymer micelles for drug delivery: loading and release of doxorubicin. J. Control. Release 48, 195–201.

La, S.B., Okano, T., Kataoka, K., 1996. Preparation and characterization of the micelle-forming polymeric drug indomethacin-incorporated poly(ethylene oxide)–poly(b-benzyl L-aspartate) block copolymer micelles. J. Pharm. Sci. 85, 85–90.

Lee, S.C., Kim, K.J., Jeong, Y.K., Chang, J.H., Choi, J., 2005. pH induced reversible complexation of poly(ethylene glycol) and poly(ε-caprolactone)-*b*-poly(methacrylic acid) copolymer micelles. Macromolecules 38, 9291–9297.

Lee, K.S., Chung, H.C., Im, S.A., Park, Y.H., Kim, C.S., Kim, S.B., et al., 2008. Multicenter phase II trial of Genexol-PM, a Cremophor-free, polymeric micelle formulation of paclitaxel, in patients with metastatic breast cancer. Breast Cancer Res. Treat. 108, 241–250.

Lee, H., Hoang, B., Fonge, H., Reilly, R.M., Allen, C., 2010. In vivo distribution of polymeric nanoparticles at the whole-body, tumor, and cellular levels. Pharm. Res. 11, 2343–2355.

Licciardi, M., Tang, Y., Billingham, N.C.S.P., 2005. Synthesis of novel folic acidfunctionalized biocompatible block copolymers by atom transfer radical polymerization for gene delivery and encapsulation of hydrophobic drugs. Biomacromolecules 6, 1085–1096.

Lieberman, H.A., Rieger, M.M., Banker, G.S., 1996. Pharmaceutical Dosage Forms: Disperse Systems, vol. 1. Marcel Dekker Inc, New York.

Liu, J., Yang, S., Lee, C.S., DeVoe, D.L., 2008. Polyacrylamide gel plugs enabling 2-D microfluidic protein separations via isoelectric focusing and multiplexed sodium dodecyl sulfate gel electrophoresis. Electrophoresis 29, 2241–2250.

Liu, G.Y., Chen, C.J., Li, D.D., Wanzg, S.S., Ji, J., 2012. Near-infrared light-sensitive micelles for enhanced intracellular drug delivery. J. Mater. Chem. 22, 16865–16871.

Lu, Y., Park, K., 2013. Polymeric micelles and alternative nanonized delivery vehicles for poorly soluble drugs. Int. J. Pharm. 453, 198–214.

Maeda, H., Seymour, L.W., Miyamoto, Y., 1992. Conjugates of anticancer agents and polymers: advantages of macromolecular therapeutics in vivo. Bioconjug. Chem. 3, 351–362.

Matsumura, Y., 2008a. Poly(amino acid) micelle nanocarriers in preclinical and clinical studies. Adv. Drug Deliv. Rev. 60, 899–914.

Matsumura, Y., 2008b. Polymeric micellar delivery systems in oncology. Jpn. J. Clin. Oncol. 38, 793–802.

Matsumura, Y., Kataoka, K., 2009. Preclinical and clinical studies of anticancer agent-incorporating polymer micelles. Cancer Sci. 100, 572–579.

Mittal, K.L., 1996. Micellization, Solubilization, and Microemulsions. Plenum Press, New York.

Miwa, A., Ishibe, A., Nakano, M., Yamahira, T., Itai, S., Jinno, S., et al., 1998. Development of novel chitosan derivatives as micellar carriers of taxol. Pharm. Res. 15, 1844–1850.

Moughton, A.O., O'Reilly, R.K., 2008. Noncovalently connected micelles, nanoparticles, and metal-functionalized nanocages using supramolecular self-assembly. J. Am. Chem. Soc. 130, 8714–8725.

Mourya, V.K., Inamdar, N., Nawale, R.B., Kulthe, S.S., 2011. Polymeric micelles: general considerations and their applications. Ind. J. Pharm. Educ. Res. 45, 128–138.

Nagasaki, Y., Okada, T., Scholz, C., Iijima, M., Kato, M., Kataoka, K., 1998. The reactive polymeric micelle based on an aldehyde-ended poly(-ethylene glycol)/poly(lactide) block copolymer. Macromolecules 31, 1473–1479.

Nakamura, N., Ryuichi, E., Takeda, M., 1976. Surface properties of styrene ethylene oxide block copolymers. J. Polym. Sci. 14, 1287–1295.

Nasongkla, N., Shuai, X., Ai, H., Weinberg, B.D., Pink, J., Boothman, D.A., et al., 2004. cRGD-functionalized polymer micelles for targeted doxorubicin delivery. Angew. Chem. Int. Ed. Engl. 43, 6323–6327.

Nijsen, J.F., Krijger, G.C., Van Het Schip, A.D., 2007. The bright future of radionuclides for cancer therapy. Anticancer Agents Med. Chem. 7, 271–290.

Nishiyama, N., Kataoka, K., 2006. Current state, achievements and future prospects of polymeric micelles as nanocarriers for drug and gene delivery. Pharmacol. Ther. 112, 630–648.

Oerlemans, C., Bult, W., Bos, M., Storm, G., Nijsen, J.F., Hennink, W.E., 2010. Polymeric micelles in anticancer therapy: targeting, imaging and triggered release. Pharm. Res. 27, 2569–2589.

Onyuksel, H., Mohanty, P.S., Rubinstein, I., 2009. VIP-grafted sterically stabilized phospholipid nanomicellar 17-allylamino-17-demethoxy geldanamycin: a novel targeted nanomedicine for breast cancer. Int. J. Pharm. 365, 157–161.

Ouchi, T., Miyazaki, H., Arimura, H., Tasaka, F., Hamada, A., Ohya, Y., 2004. The formation of biodegradable polymeric micelles from newly synthesized poly(aspartic acid)-block-polylactide AB-type diblock copolymers. Macromol. Rapid Commun. 25, 743–747.

Owen, S.C., Chan, D.P.Y., Shoichet, M.S., 2012. Polymeric micelle stability. Nano Today 7, 53–65.

Park, E.K., Lee, S.B., Lee, Y.M., 2005. Preparation and characterization of methoxy poly (ethylene glycol)/poly(ε-caprolactone) amphiphilic block copolymeric nanospheres for tumor-specific folatemediated targeting of anticancer drugs. Biomaterials 26, 1053–1061.

Peng, X., Zhang, L., 2007. Formation and morphologies of novel self-assembled micelles from chitosan derivatives. Langmuir 23, 10493–10498.

Peng, H., Blakey, I., Dargaville, B., Rasoul, F., Rose, S., Whittaker, A.K., 2009. Synthesis and evaluation of partly fluorinated block copolymers as MRI imaging agents. Biomacromolecules 10, 374–381.

Piskin, E., Kaitian, X., Denkbas, E.B., Kucukyavuz, Z., 1995. Novel PDLLA/PEG copolymer micelles as drug carriers. J. Biomater. Sci. Polym. Ed. 7, 359–373.

Podoltsev, N.A., Rubin, M., Figueroa, J., Lee, M.Y., Kwonj, J., Yu, J., et al., 2008. Phase II clinical trial of paclitaxel loaded polymeric micelle (GPM) in patients (pts) with advanced pancreatic cancer (APC). J. Clin. Oncol. 26, (May 20 suppl; abstr 4627).

Pressly, E.D., Rossin, R., Hagooly, A., Fukukawa, K., Messmore, B.W., Welch, M.J., et al., 2007. Structural effects on the biodistribution and positron emission tomography (PET) imaging of well-defined (64)Cu-labeled nanoparticles comprised of amphiphilic block graft copolymers. Biomacromolecules 8, 3126–3134.

Price, C., 1983. Micelle formation by block copolymer in organic solvents. Pure Appl. Chem. 55, 1563–1572.

Prochazka, K., Martin, T.J., Munk, P., Webber, S.E., 1996. Polyelectrolyte poly(tert-butyl acrylate)-block-poly(2-vinylpyridine) micelles in aqueous media. Macromolecules 29, 6518–6525.

Rangel-Yagui, C.O., Pessoa Junior, A., Taveres, L.C., 2005. Micellar solubilization of drugs. Can. J. Pharm. Sci. 8, 147–163.

Ranger, M., Jones, M.C., Yessine, M.A., Leroux, J.C., 2001. Well-defined diblock copolymers prepared by a versatile atom transfer radical polymerization method to supra molecular assemblies. J. Polym. Sci. A: Polym. Chem. 39, 3861–3874.

Riess, G., Hurtrez, G., Bahadur, P., 1985. Block copolymers. In: Mark, H.S., Kroschwitz, J.I. (Eds.), Encyclopedia of Polymer Science and Engineering. Wiley, New York, pp. 324–334.

Saif, M.W., Rubin, M., Figueroa, J., Kerr, R.O., 2008. Multicenter phase II trial of Genexol-PM (GPM), a novel Cremophor-free, polymeric micelle formulation of paclitaxel in patients with advanced pancreatic cancer (APC): final results. Gastrointestinal Cancers Symposium.

Schmeidl, U.P., Nelson, J.A., Teng, L., Starr, F., Malek, R., Ho, R.J., 1995. Magnetic resonance imaging of the hepatobiliary system: intestinal absorption studies of manganese mesoporphyrin. Acad. Radiol. 2, 994–999.

Seevinck, P.R., Seppenwoolde, J.H., de Wit, T.C., Nijsen, J.F., Beekman, F.J., van het Schip, A.D., 2007. Factors affecting the sensitivity and detection limits of MRI, CT, and SPECT for multimodal diagnostic and therapeutic agents. Anticancer Agents Med. Chem. 7, 317–334.

Shin, I.L., Kim, S.Y., Lee, Y.M., Cho, C.S., Sung, Y.K., 1998. Methoxy poly(ethylene glycol)/e-caprolactone amphiphilic block copolymeric micelle containing indomethacin, I. Preparation and characterization. J. Control. Release 51, 1–11.

Shiraishi, K., Kawano, K., Minowa, T., Maitani, Y., Yokoyama, M., 2009. Preparation and in vivo imaging of PEG–poly(L-lysine)-based polymeric micelle MRI contrast agents. J. Control. Release 136, 14–20.

Shokeen, M., Fettig, N.M., Rossin, R., 2008. Synthesis, in vitro and in vivo evaluation of radiolabeled nanoparticles. Quart. J. Nucl. Med. Mol. Imaging 52, 267–277.

Soman, N.R., Lanza, G.M., Heuser, J.M., Schlesinger, P.H., Wickline, S.A., 2008. Synthesis and characterization of stable fluorocarbon nanostructures as drug delivery vehicles for cytolytic peptides. Nano Lett. 8, 1131–1136.

Sun, H.L., Guo, B.N., Li, X.Q., Cheng, R., Meng, F.H., Liu, H.Y., 2010. Shell-sheddable micelles based on dextran-*ss*-poly(ε-caprolactone) diblock copolymer for efficient intracellular release of doxorubicin. Biomacromolecules 11, 848–854.

Sutton, D., Nasongkla, N., Blanco, E., Gao, J., 2007. Functionalized micellar systems for cancer targeted drug delivery. Pharm. Res. 24, 1029–1046.

Talelli, M., Rijcken, C.J., Lammers, T., Seevinck, P.R., Storm, G., van Nostrum, C.F., et al., 2009. Superparamagnetic iron oxide nanoparticles encapsulated in biodegradable thermosensitive polymeric micelles: toward a targeted nanomedicine suitable for image-guided drug delivery. Langmuir 25, 2060–2067.

Talelli, M., Rijcken, C.J.F., Hennink, W.E., Lammers, T., 2012. Polymeric micelles for cancer therapy: 3C's to enhance efficacy. Curr. Opin. Solid State Mater. Sci. 16 (6), 302–309.

Torchilin, V.P., 2002. PEG-based micelles as carriers of contrast agents for different imaging modalities. Adv. Drug Deliv. Rev 54, 235–252.

Torchilin, V.P., 2004a. Polymeric immunomicelles: carriers of choice for targeted delivery of water insoluble pharmaceuticals. J. Drug Deliv. Technol. 4, 251–261.

Torchilin, V.P., 2004b. Targeted polymeric micelles for delivery of poorly soluble drugs. Cell. Mol. Life Sci. 61, 2549–2559.

Torchilin, V.P., 2005. Fluorescence microscopy to follow the targeting of liposomes and micelles to cells and their intracellular fate. Adv. Drug Del. Rev. 57, 95–109.

Torchilin, V.P., 2007. Micellar nanocarriers: pharmaceutical perspectives. Pharm. Res. 24, 1–16.

Torchilin, V.P., Frank-Kamenetsky, M.D., Wolf, G.L., 1999. CT visualization of blood pool in rats by using long-circulating, iodine-containing micelles. Acad. Radiol. 6, 61–65.

Torchilin, V.P., Lukyanov, A.N., Gao, Z., Papahadjopoulos-Sternberg, B., 2003a. Immunomicelles: targeted pharmaceutical carriers for poorly soluble drugs. Proc. Natl. Acad. Sci. U.S.A. 100, 6039–6044.

Torchillin, V.P., Iakaubov, L.Z., Estrov, Z., 2003b. Therapeutic potential of antinuclear autoantibodies in cancer. Cancer Ther. 1, 179–190.

Trubetskoy, V.S., Torchilin, V.P., 1995. Use of polyethylene–lipid conjugates as long-circulating carriers for delivery of therapeutic and diagnostic agents. Adv. Drug Deliv. Rev. 16, 311–320.

Trubetskoy, V.S., Torchilin, V.P., 1996. Polyethyleneglycol based micelles as carriers of therapeutic and diagnostic agents. S.T.P. Pharma Sci. 6, 79–86.

Trubetskoy, V.S., Gazelle, G., Wolf, G.L., Torchilin, P., 1997. Block–copolymer of polyethylene glycol and polylysine as a carrier of organic iodine: design of a long circulating particulate contrast medium for X-ray computed tomography. J. Drug Target. 4, 381–388.

Turro, N.J., Chung, C.J., 1984. Photoluminescent probes for water–water soluble polymers, Pressure and temperature effect on a polyol surfactant. Macromolecules 17, 2123–2126.

Ulbrich, K., Konak, C., Tuzar, Z., Kopecek, J., 1987. Solution of drug carriers based on poly[*N*-(2-hydroxypropyl)methacrylamide] containing biodegradable bonds. Makromol. Chem. 188, 1261–1272.

Vega, J., Ke, S., Fan, Z., Wallace, S., Charsangavej, C., Li, C., 2003. Targeting doxorubicin to epidermal growth factor receptors by site-specific conjugation of C225 to poly(L-glutamic acid) through a polyethylene glycol spacer. Pharm. Res. 20, 826–832.

Vetvicka, D., Hruby, M., Hovorka, O., Etrych, T., Vetrik, M., Kovar, L., et al., 2009. Biological evaluation of polymeric micelles with covalently bound doxorubicin. Bioconjug. Chem. 20, 2090–2097.

Visentin, R., Pasut, G., Veronese, F.M., Mazzi, U., 2004. Highly efficient technetium-99m labeling procedure based on the conjugation of *N*-[*N*-(3-diphenylphosphinopropionyl) glycyl]cysteine ligand with poly(ethylene glycol). Bioconjug. Chem. 15, 1046–1054.

Wang, M., Zhang, G., Chen, D., Jiang, M., Liu, S., 2001. Noncovalently connected polymeric micelles based on a homopolymer pair in solutions. Macromolecules 34, 7172–7178.

Wang, Y., Yu, L., Han, L., Sha, X., Fang, X., 2007. Difunctional Pluronic copolymer micelles for paclitaxel delivery: synergistic effect of folatemediated targeting and Pluronic-mediated overcoming multidrug resistance in tumor cell lines. Int. J. Pharm. 337, 63–73.

Webber, S.E., 1998. Polymer micelles: an example of selfassembling polymers. J. Phys. Chem. B 102, 2618–2626.

Weissig, V., Lizano, C., Torchilin, V.P., 1998a. Micellar delivery system for dequalinium. Proc. Int. Symp. Control. Release Bioact. Mater. 25, 415–416.

Weissig, V., Whiteman, K.R., Torchilin, V.P., 1998b. Accumulation of protein-loaded long-circulating micelles and liposomes in subcutaneous Lewis lung carcinoma in mice. Pharm. Res. 15, 1552–1556.

Wilhelm, M., Zhao, C.L., Wang, Y., Xu, R., Winnik, M.A., Mura, J.L., et al., 1991. Poly (styrene-ethylene oxide) block copolymer micelle formation in water: a fuorescence probe study. Macromolecules 24, 1033–1040.

Wilson, R.H., Plummer, R., Adam, J., Eatock, M., Boddy, A.V., Griffin, M., et al., 2008. Phase I and pharmacokinetic study of NC-6004, a new platinum entity of cisplatin-conjugated polymer forming micelles. J. Clin. Oncol. 26, (May 20 suppl.; abstr. 2573).

Xu, B., Yuan, J., Ding, T., Gao, Q., 2010. Amphiphilic biodegradable poly(ε-caprolactone)-poly(ethylene glycol)-poly(ε-caprolactone) triblock copolymers: synthesis, characterization and their use as drug carriers for folic acid. Polym. Bull. 64, 537–551.

Xu, W., Ling, P., Zhang, T., 2013. Polymeric micelles, a promising drug delivery system to enhance bioavailability of poorly water-soluble drugs. J. Drug Deliv. 2013, 15 (Article ID 340315).

Yokoyama, M., Miyauchi, M., Yamada, N., Okano, T., Sakurai, Y., Kataoka, K., et al., 1990. Characterization and anticancer activity of the micelle-forming polymeric anticancer drug adryamicin-conjugated poly(ethylene glycol)−poly(aspartic acid) block copolymer. Cancer Res. 50, 1700−1993.

Yokoyama, M., Kwon, G.S., Okano, T., Sakurai, Y., Seto, T., Kataoka, K., 1992. Preparation of micelle-forming polymer−drug conjugates. Bioconjug. Chem. 3, 295−301.

Yokoyama, M., Okano, T., Sakurai, Y., Kataoka, K., 1994. Improved synthesis of adriamycin-conjugated poly(ethylene oxide)−poly(aspartic acid) block copolymer and formation of unimodal micellar structure with controlled amount of physically entrapped adriamycin. J. Control. Release 32, 269−277.

Yokoyama, M., Okano, T., Sakurai, Y., Suwa, S., Kataoka, K., 1996. Introduction of cisplatin into polymeric micelles. J. Control. Release 39, 351−356.

Yokoyama, M., Fukushima, S., Uehara, R., Okamoto, K., Kataoka, K., Sakurai, Y., et al., 1998a. Characterization of physical entrapment and chemical conjugation of adriamycin in polymeric micelles and their design for in vivo delivery to a solid tumor. J. Control. Release 50, 79−92.

Yokoyama, M., Satoh, A., Sakurai, Y., Okano, T., Matsumara, Y., Kakizoe, T., et al., 1998b. Incorporation of water-insoluble anticancer drug into polymeric micelles and control of their particle size. J. Control. Release 55, 219−229.

Yoo, H.S., Park, T.G., 2004. Folate receptor targeted biodegradable polymeric doxorubicin-micelles. J. Control. Release 96, 273−283.

Yu, B.G., Okano, T., Kataoka, K., Kwon, G.S., 1998. Polymeric micelles for drug delivery: solubilization and haemolytic activity of amphotericin B. J. Control. Release 53, 131−136.

Yuan, F., Leunig, M., Huang, S.K., Berk, D.A., Papahadjopoulos, D., Jain, R.K., 1994. Microvascular permeability and interstitial penetration of sterically stabilized (Stealth) liposomes in a human tumor xenograft. Cancer Res. 54, 3352−3356.

Yuan, F., Dellian, M., Fukumura, D., Leunig, M., Berk, D.A., Torchilin, V.P., et al., 1995. Vascular permeability in a human tumor xenograft: molecular size dependence and cutoff size. Cancer Res. 55, 3752−3756.

Zeng, Y., Pitt, W.G., 2006. A polymeric micelle system with a hydrolysable segment for drug delivery. J. Biomater. Sci. Polym. 17, 591−604.

Zhang, J.X., Ma, P.X., 2009. Polymeric coreshell assemblies mediated by host−guest interactions: versatile nanocarriers for drug delivery. Angew. Chem. Int. Ed. 48, 964−968.

Zhang, X., Jackson, J.K., Burt, H.M., 1996. Development of amphiphilic diblock copolymers as micellar carriers of taxol. Int. J. Pharm. 132, 195−206.

Zhang, X., Zhuo, R., Yang, Y., 2002. Using mixed solvent to synthesize temperature sensitive poly(*N*-isopropylacrylamide) gel with rapid dynamic properties. Biomaterials 23, 1313−1318.

Zhanga, J., Lia, S., Li, X., 2009. Polymeric nanoassemblies as emerging delivery carriers for therapeutic applications: a review of recent patents. Recent Pat. Nanotechnol. 23, 225−231.

Zielhuis, S.W., Nijsen, J.F.W., Seppenwoolde, J.H., Zonnenberg, B.A., Bakker, C.J.G., Hennink, W.E., et al., 2005. Lanthanide bearing microparticulate systems for multi-modality imaging and targeted therapy of cancer. Curr. Med. Chem. Anti-Cancer Agents 5, 303–313.

FURTHER READING

Kabanov, A.V., Batrakova, E.V., Alakhov, V.Y., 2002. Pluronic block copolymers as novel polymer therapeutics for drug and gene delivery. J. Control. Release 82, 189–212.

CHAPTER

Current trends in site and target specific delivery of nanomedicine for gene therapy

4

Arunachalam Muthuraman[1,2], Seema Mehdi[1] and Narahari Rishitha[1]

[1]Department of Pharmacology, JSS College of Pharmacy, JSS Academy of Higher Education and Research, Mysuru, India [2]Pharmacology Unit, Faculty of Pharmacy, AIMST University, Semeling, Malaysia

CHAPTER OUTLINE

Nanoparticles in Pharmacotherapy. DOI: https://doi.org/10.1016/B978-0-12-816504-1.00010-7

4.1 INTRODUCTION

Gene therapy is one of the newer approaches in the treatment of various genetic disorders. It involves the nucleic acids, that is, DNA or RNA (Kaufmann et al., 2013). However, the effective delivery of genes to a specific site and cellular target remains a challenging task (Conde et al., 2016). The delivery of genes to a specific cell depends upon the cellular environment of the specific disease. The gene delivery for the specific disease is based on the level of gene sequence in the host cells in the pathological condition (Schaefer and Serrano, 2016). It may be a suppressive gene for specific proteins like cytokines and chemokines; downregulating gene for pro-apoptotic proteins; or upregulating gene for antiapoptotic and antiinflammatory proteins (Plaza-Diaz et al., 2014). Overall, gene delivery makes the substitution of defective genes or missing endogenous counterpart actions or by reducing the abnormal expression of gene product. Generally, the method of gene delivery undergoes both viral and nonviral mechanisms. Sometimes, the genetic manipulation is done via oligonucleotides therapy (Maeder and Gersbach, 2016). The biomedical application of genes is very commonly used for the immunological diseases like immunodeficiency (Ott de Bruin et al., 2015). Nowadays, the approach of gene therapy is focused on cancer, cardiac, hepatic, renal disorders, and infectious disorders, including neurodegenerative disorders like Alzheimer's and Parkinson's disease (De Toro et al., 2015).

The first gene based drug, gendicine, was introduced in the Chinese market in 2004. It is an adenovirus-p53 based gene used for head and neck squamous cell carcinoma (Chen et al., 2014). The treatment of gendicine is not documented for any serious side effects; however, the therapeutic potential of this drug remains to be investigated. In European countries, the alipogene based drug tiparvovec was launched for the treatment of familial lipoprotein lipase deficiency in 2012 (Watanabe et al., 2014). This drug is known to produce the potential therapeutic benefit in lipoprotein lipase deficiency patients. Therefore, it was a milestone in the discovery of gene therapy (Coulie et al., 2014). However, the complete utilization of genetic drugs remains limited due to multiple issues raised for clinical utilization like safety, therapeutic role, pharmacological action, mechanism, interaction of other genes, legal issues, and genomic and proteomic action (Mishra and Shukla, 2014). Similar issues are also raised in the discovery process of nanotechnology. Now, nanotechnology is developed in multiple fields including nanomedicine and nanoparticle discovery (Shi et al., 2010). The various limitations and hazards have been overcome by different methods and technology (Carroll and Charo, 2015). In addition, the nanoparticle effectively plays a key role for carrying small size and large quantity of bioactive molecules like drugs, proteins, enzymes, nucleic materials (i.e., DNA, RNA, and siRNA), oligonucleotides including genes (Gandhi et al., 2014). Therefore, this book chapter has focused on current development of possible gene therapy for the chronic life threatening disorders with nanotechnology. Nanomedicine associated gene therapy is also discussed in terms of site and target specific action in various pathological conditions.

The primary role of gene therapy is delivery of the gene to the targeted tissue. The gene delivery depends upon the types of application, that is, in vivo or ex vivo (Naldini, 2015). In addition, another factor also important in the site and target specific delivery of gene into the cells is vector dependence or vector independence (Kantor et al., 2014). In the in vivo method, the gene is delivered directly into the body; whereas in the ex vivo method, host cells are collected, cultured, transfected with the gene of interest, and reintroduced into the host body (Mali, 2013). The first clinical trials are attempted to in vivo vector (adenovirus) based delivery of gene for ornithine transcarbamylase deficiency (Wang et al., 2015a,b). It is also known as inborn disease of urea synthesis. However, the host cells are produced in the immune reaction against adenoviral vectors (Mingozzi and High, 2013). Moreover, the newer gene therapy with viral capsid proteins is ameliorating the cancer without interfering with the host immune system (Srivastava et al., 2014). Further, the retroviral vectors are also used for the delivery of genes in cancer proliferating cells (Pranjol and Hajitou, 2015). Ex vivo gene therapy revealed that the isolated cell from the patient and genetic modification is carried outside of the body (Kumar et al., 2016b). Subsequently the modified genes are reintroduced into the same patient. This is also known as autologous transplant of ex vivo gene therapy (Tsang and Atkins, 2015). The major advantages of ex vivo gene therapy are reduction of risk by ectopic expression and avoiding the off-target (nontargeted tissue) effects (Arruda and Samelson-Jones, 2015). Therefore it is possible to minimize genetic toxicity in the host cells (Ishida et al., 2015).

Other than in vivo and ex vivo gene therapy, additional gene therapy can be classified into two major varieties based on the location of action: somatic gene therapy and germline gene transmission therapy (Siddique et al., 2016). Somatic gene therapy is targeted to multiple cells and tissues except reproductive organ cells (e.g., sperm and oocytes) (Indu et al., 2013). Whereas germline therapy is targeted only to the reproductive organ cells. Also, ex vivo gene therapy allows the selection, expansion, and quality control of the gene modified cells in the culture media, thus reducing the toxicity and enhancing the safety and efficacy margins. For example, the hematopoietic stem cells (HSC) are easy to isolate from the blood and introduced genetically modified HSCs into patients with advanced stage melanoma (Watts et al., 2011).

4.2 FABRICATION AND MODIFICATION OF NANOPARTICLES FOR GENE DELIVERY

Viral vector-based gene delivery is widely employed in gene therapeutic application and it produces long-term stable gene expression (Ramamoorth and

Narvekar, 2015). There are many viral vectors are employed in the delivery of genes such as herpes simplex virus type 1 (HSV-1); adenovirus; adenovirus associated virus (AAV); lentiviruses like HIV-1, feline immunodeficiency virus, and equine infectious anemia virus; and SV40. AAV and lentivirus vectors play a key vector role in gene transfer for the central nervous system (Bowers et al., 2011; Kantor et al., 2014; Ramamoorth and Narvekar, 2015). Generally, this is used for nononcological applications due to prolonged gene expression capacity and no apparent toxicity in targeted tissue (Canter et al., 2016). Currently, AAV is also used for oncological applications due to transgene expression potential for brain neurogliosis (Xie et al., 2016a). Further, HSV is an ideal vectors for gene delivery due to its large genomic transfer and genetic manipulation potential in CNS disorders (Nayerossadat et al., 2012). Moreover, the recombinant SV40 vectors are shown to possess a promising level of gene transfection and transduction to neurons like astrocytes in different brain regions and the spinal cord (Kohyama et al., 2010). However, viral vector-based gene therapy has shown some unwanted side effects. This limitation is expected to overcome by nonviral vector-based gene therapy. Therefore, the fabrication and modification of nanoparticles for gene therapy is essential (Massadeh et al., 2016).

The fabrication of nanoparticles for gene therapy is considered based on carrying molecules like DNA, RNA, siRNA, and specific gene interest (Navarro et al., 2015). In addition, before the modification of nanoparticles, it is also necessary to analyze the types of vectors and routes used for the gene delivery (Massadeh et al., 2016). Generally, nonviral vectors like polymeric nanoparticles deliver the nucleic acid and have efficient modification potential in cellular process. The primary challenges are target cell and protein specificity and the duration of transient gene expression (Germershaus and Nultsch, 2015). Recently, the intraventricular administration of recombinant DNA encoding associated modified silica nanoparticles have enhanced the neurogenesis in subventricular zone in adult mice brain via activation of fibroblast growth factor receptor-1 (Holvoet et al., 2016). In addition, electroporation associated nucleic acids are transfected in prenatal and postnatal rodent brain and enhance the neuron density in living brain (Li et al., 2016e). Therefore, the newer method of nanoparticle fabrication and modification may promise the efficient delivery of gene in targeted tissue (Li et al., 2016c).

The novel nanoparticle has multiple physiochemical as well as biological interaction properties (Navya and Daima, 2016). Therefore, these nanoparticles are able to deliver the gene in targeted cells as well as in tissue. The nanoparticle-associated gene carriers are fabricated with modification and biocompatibility based changes (Ishihara et al., 2016). This modification process is achieved with consideration of pathophysiological conditions like pH, ionic strength, and osmotic strength in the biological fluid and in biological system (Coelho et al., 2010). In addition, these nanoparticles are expected to produce integrity in blood circulation and enhance the release of intracellular cargo release (Bozzuto and Molinari, 2015). However, the primary limitation is efficient gene delivery in the cells and enhancement of gene transcription as well as translation (Maeder and

Gersbach, 2016). The transcription and translational target and modification of nanocarriers are discussed in the following section of this chapter. The primary goal is efficient gene therapy, which is achieved by multicomponent-based nanoparticle modifications and it makes the expected achievement for controlled-release of genes in the targeted cells (Chen et al., 2016b). The single nanomaterials may have advantages in certain conditions of gene delivery, whereas the multiple nanomaterials will have diverse biological functions (Navya and Daima, 2016). The multiple nanomaterial modification is also known as composite of nanoparticles. The fabrication of composite nanoparticles has attracted great attention in the scientific and technological communities (Alghuthaymi et al., 2015). The most important step for fabrication of nanoparticles is the preparation of composite nanostructures leading to achieve well-structured composite nanoparticles (Sun et al., 2015). It is expected to produce the reproducible and well-controlled fabrication with specific applications.

The various fabrication methods are employed for the preparation of composite nanoparticles. The fabrication of inorganic nanoparticles like gold nanoparticles, magnetic nanoparticles, quantum dots, silica nanoparticles, and polymeric nanoparticles is used for gene and drug delivery (Heera and Shanmugam, 2015). The fabrication of composite nanoparticles involves the bulk mixing process. The advantages of bulk mixing type of composite nanoparticles include their efficient physicochemical properties due to their self-assembly mechanism (Wang and Huang, 2014). The major limitation of bulk mixing process shows the large polydispersity between the batch-to-batch preparations, which leads to critical challenges for the clinical translation of composite nanoparticles (Chan et al., 2016). These problems can be overcome by lithography technology and microfabrication, which can yield efficient gene carrying type of composite nanoparticles without the variation involved in batch-to-batch preparation (Zhou et al., 2016b).

Moreover, the self-assembly is also produced in composite nanoparticle by the application of microfluidic mixing process (Boken et al., 2016). Microfluidic mixing involves interdisciplinary technology including engineering, physics, chemistry, nanotechnology, and biotechnology. Therefore, it is expected to produce multiple therapeutic applications. In contrast, it also produces some limitations in the fabrication of composite nanoparticles (Tasoglu et al., 2015). The primary limitation of this type of nanoparticle mixing is difficulty in reproducing nanoparticles with identical properties for large-scale utilization of nanomedicine for clinical use (Tinkle et al., 2014). Now, the microfluidics process is expanded with chemical separations and application of semiconductor technology in ultralow volumes and accessing biological length of application (Wang and Huang, 2014). It is also expanded with soft lithography, which allows the rapid prototyping of microfluidic nanoparticle preparation (Valencia et al., 2012).

Furthermore, amphiphilic molecules like lipids and their copolymers self-assemble and aggregate with changes of solvent polarity (Lombardo et al., 2015). The increase in mixing time duration can cause aggregation of lipids and expose the heterogeneous solvent environment and prevent the stabilization of the nanoparticles in the hydrophilic portion. This is also known as microscale mixing (Wang et al., 2015a). In comparison, macroscale mixing with turbulent flow increases the diffusion and advection properties of the nanoparticle. This microfluidic mixing devices are classified as active and passive mixing (Ward and Fan, 2015). The active micromixer enhances the usage of external energy input by nanoparticles like pressure field, acoustics, and temperature. Whereas, the passive mixing devices enhance the nanoparticle aggregation by complete utilization of pumping energy (Wang and Huang, 2014). They are relatively inexpensive compared with active micromixer devices. The recent advances of nanoparticle processing have improved using digital droplet generators for fabrication of nanoparticles (Nge et al., 2013). This type of fabrication involves various building blocks like 1-adamantanamine (Ad)-polyamidoaminedendrimer, Ad-PEG (polyethylene glycol), β-cyclodextrin-polyethylenimine (PEI), and Ad-Arg-Gly-Asp (RGD)-PEG conjugates. Sometimes the building blocks are introduced into the micromixing device in sequential manner (Xu et al., 2015). Therefore, the unique properties of composite nanoparticles are prepared. In addition, rapid mixing of PLGA with acetonitrile and lipid/lipid-PEG micelles in water by using hydrodynamic flow pattern also enhances the fabrication process for homogeneous formation of composite nanoparticles with fine nanosize of the particle (Wang and Huang, 2014). It also mimics the physicochemical properties of the conventional nanoparticles like zeta potential, size, and surface functionalization (d'Amora et al., 2016). The most feasible approach by which to achieve composite nanoparticle is self-assembled nanomedicine. This involves two major methods: (1) layer-by-layer self-assembly, which is a method of constructing the composite nanoparticles with multilayer structures; and (2) imprint lithography, which is also known as particle replication in nonwetting templates technology (PRINT) (Gai et al., 2016).

The composite nanoparticles promise to deliver the targeted gene in the targeted cells. This nanoparticle fabrication and modification effectively alters the gene transcription and its translation leads to ameliorate the various disorders via genetic manipulation of targeted tissue (Cox et al., 2015). The composite nanoparticle overcomes critical barriers involved in the gene delivery process such as (1) the reduction of colloidal stability of nanoparticles after systemic administration with adsorption of serum protein compromising action, (2) the rapid reduction of nanoparticle levels in the blood circulation via active reticuloendothelial uptake process, (3) the restriction of cytoplasmic membrane for the binding and entry of nanoparticles into the cells, (4) the endosomal and lysosomal trapping of nanoparticles in cytosolic regions, and (5) the blocking action of nuclear envelope for DNA transcription with targeted nuclear materials.

4.3 PROPERTIES OF NANOPARTICLES FOR GENE DELIVERY

Various nanoparticles are reported to produce efficient carriers for genes. However, the modified nanoparticles such as liposomes, solid lipid nanoparticles, and nanostructure lipid carriers including micelles, nanotubes, and dendrimers have potential loading and carrying capacity of gene towards the specific site and cell signaling target of the cells (Dolatabadi and Omidi, 2016; Prabha et al., 2016). Moreover, the modified particles may produce the efficient therapeutic action based on the following biomedical actions: (1) gene loaded nanoparticle recognition of target cell signals, (2) bioadhesion and endocytosis into the targeted cells, (3) cytosolic entry and identifying the targeted biomolecules; (4) compatible with recognized target site and releasing the gene of interest, (5) interaction and exchange of genetic materials into the host cells, and (6) elimination of free nanoparticles (Díaz-Torres et al., 2016; Yudina and Puntes, 2016). The above process results in successful gene therapy for specific genetic and metabolic disorders (Koga et al., 2016). Currently, some stem cells are also known to play a role in gene therapy by altering the abnormal genetic sequence and signals due to their pluripotency (Fernandes and Chari, 2016).

Stem cells may differentiate the host cell's genetic materials and alter gene expression. Stem cells have three germ layers: (1) ectoderm, which is similar to neurons and epidermal cells; (2) mesoderm, which is similar to blood, cardiac, and muscle cells; and (3) endoderm, which is similar to pancreatic and liver cells (Dayem et al., 2016; Perez et al., 2016). Therefore, these potent stem cells are able to generate the disease modifying action with combined action of gene and cell therapy. Based on this understanding, the stem cells can be utilized as a gene modifying agents in diseased cells (Chen et al., 2016a). In addition, these stem cells can also modify the somatic as well as germ cells via nuclear transfer and reprogramming of oocytes via pluripotency factors like Oct4, Sox2, Klf4, and c-Myc (Hu and Li, 2016). These are essential factors for the efficient reprogramming of matured somatic cells. Therefore, the pluripotent stem cell plays a key role in the gene therapy. The current scenario of gene therapy is focused on the stem cell and gene combination therapy due to applicability in multiple pathological conditions (Liu et al., 2016c; Yin et al., 2016b). Hence, current growing nanotechnology also supports specific site and targeted gene therapy for various metabolic and immunological disorders (Gill et al., 2016). Therefore, this section is focused on the correlation of nanoparticles and gene therapy with their cellular and subcellular actions.

The molecular mechanism of chronic life threatening disorders originates from the changes of genetic material (del Pozo-Rodríguez et al., 2016). The modification of genetic materials has been identified in various chronic disorders such as cancer, diabetes, heart failure, heart attack, and neurodegenerative disorders like Alzheimer's and Parkinson's diseases (Wen et al., 2016). The pathophysiological mechanism involved in the changes in these genes is identified in patients. Gene therapy is a promising option for the successful treatment of certain disorders. Nanotechnology is used to reach and release the genetic materials in targeted tissue compartment (Jhamb et al., 2016). Earlier, the gene therapy was achieved by various viral vectors. In cancer, the nucleic acid changes are well defined and this alters and induces the expression of various abnormal cell cycle regulating proteins (Vercauteren et al., 2016; Wicki et al., 2016). Therefore, gene therapy is a future medicine and its limitations can be overcome with nanotechnological approach. The tumor suppressor genes are predominantly upregulated and/or mutated oncogenes are downregulated in cancer cells (Jeggo et al., 2016). Whereas, in the initial stage, the host cells are ready to fight against cancer cell reaction by inducing the expression of suicide genes causing cancer cell death, and immune related genes (Castro and Kipps, 2016). The administration of DNA vaccines is known to reduce the cancer cell proliferation via activation of immunological reaction. The efficiency is improved with biological vectors for gene delivery such as adenovirus and lentivirus (van der Burg et al., 2016). However, these viruses are able to cause immunogenicity and/or tumorigenicity actions with alteration of host cell gene sequence. Therefore, the safer delivery of the gene to the cancer and diseased cells remains challenging (Hill et al., 2016). This limitation is partially removed by the usage of nonviral method of gene delivery. It is expected to produce a safe and cost effective method of gene delivery. However the efficacy of gene therapy remains lower than viral based gene delivery (Lambricht et al., 2016; Lundstrom, 2016). Therefore, the current nanotechnology with specific nanoparticles such as polymers, micelles, and dendrimers is expected to produce new milestones in gene therapy for the amelioration of various pathological disorders.

The nanoparticle associated gene delivery is based on the recognition of the gene complex by targeted cells. The physiochemical properties such as ultrasound, heat, light, and applied magnetic and electric fields of nanoparticle–gene complex support the targeted cell in recognition for gene delivery (Karimi et al., 2016). These actions are more precise in the tumor and cancer cells. The known and unknown target locations of cancer cells are able to response with gene–nanoparticle complex. The site and target specific action of this complex can modify the surface property, which enhances the active and passive process of cancer-specific transfection (Li et al., 2016a). However, the small nanoparticles like polyplexes with neutral surface potential and stearic action are also responsible for the recognition and target specific action of gene therapy for tumor cells. In addition, polymeric biomaterials (with or without) target coating ligands help in enhancing the cell-specific uptake and tissue-specific accumulation of modified genes into the diseased cells (Liu et al., 2016a; Tzeng et al., 2016). Therefore, nanotechnological and nucleic acid engineering are promising for the targeted cell specific transcription and translation of genes, leading to ameliorate various metabolic and genetic disorders (André et al., 2016). The delivery and action of the gene undergoes the following targeting processes: (1) biophysical targeting, (2) biomaterial-mediated targeting, (3) ligand-mediated targeting, (4) transcriptional targeting, and (5) posttranscriptional targeting (Sánchez-Moreno et al., 2016; Barto et al., 2016).

4.3.1 BIOPHYSICAL PROPERTIES BASED GENE DELIVERY

The physicochemical properties of nanoparticle–gene complex contribute to the recognition of tumor and cancer cells. The major physiochemical modulating factors are particle size, surface charge, and functional (chemical) interacting groups, which can result in efficient cellular uptake, endocytosis, and prevent nontargeted biomolecular interaction (Saleh and Shojaosadati, 2016). Whereas, the tumor and cancer cell specific actions are increased with these modifications. Furthermore, the nanotechnology enhances the passive targeting for the nucleus of cancer cells and it expands the design of polymeric-gene delivery via suitable viral and nonviral vectors (Cheng et al., 2016). The passive targeting is increased with particle size. In addition, the very small size of the particles (i.e., 5 nm) cause them to be easily excreted via the kidneys (Karjoo et al., 2016). However, the permeation and retention efficiency of the gene complex may be lacking, due to the action of lymphatic drainage in the surrounding tumor and cancer cells. Whereas, the larger-sized nanoparticles are eliminated via activation of reticulo-endothelial system (Padera et al., 2016; Zheng et al., 2016). Therefore, the appropriate sizes of nanoparticles are essential for the best action of gene therapy. In addition, the nanoparticles coated with hydrophilic molecules like polyethylene glycol prevent the neutral surface charge and steric hindrance (Adjei et al., 2016). Many nanoparticles are prepared via electrostatic interactions between cationic polymers and anionic DNA. The net effects of biopotential changes enhance the efficiency of cellular membrane interactions as well as cellular uptake (Black et al., 2016). In addition, the surface property also changes between nonspecific anionic serum proteins and cationic charged surfaces of polymeric nanoparticles. This biopotential interaction and cellular uptake occurs in the membrane proteins. Some proteins act as a receptor. This kind of cellular entry is also known as receptor-mediated endocytosis (Majzoub et al., 2016; Thomas et al., 2016).

In contrast, some studies reported that serum protein-coated nanoparticles reduce the cellular uptake. In addition, the nonspecific protein interactions of albumin loaded nanoparticles also enhance the particle aggregation and opsonization leading to reducing the bioavailability, permeation, and cancer therapeutic potential (He et al., 2016; Larsen et al., 2016). This kind of undesirable effect is prevented by shielding of the polymeric nanoparticle surface. In this condition, the neutral charged PEG molecules are used for gene therapy (Thomas et al., 2016). The PEG coated nanoparticle readily carries the stimuli-responsive linker like metalloproteinase-cleavable linker and it has close similarity to cancer cells. Therefore, the PEG coated nanoparticles are able to interact with targeted cells and provide the efficient gene deliverable action (Ke et al., 2016). The surface shielding action on the nanoparticle is achieved by polyanionic nanoparticles like poly(glutamic acid), carboxymethyl poly(L-histidine), and hyaluronic acid (HA) (Gu et al., 2016). This formulation generates the positive coating environment and promotes the colloidal stability, thus facilitating the efficient gene delivery in vitro and in vivo (Ewe et al., 2016). Furthermore, the cationic polymer is mainly interacts with DNA by electrostatic interaction; and hydrophobic polymers like poly(lactic-co-glycolic acid) are used for encapsulation of genetic materials via emulsification process (Makita-Chingombe et al., 2016; Steinbach et al., 2016). The microspheres, a type of nanoparticle, enhance delivery of DNA to macrophages and antigen presenting cells (Chattopadhyay et al., 2016).

4.3.2 BIOMATERIAL BASED GENE DELIVERY

Different varieties of biomaterials are used for the delivery of genetic materials. The polymeric vectors with specific biomaterials are able to deliver the large quantity of genetic materials into the cells (Suk et al., 2016). The specialized biomaterials such as polypeptides like poly-L-lysine; natural polymers like chitosan, dextran, and HA (Mokhtarzadeh et al., 2016); and synthetic polymers like PEI, polyamidoamine (PAMAM) and poly(β-amino ester) (PBAE) (Gao et al., 2016). In addition, these biomaterials have their own tissue and cell specific interactive properties; one type interacts with a specific kind of cells whereas another avoids the specific cells for transfection (Ross and Sullivan, 2016). Therefore, this tissue targeting and biodistribution of polymeric nanoparticles may affect the key process of pharmacokinetics properties (Lee et al., 2016). In addition, the polymeric nanoparticles may vary the intrinsic action of particular organs or tissues. The dextran sulfate binds to more specific receptors in liver sinusoidal endothelial cells and accumulates in the liver tissue (Park et al., 2016; Zhang et al., 2016b). The PBAE nanoparticles coated with anionic poly(glutamic acid)-based peptides enhance the changes of biophysical properties and tissue-specificity for spleen and bone marrow, leading to produce better antitumor action (Wang et al., 2016a; Zhou et al., 2016a). Therefore, the biomaterial and targeted cell interactions may ameliorate the cancer cell proliferation via efficient gene therapeutic actions.

4.3.3 RECEPTOR–LIGAND TARGETED GENE DELIVERY

The gene-loaded nanoparticles are designed to target multiple targets on the cell surface, cytotosolic proteins, and nuclear materials. This property enhances the therapeutic efficacy of the nanoparticle–gene complex via gene delivery in the targeted cells (Kim et al., 2015). In cancer cells, the variety of gene dysregulation is observed along with modification of cell surface receptors like transferrin, folate, epidermal growth factor (EGF), Arg-Gly-Asp peptides (RGD), HA, and some types of specific carbohydrates (Rivoltini et al., 2016). The microenvironmental changes are abnormal and it is less or absence in the normal cells. Therefore, the receptor and ligand targeted selective nanoparticle is useful for the gene delivery into the targeted cells (Ma et al., 2016). Further, antibodies and antibody fragments loaded nanoparticle also produce the efficient gene therapy for cancer disorders due to expression of specific antigens like HER2 and prostate specific antigen by cancer cells (Gupta and Shukla, 2016; Kitano et al., 2016). The liposomal type of nanoparticles is also ready to deliver the gene in the targeted cells leading to attenuate the cancer cells' progress. The ligand targeted action of transfection varies depending upon the density of ligands, affinity, effect on zeta potential, and nanoparticle stability (Malhotra et al., 2013). The higher ligand density is raising the affinity to targeted cell receptors and enhances the cellular uptake (Elias et al., 2013). The details of specific receptor–ligand targeted gene delivery are explored in the following section.

Transferrin receptor: Transferrin (Tf) is an 80 kDa molecular weight glycoprotein. It is able to conjugate with specialized nanoparticles such as poly(L-lysine) (PLL), PEI, cyclodextrin, and PAMAM (Iyer et al., 2014; Kang et al., 2015). This overexpression of Tf receptor is identified in tumor cells; whereas, the transferrin conjugated nanoparticle (i.e., PEI and PEI–PEG nanoparticles) surface carries the target specific genes and improves antitumor action and also decreases the off-target transfection (Kozielski et al., 2016). In addition, cyclodextrin–PEG–Tf complex type nanoparticles (CALAA-01) deliver the small interfering RNA (siRNA) in solid tumor cells (Jing et al., 2014; Pradhan et al., 2014). *Folate receptor*: Folic acid is one of the essential nutrients in cancer and tumor cell growth and cell signaling process. The upregulation of folate receptor (FR) is identified in rapid dividing cells like tumor and cancer cells (Hijaz et al., 2016). The delivery of the targeted gene in the cancer cells is achieved via folate coated bPEI–PEG nanoparticles leading to increase in the cellular uptake and transfection in vitro and in vivo (Zhang et al., 2016a). The folate receptors are widely expressed on the surface of proliferating cells and it is increased 3.6-fold compared with normal cells (Senol et al., 2015). Furthermore, the luciferase gene knockdown loaded oligonucleotide-based nanoparticles are documented to produce the effective target delivering of gene into the cancer cells (Jing et al., 2014).

Epidermal growth factor associated receptor: The epidermal growth factor receptor (EGFR) is a key molecular target for the cancer treatment. This EGFR is expressed around 30% in the surface of solid tumors (Xu et al., 2016). The EGF conjugated PEI–PEG nanoparticle improves the transfection rate in the cancer cells in vitro (Williford et al., 2016). In addition, the sodium iodine symporter gene loaded synthetic peptides are conjugates with PEI–PEG nanoparticles; this formulation potentially treats the liver cancer via regulation of EGFR and GE11 integrin receptor expression (Fujita et al., 2015; Uusi-Kerttula et al., 2015). *$\alpha v\beta 3$ integrin receptor (RGD peptide)*: The RGD is a three amino acid containing peptides, that is, arginine, glycine, and aspartic acid. It is reported to produce the strong activation of $\alpha v\beta 3$ integrin receptor in tumor cells (Shan et al., 2015). This receptor expression is higher in the artery of cancer and tumor cells. It leads to enhance the angiogenesis process for the essential requirement of major nutrients (Cayrol et al., 2015). The RGD peptide has two major varieties: (1) RGD peptide and (2) cyclic-RGD peptides. Both types of peptides conjugate with $\alpha v\beta 3$ integrin receptor targeted nanoparticles and regulate the tumor cell growth in the lungs and liver (Wang et al., 2013). Therefore, the gene therapy for integrin receptor targeted nanoparticles may attenuate the severity of cancer and tumor cell proliferation (Chen et al., 2016d).

HA: The HA levels were raised in the tumor cells with expression of corresponding gene activation. In this condition, cluster of domain (CD_{44}) of the T cell receptor has higher affinity towards the HA of tumor cells, altering the tumor growth (Trivedi et al., 2016). In cancer cells, it has a putative role for the initiation of cancer and stem cells. In addition, HA is able to alter the zeta potential at physiological pH due to its negative charge potential (Wang et al., 2016b). *Antibodies:* the cancer and tumor cells actively express a specific antigen such as tumor specific antigen (HER2) and prostate specific antigen. Therefore, antibodies-loaded nanoparticles will efficiently treat the cancer cells via releasing of HRE2 gene (Barve et al., 2014). This helps in reducing the expression of a specific antigen in tumor cells. However, the use of monoclonal antibodies and antibody fragments is more expensive than other ligands (Smyth et al., 2014).

4.3.4 TRANSCRIPTION TARGETED GENE DELIVERY

Transcription is one of the common cellular processes in normal cells as well as cancer cells. In gene therapy, specialized DNA delivery is one of the key processes for anticancer activity; the targeting action in cancer cells is achieved by nanoparticle conjugation (Giang et al., 2014; Jang et al., 2015). The therapeutic expression of a specific gene in the targeted cancer cells has shown to produce beneficial effects; whereas, the expression in healthy cells causes undesirable effects via abnormal transcriptional process (Haile et al., 2016). The transcriptional targeted gene delivery can be achieved by reduction of transgene expression in tumors. In addition, the particular tissues are also used the specific promoters (Danda et al., 2013). The mechanism of transcription targeted gene delivery undergoes multiple steps: (1) transcriptional control of gene expression and (2) promoters associated transcriptional action (Sachdeva et al., 2015).

4.3.4.1 Transcriptional control of gene expression

The transcription process of eukaryotic cells is based on protein-coded genes. These interact with RNA polymerase II, *cis*-regulatory DNA elements, transcription factors, cofactors, and chromatin structure (Brophy and Voigt, 2016; Kathiriya et al., 2015). The *cis*-regulatory DNA elements are known as the promoters and recognition sites for the various transcription factors. They are also known as *trans*-acting elements (Metzger et al., 2015). The major function of promoters is upstream of gene transcription at the starting stage. They completely bind with the core promoter and proximal regulatory elements. The RNA polymerase II with core promoter complex initiates the gene transcription (Jonkers and Lis, 2015). The basic transcriptional action is raised by the RNA polymerase II binding with another set of proteins, site-specific transcription factors, and proximal and distal regulatory elements (Quaresma et al., 2016). Moreover, the *cis*-regulatory elements drive the gene transcription with exogenous therapeutic gene in specialized tissue and/or tumor specific cells (Kafshdooz et al., 2016). The major limitation of this transcriptional targeted gene therapy is accurate delivery of DNA interest and control of a specific promoter (Gottfried et al., 2016). Furthermore, cell surface binding, endosomal escape, translocation, and nucleic acid release also causes the barrier action (Yao et al., 2013).

4.3.4.2 Promoters associated transcriptional action

The control of gene transcription may be possible by multiple ways such as (1) tissue-specific promoters and (2) tumor-specific promoters (Rama et al., 2015). *Tissue-specific promoters:* The primary action of tissue-specific promoters is in regulation of transcriptional factors. This may vary from one tissue to another tissue. The tissue specific promoters target specific genes that are specifically expressed in a particular tissue (Rama et al., 2015). For example, when the tissue specific promoter is active in healthy tissues and tumor cells, the ability of cancer cell proliferation is reduced. The melanocyte specific gene, that is, tyrosinase gene, drives the expression of reporter gene in humans well as in murine melanoma cells (van der Weyden et al., 2016). But it is not affected in other cells. Hence, tyrosinase promoter is widely used in gene therapy for melanoma (Viola et al., 2013). Similarly, the prostate-specific antigen, prostate-specific membrane antigen, and probasin are also used for driving the siRNA expression in prostatic cells leading to gene silencing in (androgen-sensitive) prostate cancer (Chavan et al., 2014).

Tumor-specific promoters: The tumor-specific promoters are employed in the regulation of gene transcription in tumor cells. However, they do not affect any normal cells. In addition, the cancer-specific promoters bind to specific type of particular cancer or multiple (almost all) types of cancer. It controls the cancer cell proliferation (Lai et al., 2015). The controlling action of multiple cancer cells is due to the availability of common promoter binding site and lack of specificity for particular cancer tissue. The progression-elevated gene-3 (PEG-3) is overexpressed during the tumor cell growth and different cancer types leading to drive the cancer-specific gene expression (Das et al., 2015). The jetPEI promoter drives the firefly luciferase (PEG-Luc) expression, which specifically mimics the metastasis of human melanoma and breast cancer (Bhatnagar et al., 2014). In addition, other cancer-specific promoters have also been identified such as human telomerase reverse transcriptase (hTERT), survivin, and astrocyte elevated gene-1 (AEG-1) promoters (Frau et al., 2014). Therefore, the control of gene transcriptional activity with tumor specific promoters may serve as effective management of cancer and tumor cell growth.

Inducible promoters: The inducible promoters potentially increase the gene transcription and enhance the cell growth. Therefore, constitutive transgene delivery in inducible promoters in targeted tumor tissue enhances the expression of constitutive transgene leading to achieve efficient management of cell growth (Dogbevia et al., 2015). The inducible promoters are also regulated by various biomodulating agents like drugs, radiation, and heat. In addition, the early growth response gene-1 (Egr-1) promoter is increased with radiation and is used for the control of gene expression in different tissues (Kaliberov and Buchsbaum, 2012).

Further, the inducible promoters drive the ubiquitous transgenic expression (Powell et al., 2015). Therefore, the strategies of natural promoters overcome the weak transcriptional activity with exact specificity for the targeted tissues.

4.3.5 POSTTRANSCRIPTIONAL TARGETING FOR GENE DELIVERY

The expression levels of tissue specific genes are able to control during and after posttranscriptional activity. The exogenous administration of DNA regulates the posttranscriptional targeted action of cancer cells (Kaucsar et al., 2010). In addition, the expression of transgenes is regulated at posttranscriptional levels via controlling of RNA splicing, RNA stability, and translation initiation process (Xie et al., 2016b). *RNA splicing:* RNA splicing is a process of introns removal and coupling of exons in pre-mRNA transcripts leading to form mRNA in the nuclear site. The modulation of RNA splicing within a transcript (known as alternative RNA splicing) generates multiple mRNA isoforms in multiple cell types and tissues (Paul and Montpetit, 2016). This alternative RNA splicing is observed in malignancies due to the expression of the CD_{44} gene. However, the expression of a transgene is dependent upon the removal of the intron from the mRNA sequence (da Cunha et al., 2016; Siahpirani and Roy, 2016). *RNA stability:* The mRNA stability and their decay levels play a role in the posttranscriptional regulation by interaction with noncoding RNAs and it enhances the selective gene silencing (Incarnato and Oliviero, 2016). Therefore such kind of RNA is also known as RNA interference (RNAi) messenger or RNA stabilizer (Bobbin and Rossi, 2016).

The siRNAs enhance the RNAi process via degradation of mRNAs sequences via translation repression (Bobbin and Rossi, 2016; Zhao et al., 2016). *RNA translation:* The initiation process of translation recruits and assembles the initiator tRNA. It also regulates the start codon of the mRNA molecule in 40S and 60S ribosomal subunits (Zeng et al., 2016). In this condition, ribosomes inefficiently translate the proteins due to excessive availability of 5′ untranslated region (UTR) of the mRNA. The eIF4E factor is noted to be a rate-limiting component of translation initiation (Smirnova et al., 2016). The overexpression of the eIF4E factor enhances the translation of 5′ UTR mRNAs, which leads to reduce the ribosomal translation of proteins. These factors are raised in solid tumors and this stimulates the growth of blood vessels via fibroblast growth factor-2 (FGF-2) (Li et al., 2016b; Malka-Mahieu et al., 2016). The herpes simplex virus type 1 thymidine kinase (HTK) gene and ganciclovir conjugation with 5′ UTR of FGF-2 mRNA in nanoparticles cause the restriction of HTK mRNA translation leading to susceptibility in cancer cells (Kim et al., 2015).

4.3.6 MISCELLANEOUS TARGETS FOR GENE DELIVERY

The tumor suppressive genes like P^{53} are inactivated during the growth in wild type tumor cells. The TNF-related apoptosis-inducing ligand (TRAIL) targeted gene delivery regulates the cancer cell growth (Choonara et al., 2016). TRAIL gene has the unique capacity to induce apoptosis in specific cancer cells via activation of pro-apoptotic receptors like death receptor, that is, DR_4 and DR_5, in tumor cells and cancer cells (Chen et al., 2016e; Milutinovic et al., 2016). In addition, the mesenchymal stem cells (MSCs) like human adipose-derived MSCs (hAMSCs) migrate to tumor cells (Mangraviti et al., 2016). Therefore, the MSCs cell-based therapeutic gene delivery with DNA-encoded anticancer molecules reduces the cancer cells with bone morphogenetic protein-4 (BMP-4) (Cyr-Depauw et al., 2016). It differentiates the putative cancer stem cells and facilitates the tumor eradication (Sultan et al., 2016). This MSC plus gene therapy is proved to treat glioblastoma (GBM) via induction of brain tumor-initiating cells (BTICs) in mice (Mangraviti et al., 2016).

4.4 APPLICATION OF GENE THERAPY FOR GENETIC DISORDERS

The nanoparticle assisted gene therapy is widely used for the various genetic disorders like cancer as well as infectious, retinal, metabolic, neurodegenerative, and cardiovascular disease. The details are discussed in the following sections. This section covers the major the therapeutic options for the treatment of genetic disorders with gene delivery.

4.4.1 APPLICATION OF GENE THERAPY FOR HEMOPHILIA

Gene therapy is applied in a wide variety of genetic disorders. The effective in vivo gene delivery is targeted for postmitotic cells or tissues; whereas, ex vivo gene delivery is effective in autologous cells. The viral based vector systems such as adeno-associated virus vectors are clinically successful for in vivo delivery of genes (Arruda and Samelson-Jones, 2016). In addition, the serotypes and capsid variants—associated toxicity is also avoided by nanoparticle modification. Clinical application of ex vivo gene therapy is effective in autologous HSC for hematological disorders' connection with T lymphocytes (Adair et al., 2016). In addition, the retroviral vectors, that is, γ-retroviral or lentivirus, are reported to integrate their therapeutic action in targeted cells. Initially, it also produces adverse effects; later it promotes the therapeutic action with lentivirus and it shows better safety profile and efficient gene delivery action in nondividing cells (Brimble et al., 2016). Furthermore, rAAV vector based gene therapy has also been documented to produce an efficient ameliorative effect against lipoprotein lipase deficiency disorders (Gaudet et al., 2016). Also, the monogenic diseases such as primary immune deficiencies, hemoglobinopathies, hemophilia B, neurological diseases, ocular diseases, and cancer immunotherapies are also reported to be managed by gene therapy (Gessler and Gao, 2016).

Hemophilia is a coagulative hematological disorder caused by the mutation of an X-linked gene. The mutation occurs in the coagulation factor VIII (hemophilia A) or IX (hemophilia B) encoded X-gene chromosome (Arruda and Samelson-Jones, 2016; Sun et al., 2016). The coagulation factor of X (FIX) gene loaded AAV2 vector based gene transfer via hepatic artery attenuates the hemophilia B disorders. The inherited neurological disorders are also due to the changes of neurological genes (Yin et al., 2016a). The approach of gene therapy is documented to ameliorate the neurogenetic disorders like adrenoleukodystrophy (ALD), metachromatic leukodystrophy, and aromatic L-amino acid decarboxylase (AADC) deficiency (Kumar et al., 2016b). The delivery of genes for neurological disorders involves the both viral vectors, that is, integrating lentivirus and nonintegrating AAV vectors. The LV carries the ABCD1 gene for the gene delivery in the neurological tissue to attenuate the demyelination of CNS and adrenal cortex (Maguire et al., 2014). In addition, arylsulfatase A (ARSA) gene is used for the amelioration of late infantile meta chromatic leukodystrophy associated bone marrow damage and demyelination (Meneghini et al., 2016).

4.4.2 APPLICATION OF GENE THERAPY FOR RETINAL DISEASES

The inherited retinal diseases are also treated with AAV vectors. The gene alterations of specific retinal locations cause blindness; sometimes these are inherited forms of disease (Petit et al., 2016). Leber's congenital amaurosis type 2 (LCA2) is one of the inherited retinal diseases. The mutant retinal pigment epithelium 65 kilodalton protein (RPE65) gene reduces the expression of RPE65 leading to impaired vision and blindness (Pensado et al., 2016). The subretinal injection of AAV2 vector loaded RPE65 gene improves the vision. The chimeric antigen receptor-based immunotherapy is also documented to produce the efficient amelioration of cancer with genetic modification of autologous T cells (Chmielewski et al., 2011). The autologous CD^{8+} T cells with genetic manipulation recognize the killer cells with tumor-specific antigens (Gross and Eshhar, 2016). The T cell based cancer cells' immune reaction is also ameliorated with TCR gene delivery. The TCR genes are isolated from tumor infiltrating T cells and express the efficient management of T-cell-based cancer immunotherapy via interaction of tumor antigens (Qin et al., 2016). This genetically modified TCR gene also treats a wide variety of cancers like synovial cell sarcoma, neuroblastoma, melanoma, and colorectal cancer with chronic tumor regressive action (Klebanoff et al., 2016; Kumar et al., 2016b).

4.4.3 APPLICATION OF GENE THERAPY FOR INFECTIOUS DISEASE

Infectious disease is also effectively treated with gene delivery; the primary immunodeficiencies (PIDs) are one of the genetic infectious diseases. The heterogeneous changes of monogenic conditions alter the immune (innate and/or adaptive immune) responses. More than 300 genes are mutated in PID disorders. The genetic alteration with new pathogenic infection is widely observed in humans (Fodil et al., 2016; Jackman et al., 2016). It dysregulates the immune response with compromising of autoimmunity and causing susceptibility to lymphatic reticular malignancy. In some conditions, the efficacy of gene therapy is counterbalanced with insertional oncogenesis (Lund et al., 2016). The molecular events of oncogenesis are similar to those of immunological reaction. Advanced vector technology has resulted in efficient treatment of infectious disorders like X-linked SCID (SCID-X1), ADA-SCID, Wiskott–Aldrich syndrome (WAS), and chronic granulomatous disease (CGD) (Booth et al., 2016).

X-linked SCID is an infectious disorder due to mutations of IL2RG gene leading to lack of common gamma chain (γc) expression. This gene is also responsible for the host of cytokine receptors; interleukin (IL)-2, 4, 7, 9, 15, and 21 receptor actions; and lymphocyte development and function (Kuo and Kohn, 2016; Di Costanzo et al., 2016). The treatment of SCID-X1 with spontaneous reversion of mutated γc-encoded IL2RG gene is documented to restore immunological competence. CGD is a genetically mutated infectious disorder (Pham et al., 2016). These specific gene (i.e., gp91phox, p22phox, p47pox, and p67phox) mutations impair superoxide production in phagocytic cells leading to life-threatening abscess formation (sepsis) in skin, liver, lung, bone granuloma, and muscular inflammatory cells (Cicalese and Aiuti, 2015; Tani, 2016). In addition, IFN-γ based gene therapy shows higher success rate in recovery from various infectious disorders like severe invasive fungal infection, organ abscesses, inflammatory reaction, and autoimmune actions via reduction of superoxide anion production (Xie et al., 2016c). The administration of γRV gene-loaded vector ameliorates X-CGD via regulating action of CD34 + cells and nonmyeloablative condition (Liu et al., 2016b).

4.4.4 APPLICATION OF GENE THERAPY FOR METABOLIC DISORDERS

Genetic changes alter various metabolic processes and can cause serious complications. Niemann–Pick disease type A (NPA) is one of these metabolic disorders, which alters the sphingomyelin (SPM) protein, function of organomegaly and central nervous system (CNS) action via endocytosed cholesterol trafficking (Kaplitt and During, 2016; Sasset et al., 2016). Acid sphingomyelinase (ASM) is a primary target for the changes of metabolic activity. This enzyme is predominantly present in the lysosomes and it enhances the metabolism of membrane bound SPM to ceramide (Gessler and Gao, 2016). The primary symptoms of Niemann–Pick disease type A are hepatosplenomegaly, poor feeding behavior, reduction of motor function and deterioration of neurological tissue, and sometimes death (Chandler et al., 2016; Li et al., 2016d). In addition, low levels of HDL and hypertriglyceridemia are also observed. The major genetic change, that is, deficiency of ASM enzyme due to mutation of SMPD1 gene, leads to alteration of the mannose and mannose-6 phosphate receptors in rough endoplasmic reticulum (rRER), Golgi apparatus, and lysosomes (Acuña et al., 2016; Koehne et al., 2016).

The genetic engineered genes of NPD type B and C phenotype are targeted for the gene delivery. The ASM regulating gene is documented to ameliorate the metabolic changes and can help prevent liver transplantation, amniotic membrane transplantation, bone marrow transplantation, and enzyme replacement therapy associated with metabolic complications (Iraci et al., 2016). The administration of ASM cDNA gene shows the efficient ameliorative potential in NPA patients. Phenylketonuria (PKU) is one of the genetically modified amino acid metabolic disorders. PKU is a rare monogenetic disease that occurs due to different mutations of the phenylalanine hydroxylase (PAH) gene (Harding, 2017; Matern and Rinaldo, 2017; Tsang and Atkins, 2015). The primary manifestation of PKU is severe mental retardation, developmental impairment, seizures, eczema, hair loss, skin and iris pigmentation, and psychosocial problems. The major changes of PKU associated genes are PAH enzyme responsive genes and it is coded in chromosome 12. This gene is comprised of 13 exons and 79 kilobases (kb) length (Matern and Rinaldo, 2017; Trunzo et al., 2016; Tsang and Atkins, 2015).

PAH is abundantly expressed in liver and in other organs like kidney, pancreas, and brain. Three isozymes are identified in rat and two isoforms are detected in human liver. The 531 mutations are reported to produce the PAH gene (Manfredsson, 2016; Trunzo et al., 2016). The primary mechanism of PAH enzymes is to convert phenylalanine to tyrosine. This reaction is irreversible and tetrahydrobyopterin acts as cofactor. In addition, it is synthesized from guanosine triphosphate (GTP) with the help of GTP cyclohydrolase I (GTPCH), 6-pyruvolytetrahydrobiopterin synthase (PTPS), and sepiapterin reductase (SR) enzyme (Gessler and Gao, 2016). The delivery of phenotypic PKU gene enhances the levels of pheylalanine with reduced glutathione peroxidase and it also inhibits the catalase action (Mazzola et al., 2013).

4.4.5 APPLICATION OF GENE THERAPY FOR NEURODEGENERATIVE DISEASE

The genetic alteration is also identified in various neurological disorders. The neurodegeneration is a primary event in the progress of various neurological complications such as Alzheimer's disease (AD), Parkinson's disease (PD), amylotrophic lateral sclerosis (ALS), multiple sclerosis (MS), brain tumor, and neuropathic pain (Haston and Finkbeiner, 2016; Wolfe et al., 2016). AD is one of the common neurodegenerative disorders and it causes severe memory loss and cognitive impairment in the elderly population (Gessler and Gao, 2016). The genetic mutations are targeted in the gene of amyloid precursor protein causing the excess production of amyloid-β peptide and deposition; further it enhances the intracellular neurofibrillary tangle formation, decreased synaptic integrity, and neuronal loss overall in the brain (Yuan and Grutzendler, 2016; Yadav et al.,

2016). In addition, the acetylcholine nerve growth factor (NGF) and neurotrophin associated neuroprotective gene abnormalities also contribute to the progress of AD with neurodegeneration. The fibroblast grafts-loaded retrovirus vector is documented to restore the survival rate of cholinergic neurons (Loh et al., 2016). In addition, the NGF gene therapy also improves AD via reduction of neurotoxic Aβ peptide and removal of pathogenic Aβ peptides and allows the halting of proteolytic enzymes (Yuan and Grutzendler, 2016). Furthermore, the viral vector based active and passive vaccines and small inhibitory RNAs (siRNA) are also documented to protect the cholinergic neuron form pathogenic Aβ peptides, Aβ proteases, and suppression of APP processing genes (Snow and Albensi, 2016).

PD is also a chronic progressive neurodegenerative disease and it affects the different areas of the brain, inducing abnormal motor symptoms known as extrapyramidal syndrome (Wu et al., 2016). The gene mutation alters the dopaminergic neuron in the brain leading to reduction of dopamine levels in the substantia nigra pars compacta, striatum, and basal ganglia. In addition it also alters the mitochondrial associated metabolic changes, causing oxidative stress and neuronal excitability (Bose and Beal, 2016; Valdés and Schneider, 2016). Gene manipulation with specific gene therapy and cell therapy has ameliorated PD's progress (Visanji et al., 2016). The basic mechanism of gene therapy for PD is expected for the following actions: (1) upregulation of neurotrophic factor levels, (2) enhancing the endogenous dopamine production, and (3) improvement of neuronal circuitry (Pramanik et al., 2016).

The striatal injection of specific PD targeted gene, that is, glial-derived neurotrophic factor (GDNF) and neurturin, are documented to produce ameliorative effects (Kirik et al., 2016). Amyotrophic lateral sclerosis (ALS) is a progressive muscular disorder due to loss of motor neurons in the brain and spinal cord. The mutations of SOD1 gene have been identified for the changes of cytosolic Cu/Zn dependent superoxide dismutase (SOD) actions leading to alteration in the muscle function and generating muscle weakness (McEachin et al., 2016). In addition, the angiotensinogen (ANG) encoded gene, transactive response (TAR) DNA-binding protein TDP-43, FUS encoding fused in sarcoma protein, and optineurin genes are also involved in the progress of ALS (Kovacs, 2016; Morello and Cavallaro, 2015).

The gene therapy with fALS-associated mutant SOD1 is found to protect motor neurons in the spinal cord, thus attenuating the symptoms of sALS patients. Sometimes, the brain tumor's progress is increased with specific genetic mutation (Nizzardo et al., 2016). Malignant brain tumors, also known as GBMs, are one of the most common brain tumors (Woolf et al., 2016). In addition, some neurological tumors are also identified in the central as well as peripheral nervous system, such as vestibular schwannomas, which enhance the progress via angiogenesis process (Vijapura et al., 2016). The basic strategy of gene therapy is to remove the tumor mass from the body. The viral vector injection kills the tumor cells via activation of suicide genes (Karjoo et al., 2016). The viral vectors and targeted cells-loaded genes express various enzymes like viral thymidine kinase, bacterial cytosine deaminase/uracil phosphoribosyl transferase, and mammalian cytochrome P450/carboxyesterase for reduction of tumor cell growth (Malekshah et al., 2016). The HSV-1 and adenovirus are also used for the gene therapy to ameliorate the brain tumor (Fukuhara et al., 2016).

4.4.6 APPLICATION OF GENE THERAPY FOR CARDIOVASCULAR DISEASE

Some of the cardiovascular diseases are also identified as being due to genetic changes. The various cardiovascular genes conduct the cardiovascular functions. However, abnormal gene expression is also documented to produce serious complications (Castle and Feinstein, 2016; Kathiriya et al., 2015). The cardiovascular functional regulatory genes include tumor-suppressor p53 gene, which employs an antiproliferative action (Fishbein et al., 2010; Poller et al., 2012); metalloprotease inhibitor 3 (TIMP-3) for antimigratory action (Poller et al., 2012); hepatocyte growth factor (HGF) for antifibrotic action (Chen et al., 2016a); SOD for antioxidative action (Chen et al., 2016c); tissue factor pathway inhibitor (TFPI) for antithrombotic action (Zhang et al., 2015); dominant negative monocyte chemoattractant protein-1 (dnMCP-1) for antiinflammatory action (Paneni et al., 2016); NFκB-dependent gene (A20) for antiapoptotic (Samse et al., 2016); cardiac potassium channel missense mutation (Q9E-hMiRP1) for antiarrhythmic action (Wiedmann et al., 2016); sarcoplasmic and endoplasmic reticulum calcium ATPase 2 (SERCA-2) for pro-contractile action (Lymperopoulos et al., 2016); vascular endothelial growth factor (VEGF) for angiogenic action; and inducible nitric oxide synthase (iNOS) for pleiotropic action. The multiple-gene variable for different pathophysiological conditions in cardiovascular disease is due to mode of action, activity, and specificity of transgenes (Ng et al., 2016). The specific genes such as p53, retinoblastoma tumor suppressor (Rb), PDGF, VEGF, and other growth factors are involved in the pathological alterations of the cardiac system (Wang et al., 2016c). Genetic treatment with cardioprotective genes effectively produces the ameliorative effect with heterologous viral promoters like respiratory syncytial virus (RSV) and cytomegalovirus (CMV) promoters (Castle and Feinstein, 2016; Powell et al., 2015). In addition, the cardiac troponin T promoter associated gene therapy is documented to produce a beneficial effect in cardiovascular functions (Heckmann et al., 2016). Fig. 4.1 expresses the relationships between genetic disorders and their responsible genes.

4.5 LIMITATIONS OF GENE DELIVERY

Gene delivery and the achievement of efficient gene therapy remain limited due to various physiochemical and biological obstacles. The primary choice of gene delivery in the field of gene therapy is viral vector (Hill et al., 2016; Nam and

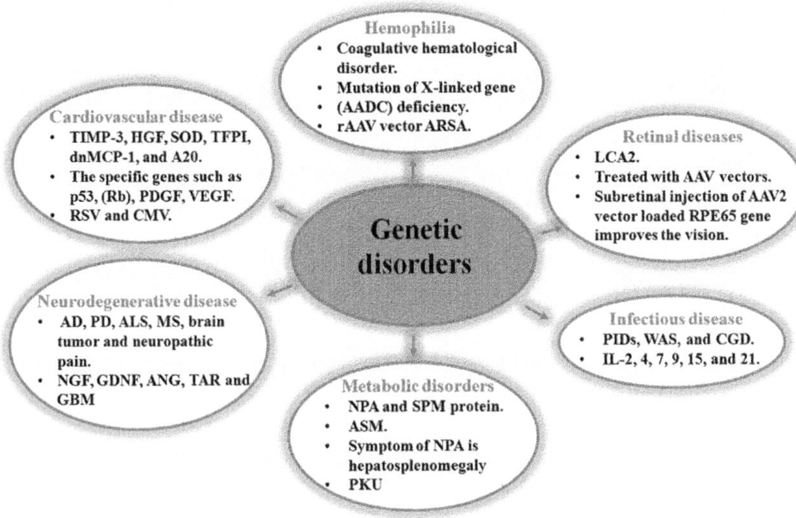

FIGURE 4.1

The genetic disorders and their responsible genes. This illustration is expressed the major genetic alteration associated multiple disorders. The details of abbreviations are AADC, aromatic L-amino acid decarboxylase; ARSA, arylsulfatase A; LCA2, Leber's congenital amaurosis type 2; RPE65, retinal pigment epithelium 65 kilodalton protein; PIDs, primary immune deficiencies; WAS, Wiskott–Aldrich syndrome; CGD, chronic granulomatous disease; IL, interleukin; ASM, acid sphingomyelinase; NPA, Niemann–Pick disease type A; PKU, phenylketonuria; AD, Alzheimer's disease; PD, Parkinson disease; ALS, amylotrophic lateral sclerosis; MS, multiple sclerosis; NGF, nerve growth factor; GDNF, glial-derived neurotrophic factor; ANG, angiotensinogen; TAR, transactive response; GBM, glioblastoma; TIMP-3, metalloprotease inhibitor 3; HGF, hepatocyte growth factor; SOD, superoxide dismutase; RSV, respiratory syncytial virus; CMV, cytomegalovirus; and VEGF, vascular endothelial growth factor.

Nah, 2016). The recombinant adeno-associated virus (rAAV) is widely used for the carrying gene for several therapeutic applications. It has specialized properties like erotype dependent tropism, nonpathogenicity, and low immunogenicity (Gessler and Gao, 2016; Kumar et al., 2016a). However, the tissue restriction of rAAV based transgene expression is higher due to physiological mimicking action, which produces undesirable side effects like transgenetoxicity (genotoxicity) and immune response (allergic and anaphylactic reactions) (Arruda and Samelson-Jones, 2016).

The immune reaction is the primary event in the nanoparticle-based gene delivery process. The severity of immune reaction depends upon the type of vector, transgene and transgene products involved in the immune response, and the targeted as well as host cells (Snyder, 2016). The humeral immune response plays a first-line defense action against systemic entry of rAAV particles leading to preexist the neutralizing antibodies (NAb). The adaptive immune system is observed in higher-level immunized patients with rAAV serotypes. The cell-mediated immune response is another obstacle for the gene delivery in the targeted tissue (Snyder, 2016). The cell mediated immunity resists the AAV-based gene delivery by specialized detection by infected cells. The immune response is activated by multiple varieties of cells such as natural killer cells (innate response), phagocytic cells (innate and adaptive response), and cytotoxic T cells (adaptive response) against viral capsids and/or transgenes (Galluzzi et al., 2016; Rey-Rico and Cucchiarini, 2016). Furthermore, there is the activation of toll-like receptor (TLR) 9 carrying cells, that is, monocyte by rAAV vector genome. The activation of CD^{8+} cytotoxic T cells by rAAV capsid infected cells via major histocompatibility complex-induced cells (Samaranch et al., 2016). Moreover, the antigen-presenting cells (APC) also restrict delivery of transgenes (Ju et al., 2016). The selection of relevant cells and organs with specific route of administration-based fabrication of nanoparticles helps to avoid the transgene toxicity or immunological reactions. The most common rAAV serotypes are identified for the better therapeutic outcome for gene therapy such as intracranial injection rAAV2 for phenylketonuria and neuronal Parkinson disease; intravenous administration of rAAV9, rh.8, and rh.10 for Canavan disease (Al Hafid and Christodoulou, 2015; Craig and Housley, 2016). Fig. 4.2 provides a summary of applications and limitations of gene therapy.

Gene therapy	
Applications	**Limitations**
1. Genetic disease	<u>Retinal diseases</u> • Visual defects and blindness. <u>Metabolic disorders</u> • Phenylketonuria. • Hepatosplenomegaly. • Mental retardation, developmental impairment, seizures, eczema, hair loss, skin and iris pigmentation.
2. Infectious disease	• Susceptible to lymphatic reticular malignancy. • Life-threatening abscess formation (sepsis) in skin, liver, lung, bone granuloma and muscular inflammatory cells.
3. Inflammatory disease	Lymphatic reticular malignancy.
4. Cancer therapy	Glioblastoma.
5. Vascular disease	Antiproliferative action. Metalloprotease inhibitor 3.
6. Neurodegenerative disease	Mictochondrial associated metabolic changes. Oxidative stress and neuronal excitability.

FIGURE 4.2

The applications and limitations of gene therapy. This figure expresses the possible applications and major hazards of gene therapy in the biological system.

4.6 FUTURE DIRECTIONS

Based on this literature and a basic understanding of the current concepts of site and target specific action, nanomedicine may be a future goal to explore the better medicine for chronic life threatening genetic disorders via gene therapeutic approach (Koushik et al., 2016; Merchant et al., 2016). Clinically, the rate of success for nanomedicine assisted gene delivery is limited; due to multiple challenges, hurdles, and ethical issues (Decuzzi, 2016; Landesman-Milo and Peer, 2016). This book chapter has also discussed the limitations and challenges of oncogenic and nononcogenic based nanomedicine and gene delivery (viral and nonviral vectors). The possibilities to overcome these challenges have also been expressed in this chapter. Therefore, new technology and genetic manipulation methods are useful for various genetic and metabolic disorders. A multidisciplinary approach including biotechnology, nanotechnology, biophysics, and biochemistry will open a new way to generate potential nanogene therapies and can reach new milestones for the treatment of genetic disorders.

ABBREVIATIONS

AADC	aromatic L-amino acid decarboxylase
AAV	adenovirus; adenovirus associated virus
AD	Alzheimer's disease
AEG-1	astrocyte elevated gene-1
ALD	adrenoleukodystrophy
ALS	amylotrophic lateral sclerosis
ANG	angiotensinogen
ARSA	arylsulfatase A
ASM	acid sphingomyelinase
BMP-4	bone morphogenetic protein-4
BTIC	brain tumor-initiating cell
CALA-01	cyclodextrin–PEG–Tf complex type nanoparticles
CD_{44}	cluster of domain
CGD	chronic granulomatous disease
CMV	cytomegalovirus
EGF	epidermal growth factor
Egr-1	early growth response gene-1
FIX	coagulation factor of X gene
FR	folate receptor
GBM	glioblastoma
GDNF	glial-derived neurotrophic factor
GTP	guanosine triphosphate
GTPCH	GTP cyclohydrolase I
HA	hyaluronic acid
hAMSCs	human adipose-derived MSCs

HER2	tumor specific antigen
HGF	hepatocyte growth factor
HSC	hematopoietic stem cells
HSV-1	herpes simplex virus type 1
hTERT	human telomerase reverse transcriptase
HTK	herpes simplex virus type 1 thymidine kinase
IL	interleukin
iNOS	inducible nitric oxide synthase
LCA2	Leber's congenital amaurosis type 2
MS	multiple sclerosis
MSCs	mesenchymal stem cells
Nab	neutralizing antibodies
NGF	nerve growth factor
NPA	Niemann–Pick disease type A
PABE	poly(β-amino ester)
PAH	phenylalanine hydroxylase gene
PAMAM	polyamidoamine
PD	Parkinson disease
PEI	polyethylenimine
PIDs	primary immunodeficiencies
PKU	phenylketonuria
PLI	poly(L-lysine)
PRINT	particle replication in nonwetting templates technology
PTPS	6-pyruvolytetrahydrobiopterin synthase
rAAV	recombinant adeno-associated virus
RGD peptide	arginine, glycine and aspartic acid peptides
RNAi	interference RNA
RPE65	retinal pigment epithelium 65 kilodalton protein
rRER	rough endoplasmatic reticulum
RSV	respiratory syncytial virus
SERCA-2	sarcoplasmic and endoplasmic reticulum calcium ATPase 2
siRNA	small interfering RNA
SOD	superoxide dismutase
SR	sepiapterin reductase
TAR	transactive response
TIMP-3	Metalloprotease inhibitor 3
TRAIL	TNF-related apoptosis-inducing ligand
UTR	5′-untranslated region
VEGF	vascular endothelial growth factor
WAS	Wiskott–Aldrich syndrome

ACKNOWLEDGMENT

The authors are thankful to Department of Pharmacology, JSS College of Pharmacy, Jagadgury Sri Shivarathreeswara University, Mysuru 570 015, Karnataka, India for their unconditional support and providing the technical facilities for the preparation of this book chapter.

REFERENCES

Acuña, M., Castro-Fernández, V., Latorre, M., Castro, J., Schuchman, E.H., Guixé, V., et al., 2016. Structural and functional analysis of the ASM p. Ala359Asp mutant that causes acid sphingomyelinase deficiency. Biochem. Biophys. Res. Commun. 479 (3), 496–501.

Adair, J.E., Waters, T., Haworth, K.G., Kubek, S.P., Trobridge, G.D., Hocum, J.D., et al., 2016. Semi-automated closed system manufacturing of lentivirus gene-modified hematopoietic stem cells for gene therapy. Nat. Commun. 7, 13173.

Adjei, I.M., Sharma, B., Peetla, C., Labhasetwar, V., 2016. Inhibition of bone loss with surface-modulated, drug-loaded nanoparticles in an intraosseous model of prostate cancer. J. Control. Release 232, 83–92.

Alghuthaymi, M.A., Almoammar, H., Rai, M., Said-Galiev, E., Abd-Elsalam, K.A., 2015. Myconanoparticles: synthesis and their role in phytopathogens management. Biotechnol. Biotechnol. Equip. 29 (2), 221–236.

Al Hafid, N., Christodoulou, J., 2015. Phenylketonuria: a review of current and future treatments. Transl. Pediatr. 4 (4), 304–317.

André, E.M., Passirani, C., Seijo, B., Sanchez, A., Montero-Menei, C.N., 2016. Nano and microcarriers to improve stem cell behaviour for neuroregenerative medicine strategies: Application to Huntington's disease. Biomaterials 83, 347–362.

Arruda, V.R., Samelson-Jones, B.J., 2015. Obstacles and future of gene therapy for hemophilia. Expert Opin. Orphan Drugs 3 (9), 997–1010.

Arruda, V.R., Samelson-Jones, B.J., 2016. Gene therapy for immune tolerance induction in hemophilia with inhibitors. J. Thromb. Haemost. 14 (6), 1121–1134.

Barto , D., Barto , A., Irimie, A., Szabo, B., Breazu, C., Tăbăran, F., et al., 2016. Ex vivo model of colon cancer in normothermic conditions: applications in nanomedicine. Particul. Sci. Technol. 34, 1–6.

Barve, A., Jin, W., Cheng, K., 2014. Prostate cancer relevant antigens and enzymes for targeted drug delivery. J. Control. Release 187, 118–132.

Bhatnagar, A., Wang, Y., Mease, R.C., Gabrielson, M., Sysa, P., Minn, I., et al., 2014. AEG-1 promoter-mediated imaging of prostate cancer. Cancer Res. 74 (20), 5772–5781.

Black, K.C., Ibricevic, A., Gunsten, S.P., Flores, J.A., Gustafson, T.P., Raymond, J.E., et al., 2016. In vivo fate tracking of degradable nanoparticles for lung gene transfer using PET and Ĉerenkov imaging. Biomaterials 98, 53–63.

Bobbin, M.L., Rossi, J.J., 2016. RNA interference (RNAi)-based therapeutics: delivering on the promise? Annu. Rev. Pharmacol. Toxicol. 56, 103–122.

Boken, J., Soni, S.K., Kumar, D., 2016. Microfluidic synthesis of nanoparticles and their bio-sensing applications. Crit. Rev. Anal. Chem. 46 (6), 538–561.

Booth, C., Gaspar, H.B., Thrasher, A.J., 2016. Treating Immunodeficiency through HSC gene therapy. Trends Mol. Med. 22 (4), 317–327.

Bose, A., Beal, M.F., 2016. Mitochondrial dysfunction in Parkinson's disease. J. Neurochem. 139 (Suppl. 1), 216–231.

Bowers, W.J., Breakefield, X.O., Sena-Esteves, M., 2011. Genetic therapy for the nervous system. Hum. Mol. Genet. 20 (R1), R28–41.

Bozzuto, G., Molinari, A., 2015. Liposomes as nanomedical devices. Int. J. Nanomed. 10, 975–999.

Brimble, M.A., Reiss, U.M., Nathwani, A.C., Davidoff, A.M., 2016. New and improved AAVenues: current status of hemophilia B gene therapy. Expert Opin. Biol. Ther. 16 (1), 79–92.

Brophy, J.A., Voigt, C.A., 2016. Antisense transcription as a tool to tune gene expression. Mol. Syst. Biol. 12 (1), 854.

Canter, R.J., Grossenbacher, S.K., Ames, E., Murphy, W.J., 2016. Immune targeting of cancer stem cells in gastrointestinal oncology. J. Gastrointest. Oncol. 7 (Suppl. 1), S1–S10.

Carroll, D., Charo, R.A., 2015. The societal opportunities and challenges of genome editing. Genome. Biol. 16, 242.

Castle, J., Feinstein, S.B., 2016. Drug and Gene Delivery Using Sonoporation for Cardiovascular Disease, Therapeutic Ultrasound. Springer, pp. 331–338.

Castro, J.E., Kipps, T.J., 2016. Adoptive cellular therapy for chronic lymphocytic leukemia and B cell malignancies. CARs and more. Best Pract. Res. Clin. Haematol. 29 (1), 15–29.

Cayrol, F., Diaz Flaque, M.C., Fernando, T., Yang, S.N., Sterle, H.A., Bolontrade, M., et al., 2015. Integrin alphavbeta3 acting as membrane receptor for thyroid hormones mediates angiogenesis in malignant T cells. Blood 125 (5), 841–851.

Chan, H.F., Ma, S., Leong, K.W., 2016. Can microfluidics address biomanufacturing challenges in drug/gene/cell therapies? Regen. Biomater. 3 (2), 87–98.

Chandler, R.J., Williams, I.M., Gibson, A.L., Davidson, C.D., Incao, A.A., Hubbard, B.T., et al., 2016. Systemic AAV9 gene therapy improves the lifespan of mice with Niemann–Pick disease, type C1. Hum. Mol. Genet. 26, ddw367.

Chattopadhyay, S., Dash, S.K., Mandal, D., Das, B., Tripathy, S., Dey, A., et al., 2016. Metal based nanoparticles as cancer antigen delivery vehicles for macrophage based antitumor vaccine. Vaccine 34 (7), 957–967.

Chavan, S.V., Maitra, A., Roy, N., Chavan, P.R., 2014. Contribution of allelic variability in prostate specific antigen (PSA) and androgen receptor (AR) genes to serum PSA levels in men with prostate cancer. Indian J. Med. Res. 139 (3), 371–378.

Chen, G.X., Zhang, S., He, X.H., Liu, S.Y., Ma, C., Zou, X.P., 2014. Clinical utility of recombinant adenoviral human p53 gene therapy: current perspectives. Onco Targets Ther. 7, 1901–1909.

Chen, H., Xia, R., Li, Z., Zhang, L., Xia, C., Ai, H., et al., 2016a. Mesenchymal stem cells combined with hepatocyte growth factor therapy for attenuating ischaemic myocardial fibrosis: assessment using multimodal molecular imaging. Sci. Rep. 6, 33700.

Chen, J., Guo, Z., Tian, H., Chen, X., 2016b. Production and clinical development of nanoparticles for gene delivery. Mol. Ther. Methods Clin. Dev. 3, 16023.

Chen, O., Ye, Z., Cao, Z., Manaenko, A., Ning, K., Zhai, X., et al., 2016c. Methane attenuates myocardial ischemia injury in rats through anti-oxidative, anti-apoptotic and anti-inflammatory actions. Free Radic. Biol. Med. 90, 1–11.

Chen, R., Cohen, A.L., Colman, H., 2016d. Targeted therapeutics in patients with high-grade gliomas: past, present, and future. Curr. Treat. Options Oncol. 17 (8), 42.

Chen, Z., Penet, M.-F., Krishnamachary, B., Banerjee, S.R., Pomper, M.G., Bhujwalla, Z. M., 2016e. PSMA-specific theranostic nanoplex for combination of TRAIL gene and 5-FC prodrug therapy of prostate cancer. Biomaterials 80, 57–67.

Cheng, J., Tang, X., Zhao, J., Shi, T., Zhao, P., Lin, C., 2016. Multifunctional cationic polyurethanes designed for non-viral cancer gene therapy. Acta Biomater. 30, 155–167.

Chmielewski, M., Kopecky, C., Hombach, A.A., Abken, H., 2011. IL-12 release by engineered T cells expressing chimeric antigen receptors can effectively Muster an antigen-independent macrophage response on tumor cells that have shut down tumor antigen expression. Cancer Res. 71 (17), 5697–5706.

Choonara, Y.E., Kumar, P., Modi, G., Pillay, V., 2016. Improving drug delivery technology for treating neurodegenerative diseases. Expert Opin. Drug. Deliv. 13 (7), 1029–1043.

Cicalese, M.P., Aiuti, A., 2015. Clinical applications of gene therapy for primary immunodeficiencies. Hum. Gene Ther. 26 (4), 210–219.

Coelho, J.F., Ferreira, P.C., Alves, P., Cordeiro, R., Fonseca, A.C., Gois, J.R., et al., 2010. Drug delivery systems: advanced technologies potentially applicable in personalized treatments. EPMA J. 1 (1), 164–209.

Conde, J., Arnold, C.E., Tian, F., Artzi, N., 2016. RNAi nanomaterials targeting immune cells as an anti-tumor therapy: the missing link in cancer treatment? Mater. Today 19 (1), 29–43.

Coulie, P.G., Van den Eynde, B.J., van der Bruggen, P., Boon, T., 2014. Tumour antigens recognized by T lymphocytes: at the core of cancer immunotherapy. Nat. Rev. Cancer 14 (2), 135–146.

Cox, D.B.T., Platt, R.J., Zhang, F., 2015. Therapeutic genome editing: prospects and challenges. Nat. Med. 21 (2), 121–131.

Craig, A.J., Housley, G.D., 2016. Evaluation of gene therapy as an intervention strategy to treat brain injury from stroke. Front. Mol. Neurosci. 9.

Cyr-Depauw, C., Northey, J.J., Tabariès, S., Annis, M.G., Dong, Z., Cory, S., et al., 2016. Chordin-like 1 suppresses bone morphogenetic protein 4-induced breast cancer cell migration and invasion. Mol. Cell Biol. 36 (10), 1509–1525.

da Cunha, C.B., Klumpers, D.D., Koshy, S.T., Weaver, J.C., Chaudhuri, O., Seruca, R., et al., 2016. CD44 alternative splicing in gastric cancer cells is regulated by culture dimensionality and matrix stiffness. Biomaterials 98, 152–162.

d'Amora, M., Rodio, M., Bartelmess, J., Sancataldo, G., Brescia, R., Zanacchi, F.C., et al., 2016. Biocompatibility and biodistribution of functionalized carbon nano-onions (f-CNOs) in a vertebrate model. Sci. Rep. 6.

Danda, R., Krishnan, G., Ganapathy, K., Krishnan, U.M., Vikas, K., Elchuri, S., et al., 2013. Targeted expression of suicide gene by tissue-specific promoter and microRNA regulation for cancer gene therapy. PLoS One 8 (12), e83398.

Das, S.K., Menezes, M.E., Bhatia, S., Wang, X.Y., Emdad, L., Sarkar, D., et al., 2015. Gene therapies for cancer: strategies, challenges and successes. J. Cell. Physiol. 230 (2), 259–271.

Dayem, A.A., Choi, H.Y., Yang, G.M., Kim, K., Saha, S.K., Kim, J.H., et al., 2016. The potential of nanoparticles in stem cell differentiation and further therapeutic applications. Biotechnol. J. 11 (12), 1550–1560.

Decuzzi, P., 2016. Facilitating the clinical integration of nanomedicines: the roles of theoretical and computational scientists. ACS Nano 10 (9), 8133–8138.

del Pozo-Rodríguez, A., Solinís, M.Á., Rodríguez-Gascón, A., 2016. Applications of lipid nanoparticles in gene therapy. Eur. J. Pharm. Biopharm. 109, 184–193.

De Toro, J., Herschlik, L., Waldner, C., Mongini, C., 2015. Emerging roles of exosomes in normal and pathological conditions: new insights for diagnosis and therapeutic applications. Novel clinical applications of extracellular vesicles. Front. Immunol. 6, 203.

Díaz-Torres, R., López-Arellano, R., Escobar-Chávez, J.J., García-García, E., Domínguez-Delgado, C.L., Ramírez-Noguera, P., 2016. Effect of size and functionalization of pharmaceutical nanoparticles and their interaction with biological systems. Handbook of Nanoparticles. Springer, pp. 1041–1060.

Di Costanzo, M., Morelli, M., Malamisura, M., Puoti, M.G., Canani, R.B., 2016. Immunodeficiency disorders resulting in malabsorption. Textbook of Pediatric Gastroenterology, Hepatology and Nutrition. Springer, pp. 425–436.

Dogbevia, G.K., Marticorena-Alvarez, R., Bausen, M., Sprengel, R., Hasan, M.T., 2015. Inducible and combinatorial gene manipulation in mouse brain. Front. Cell. Neurosci. 9, 142.

Dolatabadi, J.E.N., Omidi, Y., 2016. Solid lipid-based nanocarriers as efficient targeted drug and gene delivery systems. TrAC Trends Anal. Chem. 77, 100–108.

Elias, D.R., Poloukhtine, A., Popik, V., Tsourkas, A., 2013. Effect of ligand density, receptor density, and nanoparticle size on cell targeting. Nanomedicine 9 (2), 194–201.

Ewe, A., Przybylski, S., Burkhardt, J., Janke, A., Appelhans, D., Aigner, A., 2016. A novel tyrosine-modified low molecular weight polyethylenimine (P10Y) for efficient siRNA delivery in vitro and in vivo. J. Control. Release 230, 13–25.

Fernandes, A.R., Chari, D.M., 2016. Part I: Minicircle vector technology limits DNA size restrictions on ex vivo gene delivery using nanoparticle vectors: overcoming a translational barrier in neural stem cell therapy. J. Control. Release 238, 289–299.

Fishbein, I., Chorny, M., Levy, R.J., 2010. Site-specific gene therapy for cardiovascular disease. Curr. Opin. Drug. Discov. Dev. 13 (2), 203–213.

Fodil, N., Langlais, D., Gros, P., 2016. Primary immunodeficiencies and inflammatory disease: a growing genetic intersection. Trends Immunol. 37 (2), 126–140.

Frau, M., Feo, C.F., Feo, F., Pascale, R.M., 2014. New insights on the role of epigenetic alterations in hepatocellular carcinoma. J. Hepatocell. Carcinoma 1, 65–83.

Fujita, Y., Kuwano, K., Ochiya, T., 2015. Development of small RNA delivery systems for lung cancer therapy. Int. J. Mol. Sci. 16 (3), 5254–5270.

Fukuhara, H., Ino, Y., Todo, T., 2016. Oncolytic virus therapy: a new era of cancer treatment at dawn. Cancer Sci. 107 (10), 1373–1379.

Gai, M., Frueh, J., Kudryavtseva, V.L., Mao, R., Kiryukhin, M.V., Sukhorukov, G.B., 2016. Patterned microstructure fabrication: polyelectrolyte complexes vs polyelectrolyte multilayers. Sci. Rep. 6.

Galluzzi, L., Buqué, A., Kepp, O., Zitvogel, L., Kroemer, G., 2016. Immunogenic cell death in cancer and infectious disease. Nat. Rev. Immunol. 17 (2), 97–111.

Gandhi, N.S., Tekade, R.K., Chougule, M.B., 2014. Nanocarrier mediated delivery of siRNA/miRNA in combination with chemotherapeutic agents for cancer therapy: current progress and advances. J. Control. Release 194, 238–256.

Gao, Y., Huang, J.-Y., O'Keeffe Ahern, J., Cutlar, L., Zhou, D., Lin, F.-H., et al., 2016. Highly Branched Poly (β-amino esters) for Non-Viral Gene Delivery: High Transfection Efficiency and Low Toxicity Achieved by Increasing Molecular Weight. Biomacromolecules 17 (11), 3640–3647.

Gaudet, D., Stroes, E.S., Methot, J., Brisson, D., Tremblay, K., Bernelot Moens, S.J., et al., 2016. Long-term retrospective analysis of gene therapy with alipogene tiparvovec and its effect on lipoprotein lipase deficiency-induced pancreatitis. Hum. Gene Ther. 27 (11), 916–925.

Germershaus, O., Nultsch, K., 2015. Localized, non-viral delivery of nucleic acids: opportunities, challenges and current strategies. Asian J. Pharm. Sci. 10 (3), 159–175.

Gessler, D.J., Gao, G., 2016. Gene therapy for the treatment of neurological disorders: metabolic disorders. Gene therapy for neurological disorders: methods and protocols. Methods Mol. Biol. 1382, 429–465.

Giang, I., Boland, E.L., Poon, G.M., 2014. Prodrug applications for targeted cancer therapy. AAPS J. 16 (5), 899–913.

Gill, K.S., Fernandes, P., O'Donovan, T.R., McKenna, S.L., Doddakula, K.K., Power, D. G., et al., 2016. Glycolysis inhibition as a cancer treatment and its role in an anti-tumour immune response. Biochim. Biophys. Acta (BBA)—Rev. Cancer 1866 (1), 87–105.

Gottfried, L., Lin, X., Barravecchia, M., Dean, D.A., 2016. Identification of an alveolar type I epithelial cell-specific DNA nuclear import sequence for gene delivery. Gene Ther. 23 (10), 734–742.

Gross, G., Eshhar, Z., 2016. Therapeutic potential of T cell chimeric antigen receptors (CARs) in cancer treatment: counteracting off-tumor toxicities for safe car T cell therapy. Annu. Rev. Pharmacol. Toxicol. 56, 59–83.

Gu, J., Chen, X., Ren, X., Zhang, X., Fang, X., Sha, X., 2016. CD44-targeted hyaluronic acid-coated redox-responsive hyperbranched poly(amido amine)/plasmid DNA ternary nanoassemblies for efficient gene delivery. Bioconjug. Chem. 27 (7), 1723–1736.

Gupta, S.K., Shukla, P., 2016. Microbial platform technology for recombinant antibody fragment production: a review. Crit. Rev. Microbiol. 1–12.

Haile, S.T., Conner, J., Mackall, C., 2016. Attenuated oncolytic virus HSV1716 increases in vivo expansion of GD2-targeting CAR T cells in murine solid tumor models. Cancer Res. 76 (Suppl. 14), Abstract No. 2304.

Harding, C.O., 2017. Gene and cell therapy for inborn errors of metabolism. Inherited Metabolic Diseases. Springer, pp. 155–171.

Haston, K.M., Finkbeiner, S., 2016. Clinical trials in a dish: the potential of pluripotent stem cells to develop therapies for neurodegenerative diseases. Ann. Rev. Pharmacol. Toxicol. 56, 489.

He, C., Tang, Z., Tian, H., Chen, X., 2016. Co-delivery of chemotherapeutics and proteins for synergistic therapy. Adv. Drug Deliv. Rev. 98, 64–76.

Heckmann, M., Bauer, R., Jungmann, A., Winter, L., Rapti, K., Strucksberg, K., et al., 2016. AAV9-mediated gene transfer of desmin ameliorates cardiomyopathy in desmin-deficient mice. Gene Ther. 23 (8–9), 673.

Heera, P., Shanmugam, S., 2015. Nanoparticle characterization and application: an overview. Int. J. Curr. Microbiol. Appl. Sci. 4, 379–386.

Hijaz, M., Das, S., Mert, I., Gupta, A., Al-Wahab, Z., Tebbe, C., et al., 2016. Folic acid tagged nanoceria as a novel therapeutic agent in ovarian cancer. BMC Cancer 16, 220.

Hill, A.B., Chen, M., Chen, C.-K., Pfeifer, B.A., Jones, C.H., 2016. Overcoming gene-delivery hurdles: physiological considerations for nonviral vectors. Trends Biotechnol. 34 (2), 91–105.

Holvoet, B., De Waele, L., Quattrocelli, M., Gheysens, O., Sampaolesi, M., Verfaillie, C. M., et al., 2016. Increased understanding of stem cell behavior in neurodegenerative and neuromuscular disorders by use of noninvasive cell imaging. Stem Cells Int. 2016.

Hu, C., Li, L., 2016. Current reprogramming systems in regenerative medicine: from somatic cells to induced pluripotent stem cells. Regen. Med. 11 (91), 105–132.

Incarnato, D., Oliviero, S., 2016. The RNA epistructurome: uncovering RNA function by studying structure and post-transcriptional modifications. Trends Biotechnol. 35 (4), 318–333.

Indu, S., Ramesh, V., Oza, N., Prashad, K.V., 2013. Gene therapy: an overview. J. Orofac. Sci. 5 (2), 83.

Iraci, N., Leonardi, T., Gessler, F., Vega, B., Pluchino, S., 2016. Focus on extracellular vesicles: physiological role and signalling properties of extracellular membrane vesicles. Int. J. Mol. Sci. 17 (2), 171.

Ishida, K., Gee, P., Hotta, A., 2015. Minimizing off-target mutagenesis risks caused by programmable nucleases. Int. J. Mol. Sci. 16 (10), 24751–24771.

Ishihara, K., Chen, W., Liu, Y., Tsukamoto, Y., Inoue, Y., 2016. Cytocompatible and multifunctional polymeric nanoparticles for transportation of bioactive molecules into and within cells. Sci. Technol. Adv. Mater. 17 (1), 300–312.

Iyer, A.K., Duan, Z., Amiji, M.M., 2014. Nanodelivery systems for nucleic acid therapeutics in drug resistant tumors. Mol. Pharm. 11 (8), 2511–2526.

Jackman, J.A., Lee, J., Cho, N.J., 2016. nanomedicine for infectious disease applications: innovation towards broad-spectrum treatment of viral infections. Small 12 (9), 1133–1139.

Jang, M., Kim, J.H., Nam, H.Y., Kwon, I.C., Ahn, H.J., 2015. Design of a platform technology for systemic delivery of siRNA to tumours using rolling circle transcription. Nat. Commun. 6, 7930.

Jeggo, P.A., Pearl, L.H., Carr, A.M., 2016. DNA repair, genome stability and cancer: a historical perspective. Nat. Rev. Cancer 16 (1), 35–42.

Jhamb, S., Vangaveti, V.N., Malabu, U.H., 2016. Genetic and molecular basis of diabetic foot ulcers: clinical review. J. Tissue Viability 25 (4), 229–236.

Jing, F., Li, D., Xu, W., Liu, Y., Wang, K., Sui, Z., 2014. Transferrin- and folate-modified, double-targeted nanocarriers for gene delivery. Pharm. Biol. 52 (5), 570–574.

Jonkers, I., Lis, J.T., 2015. Getting up to speed with transcription elongation by RNA polymerase II. Nat. Rev. Mol. Cell Biol. 16 (3), 167–177.

Ju, H.L., Han, K.H., Lee, J.D., Ro, S.W., 2016. Transgenic mouse models generated by hydrodynamic transfection for genetic studies of liver cancer and preclinical testing of anti-cancer therapy. Int. J. Cancer 138 (7), 1601–1608.

Kafshdooz, T., Kafshdooz, L., Akbarzadeh, A., Hanifehpour, Y., Joo, S.W., 2016. Applications of nanoparticle systems in gene delivery and gene therapy. Artif. Cells Nanomed. Biotechnol. 44 (2), 581–587.

Kaliberov, S.A., Buchsbaum, D.J., 2012. Chapter seven—Cancer treatment with gene therapy and radiation therapy. Adv. Cancer Res. 115, 221–263.

Kang, L., Gao, Z., Huang, W., Jin, M., Wang, Q., 2015. Nanocarrier-mediated co-delivery of chemotherapeutic drugs and gene agents for cancer treatment. Acta Pharm. Sin. B 5 (3), 169–175.

Kantor, B., Bailey, R.M., Wimberly, K., Kalburgi, S.N., Gray, S.J., 2014. Methods for gene transfer to the central nervous system. Adv. Genet. 87, 125–197.

Kaplitt, M.G., During, M.J., 2016. GAD gene therapy for parkinson's disease. Translational Neuroscience. Springer, pp. 89–98.

Karimi, M., Eslami, M., Sahandi-Zangabad, P., Mirab, F., Farajisafiloo, N., Shafaei, Z., et al., 2016. pH-sensitive stimulus-responsive nanocarriers for targeted delivery of therapeutic agents. Wiley Interdiscip. Rev.: Nanomed. Nanobiotechnol. 8 (5), 696–716.

Karjoo, Z., Chen, X., Hatefi, A., 2016. Progress and problems with the use of suicide genes for targeted cancer therapy. Adv. Drug Deliv. Rev. 99, 113–128.

Kathiriya, I.S., Nora, E.P., Bruneau, B.G., 2015. Investigating the transcriptional control of cardiovascular development. Circ. Res. 116 (4), 700–714.

Kaucsar, T., Racz, Z., Hamar, P., 2010. Post-transcriptional gene-expression regulation by micro RNA (miRNA) network in renal disease. Adv. Drug Deliv. Rev. 62 (14), 1390–1401.

Kaufmann, K.B., Buning, H., Galy, A., Schambach, A., Grez, M., 2013. Gene therapy on the move. EMBO Mol. Med. 5 (11), 1642–1661.

Ke, W., Li, J., Zhao, K., Zha, Z., Han, Y., Wang, Y., et al., 2016. Modular design and facile synthesis of enzyme-responsive peptide-linked block copolymers for efficient delivery of doxorubicin. Biomacromolecules 17 (10), 3268–3276.

Kim, J., Wilson, D.R., Zamboni, C.G., Green, J.J., 2015. Targeted polymeric nanoparticles for cancer gene therapy. J. Drug Target. 23 (7–8), 627–641.

Kirik, D., Cederfjäll, E., Halliday, G., Petersén, Å., 2016. Gene therapy for Parkinson's disease: disease modification by GDNF family of ligands. Neurobiol. Dis. 97 (Pt B), 179–188.

Kitano, H., Mamiya, A., Ishikawa, T., Egoshi, K., Kokubun, S., Hidai, C., 2016. Long-term gene therapy with Del1 fragment using nonviral vectors in mice with explanted tumors. Onco Targets Ther. 9, 503–516.

Klebanoff, C.A., Rosenberg, S.A., Restifo, N.P., 2016. Prospects for gene-engineered T cell immunotherapy for solid cancers. Nat. Med. 22 (1), 26–36.

Koehne, T., Markmann, S., Schweizer, M., Muschol, N., Friedrich, R.E., Hagel, C., et al., 2016. Mannose 6-phosphate-dependent targeting of lysosomal enzymes is required for normal craniofacial and dental development. Biochim. Biophys. Acta (BBA): Mol. Basis Disease 1862 (9), 1570–1580.

Koga, J.-I., Matoba, T., Egashira, K., 2016. Anti-inflammatory nanoparticle for prevention of atherosclerotic vascular diseases. J. Atheroscler. Thromb. 23 (7), 757–765.

Kohyama, J., Sanosaka, T., Tokunaga, A., Takatsuka, E., Tsujimura, K., Okano, H., et al., 2010. BMP-induced REST regulates the establishment and maintenance of astrocytic identity. J. Cell. Biol. 189 (1), 159–170.

Koushik, O., Rao, Y., Kumar, P., Karthikeyan, R., 2016. Nano drug delivery systems to overcome cancer drug resistance—a review. J. Nanomed. Nanotechnol. 7 (378), 2.

Kovacs, G.G., 2016. Molecular pathological classification of neurodegenerative diseases: turning towards precision medicine. Int. J. Mol. Sci. 17 (2), 189.

Kozielski, K.L., Rui, Y., Green, J.J., 2016. Non-viral nucleic acid containing nanoparticles as cancer therapeutics. Expert Opin. Drug Deliv. 13 (10), 1475–1487.

Kumar, M., Dravid, A., Kumar, A., Sen, D., 2016a. Gene therapy as a potential tool for treating neuroblastoma—a focused review. Cancer Gene Ther. 23 (5), 115–124.

Kumar, S.R., Markusic, D.M., Biswas, M., High, K.A., Herzog, R.W., 2016b. Clinical development of gene therapy: results and lessons from recent successes. Mol. Ther. Methods Clin. Dev. 3, 16034.

Kuo, C.Y., Kohn, D.B., 2016. Gene therapy for the treatment of primary immune deficiencies. Curr. Allergy Asthma Rep. 16 (5), 1–8.

Lai, Y.H., Lin, C.C., Chen, S.H., Tai, C.K., 2015. Tumor-specific suicide gene therapy for hepatocellular carcinoma by transcriptionally targeted retroviral replicating vectors. Gene Ther. 22 (2), 155–162.

Lambricht, L., Lopes, A., Kos, S., Sersa, G., Préat, V., Vandermeulen, G., 2016. Clinical potential of electroporation for gene therapy and DNA vaccine delivery. Expert Opin. Drug Deliv. 13 (2), 295–310.

Landesman-Milo, D., Peer, D., 2016. Transforming nanomedicines from lab scale production to novel clinical modality. Bioconjug. Chem. 27 (4), 855–862.

Larsen, M.T., Kuhlmann, M., Hvam, M.L., Howard, K.A., 2016. Albumin-based drug delivery: harnessing nature to cure disease. Mol. Cell. Ther. 4 (1), 1.

Lee, B.K., Yun, Y., Park, K., 2016. PLA micro- and nano-particles. Adv. Drug Deliv. Rev. 107, 176–191.

Li, B.-C., Chang, H., Ren, K.-f, Ji, J., 2016a. Substrate-mediated delivery of gene complex nanoparticles via polydopamine coating for enhancing competitiveness of endothelial cells. Colloids Surf. B: Biointerfaces 147, 172–179.

Li, D., Xie, K., Zhang, L., Yao, X., Li, H., Xu, Q., et al., 2016b. Dual blockade of vascular endothelial growth factor (VEGF) and basic fibroblast growth factor (FGF-2) exhibits potent anti-angiogenic effects. Cancer Lett. 377 (2), 164–173.

Li, J., Xue, S., Mao, Z.-W., 2016c. Nanoparticle delivery systems for siRNA-based therapeutics. J. Mater. Chem. B 4 (41), 6620–6639.

Li, L., Niu, H., Sun, B., Xiao, Y., Li, W., Yuan, H., et al., 2016d. Riccardin DN induces lysosomal membrane permeabilization by inhibiting acid sphingomyelinase and interfering with sphingomyelin metabolism in vivo. Toxicol. Appl. Pharmacol. 310, 175–184.

Li, S.S., Qu, Z., Haas, M., Ngo, L., Heo, Y.J., Kang, H.J., et al., 2016e. The HSA21 gene EURL/C21ORF91 controls neurogenesis within the cerebral cortex and is implicated in the pathogenesis of down syndrome. Sci. Rep. 6, 29514.

Liu, J., Wei, T., Zhao, J., Huang, Y., Deng, H., Kumar, A., et al., 2016a. Multifunctional aptamer-based nanoparticles for targeted drug delivery to circumvent cancer resistance. Biomaterials 91, 44–56.

Liu, X.S., Wu, H., Ji, X., Stelzer, Y., Wu, X., Czauderna, S., et al., 2016b. Editing DNA methylation in the mammalian genome. Cell 167 (1), 233–247. e217.

Liu, Y., Yang, M., Zhang, J., Zhi, X., Li, C., Zhang, C., et al., 2016c. Human induced pluripotent stem cells for tumor targeted delivery of gold nanorods and enhanced photothermal therapy. ACS Nano 10 (2), 2375–2385.

Loh, C.Y.Y., Wang, A.Y.L., Kao, H.-K., Cardona, E., Chuang, S.-H., Wei, F.-C., 2016. Episomal induced pluripotent stem cells promote functional recovery of transected murine peripheral nerve. PLoS ONE 11 (10), e0164696.

Lombardo, D., Kiselev, M.A., Magazù, S., Calandra, P., 2015. Amphiphiles self-assembly: basic concepts and future perspectives of supramolecular approaches. Adv. Condens. Matter Phys. 2015.

Lund, A.W., Medler, T.R., Leachman, S.A., Coussens, L.M., 2016. Lymphatic vessels, inflammation, and immunity in skin cancer. Cancer Discov. 6 (1), 22–35.

Lundstrom, K., 2016. Self-replicating RNA viral vectors in vaccine development and gene therapy. Fut. Virol. 11 (5), 345–356.

Lymperopoulos, A., McCrink, K., Brill, A., 2016. Impact of CYP2D6 genetic variation on the response of the cardiovascular patient to carvedilol and metoprolol. Curr. Drug Metab. 17 (1), 30–36.

Ma, Y., Yang, H., Pitt, J.M., Kroemer, G., Zitvogel, L., 2016. Therapy-induced microenvironmental changes in cancer. J. Mol. Med. (Berl.) 94 (5), 497–508.

Maeder, M.L., Gersbach, C.A., 2016. Genome-editing technologies for gene and cell therapy. Mol. Ther. 24 (3), 430–446.

Maguire, C.A., Ramirez, S.H., Merkel, S.F., Sena-Esteves, M., Breakefield, X.O., 2014. Gene therapy for the nervous system: challenges and new strategies. Neurotherapeutics 11 (4), 817–839.

Majzoub, R.N., Ewert, K.K., Safinya, C.R., 2016. Cationic liposome–nucleic acid nanoparticle assemblies with applications in gene delivery and gene silencing. Philos. Trans. R. Soc. A 374 (2072), 20150129.

Makita-Chingombe, F., Kutscher, H.L., DiTursi, S.L., Morse, G.D., Maponga, C.C., 2016. Poly(lactic-co-glycolic) acid-chitosan dual loaded nanoparticles for antiretroviral nanoformulations. J. Drug Deliv. 2016.

Malekshah, O.M., Chen, X., Nomani, A., Sarkar, S., Hatefi, A., 2016. Enzyme/prodrug systems for cancer gene therapy. Curr. Pharmacol. Rep. 2 (6), 299–308.

Malhotra, M., Tomaro-Duchesneau, C., Saha, S., Prakash, S., 2013. Systemic siRNA delivery via peptide-tagged polymeric nanoparticles, targeting PLK1 gene in a mouse xenograft model of colorectal cancer. Int. J. Biomater. 2013, Article ID 252531.

Mali, S., 2013. Delivery systems for gene therapy. Indian J. Hum. Genet. 19 (1), 3–8.

Malka-Mahieu, H., Newman, M., Désaubry, L., Robert, C., Vagner, S., 2016. Molecular pathways: the eIF4F translation initiation complex—new opportunities for cancer treatment. Clin. Cancer Res. 23 (1), 21–25.

Manfredsson, F.P., 2016. Gene therapy for neurological disorders. Methods Mol. Biol. 1382.

Mangraviti, A., Tzeng, S.Y., Gullotti, D., Kozielski, K.L., Kim, J.E., Seng, M., et al., 2016. Non-virally engineered human adipose mesenchymal stem cells produce BMP4, target brain tumors, and extend survival. Biomaterials 100, 53–66.

Massadeh, S., Al-Aamery, M., Bawazeer, S., AlAhmad, O., AlSubai, R., Barker, S., et al., 2016. Nano-materials for gene therapy: an efficient way in overcoming challenges of gene delivery. J. Biosens. Bioelectron. 2016.

Matern, D., Rinaldo, P., 2017. Newborn screening for inherited metabolic disease. Inherited Metabolic Diseases. Springer, pp. 421–437.

Mazzola, P.N., Karikas, G.A., Schulpis, K.H., Dutra-Filho, C.S., 2013. Antioxidant treatment strategies for hyperphenylalaninemia. Metab. Brain Dis. 28 (4), 541–550.

McEachin, Z.T., Donsante, A., Boulis, N., 2016. Gene therapy for the treatment of neurological disorders: amyotrophic lateral sclerosis. Gene Ther. Neurol. Disord.: Methods Protoc. 399–408.

Meneghini, V., Lattanzi, A., Tiradani, L., Bravo, G., Morena, F., Sanvito, F., et al., 2016. Pervasive supply of therapeutic lysosomal enzymes in the CNS of normal and Krabbe-affected non-human primates by intracerebral lentiviral gene therapy. EMBO Mol. Med. 8 (5), 489–510.

Merchant, Z., Buckton, G., Taylor, K.M., Stapleton, P., Saleem, I.Y., Zariwala, M.G., et al., 2016. A new era of pulmonary delivery of nano-antimicrobial therapeutics to treat chronic pulmonary infections. Curr. Pharm. Des. 22 (17), 2577–2598.

Metzger, C.S., Koutsimpelas, D., Brieger, J., 2015. Transcriptional regulation of the VEGF gene in dependence of individual genomic variations. Cytokine 76 (2), 519–526.

Milutinovic, S., Kashyap, A.K., Yanagi, T., Wimer, C., Zhou, S., O'Neil, R., et al., 2016. Dual agonist surrobody simultaneously activates death receptors DR4 and DR5 to induce cancer cell death. Mol. Cancer Ther. 15 (1), 114–124.

Mingozzi, F., High, K.A., 2013. Immune responses to AAV vectors: overcoming barriers to successful gene therapy. Blood 122 (1), 23–36.

Mishra, N.K., Shukla, M., 2014. Application of computational proteomics and lipidomics in drug discovery. J. Theor. Comput. Sci. 2014.

Mokhtarzadeh, A., Alibakhshi, A., Hashemi, M., Hejazi, M., Hosseini, V., de la Guardia, M., et al., 2016. Biodegradable nano-polymers as delivery vehicles for therapeutic small non-coding ribonucleic acids. J. Control. Release 245, 116–126.

Morello, G., Cavallaro, S., 2015. Transcriptional analysis reveals distinct subtypes in amyotrophic lateral sclerosis: implications for personalized therapy. Fut. Med. Chem. 7 (10), 1335–1359.

Naldini, L., 2015. Gene therapy returns to centre stage. Nature 526 (7573), 351–360.

Nam, J.-P., Nah, J.-W., 2016. Target gene delivery from targeting ligand conjugated chitosan–PEI copolymer for cancer therapy. Carbohydr. Polym. 135, 153–161.

Navarro, G., Pan, J., Torchilin, V.P., 2015. Micelle-like nanoparticles as carriers for DNA and siRNA. Mol. Pharm. 12 (2), 301–313.

Navya, P., Daima, H.K., 2016. Rational engineering of physicochemical properties of nanomaterials for biomedical applications with nanotoxicological perspectives. Nano Converg. 3 (1), 1.

Nayerossadat, N., Maedeh, T., Ali, P.A., 2012. Viral and nonviral delivery systems for gene delivery. Adv. Biomed. Res. 1, 27.

Ng, T.M., Goland, S., Elkayam, U., 2016. Relaxin for the treatment of acute decompensated heart failure: pharmacology, mechanisms of action, and clinical evidence. Cardiol. Rev. 24 (4), 194–204.

Nge, P.N., Rogers, C.I., Woolley, A.T., 2013. Advances in microfluidic materials, functions, integration, and applications. Chem. Rev. 113 (4), 2550–2583.

Nizzardo, M., Simone, C., Rizzo, F., Ulzi, G., Ramirez, A., Rizzuti, M., et al., 2016. Morpholino-mediated SOD1 reduction ameliorates an amyotrophic lateral sclerosis disease phenotype. Sci. Rep. 6, 21301.

Ott de Bruin, L.M., Volpi, S., Musunuru, K., 2015. Novel genome-editing tools to model and correct primary immunodeficiencies. Front. Immunol. 6, 250.

Padera, T.P., Meijer, E.F., Munn, L.L., 2016. The lymphatic system in disease processes and cancer progression. Annu. Rev. Biomed. Eng. 18, 125–158.

Paneni, F., Costantino, S., Kränkel, N., Cosentino, F., Lüscher, T.F., 2016. Reprogramming ageing and longevity genes restores paracrine angiogenic properties of early outgrowth cells. Eur. Heart J. 37 (22), 1733–1737.

Park, J.-K., Utsumi, T., Seo, Y.-E., Deng, Y., Satoh, A., Saltzman, W.M., et al., 2016. Cellular distribution of injected PLGA–nanoparticles in the liver. Nanomed.: Nanotechnol. Biol. Med. 12 (5), 1365–1374.

Paul, B., Montpetit, B., 2016. Altered RNA processing and export lead to retention of mRNAs near transcription sites and nuclear pore complexes or within the nucleolus. Mol. Biol. Cell. 27 (17), 2742–2756.

Pensado, A., Diaz-Corrales, F.J., De la Cerda, B., Valdés-Sánchez, L., del Boz, A.A., Rodriguez-Martinez, D., et al., 2016. Span poly-L-arginine nanoparticles are efficient non-viral vectors for PRPF31 gene delivery: an approach of gene therapy to treat retinitis pigmentosa. Nanomed.: Nanotechnol. Biol. Med. 12 (8), 2251–2260.

Perez, R.A., Choi, S.-J., Han, C.-M., Kim, J.-J., Shim, H., Leong, K.W., et al., 2016. Biomaterials control of pluripotent stem cell fate for regenerative therapy. Prog. Mater. Sci. 82, 234–293.

Petit, L., Khanna, H., Punzo, C., 2016. Advances in gene therapy for diseases of the eye. Hum. Gene Ther. 27 (8), 563–579

Pham, C., Martin, A., Antony, J., MacLeod, D., Brown, A., Nicholson, M., et al., 2016. 200. Generation of CAR-T cells lacking T cell receptor and human leukocyte antigen using engineered meganucleases. Mol. Ther. 24 (1).

Plaza-Diaz, J., Gomez-Llorente, C., Fontana, L., Gil, A., 2014. Modulation of immunity and inflammatory gene expression in the gut, in inflammatory diseases of the gut and in the liver by probiotics. World J. Gastroenterol. 20 (42), 15632–15649.

Poller, W., Rother, M., Skurk, C., Scheibenbogen, C., 2012. Endogenous migration modulators as parent compounds for the development of novel cardiovascular and anti-inflammatory drugs. Br. J. Pharmacol. 165 (7), 2044–2058.

Powell, S.K., Rivera-Soto, R., Gray, S.J., 2015. Viral expression cassette elements to enhance transgene target specificity and expression in gene therapy. Discov. Med. 19 (102), 49–57.

Prabha, S., Arya, G., Chandra, R., Ahmed, B., Nimesh, S., 2016. Effect of size on biological properties of nanoparticles employed in gene delivery. Artif. Cells Nanomed. Biotechnol. 44 (1), 83–91.

Pradhan, A., Pinheiro, J.P., Seena, S., Pascoal, C., Cassio, F., 2014. Polyhydroxyfullerene binds cadmium ions and alleviates metal-induced oxidative stress in *Saccharomyces cerevisiae*. Appl. Environ. Microbiol. 80 (18), 5874–5881.

Pramanik, S., Sulistio, Y.A., Heese, K., 2016. Neurotrophin signaling and stem cells—implications for neurodegenerative diseases and stem cell therapy. Mol. Neurobiol. 1–59.

Pranjol, M.Z., Hajitou, A., 2015. Bacteriophage-derived vectors for targeted cancer gene therapy. Viruses 7 (1), 268–284.

Quaresma, A.J.C., Bugai, A., Barboric, M., 2016. Cracking the control of RNA polymerase II elongation by 7SK snRNP and P-TEFb. Nucleic Acids Res. 44 (16), 7527–7539.

Qin, A., Coffey, D.G., Warren, E.H., Ramnath, N., 2016. Mechanisms of immune evasion and current status of checkpoint inhibitors in non-small cell lung cancer. Cancer Med. 5 (9), 2567–2578.

Rama, A.R., Aguilera, A., Melguizo, C., Caba, O., Prados, J., 2015. Tissue specific promoters in colorectal cancer. Dis. Markers 2015, 390161.

Ramamoorth, M., Narvekar, A., 2015. Non viral vectors in gene therapy- an overview. J. Clin. Diagn. Res. 9 (1), GE01-06.

Rey-Rico, A., Cucchiarini, M., 2016. Controlled release strategies for rAAV-mediated gene delivery. Acta Biomater. 29, 1–10.

Rivoltini, L., Chiodoni, C., Squarcina, P., Tortoreto, M., Villa, A., Vergani, B., et al., 2016. TNF-related apoptosis-inducing ligand (TRAIL)-armed exosomes deliver proapoptotic signals to tumor site. Clin. Cancer Res. 22 (14), 3499–3512.

Ross, N.L., Sullivan, M.O., 2016. Overexpression of caveolin-1 in inflammatory breast cancer cells enables IBC-specific gene delivery and prodrug conversion using histone-targeted polyplexes. Biotechnol. Bioeng.

Sachdeva, M., Sachdeva, N., Pal, M., Gupta, N., Khan, I.A., Majumdar, M., et al., 2015. CRISPR/Cas9: molecular tool for gene therapy to target genome and epigenome in the treatment of lung cancer. Cancer Gene Ther. 22 (11), 509–517.

Saleh, T., Shojaosadati, S.A., 2016. Multifunctional nanoparticles for cancer immunotherapy. Hum. Vac. Immunother. 1–13.

Samaranch, L., Hadaczek, P., Kells, A.P., Bringas, J.R., Stockinger, D., San Sebastian, W., et al., 2016. Slow AAV2 clearance from the brain of nonhuman primates and anti-capsid immune response. Gene Ther. 23 (4), 393–398.

Samse, K., Hariharan, N., Sussman, M.A., 2016. Personalizing cardiac regenerative therapy: at the heart of Pim1 kinase. Pharmacol. Res. 103, 13–16.

Sánchez-Moreno, P., Ortega-Vinuesa, J., Peula-García, J., Marchal, J., Boulaiz, H., 2016. Smart drug-delivery systems for cancer nanotherapy. Curr. Drug Targets 17.

Sasset, L., Zhang, Y., Dunn, T.M., Di Lorenzo, A., 2016. Sphingolipid de novo biosynthesis: a rheostat of cardiovascular homeostasis. Trends Endocrinol. Metabol. 27 (11), 807–819.

Schaefer, M.H., Serrano, L., 2016. Cell type-specific properties and environment shape tissue specificity of cancer genes. Sci. Rep. 6, 20707.

Senol, S., Ceyran, A.B., Aydin, A., Zemheri, E., Ozkanli, S., Kosemetin, D., et al., 2015. Folate receptor alpha expression and significance in endometrioid endometrium carcinoma and endometrial hyperplasia. Int. J. Clin. Exp. Pathol. 8 (5), 5633–5641.

Shan, D., Li, J., Cai, P., Prasad, P., Liu, F., Rauth, A.M., et al., 2015. RGD-conjugated solid lipid nanoparticles inhibit adhesion and invasion of alphavbeta3 integrin-overexpressing breast cancer cells. Drug Deliv. Transl. Res. 5 (1), 15–26.

Shi, J., Votruba, A.R., Farokhzad, O.C., Langer, R., 2010. Nanotechnology in drug delivery and tissue engineering: from discovery to applications. Nano Lett. 10 (9), 3223–3230.

Siahpirani, A.F., Roy, S., 2016. A prior-based integrative framework for functional transcriptional regulatory network inference. Nucleic Acids Res. 45 (4), gkw963.

Siddique, N., Raza, H., Ahmed, S., Khurshid, Z., Zafar, M.S., 2016. Gene therapy: a paradigm shift in dentistry. Genes (Basel) 7 (11).

Smirnova, V.V., Terenin, I.M., Khutornenko, A.A., Andreev, D.E., Dmitriev, S.E., Shatsky, I.N., 2016. Does HIV-1 mRNA 5′-untranslated region bear an internal ribosome entry site? Biochimie 121, 228–237.

Smyth, E.C., Sclafani, F., Cunningham, D., 2014. Emerging molecular targets in oncology: clinical potential of MET/hepatocyte growth-factor inhibitors. Onco Targets Ther. 7, 1001–1014.

Snow, W.M., Albensi, B.C., 2016. Neuronal gene targets of NF-κB and their dysregulation in Alzheimer's disease. Front. Mol. Neurosci. 9.

Snyder, R.O., 2016. An overview of rAAV vector product development for gene therapy. Regenerative Medicine for Degenerative Muscle Diseases. Springer, pp. 21–37.

Srivastava, A., Zhong, L., Zolotukhin, S., Aslanidi, G.V., Agbandje-McKenna, M., Van Vliet, K.M., 2014. Capsid-modified, raav3 vector compositions and uses in gene therapy of human liver cancer. Google Patents.

Steinbach, J.M., Seo, Y.-E., Saltzman, W.M., 2016. Cell penetrating peptide-modified poly (lactic-co-glycolic acid) nanoparticles with enhanced cell internalization. Acta Biomater. 30, 49–61.

Suk, J.S., Xu, Q., Kim, N., Hanes, J., Ensign, L.M., 2016. PEGylation as a strategy for improving nanoparticle-based drug and gene delivery. Adv. Drug Deliv. Rev. 99 (Pt A), 28–51.

Sultan, M., Coyle, K.M., Vidovic, D., Thomas, M.L., Gujar, S., Marcato, P., 2016. Hide-and-seek: the interplay between cancer stem cells and the immune system. Carcinogenesis 38 (2), 107–118. bgw115.

Sun, Q., Zhang, X.-Q., Wang, Y., Lu, A.-H., 2015. Recent progress on core-shell nanocatalysts. Chin. J. Catal. 36 (5), 683–691.

Sun, N., Chen, Y., Peng, H., Luo, Y., Zhang, G., 2016. A novel Ala275Val mutation in factor X gene influences its structural compatibility and impairs intracellular trafficking and coagulant activity. Thromb. Res. 138, 108–113.

Tani, K., 2016. Current status of ex vivo gene therapy for hematological disorders: a review of clinical trials in Japan around the world. Int. J. Hematol. 104 (1), 42–72.

Tasoglu, S., Tekin, H.C., Inci, F., Knowlton, S., Wang, S., Wang-Johanning, F., et al., 2015. Advances in nanotechnology and microfluidics for human papillomavirus diagnostics. Proc. IEEE 103 (2), 161–178.

Thomas, T., Tajmir-Riahi, H., Thomas, T., 2016. Polyamine–DNA interactions and development of gene delivery vehicles. Amino Acids. 48 (10), 2423–2431.

Tinkle, S., McNeil, S.E., Mühlebach, S., Bawa, R., Borchard, G., Barenholz, Y.C., et al., 2014. Nanomedicines: addressing the scientific and regulatory gap. Ann. N. Y. Acad. Sci. 1313 (1), 35–56.

Trivedi, M., Talekar, M., Shah, P., Ouyang, Q., Amiji, M., 2016. Modification of tumor cell exosome content by transfection with wt-p53 and microRNA-125b expressing plasmid DNA and its effect on macrophage polarization. Oncogenesis 5 (8), e250.

Trunzo, R., Santacroce, R., Shen, N., Jung-Klawitter, S., Leccese, A., De Girolamo, G., et al., 2016. In vitro residual activity of phenylalanine hydroxylase variants and correlation with metabolic phenotypes in PKU. Gene 594 (1), 138–143.

Tsang, J.J., Atkins, H.L., 2015. The ex vivo purge of cancer cells using oncolytic viruses: recent advances and clinical implications. Oncol. Virother. 4, 13–23.

Tzeng, S.Y., Wilson, D.R., Hansen, S.K., Quiñones-Hinojosa, A., Green, J.J., 2016. Polymeric nanoparticle-based delivery of TRAIL DNA for cancer-specific killing. Bioeng. Transl. Med. 1 (2), 149–159.

Uusi-Kerttula, H., Legut, M., Davies, J., Jones, R., Hudson, E., Hanna, L., et al., 2015. Incorporation of peptides targeting EGFR and FGFR1 into the adenoviral fiber knob domain and their evaluation as targeted cancer therapies. Hum. Gene Ther. 26 (5), 320–329.

Valdés, P., Schneider, B.L., 2016. Gene therapy: a promising approach for neuroprotection in Parkinson's disease? Front. Neuroanat. 10.

Valencia, P.M., Farokhzad, O.C., Karnik, R., Langer, R., 2012. Microfluidic technologies for accelerating the clinical translation of nanoparticles. Nat. Nanotechnol. 7 (10), 623–629.

van der Burg, S.H., Arens, R., Ossendorp, F., van Hall, T., Melief, C.J., 2016. Vaccines for established cancer: overcoming the challenges posed by immune evasion. Nat. Rev. Cancer 16 (4), 219–233.

van der Weyden, L., Patton, E.E., Wood, G.A., Foote, A.K., Brenn, T., Arends, M.J., et al., 2016. Cross-species models of human melanoma. J. Pathol. 238 (2), 152–165.

Vercauteren, K., Hoffman, B.E., Zolotukhin, I., Keeler, G.D., Xiao, J.W., Basner-Tschakarjan, E., et al., 2016. Superior in vivo transduction of human hepatocytes using engineered AAV3 capsid. Mol. Ther. 24 (6), 1042–1049.

Vijapura, C., Saad Aldin, E., Capizzano, A.A., Policeni, B., Sato, Y., Moritani, T., 2016. Genetic syndromes associated with central nervous system tumors. Radiographics 37 (1), 160057.

Viola, J.R., Rafael, D.F., Wagner, E., Besch, R., Ogris, M., 2013. Gene therapy for advanced melanoma: selective targeting and therapeutic nucleic acids. J. Drug Deliv. 2013, 897348.

Visanji, N.P., Brotchie, J.M., Kalia, L.V., Koprich, J.B., Tandon, A., Watts, J.C., et al., 2016. alpha-Synuclein-based animal models of Parkinson's Disease: challenges and opportunities in a new era. Trends Neurosci. 39 (11), 750–762.

Wang, Y., Huang, L., 2014. Composite nanoparticles for gene delivery. Adv. Genet. 88, 111–137.

Wang, F., Li, Y., Shen, Y., Wang, A., Wang, S., Xie, T., 2013. The functions and applications of RGD in tumor therapy and tissue engineering. Int. J. Mol. Sci. 14 (7), 13447–13462.

Wang, Y., Miao, L., Satterlee, A., Huang, L., 2015a. Delivery of oligonucleotides with lipid nanoparticles. Adv. Drug Deliv. Rev. 87, 68–80.

Wang, L., Bell, P., Somanathan, S., Wang, Q., He, Z., Yu, H., et al., 2015b. Comparative study of liver gene transfer with AAV vectors based on natural and engineered AAV capsids. Mol. Ther. 23 (12), 1877–1887.

Wang, H., Wan, G., Liu, Y., Chen, B., Chen, H., Zhang, S., et al., 2016a. Dual-responsive nanoparticles based on oxidized pullulan and a disulfide-containing poly(β-amino) ester for efficient delivery of genes and chemotherapeutic agents targeting hepatoma. Polym. Chem. 7 (41), 6340–6353.

Wang, S., Tian, Y., Tian, W., Sun, J., Zhao, S., Liu, Y., et al., 2016b. Selectively sensitizing malignant cells to photothermal therapy using a CD44-targeting heat shock protein 72 depletion nanosystem. ACS Nano 10 (9), 8578–8590.

Wang, Z., Wang, Y., Chen, Y., Lv, J., 2016c. The IL-24 gene protects human umbilical vein endothelial cells against H_2O_2-induced injury and may be useful as a treatment for cardiovascular disease. Int. J. Mol. Med. 37 (3), 581–592.

Ward, K., Fan, Z.H., 2015. Mixing in microfluidic devices and enhancement methods. J. Micromech. Microeng. 25 (9).

Watanabe, N., Yano, K., Tsuyuki, K., Okano, T., Yamato, M., 2014. Re-examination of regulatory opinions in Europe: possible contribution for the approval of the first gene therapy product Glybera. Mol. Ther. Methods Clin. Dev. 2, 14066.

Watts, K.L., Adair, J., Kiem, H.P., 2011. Hematopoietic stem cell expansion and gene therapy. Cytotherapy 13 (10), 1164–1171.

Wen, R., Banik, B., Pathak, R.K., Kumar, A., Kolishetti, N., Dhar, S., 2016. Nanotechnology inspired tools for mitochondrial dysfunction related diseases. Adv. Drug Deliv. Rev. 99, 52–69.

Wicki, A., Mandalà, M., Massi, D., Taverna, D., Tang, H., Hemmings, B.A., et al., 2016. Acquired resistance to clinical cancer therapy: a twist in physiological signaling. Physiol. Rev. 96 (3), 805–829.

Wiedmann, F., Schmidt, C., Lugenbiel, P., Staudacher, I., Rahm, A.-K., Seyler, C., et al., 2016. Therapeutic targeting of two-pore-domain potassium (K2P) channels in the cardiovascular system. Clin. Sci. 130 (9), 643–650.

Williford, J.M., Archang, M.M., Minn, I., Ren, Y., Wo, M., Vandermark, J., et al., 2016. Critical length of PEG grafts on lPEI/DNA nanoparticles for efficient in vivo delivery. ACS Biomater. Sci. Eng. 2 (4), 567–578.

Wolfe, D., Krisky, D., Goss, J., Wechuck, J., Mata, M., Fink, D.J., 2016. Translating gene therapy for pain from animal studies to the clinic. Translational Neuroscience. Springer, pp. 167–183.

Woolf, E.C., Syed, N., Scheck, A.C., 2016. Tumor Metabolism, the Ketogenic Diet and β-Hydroxybutyrate: Novel Approaches to Adjuvant Brain Tumor Therapy. Front. Mol. Neurosci. 9.

Wu, Q., Yang, X., Zhang, Y., Zhang, L., Feng, L., 2016. Chronic mild stress accelerates the progression of Parkinson's disease in A53T α-synuclein transgenic mice. Exp. Neurol. 285, 61–71.

Xie, A., Petravicz, J., McCarthy, K., 2016a. Molecular approaches for manipulating astrocytic signaling in vivo. The Role of Glia in Plasticity and Behavior .

Xie, H.Q., Xu, T., Chen, Y., Li, Y., Xia, Y., Xu, S.L., et al., 2016b. New perspectives for multi-level regulations of neuronal acetylcholinesterase by dioxins. Chem. Biol. Interact. 259, 286–290.

Xie, Y.L., Rosen, L.B., Sereti, I., Barber, D.L., Chen, R.Y., Hsu, D.C., et al., 2016c. Severe paradoxical reaction during treatment of disseminated tuberculosis in a patient with neutralizing anti-IFNγ autoantibodies. Clin. Infect. Dis. 62 (6), 770–773.

Xu, J., Yang, J., Miao, Y., 2015. Dual receptor-targeting (9)(9)mTc-labeled Arg-Gly-Asp-conjugated alpha-melanocyte stimulating hormone hybrid peptides for human melanoma imaging. Nucl. Med. Biol. 42 (4), 369–374.

Xu, W., Yang, Z., Lu, N., 2016. Molecular targeted therapy for the treatment of gastric cancer. J. Exp. Clin. Cancer Res. 35, 1.

Yadav, V.N., Castro, M.G., Lowenstein, P.R., 2016. Viral gene therapy for central nervous system diseases. Neurotropic Viral Infections. Springer, pp. 519–544.

Yao, J., Fan, Y., Li, Y., Huang, L., 2013. Strategies on the nuclear-targeted delivery of genes. J. Drug Target. 21 (10), 926–939.

Yin, P.T., Han, E., Lee, K.B., 2016a. Engineering stem cells for biomedical applications. Adv. Healthcare Mater. 5 (1), 10–55.

Yin, P.T., Shah, S., Pasquale, N.J., Garbuzenko, O.B., Minko, T., Lee, K.-B., 2016b. Stem cell-based gene therapy activated using magnetic hyperthermia to enhance the treatment of cancer. Biomaterials 81, 46–57.

Yuan, P., Grutzendler, J., 2016. Attenuation of β-amyloid deposition and neurotoxicity by chemogenetic modulation of neural activity. J. Neurosci. 36 (2), 632–641.

Yudina, T., Puntes, V., 2016. Synthesis, kinetic control and properties engineering of cerium oxide nanoparticles for biomedical applications. PhD Thesis, Departament de Bioquímica i Biologia Molecular, Universitat Autònoma de, Barcelona, ISBN: 9788449064623.

Zeng, Y., Chen, P., Chen, Y., Li, M., 2016. β-Thalassemia for translating human β-globin gene and RNA interference of α-globin gene expression by lentivirus vectors. Int. J. Clin. Exp. Pathol. 9 (10), 9898–9906.

Zhang, Y., Wang, H., Wen, D., Zhang, J., Zheng, F., Jiang, N., et al., 2015. Tissue factor pathway inhibitor-2 is critical in zebrafish cardiogenesis. Biochem. Biophys. Res. Commun. 456 (3), 827–833.

Zhang, M., Kim, Y.K., Cui, P., Zhang, J., Qiao, J., He, Y., et al., 2016a. Folate-conjugated polyspermine for lung cancer-targeted gene therapy. Acta Pharm. Sin. B 6 (4), 336–343.

Zhang, Y.N., Poon, W., Tavares, A.J., McGilvray, I.D., Chan, W.C., 2016b. Nanoparticle-liver interactions: cellular uptake and hepatobiliary elimination. J. Control. Release 240, 332–348.

Zhao, B.S., Roundtree, I.A., He, C., 2016. Post-transcriptional gene regulation by mRNA modifications. Nat. Rev. Mol. Cell Biol.

Zheng, M., Kimura, S., Junko, N.-K., Iwanaga, T., 2016. The selective distribution of LYVE-1-expressing endothelial cells and reticular cells in the reticulo-endothelial system (RES). Biomed. Res. 37 (3), 187–198.

Zhou, A., Li, M., He, B., Feng, W., Huang, F., Xu, B., et al., 2016a. Lipopolysaccharide treatment induces genome-wide pre-mRNA splicing pattern changes in mouse bone marrow stromal stem cells. BMC Genom. 17 (7), 509.

Zhou, Y., Basu, S., Wohlfahrt, K.J., Lee, S.F., Klenerman, D., Laue, E.D., et al., 2016b. A microfluidic platform for trapping, releasing and super-resolution imaging of single cells. Sens. Actuators B. Chem. 232, 680–691.

CHAPTER

Nanoparticles for cosmetic use and its application

5

Bruna Galdorfini Chiari-Andréo[1,2], Maria Gabriela José de Almeida-Cincotto[1], João Augusto Oshiro Jr[1], Cibele Yumi Yamada Taniguchi[1,2], Leila Aparecida Chiavacci[1] and Vera Lucia Borges Isaac[1]

[1]São Paulo State University (UNESP), School of Pharmaceutical Sciences, Araraquara, São Paulo, Brazil [2]Araraquara University (UNIARA), Araraquara, São Paulo, Brazil

CHAPTER OUTLINE

Nanoparticles in Pharmacotherapy. DOI: https://doi.org/10.1016/B978-0-12-816504-1.00013-2

5.1 INTRODUCTION

The cosmetic market has exhibited promising and meaningful growth for years. To support this advance, innovative formulations and products are needed, as well as the improvement of its formulations, offering real benefits to the users. Nanoparticles have been cited as a tool to solve many pharmaceutical problems, as well as problems in the cosmetic field.

Some applications of nanoparticles are to vectorize the release and action of a cosmetic active substance; to retard the release of it; to keep it chemically stable, effective, and safe; to improve drug solubility; film formation; controlled occlusion; skin hydration; enhanced skin bioavailability; the combination of two or more active substances, among other uses (Brigger et al., 2012; Da Costa et al., 2012; Muller et al., 2007).

For this end, many kinds of nanoparticles could be employed, such as nanospheres, nanocapsules, solid lipid nanoparticles (SLNPs), liposomes, cyclodextrins, and micelles (Janib et al., 2010). Each one of them shows peculiarities that should be evaluated before the choice of the most suitable for each case. Al-Jamal and Kostarelos (2011), for example, affirm that liposomes, earlier used only to encapsulate drugs, could improve the therapeutics and even the diagnostic efficacy (Al-Jamal and Kostarelos, 2011).

However, some challenges must be overcome in their cosmetic use. One of them is the difficulty to scale up the technology developed, and another is their safety assessment.

Cosmetics, according to their definition, should not reach the bloodstream, acting only on the skin, which means that while penetration is permitted, permeation is not. Thus, the extensive study of its route on the skin should be assessed and well defined. In this vein, Tinkle (2010) explains that the application of nanoparticles could cause adverse consequences, and so a detailed assessment of their safety is indispensable (Tinkle, 2010).

Nohynek and Dufour (2012) discussed the safety in using nanoparticles in sunscreens (Nohynek and Dufour, 2012).

Nazarenko et al. (2012) reported the risks of inhalation exposure to engineered nanoparticles from nanotechnology-based cosmetic powders (Nazarenko et al., 2012).

Considering the above, this chapter aimed to show the main kinds of nanoparticles employed in the cosmetic field, highlighting the advantages obtained with their use. Also, it discusses the scale-up and the risks, mainly in relation to the pathway followed by the nanoparticle when applied on the skin.

To complete this discussion, an extensive review of the literature was carried out to show the state of art of this issue.

Papers previously published in renowned journals have been used, exploring the earliest and the most recent publications about the subject. Also, the group's experience in this subject was used to elaborate the manuscript.

The authors hope that this chapter helps researchers and others interested in the matter to better understand and update their knowledge about the use of nanoparticles in cosmetics.

5.2 THE COSMETIC MARKET

In recent decades the cosmetics sector has shown steady growth, moving billions of dollars around the world. On the world market ranking of Toiletry products, Perfumery and Cosmetics, the first five listed countries alone (United States, China, Japan, Brazil, and the United Kingdom) moved more than 210 billion dollars in 2015, according to data from the Brazilian Association of Toiletries, Perfumes & Cosmetics (ABIHPEC, 2016).

This growth is driven by several factors including, for example, the increase in the elderly population, the anticipation of the population in the labor market, the increase in global propagandist placement through websites and exaltation of the image by social media, the appreciation of male esthetics and, also, the technological advancement of products and features that enable the maintenance of cosmetic beauty.

Cosmetics, according to the Federal Food, Drug, and Cosmetic Act (FD&C Act), are defined as articles intended to be rubbed, poured, sprinkled, or sprayed on, introduced into, or otherwise applied to the human body for cleansing, beautifying, promoting attractiveness, or altering the appearance, such as skin moisturizers, perfumes, lipsticks, fingernail polishes, eye and facial makeup preparations, cleansing shampoos, permanent waves, hair colors, and deodorants, as well as any substance intended for use as a component of the cosmetic product.

In 1984, Dr. Albert M. Kligman, during a conference, first used the term *cosmeceutical* to designate a class of products with intermediate action between cosmetics and drugs. Cosmeceuticals were therefore a class not only restricted to embellishment, but defined as topical products that could promote structural and/or functional changes in the skin, mucous membrane, and its attachments without therapeutic claim, but with possible preventive action (Kligman, 1998).

Currently several other synonyms are used to designate products for this purpose, such as skin care products, functional cosmetics, active cosmetics, dermocosmetics, among others. These terms generally become common among professionals who work with the skin, however, such nomenclatures, trying to exalt the action and function of the product in terms of its bioactivity, are still controversial and not recognized by most countries.

The governmental bodies that regulate cosmetic products in the United States (FD&C Act), Europe (EU Council Directive), Japan (Ministry of Health, Labor and Welfare), and Brazil (ANVISA, National Health Surveillance Agency) do not recognize these new terms, neither does the International Cooperation on Cosmetics Regulation (ICCR), an international group that includes regulators from the United States, European Union, Japanese, Canadian, and Brazilian cosmetics industries. Although the goal of this group is trying to propose rules governing cosmetics between countries to minimize barriers to international trade and ensure global consumer protection, the recognition of these nomenclatures is still far from a consensus, since the classification of cosmetics and drugs from one country to another also varies depending on the product.

The issue is that there are cosmetics on the market that act more broadly and effectively than the functions of beautifying and maintaining good skin conditions. This is the result of major technological breakthroughs regarding the development of these products. One such technology is coming from nanoscience, that is, studying materials whose dimensions are on the nanometer order ($1\ nm = 10^{-9}\ m$).

The use of nanoscale materials allowed the development of cosmetic products of better quality, with different properties and better performance due to the small size of their particles. Since the last decade, nanotechnology has grown substantially in various fields of science. The National Science Foundation has estimated that by 2015 nanotechnology would have an overall impact of about $1 trillion into the economy (Melo et al., 2015). In the United States alone, investment in the nanotechnology sector, called the National Nanotechnology Initiative, was $494 million in fiscal year 2001, when the project began. By 2016, the proportion had increased to $1.5 billion, reflecting the strength of this sector (White-House, 2014).

Cosmetic products followed the technological evolution of this market through nanocosmetics. These products are defined as cosmetic formulations containing nanoscaled materials and can provide improved aspects, due to the properties of nanomaterials, compared with formulas that contain only macroscopic materials (Guterres et al., 2012).

Humans have used nanomaterials for a long time. According to Landriscina et al. (2016), the cosmetic application of nanomaterials dates back to the Greco-Roman period, when a mixture of lime and water was used to color dark hair. Further studies showed that the mixture contained 5-nm nanocrystals of lead selenide (Landriscina et al., 2016). However, the use of nanomaterials as topical therapeutic agents has been considered recently. Nanocosmetics gained prominence in the global cosmetology market in the 1990s, when Lancôme, a luxury division of L'Oreal, launched in 1995 the first nanotechnology-based cosmetic product, a facial cream composed of vitamin E nanocapsules to combat signs of aging. Since then, several other major international companies have followed suit (Melo et al., 2015).

The nanocosmetics present advantages in relation to traditional cosmetics, such as better penetration of actives into the skin, the possibility of transdermal controlled release, protection against degradation of labile active substances, better sensory properties, better stabilization of the formulation, and low toxicity (Duarah et al., 2016; Durán et al., 2011).

5.3 NANOCOSMETICS AND SKIN PENETRATION

It is known that the skin is the first protective barrier of the organism to the external environment. It protects, in a way, against mechanical and electrical trauma, against the invasion of pathogenic microorganisms, against excessive ultraviolet (UV) radiation and, also, against loss of water and electrolytes by the transepidermal route.

However, the cosmetic application occurs primarily on the surface of the skin and, therefore, the clinical efficacy of the product is based often on the diffusion of the active substances through the epidermis layer. The major barrier to penetration of substances through the skin is the stratum corneum, the outermost layer of the epidermis, which is in direct contact with the environment.

The stratum corneum is formed by a set of keratinized and anucleated cells arranged flat manner against each other. These cells are collected into 10–15 cell layers surrounded by extracellular lipid material. This gives to the skin surface an important structural and organizational characteristic, as well as a hydrophobic property, being these the key factors for its protective function and, also, for the selective control of cutaneous permeability.

Due to these features, some substances hardly penetrate through the skin by passive diffusion in amounts sufficient to achieve therapeutic levels.

In this context, nanotechnology was instrumental in the development of strategies that would enable increased penetration of actives via transepidermal route. Several studies of cutaneous transport in vitro and in vivo have been published since 1999. Most of these studies include the investigation of active substances in nanocapsules (Guterres et al., 2012).

The nanoencapsulation refers to a type of carrier system having the ability to enhance penetration of active cosmetics due to their small size nanocapsules (20–1000 nm diameter) (Cornell et al., 2012). Schmaltz et al. (2005) explains that instead of being directly added to the cosmetic vehicle in free form, the active substance is encapsulated in nanometer vesicles called nanocapsules or nanospheres, which, in addition to improving the penetration of active substance, also control its release (Schmaltz et al., 2005).

Thus, the nanocapsules provide a better performance of this active on the skin, allow a gradual release of the substance, which reduces potential irritation reactions caused by a sudden release, and moreover, protect the active substance from degradation by reducing its contact with the remaining of the formulation. Some physicochemical characteristics of the nanocarriers are important in the permeation mechanism, such as stiffness, hydrophobicity, charge, and size of molecule (Durán et al., 2011).

Besides the nanocapsules there are several other methods and nanoscaled materials based on the use of technology. These different methods will be seen below.

Despite the increase in the placement of active substances through the skin, studies with different scaffolds demonstrated that there is penetration of particles to the viable skin layers, which could result in low risk of systemic absorption (Guterres et al., 2012). Contri et al. (2011) explains that, in general, nanoparticles cannot cross the skin barrier due to cell cohesion and lipids present in the extra-cellular matrix, but the use of nanostructured substances can modulate this penetration by controlling the release of active substances and also by increasing the time of permanence in the skin (Contri et al., 2011).

5.4 NANOPARTICLES OF CONTROLLED RELEASE

The advent of nanotechnology promoted a major revolution in the clinical and scientific field of medicine. As cites Couvreur (2013), until the 1970s, intravenous administration pharmaceutical suspensions (dispersion of solid particles in liquid) was considered impossible due to the risk of causing embolism. But, nanotechnology has enabled the development of very small solid particles that reduced not only the risks but also the levels of toxicity, since they can direct their actions to the tissue/target cell (Couvreur, 2013).

The nanoparticles act as carrier compartments for cosmetic active substances. Thus, the structure functions as a reservoir or barrier that controls the delivery of the cosmetic substances, allowing a more controlled delivery and targeted to desired local of treatment. This adjustment, which is more precise and sustained, in the delivery system reduces the irritation potential of the cosmetic active substance, since it decreases the contact with the irritants and allows the use of these in lower concentrations (Guterres et al., 2012).

The choice of the type of nanoparticle used is one of the factors that determines the release to a specific target tissue. It is known that physical and chemical characteristics of both active substance and cosmetic formulation influence the depth of penetration of the substances into the skin, and also influence the affinity of the nanosized system to different structures. The transepidermal flow is influenced not only by the particle size but also by its nature, solubility, and polarity (Wokovich et al., 2006; Silva et al., 2010).

The fact that the stratum corneum shows characteristics of impermeability makes it the main barrier to the penetration of topically applied substances. The main routes of substances' penetration through the skin are transcellularly (through cells diffusion), intercellularly (diffusion between cells), or transanexially (diffusion between the annexes through the hair follicles and glands) (Fig. 5.1).

All routes offer limiting factors that affect the success of active substance penetration. The transanexial route, for example, depends on the surface area covered

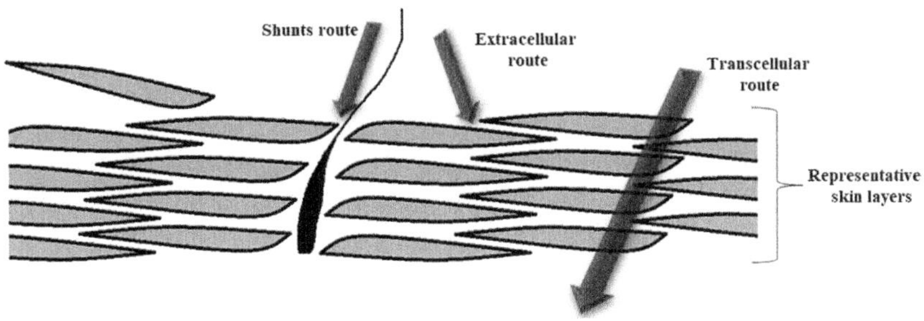

FIGURE 5.1

Potential penetration pathways through the stratum corneum. Topically applied compounds can enter the skin via shunts (e.g., the pilososebaceous units and sweat glands), via the transcellular route over the corneocytes or via the tortuous extracellular pathway in the lipid matrix.

by skin annexes, which is generally small (about 0.1% of the total skin surface); in addition, active substances are conveyed through the ducts of glands and follicles, which release their products to the external environment; thus, these active substances need to go against the spread route; depending on the nature of the substance it can be easier to penetrate the sweat ducts (when hydrophilic) or the sebaceous ducts (when lipophilic). This pathway facilitates the transport of large and polar molecules (Morrow et al., 2007).

However, by the intracellular route, active substances need to cross through the corneocytes, which are formed by keratin. The proteins facilitate the passage of small, hydrophilic molecules, but hinder the passage of lipophilic molecules. Conversely, the intercellular route facilitates the passage of small molecules due to their lipophilicity dissolving the intercellular lipids of the stratum corneum (Morrow et al., 2007).

Thus, some authors affirm that if an active substance more easily penetrates through the skin it cannot be very hydrophilic or very lipophilic, since if it happens, it may be retained on the skin surface (Oliveira, 2008). Some evidence indicates that the nanostructured materials tend to have a higher degree of penetration if shown affinity by the lipids of the stratum corneum (Guterres et al., 2012), thus, SLNPs and liposomes are often used.

5.5 NANOPARTICLES AND COSMETIC SENSORY ASPECTS

Besides enabling modulation of cosmetic active substances penetration through the skin, nanoparticles can bind more efficiently on the skin surface depending on its characteristics, increasing its adhesion area of the stratum corneum. This enables the formation of a more homogeneous protective film and increases skin retention time of lipophilic substances (Alvarez-Roman et al., 2004; Külkamp-Guerreiro et al., 2013).

The addition of nanoparticle materials to the sunscreens, for example, enabled the improvement not only of the spreadability of the products, but also of the visual aspects after application on the skin. It was common, for example, for sunscreens to provide a whitish appearance of the product on the skin surface. However, the use of nanoparticles in the preparation of such products secured a more esthetic visual appearance after application. Jansen (2010) demonstrated the influence of titanium dioxide used in different nanometric particle sizes. The conclusion is that the smaller are the titanium dioxide particles, the better is the visual appearance of the product when applied to the skin.

The use of nanoparticles also promoted a major breakthrough in quality of the makeup and hair dyes. Due to the small size of its particles they ensure better coverage and pleasant texture on the skin. Furthermore, the use of nanoparticulated dye propitiates the penetration of the coloring agent through the hair cuticles, causing less damage to the hair fibers, contributing to the development of products that offer improved esthetic results (Jansen, 2010).

In conclusion, innumerous advantages could be obtained employing nanosystems in cosmetics, including the preference of the potential consumers, which suggests nanotechnology as being synonymous with good quality products. Thus, the following sections include details about various nanosystems that could be used in cosmetics.

5.6 CYCLODEXTRINS

Cyclodextrins (CDs), also known as cycloamyloses, cycloglucoamyloses, or Schardinger's γ-dextrin (Ghosh et al., 2011), are cyclic oligosaccharides of natural origin, obtained from renewable sources, composed of D-(+)-glucopyranose units linked by α-1,4 bonds. Villers, in 1891, was the researcher who discovered them (Andreaus et al., 2010).

There are naturally three types of CD: α, β, and γ where the cavities are formed by 6, 7, and 8 units of glucopyranose, respectively. These types of CDs are obtained from the enzymatic degradation of starch by the action of bacterial glycosyl transferase (Barreto and Cunha, 2008; Rasheed et al., 2008). α, β, and γ CDs, due to their shape (cone), have a hollow tapered cavity with 0.79 nm of depth and 1.37, 1.53, 1.69 nm of width at the top, respectively (Zhang and Ma, 2013).

The shape is similar to cones, where the outer layer is hydrophilic due to the hydroxylic groups guided in this sense and the inner cavity is structured by the carbons and ethers bonds of the glucose residues conferring a high hydrophobicity. Due to this conformation, CDs are soluble in water and able to bind selectively to the different active substances (organic, inorganic, or biological molecules) in the cavity (Alvarez-Dorta et al., 2015; Guo et al., 2011). Besides, they are able to protect the active substances of the external environment and are extremely attractive due to their toxicological profile (Moya-Ortega et al., 2012).

These characteristics lead the CDs to be widely employed in the food, pharmaceutical, and cosmetic industries. In the cosmetic industry, CDs are used in the stabilization (preventing decomposition of substances by the light, temperature, and oxygen), solubilization, and to decrease or prevent the skin irritation after topical application (Koontz et al., 2003; Vilanova and Solans, 2015; Zaibunnisa et al., 2011).

For instance, a CDs-based cosmetic formulation has been developed to increase the solubility and biological activity of some compounds, such as retinol, ubiquinone, and one nitrogen-containing agent (biotin or carnitine). This formulation is able to protect the skin against environmental (extrinsic) factors that enhance skin aging, such as, UV light, ozone, cigarette smoke, and others. Another advantage is the reduction of fine lines and wrinkles and/or to improve the elasticity of the skin (Max et al., 2004).

Cosmetic formulation for skin rejuvenation based on CDs was developed by Gnain in 2008 (Gnaim, 2008). The formulation can host ascorbic acid (vitamin C), which possesses antioxidant active, protecting the skin against UV light. However, the advantage obtained by CDs it the prevention of the easy oxidation of ascorbic acid. This same system also is used to host retinol palmitate (vitamin A) avoiding its chemical degradation/oxidation, making this formulation a powerful system for the treatment of common acne, and as antiaging and rejuvenation formulation (Gnaim, 2008).

The number of related publications with CDs have increased significantly, thus some examples of applications are listed in Table 5.1.

It can be seen from Table 5.1 that β-CDs followed by α-CDs are the most used in comparison with γ-CD. This fact is based on the more suitable characteristics with respect to size, hollowness, water solubility, bioavailability, and also, low yield and the high price of the γ-CD (Li et al., 2007).

5.6.1 CYCLODEXTRINS: METHODS FOR PRODUCTION AND CHARACTERIZATION

According to Ban et al. (2015) there is a difficulty in the industrial scale production of CDs, since the methods result in mixtures of α-, β-, and γ-CDs (Ban et al., 2015).

In general, there are two options of process that can be used to produce CDs: the solvent and nonsolvent processes. The main difference between these methods is that the solvent process requires an organic complexing agent, which directs the enzymatic reaction to obtain the cyclodextrin of interest (Ban et al., 2015; Li et al., 2007).

The nanoprecipitation or nonsolvent method is used to incorporate lipophilic active substances in the CDs due to the miscibility of the solvent with the aqueous

Table 5.1 Some Examples of CDs Application in the Cosmetic Industry

CDs	Active Substance	Effect	Comments	References
β-CD	Diosgenin	Recovers skin thickness and hyperlipidemia	The bind β-CD-Diosgenin improved the skin distribution and oral bioavailability	Okawara et al. (2013)
β-CD	Retinoic acid	Treatment of acne vulgaris	The retinoic acid hosted in CD is more efficient than commercial product.	Anadolu et al. (2004)
β-CD	Phytochemical	Antibacterial agent	Phytochemical were hosted in CD to the development of cosmetic formulations (cream, shampoo, or hair tonic) against Gram (+) and *Malassezia furfur*, which cause dandruff.	Yoo et al. (2011)
β-CD	Milk thistle (MT)	Antiinflammatory, and antioxidant activity	Results reveal that emulsion with MT linked to CD improves the penetration through the skin and the control release of MT.	Spada et al. (2013)
β-CD	Methyl -β-CD (MβCD)	Anticarveolins activity in the skin	Results of the MβCD application revealed that they can be a potent antiaging system. Its mechanism of action is based on the inhibition of Cav-1.	Lee et al. (2015)
α-CD	Dehydroepiandrosterone (DHEA)	Age-related illnesses	Transdermal gel with CD reveals that solubility, dissolution properties, and partition coefficient of DHEA was improved.	Ceschel et al. (2002)
α-CD, β-CD, and γ-CD	Vitamin F (Vit-F)	Attenuate the dry, rough, irritated skin	Vit-F is a substance that undergoes oxidation with oxygen. The bind with CD stabilized Vit-F against oxidative decomposition.	Regiert and Kupka (2010)

phase. This method consists of precipitation or interfacial deposition of a preformed polymer, since it occurs at the interface of an O/W emulsion. The nonsolvent method also allows the achievement of nanospheres or nanocapsules. Nanospheres are obtained when the active substance is dissolved or dispersed in the polymer organic solution. Nanocapsules are obtained when the active ingredient is previously dissolved in an oil phase, and then, it is emulsified in the organic polymer solution before dispersing the internal phase (Nagavarma et al., 2012; Souto et al., 2012).

Inclusion complexes formed between the active substance and cyclodextrin can be characterized and studied in both solids or solutions. Fourier transform infrared spectroscopy and X-ray powder diffraction are techniques used to guarantee the inclusion complexes' formation and to analyze the type and the degree of crystalline phases of these materials. Differential scanning calorimetry (DSC) reveals some information on solid-state interactions between active substance and cyclodextrin (Kacso et al., 2010). Singh et al. (2010) affirm that several techniques may be used, such as scanning electron microscopy, nuclear magnetic resonance (NMR) spectroscopy, ultraviolet/visible (UV/vis) spectroscopy, electrochemistry, and fluorescence spectroscopy. In this chapter more details about these techniques are described (Singh et al., 2010).

5.7 NANOEMULSIONS

Emulsions are heterogeneous dispersed systems consisting of two or more immiscible liquids. They are composed of water phase, oily phase, and surfactant, in which one of these phases is called the disperse phase (droplets) and the other is the continuous phase (Foley, 2011). The type of surfactant determines if the continuous phase of the emulsion will be the water or the oily phase. These systems can be classified into macroemulsions/emulsions, microemulsions, and nanoemulsions (NE).

The NE, depending on the composition of the droplets, may exist as oil-in-water or water-in-oil forms, where the core of the droplet is oil or water, respectively (Sharma and Sarangdevot, 2012).

Among these, the NEs have been highlighted due to their characteristics, such as stability without showing flocculation or coalescence during long-term storage (this is due to the size of the droplets which minimizes the action of gravitational force), ability to incorporate and increase the bioavailability of lipophilic substances, and because they are translucent, exhibit low viscosity with Newtonian behavior, large surface area, and better penetration into the skin and mucous membrane (Gao et al., 2011; Qian et al., 2012; Singh and Vingkar, 2008). The dispersed phase can act as a reservoir that releases active substances in a controlled manner (Kong et al., 2011).

The surface charge of the nanodroplets is important because it contributes to the stabilization of these systems by creating electrostatic repulsion between them. This data is defined by zeta potential (ZP). NE with ZPs less than +10 mV are more susceptible to destabilization, especially during a sterilization process using higher temperatures, being +20 to +40 mV the better range for ZP (Lallemand et al., 2012).

NE consists of small size droplet with radius ranging from 50 to 200 nm (Kong and Park, 2011), however this size has no consensus among researchers. Table 5.2 shows a compilation of definitions proposed by several researchers in relation to the size of the NE.

Because of these discrepancies found, we can assume droplets of 10−1000 nm are named NE. However, it is important to clarify that droplets of size less than 200 nm are more stable and transparent, while that droplets with size around to 500 nm are milky (Jafari et al., 2007). This classification is not a consensus between authors.

The NE may be administered by various routes such as ocular, oral, parenteral, nasal, and transdermal (Ammar et al., 2009; Ko et al., 1998; Kong et al., 2011; Wulff-Perez et al., 2009; Yu and Huang, 2012).

The NEs have been introduced to cosmetic use due to their greater power of skin moisturization and ability to penetrate through the skin barriers. They also offer uniform product distribution on the skin, pleasant sensory aspect, transparency, fluidity, and can be presented in several forms, such as foams, creams, liquids, and sprays (Sharma and Sarangdevot, 2012). Thus, cosmetic products can be found in the market based on NEs. For instance, Kerastase Nutritive Aqua-Oléum (L'Óreal) is a NE of intense moisture for dry hair. Also, the Active Dermato, according to the manufacturer (Boticário), acts in the treatment of hyperchromias caused by various factors such as aging, exposure to sunlight, and the artificial light. Skin Caviar, by La Prairie, is another product found on the market that is a NE to deliver the active substances into the skin aimed at treatment of aging. Bruma de leite (Natura) is a moisturizer for the body using NEs, with droplets of approximately 150 nm (Melo et al., 2015).

Further products include All Soft and Lancôme Primordiale Nanolotion (L'Óreal), Vitactive (Boticário), Bepanthol Facial Cream Ultra Protect (Bayer HealthCare), and Gãrshan slim (Lakshmi).

Table 5.2 Some Definitions Found in Literature About Size of NE

Definition	Reference
Droplet ranging from 10 to 200 nm	Singh and Vingkar (2008)
Droplet with radius ranging from 50 to 200 nm	Kong et al. (2011)
Droplet diameter ranging from 10 to 100 nm	McClements (2011)
Droplet size usually below 500 nm	Tang et al. (2012)
Particle sizes between 100 and 500 nm	Khurana et al. (2013)
Droplet size ranging from 100 to 600 nm	Sharma et al. (2010)
Droplet diameters ranging from 50 to 1000 nm	Sharma and Sarangdevot (2012)

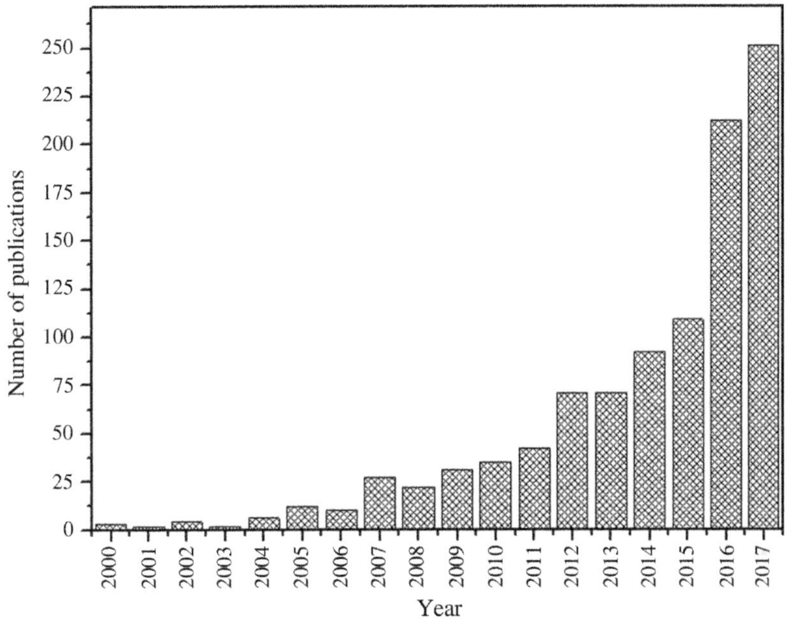

FIGURE 5.2

Number of publications that contain the terms "nanoemulsion" and "cosmetic" (Science Direct, accessed February 2018).

Besides the high number of products developed by industries, the number of scientific papers using NEs for cosmetic application have increased considerably over the last 10 years. Fig. 5.2 presents the number of publications related to the terms "nanoemulsion" and "cosmetic," revealing the importance of these NEs nowadays.

5.7.1 NANOEMULSIONS: METHODS FOR PRODUCTION AND CHARACTERIZATION

The methods for preparation of NEs aim to achieve droplet size ranging from 10 to 1000 nm, as exposed above. Thus, a great power for droplets dispersion is required. It is provided by mechanical devices or chemical potential inherent within the component (Sharma et al., 2010). These methods can be classified as low energy and high energy methods. Methods such as sonication, ultrasonication, and high pressure homogenization are approaches that utilize high energy from mechanical devices (Qian and McClements, 2011), while, phase inversion temperature and emulsion inversion point are methods of NE spontaneous formation using the rapid diffusion of a solvent soluble in water or use polyethoxylated surfactants to modify their partitioning coefficient as a function of the temperature (Anton et al., 2008).

In the high pressure homogenization the operating pressure is up to 150 MPa to produce NEs with smaller size droplets. This method has the advantage of easy scalability in industrial production (Lee and Norton, 2013).

Direct membrane emulsification technique using standard Shirasu porous glass membranes is used to prepared NE with uniform size. In this method, the disperse phase is forced through the membrane into the continuous phase (Oh et al., 2011).

Ultrasonic emulsification, also known as acoustic emulsification, is a high energy method that uses sound waves in a frequency larger than 20 kHz. This method results in NE with low size and low polydispersity, besides being fast (Ghosh et al., 2013).

In relation to methods of characterization, various studies in the literature utilize dynamic light scattering (DLS), ZP, and transmission electron microscopy (TEM) to check the size of the particles, the polydispersibility index, the electric charge on its surface, and the morphology of the particles (Karthikeyan et al., 2012; Wong et al., 2014; Yu et al., 2014b).

5.8 MICELLES

The surfactants are indispensable raw materials for the formation of micelles. They are amphipathic molecules constituted of a polar or hydrophilic part and a hydrophobic carbonic chain or nonpolar part, wherein the polar part can be positively or negatively charged (cationic or anionic), dipolar (zwitterionic), or noncharged (nonionic) (Singh et al., 2007). Surfactants can promote solubilization and emulsification of insoluble materials and can reduce the surface tension of the solution (Griffiths and Cheung, 2002). A fundamental characteristic that is directly linked to the action of the surfactants is that in an aqueous medium, when a certain concentration is reached, the surfactants are able to form micelles and other aggregates in which the lipophilic portion of the molecule is oriented to the inner part of the micelle while only the polar groups remain on the outer part of the micelle in contact with water. The concentration in which these micelles are formed is the critical micelle concentration (CMC), which corresponds to the minimum surfactant concentration necessary to reduce the surface tension to a maximum. When the CMC is reached several micelles are formed obtaining a hydrophobic core and a hydrophilic shell (Cortis and Ghezzehei, 2007).

Micelles are "particles" dispersed in water, with a size ranging from 10 to 100 nm and composed of phospholipids or prepared from synthetic polymers that contain hydrophilic and hydrophobic moieties. The hydrophilic part is generally composed of polyethylene oxide units and its hydrophobic portion is composed of amino acids (Adams et al., 2003).

The attraction forces in the inner core utilized in micelle formation include hydrophobic, electrostatic, $\pi-\pi$ interactions, and hydrogen bonds (Kwon, 2005).

The spherical shape is the type of micelle most commonly used. However, various morphologies depending on the characteristics of the copolymer are possible, thereby, a transition between micelles and vesicles can occur (Choucair and Eisenberg, 2003; Zhang et al., 1996). The micelle structure allows the incorporation of low solubility substances on their hydrophobic core, thereby improving its stability and bioavailability. In addition, its outer layer has an important role in the in vivo behavior due to its steric stabilization and its ability to interact with cells. Polymeric micelles also facilitate the transport of drugs through endocytosis (Yoncheva et al., 2012).

For the formation of spherical micelles the hydrophilic part of the polymer should not overlap the hydrophobic moiety (Sidim and Acar, 2013). The properties that define the micelles are CMC, aggregation number, size, and shape.

The purpose of using micelles is to minimize problems related to the solubility of hydrophobic substances, pharmacokinetics, and toxicity (Hamaguchi et al., 2005; Matsumura et al., 2004). Also, due to the small size, micelles improve the penetration of the substances into the deeper skin layers in topical systems.

Thereby, different uses of micelles in cosmetic application have been introduced. For instance, a cosmetic topical formulation (gel) was developed using cylindrical micelles based on lecithin for antiaging treatment because itself offers protection against UV. Furthermore, a study where drugs were hosted in lecithin can be found in the literature. The aim of this approach is to promote better protection against aging (for more information see review done by Raut et al., 2012).

Another example of cosmetic product that has been found in the market is micellar water (Garnier, L'Oreal, Avène, La Roche-Posay, and others). This product, according to the manufacturers, should be applied to remove the makeup and impurities without stripping skin. There is still the possibility of host substances for improving its properties, such as selenium, an antioxidant known for its antiirritating properties.

5.8.1 MICELLES: METHODS OF PRODUCTION AND CHARACTERIZATION

Several methods can be used for micelles production, such as direct dissolution, solid dispersion, oil-in-water emulsion method, dialysis, cosolvent evaporation, and others.

The direct dissolution method has been used for preparation of micelles when the copolymer used is very hydrophobic and requires heating to start micellization. In this method is used the aqueous solvent to dissolve the copolymer and the active substance (Ding et al., 2013; Yu et al., 2014a).

Solid dispersion method is used in the preparation of micelles because it is simple, allows the reducing of particle size, and is easy to scale up (Li et al., 2013;

Liu et al., 2013). Vo and collaborators defined the solid dispersion, as "a dispersion of one or more active substances in an inert matrix at the solid state prepared by solvent, melting or solvent–melting method. The active substance in solid dispersions can be dispersed as separate molecules, amorphous particles, or crystalline particles while the carrier can be in the crystalline or amorphous state" (Vo et al., 2013).

Incorporation of an active substance (hydrophobic) can be done through the oil-in-water emulsion method, which uses an organic solvent immiscible with water (Zhang et al., 2015). Another method is dialysis, which results in systems with homogeneous sizes in the range of 50–100 nm. The principle is based on dissolving lipids in the presence of a surfactant above its CMC, to form mixed micelles (Holzer et al., 2009).

Teagarden and Baker (2002) described a process using just one step in the preparation of micellar systems incorporated with active substances, which is based on dissolving the active substance with the polymer in water/tert-butanol and then lyophilization. Micelles loaded with active substance are formed when there is a reconstitution of this substance–polymer complex in an injectable vehicle. This process allows the production of stable spherical micelles, with a monodisperse size distribution and an average diameter of 30–60 nm.

In the literature some methods are utilized by characterizing the micelles. TEM is used by observe the morphology (shape and diameter) and the DLS can be used to confirm the micelles size (Hong et al., 2014). Small angle X-ray scattering (SAXS) allows to investigate the structure (shape) of micelles by mean of the gyration radius as can be determined using small angle neutron scattering (SANS). However, using SANS a more accurate information than using SAXS can be obtained about the micelles' internal structure (Chamieh et al., 2015). The NMR can be used to determine the molecular weight and to analyze possible changes in the micelles structure (Kwon et al., 2014).

5.9 NANOCAPSULES

The nanocapsules are systems that can be characterized as nanovesicles because they are composed of a polymeric matrix, which forms its shell, and an oily core connected with surfactants, which can be more hydrophilic or more lipophilic. The active substances to be incorporated in this system can be solid or liquid (Bizerra and Silva, 2016; Daudt et al., 2013; Devissaguet and Fessi, 1991).

Moreover, the active substances may be solubilized in the core if it exhibits lipophilic characteristics, or may be adsorbed on the polymeric matrix depending on its solubility. Liposoluble substances are easily dissolved in the core, while the amphiphilics will have greater affinity for the polymeric matrix (Danckwerts and Fassihi, 1991; Dash and Cudworth, 1998; Kimura and Ogura, 2001).

Because of these characteristics, this system can be characterized as a reservoir-type system and the release of the active substance depends on the affinity between it and the system components (Danckwerts and Fassihi, 1991; Daudt et al., 2013; Kimura and Ogura, 2001; Park and Park, 1996).

The nanocapsules differ from microcapsules in relation to the particle size. The nanocapsules have a diameter smaller than 1 μm, with the ideal diameters between 100 and 500 nm (Daudt et al., 2013; Morganti, 2010).

The size obtained during synthesis is directly related to the process and to the materials used, such as the type of polymer, the amount of surfactant, and the degree of hydro/lipophilicity of the same. Considering the process, the interference should be attributed from the speed of stirring, the type of solvent used to solubilize the polymer, etc. (Moinard-Checot et al., 2008; Esmaeili et al., 2011).

The kind of polymer used in the development of the nanocapsules can directly affect the size. In addition, these materials must be biocompatible and biodegradable, providing compatibility with the biological fluid and being degraded in vivo into carbon dioxide and water, which can be easily excreted by the body. Studies have been conducted with polymers used to increase release providing increased activity/efficacy and, at the same time, reducing side effects caused by the polymer (Pillai and Panchagnula, 2001; Schaffazick et al., 2003; Esmaeili et al., 2011).

The polymers used for the synthesis of nanocapsules can be of various origins, such as natural, synthetic, or semisynthetic. The most commonly used are polylactic acid, polyglycolic acid, PLGA (poly(lactic-*co*-glycolic acid)), cellulose derivatives such as hydroxypropylmethyl cellulose, chitosan, etc. (Pillai and Panchagnula, 2001; Schaffazick et al., 2003).

The nanocapsules are a kind of nanoparticle that have been applied in cosmetic and dermocosmetic formulations to act as a reservoir for prolonged release of active substances, to protect them from oxidative processes, to reduce unpleasant odors, among others (Daudt et al., 2013).

Wu and Guy (2009) reported in a review paper that sunscreen formulations may use nanocapsules containing substances such as octyl methoxycinnamate and benzophenone-3, because this system can form a film on the skin and slowly release these substances promoting an extended protection (Wu and Guy, 2009).

Kaur and Agrawa (2007) used nanocapsules as a system for carrier vitamin A in emulsified formulation with antiaging purpose. These nanocapsules acted as a reservoir system releasing vitamin A, and its action was prolonged (Kaur and Agrawal, 2007).

Wu et al. (2013) developed a nanocapsule of γ-polyglutamic acid/quitosans containing jasmine essence and detected that there was a sustained release of the essence at room temperature (Wu et al., 2013). Thus, it is possible to say that the nanoencapsulation of fragrances can also be used to prolong the pleasant odor of cosmetic products.

5.10 NANOSPHERES

The nanospheres are nanoparticles that as well as the nanocapsules are smaller than 1 μm. However, they differ in their structural characteristics. The nanospheres are matrix systems being formed of a polymeric matrix, which can contain a polymer or a mixture of it to obtain different properties, where the active substance is incorporated or adsorbed. Thus, no oil component is employed on its composition (Bizerra and Silva, 2016; Daudt et al., 2013; Dos Santos et al., 2015; Morganti, 2010).

This type of nanoparticle may be used in the cosmetic industry for the purpose of prolonged release of active substances that provide benefits to the skin, such as vitamins, or to prolong the release of fragrances (Shefer et al., 2004; Tree-udom et al., 2011; Tsujimoto et al., 2004). In the case of fragrances it could be used in products for consumer use or in the catalogs of direct selling companies, in which the particles are placed on a sheet of paper and after being rubbed release a pleasant odor of the fragrance.

The same polymers employed in the preparation of nanocapsules should be used for nanospheres.

The release of active substances from nanospheres happens slowly, since the dissolution of the content occurs in the biological fluid (Dos Santos et al., 2015).

Tsujimoto et al. (2004) developed nanospheres containing vitamin A, C, and E using PLGA as polymer and related positive results to the clinical efficacy as antiaging and for use in skin whitening.

5.10.1 NANOCAPSULES AND NANOSPHERES: METHODS OF PRODUCTION AND CHARACTERIZATION

The preparation and characterization of nanocapsules and nanospheres are made by the same processes, since both are polymeric nanoparticles.

There are several methods for obtaining nanocapsules and nanospheres. They are mostly based on in situ polymerization, nanoprecipitation of preformed polymers, emulsification–diffusion, and emulsification–solvent evaporation (Fessi et al., 1989; Guterres et al., 2007; Lima and Reis, 2015; Schaffazick et al., 2003; Dimer et al., 2013; Mora-Huertas et al., 2010; Carneiro et al., 2015). The main difference between them is that for obtaining nanocapsules, an oily substance is used and in the preparation of nanospheres, such a substance is absent (Guterres et al., 2007).

The characterization of nanocapsules and nanospheres is also carried out by methods that serve both types of polymeric nanoparticles. There are several methods that can be used, among them the most applied are chromatography size exclusion, liquid chromatography, ultracentrifugation and ultrafiltration, potentiometric, dynamic and static light scattering, SAXS, atomic force microscopy (AFM), and TEM (Guterres et al., 2007; Silva et al., 2013; Pan et al., 2013; Chen et al., 2013).

5.11 SOLID LIPID NANOPARTICLES

As can be understood from its denomination, SLNPs are systems formed from solid lipids (Daudt et al., 2013; Dos Santos et al., 2015), exhibiting diameters less than 1 μm. Among their main features are physical stability, protection of unstable substances, control of the active substance release, and film forming ability on the skin (Dos Santos et al., 2015; Garud et al., 2012; Mihranyan et al., 2012; Sarathchandiran, 2012). Thus, SLNPs are an important alternative to improve the cosmetic products effectiveness.

This type of nanoparticle is used for incorporation of lipophilic substances, even when a water-based vehicle is required. This application can also be useful when the objective is to incorporate sunscreens and labile chemicals, those that dissolve by chemical or physical means when in contact with skin, protecting them from degradation and favoring the controlled release (Mihranyan et al., 2012; Wu and Guy, 2009; Daudt et al., 2013; Dos Santos et al., 2015; Ereno, 2008; Garud et al., 2012).

In addition to the advantages already discussed, this type of nanoparticle is easy to produce because it does not present difficulties in the scale up and even with the sterilization process (Daudt et al., 2013; Mihranyan et al., 2012; Nemen and Lemos-Senna, 2011; Wu and Guy, 2009; Garud et al., 2012).

To obtain SLNPs, the raw materials used are lubricants, such as waxes, steroids, triglycerides, fatty acids, etc. (Daudt et al., 2013; Tiarks et al., 2001).

For use in cosmetics, SLNPs are advantageous for delivery of chemically labile active substances, such as retinol (Wu and Guy, 2009).

Another advantage observed for cosmetic use is the possibility of sunscreens incorporation in the lipid matrix, delaying the penetration of the active, keeping it above the stratum corneum. This fact is extremely important, since this product to be effective should be in the skin surface and, also, it could reduce the toxic potential when compared with a conventional product (Daudt et al., 2013; Wu and Guy, 2009).

Despite the range of advantages offered by the use of SLNPs, there are also disadvantages that can affect system stability, such as difficulty maintaining active encapsulation when compared with systems such as micelles and liposomes, and physical instability during administration in the storage period (Daudt et al., 2013).

5.11.1 SOLID LIPID NANOPARTICLE: METHODS OF PRODUCTION AND CHARACTERIZATION

For the preparation of SLNPs various techniques are described in the literature. Among these techniques, ultrasonication is the one that can easily be scaled up, however, it presents problems such as particle size distribution and physical instability (Eldem et al., 1991; Garud et al., 2012).

Another technique that exhibits advantage is the use of supercritical fluid because there is no use of toxic solvents and, also, because the process results in powdered SLNPs (Chen et al., 2006).

The use of spray drying is also an alternative. This technique also results in a powdered SLNP and the method offers a higher yield than lyophilization. A limitation is that it should be used for solids that exhibit melting point higher than 70°C to prevent aggregation of the particles at this temperature (Garud et al., 2012).

SLNPs can be also be produced by high pressure homogenization in cold and hot emulsification, solvent evaporation, double emulsion, among others (Garud et al., 2012; Jenning et al., 2000; Mehnert and Mäder, 2001; Müller et al., 2002).

The characterization of SLNPs is important for the quality control of the same, so the particle size, polymorphism, and coexistence of other colloidal structures such as liposomes, micelles, among others should be verified (Garud et al., 2012). In these cases, ZP, AFM, TEM, DSC, and NMR are useful techniques (Garud et al., 2012; Mukherjee et al., 2009; Pandey et al., 2005).

5.12 LIPOSOMES

Liposomes as well as nanocapsules are reservoir systems. They are spherical vesicles that can be composed of a phospholipidic membrane having inside an aqueous core (Daudt et al., 2013; Dos Santos et al., 2015; Mihranyan et al., 2012; Wu and Guy, 2009).

Liposomes may contain one or more membranes; in this manner they are classified as unilamellar or multilamellar, having particle diameter from 20 to 3500 nm. Liposome with a single membrane and with a small aqueous compartment are called small unilamellar liposomes, and their diameter varies between 20 and 50 nm; however, when the aqueous cavity is large, but there is only one lipid bilayer membrane, vesicles obtained vary from 200 to 1000 nm. The multilamellar liposomes are those in which there are several lipid bilayers and aqueous spaces between them; this fact causes the diameter increase and structure varies from 400 to 3500 nm (Mozafari, 2005).

Considering the diameter of the various liposome classifications it is easy to understand why not all liposomes may be administered by any route. For example, for intravenous use, particles may have a size smaller than 100 nm, in addition, the use of very large size particles can harm the skin penetration and may be used when the product should remain longer on the skin releasing its content and preventing permeation, which could lead to toxicity.

Liposomes are formed when phospholipid molecules, usually phosphatidylcholine derived from egg or soy lecithin, are put into excess water and form, spontaneously, lipid aggregates through the interaction of the lipophilic portions of the phospholipid. To stabilize this link, the formed membrane closes forming a vesicle containing water inside (Heather, 2005; Pautot et al., 2003). When we want the liposomes to become more stable, we can include in its composition cholesterol or hydrogenated soy phosphatidylcholine generating rigid vesicle structures (Daudt et al., 2013; Pietzyk and Henschke, 2000).

Due to its structure, that is, of phospholipids forming the interface and water content, the liposomes are systems that allow carrying water-soluble substances, but also, liposoluble or amphiphilic. The liposoluble substances, due to their affinity to phospholipids, are inserted into the membrane, while hydrophilics are inserted within the vesicle (Mihranyan et al., 2012; Wu and Guy, 2009; Batista et al., 2007; Bizerra and Silva, 2016).

There are structural similarities between the cell membrane and the liposomes. In this way, an interaction between the liposomes and skin cells can occur, allowing greater control on the release (Bizerra and Silva, 2016). The liposomes release the chemical encapsulated by erosion and/or diffusion of the vesicle while the distribution is controlled by the size and composition of the membrane of the same (Bizerra and Silva, 2016; Fialho and Cunha Junior, 2007).

The amount of encapsulated active depends on the size of the vesicles, since the penetration depends on the size and lipid composition (Bizerra and Silva, 2016; Fialho and Cunha Junior, 2007).

A wide variety of applications is found for liposomes in cosmetics. First, the liposome has the ability to act directly on the skin and, because of its chemical characteristics, can replace the lipids of the stratum corneum, a fact that favors skin hydration (Wu and Guy, 2009).

Furthermore, the release of its content, the cosmetic active substance, is expected in the epidermis (Lasic, 1995; Daudt et al., 2013; Dos Santos et al., 2015).

Due to the incorporation capacity of substances with different solubility characteristics, liposomes may be used in various cosmetics, such as antiaging product, sunscreen, aftershave, moisturizers, makeup, and even as carrier vehicles for tanning agents. Some existing product can be mentioned, such as liposome carriers of vitamins A, C, and E, ceramides, tanning agents, taking as examples the L'Oreal, Lancome, and Estee Lauder products (Daudt et al., 2013; Dos Santos et al., 2015; Lasic, 1995).

5.12.1 LIPOSOMES: METHODS OF PRODUCTION AND CHARACTERIZATION

Liposomes have been extensively used in cosmetic products and their preparation is summarized in three steps. The first is the hydration of the lipid, the second is the design of the liposome, and the third is the removal of active substance that was not encapsulated. In industrial scale often steps 1 and 2 are performed simultaneously (Kulkarni et al., 2011).

After the preparation it is necessary to perform the chemical and physical characterization of liposomes, since these characteristics can affect the release and delivery of substances (Kulkarni et al., 2011; Senior, 1987). It is not necessary to wait for the liposome stabilizing to perform the characterization, thus it can be performed immediately after preparation.

The main parameters analyzed are size, measured by electron microscopy techniques, light scattering, ultracentrifugation, and Coulter counter; membrane fluidity by fluorescence microscopy; number of lamellae by SAXS and NMR spectroscopy; and internal morphology by microelectrophoresis (Kulkarni et al., 2011; Senior, 1987).

5.13 SCALE-UP OF NANOPARTICLES PROCESSING OPERATIONS: A CHALLENGE THAT NEEDS TO BE OVERCOME

The development of a new product and its insertion in the market depends on a number of factors. One of the most discussed is the design of the new drug and formulation. However, these initial steps happen at small scale and then need to be transposed to large scale (Galindo-Rodriguez et al., 2005).

Scale-up is the adaptation of a process, developed at laboratorial scale, for production at industrial scale, mainly with the aim of commercialization of the product manufactured.

The development of nanoparticles, until today, generally remains the exclusive domain of research centers or of big companies that dominate this area of study; however, the scale-up is not always successful.

It is a challenge not exclusive to nanoparticles. Klinzing and Dhodapkar (2005) reported that "the scale-up takes a long time to achieve a steady state and sometimes does not even achieve functionality." Galindo-Rodríguez et al. (2005) emphasized that there is a lack of information about the scaling-up of polymeric nanoparticles processing, as this is a limiting factor that could hinder their introduction in the pharmaceutical market (Galindo-Rodriguez et al., 2005). De Paula and Ribeiro (2001) clarify that this seems to be a tacit knowledge, transmitted from one professional to another. Thus, many times the scale-up problems are solved by many trials, until the achievement of the expected result (De Paula and Ribeiro, 2001).

This process is usually not direct, which means that the transposition happens gradually, using intermediate scales before the attainment of the final large scale (Galindo-Rodriguez et al., 2005; Shegokar et al., 2011), being possible the adaptation, arrangements in formulations, and other gradual modifications.

During the scale-up some alterations is commonly verified in the nanoparticles, such as mean size, percentage of drug load, shape, and others.

Galindo-Rodríguez et al. (2005) assessed the scale-up of ibuprofen-loaded nanoparticles and showed that it is possible. Three manufacturing processes were used for this, that is, salting-out, emulsification–diffusion, and nanoprecipitation, by increasing 20-fold the volume of the laboratory batches. These authors were concerned about the maintenance of the geometric similarities of the agitation systems used at lab and pilot scales, to obtain a similar fluid motion at both scales. They verified for both lab and pilot scale that by increasing the stirring, the mean size of nanoparticles decreased. However, in a general manner, the same rate of agitation promoted smaller nanoparticles in the pilot scale than in the lab scale. The rate of stirring should be determined according to the mean size required for each application (Galindo-Rodriguez et al., 2005).

Yin et al. (2004) also studied the large-scale production of silver nanoparticles prepared under microwave irradiation, and, according to these authors, metal nanoparticles found a range of applications (Thakkar et al., 2010; Yin et al., 2004). However, considering the difficulty in the removal of stabilizers from the nanoparticles' surface, the comprehension of large-scale synthesis of nanosized silver particles using removable reductants and stabilizers is worthy of investigation.

Marengo et al. (2000) developed an apparatus, based on a thermostated aluminum chamber and a pneumatic piston, to scale up the production of solid lipid nanospheres. The nanospheres are formed by the dispersion of a warm microemulsion through a needle in cold water. The parameters that were evaluated and that altered physical characteristics, mainly the mean size of nanoparticles, were pressure applied to the pneumatic cylinder, temperature of the aluminum chamber, needle gauge, and volume of the dispersing waters (Marengo et al., 2000).

Discussing the scale-up of nanoparticle processing operations, an alternative should be cited, that is, the biological approach for synthesis of nanoparticles. Thakkar et al. (2010) listed some options to biologically produce nanoparticles, using plant products and, also, algae, fungi, yeast, bacteria, and viruses. The nanoparticle synthesis using fungi, for example, is easily to scale up, since large-scale production of its products is easily obtained (Thakkar et al., 2010).

5.14 RISKS ASSOCIATED WITH EMPLOYMENT OF NANOPARTICLES

Despite the numerous advantages related to the efficacy of nanoparticles, many strong concerns have been raised about their safety (Moore, 2006), including for cosmetic use. The risks that should be considered include those related to consumers, but also, related to the workers who could be in daily contact with these nanosystems, or the risks related to the environmental and ecosystem contamination.

The main concern is related to the diminished size of these particles, which may make them easier to cross the barriers of human body, entering at different points such as skin, blood, and respiratory pathways (Simko and Mattsson, 2010). From one point of view this facility in surpassing these obstacles could be desired, making easier the treatment of the body. However, it may also mean the "opening" of the body to foreign particles. Also, nanomaterials exhibit peculiarities in relation to their reactivities, surface area, physicochemical and toxicological properties that differ from the larger particles (Handy et al., 2008). Simko and Mattsson (2010) elaborated a critical review showing that some nanoparticles, after reaching the body, are able to promote hazardous effects in the brain. But, these authors also reinforce that the present state of knowledge is unsatisfactory. Thus, they also suggest, as well as other researchers (Moore, 2006), that there is a need for focused research to support risk assessment, which should use suitable methods, including the right in vitro and/or in vivo model, controls, doses, characteristics of nanoparticles, etc.

Warheit (2004) reports in this paper the fact that rats, when exposed to nanoparticles by instillation, developed inflammation and subsequent tumors in lungs. However, this author also cautions that the risk is a combination of the hazard and the exposure. Considering the cosmetics users, the way of exposure, by topical application, for example, may be not able to promote this kind of exposure (inhalation) necessary to promote these damages. This situation could be clearly changed if the application happens by aerosol, for example (Warheit, 2004).

In terms of cosmetic use, the main concern would be the breaching of the skin barrier and reaching the bloodstream.

The advantages in the use of nanoparticles are often studied, however, for some nanoparticles, the risks, including the environmental ones, are largely unknown at present (Grieger et al., 2010).

Some papers already published confirm this problematic and difficult position of regulatory agencies, which need more detailed information and scientific evidence.

It is clear that it is not a simple matter, and that many factors influence the toxicological effects. Warheit (2004) highlights that the species, level of particle aggregation, and surface coatings are important variables that alter the toxicity of nanoparticles.

Nohynek et al. (2008) discuss that, more than the particle size, the chemistry of the particle is responsible for the toxicity. They based this affirmation, between other indicatives, on the evaluation of TiO_2 nano- and microparticles, showing that in both there was no verified toxic effect for humans (Nohynek et al., 2008). Also, Nohynek and Dufour (2012) highlight the studies carried out by the US FDA (United States Food and Drug Administration) proving the absence of penetration of TiO_2 nanoparticles (Nohynek and Dufour, 2012; Sadrieh et al., 2010).

Thus, a conclusion that could be obtained after this extensive discussion is that we still have a lack of knowledge about nanoparticles. Although the methods of development and likewise the studies to better understand them have been increasing, there is a lack of data and information, still to be determined. Also, it is not possible to generalize, since each nanoparticle should be assessed to obtain its toxic profile for humans, environment, etc.

Another discussion that should be evoked is that the risks and the benefits should be weighed, considering that these materials provide a wide variety of benefit to humans (Nohynek and Dufour, 2012).

Considering the cosmetic use of nanoparticles, this discussion becomes even more problematic, since many countries still have legislation and concerns related to these products being treated differently from drugs, for example. Sometimes, cosmetics are considered as safe products in all situations.

ACKNOWLEDGMENT

This work had financial support by the Coordenação de Aperfeiçoamento de Pessoal de Nível Superior – Brasil (CAPES) – Finance Code 001.

REFERENCES

ABIHPEC, 2016. Available from: <https://www.abihpec.org.br/> (accessed 15.10.16).

Adams, M.L., Lavasanifar, A., Kwon, G.S., 2003. Amphiphilic block copolymers for drug delivery. J. Pharm. Sci. 92, 1343–1355.

Al-Jamal, W.T., Kostarelos, K., 2011. Liposomes: from a clinically established drug delivery system to a nanoparticle platform for theranostic nanomedicine. Acc. Chem. Res. 44, 1094–1104.

Alvarez-Dorta, D., Leon, E.I., Kennedy, A.R., Martin, A., Perez-Martin, I., Suarez, E., 2015. Easy access to modified cyclodextrins by an intramolecular radical approach. Angew. Chem. Int. Ed. 54, 3674–3678.

Alvarez-Roman, R., Naik, A., Kalia, Y., Guy, R.H., Fessi, H., 2004. Skin penetration and distribution of polymeric nanoparticles. J. Control. Release 99, 53–62.

Ammar, H.O., Salama, H.A., Ghorab, M., Mahmoud, A.A., 2009. Nanoemulsion as a potential ophthalmic delivery system for dorzolamide hydrochloride. AAPS PharmSciTech 10, 808–819.

Anadolu, R.Y., Sen, T., Tarimci, N., Birol, A., Erdem, C., 2004. Improved efficacy and tolerability of retinoic acid in acne vulgaris: a new topical formulation with cyclodextrin complex psi. J. Eur. Acad. Dermatol. Venereol. 18, 416–421.

Andreaus, J., Dalmolin, M.C., De oliveira, I.B., Barcellos, I.O., 2010. Application of cyclodextrins in textile processes. Quim. Nova 33, 929–937.

Anton, N., Benoit, J.P., Saulnier, P., 2008. Design and production of nanoparticles formulated from nano-emulsion templates—a review. J. Control. Release 128, 185–199.

Ban, X.F., Gu, Z.B., Li, C.M., Huang, M., Cheng, L., Hong, Y., et al., 2015. Mutations at calcium binding site III in cyclodextrin glycosyltransferase improve beta-cyclodextrin specificity. Int. J. Biol. Macromol. 76, 224–229.

Barreto, L.C.L.S., Cunha, M.S.S., 2008. Cyclodextrin: important pharmaceutical functional excipient. Lat. Am. J. Pharm. 27, 629–636.

Batista, C.M., Carvalho, C.M.B., Magalhães, N.S.S., 2007. Lipossomas e suas aplicações terapêuticas: estado da arte. Revista Brasileira de Ciências Farmacêuticas 43, 167–179.

Bizerra, A., Silva, V., 2016. Sistemas de Liberação Controlada: Mecanismos e aplicações. Revista Saúde e Meio Ambiente 3, 1–12.

Brigger, I., Dubernet, C., Couvreur, P., 2012. Nanoparticles in cancer therapy and diagnosis. Adv. Drug Deliv. Rev. 64, 24–36.

Carneiro, S.P., Cruz, R.M., Dos Santos, O.D.H., 2015. Evaluation of short cycles of ultrasound application in nanoemulsions to obtain nanocapsules. Ultrason. Sonochem. 27, 536–542.

Ceschel, G.C., Mora, P.C., Borgia, S.L., Maffei, P., Ronchi, C., 2002. Skin permeation study of dehydroepiandrosterone (DHEA) compared with its alpha-cyclodextrin complex form. J. Pharm. Sci. 91, 2399–2407.

Chamieh, J., Davanier, F., Jannin, V., Demarne, F., Cottet, H., 2015. Size characterization of commercial micelles and microemulsions by Taylor dispersion analysis. Int. J. Pharm. 492, 46–54.

Chen, Y.-J., Jin, R.-X., Zhou, Y.-Q., Zeng, J., Zhang, H., Feng, Q.-R., 2006. Preparation of solid lipid nanoparticles loaded with Xionggui powder-supercritical carbon dioxide fluid extraction and their evaluation in vitro release. China J. Chin. Mater. Med. 31, 376–379.

Chen, T.Y., Du, B.Y., Zhang, X.H., Fan, Z.Q., 2013. Fabrication of polymer nanocapsules with controllable oligo(ethylene glycol) densities, permeation properties and robustly crosslinked walls. ACS Appl. Mater. Interfaces 5, 3748–3756.

Choucair, A., Eisenberg, A., 2003. Control of amphiphilic block copolymer morphologies using solution conditions. Eur. Phys. J. E. 10, 37–44.

Contri, R.V., Fiel, L.A., Pohlmann, A.R., Guterres, S.S., Beck, R.C., 2011. Transport of substances and nanoparticles across the skin and in vitro models to evaluate skin permeation and/or penetration. Nanocosmetics and Nanomedicines. Springer, pp. 3–35.

Cornell, M., Pillai, S., Oresajo, C., 2012. Aplicação de ativos cosméticos na pele. In: Draelos, Z.D. (Ed.), Dermatologia Cosmética—Produtos e Procedimentos. Santos, SP, pp. 62–70.

Cortis, A., Ghezzehei, T.A., 2007. On the transport of emulsions in porous media. J. Colloid Interface Sci. 313, 1–4.

Couvreur, P., 2013. Nanoparticles in drug delivery: past, present and future. Adv. Drug Deliv. Rev. 65, 21–23.

Da Costa, P.R., Perez-Rebolledo, A., Costa, D.A.S., Gomes, D.A., Rodrigues, M.A., Serakides, R., 2012. Nanopartículas de ouro: uma ferramenta diagnóstica e terapêutica. 7° COLAOB—Congresso Latino Americano de Órgãos Artificiais e Biomateriais. Natal, RN. SLABO.

Danckwerts, M., Fassihi, A., 1991. Implantable controlled release drug delivery systems—a review. Drug Dev. Ind. Pharm. 17, 1465–1502.

Dash, A.K., Cudworth, G.C., 1998. Therapeutic applications of implantable drug delivery systems. J. Pharmacol. Toxicol. Methods 40, 1–12.

Daudt, R.M., Emanuelli, J., Kulkap-Guerreiro, I.C., Pohlmann, A.R., Guterres, S.S., 2013. A nanotecnologia como estratégia para o desenvolvimento de cosméticos. Ciência e Cultura 65, 28–31.

De Paula, I.C., Ribeiro, J.L.D., 2001. Problemas de scaling up no desenvovimento de produtos farmacêuticos em empresas brasileiras. Produto Produção 5, 17–32.

Devissaguet, J.P., Fessi, H., 1991. Process for the preparation of dispersible colloidal system of a substance in the form of nanocapsules. France Patent Application.

Dimer, F.A., Friedrich, R.B., Beck, R.C.R., Guterres, S.S., Pohlmann, A.R., 2013. Impact of nanotechnology on public health: production of medicines. Quim. Nova 36, 1520–1526.

Ding, J.X., Chen, J.J., Li, D., Xiao, C.S., Zhang, J.C., He, C.L., et al., 2013. Biocompatible reduction-responsive polypeptide micelles as nanocarriers for enhanced chemotherapy efficacy in vitro. J. Mater. Chem. B 1, 69–81.

Dos Santos, P.O., Miyashiro, P.Y., Da silva, V.A., 2015. A Nanotecnologia Em Formulação Cosmética. Revista Eletrônica—Belezain

Duarah, S., Pujari, K., Durai, R.D., Narayanan, V.H.B., 2016. Nanotechnology-based cosmeceuticals: a review. Int. J. Appl. Pharm. 8, 8–12.

Durán, N., Teixeira, Z., Marcato, P.D., 2011. Topical application of nanostructures: solid lipid, polymeric and metallic nanoparticles. Nanocosmetics and Nanomedicines. Springer, pp. 69–99.

Eldem, T., Speiser, P., Hincal, A., 1991. Optimization of spray-dried and -congealed lipid micropellets and characterization of their surface morphology by scanning electron microscopy. Pharm. Res. 8, 47–54.

Ereno, D., 2008. Beleza fundamentada. Pesquisa Fapesp 146, 80–85.

Esmaeili, A., Saremnia, B., Koohian, A., Rezazadeh, S., 2011. Mechanism of nanocapsules of *Matricaria recutita* L. extract formation by the emulsion–diffusion process. Superlattices Microstruct. 50, 340–349.

Fessi, H., Puisieux, F., Devissaguet, J.P., Ammoury, N., Benita, S., 1989. Nanocapsule formation by interfacial polymer deposition following solvent displacement. Int. J. Pharm. 55, R1–R4.

Fialho, S.L., Cunha Junior, A.S., 2007. Sistemas de transporte de drogas para o segmenti posterior do olho: bases fundamentais e aplicações. Arquivo Brasileiro De Oftalmologia 70, 1465–1502.

Foley, M., 2011. Utilizing nanoscale materials as dispersants, surfactants or stabilizing molecules, methods of making the same, and products produced therefrom. Google Patents.

Galindo-Rodriguez, S.A., Puel, F., Briancon, S., Allemann, E., Doelker, E., Fessi, H., 2005. Comparative scale-up of three methods for producing ibuprofen-loaded nanoparticles. Eur. J. Pharm. Sci. 25, 357–367.

Gao, F., Zhang, Z.W., Bu, H.H., Huang, Y., Gao, Z.W., Shen, J.N., et al., 2011. Nanoemulsion improves the oral absorption of candesartan cilexetil in rats: performance and mechanism. J. Control. Release 149, 168–174.

Garud, A., Deepti, S., Garud, N., 2012. Solid lipid nanoparticles (SLN): method, characterization and applications. Int. J. Curr. Pharm. J. 1, 384–393.

Ghosh, S., Badruddoza, A.Z.M., Uddin, M.S., Hidajat, K., 2011. Adsorption of chiral aromatic amino acids onto carboxymethyl-beta-cyclodextrin bonded Fe_3O_4/SiO_2 core–shell nanoparticles. J. Colloid Interface Sci. 354, 483–492.

Ghosh, V., Mukherjee, A., Chandrasekaran, N., 2013. Ultrasonic emulsification of food-grade nanoemulsion formulation and evaluation of its bactericidal activity. Ultrason. Sonochem. 20, 338–344.

Gnaim, J.M., 2008. Cyclodextrin-containing polymers and uses thereof. Google Patents.

Grieger, K.D., Fjordboge, A., Hartmann, N.B., Eriksson, E., Bjerg, P.L., Baun, A., 2010. Environmental benefits and risks of zero-valent iron nanoparticles (nZVI) for in situ remediation: risk mitigation or trade-off? J. Contam. Hydrol. 118, 165–183.

Griffiths, P.C., Cheung, A.Y.F., 2002. Interaction between surfactants and gelatin in aqueous solutions. Mater. Sci. Technol. 18, 591–599.

Guo, Y.J., Guo, S.J., Li, J., Wang, E.K., Dong, S.J., 2011. Cyclodextrin–graphene hybrid nanosheets as enhanced sensing platform for ultrasensitive determination of carbendazim. Talanta 84, 60–64.

Guterres, S.S., Alves, M.P., Pohlmann, A.R., 2007. Polymeric nanoparticles, nanospheres and nanocapsules for cutaneous apllications. Drug Target Insights 2, 148–157.

Guterres, S.S., Beck, R.C.R., Pohlmann, A.R., 2012. Nanocosmecêuticos. In: Costa, A. (Ed.), Tratado Internacional de Cosmecêuticos. Guanabara Koogan, Rio de Janeiro, RJ, pp. 428–435.

Hamaguchi, T., Matsumura, Y., Suzuki, M., Shimizu, K., Goda, R., Nakamura, I., et al., 2005. NK105, a paclitaxel-incorporating micellar nanoparticle formulation, can extend in vivo antitumour activity and reduce the neurotoxicity of paclitaxel. Br. J. Cancer 92, 1240–1246.

Handy, R.D., von der Kammer, F., Lead, J.R., Hassellov, M., Owen, R., Crane, M., 2008. The ecotoxicology and chemistry of manufactured nanoparticles. Ecotoxicology 17, 287–314.

Heather, A.E.B., 2005. Transdermal drug delivery: penetration enhancement techniques. Curr. Drug Deliv. 2, 23–33.

Holzer, M., Barnert, S., Momm, J., Schubert, R., 2009. Preparative size exclusion chromatography combined with detergent removal as a versatile tool to prepare unilamellar and spherical liposomes of highly uniform size distribution. J. Chromatogr. A 1216, 5838–5848.

Hong, W., Chen, D.W., Jia, L., Gu, J.C., Hu, H.Y., Zhao, X.L., et al., 2014. Thermo- and pH-responsive copolymers based on PLGA–PEG–PLGA and poly(L-histidine): synthesis and in vitro characterization of copolymer micelles. Acta Biomater. 10, 1259–1271.

Jafari, S.M., He, Y.H., Bhandari, B., 2007. Production of sub-micron emulsions by ultrasound and microfluidization techniques. J. Food Eng. 82, 478–488.

Janib, S.M., Moses, A.S., Mackay, J.A., 2010. Imaging and drug delivery using theranostic nanoparticles. Adv. Drug Deliv. Rev. 62, 1052–1063.

Jansen, J., 2010. Nanocosméticos & Absorção Percutânea. Seminário de Nanotecnologia—Aplicação em Cosméticos. Curitiba. Available from: <https://www.abihpec.org.br/conteudo/ivrodadatecnologica/04JOCELIA-UEPG.pdf> (accessed 14.10.16).

Jenning, V., Gysler, A., Schäfer-Korting, M., Gohla, S.H., 2000. Vitamin A loaded solid lipid nanoparticles for topical use: occlusive properties and drug targeting to the upper skin. Eur. J. Pharm. Biopharm. 49, 211–218.

Kacso, I., Borodi, G., Farcas, S.I., Hernanz, A., Bratu, I., 2010. Host–guest system of Vitamin B10 in beta-cyclodextrin: characterization of the interaction in solution and in solid state. J. Incl. Phenom. Macrocycl. Chem. 68, 175–182.

Karthikeyan, S., Jeeva, P., Jerobin, J., Mukherjee amitava, C.N., 2012. Formulation and characterization of nanoemulsion coatings from *Azadirachta indica*. Int. J. ChemTech. Res. 4, 1566–1570.

Kaur, I.P., Agrawal, R., 2007. Nanotechnology: a new paradigm in cosmecauticals. Recent Pat. Drug Deliv. Formul. 1.

Khurana, S., Jain, N.K., Bedi, P.M.S., 2013. Nanoemulsion based gel for transdermal delivery of meloxicam: physico-chemical, mechanistic investigation. Life Sci. 92, 383–392.

Kimura, H., Ogura, Y., 2001. Biodegradable polymers for ocular drug delivery. Ophthalmologica 215, 143–155.

Kligman, A.M., 1998. Cosmeceuticals as third category. Cosmet. Toilet. 113, 33–40.

Klinzing, G.E., Dhodapkar, S.V., 2005. Scale-up in particle processing. Powder Technol. 150, 59–59.

Ko, K.T., Needham, T.E., Zia, H., 1998. Emulsion formulations of testosterone for nasal administration. J. Microencapsul. 15, 197–205.

Kong, M., Park, H.J., 2011. Stability investigation of hyaluronic acid based nanoemulsion and its potential as transdermal carrier. Carbohydr. Polym. 83, 1303–1310.

Kong, M., Chen, X.G., Kweon, D.K., Park, H.J., 2011. Investigations on skin permeation of hyaluronic acid based nanoemulsion as transdermal carrier. Carbohydr. Polym. 86, 837–843.

Koontz, J.L., Marcy, J.E., Barbeau, W.E., Duncan, S.E., 2003. Stability of natamycin and its cyclodextrin inclusion complexes in aqueous solution. J. Agric. Food Chem. 51, 7111–7114.

Külkamp-Guerreiro, I., Berlitz, S., Contri, R., Alves, L., Henrique, E., Barreiros, V., et al., 2013. Influence of nanoencapsulation on the sensory properties of cosmetic formulations containing lipoic acid. Int. J. Cosmetic. Sci. 35, 105–111.

Kulkarni, P.R., Yadav, J.D., Vaidya, K.A., 2011. Lipossomes: a novel drug delivery system. Int. J. Curr. Pharm. Res. 3, 11–18.

Kwon, D.Y., Tai, G.Z., Park, J.H., Lee, B.K., Lee, J.H., Kim, J.H., et al., 2014. Preparation of methoxy poly(ethyleneglycol)-*b*-poly(epsilon-caprolactone-*co*-L-lactide) and characterization as biodegradable micelles. J. Polym. Res. 21.

Kwon, G.L., 2005. Polymeric Drug Delivery Systems. Taylor & Francis Group, Boca Raton, FL.

Lallemand, F., Daull, P., Benita, S., Buggage, R., Garrigue, J.-S., 2012. Successfully improving ocular drug delivery using the cationic nanoemulsion, novasorb. J. Drug Deliv. 2012.

Landriscina, A., Rosen, F., Adam, J., 2016. Nanodermatology: nanoescale therapeutics for topical application. In: Quan, T. (Ed.), Molecular Mechanisms of Skin Aging and Age-Related Diseases. CRC Press, Taylor & Francis Group, New York.

Lasic, D.D., 1995. Applications of lipossomes. In: Lipowsky, R., Sackmann, E. (Eds.), Handbook of Biological Physics. Elsevier Science B.V., Menlo Park, CA.

Lee, L., Norton, I.T., 2013. Comparing droplet breakup for a high-pressure valve homogeniser and a microfluidizer for the potential production of food-grade nanoemulsions. J. Food Eng. 114, 158–163.

Lee, J.A., Choi, D.I., Choi, J.Y., Kim, S.O., Cho, K.A., Lee, J.B., et al., 2015. Methyl-beta-cyclodextrin up-regulates collagen I expression in chronologically-aged skin via its anti-caveolin-1 activity. Oncotarget 6, 1942–1953.

Li, Z.F., Wang, M., Wang, F., Gu, Z.B., Du, G.C., Wu, J., et al., 2007. gamma-Cyclodextrin: a review on enzymatic production and applications. Appl. Microbiol. Biotechnol. 77, 245–255.

Li, Y., Xu, X., Shen, Y., Qian, C., Lu, F., Guo, S., 2013. Preparation and evaluation of copolymeric micelles with high paclitaxel contents and sustained drug release. Colloids Surf. A: Physicochem. Eng. Asp. 429, 12–18.

Lima, S.A.C., Reis, S., 2015. Temperature-responsive polymeric nanospheres containing methotrexate and gold nanoparticles: a multi-drug system for theranostic in rheumatoid arthritis. Colloids Surf. B: Biointerfaces 133, 378–387.

Liu, L., Sun, L., Wu, Q., Guo, W., Li, L., Chen, Y., et al., 2013. Curcumin loaded polymeric micelles inhibit breast tumor growth and spontaneous pulmonary metastasis. Int. J. Pharm. 443, 175–182.

Marengo, E., Cavalli, R., Caputo, O., Rodriguez, L., Gasco, M.R., 2000. Scale-up of the preparation process of solid lipid nanospheres. Part I. Int. J. Pharm. 205, 3–13.

Matsumura, Y., Hamaguchi, T., Ura, T., Muro, K., Yamada, Y., Shimada, Y., et al., 2004. Phase I clinical trial and pharmacokinetic evaluation of NK911, a micelle-encapsulated doxorubicin. Br. J. Cancer 91, 1775–1781.

Max, H.D., Raschke, T.D., Schimpf, R.D., 2004. Skin care compositions with retinoids, ubiquinones, and biotin or carnitine. Google Patents.

Mcclements, D.J., 2011. Edible nanoemulsions: fabrication, properties, and functional performance. Soft Matter 7, 2297–2316.

Mehnert, W., Mäder, K., 2001. Solid lipid nanoparticles: production, characterization and applications. Adv. Drug Deliv. Rev. 47, 165–196.

Melo, A., Amadeu, M.S., Lancellotti, M., De hollanda, L.M., Machado, D., 2015. The role of nanomaterials in cosmetics: national and international legislative aspects. Quim. Nova 38, 599–603.

Mihranyan, A., Ferraz, N., Stromme, M., 2012. Current status and future prospects of nanotechnology in cosmetics. Prog. Mater. Sci. 57, 875–910.

Moinard-Checot, D., Chevalier, Y., Briancon, S., Beney, L., Fessi, H., 2008. Mechanism of nanocapsules formation by the emulsion-diffusion process. J. Colloid Interface Sci. 317, 458–468.

Moore, M.N., 2006. Do nanoparticles present ecotoxicological risks for the health of the aquatic environment? Environ. Int. 32, 967–976.

Mora-Huertas, C.E., Fessi, H., Elaissari, A., 2010. Polymer-based nanocapsules for drug delivery. Int. J. Pharm. 385, 113–142.

Morganti, P., 2010. Use and potential of nanotechnology in cosmetic dermatology. Clin. Cosmet. Investig. Dermatol. 3, 5–13.

Morrow, D., Mccarron, P., Woolfson, A., Donnelly, R., 2007. Innovative strategies for enhancing topical and transdermal drug delivery. Open Drug Deliv. J. 1, 36–59.

Moya-Ortega, M.D., Alvarez-Lorenzo, C., Concheiro, A., Loftsson, T., 2012. Cyclodextrin-based nanogels for pharmaceutical and biomedical applications. Int. J. Pharm. 428, 152–163.

Mozafari, M.R., 2005. Liposomes: an overview of manufacturing techniques. Cell. Mol. Biol. Lett. 10, 711–719.

Mukherjee, S., Ray, S., Thakur, R.S., 2009. Solid lipid nanoparticles: a modern formulation approach in drug delivery system. Indian J. Pharm. Sci. 71, 349–358.

Müller, R.H., Radtke, M., Wissing, S.A., 2002. Solid lipid nanoparticles (SLN) and nanostructured lipid carriers (NLC) in cosmetic and dermatological preparations. Adv. Drug Deliv. Rev. 54 (Suppl.), S131–S155.

Muller, R.H., Petersen, R.D., Hornmoss, A., Pardeike, J., 2007. Nanostructured lipid carriers (NLC) in cosmetic dermal products. Adv. Drug Deliv. Rev. 59, 522–530.

Nagavarma, B., Yadav, H.K., Ayaz, A., Vasudha, L., Shivakumar, H., 2012. Different techniques for preparation of polymeric nanoparticles—a review. Asian J. Pharm. Clin. Res. 5, 16–23.

Nazarenko, Y., Zhen, H.J., Han, T., Lioy, P.J., Mainelis, G., 2012. Potential for inhalation exposure to engineered nanoparticles from nanotechnology-based cosmetic powders. Environ. Health Perspect. 120, 885–892.

Nemen, D., Lemos-Senna, E., 2011. Preparation and characterization of resveratrol-loaded lipid-based nanocarriers for cutaneous administration. Quim. Nova 34, 408–413.

Nohynek, G.J., Dufour, E.K., 2012. Nano-sized cosmetic formulations or solid nanoparticles in sunscreens: a risk to human health? Arch. Toxicol. 86, 1063–1075.

Nohynek, G.J., Dufour, E.K., Roberts, M.S., 2008. Nanotechnology, cosmetics and the skin: is there a health risk? Skin Pharmacol. Physiol. 21, 136–149.

Oh, D.H., Balakrishnan, P., Oh, Y.K., Kim, D.D., Yong, C.S., Choi, H.G., 2011. Effect of process parameters on nanoemulsion droplet size and distribution in SPG membrane emulsification. Int. J. Pharm. 404, 191–197.

Okawara, M., Tokudome, Y., Todo, H., Sugibayashi, K., Hashimoto, F., 2013. Enhancement of diosgenin distribution in the skin by cyclodextrin complexation following oral administration. Biol. Pharm. Bull. 36, 36–40.

Oliveira, R.C.S., 2008. Desenvolvimento, formulação e avaliação de sistemas de libertação transdérmica incorporando sistemas ternários de complexação. Tese de Doutorado, Universidade do Porto.

Pan, G.H., Barras, A., Boussekey, L., Qu, X.S., Addad, A., Boukherroub, R., 2013. Preparation and characterization of decyl-terminated silicon nanoparticles encapsulated in lipid nanocapsules. Langmuir 29, 12688–12696.

Pandey, R., Sharma, S., Khuller, G.K., 2005. Oral solid lipid nanoparticle-based antitubercular chemotherapy. Tuberculosis 85, 415–420.

Park, H., Park, K., 1996. Biocompatibility issues of implantable drug delivery systems. Pharm. Res. 13, 1770–1776.

Pautot, S., Frisken, B.J., Cheng, J.X., Xie, X.S., Weitz, D.A., 2003. Spontaneous formation of lipid structures at oil/water lipid interfaces. Langmuir 19, 10281–10287.

Pietzyk, B., Henschke, K., 2000. Degradation of phosphatidylcholine in liposomes containing carboplatin in dependence on composition and storage conditions. Int. J. Pharm. 196, 215–218.

Pillai, O., Panchagnula, R., 2001. Polymers in drug delivery. Curr. Opin. Chem. Biol. 5, 447–451.

Qian, C., Mcclements, D.J., 2011. Formation of nanoemulsions stabilized by model food-grade emulsifiers using high-pressure homogenization: factors affecting particle size. Food Hydrocoll. 25, 1000–1008.

Qian, C., Decker, E.A., Xiao, H., Mcclements, D.J., 2012. Nanoemulsion delivery systems: influence of carrier oil on beta-carotene bioaccessibility. Food Chem. 135, 1440–1447.

Rasheed, A., Kumar, C.K., Sravanthi, V.V.N.S.S., 2008. Cyclodextrins as drug carrier molecule: a review. Sci. Pharm. 76, 567–598.

Raut, S., Bhadoriya, S.S., Uplanchiwar, V., Mishra, V., Gahane, A., Jain, S.K., 2012. Lecithin organogel: a unique micellar system for the delivery of bioactive agents in the treatment of skin aging. Acta Pharm. Sin. B 2, 8–15.

Regiert, M., Kupka, M., 2010. Cosmetic composition comprising a complex of cyclodextrin and vitamin F. Google Patents.

Sadrieh, N., Wokovich, A.M., Gopee, N.V., Zheng, J.W., Haines, D., Parmiter, D., et al., 2010. Lack of significant dermal penetration of titanium dioxide from sunscreen formulations containing nano- and submicron-size TiO_2 particles. Toxicol. Sci. 115, 156–166.

Sarathchandiran, I., 2012. A review on nanotechnology in solid lipid nanoparticles. Int. J. Pharm. Dev. Technol. 2, 45–61.

Schaffazick, S.R., Guterres, S.S.U., Freitas, L.D., Pohlmann, A.R., 2003. Physicochemical characterization and stability of the polymeric nanoparticle systems for drug administration. Quim. Nova 26, 726–737.

Schmaltz, C., Santos, J.V., Guterres, S.S., 2005. Nanocápsulas como uma tendência promissora na área cosmética: a imenda potencialidade deste pequeno grande recurso. Infarma 16, 80–85.

Senior, J.H., 1987. Fate and behavior of liposomes in vivo: a review of controlling factors. Crit. Rev. Ther. Drug Carrier Syst. 3, 123–193.

Sharma, S., Sarangdevot, K., 2012. Nanoemulsions for cosmetics. IJARPB 1, 408–415.

Sharma, N., Bansal, M., Visht, S., Sharma, P., Kulkarni, G., 2010. Nanoemulsion: a new concept of delivery system. Chron. Young Sci. 1, 2.

Shefer, A., Ng, C., Shefer, S., 2004. Nanotechnology enhances bio-adhesion and release rinse-off. Cosmet. Toilet. 119.

Shegokar, R., Singh, K.K., Muller, R.H., 2011. Production and stability of stavudine solid lipid nanoparticles—from lab to industrial scale. Int. J. Pharm. 416, 461–470.

Sidim, T., Acar, G., 2013. Alcohols effect on critic micelle concentration of polysorbate 20 and cetyl trimethyl ammonium bromine mixed solutions. J. Surf. Deterg. 16, 601–607.

Silva, J.A., Apolinário, A.C., Souza, M.S.R., Damanesceno, B.P.G.L., Medeiros, A.C.D., 2010. Administração cutâena de fármacos: desafios e estratégias para o desenvolvimento de formulações transdérmicas. Revista de Ciências Farmacêuticas Básica e Aplicada 31, 125–131.

Silva, A.L.M., Contri, R.V., Jornada, D.S., Pohlmann, A.R., Guterres, S.S., 2013. Vitamin K1-loaded lipid-core nanocapsules: physicochemical characterization and *in vitro* skin permeation. Skin Res. Technol. 19, e223–e230.

Simko, M., Mattsson, M.O., 2010. Risks from accidental exposures to engineered nanoparticles and neurological health effects: a critical review. Part. Fibre Toxicol. 7.

Singh, K.K., Vingkar, S.K., 2008. Formulation, antimalarial activity and biodistribution of oral lipid nanoemulsion of primaquine. Int. J. Pharm. 347, 136–143.

Singh, A., Van hamme, J.D., Ward, O.P., 2007. Surfactants in microbiology and biotechnology. Part II. Application aspects. Biotechnol. Adv. 25, 99–121.

Singh, R., Bharti, N., Madan, J., Hiremath, S.N., 2010. Caracterization of cyclodextrin inclusion complexes—a review. J. Pharm. Sci. Technol. 2, 171–183.

Souto, E.B., Severino, P., Santana, M.H.A., 2012. Preparation of polymeric nanoparticles from pre-formedpolymers. Part II. Polimeros-Ciencia E Tecnologia 22, 101–106.

Spada, G., Gavini, E., Cossu, M., Rassu, G., Carta, A., Giunchedi, P., 2013. Evaluation of the effect of hydroxypropyl-β-cyclodextrin on topical administration of milk thistle extract. Carbohydr. Polym. 92, 40–47.

Tang, S.Y., Manickam, S., Wei, T.K., Nashiru, B., 2012. Formulation development and optimization of a novel Cremophore EL-based nanoemulsion using ultrasound cavitation. Ultrason. Sonochem. 19, 330–345.

Teagarden, D.L., Baker, D.S., 2002. Practical aspects of lyophilization using non-aqueous co-solvent systems. Eur. J. Pharm. Sci. 15, 115–133.

Thakkar, K.N., Mhatre, S.S., Parikh, R.Y., 2010. Biological synthesis of metallic nanoparticles. Nanomed. Nanotechnol. Biol. Med. 6, 257–262.

Tiarks, F., Landfester, K., Antonietti, M., 2001. Preparation of polymeric nanocapsules by miniemulsion polymerization. Langmuir 17, 908–918.

Tinkle, S.S., 2010. Maximizing safe design of engineered nanomaterials: the NIH and NIEHS research perspective. Wiley Interdiscip. Rev. Nanomed. Nanobiotechnol. 2, 88–98.

Tree-Udom, T., Wanichwecharungruang, S.P., Seemork, J., Arayachukeat, S., 2011. Fragrant chitosan nanospheres: controlled release systems with physical and chemical barriers. Carbohydr. Polym. 86, 1602–1609.

Tsujimoto, H., Hara, K., Yokoyama, T., Yamamoto, H., Takeuchi, H., Kawashima, Y., et al., 2004. Percutaneous absorption study of biodegradable PLGA nano-spheres via human skin biopsies. J. Soc. Powder Technol. 41.

Vilanova, N., Solans, C., 2015. Vitamin A palmitate-beta-cyclodextrin inclusion complexes: characterization, protection and emulsification properties. Food Chem. 175, 529–535.

Vo, C.L.N., Park, C., Lee, B.J., 2013. Current trends and future perspectives of solid dispersions containing poorly water-soluble drugs. Eur. J. Pharm. Biopharm. 85, 799–813.

Warheit, D.B., 2004. Nanoparticles: health impacts? Mater. Today 7, 32–35.

White-House, 2014. Available from: <www.whitehouse.gov> (accessed 14.10.16).

Wokovich, A.M., Prodduturi, S., Doub, W.H., Hussain, A.S., Buhse, L.F., 2006. Transdermal drug delivery system (TDDS) adhesion as a critical safety, efficacy and quality attribute. Eur. J. Pharm. Biopharm. 64, 1–8.

Wong, P.T., Wang, S.H., Ciotti, S., Makidon, P.E., Smith, D.M., Fan, Y.Y., et al., 2014. Formulation and characterization of nanoemulsion intranasal adjuvants: effects of surfactant composition on mucoadhesion and immunogenicity. Mol. Pharm. 11, 531–544.

Wu, X., Guy, R.H., 2009. Applications of nanoparticles in topical drug delivery and in cosmetics. J. Drug Deliv. Sci. Technol. 19, 371–384.

Wu, Y., Ju, L., Ma, X., 2013. Application of γ-polyglutamic acid/chitosan nanocapsule in delayed release of essence. Flav. Frag. Cosm. 1.

Wulff-Perez, M., Torcello-Gomez, A., Galvez-Ruiz, M.J., Martin-Rodriguez, A., 2009. Stability of emulsions for parenteral feeding: preparation and characterization of o/w nanoemulsions with natural oils and Pluronic f68 as surfactant. Food Hydrocoll. 23, 1096–1102.

Yin, H.B., Yamamoto, T., Wada, Y., Yanagida, S., 2004. Large-scale and size-controlled synthesis of silver nanoparticles under microwave irradiation. Mater. Chem. Phys. 83, 66–70.

Yoncheva, K., Calleja, P., Agueros, M., Petrov, P., Miladinova, I., Tsvetanov, C., et al., 2012. Stabilized micelles as delivery vehicles for paclitaxel. Int. J. Pharm. 436, 258–264.

Yoo, D.-C., Cho, H.-N., Kim, K.-R., Byun, H.-J., Kim, J.-H., Park, H.-B., et al., 2011. The anti-bacterial activity of supramolecule containing cosmetic materials. J. Soc. Cosm. Sci. Korea 37, 337–345.

Yu, C., Gao, C., Lü, S., Chen, C., Yang, J., Di, X., et al., 2014a. Facile preparation of pH-sensitive micelles self-assembled from amphiphilic chondroitin sulfate–histamine conjugate for triggered intracellular drug release. Colloids Surf. B: Biointerfaces 115, 331–339.

Yu, H.L., Huang, Q.R., 2012. Improving the oral bioavailability of curcumin using novel organogel-based nanoemulsions. J. Agric. Food Chem. 60, 5373–5379.

Yu, M., Ma, H.X., Lei, M.Z., Li, N., Tan, F.P., 2014b. In vitro/in vivo characterization of nanoemulsion formulation of metronidazole with improved skin targeting and anti-rosacea properties. Eur. J. Pharm. Biopharm. 88, 92–103.

Zaibunnisa, A.H., Aini marhanna, M.N.A., Ainun atirah, M., 2011. Characterization and solubility study of γ-cyclodextrin and β-carotene complex. Int. Food Res. J. 18, 1061–1065.

Zhang, J.X., Ma, P.X., 2013. Cyclodextrin-based supramolecular systems for drug delivery: recent progress and future perspective. Adv. Drug Deliv. Rev. 65, 1215–1233.

Zhang, L.F., Yu, K., Eisenberg, A., 1996. Ion-induced morphological changes in ‘‘crew-cut’’ aggregates of amphiphilic block copolymers. Science 272, 1777–1779.

Zhang, H., Zeeb, B., Salminen, H., Weiss, J., 2015. Isothermal titration calorimetric analysis on solubilization of an octane oil-in-water emulsion in surfactant micelles and surfactant–anionic polymer complexes (vol. 438, p. 7, 2015). J. Colloid Interface Sci. 452, 1–1.

FURTHER READING

SCIENCEDIRECT, 2018. Available from: <https://www.sciencedirect.com/search?qs = nanoemulsion%20cosmetics&authors = &pub = &volume = &issue = &page = &origin = home&zone = qSearch&sortBy = date> (accessed 07.02.18).

Simonsson, C., 2011. New insights in contact allergy and drug delivery: a study of formulation effects and hapten targets in skin using two-photon fluorescence microscopy (Doctoral Thesis) (Philosophy in Chemistry). University of Gothenburg.

Wichlab, 2016. Available from: <http://www.wichlab.com/research/> (accessed 14.10.16).

CHAPTER

6

Nanoparticulate drug delivery systems for the oral administration of macromolecular drugs

Javed Iqbal*[,1], Gautam Behl*[,2], Gavin Walker[1], Chris Edin[1], Parveen Kumar[3], Niall O'Reilly[2], Rohit Bhatia[4], Laurence Fitzhenry[2] and Taruna Arora[5]

[1]Pharmaceutical Manufacturing Technology Centre, Bernal Institute, University of Limerick, Limerick, Ireland [2]Pharmaceutical and Molecular Biotechnology Research Centre, Department of Science, Waterford Institute of Technology, Waterford, Ireland [3]Department of Chemistry, Dyal Singh College, University of Delhi, New Delhi, India [4]Division of Sustainable Chemistry, Rudraksh Proudhyogiki Sangathan, Delhi, India [5]Department of Biochemistry, Institute of Home Economics, University of Delhi, New Delhi, India

CHAPTER OUTLINE

*Both authors have contributed equally to this chapter.

Nanoparticles in Pharmacotherapy. DOI: https://doi.org/10.1016/B978-0-12-816504-1.00017-X

6.1 INTRODUCTION

Recent advances in pharmaceutical biotechnology promote macromolecules as a front line therapeutic agent for the treatment of several major disorders, including AIDS, hepatitis, cardiovascular disease, and cancer (Brannon-Peppas and Blanchette, 2004; Kawasaki and Player, 2005). Most macromolecular drugs are peptides and proteins and are usually administered as a solution or suspension, through what can be painful and often unacceptable invasive routes, such as intravenous or subcutaneous injections (Jitendra et al., 2011). As most macromolecular drugs are intended for the long-term treatment of nonacute, life-threatening diseases, easier routes of administration are necessary.

Oral administration of macromolecular drug is hindered by physical and biochemical barrier offered by gastrointestinal (GI) epithelium. The physical barrier is primarily due to the impermeable GI epithelium, while the biochemical barrier mainly involves enzymatic degradation of the drug. The basic understanding of the barrier function of GI tract is therefore necessarily required for designing optimal deliver approaches for macromolecular drugs (Vilar et al., 2012). Thus, enthusiastic efforts have been made over the last two decades in the development of safe and effective formulations for the oral administration of these therapeutic agents. Innovations in the field of drug delivery, such as enteric coating, encapsulation with polymeric materials, coadministration of enzyme inhibitors, reducing the number of amino acids to achieve the same effect, the use of permeation enhancers, mucoadhesive materials, and nanoparticles has made it possible to overcome these absorption barriers to some extent and has led to enhanced oral bioavailability of macromolecular therapeutic agents (Hamman et al., 2005; Renukuntla et al., 2013; Smart et al., 2014).

In general, *nanoparticles*, *nanoparticulate*, *nanocarrier*, *nanospheres*, and *nanocapsules* are some of the most common terms used to define solid particles based on natural or synthetic biodegradable or nonbiodegradable polymers, with diameters ranging from 1 to 1000 nm (Couvreur, 1988). Nanospheres themselves are comprised of a matrix-type structure. Drugs may be adsorbed at the sphere surface or encapsulated within the particle. In contrast, nanocapsules are vesicular systems in which the drug is restrained to a cavity consisting of an inner liquid core surrounded by a polymeric membrane (Couvreur et al., 1995). In this case the active substances are usually dissolved in the inner core but may also be adsorbed to the capsule surface (Allemann et al., 1993). Due to their extremely small size, nanoparticulate drug delivery systems (DDS) permit association of therapeutic agent inside their polymeric matrix/core. Thus, they provide a protecting shield to the encapsulated drug against harsh environment, as well as hydrolytic and enzymatic degradation.

This chapter addresses the developments in the use of nanoparticulate DDS for the oral administration of macromolecular drugs, which include proteins and peptides. The materials and techniques used in the development of nanoparticulate DDS are also described in detail.

6.2 MODE OF TRANSPORT

The anatomy and physiology of intestinal epithelium is constructed in such a way that it can efficiently prevent uptake of particulate matter into the internal milieu. The absorption of orally presented macromolecules is hindered by a series of physiological and morphological obstructions in GI tract, such as (1) enzymatic barrier in gut lumen (e.g., trypsin, chymotrypsin and pepsin), (2) brush border membrane enzymes (endopeptidases), (3) mucus gel layer, (4) bacterial gut flora, and (5) epithelial cell lining itself (tight junctions). Theoretically, molecules can be transported across intestinal mucosa via intercellular uptake (paracellular or persorption) or transcellular, depending on their chemical and physical properties (Pade and Stavchansky, 1997). However, hydrophilic nature, large molecular mass, and charged structure of macromolecules make it difficult for them to access the transcellular route. Thus paracellular pathway is described to be mainly involved in the transport of hydrophilic macromolecules (Lee et al., 1991). Strategies have been designed to bypass the intestinal barriers faced in oral protein delivery. Nanoparticles have proven to efficiently absorb in the intestine due to their small size and ease of desirable modification with absorption enhancers (Kreuter, 1996). The absorption and pathway of nanoparticle uptake is diverse in different parts of the intestine (Fasano, 1998). The M-cells of the Peyer's patches as mentioned in Fig. 6.1 symbolize a sort of lymphatic island within the intestinal mucosa and, presumably, the major gateway through which particles can be absorbed. Cell-specific carbohydrate is present on the surface of M-cells serving as the binding site of specific ligands. Certain type of glycoprotein and lectins are known to bind specifically to these surface entities by receptor-mediated mechanism (Cai et al., 2010; Powell et al., 2010). Vitamin B_{12} absorption from the gut under physiological conditions occurs via receptor-mediated endocytosis. The applicability of vitamin B_{12} in increasing the oral bioavailability of various peptides and proteins has also been evaluated and appreciable results have been observed (Damgé et al., 1990; Brandtzaeg et al., 1997; Florence, 2004). The internalization of the nanoparticle occurs by either receptor-mediated endocytosis, an active mode of targeting, or simply by adsorption endocytosis, which does not require any ligand. This process is initiated by an unspecific physical adsorption of material to the cell surface by electrostatic forces such as hydrogen bonding or

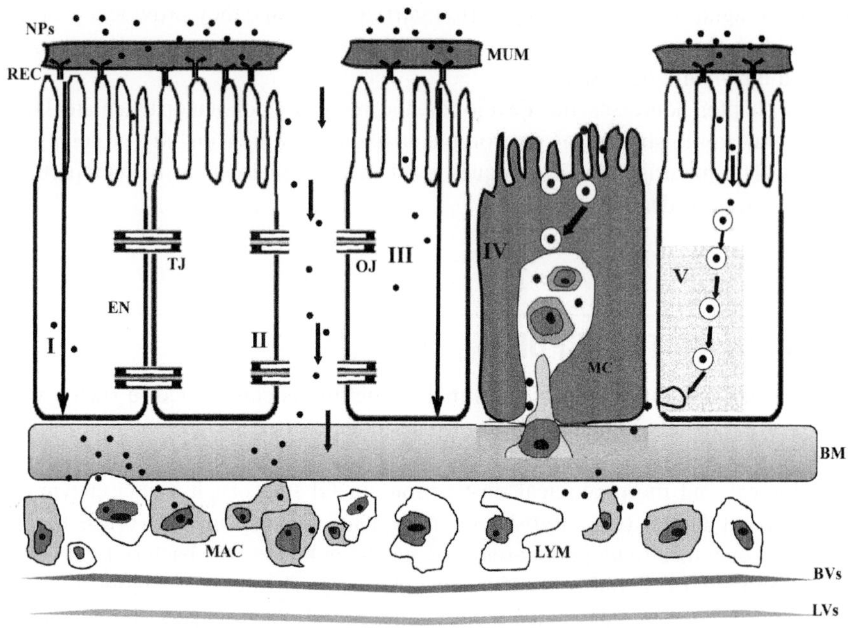

FIGURE 6.1

Schematic presentation describing the transport mechanisms of nanoparticles for oral drug delivery: (I) receptor-mediated endocytosis, (II) paracellular transport, (III) nonspecific transcellular transport, (IV) M cells mediated transport, (V) endocytotic uptake by enterocytes. *Abbreviations*: NPs, nanoparticles; MUM, mucus membrane; REC, receptors; TJ, tight junction; OJ, open junction; EN, enterocytes; MC, M-cells; BM, basal membrane; LYM, lymphocytes; MAC, macrophages; BVs, blood vessels; LVs, lymphatic vessels.

hydrophobic interactions. The demonstration of all endocytosis mechanisms that are restricted to M-cells and phagocytic immune cells and pinocytosis shown in Fig. 6.1 has been recently reviewed (Conner and Schmid, 2003; Bareford and Swaan, 2007; Sahay et al., 2010).

6.3 POTENTIAL ADVANTAGES

Regarding the drug delivery perspective, the use of nanoparticles has been associated with numerous potential rewards, such as that they (1) enhance the solubility of poorly water-soluble drugs, (2) reduce immunogenicity, (3) are able to extend the half-life of drug systemic circulation, (4) release drugs at sustained rate, (5) reduce the dosing frequency, (6) have low therapeutic toxicity, (7) provide more convenient route of drug administration, (8) reduce healthcare costs and finally (9) improve patient compliance (El-Shabouri, 2002; Schiffelers, 2004; Connor et al., 2005; Ferrari, 2005; Gelperina et al., 2005; Kesisoglou et al., 2007; Sung et al., 2007; Iqbal et al., 2011; Cheng et al., 2012). Nanoparticles fabricated with targeting moieties specific to certain tissues can be developed to release the drug at the site of action (Yun et al., 2013; Li et al., 2015; Masood, 2016). Folate conjugated PEG–PLGA nanoparticles have been studied for the oral delivery of insulin showed a twofold increase in oral bioavailability without hypoglycemic shock as compared to subcutaneous injection of regular insulin solution. Further the targeted nanoparticles maintained the glucose levels for 24 h, as compared to regular insulin solution, which resulted in hypoglycemic shock following a transient effect in less than 8 h (Jain et al., 2012). Furthermore, multitasking, encompassing diagnostic agents (Merisko-Liversidge and Liversidge, 2011), DNA manipulation (Kawakami, 2012), molecular medicine (Galindo-Rodriguez et al., 2005), and follow-up monitoring (Patel et al., 2011; Desai et al., 2012), is promising by single nanoparticle-based agents owing to their designable multifunctionality as shown in Fig. 6.2.

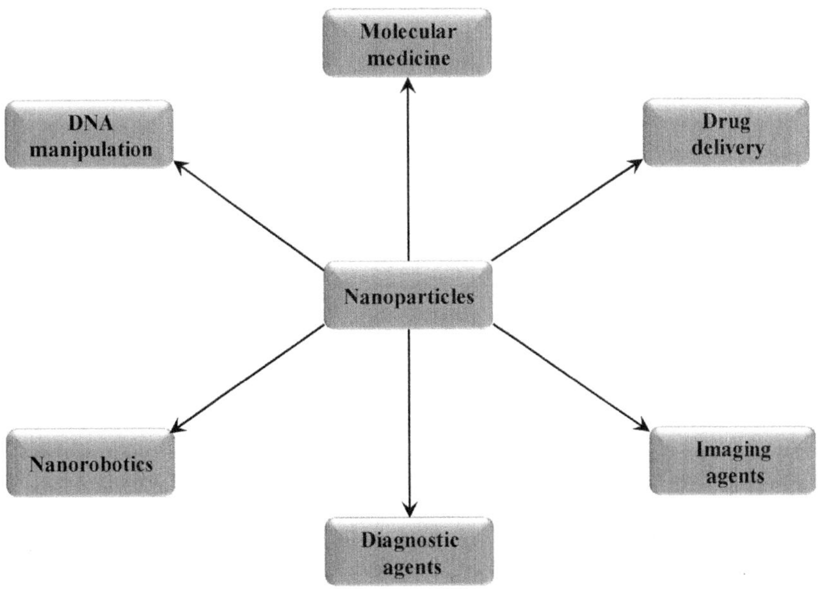

FIGURE 6.2

Multitasking advantages of nanoparticles.

6.4 PRODUCTION TECHNIQUES

As the figure of diverse type of nanoparticles is increasing rapidly, the majority can be classified according to their size, production processes and type of material (Fig. 6.3). Liposomes, dendrimers, carbon nanotubes, emulsions, and other polymers are all well-established types of organic particles.

Use of these organic nanoparticles has previously produced exciting results. An inorganic (ceramics) nanoparticle consists of a central core, usually metals that define the optical, magnetic, fluorescence, and electronic properties of the particle, with a protective organic coating on the surface (Yezhelyev et al., 2006). Nanoparticles can be generated by adapting numerous techniques regarding the type of material (organic or inorganic) employed for formulations. Accordingly, production techniques for organic and inorganic (ceramics) nanoparticles are provided in Fig. 6.4.

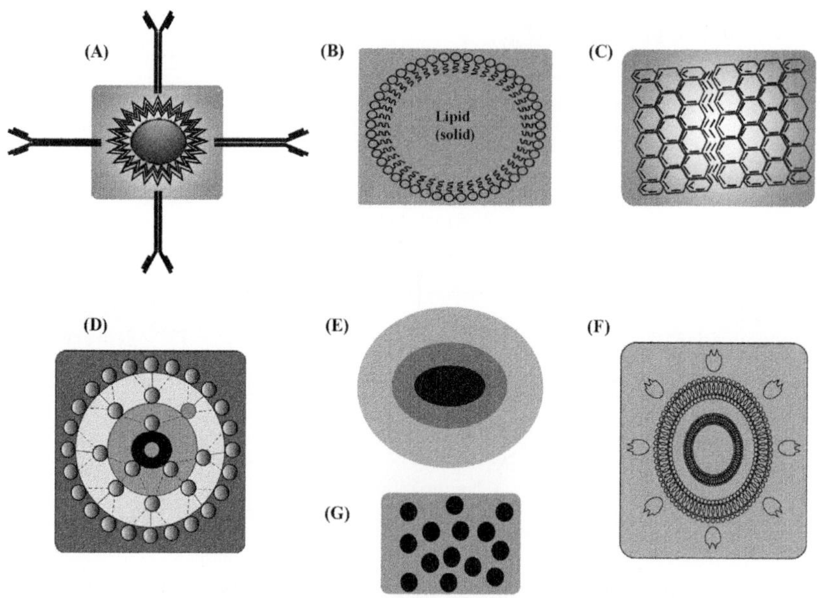

FIGURE 6.3

Different types of nanoparticles. (A) Functionalized nanoparticles: uniform shape with pharmacological agents and targeting molecules; (B) solid lipid nanoparticles (SLN): colloids based on solid lipid; (C) nanotubes: self-assembled lipid tubes; (D) dendrimers: series of branches around an inner core for aided visualization of various pathological processes; (E) nanocapsules: drug is encapsulated within polymeric membrane for modified release; (F) liposomes: bilayered vesicles entirely enclosed by a membranous lipid bilayer with targeted ligand; (G) nanospheres: drug is dispersed throughout the polymeric matrix for modified release.

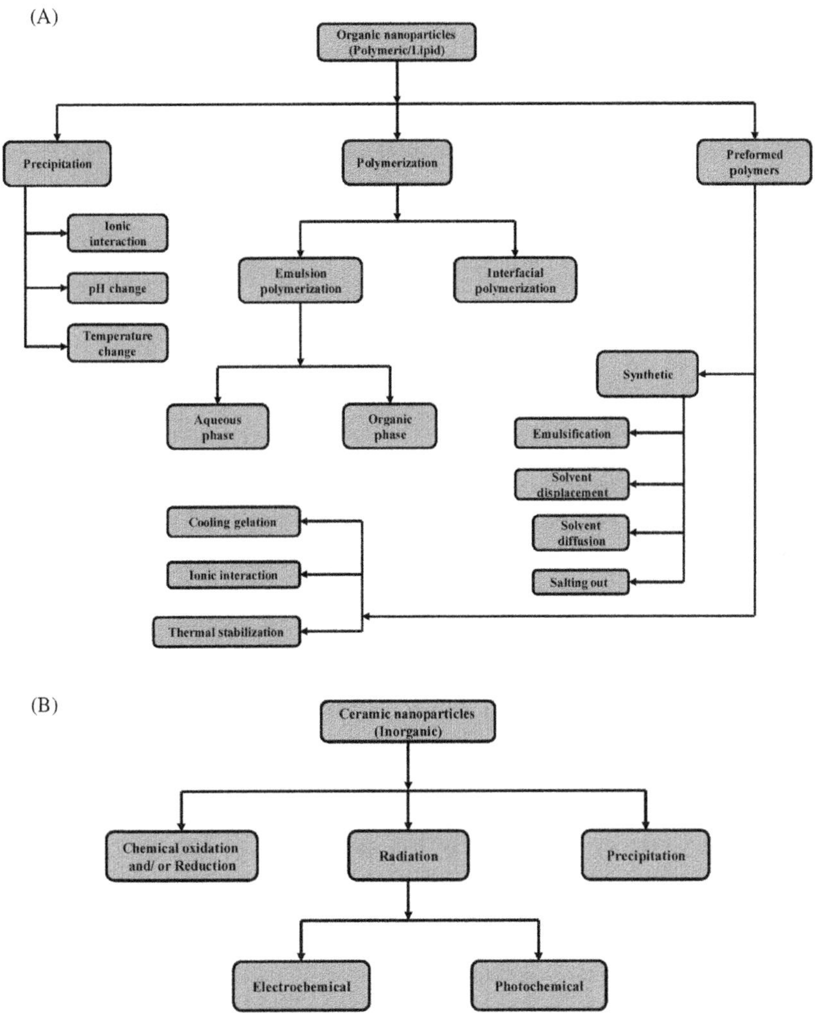

FIGURE 6.4

Schematic presentation of production techniques used for the development of organic (A) and inorganic/ceramics (B) nanoparticles.

6.5 MACROMOLECULAR DELIVERY: STATE OF THE ART

The ease and suitability of the oral route for drug administration dictates that it has received much consideration for the delivery of macromolecules. Various advances have been demonstrated to overcome the delivery issues of macromolecules by employing nanoparticles. The potential of nanoparticles to enhance the transport of the associated macromolecules has been attributed to different mechanisms depending on the nanoparticles composition. These mechanisms are (1) mucoadhesion, (2) particle internalization phenomenon, and (3) permeation enhancing effect (Renukuntla et al., 2013; Date et al., 2016). Various techniques have been used in an endeavor to overcome these barriers and to increase the oral bioavailability of such macromolecules including the use of lipid-based carrier systems, inorganic ceramic-based carrier systems, and polymeric-based carrier systems. Also the most important part is played by the approach adopted for formulation development. Different strategies have been studied to develop stable delivery carriers for macromolecules protecting them from degradation in GI tract besides enhancing permeation. Therefore, before going to the discussion of nanoparticles being used, a brief account of formulation strategies would give a better understanding of the concept. Being hydrophilic, large size and charged skeleton proteins and peptides have very poor permeation and thus suffer from very low bioavailability. Further they are also susceptible to degradation by proteolytic enzymes in the GI tract. The answer to these two issues have been devised in terms of application of absorption enhancers and enzyme inhibitors for successful oral delivery of macromolecules (Bruno et al., 2013; Renukuntla et al., 2013; Date et al., 2016).

Permeation enhancers either increase paracellular permeability by slight modification of tight junction (TJ) functional properties or increase transcellular permeation by membrane perturbation (Tscheik et al., 2013). Enhancers such as fatty acid, bile salts, chitosan, polymeric thiomers, and Pz-peptide have been used for the oral delivery of insulin resulting in the reduction in blood glucose levels. Surfactants have also been utilized for the delivery of macromolecules as they enhance transcellular transport by disrupting the lipid bilayer thus making it more permeable to drugs. However, beside the advantages prolonged use of permeation enhancers is also associated with membrane damage causing local inflammation (Renukuntla et al., 2013; Muheem et al., 2016).

A number of enzymes are involved in cleavage of amino acid chains in peptides and proteins, that is, trypsin, chymotrypsin, elastase, pepsin, carboxypeptidases, etc. Therefore formulating enzyme inhibitors is a viable approach to keep therapeutic macromolecules intact and preserve their activity. Recently, chicken and duck ovomucoids have been studied for the delivery of insulin where about 100% protection was observed against trypsin and α-chymotrypsin. Carboxymethyl cellulose–Elastinal is an example of polymer-inhibitor conjugate studied that has shown protection from trypsin, α-chymotrypsin, and elastase. About 30% of the therapeutic was found to be active even after 4 h of incubation with elastase. Members from serine protease inhibitor group are known to form covalent complexes with protease by undergoing conformational changes thus protecting the therapeutic protein from degradation (Renukuntla et al., 2013; Muheem et al., 2016).

6.6 LIPID-BASED CARRIER SYSTEMS

6.6.1 LIPOSOMES

Liposomes were introduced in pharmaceutical literature as a drug carrier in the early 1970s (Gregoriadis and Ryman, 1972). Liposomes are broadly defined as lyotropic liquid crystals (vesicles) with a central aqueous space encapsulated by natural or synthetic phospholipid bilayer(s), as illustrated in Fig. 6.3. The hydrophilic heads of the phospholipid layer, pointing in an outward direction, make liposomes water soluble. Unilamellar vesicles (ULVs) are surrounded by single lipid layer with a diameter of 25–100 nm, whereas, multilamellar vesicles (MLVs) are composed of several (up to 14) lipid layers separated from each other by aqueous layers (Kozubek et al., 2001). Due to their amphiphilic nature, liposomes can entrap hydrophilic as well as hydrophobic molecules easily. Liposomes transport the drugs into cells by fusion or endocytosis mechanisms. Liposome properties differ substantially with lipid composition, size, surface charge, and the method of preparation (Makino and Shibata, 2006). The natural lipid constituents of liposomes make them biodegradable, biologically inert, nontoxic and nonimmunogenic, further they also protect the drug from enzymatic degradation. Thus, the encapsulated drug can be transported securely to the target site without inactivation and/or degradation (Allen and Moase, 1996; Allen, 1997). Liposomal formulations have shown a potential to improve the pharmacodynamic and pharmacokinetic profile of various therapeutic agents especially peptides and proteins (Gabizon et al., 1998), (Table 6.1). Parmentier et al. (2010) reported three- to fourfold improved permeation of dextran entrapped in egg phosphatidylcholine/cholesterol liposomes through Caco-2 cells. Moreover, liposome surfaces can be easily modified by the attachment of PEGs or mucin, leading to the formation of stealth liposomes. For instance, long circulating PEG-coated and mucin-coated liposomal formulation of insulin leading to the enhanced and prolonged hypoglycemic effects of insulin relative to the free drug in rats (Iwanaga et al., 1997, 1999). However, use of liposomes may be restricted due to challenges posed by their leakage of hydrophilic drug during their passage through the GI tract, heterogeneous size, and inadequate in vivo stability.

To address these issues archaeosomes have been devised, which are liposomes prepared from the lipids that are unique to archaeobacterial membrane. Archaeosomes have been found to withstand extreme pH and oxidative conditions and are also reported to be stable in bile salts and lipases (Barbeau et al., 2011). Archaeosomes prepared from polar lipid fraction E were evaluated for oral insulin delivery in diabetic rats. The formulation was found to be stable in GI tract and better control over blood glucose level was observed as compared to conventional liposomes (Li et al., 2010). The high stability of archaesomes thus makes them suitable candidates for the oral delivery of peptide and proteins.

Self-assembling pectin–liposome nanocomplexes and silica–liposome nanocapsules are some of the other advanced variant liposomes devised for improved

Table 6.1 Lipid Microparticles, Solid Lipid Nanoparticles, and Liposomes as a Drug Delivery System With Production Procedure and the Association Efficiency for Peptides and Proteins

Protein/Peptide	Formulation Technique	Carrier System	Association Efficiency (%)	Cumulative Release (%)	Protein Stability/ Biological Activity	References
Adamantyltripeptides	Dry lipid hydration		NA	NA	NA	Frkanec et al. (2003)
Antide	Milling of drug–lipid solid solutions (comelting)	LM	>85	≤60/24 h	100% BA	Del Curto et al. (2003)
Antide	Milling of drug–lipid solid solutions (solvent stripping)	LM	>80	≤40/24 h	100% BA	Del Curto et al. (2003)
Antiovalbumin antibodies	Dry lipid hydration	Liposomes	70–80	NA	100% BA	Brgles et al. (2007)
Basic fibroblast growth factor	Freeze–thawing extrusion	Liposomes	75–80	20 ng cm^{-2} uptake	BA proved in vivo	Plum (2000), Luo et al. (2003)
BSA	Reversal evaporation	Liposomes	25–71	80/15 days	100% BA	Dai et al. (2005, 2006)
BSA	Double emulsification	Liposomes	43–71	NA	NA	Murakami et al. (2006)
BSA	Freeze–thawing		20–45	NA	NA	Murakami et al. (2006)
BSA	Solvent evaporation (w/o/w)	LM	58.1–69.5	≈30/24 h	>98% intact, BA proved in vivo	Saraf et al. (2006)
BSA	Coating with lipid using SCF	LM	13–62	≈80/24 h	100% intact	Ribeiro Dos Santos et al. (2002)
BSA	Adsorption onto SLN	SLN	NA	NA	100% intact	Gualbert et al. (2003)
BSA-FITC	Adsorption onto LM	LM	NA	NA	100% intact	Erni et al. (2002)
Calcitonin	Dry lipid hydration		20			Takeuchi et al. (2003)
Calcitonin	Dry lipid hydration	SLN	>90	4/6 h	BA proved in vivo	Takeuchi et al. (2005)
Cyc A	HPH hot dispersion HPH cold dispersion	SLN	95.44–97.878.5–93.9	NA	100% intact.	Mueller et al. (1998), Penkler et al. (1999)
Cyc A	HPH hot dispersion	SLN	96.6–97.8	NA	100% intact	Müller et al. (2007)

Cyc A	HPH hot dispersion	SLN	NA	NA	100% intact	Mudshinge et al. (2011), Selvamuthukumar and Velmurugan (2012)
Cyc A	HPH hot dispersion	SLN	96.1	NA	BA proved in vivo	Müller et al. (2006)
Cyc A	Warm microemulsion (o/w)	SLN	13	<5/2 h	100% intact	Ugazio et al. (2002)
Cyc A	HPH hot dispersion	SLN	88.4	NA	BA proved in vivo	Zhang et al. (2000)
Cyc A	Self-emulsification	Lipospheres	NA	NA	BA proved in vivo	Bekerman et al. (2004)
Enkephalin	Double emulsification	Liposomes	50–85	60/3 days	100% intact	Ye et al. (2000)
Epidermal growth factor receptor	Freeze–thawing extrusion	Liposomes	20–30	NA	100% intact	Kullberg et al. (2005)
Gonadorelin	Solvent displacement	SLN	50.4–69.4	80/14 days	100% intact	Hu (2004)
Haptides		Liposomes	~100	NA	100% intact	Gorodetsky et al. (2004)
Hemoglobin	Dry lipid hydration extrusion	Liposomes	37–62	NA	100% intact	Arifin and Palmer (2003), Oda et al. (2005), Patton and Palmer (2005), Xi and Guo (2007)
Horseradish peroxidase	Extrusion	2–5	70–80	NA	Intact protein	Visser et al. (2005)
Human gamma-globulin	Dehydration–rehydration	30–31	75–85	NA	BA proved in vivo	García-Santana et al. (2006)
HSA	Adsorption onto SLN	SLN	12.4–32.4 (stealth) 7–43.5 (nonstealth)	NA	NA	Cavalli et al. (1999)
Human recombinant epidermal growth factor	Microemulsion	SLN	100	NA	NA	Pedersen et al. (2006)
Insulin	w/o/w double emulsion	SLN	27–6835–45	100/24 h	Intact protein	Das and Chaudhury (2010)
Insulin	Reverse phase evaporation	Liposomes	30–82	NA	Intact protein, BA proved in vivo	Zhang et al. (2005), Goto et al. (2006)

(Continued)

Table 6.1 Lipid Microparticles, Solid Lipid Nanoparticles, and Liposomes as a Drug Delivery System With Production Procedure and the Association Efficiency for Peptides and Proteins *Continued*

Protein/Peptide	Formulation Technique	Carrier System	Association Efficiency (%)	Cumulative Release (%)	Protein Stability/ Biological Activity	References
Insulin	Solvent evaporation (o/w)	LM	45–94	<60/3 days	Some aggregation	Reithmeier et al. (2001b)
Insulin	Melt-dispersion (o/w or w/o/w)	LM	≈50 or >80	<60/3 days	Some aggregation	García-Fuentes et al. (2003)
Insulin	Solvent evaporation (w/o/w)	SLN	NA	45/1 h	NA	Trotta et al. (2005)
Insulin	Solvent diffusion (w/o/w)	LM	78–84	≈30/24 h	Intact protein	Zhang et al. (2006)
Insulin	Warm microemulsion (w/o/w)	SLN	37.8	NA	NA	Caliceti et al. (2006)
Insulin	Solvent evaporation (w/o/w)	SLN	67.9	NA	BA proved in vivo	Caliceti et al. (2006)
Insulin	Solvent displacement	SLN	26.8	NA	NA	Caliceti et al. (2006)
Insulin	Supercritical CO_2 (PGSS)	SLN	75	100/4 days	BA proved in vivo	Morel et al. (1995)
J E Ag	Solvent evaporation (w/o/w)	LM	73.8	≈20/10 days	BA proved in vivo	Pichayakorn (2006)
Leishmania antigen	Freeze–thawing extrusion	Liposomes	>80	NA	BA proved in vivo	Badiee et al. (2007)
Leridistim	Double emulsification	Liposomes	>70	100/10 days	BA proved in vivo	Langston (2003)
Leuprolide [D-Trp-6] LHRH	Warm microemulsion (w/o/w)	SLN	90	≈10/8 h	NA	Morel et al. (1995)
Leuprolide [D-Trp-6] LHRH	Dry lipid hydration	Liposomes	>70	NA	NA	Guo (2003)
Lysozyme	HPH cold dispersion	SLN	43.2–59.2	NA	≈100% BA	Almeida et al. (1997)
M Ag R32NS1 (Recombinant malaria protein antigens)	Melt-dispersion (o/w)	Lipospheres	>80	NA	BA proved in vivo	Amsteel et al. (1992)

Nerve growth factor	Reverse phase evaporation	Liposomes	24–34	NA	BA proved in vivo	Xie et al. (2005)
Octreotide	Double emulsification	Liposomes	50–85	80/8 days	100% intact	Ye et al. (2000)
Ovalbumin	Melt-dispersion (o/w)	SLN	>80	NA	Intact protein	Reithmeier et al. (2001a)
Ovalbumin	Adsorption onto SLN	SLN	70–97	53/24 h	Intact protein	Videira (2002)
Progenipoietin	Double emulsification		80–90	NA	BA proved in vivo	Ramprasad et al. (2003)
Somatostatin	Solvent evaporation (o/w or w/o/w)	LM	<75%	NA	NA	Reithmeier et al. (2001a)
Somatostatin	Melt-dispersion (o/w)	LM	65–97	70–80/14 days	Intact protein	Reithmeier et al. (2001a)
Streptavidin	Microemulsion technique	SLN	~100	NA	NA	Pedersen et al. (2006)
Superoxide dismutase	Dry lipid hydration		1–13			
	Dehydration–rehydration		2–3			
	Pro-liposome		39–65	NA	NA	Galović Rengel et al. (2002)
TAT peptide	Extrusion		~100	NA	NA	Torchilin et al. (2001)
Thymopentin	Warm microemulsion (o/w with ionic pair or w/o/w)	Liposomes	5.2 or 1.7		Intact peptide	Morel et al. (1996)
Thymocartin	Solvent evaporation (o/w or w/o/w)	LM	<10 or <50	65–90/5 days	Intact peptide	Reithmeier et al. (2001b)
Thymocartin	Melt-dispersion (o/w or w/o/w)	LM			Intact peptide	Reithmeier et al. (2001b)

BA, biological activity; BSA, bovine serum albumin; Cyc A, cyclosporine A; FITC, flourescein-5-isothiocyanate; has, human serum albumin; HPH, high pressure homogenization; J E Ag, Japanese encephalitis antigen; LHRH, luteinizing hormone–releasing hormone; LM, lipid microparticles; M Ag R32NS1, malaria antigen R32NS1; NA, nonavailable; PGSS, particles from gas saturated solution technique.

intestinal absorption and stability, where in the former case pectin played a role of enhancing the intestinal absorption of calcitonin and in latter case silica nanoparticles coating on liposomes protected the liposomes and loaded insulin from enzymatic degradation (Thirawong et al., 2008; Mohanraj et al., 2010).

6.6.2 SOLID LIPID NANOPARTICLES

Solid lipid nanoparticles (SLNs) are composed of solid triglycerides. They were first developed in the early 1990s as a safe alternative drug delivery carrier to emulsions, liposomes, and polymeric nanoparticles (Muller, 1996). SLNs are usually prepared by the dispersion of solid lipids in water or in an aqueous surfactant solution (Rawat et al., 2006). They consist of a solid hydrophobic core surrounded by a monolayer of phospholipid coating. The entrapped drug is dispersed or dissolved in the solid, high-melting lipid matrix to form the solid hydrophobic core. The phospholipid chains are rooted in the fat matrix. Emulsifiers (surfactants) are usually employed to provide stability to the SLN system. Owing to the advantage of being in the solid state, the lipid components of SLN are supposed to degrade slowly and thus provide more stability and long circulation time to the SLN (Speiser, 1990). Compared to other particulate transporters, SLNs have certain merits as DDS, such as nontoxicity (Müller, 1996), superior permissibility (Gide et al., 2013), biodegradation (Almeida and Souto, 2007), and the possibility of production at commercial scale (Muchow et al., 2008). Further SLN preparation does not require use of toxic organic solvents, and therefore no associated issue of protein stability during formulation. It has also been reported that nanoencapsulation of peptides and proteins in SLNs resulted in improved bioavailability, prolonged blood residence time, and/or customized biodistribution (Garcia-Fuentes et al., 2005). For instance, Sarmento et al. (2007) reported that oral administration of insulin containing cetyl palmitate-based SLN revealed a significant and prolonged hypoglycemic effect in rats in comparison to oral insulin solution. Moreover, coating of SLN with hydrophilic substances like PEGs provide protection to SLNs from enzymatic degradation and aggregation, support the interaction with epithelia, and enhance blood circulation time of particles thus prolonging the efficiency of drugs. Alonso et al. examined that insulin containing PEG-stearate coated SLNs were found to be more stable in the simulated gastric and intestinal fluids in comparison to noncoated lipid nanoparticles (García-Fuentes et al., 2003). Further, chitosan-coated lipid nanostructures have also been evaluated for the oral delivery of salmon calcitonin in rats, where a significant reduction in serum calcium was observed as compared to the study performed with calcitonin solution (Garcia-Fuentes et al., 2005). All these observations clearly indicate the ability of lipid nanoparticle in enhancing the bioavailability of orally delivered macromolecules besides protecting them from harsh environment in the GI tract.

6.6.3 NANOSTRUCTURED LIPID CARRIERS

Nanostructured lipid carriers (NLC) are second generation lipid-based carrier system introduced for delivery of therapeutics (Lück et al., 1998). NLC are formulated by utilization of both solid lipids and liquid lipids (oils) in a proportion that the mixture remains solid at 40°C. The concept of the NLC is executed by nanostructuring the lipid matrix to provide more flexibility for modulation of drug release, to increase the drug payload and to prevent drug expulsion. This could be comprehended in three ways: (1) the imperfect type, (2) the multiple type, and (3) the amorphous type. Most of the drugs utilized in pharmaceutical science possess a higher solubility in oils and hence the drugs dissolved in the oil phase can be protected from degradation by surrounding solid lipids. Examples of macromolecules associated to NLC for oral delivery are the calcitonin and cyclosporine A. Calcitonin loaded NLC have been fabricated by w/o/w double emulsion, exhibiting association efficiency higher than 90% (Müller et al., 2002; Morishita and Peppas, 2006). NLC with cyclosporine A have been generated by hot HPH (Patlolla et al., 2010).

6.7 INORGANIC CERAMIC-BASED CARRIER SYSTEMS AND GOLD NANOPARTICLES

Inorganic ceramic nanoparticles represent a novel generation of nanoparticles made from naturally occurring inorganic ceramic compounds like silica (SiO_2), titania (TiO_2), alumina (Al_2O_3), iron oxide (Fe_3O_4, Fe_2O_3), zinc oxide (ZnO), ceria (CeO_2), and zirconia (ZrO_2) (Jain et al., 1998; Lal et al., 2000). These nanoparticulate systems exhibit numerous advantageous properties over traditional organic polymeric nanocarriers, making them a suitable candidate for oral drug delivery of peptides and proteins. Inorganic ceramic nanoparticles can easily avoid the immune system (monocytes and macrophages) because of their extremely small size (typically less than 50 nm). These materials can be prepared at low temperature so that the enzyme activity can be retained and they are not subjected to microbial attack (Weetall, 1970). Moreover, they display high entrapment efficiencies, stability over a broad pH range, thermal stability, biocompatibility, and flexibility in terms of ease of surface modification (Dey, 2006; Brinker et al., 1990). The effectiveness of inorganic ceramic-based nanoparticulate systems in the delivery of proteins and genes has been demonstrated by numerous research groups (Cherian et al., 2000; Roy et al., 2003). For instance, Jain and coworkers reported better therapeutic effect of insulin entrapped in self-assembled carbohydrate-stabilized ceramic nanoparticles in comparison to standard porcine insulin solution. Unfortunately, inorganic ceramic is nonbiodegradable, slow dissolving, and nevertheless must be somehow eliminated from the body. Another technological problem limiting the use of metal nanoparticles in some applications is their high reactivity, which makes them difficult to produce, transfer, and store without particle contamination (Cherian et al., 2000).

Due to their versatility to fabricate and easily modifiable surface, gold nanoparticles (AuNPs) have been investigated for the oral delivery of macromolecules (Diab et al., 2012). Chitosan-coated AuNPs were synthesized and evaluated for the delivery of insulin. A significant reduction in blood glucose level was observed after 2 h of oral administration in diabetic rats (Bhumkar et al., 2007). Although a very limited studies have been performed with AuNPs for the delivery of macromolecules, but ease of their modification with polymers and other biomolecules makes them a perfect candidate for oral delivery of macromolecules in the near future.

6.8 POLYMERIC-BASED CARRIER SYSTEMS

Polymeric nanoparticles are colloidal carriers based on biodegradable and biocompatible, natural or synthetic polymers that vary in size ranging from 10 to 1000 nm. Based on the method of preparation, polymeric nanocarriers can be classified into nanospheres or nanocapsules in which the drug is either encapsulated or dissolved inside the inner core surrounded by polymeric shell or it can be attached directly to the polymeric matrix. Polymeric nanoparticles exhibit several advantageous properties, including biodegradability, biocompatibility, ease of surface modification, and functionalization of polymers. Moreover, polymeric materials provide a shielding effect to the associated drugs against enzymatic attacks. Thus, an improved bioavailability, prolonged blood residence time, and/or steady biodistribution are expected in the presence of polymeric nanocarriers. These features suggest that they are suitable carriers for controlled drug delivery for peptides and proteins (Duncan, 2003; Behl et al., 2013, 2014; Kumar et al., 2016).

Over the past three decades, research in the field of nanobiotechnology has explored several interesting nanoparticles based on functional polymers for the oral delivery of therapeutic peptides and proteins (Table 6.2).

6.8.1 STIMULI-RESPONSIVE POLYMERIC NANOPARTICLES

Stimuli-responsive polymeric nanoparticles are those that are capable of conformational and chemical changes on receiving an external signal. The external signal is the result of changes in the polymer environment such as the pH, temperature, ionic strength of the surrounding medium, the quality of the solvent, and the action of the external electromagnetic field. Ionizable groups in the backbone of stimuli-responsive polymers can accept or donate protons in response to the environmental change in pH (Stuart et al., 2010). Hydrogel nanoparticles or so-called nanogels have received considerable interest in recent years as a promising drug carrier, due to certain interesting properties, like the enormous water

Table 6.2 Polymeric-Based Carrier as a Drug Delivery System With Production Technique and the Particle Size for Peptides and Proteins

Protein/ Peptide	Formulation Technique	Carrier System	Particle Size (nm)	Reference
Albumin	Ionic gelation	–		Lochmann et al. (2005)
Antisense oligonucleotide	Ionic gelation	–	230–320	Zimmer (1999)
Antisense oligonucleotide	Emulsion polymerization	Poly(hexylcyanoacrylate)/DEAE-dextran/CTAB	250	Zimmer (1999)
Hydrocortisone	Inclusion complexation	β-Cyclodextrin/2-hydroxypropyl-β-cyclodextrin	<100	Cavalli (1999)
Insulin	Antisolvent coprecipitation	Polymethacrylate derivative (ERL)/ERL + hydroxypropyl methylcellulose	315–331	Morales et al. (2014)
Insulin	Interfacial polymerization	Poly(ethylcyanoacrylate)	130–180	Watnasirichaikul et al. (2002)
Insulin	Interfacial polymerization	Poly(isobutylcyanoacrylate)	150–300	Aboubakar et al. (1999)
Insulin	Ionotropic gelation	Chitosan/poly(γ-glutamic acid)	245–260	Sonaje et al. (2010a,b)
Insulin	Polymerization	Poly(alkyl cyanoacrylate)	128–524	Damgé et al. (1997)
Insulin	Polymerization	Poly(ethylcyanoacrylate)	500	Radwan and Aboul-Enein (2002)
Insulin	Ionotropic gelation	N-trimethyl chitosan	220	Sandri et al. (2010)
Insulin	Ionotropic gelation	Anacardoylated chitosan	214	Shelma et al. (2010)
Insulin	Electrostatic attraction	Alginate/chitosan/β-cyclodextrin	<350	Zhang et al. (2010)
Insulin	Complex coacervation method	Chitosan/Eudragit L 100-55	199	Jelvehgari et al. (2010)

(Continued)

Table 6.2 Polymeric-Based Carrier as a Drug Delivery System With Production Technique and the Particle Size for Peptides and Proteins *Continued*

Protein/ Peptide	Formulation Technique	Carrier System	Particle Size (nm)	Reference
Insulin	Double-emulsion method (w/o/w)	PLGA/PEG	170–220	Hosseininasab et al. (2014)
Plasmid DNA	Water-in-oil reverse microemulsion	Chitosan	<100	Lotfipour et al. (2011)
Progesterone	Aerobic polymerization	Polyethylene oxide–polypropylene oxide (Pluronic F-68)/β-cyclodextrin	85–95	Rohani et al. (2012)
Progesterone	Emulsification solvent evaporation	Amphiphilic cyclodextrin	100–300	Lemos-Senna et al. (1998a,b)
Progesterone	Emulsion polymerization	Polybutylcyanoacrylate	250	Li et al. (1986)
Progesterone	Inclusion complexation	β-Cyclodextrin/2-hydroxypropyl-β-cyclodextrin	<100	Cavalli (1999)
Protamine	Ionic gelation	–	230–320	Lochmann et al. (2005)
Salmon calcitonin	Emulsion polymerization	Trimyristin/Poloxamer 407	200	Martins et al. (2009)
Salmon calcitonin	Interfacial polymerization	Eudragit RSPO/Eudragit L100/Eudragit–poly(lactic-co-glycolic acid)/sodium taurodeoxycholate	179.7–308.9	Cetin et al. (2011)
Salmon calcitonin	Free radical copolymerization	p-Chloromethyl styrene/*N*-vinylacetamide/*N*-isopropylacrylamide/t-butyl methacrylate	148–895	Sakuma et al. (2002)
Salmon calcitonin	Double emulsion solvent evaporation	Eudragit RS/PLGA	200	Glowka et al. (2009)
Octreotide	Solvent evaporation	Poly(ε-caprolactone) methyl ether/poly(ethylene glycol)	130–195	Dubey et al. (2012)

absorptive capacity, hydrophilicity, flexibility, versatility, biocompatibility, long circulation life span, and sustained release behavior (Xia et al., 2013). Due to the presence of certain pH-responsive acidic or basic pendant groups on hydrogels such as −OH, −CONH−, $-CONH_2-$, and $-SO_3H$, which change ionization in response to change in the pH, these have become the subject matter of major interest for use as carriers in oral drug delivery research (Bajpai et al., 2008). Weak polyacids (or polybases), which exhibit an ionization/deionization switch from pH four to eight, are employed as pH-responsive polymers, for example, poly(acrylic acid) (PAA) and poly(methacrylic acid) (PMAA). Their carboxylic pendant groups accept protons at low pH, while releasing them at high pH. This provides a momentum along with the hydrophobic interaction to govern the precipitation or solubilization of molecular chains, deswelling/swelling, or hydrophobic/hydrophilic characteristics of surfaces. Consequently, at low pH (1−2) these polymers shrink and the encapsulated drug can be shielded from the acidic environment of the stomach due to its limited release. After transport into the small intestine, the pH changes from acidic to neutral (6−7.4) leading the matrix to swell and to release the encapsulated drug (Cheng et al., 2013). Poly(*N*-isopropylacrylamide) (PNIPAAM) is one of the most extensively studied temperature responsive polymers with a low critical solution temperature (LCST) around 32°C. The copolymer of PNIPAAM with methacrylic acid (PNIPAAM−MAA) has been a material of interest due to the added advantage of ionizable groups at a particular biological pH range (Xia et al., 2005). The copolymer was utilized for developing pH and thermoresponsive composite membranes and investigated for their application in delivery of peptides and protein in response to external stimuli. It was demonstrated that the permeability of the solutes across the membranes increased with increasing temperature and decreased with increasing pH (Zhang and Wu, 2004; Lokuge et al., 2007). Recently, stimuli-responsive nanoparticles and nanocapsules have achieved great significance because of the extensive opportunities for in vivo applications (Gao et al., 2010; Lu et al., 2014). Among the various advances used to improve the efficacy of therapy is the employment of carrier systems that release a drug in response to stimuli, such as changes in pH, the existence of certain enzymes, or glutathione concentration, that are selectively encountered in relevant cell organelles. Drug release can be activated on demand by local transitions in pH or by distant physical stimuli (Gao et al., 2010; Lu et al., 2014). For instance, the selective addressing of intracellular layer-by-layer (LbL) microcapsules was demonstrated with laser light. Moreover, LbL microcapsule tracking inside cells was confirmed through the encapsulation of pH-sensitive dyes (Monge et al., 2015). A pH-sensitive hydrogel was synthesized by Gao et al. and utilized for the oral delivery of insulin. The hydrogels were composed of four different types of polyacrylic acid derivatives cross-linked by biodegradable poly(L-glutamic acid) (PGA) cross-linker. The hydrogels exhibited lower swelling and biodegradation in the low pH of stomach, while a higher swelling and degradation was observed in the intestinal environment. The formulation resulted in significant hypoglycemic effect in diabetic rats with 7 h of oral administration (Gao et al., 2010). In another study pH-responsive nanoparticles were synthesized from chitosan and PGA cross-linked with a blend of sodium tripolyphosphate (TPP) and $MgSO_4$. The nanoparticles were evaluated for oral delivery of insulin, where release was observed at higher pH by the deprotonation of chitosan (Lin et al., 2008).

6.8.2 MUCOADHESIVE POLYMERIC NANOPARTICLES

Mucoadhesive nanoparticles have been used widely for the oral delivery of macromolecules (Maurya et al., 2010; Shaikh et al., 2011; Bagan et al., 2012). Mucoadhesive nanoparticles are based on synthetic or natural polymers that interact with the mucus layer covering the mucosal epithelial surface and main molecules constituting a major part of mucus. Mucoadhesive nanoparticles especially increase the residence time of drug at the absorption site and facilitate enhanced duration of contact of drug with underlying absorptive surface thereby resulting in improved therapeutic efficiency (Chaturvedi et al., 2011; Singh et al., 2012). The mechanism of mucoadhesion is described to involve physical interactions, that is, hydrogen bonding, van der Waals forces, and covalent chemical interactions. Various polymers such as sodium alginate, chitosan, sodium carboxymethylcellulose, guar gum, hydroxyethylcellulose, karya gum, methylcellulose, polyethylene glycol (PEG), retene, tragacanth, and poly(acrylic acid) were extensively explored for the development of formulations having mucoadhesive properties (Renukuntla et al., 2013). Polycarbophil, a polyacrylate derivative known for its mucoadhesive properties, has been evaluated for oral delivery of 9-desglycinamide, 8-arginine vasopressin (DGAVP). The in vivo results showed improved intestinal absorption in rats (Lehr and Haas, 2002). Apart from this study, significantly enhanced reduction in blood calcium level has been observed utilizing calcitonin loaded chitosan-coated mucoadhesive nanospheres (Prego et al., 2005). Das and Lin (Das and Lin, 2005) developed double-coated (Tween 80 and PEG 20,000) poly(butylcyanoacrylate) nanoparticulate delivery systems (PBCA) for oral delivery of dalargin. Previous studies illustrated similar approaches to protect and orally deliver vasopressin (Lehr et al., 1992), peptidomimetics (Leroux et al., 1995), luteinizing hormone–releasing hormone (LHRH) (Hillery et al., 1996), interferon (Eyles et al., 1997), and cyclosporine A (Bonduelle et al., 1991, 1995, 1996).

A very recent study has revealed that mucoadhesion not only increase the absorption but also affects the distribution as well (Reineke et al., 2013). A remarkably enhanced uptake (from 5.8 ± 1.9 to 66.9 ± 12.9 SEM%) of mucoadhesive poly(butadiene-maleic anhydride-co-L-dopa)-coated polystyrene nanoparticles was observed in an in vivo isolated rat jejunal loop as compared to uncoated polystyrene beads. Further variation in organ distribution was also observed, where polystyrene beads were mainly taken up by liver and a negligible uptake was observed for mucoadhesive nanoparticles, which were mainly found in blood but relatively high in the blood, heart, lungs, and spleen (Reineke et al., 2013;

Griffin et al., 2016). The observation indicates that the surface modification may play an important role in biodistribution and can be useful tool for passive targeting.

6.8.3 BIODEGRADABLE POLYMERIC NANOPARTICLES

Numerals of different polymers, both synthetic and natural, have been utilized in formulating biodegradable nanoparticles (Jung, 2000; Soppimath et al., 2001; Kumari and Yadav, 2010; Behl et al., 2011). Biodegradable polymers acquire chemical functionalities that are unstable within living environments, for example, anhydride, ester, or amide bonds. The most frequent routes of biodegradation in vivo are hydrolysis and enzymatic cleavage resulting in scission of the polymer backbone (Nair and Laurencin, 2007). The drug entrapped in biodegradable matrix is released at a sustained rate through diffusion of the drug in the polymer matrix and by degradation of the polymer matrix. The polymers used for the formulation of nanoparticles include synthetic polymers such as polylactide-polyglycolide copolymers, polyacrylates, and polycaprolactone (PCL) or natural polymers such as gelatin, alginate, collagen, and chitosan (Soppimath et al., 2001). Of these polymers, polylactides (PLA) and poly(D,L-lactide-co-glycolide) (PLGA) have been the most extensively investigated for drug delivery (Jain, 2000; Panyam and Labhasetwar, 2003; Panyam et al., 2003). Woitiski et al. studied alginate–dextran sulfate nanoparticles coated with albumin for the delivery of insulin. The albumin coated nanoparticles resulted in enhanced intestinal uptake, stabilizing the insulin from intestinal condition thus protecting it proteolytic enzymes. Appreciable results were observed in in vitro and ex vivo studies with Caco-2 cell monolayer and excised intestinal mucosa of Wistar rats. Further in vivo studies revealed significant reduction blood glucose levels with a sustained hypoglycemic affect over 24 h (Woitiski et al., 2010, 2011).

Xiong et al. developed PLA-b-pluronic-b-PLA vesicles for oral delivery of insulin. A sustained hypoglycemic effect was observed when administered orally in diabetic mice with significant reduction in blood glucose levels from 18.5 to 5.3 mmol/L in first 4.5 h, followed by its lowest level in just 5 h, with same level maintained up to 18.5 h (Xiong et al., 2007, 2013).

As mentioned earlier PLGA has been studied extensively for the oral delivery of insulin owing to its biodegradability and biocompatibility (Zhang et al., 2012; Balasubramanian et al., 2015). A recent study by Reix et al. revealed insulin loaded PLGA nanoparticles followed a time dependent clathrin-mediated endocytosis on Caco-2 cell monolayer models. Further the nanoparticles were found to be stable after intraduodenal administration in diabetic rats for a sufficient period of time to allow its uptake into the bloodstream. The formulation was found to show similar effect compared to long-acting insulin (Reix et al., 2012). In another study, Yang et al. reported a pH release of insulin from PLGA nanoparticles prepared by double emulsion solvent evaporation technique. A very slow release rate was observed at pH 1.0, reaching 90% in 11 days, whereas a faster release was observed at pH 7.8 reaching 90% in 3 days. The formulation was further evaluated by oral administration in diabetic rats resulting in decrease in blood glucose level as the level of insulin increased in bloodstream (Yang et al., 2012). These results confirm the applicability of PLGA in preparation of sustained release formulation and protection of insulin. Although appreciable results have been observed with PLGA, the encapsulation of hydrophilic insulin has always been a challenging task, leading to the design of different strategies to enhance the lipid solubility and loading efficiency of insulin. One of the strategies involved complexation of insulin with soybean phosphatidylcholine to enhance lipid solubility leading to an association efficiency of about 90%. The nanoparticles showed a decrease in plasma glucose level to 57% in first 8 h, and prolonged effect up to 12 h when administered orally in diabetic rats (Cui et al., 2006). A complex of insulin and sodium oleate was formulated into PLGA nanoparticles by double emulsion solvent evaporation method, resulting in an association efficiency of 91%. The nanoparticles resulted in a reduction in blood glucose levels to about 23.9% in diabetic rats after 12 h after administration. The low blood glucose levels were even maintained up to 24 h (Sun et al., 2010).

Apart from all these studies, the major concern for macromolecular delivery is their stability in acidic gastric pH and protection from the proteolytic enzymes present in the GI tract. A number of studies to address this issue have been described here. PLGA nanoparticles blended with poloxamer (Pluronic F-68) and poloxamine (Tetronic T904) respectively were assessed for their stability in gastric and intestinal fluids with an aim of developing oral insulin delivery system. The blended particles were found to quite stable as compared to unblended particles with no observed interaction with fluids (Santander-Ortega et al., 2009). Another study involved the loading of insulin along with an antacid (2% magnesium hydroxide or zinc carbonate) into the PLGA nanoparticles. Structural integrity of the insulin was found to be maintained as confirmed by Fourier transform infrared spectroscopy, circular dichroism (CD), and fluorescence spectroscopy. The in vivo study in healthy rats revealed 6 times enhancement in oral bioavailability compared to when an insulin solution was used. Further the formulation resulted in significant hypoglycemic effect when administered orally in diabetic rats (Sharma et al., 2013).

Another approach tried by Cui et al. is the enteric coating of PLGA nanoparticles with hydroxypropyl methylcellulose phthalate (HPMCP) to prevent the initial burst of insulin stomach and deliver insulin specifically in small intestine. The enteric coating significantly reduced the burst phase, with only 20% of insulin was released in first hour, as compared to 50% release in PLGA nanoparticles without coating. Further the relative bioavailability of HPMCP coated and uncoated PLGA nanoparticles as compared with subcutaneous insulin injection was 6.3% and 3.7%, respectively, when studied in diabetic rats (Cui et al., 2007).

The surface modification of PLGA nanoparticles has also been carried out to improve formulation properties, stability, and enhance permeation across intestinal membrane (Paolicelli et al., 2010; Zhu et al., 2015). Chitosan-coated PLGA nanoparticles were developed by Zhang et al. where significantly higher bioadhesive properties were observed compared to uncoated particles. Further a relatively higher bioavailability of insulin was observed compared to insulin solution (Zhang et al., 2012). Pegylation of PLGA has also been tried for enhancing the bioavailability of insulin. Jain et al. utilized folate conjugated pegylated PLGA nanoparticles for oral delivery of insulin. About two times enhancement in bioavailability was observed as compared to subcutaneous administration of insulin when studied in diabetic rats (Jain et al., 2012). Liu et al. studied cell penetrating peptides (poly[arginine]8 enantiomers, L-R8 and D-R8) functionalized PEG–PLGA nanoparticles for insulin delivery. Enhanced permeation of insulin was observed across Caco-2 cells monolayer when studied and an improved bioavailability along with significant hypoglycemic effect was observed after oral administration in diabetic rats (Liu et al., 2013).

PCL is another example of biodegradable polymer. Its slower degradation rate as compared to PLGA makes PCL preferable for developing prolonged release formulation. Damge et al. prepared a formulation of insulin from PCL and Eudragit RS blend, where a dose dependent effect involving decrease in fasting glycemia was observed. The nanoparticles showed enhanced mucoadhesive properties as observed from strong adherence of nanoparticles containing fluorescein isothiocyanate labeled insulin to intestinal mucosa leading to the insulin delivery mainly by Peyer's patches (Damgé et al., 2010). The same group further utilized the nanoparticles for the delivery of commercial insulins, Actrapid and Novorapid, and an enhanced glycemic response was observed in dose dependent manner (Damgé et al., 2010).

Polyacrylate based polymers have also been utilized for insulin delivery because of their proteolytic protection, strong mucoadhesive properties, and ability to alter cell TJs to improve intestinal uptake. Thiomer nanoparticles of cysteine derivatized poly(acrylic acid) were utilized for oral delivery of insulin in diabetic rats. An overall protection form enzymes and enhanced permeation was observed due to increase mucoadhesive properties as a result of dervatization with cysteine (Deutel et al., 2008). Foss et al. synthesized PEG grafted cross-linked matrix of acrylic acid and methacrylic acid nanosphere and used them for the delivery of insulin. The nanospheres released preferably higher amount insulin at pH 7.0 and resulted in significant reduction in blood glucose levels lasting for 6 h, when studied in vivo in diabetic rats (Foss et al., 2004). Methacrylates are described to be useful for opening TJs thus enhancing intestinal permeation and PEG plays the role of stabilizing the protein and increased mucoadhesion. Li et al. explored the pH-sensitive property of methacrylic acid and chitosan by preparing nanoparticles from a blend of both the polymers and studied their applicability for the delivery of insulin. The nanoparticles were found to protect insulin from harsh environment of stomach and increased the interaction with intestinal mucosa (Li et al., 2007). Among acrylates the most popular polymer used as excipient in pharmaceutical products is Eudragit, which is a copolymer of ethyl acrylate, methyl methacrylate, and methacrylic acid ester with quaternary ammonium groups (Thakral et al., 2012; Moustafine et al., 2013). Zhang et al. prepared nanoparticles of thiolated Eudragit L100 (prepared by cysteine derivatization) by a precipitation method. The nanoparticles showed a pH dependent release of insulin in vitro and structural integrity of insulin was found to be maintained as revealed by CD measurements. Further enhanced mucoadhesion was observed in rat jejunum and ileum. Oral administration of the nanoparticles in diabetic rats resulted in prolonged hypoglycemic effect, and a 2.8-fold increase in insulin bioavailability was observed as compared to nonthiolated nanoparticles. The enhancement in bioavailability of insulin was attributed to the thiol group immobilized on Eudragit L100 resulting in enhanced mucoadhesion and permeation across intestinal mucosa (Zhang et al., 2014).

6.8.4 CHITOSAN BASED COPOLYMER NANOPARTICLES

Chitosan is a natural polysaccharide that is mainly obtained by enzymatic or alkaline deacetylation of chitin, a kind of seafood waste. Chitosan has been extensively used in numerous clinical applications due to its nontoxic, biocompatible, and nonantigenic nature (Arbia et al., 2013). Moreover, chitosan is capable to adhere to the mucosal surface and transiently open the TJs between epithelial cells (Sung et al., 2012; Ahmed and Aljaeid, 2016; Suh et al., 2016). Being polycationic it easily forms nanoparticles with sodium TPP or simply undergoes polyelectrolytic complexation with insulin to form nanoparticles (Nam et al., 2010; Huang and Lapitsky, 2012; Mukhopadhyay et al., 2013; Al-Kurdi et al., 2015; Behl et al., 2016). Pan et al. reported enhanced intestinal absorption of insulin loaded in chitosan nanoparticles. The in vivo study revealed that the nanoparticles resulted in hypoglycemic effect for 15 h and higher relative bioavailability of up to 14.9% was observed as compared to subcutaneous injection of insulin solution (Pan et al., 2002).

Despite these advantages, a disadvantage of chitosan is that it lacks good solubility at physiological pH values ($pKa = 5.5$). Dedicated effort has been applied in the past to overcome this drawback. These efforts have taken the direction of the synthesis of novel chitosan graft copolymers and nanoparticles (Qian et al., 2006; Riva et al., 2011). For instance, Thanou et al. reported that nanoparticles consisting of a quaternized chitosan derivative were able to enhance the intestinal permeation of hydrophilic macromolecular drugs (Thanou et al., 2007). Benediktsdóttir et al. utilized *N,N,N*-trimethyl chitosan (TMC) and *N*-propyl-(quatpropyl), *N*-butyl-(quatbutyl), and *N*-hexyl-(quathexyl)-*N,N*-dimethyl chitosan derivatives for delivery of model macromolecule payload FITC–dextran 4 kDa (FD4) and an enhanced permeation (two- to fivefold) was observed (Benediktsdóttir et al., 2014). Ionic interaction derived trimethyl chitosan–enoxaparin nanoparticles were developed by Dadras and coworkers. The nanoparticles were found to be stable at varying pH with high encapsulation efficiency of enoxaparin together with sustained release (Deadras et al., 2014). Jin et al. further demonstrated that peptide functionalization led to enhanced nanoparticle uptake and permeation of drug across epithelium. CSK peptide (CSKSSDYQC) functionalized TMC nanoparticles loaded with insulin were developed for targeting goblet cells. A higher insulin permeation of insulin across the epithelium was observed in peptide conjugated nanoparticles as a result of higher internalization via clathrin and caveolae mediated endocytosis in goblet cell-like HT29-MTX cells. Apart from this a higher bioavailability of insulin and enhanced hypoglycemic effect (1.5 times higher) was observed as compared to nanoparticles not functionalized with peptide (Jin et al., 2012). In addition to TMC, diethylmethyl chitosan (DEMC), triethyl chitosan (TEC), and dimethylethyl chitosan (DMEC) derivatives have also been developed and utilized for the delivery proteins and peptides at varying intestinal pH. The preparations have showed sustained release of insulin with minimum burst effect. The in vivo studies further revealed an enhanced intestinal absorption of insulin from the nanoparticles as compared to the subcutaneously administered insulin solution (Fonte et al., 2015).

6.8.4.1 Chitosan–cyclodextrin nanoparticles

Cyclodextrins (CDs) are cyclic oligosaccharides containing α-D-glucosyl units linked by α-(1,4)-glucosidic bonds (Aachmann et al., 2003). Macromolecular drugs are large molecules that can often partially interact with CDs via their hydrophobic side chains (Irie, 1999). CDs are known to enhance the stability (Dotsikas and Loukas, 2002) and solubility of such incorporated drugs (Filipović-Grčić et al., 2000). Moreover, they can also function as permeation enhancers (Merkus et al., 1991; Sakr, 1996; Yang et al., 2004) and prevent macromolecular drugs from enzymatic degradation (Matsubara et al., 1997). Chitosan–cyclodextrin represents a new generation of hybrid polysaccharide nanocarriers. Alonso et al. prepared chitosan/carboxymethyl-β-cyclodextrin (CM-β-CD) nanoparticles via ionotropic gelation technique. The resulting nanoparticles showed a great capacity to associate insulin and heparin.

6.8.4.2 Dextran nanoparticles

Dextran is another biocompatible material composed of α-D-glucose units that bind to each other through glycosidic bonds that has also been utilized for the oral delivery of insulin (Varshosaz, 2012; Fonte et al., 2015; Alibolandi et al., 2016; Ishak et al., 2016; Wasiak et al., 2016). Chalasani et al. studied vitamin B_{12} surface modified dextran nanoparticles for insulin delivery. The nanoparticles showed an initial burst followed by a sustained release phase with more than 90% in 48 h. The in vivo results also showed a reduction in blood glucose levels (70%–75%) with 5 h of oral administration in diabetic rats and the effect was quite prolonged for 54 h. The relative bioavailability of insulin from vitamin B12 modified nanoparticles was significantly higher as compared to unmodified nanoparticles; further the relative bioavailability was 29.4% higher as compared with subcutaneously administered insulin solution (Chalasani et al., 2007a,b).

Sarmento et al. prepared insulin loaded nanoparticles by complexing dextran sulfate with chitosan in an aqueous solution. The nanoparticles retained insulin in gastric fluid with almost no release and a sustained release was observed in simulated intestinal fluid for about 24 h. Appreciable results were observed when studied in vivo in diabetic rats with about 35% reduction in serum glucose levels maintained for 24 h. The overall bioavailability was also increased significantly as compared to orally administered insulin solution (Fonte et al., 2015).

6.8.4.3 Hyaluronic acid nanoparticles

Hyaluronic acid is a biodegradable anionic linear polymer of nonsulfated glycosaminoglycan subunits. The polymer has been used widely in drug and gene delivery, especially in cancer research because of it specific targeting capability to C44 receptor overexpressed in various types of cancers (Widjaja et al., 2013; Miller et al., 2014; Tripodo et al., 2015; Jiao et al., 2016; Dosio et al., 2016). Some of the studies carried out for oral delivery utilizing hyaluronic acid nanoparticles have been described here. Han et al. prepared insulin loaded hyaluronic acid nanoparticles by reverse emulsion freeze drying method for possible oral delivery. The nanoparticles were able to protect insulin from degradation at low gastric pH and released insulin preferably at intestinal environment. The intestinal uptake studies revealed a twofold increase in insulin uptake as compared to insulin solution. Further, strong hypoglycemic effect was observed as compared to insulin solution when administered orally. The mechanistic studies also revealed that insulin uptake occurred through active transport (Han et al., 2012). Liu et al explored the potential application oleoyl-carboxymethyl-chitosan (OCMCS)/hyaluronic acid (OCMCS−HA) nanopolyplexes for oral gene vaccine delivery. The nanopolyplexes were prepared by loading microbial antigen (aerA) aerolysin gene from *Aeromonas hydrophila* into the OCMCS−HA. The in vitro studies with Caco-2 cells revealed a 2.5-fold increase in permeation as compared to nanoparticles prepared without hyaluronic acid. The in vivo evaluation carried out with carp further revealed significantly higher ($P < .05$) antigen-specific antibodies in serum after oral administration (Liu et al., 2012).

6.8.5 HYDROGELS NANOPARTICLES

Hydrogels are natural (chitosan, pectin, collagen, and dextran) or synthetic polymeric (PVA, PEO, PEI, PVP, PEG, PLA, PLGA, PCL, and PHB) systems with three-dimensional arrangements (Hamidi et al., 2008). Due to the presence of certain hydrophilic groups such as −OH, −CONH−, $-CONH_2-$, and $-SO_3H$, hydrogels are capable of engrossing elevated amounts of water and/or biological fluids from the surroundings (Peppas, 1986). Hydrogel nanoparticles (or designated nanogels) have received noticeable attention in recent years as a promising drug carrier due to certain interesting properties such as enormous water absorption capacity, hydrophilicity, flexibility, versatility, biocompatibility, long circulation life span, and sustained release. In spite of the type of polymeric system used, the multifaceted release mechanism of the incorporated therapeutic agent from hydrogel nanoparticles is consequential from three key phenomena, that is, passive diffusion of drug molecules, swelling of the hydrogel matrix, and chemical reactivity of the drug/matrix (Amsden, 1998).

6.8.6 THIOMER NANOPARTICLES

Thiolated polymers (thiomers) have been developed as another promising class of mucoadhesive polymers, with reactive thiol groups immobilized on the polymeric backbone (Iqbal et al., 2011; Hauptstein and Bernkop-Schnürch, 2012). They can strongly adhere to the intestinal mucus layer for a prolonged time, through covalent bonding with mucin glycoproteins via thiol-disulfide exchange reactions, consequently providing a steep drug concentration gradient at the absorption sites (Iqbal et al., 2010). Recently, the permeation and mucoadhesive properties of the nanoparticles based on thiolated poly (acrylic acid) were increased up to 2.0- and 1.9-fold, respectively, due to immobilization of the thiol groups in comparison to an unmodified control (Köllner et al., 2015). Similarly, 2.1- to 4.7-fold increase in mucoadhesion was observed for trimethyl chitosan–cysteine/insulin nanoparticles (TMC–Cys NP) as compared to TMC/insulin nanoparticles (TMC NP), which might be partly due to disulfide formation between TMC–Cys and mucin. Moreover, the AUC of human insulin after oral administration of thiolated chitosan nanoparticles had a fourfold improvement compared to unmodified chitosan nanoparticles (Yin et al., 2009). In another study, relative oral bioavailability of peptide drug leuprolide has been shown to improve up to 4.2-fold in the presence of tablets containing thiomer nanoparticles (Iqbal et al., 2011).

6.9 TARGETED NANOPARTICLES FOR ORAL PROTEIN DELIVERY

Surface modification of nanoparticles with targeting moieties has been described to be the most efficient way to ensure optimum contact between the biological surface under study and the nanoparticle to enhance drug absorption. Generally, a strong interaction between the nanoparticle and epithelial surface is necessary for the increased uptake of drug. A majority of the research has been devoted to the development of targeted nanoparticles especially by active means (active targeting) to deliver drugs at the actual site of action (Yun et al., 2013). Surface modified nanoparticles with targeting moieties alter the mode of cellular internalization and maximize the uptake (Kamaly et al., 2012). Small targeting ligands namely folic acid, albumin, and cholesterol involve the cellular uptake by caveolin mediated endocytosis and besides glycol receptor specific ligands undergo clathrin-mediated endocytosis (Ernsting et al., 2013). Peptide ligands have also been used as targeting entity either by adsorption or covalent attachment to the nanoparticle surface. The advantage of using peptide ligands is their ease of synthesis and advancement of conjugation techniques to nanoparticles' surface, which led to improved application in diagnosis and drug delivery (Veiseh et al., 2010; Fan et al., 2014; Li et al., 2014; Hwang and Byun, 2014; Hua, 2014; Tian et al., 2016). Some of the important peptide ligands utilized with nanoparticles for oral delivery have been described here. The peptide CKS9 (CKSTHPLSC) was explored for specifically target follicle-associated epithelium region of Peyer's patch for site specific delivery of vaccine (Yoo et al., 2010). Chitosan nanoparticles were surface modified with peptide and evaluated. The peptide functionalized nanoparticles were found to spread more effectively across M cell model and preferably accumulated in the Peyer's patch regions as compared to unfunctionalized nanoparticles. Fievez et al. utilized two different peptides (LRVG and CTGKSC) functionalized nanoparticles and reported improved (up to 8 times) transport of vaccine across M-cells layers as compared to unmodified nanoparticles (Fievez et al., 2010). Among different types of targeting molecules lectins have been studied extensively (Gavrovic-Jankulovic, 2011; Diesner et al., 2012; Kumar et al., 2012; Yun et al., 2013). Lectins are a family of proteins that bind specifically to carbohydrates present on glycocalyx of the intestinal enterocytes and the mucus layer (Pelaseyed et al., 2014). Therefore, conjugation of lectins to polymeric nanoparticles significantly enhance the transport of nanoparticles across intestinal mucosa by significantly increased interaction with mucus (Yun et al., 2013; Zhang and Wu, 2014). Oral delivery insulin has been tried via lectin conjugated nanoparticles. Appreciable results were observed with good physical stability and sustained release behavior of insulin. Intestinal uptake studies carried out by YaShu et al. revealed most of the lectins conjugated nanoparticles were accumulated in small intestine, showing enhanced bioadhesion and endocytosis. There was almost threefold increase in uptake of lectins conjugated nanoparticles observed as compared with unconjugated nanoparticles, which showed less than 4.9% uptake only (Yun et al., 2013). The well-known cell penetrating peptide (CPP) RGD (arginine–glycine–aspartic acid) has been used popularly for conjugation to the nanoparticles for targeted delivery (des Rieux et al., 2013; Kong et al., 2015; Zhang et al., 2016). A 50-fold increased uptake of RGD conjugated polystyrene nanoparticles was observed across human intestinal epithelial cells as compared to unmodified nanoparticles (Yun et al., 2013). Despite the advancements in research an exact mechanism of the nanoparticle uptake and clearance from the body is yet to be determined. The potential of targeting the specific site and modulation of transit through GI tract by the application of surface modified nanoparticles has been extensively studied and appears to be a promising approach; however the development of oral protein formulation is a big challenge and requires rigorous research in the field, and so the quest to overcome the barriers is still ongoing.

6.10 MACROMOLECULE DELIVERY TECHNOLOGY AT CLINICAL STAGE

As evident from the extensive literature survey in this chapter, an extensive research effort has been devoted to the development of delivery carriers for oral macromolecule delivery utilizing tools and techniques of nanotechnology. Though the field is growing at great pace and appreciable results have been observed at preliminary stages, most of the technologies remain at developmental stages without progressing to clinical evaluation. The main focus of the pharmaceutical companies is on the strategies for enhanced permeation, and protection for degradation of macromolecules by use of protease inhibitors (Park and Kwon, 2011). Some of the technologies that have reached the clinical stage of development are described in this section. CobOral technology developed by Access Pharmaceutical, Inc. (Dallas, TX, United States) is a polymeric nanoparticulate technology with vitamin B12 attached on the surface. The technology utilizes the advantage of oral uptake of vitamin B12 to enhance the intestinal absorption of insulin. Following oral administration the vitamin B12 present on nanoparticles' surface binds to heptocorrin in stomach and the complex so formed migrates to duodenum followed by ileum, finally leading to the endocytosis of the nanoparticles reaching the bloodstream (Fonte et al., 2015). Chalasani et al. developed dextran nanoparticles surface functionalized with vitamin B12 for oral delivery of insulin. The in vitro release profile suggested a burst release phase initially followed by sustained release phase over a period of 48 h. the in vivo results with diabetic rats resulted in significant reduction (70%–75%) in plasma glucose levels within 5 h of oral administration. Further molecular weight of dextran was also found to play an important role where 1.4% higher pharmacological activity was observed with nanoparticles prepared from 70,000 molecular weight dextran as compared with Mw 10,000 (Chalasani et al., 2007a,b). A novel nanoparticulate delivery carrier based on chitosan-coated γ-polyglutamic acid has been developed by NanoMega Medical Corporation (Lake Forest, CA, United States). The positively charged chitosan shell plays an important function of opening the TJ of the cell for enhanced uptake of insulin along with promoting mucoadhesion and γ-polyglutamic acid core of the nanoparticles stabilizes the nanoparticles. The oral administration of the insulin loaded nanoparticles to diabetic rats resulted in reduction in blood glucose level in a sustained manner over 8 h, whereas subcutaneously administered insulin solution decreased the blood glucose level faster initially, but the effect became weak after 3 h. Further the orally administered insulin solution was found to be ineffective in reducing blood glucose levels, proving the applicability of the nanoparticles in increasing intestinal uptake of the insulin (NanoMega, 2014). Pharmaceutical companies are putting continuous efforts in developing oral delivery systems for insulin and quite a few results have been revealed till now and the most relevant results have been described in this section. Although positive results have been observed in initial studies, extensive research is still required for the progress of the field. Companies like Aphios Corp. (Woburn, MA, United States) and NOD Pharmaceuticals, Inc. (San Diego, CA, United States) are doing rigorous research in the development of oral delivery carriers for insulin, however limited information about the polymers used and products under development has been disclosed. APH-0907 is a product in the preclinical stage of development by Aphios Corp. for insulin delivery. Nodlin is a product at clinical phase trial I prepared from bioadhesive nanoparticles for insulin delivery for basal supplementation (NOD: Emerging leader in oral delivery of Biopharmaceuticals, 2014; Fonte et al., 2015). A number of pharmaceutical companies have claimed the development of an oral delivery system for insulin, however quite few results have been presented so far, suggesting a gap in the field requiring more focused research.

6.11 OUTLOOK AND CONCLUSION

This chapter has outlined a range of nanoparticulate DDS suitable for improving the oral availability of macromolecular drugs. Throughout this article, the advantages and disadvantages of inorganic and organic nanoparticulate systems have been highlighted. In the former, advantages such as a high degree of biocompatibility and small size are somewhat offset by issues relating to their lack of biodegradability and poor elimination profiles. First generation lipid-based carriers, liposomes, have distinct advantages, such as biodegradability and biological inertness but are problematic in terms of potential issues with in vivo stability. However, this may be overcome using second or third generation lipid carriers. It is perhaps the use of polymer-based carrier systems that holds the key to fully realizing the potential of macromolecular drugs by oral administration. With the capacity to develop, smart, stimuli-responsive materials that can trigger release based on pH, temperature, or in response to certain biomarkers, for example, there are myriad polymer combinations and permutations that can be developed to suit a particular macromolecule for a specific indication. The use of biodegradable and biocompatible materials, such as the synthetic PEG or the natural chitosan or alginate, further points to the almost limitless usefulness of these materials. Coupled with these factors is the ability for these polymers to be prepared as mucodhesive materials, further expanding the sites of delivery and modes of transport that can be applied to release throughout the body. The future for nanoparticulate delivery of macromolecules is sure to be an interesting and ever-expanding one. It would seem that with an increasing knowledge of the physiological factors associated with disease management, the ability to tailor polymer-based nanocarriers will direct future research to tailoring the behavior of these materials to apply to the delivery of a broad spectrum of drugs for a range of indications. However, the use of inorganic and lipid-based carriers holds many advantages and must not be forgotten. As such, the combination of these approaches, or the use of ceramics and lipids as models for which to design the properties of polymer-based materials, will be at the forefront of developing nanoparticulate delivery systems.

REFERENCES

Aachmann, F.L., Otzen, D.E., Larsen, K.L., Wimmer, R., 2003. Structural background of cyclodextrin-protein interactions. Protein Eng. Des. Sel. 16 (12), 905–912.

Aboubakar, M., Puisieux, F., Couvreur, P., Deyme, M., Vauthier, C., 1999. Study of the mechanism of insulin encapsulation in poly(isobutylcyanoacrylate) nanocapsules obtained by interfacial polymerization. J. Biomed. Mater. Res. 47 (4), 568–576.

Ahmed, T., Aljaeid, B., 2016. Preparation, characterization, and potential application of chitosan, chitosan derivatives, and chitosan metal nanoparticles in pharmaceutical drug delivery. Drug Des. Devel. Ther. 10, 483–507.

Alibolandi, M., Alabdollah, F., Sadeghi, F., Mohammadi, M., Abnous, K., Ramezani, M., et al., 2016. Dextran-b-poly(lactide-co-glycolide) polymersome for oral delivery of insulin: in vitro and in vivo evaluation. J. Control. Release 227, 58–70.

Al-Kurdi, Z., Chowdhry, B., Leharne, S., Omari, M., Badwan, A., 2015. Low molecular weight chitosan–insulin polyelectrolyte complex: characterization and stability studies. Mar. Drugs 13 (4), 1765–1784.

Allemann, E., Gurny, R., Doelker, E., 1993. Drug-loaded nanoparticles: preparation methods and drug targeting issues. Eur. J. Pharm. Biopharm. 39 (5), 173–191.

Allen, T., 1997. Liposomes. Opportunities in drug delivery. Drugs 54, 8–14.

Allen, T.M., Moase, E.H., 1996. Therapeutic opportunities for targeted liposomal drug delivery. Adv. Drug Deliv. Rev. 21 (2), 117–133.

Almeida, A., Souto, E., 2007. Solid lipid nanoparticles as a drug delivery system for peptides and proteins. Adv. Drug Deliv. Rev. 59 (6), 478–490.

Almeida, A.J., Runge, S., Müller, R.H., 1997. Peptide-loaded solid lipid nanoparticles (SLN): influence of production parameters. Int. J. Pharm. 149 (2), 255–265.

Amsden, B., 1998. Solute diffusion within hydrogels. Mechanisms and models. Macromolecules 31 (23), 8382–8395.

Amsteel, S., Domb, A., Laving, C., 1992. Lipospheres as a vaccine carrier system: effects of size, charge, and phospholipids composition. Vaccine Res. 1 (4), 383–395.

Arbia, W., Arbia, L., Adour, L., Amrane, L., 2013. Chitin extraction from crustacean shells using biological methods—a review. Food Technol. Biotechnol. 51 (1), 12–25.

Arifin, D.R., Palmer, A.F., 2003. Determination of size distribution and encapsulation efficiency of liposome-encapsulated hemoglobin blood substitutes using asymmetric flow field-flow fractionation coupled with multi-angle static light scattering. Biotechnol. Prog. 19 (6), 1798–1811.

Badiee, A., Jaafari, M.R., Khamesipour, A., 2007. Leishmania major: immune response in BALB/c mice immunized with stress-inducible protein 1 encapsulated in liposomes. Exp. Parasitol. 115 (2), 127–134.

Bagan, J., Paderni, C., Termine, N., Campisi, G., Lo Russo, L., Compilato, D., et al., 2012. Mucoadhesive polymers for oral transmucosal drug delivery: a review. Curr. Pharm. Design 18 (34), 5497–5514.

Bajpai, A.K., Shukla, S.K., Bhanu, S., Kankane, S., 2008. Responsive polymers in controlled drug delivery. Prog. Polym. Sci. 33 (11), 1088–1118.

Balasubramanian, S., Sampath, M., Perumal, N., Pandiyan, V., Webster, T.J., 2015. Novel PLGA-based nanoparticles for the oral delivery of insulin. Int. J. Nanomed. 10, 2207–2218.

Barbeau, J., Cammas-Marion, S., Auvray, P., Benvegnu, T., 2011. Preparation and characterization of stealth archaeosomes based on a synthetic PEGylated archaeal tetraether lipid. J. Drug Deliv. 2011, 1–11.

Bareford, L., Swaan, P., 2007. Endocytic mechanisms for targeted drug delivery. Adv. Drug Deliv. Rev. 59 (8), 748–758.

Behl, G., Sharma, M., Dahiya, S., Chhikara, A., Chopra, M., 2011. Synthesis, characterization, and evaluation of radical scavenging ability of ellagic acid-loaded nanogels. J. Nanomater. 2011, 1–9.

Behl, G., Sharma, M., Sikka, M., Dahiya, S., Chhikara, A., Chopra, M., 2013. Gallic acid loaded disulfide cross-linked biocompatible polymeric nanogels as controlled release system: synthesis, characterization, and antioxidant activity. J. Biomater. Sci., Polym. Ed. 24 (7), 865–881.

Behl, G., Sikka, M., Chhikara, A., Chopra, M., 2014. PEG-coumarin based biocompatible self-assembled fluorescent nanoaggregates synthesized via click reactions and studies of aggregation behavior. J. Colloid Interface Sci. 416, 151–160.

Behl, G., Iqbal, J., O'Reilly, N.J., McLoughlin, P., Fitzhenry, L., 2016. Synthesis and characterization of poly(2-hydroxyethylmethacrylate) contact lenses containing chitosan nanoparticles as an ocular delivery system for dexamethasone sodium phosphate. Pharm. Res. 33 (7), 1638–1648.

Bekerman, T., Golenser, J., Domb, A., 2004. Cyclosporin nanoparticulate lipospheres for oral administration. J. Pharm. Sci. 93 (5), 1264–1270.

Benediktsdóttir, B.E., Gudjónsson, T., Baldursson, Ó., Másson, M., 2014. N-alkylation of highly quaternized chitosan derivatives affects the paracellular permeation enhancement in bronchial epithelia in vitro. Eur. J. Pharm. Biopharm. 86 (1), 55–63.

Bhumkar, D.R., Joshi, H.M., Sastry, M., Pokharkar, V.B., 2007. Chitosan reduced gold nanoparticles as novel carriers for transmucosal delivery of insulin. Pharm. Res. 24 (8), 1415–1426.

Bonduelle, S., Foucher, C., Leroux, J., Chouinard, F., Cadieux, C., Lenaerts, V., 1991. Association of cyclosporin to isohexylcyanoacrylate nanospheres and subsequent release in human plasma in vitro. J. Microencapsul. 9 (2), 173–182.

Bonduelle, S., Pimenta, C., Benoit, J.-P., Lenearts, V., 1995. Body distribution in mice of intravenously injected radiolabelled cyclosporin associated with polyisohexylcyanoacrylate nanocapsules or nanospheres. Eur. J. Pharm. Biopharm. 41 (1), 27–30.

Bonduelle, S., Carrier, M., Pimenta, C., Benoit, J.-P., Lenaerts, V., 1996. Tissue concentration of nanoencapsulated radio-labelled cyclosporin following peroral delivery in mice or ophthalmic application in rabbits. Eur. J. Pharm. Biopharm. 42 (5), 313–319.

Brandtzaeg, P., Berstad, A., Farstad, I., Haraldsen, G., Helgeland, L., Jahnsen, F., et al., 1997. Mucosal immunity—a major adaptive defence mechanism. Behring Inst. Mitt. (98), 1–23.

Brannon-Peppas, L., Blanchette, J.O., 2004. Nanoparticle and targeted systems for cancer therapy. Adv. Drug Deliv. Rev. 56 (11), 1649–1659. Available from: https://doi.org/10.1016/j.addr.2004.02.014.

Brgles, M., Mirosavljević, K., Noethig-Laslo, V., Frkanec, R., Tomašić, J., 2007. Spin-labelling study of interactions of ovalbumin with multilamellar liposomes and specific anti-ovalbumin antibodies. Int. J. Biol. Macromol. 40 (4), 312–318.

Brinker, J.C., Scherer, G.W., Brinker, C.J., 1990. Sol–Gel Science: The Physics and Chemistry of Sol–Gel Processing, fifth ed Academic Press, Boston.

Bruno, B.J., Miller, G.D., Lim, C.S., 2013. Basics and recent advances in peptide and protein drug delivery. Ther. Deliv. 4 (11), 1443–1467.

Cai, Z., Wang, Y., Zhu, L.-J., Liu, Z.-Q., 2010. Nanocarriers: a general strategy for enhancement of oral bioavailability of poorly absorbed or pre-systemically metabolized drugs. Curr. Drug Metabol. 11 (2), 197–207.

Caliceti, P., Brossa, A., Salmaso, S., Bersani, S., Elvassore, N., Bertucco, A., 2006. Preparation of protein loaded solid lipid nano-particles by compressed fluid process. Proc. Int. Symp. Control. Rel. Bioact. Mater. 33 (1), 383.

Cavalli, R., 1999. Solid lipid nanoparticles as carriers of hydrocortisone and progesterone complexes with β-cyclodextrins. Int. J. Pharm. 182 (1), 59–69.

Cavalli, R., Bocca, C., Miglietta, A., Caputo, O., Gasco, M.R., 1999. Albumin adsorption on stealth and non-stealth solid lipid nanoparticles. STP Pharma. Sci. 9 (2), 183–189.

Cetin, M., Aktas, M.S., Vural, I., Ozturk, M., 2011. Salmon calcitonin-loaded Eudragit® and Eudragit®–PLGA nanoparticles: in vitroandin vivoevaluation. J. Microencapsul. 29 (2), 156–166.

Chalasani, K.B., Russell-Jones, G.J., Jain, A.K., Diwan, P.V., Jain, S.K., 2007a. Effective oral delivery of insulin in animal models using vitamin B12-coated dextran nanoparticles. J. Control. Release 122 (2), 141–150.

Chalasani, K.B., Russell-Jones, G.J., Yandrapu, S.K., Diwan, P.V., Jain, S.K., 2007b. A novel vitamin B12-nanosphere conjugate carrier system for peroral delivery of insulin. J. Control. Release 117 (3), 421–429.

Chaturvedi, M., Kumar, M., Pathak, K., 2011. A review on mucoadhesive polymer used in nasal drug delivery system. J. Adv. Pharm. Technol. Res. 2 (4), 215.

Cheng, R., Meng, F., Deng, C., Klok, H.-A., Zhong, Z., 2013. Dual and multi-stimuli responsive polymeric nanoparticles for programmed site-specific drug delivery. Biomaterials 34 (14), 3647–3657.

Cheng, Z., Al Zaki, A., Hui, J.Z., Muzykantov, V.R., Tsourkas, A., 2012. Multifunctional nanoparticles: cost versus benefit of adding targeting and imaging capabilities. Science 338 (6109), 903–910.

Cherian, A.K., Rana, A.C., Jain, S.K., 2000. Self-assembled carbohydrate-stabilized ceramic nanoparticles for the parenteral delivery of insulin. Drug Dev. Ind. Pharm. 26 (4), 459–463.

Conner, S.D., Schmid, S.L., 2003. Regulated portals of entry into the cell. Nature 422 (6927), 37–44.

Connor, E.E., Mwamuka, J., Gole, A., Murphy, C.J., Wyatt, M.D., 2005. Gold nanoparticles are taken up by human cells but do not cause acute cytotoxicity. Small 1 (3), 325–327.

Couvreur, P., 1988. Polyalkylcyanoacrylates as colloidal drug carriers. Crit. Rev. Ther. Drug Carrier Syst. 5 (1), 1–20.

Couvreur, P., Dubernet, C., Puisieux, F., 1995. Controlled drug delivery with nanoparticles: current possibilities and future trends. Eur. J. Pharm. Biopharm. 41 (1), 2–13.

Cui, F., Shi, K., Zhang, L., Tao, A., Kawashima, Y., 2006. Biodegradable nanoparticles loaded with insulin–phospholipid complex for oral delivery: preparation, in vitro characterization and in vivo evaluation. J. Control. Release 114 (2), 242–250.

Cui, F., Tao, A., Cun, D., Zhang, L., Shi, K., 2007. Preparation of insulin loaded PLGA–Hp55 nanoparticles for oral delivery. J. Pharm. Sci. 96 (2), 421–427.

Dai, C., Wang, B., Zhao, H., Li, B., 2005. Factors affecting protein release from microcapsule prepared by liposome in alginate. Colloids Surf. B: Biointerfaces 42 (3–4), 253–258.

Dai, C., Wang, B., Zhao, H., Li, B., Wang, J., 2006. Preparation and characterization of liposomes-in-alginate (LIA) for protein delivery system. Colloids Surf. B: Biointerfaces 47 (2), 205–210.

Damgé, C., Michel, C., Aprahamian, M., Couvreur, P., Devissaguet, J.P., 1990. Nanocapsules as carriers for oral peptide delivery. J. Control. Release 13 (2–3), 233–239.

Damgé, C., Vranckx, H., Balschmidt, P., Couvreur, P., 1997. Poly(alkyl cyanoacrylate) nanospheres for oral administration of insulin. J. Pharm. Sci. 86 (12), 1403–1409.

Damgé, C., Socha, M., Ubrich, N., Maincent, P., 2010. Poly(ε-caprolactone)/Eudragit nanoparticles for oral delivery of aspart-insulin in the treatment of diabetes. J. Pharm. Sci. 99 (2), 879–889.

Das, D., Lin, S., 2005. Double-coated poly(butylcynanoacrylate) nanoparticulate delivery systems for brain targeting of dalargin via oral administration. J. Pharm. Sci. 94 (6), 1343–1353.

Das, S., Chaudhury, A., 2010. Recent advances in lipid nanoparticle formulations with solid matrix for oral drug delivery. AAPS PharmSciTech 12 (1), 62–76.

Date, A.A., Hanes, J., Ensign, L.M., 2016. Nanoparticles for oral delivery: design, evaluation and state-of-the-art. J. Control. Release 240, 504–526.

Deadras, O., Goharzadeh, M., Avadi, M., Ghassemi, Z., Sadeghi, A., 2014. In vitro release studies of enoxaparin in nanoparticle form and enterically coated tablets containing surfactants. J. Appl. Chem. Res. 8 (3), 7–16.

Del Curto, M.D., Chicco, D., D'Antonio, M., Ciolli, V., Dannan, H., D'Urso, S., et al., 2003. Lipid microparticles as sustained release system for a GnRH antagonist (Antide). J. Control. Release 89 (2), 297–310.

Desai, P.P., Date, A.A., Patravale, V.B., 2012. Overcoming poor oral bioavailability using nanoparticle formulations—opportunities and limitations. Drug Discov. Today: Technol. 9 (2), e87–e95.

Deutel, B., Greindl, M., Thaurer, M., Bernkop-Schnürch, A., 2008. Novel insulin thiomer nanoparticles: in vivo evaluation of an oral drug delivery system. Biomacromolecules 9 (1), 278–285.

Dey, S., 2006. Inorganic biohybrid nanoparticles for targeted drug delivery. In: Proc 6th Annual: Cambridge Health Institute.

Diab, R., Jaafar-Maalej, C., Fessi, H., Maincent, P., 2012. Engineered nanoparticulate drug delivery systems: the next frontier for oral administration? AAPS J. 14 (4), 688–702.

Diesner, S.C., Wang, X.-Y., Jensen-Jarolim, E., Untersmayr, E., Gabor, F., 2012. Use of lectin-functionalized particles for oral immunotherapy. Ther. Deliv. 3 (2), 277–290.

Dosio, F., Arpicco, S., Stella, B., Fattal, E., 2016. Hyaluronic acid for anticancer drug and nucleic acid delivery. Adv. Drug Deliv. Rev. 97, 204–236.

Dotsikas, Y., Loukas, Y.L., 2002. Kinetic degradation study of insulin complexed with methyl-beta cyclodextrin. Confirmation of complexation with electrospray mass spectrometry and ^{1}H NMR. J. Pharm. Biomed. Anal. 29 (3), 487–494.

Dubey, N., Varshney, R., Shukla, J., Ganeshpurkar, A., Hazari, P.P., Bandopadhaya, G.P., et al., 2012. Synthesis and evaluation of biodegradable PCL/PEG nanoparticles for neuroendocrine tumor targeted delivery of somatostatin analog. Drug Deliv. 19 (3), 132–142.

Duncan, R., 2003. The dawning era of polymer therapeutics. Nat. Rev. Drug Discov. 2 (5), 347–360.

El-Shabouri, M.H., 2002. Positively charged nanoparticles for improving the oral bioavailability of cyclosporin-a. Int. J. Pharm. 249 (1–2), 101–108.

Erni, C., Suard, C., Freitas, S., Dreher, D., Merkle, H.P., Walter, E., 2002. Evaluation of cationic solid lipid microparticles as synthetic carriers for the targeted delivery of macromolecules to phagocytic antigen-presenting cells. Biomaterials 23 (23), 4667–4676.

Ernsting, M.J., Murakami, M., Roy, A., Li, S.-D., 2013. Factors controlling the pharmacokinetics, biodistribution and intratumoral penetration of nanoparticles. J. Control. Release 172 (3), 782–794.
Eyles, J.E., Alpar, H.O., Conway, B.R., Keswick, M., 1997. Oral delivery and fate of poly (lactic acid) microsphere-encapsulated interferon in rats. J. Pharm. Pharm. 49 (7), 669–674.
Fan, T., Chen, C., Guo, H., Xu, J., Zhang, J., Zhu, X., et al., 2014. Design and evaluation of solid lipid nanoparticles modified with peptide ligand for oral delivery of protein drugs. Eur. J. Pharm. Biopharm. 88 (2), 518–528.
Fasano, A., 1998. Novel approaches for oral delivery of macromolecules. J. Pharm. Sci. 87 (11), 1351–1356.
Ferrari, M., 2005. Cancer nanotechnology: opportunities and challenges. Nat. Rev. Cancer 5 (3), 161–171.
Fievez, V., Plapied, L., Plaideau, C., Legendre, D., des Rieux, A., Pourcelle, V., et al., 2010. In vitro identification of targeting ligands of human M cells by phage display. Int. J. Pharm. 394 (1–2), 35–42.
Filipović-Grčić, J., Voinovich, D., Moneghini, M., Bećirević-Laćan, M., Magarotto, L., Jalšenjak, I., 2000. Chitosan microspheres with hydrocortisone and hydrocortisone–hydroxypropyl-β-cyclodextrin inclusion complex. Eur. J. Pharm. Sci. 9 (4), 373–379.
Florence, A.T., 2004. Issues in oral nanoparticle drug carrier uptake and targeting. J. Drug Target. 12 (2), 65–70.
Fonte, P., Araújo, F., Silva, C., Pereira, C., Reis, S., Santos, H.A., et al., 2015. Polymer-based nanoparticles for oral insulin delivery: revisited approaches. Biotechnol. Adv. 33 (6), 1342–1354.
Foss, A.C., Goto, T., Morishita, M., Peppas, N.A., 2004. Development of acrylic-based copolymers for oral insulin delivery. Eur. J. Pharm. Biopharm. 57 (2), 163–169.
Frkanec, R., Noethig-Laslo, V., Vranešić, B., Mirosavljević, K., Tomašić, J., 2003. A spin labelling study of immunomodulating peptidoglycan monomer and adamantyltripeptides entrapped into liposomes. Biochim. Biophys. Acta (BBA)—Biomembr. 1611 (1–2), 187–196.
Gabizon, A., Goren, D., Cohen, R., Barenholz, Y., 1998. Development of liposomal anthracyclines: from basics to clinical applications: 1. This paper is based on a lecture presented at the 8th international symposium on recent advances in drug delivery systems (Salt Lake City, UT, USA, 1997). J. Control. Release 53 (1–3), 275–279.
Galindo-Rodriguez, S.A., Allemann, E., Fessi, H., Doelker, E., 2005. Polymeric nanoparticles for oral delivery of drugs and vaccines: a critical evaluation of in vivo studies. Crit. Rev. Ther. Drug Carrier Syst. 22 (5), 419–464.
Galović Rengel, R., Barišić, K., Pavelić, Ž., Žanić Grubišić, T., Čepelak, I., Filipović-Grčić, J., 2002. High efficiency entrapment of superoxide dismutase into mucoadhesive chitosan-coated liposomes. Eur. J. Pharm. Sci. 15 (5), 441–448.
Gao, W., Chan, J.M., Farokhzad, O.C., 2010. PH-responsive nanoparticles for drug delivery. Mol. Pharm. 7 (6), 1913–1920.
García-Fuentes, M., Torres, D., Alonso, M.J., 2003. Design of lipid nanoparticles for the oral delivery of hydrophilic macromolecules. Colloids Surf. B: Biointerfaces 27 (2–3), 159–168.
Garcia-Fuentes, M., Prego, C., Torres, D., Alonso, M.J., 2005. A comparative study of the potential of solid triglyceride nanostructures coated with chitosan or poly(ethylene glycol) as carriers for oral calcitonin delivery. Eur. J. Pharm. Sci. 25 (1), 133–143.

García-Santana, M.A., Duconge, J., Sarmiento, M.E., Lanio-Ruíz, M.E., Becquer, M.A., Izquierdo, L., et al., 2006. Biodistribution of liposome-entrapped human gamma-globulin. Biopharm. Drug Dispos. 27 (6), 275–283.

Gavrovic-Jankulovic, M., 2011. Drug delivery: plant lectins as bioadhesive drug delivery systems. J. Biomater. Nanobiotechnol. 02 (05), 614–621.

Gelperina, S., Kisich, K., Iseman, M.D., Heifets, L., 2005. The potential advantages of nanoparticle drug delivery systems in chemotherapy of tuberculosis. Am. J. Respir. Crit. Care Med. 172 (12), 1487–1490.

Gide, P., Gidwani, S., Kothule, K., 2013. Enhancement of transdermal penetration and bioavailability of poorly soluble acyclovir using solid lipid nanoparticles incorporated in gel cream. Indian J. Pharm. Sci. 75 (2), 138–142.

Glowka, E., Sapin-Minet, A., Leroy, P., Lulek, J., Maincent, P., 2009. Preparation and in vitro–in vivo evaluation of salmon calcitonin-loaded polymeric nanoparticles. J. Microencapsul. 27 (1), 1–12.

Gorodetsky, R., Levdansky, L., Vexler, A., Shimeliovich, I., Kassis, I., Ben-Moshe, M., et al., 2004. Liposome transduction into cells enhanced by haptotactic peptides (haptides) homologous to fibrinogen c-termini. J. Control. Release 95 (3), 477–488.

Goto, T., Morishita, M., Nishimura, K., Nakanishi, M., Kato, A., Ehara, J., et al., 2006. Novel mucosal insulin delivery systems based on fusogenic liposomes. Pharm. Res. 23 (2), 384–391.

Gregoriadis, G., Ryman, B.E., 1972. Lysosomal localization of enzyme-containing liposomes injected into rats. Biochem. J. 128 (4), 142P–143P.

Griffin, B.T., Guo, J., Presas, E., Donovan, M.D., Alonso, M.J., O'Driscoll, C.M., 2016. Pharmacokinetic, pharmacodynamic and biodistribution following oral administration of nanocarriers containing peptide and protein drugs. Adv. Drug Deliv. Rev. Available from: https://doi.org/10.1016/j.addr.2016.06.006.

Gualbert, J., Shahgaldian, P., Coleman, A.W., 2003. Interactions of amphiphilic calix[4] arene-based solid lipid nanoparticles with bovine serum albumin. Int. J. Pharm. 257 (1–2), 69–73.

Guo, J., 2003. Chitosan-coated liposomes: characterization and interaction with leuprolide. Int. J. Pharm. 260 (2), 167–173.

Hamidi, M., Azadi, A., Rafiei, P., 2008. Hydrogel nanoparticles in drug delivery. Adv. Drug Deliv. Rev. 60 (15), 1638–1649.

Hamman, J., Enslin, G., Kotzé, A., 2005. Oral delivery of peptide drugs: barriers and developments. BioDrugs: Clin. Immunother. Biopharm. Gene Ther. 19 (3), 165–177.

Han, L., Zhao, Y., Yin, L., Li, R., Liang, Y., Huang, H., et al., 2012. Insulin-loaded pH-sensitive hyaluronic acid nanoparticles enhance transcellular delivery. AAPS PharmSciTech 13 (3), 836–845.

Hauptstein, S., Bernkop-Schnürch, A., 2012. Thiomers and thiomer-based nanoparticles in protein and DNA drug delivery. Expert Opin. Drug Deliv. 9 (9), 1069–1081.

Hillery, A.M., Toth, I., Florence, A.T., 1996. Co-polymerised peptide particles II: oral uptake of a novel co-polymeric nanoparticulate delivery system for peptides. J. Control. Release 42 (1), 65–73.

Hosseininasab, S., Pashaei-Asl, R., Khandaghi, A.A., Nasrabadi, H.T., Nejati-Koshki, K., Akbarzadeh, A., et al., 2014. Synthesis, characterization, and in vitro studies of PLGA–PEG nanoparticles for oral insulin delivery. Chem. Biol. Drug Des. 84 (3), 307–315.

Hu, F., 2004. Preparation and characterization of solid lipid nanoparticles containing peptide. Int. J. Pharm. 273 (1–2), 29–35.

Hua, S., 2014. Orally administered liposomal formulations for colon targeted drug delivery. Front. Pharmacol. 5, 138. Available from: https://doi.org/10.3389/fphar.2014.00138.

Huang, Y., Lapitsky, Y., 2012. Salt-assisted mechanistic analysis of chitosan/tripolyphosphate micro- and nanogel formation. Biomacromolecules 13 (11), 3868–3876.

Hwang, S.R., Byun, Y., 2014. Advances in oral macromolecular drug delivery. Expert Opin. Drug Deliv. 11 (12), 1955–1967.

Ishak, A.H., Osman, R., Awad, G., 2016. Dextran-based nanocarriers for delivery of bioactives. Curr. Pharm. Des. 22 (22), 3411–3428.

Iqbal, J., Hombach, J., Matuszczak, B., Bernkop-Schnürch, A., 2010. Design and in vitro evaluation of a novel polymeric P-glycoprotein (P-gp) inhibitor. J. Control. Release 147 (1), 62–69.

Iqbal, J., Vigl, C., Moser, G., Gasteiger, M., Perera, G., Bernkop-Schnürch, A., 2011. Development and in vivo evaluation of a new oral nanoparticulate dosage form for leuprolide based on polyacrylic acid. Drug Deliv. 18 (6), 432–440.

Irie, T., 1999. Cyclodextrins in peptide and protein delivery. Adv. Drug Deliv. Rev. 36 (1), 101–123.

Iwanaga, K., Ono, S., Narioka, K., Morimoto, K., Kakemi, M., Yamashita, S., et al., 1997. Oral delivery of insulin by using surface coating liposomes. Int. J. Pharm. 157 (1), 73–80.

Iwanaga, K., Ono, S., Narioka, K., Kakemi, M., Morimoto, K., Yamashita, S., et al., 1999. Application of surface-coated liposomes for oral delivery of peptide: effects of coating the liposome's surface on the GI transit of insulin. J. Pharm. Sci. 88 (2), 248–252.

Jain, R.A., 2000. The manufacturing techniques of various drug loaded biodegradable poly (lactide-co-glycolide) (PLGA) devices. Biomaterials 21 (23), 2475–2490.

Jain, S., Rathi, V.V., Jain, A.K., Das, M., Godugu, C., 2012. Folate-decorated PLGA nanoparticles as a rationally designed vehicle for the oral delivery of insulin. Nanomedicine 7 (9), 1311–1337.

Jain, T.K., Roy, I., De, T.K., Maitra, A., 1998. Nanometer silica particles encapsulating active compounds: a novel ceramic drug carrier. J. Am. Chem. Soc. 120 (43), 11092–11095.

Jelvehgari, M., Zakeri-Milani, P., Siahi-Shadbad, M.R., Loveymi, B.D., Nokhodchi, A., Azari, Z., et al., 2010. Development of pH-sensitive insulin nanoparticles using Eudragit L100-55 and chitosan with different molecular weights. AAPS PharmSciTech 11 (3), 1237–1242.

Jiao, Y., Pang, X., Zhai, G., 2016. Advances in hyaluronic acid-based drug delivery systems. Curr. Drug Targets 17 (6), 720–730.

Jin, Y., Song, Y., Zhu, X., Zhou, D., Chen, C., Zhang, Z., et al., 2012. Goblet cell-targeting nanoparticles for oral insulin delivery and the influence of mucus on insulin transport. Biomaterials 33 (5), 1573–1582.

Jitendra, Sharma, P.K., Bansal, S., Banik, A., 2011. Noninvasive routes of proteins and peptides drug delivery. Indian J. Pharm. Sci. 73 (4), 367–375.

Jung, T., 2000. Biodegradable nanoparticles for oral delivery of peptides: is there a role for polymers to affect mucosal uptake? Eur. J. Pharm. Biopharm. 50 (1), 147–160.

Kamaly, N., Xiao, Z., Valencia, P.M., Radovic-Moreno, A.F., Farokhzad, O.C., 2012. Targeted polymeric therapeutic nanoparticles: design, development and clinical translation. Chem. Soc. Rev. 41 (7), 2971.

Kawakami, K., 2012. Modification of physicochemical characteristics of active pharmaceutical ingredients and application of supersaturatable dosage forms for improving bioavailability of poorly absorbed drugs. Adv. Drug Deliv. Rev. 64 (6), 480–495.

Kawasaki, E.S., Player, A., 2005. Nanotechnology, nanomedicine, and the development of new, effective therapies for cancer. Nanomed.: Nanotechnol., Biol. Med. 1 (2), 101–109. Available from: https://doi.org/10.1016/j.nano.2005.03.002.

Kesisoglou, F., Panmai, S., Wu, Y., 2007. Nanosizing—oral formulation development and biopharmaceutical evaluation. Adv. Drug Deliv. Rev. 59 (7), 631–644.

Kong, L., Alves, C.S., Hou, W., Qiu, J., Möhwald, H., Tomás, H., et al., 2015. RGD peptide-modified dendrimer-entrapped gold nanoparticles enable highly efficient and specific gene delivery to stem cells. ACS Appl. Mater. Interfaces 7 (8), 4833–4843.

Kozubek, A., Gubernator, J., Przeworska, E., Stasiuk, M., 2001. Liposomal drug delivery, a novel approach: PLARosomes. Acta Biochim. Pol. 47 (3), 639–649.

Kreuter, J., 1996. Nanoparticles and microparticles for drug and vaccine delivery. J. Anat. 189 (Pt 3), 503–505.

Kullberg, E.B., Wei, Q., Capala, J., Giusti, V., Malmström, P.-U., Gedda, L., 2005. EGF-receptor targeted liposomes with boronated acridine: growth inhibition of cultured glioma cells after neutron irradiation. Int. J. Radiat. Biol. 81 (8), 621–629.

Kumar, K.K., Reddy, G.S., Reddy, B., Shekar, P.C., Sumanthi, J., Chandra, K.L.P., 2012. Biological role of lectins: a review. J. Orofac. Sci. 4 (1), 20.

Kumar, P., Behl, G., Sikka, M., Chhikara, A., Chopra, M., 2016. Poly(ethylene glycol)-co-methacrylamide-co-acrylic acid based nanogels for delivery of doxorubicin. J. Biomater. Sci. Polym. Ed. 27 (14), 1413–1433.

Kumari, A., Yadav, S.K., 2010. Biodegradable polymeric nanoparticles based drug delivery systems. Colloids Surf. B: Biointerfaces 75 (1), 1–18.

Köllner, S., Dünnhaupt, S., Waldner, C., Hauptstein, S., Pereira de Sousa, I., Bernkop-Schnürch, A., 2015. Mucus permeating thiomer nanoparticles. Eur. J. Pharm. Biopharm. 97, 265–272.

Lal, M., Levy, L., Kim, K.S., He, G.S., Wang, X., Min, Y.H., et al., 2000. Silica nanobubbles containing an organic dye in a multilayered organic/inorganic heterostructure with enhanced luminescence. Chem. Mater. 12 (9), 2632–2639.

Langston, M., 2003. Modulation of the sustained delivery of myelopoietin (Leridistim) encapsulated in multivesicular liposomes (DepoFoam). J. Control. Release 89 (1), 87–99.

Lee, V., Yamamoto, A., Kompella, U., 1991. Mucosal penetration enhancers for facilitation of peptide and protein drug absorption. Crit. Rev. Ther. Drug Carrier Syst. 8 (2), 91–192.

Lehr, C.-M., Haas, J., 2002. Developments in the area of bioadhesive drug delivery systems. Expert Opin. Biol. Ther. 2 (3), 287–298.

Lehr, C.-M., Bouwstra, J.A., Kok, W., Boer, A.G., Tukker, J.J., Verhoef, J.C., et al., 1992. Effects of the mucoadhesive polymer polycarbophil on the intestinal absorption of a peptide drug in the rat. J. Pharm. Pharmacol. 44 (5), 402–407.

Lemos-Senna, E., Wouessidjewe, D., Lesieur, S., Duchêne, D., 1998a. Preparation of amphiphilic cyclodextrin nanospheres using the emulsification solvent evaporation method. Influence of the surfactant on preparation and hydrophobic drug loading. Int. J. Pharm. 170 (1), 119–128.

Lemos-Senna, E., Wouessidjewe, D., Lesieur, S., Puisieux, F., Couarraze, G., Duchêne, D., 1998b. Evaluation of the hydrophobic drug loading characteristics in nanoprecipitated amphiphilic cyclodextrin nanospheres. Pharm. Dev. Technol. 3 (1), 85–94.

Leroux, J.-C., Cozens, R., Roesel, J.L., Galli, B., Kubel, F., Doelker, E., et al., 1995. Pharmacokinetics of a novel HIV-1 protease inhibitor incorporated into biodegradable or enteric nanoparticles following intravenous and oral administration to mice. J. Pharm. Sci. 84 (12), 1387–1391.

Li, J., Wang, Y., Liang, R., An, X., Wang, K., Shen, G., et al., 2015. Recent advances in targeted nanoparticles drug delivery to melanoma. Nanomed.: Nanotechnol. Biol. Med. 11 (3), 769–794.

Li, M., Lu, W., Wang, J., Zhang, X., Wang, X., Zheng, A., et al., 2007. Distribution, transition, adhesion and release of insulin loaded nanoparticles in the gut of rats. Int. J. Pharm. 329 (1–2), 182–191.

Li, V.H.K., Wood, R.W., Kreuter, J., Harmia, T., Robinson, J.R., 1986. Ocular drug delivery of progesterone using nanoparticles. J. Microencapsul. 3 (3), 213–218.

Li, X., Yu, M., Fan, W., Gan, Y., Hovgaard, L., Yang, M., 2014. Orally active-targeted drug delivery systems for proteins and peptides. Expert Opin. Drug Deliv. 11 (9), 1435–1447.

Li, Z., Chen, J., Sun, W., Xu, Y., 2010. Investigation of archaeosomes as carriers for oral delivery of peptides. Biochem. Biophys. Res. Commun. 394 (2), 412–417.

Lin, Y.-H., Sonaje, K., Lin, K.M., Juang, J.-H., Mi, F.-L., Yang, H.-W., et al., 2008. Multi-ion-crosslinked nanoparticles with pH-responsive characteristics for oral delivery of protein drugs. J. Control. Release 132 (2), 141–149.

Liu, X., Liu, C., Zhang, W., Xie, C., Wei, G., Lu, W., 2013. Oligoarginine-modified biodegradable nanoparticles improve the intestinal absorption of insulin. Int. J. Pharm. 448 (1), 159–167.

Liu, Y., Kong, M., Feng, C., Yang, K., Li, Y., Su, J., et al., 2012. Biocompatibility, cellular uptake and biodistribution of the polymeric amphiphilic nanoparticles as oral drug carriers. Colloids Surf. B: Biointerfaces 103, 345–353.

Lochmann, D., Weyermann, J., Georgens, C., Prassl, R., Zimmer, A., 2005. Albumin–protamine–oligonucleotide nanoparticles as a new antisense delivery system: part 1. Physicochemical characterization. Eur. J. Pharm. Biopharm. 59 (3), 419–429.

Lokuge, I., Wang, X., Bohn, P.W., 2007. Temperature-controlled flow switching in nanocapillary array membranes mediated by poly(N-isopropylacrylamide) polymer brushes grafted by atom transfer radical polymerization. Langmuir 23 (1), 305–311.

Lotfipour, F., Hallaj-Nezhadi, S., Valizadeh, H., Dastmalchi, S., Baradaran, B., Jalali, M. B., et al., 2011. Preparation of chitosan-plasmid DNA nanoparticles encoding interleukin-12 and their expression in CT-26 colon carcinoma cells. J. Pharm. Pharm. Sci. 14 (2), 181.

Lu, Y., Sun, W., Gu, Z., 2014. Stimuli-responsive nanomaterials for therapeutic protein delivery. J. Control. Release 194, 1–19.

Luo, X., Belcastro, R., Cabacungan, J., Hannam, V., Negus, A., Wen, Y., et al., 2003. Transfection of lung cells in vitro and in vivo: effect of antioxidants and intraliposomal bFGF. Am. J. Physiol.—Lung Cell. Mol. Physiol. 284 (5), L817–L825.

Lück, M., Pistel, K.-F., Li, Y.-X., Blunk, T., Müller, R.H., Kissel, T., 1998. Plasma protein adsorption on biodegradable microspheres consisting of poly(D,L-lactide-co-glycolide), poly(L-lactide) or ABA triblock copolymers containing poly(oxyethylene). J. Control. Release 55 (2–3), 107–120.

Makino, K., Shibata, A., 2006. Chapter 2: surface properties of liposomes depending on their composition. Adv. Planar Lipid Bilayers Liposomes 4, 49–77.

Martins, S., Silva, A.C., Ferreira, D.C., Souto, E.B., 2009. Improving oral absorption of samon calcitonin by trimyristin lipid nanoparticles. J. Biomed. Nanotechnol. 5 (1), 76–83.

Masood, F., 2016. Polymeric nanoparticles for targeted drug delivery system for cancer therapy. Mater. Sci. Eng. C 60, 569–578.

Matsubara, K., Ando, Y., Irie, T., Uekama, K., 1997. Protection afforded by maltosyl-beta-cyclodextrin against alpha-chymotrypsin-catalyzed hydrolysis of a luteinizing hormone-releasing hormone agonist, buserelin acetate. Pharm. Res. 14 (10), 1401–1405.

Maurya, S.K., Pathak, K., Bali, V., 2010. Therapeutic potential of mucoadhesive drug delivery systems—an updated patent review. Recent Pat. Drug Deliv. Formul. 4 (3), 256–265.

Merisko-Liversidge, E., Liversidge, G.G., 2011. Nanosizing for oral and parenteral drug delivery: a perspective on formulating poorly-water soluble compounds using wet media milling technology. Adv. Drug Deliv. Rev. 63 (6), 427–440.

Merkus, F., Verhoef, J., Romeijn, S., Schipper, N., 1991. Absorption enhancing effect of cyclodextrins on intranasally administered insulin in rats. Pharm. Res. 8 (5), 588–592.

Miller, T., Goude, M.C., McDevitt, T.C., Temenoff, J.S., 2014. Molecular engineering of glycosaminoglycan chemistry for biomolecule delivery. Acta Biomater. 10 (4), 1705–1719.

Mohanraj, V.J., Barnes, T.J., Prestidge, C.A., 2010. Silica nanoparticle coated liposomes: a new type of hybrid nanocapsule for proteins. Int. J. Pharm. 392 (1–2), 285–293.

Monge, C., Almodóvar, J., Boudou, T., Picart, C., 2015. Spatio-temporal control of LbL films for biomedical applications: from 2D to 3D. Adv. Healthcare Mater. 4 (6), 811–830.

Morales, J.O., Huang, S., Williams, R.O., McConville, J.T., 2014. Films loaded with insulin-coated nanoparticles (ICNP) as potential platforms for peptide buccal delivery. Colloids Surf. B: Biointerfaces 122, 38–45.

Morel, S., Gasco, M., Cavalli, R., 1995. Incorporation in lipospheres of [D-Trp-6]LHRH. Int. J. Pharm. 119 (1), 126.

Morel, S., Ugazio, E., Cavalli, R., Gasco, M.R., 1996. Thymopentin in solid lipid nanoparticles. Int. J. Pharm. 132 (1–2), 259–261.

Morishita, M., Peppas, N.A., 2006. Is the oral route possible for peptide and protein drug delivery? Drug Discov. Today 11 (19–20), 905–910.

Moustafine, R.I., Bukhovets, A.V., Sitenkov, A.Y., Kemenova, V.A., Rombaut, P., Van den Mooter, G., 2013. Eudragit E PO as a complementary material for designing oral drug delivery systems with controlled release properties: comparative evaluation of new interpolyelectrolyte complexes with countercharged Eudragit L100 copolymers. Mol. Pharm. 10 (7), 2630–2641.

Muchow, M., Maincent, P., Müller, R.H., 2008. Lipid nanoparticles with a solid matrix (SLN®, NLC®, LDC®) for oral drug delivery. Drug Dev. Ind. Pharm. 34 (12), 1394–1405.

Mudshinge, S.R., Deore, A.B., Patil, S., Bhalgat, C.M., 2011. Nanoparticles: emerging carriers for drug delivery. Saudi Pharm. J. 19 (3), 129–141.

Mueller, R., Ravelli, V., Runge, S.A., Giul, P.I.S., 1998. Patent DE19819273A1—Pharmazeutische Ciclosporin-Formulierung mit verbesserten biopharmazeutischen Eigenschaften, erhöhter physikalischer Qualität und Stabilität sowie Verfahren zur Herstellung Ciclosporin pharmaceutical formulation with improved biophar (accessed 11.10.16).

Muheem, A., Shakeel, F., Jahangir, M.A., Anwar, M., Mallick, N., Jain, G.K., et al., 2016. A review on the strategies for oral delivery of proteins and peptides and their clinical perspectives. Saudi Pharm. J. 24 (4), 413–428.

Mukhopadhyay, P., Sarkar, K., Chakraborty, M., Bhattacharya, S., Mishra, R., Kundu, P.P., 2013. Oral insulin delivery by self-assembled chitosan nanoparticles: in vitro and in vivo studies in diabetic animal model. Mater. Sci. Eng. C 33 (1), 376–382.

Muller, R.H., 1996. Azneistoffträger aus festen Lipidteilchen feste Lipid Nanosphären (SLN). European Patent 0605497.

Murakami, S., Ono, T., Sakai, S., Ijima, H., Kawakami, K., 2006. Effect of diglucosamine on the entrapment of protein into liposomes. J. Liposome Res. 16 (2), 103–112.

Müller, R., 1996. Cytotoxicity of magnetite-loaded polylactide, polylactide/glycolide particles and solid lipid nanoparticles. Int. J. Pharm. 138 (1), 85–94.

Müller, R.H., Radtke, M., Wissing, S.A., 2002. Nanostructured lipid matrices for improved microencapsulation of drugs. Int. J. Pharm. 242 (1–2), 121–128.

Müller, R.H., Runge, S., Ravelli, V., Mehnert, W., Thünemann, A.F., Souto, E.B., 2006. Oral bioavailability of cyclosporine: solid lipid nanoparticles (SLN®) versus drug nanocrystals. Int. J. Pharm. 317 (1), 82–89.

Müller, R., Runge, S., Ravelli, V., Thünemann, A., Mehnert, W., Souto, E., 2007. Cyclosporine-loaded solid lipid nanoparticles (SLN): drug-lipid physicochemical interactions and characterization of drug incorporation. Eur. J. Pharm. Biopharm.: Off. J. Arbeitsgemeinschaft für Pharmazeutische Verfahrenstechnik e.V. 68 (3), 535–544.

Nair, L.S., Laurencin, C.T., 2007. Biodegradable polymers as biomaterials. Prog. Polym. Sci. 32 (8–9), 762–798.

Nam, J.-P., Choi, C., Jang, M.-K., Jeong, Y.-I., Nah, J.-W., Kim, S.-H., et al., 2010. Insulin-incorporated chitosan nanoparticles based on polyelectrolyte complex formation. Macromol. Res. 18 (7), 630–635.

NanoMega, 2014. NanoMega Medical Corp. Available from: http://www.nanomegamedical.com.

NOD: Emerging leader in oral delivery of Biopharmaceuticals, 2014. Available from: http://www.nodpharm.com.

Oda, T., Kimura, T., Ogata, Y., Fujise, Y., 2005. Hemodilution with liposome-encapsulated low-oxygen-affinity hemoglobin does not attenuate hypothermic cerebral ischemia in rats. J. Art. Organs 8 (4), 263–269.

Pade, V., Stavchansky, S., 1997. Estimation of the relative contribution of the transcellular and paracellular pathway to the transport of passively absorbed drugs in the Caco-2 cell culture model. Pharm. Res. 14 (9), 1210–1215.

Pan, Y., Li, Y., Zhao, H., Zheng, J., Xu, H., Wei, G., et al., 2002. Bioadhesive polysaccharide in protein delivery system: chitosan nanoparticles improve the intestinal absorption of insulin in vivo. Int. J. Pharm. 249 (1–2), 139–147.

Panyam, J., Labhasetwar, V., 2003. Biodegradable nanoparticles for drug and gene delivery to cells and tissue. Adv. Drug Deliv. Rev. 55 (3), 329–347.

Panyam, J., Dali, M.M., Sahoo, S.K., Ma, W., Chakravarthi, S.S., Amidon, G.L., et al., 2003. Polymer degradation and in vitro release of a model protein from poly(D,L-lactide-co-glycolide) nano- and microparticles. J. Control. Release 92 (1–2), 173–187.

Paolicelli, P., Prego, C., Sanchez, A., Alonso, M.J., 2010. Surface-modified PLGA-based nanoparticles that can efficiently associate and deliver virus-like particles. Nanomedicine 5 (6), 843–853.

Park, K., Kwon, I.C., 2011. Oral protein delivery: current status and future prospect. React. Funct. Polym. 71 (3), 280–287.

Parmentier, J., Hartmann, F.J., Fricker, G., 2010. In vitro evaluation of liposomes containing bio-enhancers for the oral delivery of macromolecules. Eur. J. Pharm. Biopharm. 76 (3), 394–403.

Patel, V.F., Liu, F., Brown, M.B., 2011. Advances in oral transmucosal drug delivery. J. Control. Release 153 (2), 106–116.

Patlolla, R.R., Chougule, M., Patel, A.R., Jackson, T., Tata, P.N.V., Singh, M., 2010. Formulation, characterization and pulmonary deposition of nebulized celecoxib encapsulated nanostructured lipid carriers. J. Control. Release 144 (2), 233–241.

Patton, J.N., Palmer, A.F., 2005. Photopolymerization of bovine hemoglobin entrapped nanoscale hydrogel particles within liposomal reactors for use as an artificial blood substitute. Biomacromolecules 6 (1), 414–424.

Pedersen, N., Hansen, S., Heydenreich, A.V., Kristensen, H.G., Poulsen, H.S., 2006. Solid lipid nanoparticles can effectively bind DNA, streptavidin and biotinylated ligands. Eur. J. Pharm. Biopharm. 62 (2), 155–162.

Pelaseyed, T., Bergström, J.H., Gustafsson, J.K., Ermund, A., Birchenough, G.M.H., Schütte, A., et al., 2014. The mucus and mucins of the goblet cells and enterocytes provide the first defense line of the gastrointestinal tract and interact with the immune system. Immunol. Rev. 260 (1), 8–20.

Penkler, L.J., Müller, R.H., Runge, S.A., Ravelli, V., 1999. Patent US6551619—Pharmaceutical cyclosporin formulation with improved biopharmaceutical properties, improved physical quality and greater stability, and method for producing said formulation. Available from: https://www.google.com/patents/US6551619 (accessed 11.10.16).

Peppas, N., 1986. In: Mikos, A.G. (Ed.), Preparation Methods and Structure of Hydrogels, first ed. CRC Press, Boca Raton, FL.

Pichayakorn, W., 2006. Solid lipid microparticles for oral Japanese encephalitis vaccine delivery: in vitro study in Caco-2 cells and human intestinal M-cells models and in vivo immunization in mice. In: Proceedings International Symposium Control. Rel. Bioact. Mater. p. 560.

Plum, S., 2000. Administration of a liposomal FGF-2 peptide vaccine leads to abrogation of FGF-2-mediated angiogenesis and tumor development. Vaccine 19 (9–10), 1294–1303.

Powell, J.J., Faria, N., Thomas-McKay, E., Pele, L.C., 2010. Origin and fate of dietary nanoparticles and microparticles in the gastrointestinal tract. J. Autoimmun. 34 (3), J226–J233.

Prego, C., García, M., Torres, D., Alonso, M.J., 2005. Transmucosal macromolecular drug delivery. J. Control. Release 101 (1–3), 151–162.

Qian, F., Cui, F., Ding, J., Tang, C., Yin, C., 2006. Chitosan graft copolymer nanoparticles for oral protein drug delivery: preparation and characterization. Biomacromolecules 7 (10), 2722–2727.

Radwan, M.A., Aboul-Enein, H.Y., 2002. The effect of oral absorption enhancers on the in vivo performance of insulin-loaded poly(ethylcyanoacrylate) nanospheres in diabetic rats. J. Microencapsul. 19 (2), 225–235.

Ramprasad, M.P., Amini, A., Kararli, T., Katre, N.V., 2003. The sustained granulopoietic effect of progenipoietin encapsulated in multivesicular liposomes. Int. J. Pharm. 261 (1–2), 93–103.

Rawat, M., Singh, D., Saraf, S., 2006. Nanocarriers: promising vehicle for bioactive drugs. Biol. Pharm. Bull. 29 (9), 1790–1798.

Reineke, J., Cho, D.Y., Dingle, Y.L., Cheifetz, P., Laulicht, B., Lavin, D., et al., 2013. Can bioadhesive nanoparticles allow for more effective particle uptake from the small intestine? J. Control. Release 170 (3), 477–484.

Reithmeier, H., Herrmann, J., Göpferich, A., 2001a. Development and characterization of lipid microparticles as a drug carrier for somatostatin. Int. J. Pharm. 218 (1–2), 133–143.

Reithmeier, H., Herrmann, J., Göpferich, A., 2001b. Lipid microparticles as a parenteral controlled release device for peptides. J. Control. Release 73 (2–3), 339–350.

Reix, N., Parat, A., Seyfritz, E., Van Der Werf, R., Epure, V., Ebel, N., et al., 2012. In vitro uptake evaluation in Caco-2 cells and in vivo results in diabetic rats of insulin-loaded PLGA nanoparticles. Int. J. Pharm. 437 (1–2), 213–220.

Renukuntla, J., Vadlapudi, A.D., Patel, A., Boddu, S.H.S., Mitra, A.K., 2013. Approaches for enhancing oral bioavailability of peptides and proteins. Int. J. Pharm. 447 (1–2), 75–93. Available from: https://doi.org/10.1016/j.ijpharm.2013.02.030.

Ribeiro Dos Santos, I., Richard, J., Pech, B., Thies, C., Benoit, J.P., 2002. Microencapsulation of protein particles within lipids using a novel supercritical fluid process. Int. J. Pharm. 242 (1–2), 69–78.

des Rieux, A., Pourcelle, V., Cani, P.D., Marchand-Brynaert, J., Préat, V., 2013. Targeted nanoparticles with novel non-peptidic ligands for oral delivery. Adv. Drug Deliv. Rev. 65 (6), 833–844.

Riva, R., Ragelle, H., des Rieux, A., Duhem, N., Jérôme, C., Préat, V., 2011. Chitosan and chitosan derivatives in drug delivery and tissue engineering. Chitosan Biomater. II 244 (1), 19–44.

Rohani, S., Ragab, D., Costa, S., 2012. Controlled release of 5-fluorouracil and progesterone from magnetic nanoaggregates. Int. J. Nanomed. 7 (1), 3167–3189.

Roy, I., Ohulchanskyy, T.Y., Pudavar, H.E., Bergey, E.J., Oseroff, A.R., Morgan, J., et al., 2003. Ceramic-based nanoparticles entrapping water-insoluble photosensitizing anticancer drugs: a novel drug-carrier system for photodynamic therapy. J. Am. Chem. Soc. 125 (26), 7860–7865.

Sahay, G., Alakhova, D.Y., Kabanov, A.V., 2010. Endocytosis of nanomedicines. J. Control. Release 145 (3), 182–195.

Sakr, F.M., 1996. Nasal administration of glucagon combined with dimethyl-β-cyclodextrin: comparison of pharmacokinetics and pharmacodynamics of spray and powder formulations. Int. J. Pharm. 132 (1–2), 189–194.

Sakuma, S., Suzuki, N., Sudo, R., Hiwatari, K., Kishida, A., Akashi, M., 2002. Optimized chemical structure of nanoparticles as carriers for oral delivery of salmon calcitonin. Int. J. Pharm. 239 (1–2), 185–195. Available from: https://doi.org/10.1016/s0378-5173(02)00113-8.

Sandri, G., Bonferoni, M.C., Rossi, S., Ferrari, F., Boselli, C., Caramella, C., 2010. Insulin-loaded nanoparticles based on N-trimethyl chitosan: in vitro (Caco-2 model) and ex vivo (excised rat Jejunum, Duodenum, and Ileum) evaluation of penetration enhancement properties. AAPS PharmSciTech 11 (1), 362–371.

Santander-Ortega, M.J., Bastos-González, D., Ortega-Vinuesa, J.L., Alonso, M.J., 2009. Insulin-loaded PLGA nanoparticles for oral administration: an in vitro physico-chemical characterization. J. Biomed. Nanotechnol. 5 (1), 45–53.

Saraf, S., Mishra, D., Asthana, A., Jain, R., Singh, S., Jain, N.K., 2006. Lipid microparticles for mucosal immunization against hepatitis B. Vaccine 24 (1), 45–56.

Sarmento, B., Martins, S., Ferreira, D., Souto, E.B., 2007. Oral insulin delivery by means of solid lipid nanoparticles. Int. J. Nanomed. 2 (4), 743–749.

Schiffelers, R.M., 2004. Cancer siRNA therapy by tumor selective delivery with ligand-targeted sterically stabilized nanoparticle. Nucleic Acids Res. 32 (19), e149.

Selvamuthukumar, S., Velmurugan, R., 2012. Nanostructured lipid carriers: a potential drug carrier for cancer chemotherapy. Lipids Health Dis. 11, 159.

Shaikh, R., Singh, T.R.R., Garland, M.J., Woolfson, D.A., Donnelly, R.F., 2011. Mucoadhesive drug delivery systems. J. Pharm. Bioallied Sci. 3 (1), 89–100.

Sharma, G., van der Walle, C.F., Ravi Kumar, M.N.V., 2013. Antacid co-encapsulated polyester nanoparticles for peroral delivery of insulin: development, pharmacokinetics, biodistribution and pharmacodynamics. Int. J. Pharm. 440 (1), 99–110.

Shelma, R., Paul, W., Sharma, C.P., 2010. Development and characterization of self-aggregated nanoparticles from anacardoylated chitosan as a carrier for insulin. Carbohydr. Polym. 80 (1), 285–290.

Singh, R.M., Kumar, A., Pathak, K., 2012. Mucoadhesive in situ nasal gelling drug delivery systems for modulated drug delivery. Expert Opin. Drug Deliv. 10 (1), 115–130.

Smart, A.L., Gaisford, S., Basit, A.W., 2014. Oral peptide and protein delivery: intestinal obstacles and commercial prospects. Expert Opin. Drug Deliv. 11 (8), 1323–1335.

Sonaje, K., Chen, Y.-J., Chen, H.-L., Wey, S.-P., Juang, J.-H., Nguyen, H.-N., et al., 2010a. Enteric-coated capsules filled with freeze-dried chitosan/poly(γ-glutamic acid) nanoparticles for oral insulin delivery. Biomaterials 31 (12), 3384–3394.

Sonaje, K., Lin, K.-J., Wey, S.-P., Lin, C.-K., Yeh, T.-H., Nguyen, H.-N., et al., 2010b. Biodistribution, pharmacodynamics and pharmacokinetics of insulin analogues in a rat model: oral delivery using pH-responsive nanoparticles vs. subcutaneous injection. Biomaterials 31 (26), 6849–6858.

Soppimath, K.S., Aminabhavi, T.M., Kulkarni, A.R., Rudzinski, W.E., 2001. Biodegradable polymeric nanoparticles as drug delivery devices. J. Control. Release 70 (1–2), 1–20.

Speiser, P., 1990. Lipid nano pellets as drug carriers for oral administration. European Patent EP 0167825 A3, Germany.

Stuart, M.A.C., Huck, W.T.S., Genzer, J., Müller, M., Ober, C., Stamm, M., et al., 2010. Emerging applications of stimuli-responsive polymer materials. Nat. Mater. 9 (2), 101–113.

Suh, J.W., Lee, J.-S., Ko, S., Lee, H.G., 2016. Preparation and characterization of mucoadhesive buccal nanoparticles using chitosan and dextran sulfate. J. Agric. Food Chem. 64 (26), 5384–5388.

Sun, S., Liang, N., Piao, H., Yamamoto, H., Kawashima, Y., Cui, F., 2010. Insulin-s.O (sodium oleate) complex-loaded PLGA nanoparticles: formulation, characterization and in vivo evaluation. J. Microencapsul. 27 (6), 471–478.

Sung, H.-W., Sonaje, K., Liao, Z.-X., Hsu, L.-W., Chuang, E.-Y., 2012. PH-responsive nanoparticles shelled with chitosan for oral delivery of insulin: from mechanism to therapeutic applications. Accounts Chem. Res. 45 (4), 619–629.

Sung, J.C., Pulliam, B.L., Edwards, D.A., 2007. Nanoparticles for drug delivery to the lungs. Trends Biotechnol. 25 (12), 563–570.

Takeuchi, H., Matsui, Y., Yamamoto, H., Kawashima, Y., 2003. Mucoadhesive properties of carbopol or chitosan-coated liposomes and their effectiveness in the oral administration of calcitonin to rats. J. Control. Release 86 (2–3), 235–242.

Takeuchi, H., Matsui, Y., Sugihara, H., Yamamoto, H., Kawashima, Y., 2005. Effectiveness of submicron-sized, chitosan-coated liposomes in oral administration of peptide drugs. Int. J. Pharm. 303 (1–2), 160–170.

Thakral, S., Thakral, N.K., Majumdar, D.K., 2012. Eudragit®: a technology evaluation. Expert Opin. Drug Deliv. 10 (1), 131–149.

Thanou, M., Henderson, S., Kydonieus, A., Elson, C., 2007. N-sulfonato-N,O-carboxymethylchitosan: a novel polymeric absorption enhancer for the oral delivery of macromolecules. J. Control. Release 117 (2), 171–178.

Thirawong, N., Thongborisute, J., Takeuchi, H., Sriamornsak, P., 2008. Improved intestinal absorption of calcitonin by mucoadhesive delivery of novel pectin–liposome nanocomplexes. J. Control. Release 125 (3), 236–245.

Tian, Q., Ding, F., Guo, L., Wang, J., Wu, F., Yu, Y., 2016. Targeted solid lipid nanoparticles with peptide ligand for oral delivery of atorvastatin calcium. RSC Adv. 6 (42), 35901–35909.

Torchilin, V.P., Rammohan, R., Weissig, V., Levchenko, T.S., 2001. TAT peptide on the surface of liposomes affords their efficient intracellular delivery even at low temperature and in the presence of metabolic inhibitors. Proc. Nat. Acad. Sci. 98 (15), 8786–8791.

Tripodo, G., Trapani, A., Torre, M.L., Giammona, G., Trapani, G., Mandracchia, D., 2015. Hyaluronic acid and its derivatives in drug delivery and imaging: recent advances and challenges. Eur. J. Pharm. Biopharm. 97, 400–416.

Trotta, M., Cavalli, R., Carlotti, M.E., Battaglia, L., Debernardi, F., 2005. Solid lipid micro-particles carrying insulin formed by solvent-in-water emulsion–diffusion technique. Int. J. Pharm. 288 (2), 281–288.

Tscheik, C., Blasig, I.E., Winkler, L., 2013. Trends in drug delivery through tissue barriers containing tight junctions. Tissue Barriers 1 (2), e24565.

Ugazio, E., Cavalli, R., Gasco, M.R., 2002. Incorporation of cyclosporin A in solid lipid nanoparticles (SLN). Int. J. Pharm. 241 (2), 341–344.

Varshosaz, J., 2012. Dextran conjugates in drug delivery. Expert Opin. Drug Deliv. 9 (5), 509–523.

Veiseh, O., Gunn, J.W., Zhang, M., 2010. Design and fabrication of magnetic nanoparticles for targeted drug delivery and imaging. Adv. Drug Deliv. Rev. 62 (3), 284–304.

Videira, M., 2002. Preparation of solid lipid nanoparticles (SLN): a potential protein delivery system. In: V Spanish–Portuguese Conf Control Drug Deliv Seville, pp. 69–70.

Vilar, G., Tulla-Puche, J., Albericio, F., 2012. Polymers and drug delivery systems. Curr. Drug Deliv. 9 (4), 367–394.

Visser, C.C., Stevanović, S., Voorwinden, L.H., Bloois, L., van, Gaillard, P.J., Danhof, M., et al., 2005. Targeting liposomes with protein drugs to the blood–brain barrier in vitro. Eur. J. Pharm. Sci. 25 (2–3), 299–305.

Wasiak, I., Kulikowska, A., Janczewska, M., Michalak, M., Cymerman, I.A., Nagalski, A., et al., 2016. Dextran nanoparticle synthesis and properties. PLoS One 11 (1), e0146237.

Watnasirichaikul, S., Rades, T., Tucker, I.G., Davies, N.M., 2002. Effects of formulation variables on characteristics of poly(ethylcyanoacrylate) nanocapsules prepared from w/o microemulsions. Int. J. Pharm. 235 (1–2), 237–246.

Weetall, H.H., 1970. Storage stability of water-insoluble enzymes covalently coupled to organic and inorganic carriers. Biochim. Biophys. Acta (BBA)—Enzymol. 212 (1), 1–7.

Widjaja, L.K., Bora, M., Chan, P.N.P.H., Lipik, V., Wong, T.T.L., Venkatraman, S.S., 2013. Hyaluronic acid-based nanocomposite hydrogels for ocular drug delivery applications. J. Biomed. Mater. Res. A 102 (9), 3056–3065.

Woitiski, C.B., Neufeld, R.J., Veiga, F., Carvalho, R.A., Figueiredo, I.V., 2010. Pharmacological effect of orally delivered insulin facilitated by multilayered stable nanoparticles. Eur. J. Pharm. Sci. 41 (3–4), 556–563.

Woitiski, C.B., Sarmento, B., Carvalho, R.A., Neufeld, R.J., Veiga, F., 2011. Facilitated nanoscale delivery of insulin across intestinal membrane models. Int. J. Pharm. 412 (1–2), 123–131.

Xi, J., Guo, R., 2007. Interactions between flavonoids and hemoglobin in lecithin liposomes. Int. J. Biol. Macromol. 40 (4), 305–311.

Xia, L.-W., Xie, R., Ju, X.-J., Wang, W., Chen, Q., Chu, L.-Y., 2013. Nano-structured smart hydrogels with rapid response and high elasticity. Nat. Commun. 4, 2226–2236.

Xia, X., Hu, Z., Marquez, M., 2005. Physically bonded nanoparticle networks: a novel drug delivery system. J. Control. Release 103 (1), 21–30.

Xie, Y., Ye, L., Zhang, X., Cui, W., Lou, J., Nagai, T., et al., 2005. Transport of nerve growth factor encapsulated into liposomes across the blood–brain barrier: in vitro and in vivo studies. J. Control. Release 105 (1–2), 106–119.

Xiong, X.Y., Li, Y.P., Li, Z.L., Zhou, C.L., Tam, K.C., Liu, Z.Y., et al., 2007. Vesicles from pluronic/poly(lactic acid) block copolymers as new carriers for oral insulin delivery. J. Control. Release 120 (1–2), 11–17.

Xiong, X.Y., Li, Q.H., Li, Y.P., Guo, L., Li, Z.L., Gong, Y.C., 2013. Pluronic P85/poly (lactic acid) vesicles as novel carrier for oral insulin delivery. Colloids Surf. B: Biointerfaces 111, 282–288.

Yang, J., Sun, H., Song, C., 2012. Preparation, characterization and in vivo evaluation of pH-sensitive oral insulin-loaded poly(lactic-co-glycolicacid) nanoparticles. Diabetes Obes. Metabol. 14 (4), 358–364.

Yang, T., Hussain, A., Paulson, J., Abbruscato, T.J., Ahsan, F., 2004. Cyclodextrins in nasal delivery of low-molecular-weight heparins: in vivo and in vitro studies. Pharm. Res. 21 (7), 1127–1136.

Ye, Q., Asherman, J., Stevenson, M., Brownson, E., Katre, N.V., 2000. DepoFoam™ technology: a vehicle for controlled delivery of protein and peptide drugs. J. Control. Release 64 (1–3), 155–166.

Yezhelyev, M.V., Gao, X., Xing, Y., Al-Hajj, A., Nie, S., O'Regan, R.M., 2006. Emerging use of nanoparticles in diagnosis and treatment of breast cancer. Lancet Oncol. 7 (8), 657–667.

Yin, L., Ding, J., He, C., Cui, L., Tang, C., Yin, C., 2009. Drug permeability and mucoadhesion properties of thiolated trimethyl chitosan nanoparticles in oral insulin delivery. Biomaterials 30 (29), 5691–5700.

Yoo, M.-K., Kang, S.-K., Choi, J.-H., Park, I.-K., Na, H.-S., Lee, H.-C., et al., 2010. Targeted delivery of chitosan nanoparticles to Peyer's patch using M cell-homing peptide selected by phage display technique. Biomaterials 31 (30), 7738–7747.

Yun, Y., Cho, Y.W., Park, K., 2013. Nanoparticles for oral delivery: targeted nanoparticles with peptidic ligands for oral protein delivery. Adv. Drug Deliv. Rev. 65 (6), 822–832.

Zhang, D., Wang, J., Xu, D., 2016. Cell-penetrating peptides as noninvasive transmembrane vectors for the development of novel multifunctional drug-delivery systems. J. Control. Release 229, 130–139.

Zhang, K., Wu, X.Y., 2004. Temperature and pH-responsive polymeric composite membranes for controlled delivery of proteins and peptides. Biomaterials 25 (22), 5281–5291.

Zhang, N., Ping, Q.N., Huang, G.H., Xu, W.F., 2005. Investigation of lectin-modified insulin liposomes as carriers for oral administration. Int. J. Pharm. 294 (1–2), 247–259.

Zhang, N., Ping, Q., Huang, G., Xu, W., Cheng, Y., Han, X., 2006. Lectin-modified solid lipid nanoparticles as carriers for oral administration of insulin. Int. J. Pharm. 327 (1–2), 153–159.

Zhang, N., Li, J., Jiang, W., Ren, C., Xin, J., Li, K., 2010. Effective protection and controlled release of insulin by cationic β-cyclodextrin polymers from alginate/chitosan nanoparticles. Int. J. Pharm. 393 (1–2), 213–219.

Zhang, Q., Yie, G., Li, Y., Yang, Q., Nagai, T., 2000. Studies on the cyclosporin A loaded stearic acid nanoparticles. Int. J. Pharm. 200 (2), 153–159.

Zhang, X., Wu, W., 2014. Ligand-mediated active targeting for enhanced oral absorption. Drug Discov. Today 19 (7), 898–904.

Zhang, X., Sun, M., Zheng, A., Cao, D., Bi, Y., Sun, J., 2012. Preparation and characterization of insulin-loaded bioadhesive PLGA nanoparticles for oral administration. Eur. J. Pharm. Sci. 45 (5), 632–638.

Zhang, Y., Du, X., Li, G., Cai, C., Xu, J., Tang, X., 2014. Thiolated Eudragit-based nanoparticles for oral insulin delivery: preparation, characterization, and evaluation using intestinal epithelial cells in vitro. Macromol. Biosci. 14 (6), 842–852.

Zhu, S., Chen, S., Gao, Y., Guo, F., Li, F., Xie, B., et al., 2015. Enhanced oral bioavailability of insulin using PLGA nanoparticles co-modified with cell-penetrating peptides and Engrailed secretion peptide (Sec). Drug Deliv. 23 (6), 1980–1991.

Zimmer, A., 1999. Antisense oligonucleotide delivery with polyhexylcyanoacrylate nanoparticles as carriers. Methods 18 (3), 286–295.

FURTHER READING

Sarmento, B., Ribeiro, A., Veiga, F., Ferreira, D., 2006. Development and characterization of new insulin containing polysaccharide nanoparticles. Colloids Surf. B: Biointerfaces 53 (2), 193–202.

Videira, M., 1998. Entrapment of a high molecular weight protein into solid lipid nanoparticles. In: Proc. 2nd World Meeting APV/APGI Paris, pp. 629–630.

CHAPTER

Nanomaterial-based electrochemical sensors in pharmaceutical applications

7

Kotagiri Yugender Goud[1,2], Moru Satyanarayana[1,2], Akhtar Hayat[1,3], Kauveri Vengatajalabathy Gobi[2] and Jean Louis Marty[1]

[1]BAE Laboratory, Universite de Perpignan Via Domitia, Perpignan, France [2]Department of Chemistry, National Institute of Technology, Warangal, India [3]Interdisciplinary Research Centre in Biomedical Materials (IRCBM), COMSATS Institute of Information Technology (CIIT), Lahore, Pakistan

CHAPTER OUTLINE

7.1 INTRODUCTION

Sensors are devices that register a physical, chemical, or biological change and convert that into a measurable signal. The sensor contains two major components, one is a recognition element that enables the selective response to a particular analyte or a group of analytes, thus minimizing interferences from other sample components. Another main component of a sensor is the transducer or the detector device that produces a signal. A signal processor collects, amplifies, and displays the signal.

Based on the nature of the molecular recognition element, several kinds of biosensors have been constructed. Among them, the majority of the reported literature reveals that enzymes have been widely used for biosensor fabrication. Molecular recognition elements reported in the literature are antibodies (Bhatta et al., 2010), nucleic acids, aptamers (Strehlitz et al., 2012), dendrimers (Satija et al., 2014), self-assembled monolayers (Andrade et al., 2015), magnetic beads, conducting polymers (Rosy et al., 2016; Kumar et al., 2016), mesoporous

Nanoparticles in Pharmacotherapy. DOI: https://doi.org/10.1016/B978-0-12-816504-1.00015-6

materials (Walcarius, 2013; Ndamanisha and Guo, 2012; Zhao et al., 2013), sol−gel based materials (Almeida et al., 2011), molecular imprinted polymers (Kubo and Otsuka, 2016), nanomaterials (nanoparticles of noble metals, other metals and nonmetals), carbon nanotubes, their derivatives, graphene, and graphene oxide. Among the all the molecular recognition elements nanomaterials have played a very important role since the last decade in the field of biosensors.

The emergence of nanoscience and nanotechnology, which is creation and utilization of materials and tools at the nanometer scale, has been a great influence on a number of industries and particularly the pharmaceutical industry. Just as recombinant technology and biotechnology changed the landscape of the pharmaceutical industry, nanotechnology is poised to lift the pharmaceutical industry, which is at crossroads today, to new levels. However, unlike the biotechnology industry, which primarily influenced the pharmaceutical industry, nanotechnology has broader applications and therefore, the nanotechnology tools and materials developed for other industries also have potential opportunities in the pharmaceutical industry as well. In addition, there is a prevalent notion that nanotechnology tools are exotic, too futuristic, disruptive, and not suitable for quick commercialization of products. This notion is far from the truth as nanotechnology tools are being shown to add value to existing products for existing markets as well as opening opportunities in new markets. This is true not only for the pharmaceutical industry but also for other industries (Fig. 7.1).

In this present chapter, we focused on the role of carbon nanomaterials in the pharmaceutical industry to develop the biosensors for detection of various drugs by using electrochemical methods. The development of electrochemical sensors for pharmaceutical drugs detection, a brief discussion regarding cancer drugs (5-Fluorouracil and Dacarbazine) detection, and tuberculosis drug (Isoniazid) detection on carbon nanotube materials as recognition elements by using electrochemical techniques have been discussed.

7.2 FUNCTIONS OF NANOMATERIALS

The last decade has witnessed an enormous impact of nanomaterials on the field of biosensors. Significant progress has been achieved in synthetic approaches to prepare nanomaterials of desired properties such as controllable size, shape, surface charge, and physiochemical characteristics (Chen and Chatterjee, 2013). Nanomaterials can be decorated with polymers and biomolecules to achieve better biocompatibility and precise targeting. These features have made it possible to integrate nanomaterials in biosensors for any desired function, and led to an increasing use of nanomaterials in the design of electrochemical sensors. Thousands of scientific articles describing associations between nanomaterials and biomolecules have been published in the literature (Song et al., 2010). In this context, different techniques have been explored to integrate nanomaterials of

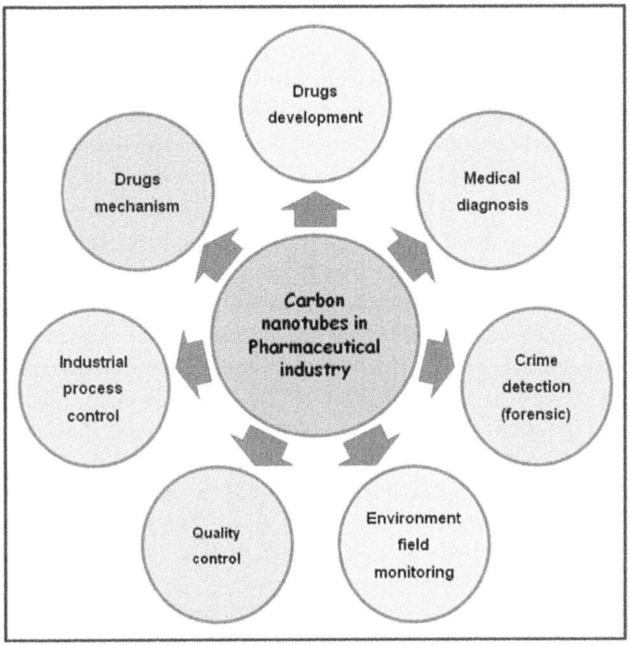

FIGURE 7.1

Carbon nanotubes in the pharmaceutical industry.

diverse nature in the electrochemical transduction platform. Recent trends have also shown the direct conjugation of nanomaterials with biomolecules or target analyte through various kinds of interactions such as physical adsorption, covalent attachment, and electrostatic interaction with weak hydrogen bonding. However, in some cases, nanomaterials are directly employed in the interacting medium on transducer surface to enhance the electroanalytical performance of the system. There is no doubt on the importance of nanostructured materials and their implications in biosensor fabrication. Here, we discuss the role of carbon nanomaterials in the field of pharmaceutical industry to develop the biosensors for detection of various drugs.

Carbon nanotubes (CNTs) have been recognized as an important material in recent years in various fields due to their unique electrical, mechanical, and structural properties. CNTs can easily promote electron transfer between the electroactive species and the electrode surface due to their unique long and tubular geometry. The remarkable property of conductivity to the surface adsorbates permits the use of CNTs in the fabrication of highly sensitive nanoscale sensors. Their use as electrode modifiers can lead to a decrease of the overpotential, a decrease in the response time, enhanced electrocatalytic activity, and an increase in available active surface area in comparison with conventional carbon electrodes. The electrocatalytic effect of CNTs has been attributed to the activity of edge-plane-like graphite sites at the CNT ends, and it can be further increased by functionalization of CNTs. CNTs also reduce the fouling of the electrode, which can greatly improve the reusability of such sensors. The low solubility of CNTs in most solvents is the major problem to overcome regarding their use as modifiers in the fabrication of chemical sensors and/or biosensors. To overcome this problem, several strategies have been proposed for effective immobilization of CNTs on electrochemical transducers, such as dispersion in different solvents or polyelectrolytes or incorporation in composite matrices using distinct binders. Recently, a number of reviews were published regarding nanomaterials importance in pharmaceutical drugs detection (Cernat et al., 2015; Dmitrienko et al., 2014; Jain, 2009; Prakash et al., 2011; Thapliyal et al., 2015; Tothill, 2009; Uslu and Ozkan, 2007; Yang et al., 2011; Yu et al., 2014; Chen and Chatterjee, 2013). Several electrochemical sensors (El-Nashar et al., 2012; Kutluay and Aslanoglu, 2014; Raoof et al., 2012; Jain and Dhanjai, 2013; Amiri et al., 2011; Mahmoud et al., 2015; Leite et al., 2012; Abbaspour and Mirzajani, 2007; Goyal et al., 2008; Shahrokhian and Asadian, 2010; Gupta et al., 2013; Arvand et al., 2011; Arvand and Palizkar, 2013; Baghayeri et al., 2016; Lu and Tsai, 2011, 2010; Abbar et al., 2012; Malode et al., 2012; Patil et al., 2011; Jain and Sharma, 2012; Jain and Rather, 2011; Keyvanfard and Alizad, 2016; Ashrafi et al., 2014; Taei et al., 2016; Ensafi et al., 2011; Sadiković et al., 2014; Baytak et al., 2016; Brahman et al., 2012; Dogan-Topal et al., 2013; Lee et al., 2016; Goyal and Bishnoi, 2009; Karadas et al., 2013; Nigović et al., 2014; Karadas and Ozkan, 2014; Silva et al., 2015; Cheraghi and Taher, 2016; Goyal et al., 2010; Khaleghi et al., 2016a) were published recently based on nanomaterials for drugs detection. These carbon nanotubes play an important role in the development of optical and electrochemical biosensors. Further, we discussed carbon nanotube-based electrochemical biosensors for pharmaceutical drugs detection.

7.3 ELECTROCHEMICAL BIOSENSORS

Various types of analytical techniques such as spectroscopic, optical, and electroanalytical techniques were employed as transduction methods for the quantification of physicochemical interactions. Spectroscopic techniques as transduction principles include UV–visible spectroscopy (Vinoth et al., 2015), Fourier transforms infrared spectroscopy (FT-IR), and Raman spectroscopy (Wang et al., 2014). Optical techniques as transducers include capillary electrophoresis (Rattanarat et al., 2016), chemiluminescence (Qi et al., 2011), electrochemiluminescence (Wei et al., 2016), fluorescence spectroscopy (Gopal et al., 2016; Nguyen et al., 2015), and surface plasmon resonance (Kinouchi et al., 2016; Altintas et al., 2016). Advanced analytical methods include chromatography techniques like gas chromatography (GC) (Afreen et al., 2015), high-performance liquid chromatography (HPLC), liquid chromatography combined with mass spectrometry (LC–MS) (Santos Pereira et al., 2016), etc. Piezoelectric techniques employed as transducers involve quartz crystal balance, either micro or nano (Feng

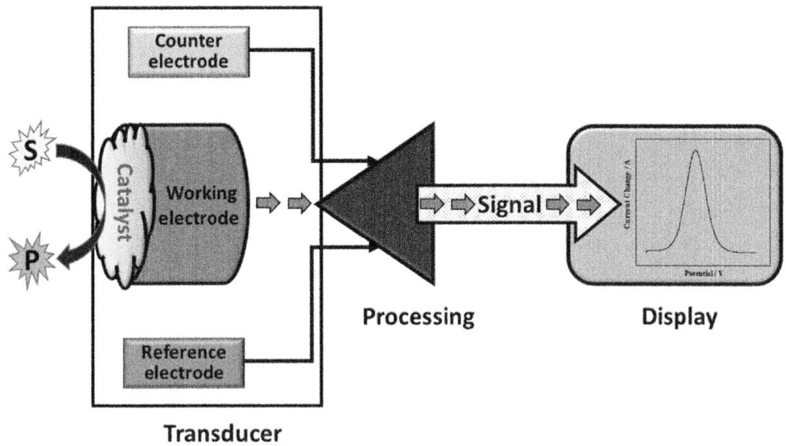

FIGURE 7.2

Electrochemical sensing principle.

and Yang, 2010). Electroanalytical techniques such as amperometry (Amatatongchai et al., 2015; Mohammadi-Behzad et al., 2016), potentiometry, voltammetry (Pisoschi et al., 2014; Kalambate et al., 2016, 2015; Khaleghi et al., 2016b), and electrochemical impedance spectroscopy (Hofmann et al., 2010; Rastogi et al., 2016) were considered as the latest, user friendly transduction methods.

Most biosensors use electrochemical detection for the transducer because of the low cost, ease of use, portability, and simplicity of construction. The reaction, which is monitored electrochemically, typically generates a measurable current (amperometry), a measurable charge accumulation or potential (potentiometry), or alters the conductive properties of the medium between electrodes (conductometry). Use of electrochemical impedance spectroscopy by monitoring both resistance and reactance in the biosensor is also becoming more common. Electrochemistry is a surface technique and offers certain advantages for detection in biosensors. It does not depend strongly on the reaction volume, and very small sample volumes can be used for measurement. Electrochemical techniques are generally organized into three main categories of measurement: current, potential, and impedance. This article focuses primarily on those techniques that measure current since they are the most commonly used in biosensors. Moreover, here we focused on carbon nanotube-based electrochemical sensors for tuberculosis and cancer drug detection (Fig. 7.2).

7.4 TUBERCULOSIS DRUGS DETECTION

Tuberculosis (TB) is an infection caused by a bacterium, *Mycobacterium tuberculosis*. TB bacteria are released into the air when someone with infectious TB coughs or sneezes. Active tuberculosis, particularly if it's a drug-resistant strain, will require several drugs at once. The most common medications used to treat tuberculosis include Isoniazid, Rifampin, Ethambutol, Pyrazinamide.

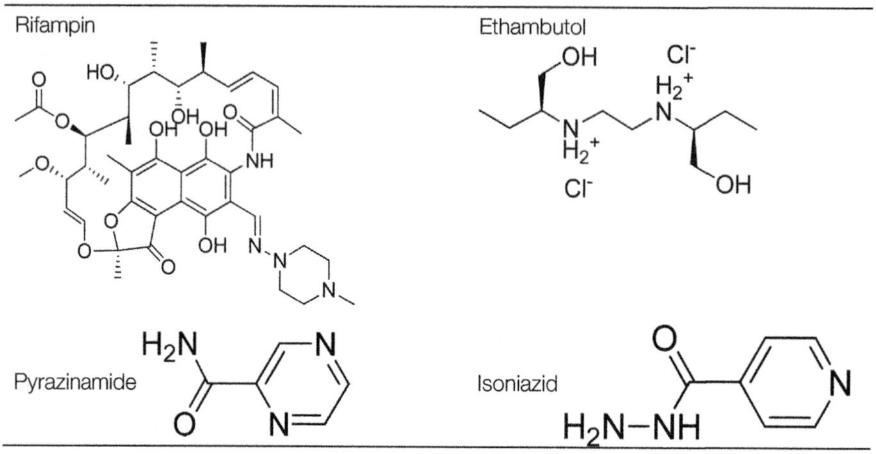

Isoniazid (isonicotinylhydrazide, INH) is one of the most widely used first-line clinical drugs for the treatment of all kinds of tuberculosis. Overdoses of the drug during chemotherapy can cause hepatotoxicity, and there has been a global increase in the prevalence of drug-resistant tuberculosis. The large-scale therapeutic use of this drug necessitated the need for the development of rapid, simple, and onsite analytical methods for determining isoniazid, in drug formulations for quality control and in biological fluids for medical diagnosis. Various analytical methods based on titrimetric, spectroscopic chromatographic, chemiluminescence, and electrochemical techniques have aimed at the quantitative determination of isoniazid. Electrochemical methods are interesting as they require inexpensive, miniaturized portable equipment, and the analytes could be detected at trace levels without any preconcentration steps and without adding any special reagents.

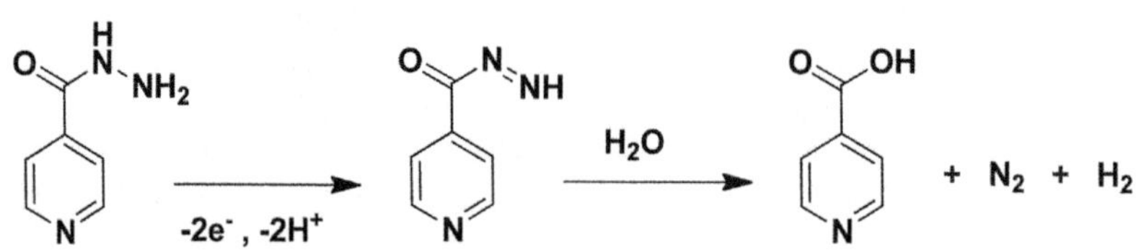

Mechanism for electrochemical oxidation of INH.

Here we explained a brief discussion of the development of an electrochemical sensor for INH detection based on the MWCNT–chitosan electrode (Satyanarayana et al., 2014). The voltammetric response of the MWCNT–Chit/GCE electrode to the presence of INH was investigated by DPV. At the MWCNT–chitosan modified electrode, the anodic peak at +0.15 V was monitored and a better voltammetric profile was obtained with a scan rate of

50 mV s^{-1}, a pulse amplitude of 50 mV, and a pulse period of 5 ms. After optimizing the operating conditions, differential pulse voltammetric measurements were carried out at the MWCNT–Chit/GCE in B–R buffer containing different concentrations of INH. The results showed that the oxidation peak current (i_p) linearly increased with the concentration of isoniazid. The linear regression equation is expressed as: i(mA) = 1.1104 + 1.673C isoniazid (mM) with the regression coefficient $R^2 = 0.9876$; the detection limit is found to be 5.5×10^{-8} M (S/N = 3), which is equivalent to ~7.5 ng mL^{-1} with a linear range of 1.0×10^{-7} M to 1.0×10^{-5} M. The performance of the present electrochemical sensor is compared with other electrochemical sensors and also with other analytical methods reported in the literature (Table 7.1).

The electrochemical response towards INH in the presence of possible electroactive physiological components such as ascorbic acid (AA), uric acid (UA), and dopamine (DA) has been investigated at the MWCNT–chitosan modified electrode using DPV. Pharmaceutical samples of isoniazid (300 mg per tablet), which were diluted appropriately, were analyzed with the proposed electrochemical method by the standard addition method, obtaining good recovery values ranging from 96.7% to 102.0% with a low RSD (1.6%). Recovery of INH in artificial urine was also examined by the direct addition of INH into undiluted artificial urine without any buffer. Isoniazid metabolizes rapidly with a half-life of 1–3 h. These results clearly indicate that the developed sensor chip could detect INH selectively from urine samples at concentrations as low as 5×10^{-7} M. This detection limit is nearly four times superior compared with the concentration of isoniazid in urine 10 h after administration, and thus diluted urine samples could also be investigated for the determination of isoniazid by the present method. From the results of the recovery analysis, we conclude that the fabricated sensor can be used efficiently for the selective determination of isoniazid from pharmaceutical formulations and from urine samples in vitro.

7.5 CANCER DRUGS DETECTION

Anticancer drugs, also called antineoplastic drugs, include any drug that is effective in the treatment of malignant, or cancerous, disease. There are several major classes of anticancer drugs; these include alkylating agents, antimetabolites, natural products, and hormones. In addition, there are a number of drugs that do not fall within those classes but that demonstrate anticancer activity and thus are used in the treatment of malignant disease.

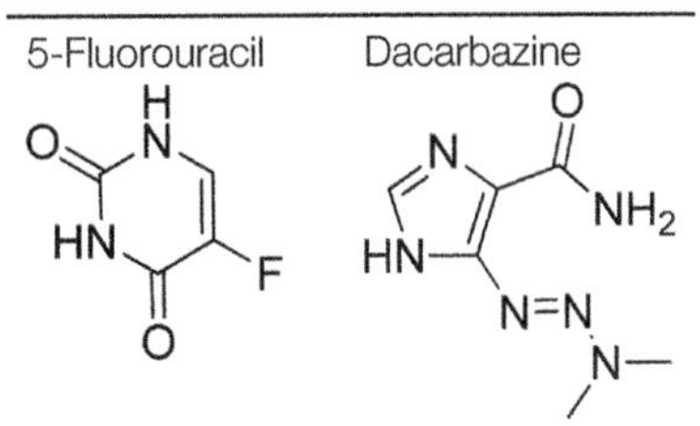

Table 7.1 Detection of INH Using Electrochemical and Various Analytical Methods

Method	Electrode	Linear Range	Limit of Detection	Reference
DPV	Carbon paste/poly-L-histidine	5.0×10^{-7}–1.1×10^{-4} M	1.7×10^{-7} M	Bergamini et al. (2010)
SWV[a]	Pt/Ferrocene mediator	5.0×10^{-5}–6.0×10^{-4} M	–	Gao et al. (2006)
DPV	MWCNT/Thionine	1.0×10^{-6}–1.0×10^{-4} M	5.0×10^{-7} M	Shahrokhian and Asadian (2010)
Amperometry	GC/polypyrrole	4.0×10^{-6}–2.0×10^{-3} M	3.2×10^{-6} M	Majidi et al. (2006)
Amperometry	GC/Nafion-ordered mesoporous carbon	1.0×10^{-7}–3.7×10^{-4} M	8.4×10^{-8} M	Yan et al. (2011)
DPV	GC/MWCNT–carbon paste	1.0×10^{-6}–1.0×10^{-3} M	5.0×10^{-7} M	Shahrokhian and Amiri (2007)
HPLC	n.a.	0.002–20 $\mu g\ mL^{-1}$	0.5 $ng\ mL^{-1}$	Zhou et al. (2009)
MEKC	n.a.	3.0–100.0 $mg\ mL^{-1}$	1.0 $mg\ mL^{-1}$	Nemutlu et al. (2007)
UV–Vis spectrophotometry	n.a.	0.3–3.5 $\mu g\ mL^{-1}$	0.26 $\mu g\ mL^{-1}$	Safavi et al. (2004)
UV–Vis spectrophotometry	n.a.	0.6–6.2 $\mu g\ mL^{-1}$	0.15 $\mu g\ mL^{-1}$	Safavi and Bagheri (2008)
DPV	MWCNT–Chit/GCE	1.0×10^{-7}–3.0×10^{-6} M (0.014–0.411 $\mu g\ mL^{-1}$)	5.5×10^{-8} M (7.5 $ng\ mL^{-1}$)	Satyanarayana et al. (2014)

[a]*SWV, square-wave voltammetry; MEKC, micellar electrokinetic capillary chromatography; n.a., not applicable.*

In recent years, various antitumor active compounds are widely used in the treatment of cancer, and one of the most widely used drugs for chemotherapy of solid tumors is 5-fluorouracil (5-FU). It is an antineoplastic agent used for the treatment of solid tumor based cancers. Generally, high serum concentrations of drugs are necessitated to effect a pharmacological activity against tumors, but 5-FU metabolizes rapidly in the body. Determining safe and effective dosage of drugs for chemotherapy is still a substantial challenge in the treatment of neoplastic diseases. Frequent measurement of the concentration of 5-FU in physiological fluids is extremely in demand to maintain an optimal concentration of 5-FU. It is noteworthy that an overdose of 5-FU may cause very high toxicity by accumulation in cancer patients. Controlling the amount of 5-FU in a given formulation and the dose to the sufferers is important in pharmaceutical quality control and for clinical diagnosis. Various analytical methods have been aimed at the detection of 5-FU based on spectroscopic, chromatographic capillary, electrophoresis, and electrochemical methods. Electrochemical transduction methodologies are advantageous because they offer miniaturization, portable analysis, rugged instrumentation with no movable parts, high sensitivity, on-site analysis, etc. However, the major problem towards the detection of 5-FU by an electrochemical method is the poor oxidation of 5-FU at bare electrodes and only a few electrochemical sensors have been reported so far for the detection of 5-FU.

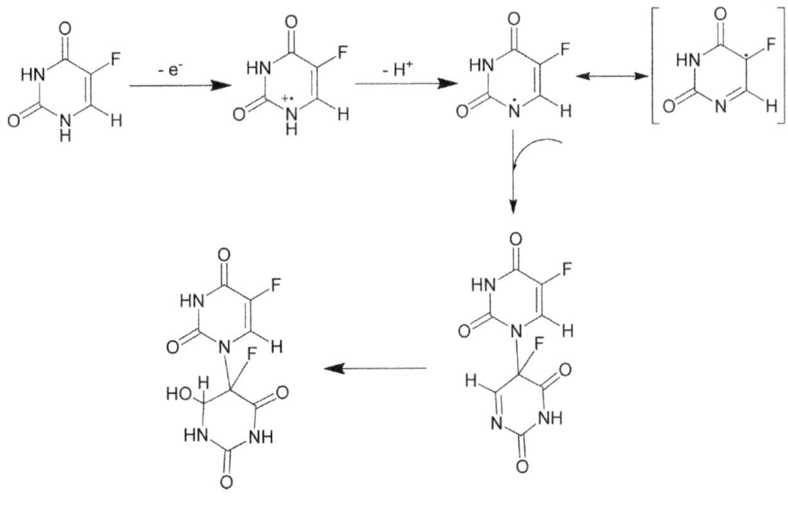

Reaction mechanism of 5-fluorouracil

Here we will discuss the development of an electrochemical sensor for 5-FU detection based on GNP–MWCNT–chitosan electrode (Satyanarayana et al., 2015). Differential pulse voltammetric analysis of 5-FU at the GNP nanocomposite electrode was carried out for quantitative analysis of 5-FU. Optimum experimental parameters for a good voltammetric profile with high peak current and

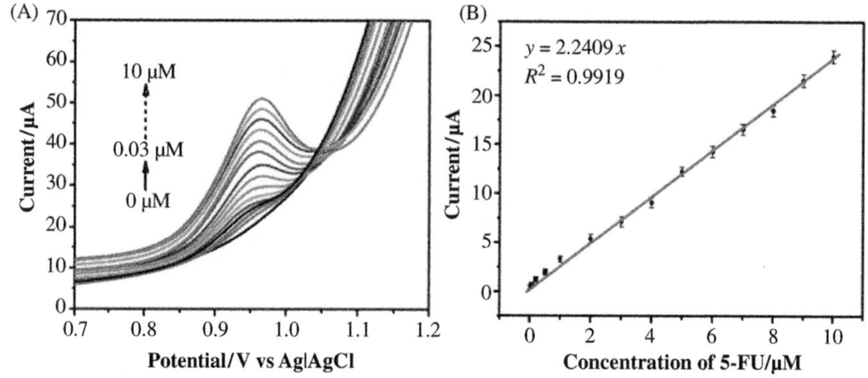

FIGURE 7.3

(A) DPVs of 5-FU in B–R buffer (pH 8.0) at different concentrations (0, 0.03, 0.06, 0.2, 0.5, 1, 2, 3, 4, 5, 6, 7, 8, 9, 10 mM) at GNP–MWCNT–Chit/GCE. (B) Plot of ip versus 5-FU concentration (Satyanarayana et al., 2015).

Reprint permission from Elsevier.

minimal background current are established to be 100 mV pulse amplitude, 100 ms pulse width, and 10 mV step increment. Differential pulse voltammograms of 5-FU at GNP–MWCNT–Chit/GCE electrode were recorded in B-R buffer (pH 8.0) over the concentration range of 0–10 mM 5-FU, and the results are shown in Fig. 7.3. A well-defined anodic peak was observed at +0.95 V. The peak current increased gradually with the increase in the concentration of 5-FU. The DPV measurements were repeated four times with independent electrodes, and the peak current was plotted against the concentration of 5-FU. A linear plot with a regression coefficient of 0.99 was obtained. Low detection limit of the sensor system is defined as the change in peak current by three relative standard deviations, and it is determined from the plot to be 20 nM. The linear determination range is 3×10^{-8}–10×10^{-6} M. The performance of the GNP nanocomposites electrode was compared with those reported previously, as shown in Table 7.2. The performance of the present GNP nanocomposites electrode for the detection of 5-FU in the presence of various potential interferents and from complex physiological and pharmaceutical samples were further investigated.

Pharmaceutical samples of 5-FU (Fluracil 250 mg/5 mL injection) were analyzed by using the fabricated nanocomposite sensor with the standard addition method. The analytical samples containing 0.5 mM of pure 5-FU together with different concentrations (0.5, 1, and 3 mM) of 5-FU of the pharmaceutical injection at the GNP–MWCNT–Chit/GCE. The experimental time for a DPV analysis is as low as 25 s. The recovery values obtained in these experiments varied from 98.3% to 102.6% and are quite satisfactory. Recovery of 5-FU from artificial urine was examined by the addition of 5-FU into the undiluted artificial urine of pH 8.0 without any buffer. The DPVs of the GNP–MWCNT–Chit/GCE

Table 7.2 Detection of 5-FU Using Electrochemical and Various Analytical Methods

Method	Electrode	Linear Range	Limit of Detection	Reference
SWV	AuNP–SPE	0.2–50 μg mL^{-1}	0.1 μg mL^{-1}	Wang et al. (2012)
Potentiometry	Membrane electrode	1.3–130 μg mL^{-1}	10 μM	Hassan et al. (1998)
DPASV	MIP–fiber electrode	9.99–426.46 ng mL^{-1}	1.30 ng mL^{-1}	Prasad et al. (2012)
DPV	IL-CPE	0.5μM–800 μM	13 nM	Zhan et al. (2011)
Cyclic voltammetry	MWNT/BTB/GCE	0.8 μM–5 mM	0.26 μM	Hua et al. (2013)
SWV	HMDE	10 pM–90 pM	7.7 pM	Mirčeski et al. (2000)
Vibrational Spectroscopy	n.a.	–	2 μg mL^{-1}	Farquharson et al. (2005)
Spectrophotometry	n.a.	30–100 μg mL^{-1}	0.5 μg mL^{-1}	Amer et al. (1998)
HPLC	n.a.	4–160 μg L^{-1}	4 μg L^{-1}	Chu et al. (2003)
HPLC	n.a.	0.2–40 μg mL^{-1}	0.6 ng mL^{-1}	Chen et al. (2012)
DPV	GNP–MWCNT–Chit/GCE	50 nM–10 μM (6.5 ng mL^{-1}–1.3 μg mL^{-1})	2×10^{-8} M (2.6 ng mL^{-1})	Satyanarayana et al. (2015)

BTB, bromothymol blue; SWV, square-wave voltammetry; HMDE, hanging mercury drop electrode; AuNP, SPE, gold nanoparticle-modified screen-printed electrode; DPASV, differential pulse adsorptive stripping voltammetry; HPLC, high-performance liquid chromatography; n.a., not applicable.

electrode recorded in the artificial urine sample in the presence of 1.0, 2.0, and 4.0 mM 5-FU. The recovery of 5-FU from undiluted artificial urine samples at different concentrations of 5-FU (1.0–4.0 mM) varied from 97.6% to 101.4%. In physiological samples of clinical patients, the mean concentration of 5-FU varies over a wide concentration range of 0.106–3.77 mg mL^{-1} (0.82–29.2 mM) in serum samples and as high as 10–60 mg mL^{-1} (77.4–464.2 mM) in urine samples. The linear determination range of the developed sensor (0.03–10 mM) along with the very good recovery limits for physiological urine and pharmaceutical samples clearly confirms that the developed nanocomposites sensor would be of biomedical interest.

Dacarbazine [5-(3,3-dimethyl-1-triazenyl)imidazole-4-carboxamide, DTIC] is also a commonly used anticancer drug to treat some malignant diseases. The drug is normally administered to the patient via veins with the dosage of 250 mg per day. The half-life of DTIC in human plasma is c.35 min and 43% of the injected DTIC dosage is excreted in urine within a period of 6 h. The major side effects are nausea and vomiting. Thus, periodical monitoring/determination of physiological concentrations of DTIC is immensely essential. Various laboratory-based analytical methods, such as UV–visible spectroscopy and HPLC, have been employed to determine DTIC from pharmaceutical formulations and physiological liquids. On the other hand, there are only a few reports available for the direct detection of DTIC using simple and rapid direct in vitro analytical methods like electrochemical technique. Most of the electrochemical studies have focused on the binding interaction of DTIC with DNA. The electrochemical detection of DTIC reported good detection limits with a typical analytical system containing HPLC integrated with amperometric methods, and reported an indirect method for the detection of DTIC via a Cu2 + complex. Here, we have demonstrated a direct electrochemical analysis for detection of DTIC, which is aimed at a reagent and bedside diagnosis for the patients in biomedical applications and also can be used for quality control and clinical applications.

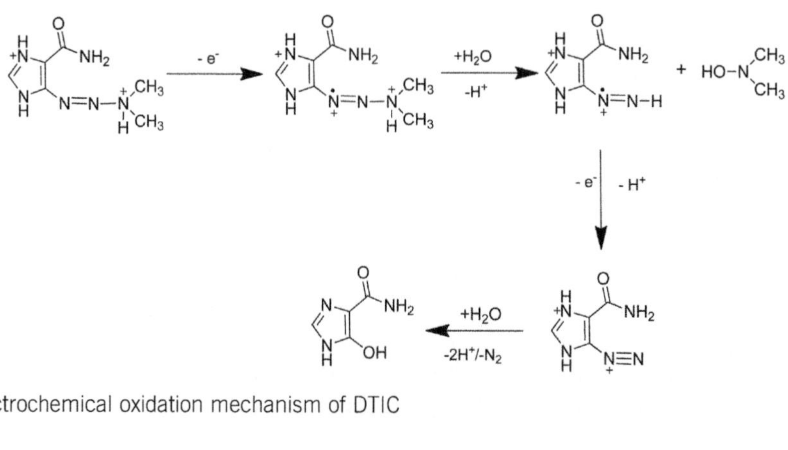

Electrochemical oxidation mechanism of DTIC

Differential pulse voltammetric analysis of DTIC at poly-ATD/CNPE was carried out for quantitative analysis of DTIC. Optimum experimental parameters for a good voltammetric profile for the oxidation of DTIC with high peak current, minimal full-width-at-half-maximum, and minimal background current are established to be 100 mV pulse amplitude, 100 ms pulse width, and 10 mV step increment. Differential pulse voltammograms of DTIC at poly-ATD/CNPE electrode were recorded in B–R buffer (pH 6.0) over the concentration range of 0–24 μM DTIC. A well-defined anodic peak was observed at +0.65 V and the peak current increased gradually with the increase in the concentration of DTIC. The DPV measurements were repeated four times with independent electrodes, and the peak current was plotted against the concentration of DTIC. A linear plot with a regression coefficient of 0.99 was obtained. Low detection limit of the sensor system is defined as the change in peak current by three relative standard deviations, and it is determined from the plot to be 35 nM. The linear determination range is 5×10^{-8}–24×10^{-6} M and the analysis time is as low as 25 s.

Pharmaceutical samples of DTIC (Decarb-200 mg injection) were analyzed by using the fabricated nanocomposite sensor by the standard addition method. Analytical samples containing 0.5 μM of pure DTIC together with different concentrations (0.5, 1 and 1.5 μM) of DTIC of the pharmaceutical injection were examined at the poly-ATD/CNPE. The recovery values obtained in these experiments varied from 99.3% to 102.6% and are quite satisfactory. Recovery of DTIC from artificial urine was examined by the addition of DTIC into the undiluted artificial urine of pH 6.0 without any buffer. The recovery of DTIC from undiluted artificial urine samples at different concentrations of DTIC (0.5–1.5 μM) varied from 98.9% to 101.6%. The observed good recovery limits of spiked DTIC concentrations from pharmaceutical formulations and artificial urine samples demonstrate that the present biosensor is highly promising for the determination of DTIC from real samples.

7.6 CONCLUSIONS

This chapter has addressed recent advances in the application of nanomaterials for electrochemical biosensing of pharmaceutical drugs. The attractive properties of materials have paved the way for the construction of a wide range of electrochemical biosensors exhibiting an attractive analytical behavior. The marked electrocatalytic activity towards the important electroactive drugs. The enhanced electrochemical reactivity is coupled to resistance to surface fouling and hence to high stability. The use of CNT and its metal composites offers great promise for achieving efficient electron transfer from electrode surfaces to the redox sites of molecules. Better control of the chemical and physical properties of CNT and understanding of their use as metal composites should lead to more efficient electrical sensing devices. Electrochemical DNA biosensors can greatly benefit from the use of CNT support platforms and from the enhanced detection of the target molecule.

Here we provided a brief survey of the enormous variety of functions of nanomaterials, which are attributed to the unique physiochemical properties of the constituent nanomaterials, that could find potential applications in electrochemical biosensors. Despite much progress, this field is still new, and there are many points in the development of nanomaterial-based electrochemical biosensors to be addressed such as (1) many complex drug molecules require specific physiological environments and a certain degree of biocompatibility, and the biosensor-integrated nanomaterial must fulfill this requirement; (2) it is highly desirable to find nanomaterials with sufficient binding sites for biomolecules; (3) the possibility to controllably tune the nanomaterials' properties for simultaneous analysis of various analytes is yet to be explored; (4) although there are reports on the joint use of nanomaterials, most of the nanomaterial combinations still need to be explored; and (5) the problem of nanomaterial aggregation in complex media should be resolved. Future research may focus on improving the above-described parameters to further improve the analytical characteristic of nanomaterial-based electrochemical biosensors for detection of pharmaceutical drugs.

SYMBOLS AND ABBREVIATIONS

cm	centimeter
s	seconds
nm	nanometer
I	current
E	potential
ν	scan rate
μA	microampere
mV	millivolt
μM	micromolar
nM	nanomolar
M	molarity
R	gas constant
T	absolute temperature
R_{CT}	charge transfer resistance
Z_W	Warburg impedance
Ω	Ohm
LDL	low-detection limit
RSD	relative standard deviation
INH	isoniazid
5-FU	5-fluorouracil
CLP	clopidogrel
DTIC	dacarbazine
B–R buffer	Britton–Rabinson buffer

MWCNT	multiwall carbon nanotubes
GNP	gold nanoparticles
SNP	silver nanoparticles
Chit	chitosan
CV	cyclic voltammetry
CA	chronoamperometry
DPV	differential pulse voltammetry
EIS	electrochemical impedance spectroscopy

REFERENCES

Abbar, J.C., Malode, S.J., Nandibewoor, S.T., 2012. Electrochemical determination of a hemorheologic drug, pentoxifylline at a multi-walled carbon nanotube paste electrode. Bioelectrochemistry 83 (1), 1–7. Available from: https://doi.org/10.1016/j.bioelechem.2011.06.008.

Abbaspour, A., Mirzajani, R., 2007. Electrochemical monitoring of piroxicam in different pharmaceutical forms with multi-walled carbon nanotubes paste electrode. J. Pharm. Biomed. Anal. 44 (1), 41–48. Available from: http://linkinghub.elsevier.com/retrieve/pii/S0731708507000684.

Afreen, S., et al., 2015. Functionalized fullerene (C60) as a potential nanomediator in the fabrication of highly sensitive biosensors. Biosens. Bioelectron. 63, 354–364. Available from: http://linkinghub.elsevier.com/retrieve/pii/S0956566314005454.

Almeida, S.A.A., et al., 2011. Sulphonamide-imprinted sol–gel materials as ionophores in potentiometric transduction. Mater. Sci. Eng. C 31 (8), 1784–1790. Available from: http://linkinghub.elsevier.com/retrieve/pii/S0928493111002360.

Altintas, Z., et al., 2016. Computationally modelled receptors for drug monitoring using an optical based biomimetic SPR sensor. Sens. Actuators B: Chem. 224, 726–737. Available from: http://linkinghub.elsevier.com/retrieve/pii/S0925400515305414.

Amatatongchai, M., et al., 2015. Amperometric flow injection analysis of glucose using immobilized glucose oxidase on nano-composite carbon nanotubes–platinum nanoparticles carbon paste electrode. Talanta 166.

Amer, M.M., et al., 1998. Spectrophotometric and spectrofluorimetric determination of fluorouracil in the presence of its degradation products. J. Pharm. Pharm. 50 (2), 133–138.

Amiri, M., et al., 2011. Electrocatalytic determination of sumatriptan on the surface of carbon-paste electrode modified with a composite of cobalt/Schiff-base complex and carbon nanotube. Bioelectrochemistry 81 (2), 81–85. Available from: https://doi.org/10.1016/j.bioelechem.2011.03.003.

Andrade, F., et al., 2015. Biological assessment of self-assembled polymeric micelles for pulmonary administration of insulin. Nanomed. Nanotechnol. Biol. Med. 11 (7), 1621–1631. Available from: http://linkinghub.elsevier.com/retrieve/pii/S1549963415001215.

Arvand, M., Ansari, R., Heydari, L., 2011. Electrocatalytic oxidation and differential pulse voltammetric determination of sulfamethoxazole using carbon nanotube paste electrode. Mater. Sci. Eng. C 31 (8), 1819–1825. Available from: https://doi.org/10.1016/j.msec.2011.08.014.

Arvand, M., Palizkar, B., 2013. Development of a modified electrode with amine-functionalized TiO_2/multi-walled carbon nanotubes nanocomposite for electrochemical sensing of the atypical neuroleptic drug olanzapine. Mater. Sci. Eng. C 33 (8), 4876–4883. Available from: https://doi.org/10.1016/j.msec.2013.08.002.

Ashrafi, A.M., et al., 2014. Electrochemical mechanism and sensitive assay of antiretroviral drug Abacavir in biological sample using multiwalled carbon nanotube modified pyrolytic graphite electrode. J. Electroanal. Chem. 712, 178–184. Available from: http://linkinghub.elsevier.com/retrieve/pii/S1572665713005092.

Baghayeri, M., et al., 2016. A novel way for detection of antiparkinsonism drug entacapone via electrodeposition of silver nanoparticles/functionalized multi-walled carbon nanotubes as an amperometric sensor. Mater. Sci. Eng. C 66, 77–83. Available from: https://doi.org/10.1016/j.msec.2016.03.077.

Baytak, A.K., et al., 2016. A sensitive determination of terbutaline in pharmaceuticals and urine samples using a composite electrode based on zirconium oxide nanoparticles. Mater. Sci. Eng. C 67, 125–131. Available from: https://doi.org/10.1016/j.msec.2016.05.008.

Bergamini, M.F., Santos, D.P., Zanoni, M.V.B., 2010. Determination of isoniazid in human urine using screen-printed carbon electrode modified with poly-L-histidine. Bioelectrochemistry (Amsterdam, Netherlands) 77 (2), 133–138.

Bhatta, D., et al., 2010. Label-free monitoring of antibody–antigen interactions using optical microchip biosensors. J. Immunol. Methods 362 (1–2), 121–126. Available from: http://linkinghub.elsevier.com/retrieve/pii/S0022175910002693.

Brahman, P.K., Dar, R.A., Pitre, K.S., 2012. Adsorptive stripping voltammetric study of vitamin B1 at multi-walled carbon nanotube paste electrode. Arab. J. Chem. 1–5. Available from: http://linkinghub.elsevier.com/retrieve/pii/S1878535212003073.

Cernat, A., et al., 2015. Electrochemical sensors based on carbon nanomaterials for acetaminophen detection: a review. Anal. Chim. Acta 886, 16–28. Available from: http://linkinghub.elsevier.com/retrieve/pii/S0003267015007941.

Chen, A., Chatterjee, S., 2013. Nanomaterials based electrochemical sensors for biomedical applications. Chem. Soc. Rev. 42 (12), 5425. Available from: http://pubs.rsc.org/en/content/articlehtml/2013/cs/c3cs35518g.

Chen, W., et al., 2012. Development and application of a validated gradient elution HPLC method for simultaneous determination of 5-fluorouracil and paclitaxel in dissolution samples of 5-fluorouracil/paclitaxel-co-eluting stents. J. Pharm. Biomed. Anal. 59 (1), 179–183.

Cheraghi, S., Taher, M.A., 2016. Fabrication of CdO/single wall carbon nanotubes modified ionic liquids carbon paste electrode as a high performance sensor in diphenhydramine analysis. J. Mol. Liq. 219, 1023–1029. Available from: https://doi.org/10.1016/j.molliq.2016.04.035.

Chu, D., et al., 2003. Sensitive liquid chromatographic assay for the simultaneous determination of 5-fluorouracil and its prodrug, tegafur, in beagle dog plasma. J. Chromatogr. B: Anal. Technol. Biomed. Life Sci. 795 (2), 377–382.

Dmitrienko, S.G., et al., 2014. Recent advances in sample preparation techniques and methods of sulfonamides detection—a review. Anal. Chim. Acta 850, 6–25. Available from: https://doi.org/10.1016/j.aca.2014.08.023.

Dogan-Topal, B., et al., 2013. Multi-walled carbon nanotube modified glassy carbon electrode as a voltammetric nanosensor for the sensitive determination of anti-viral drug valganciclovir in pharmaceuticals. Sens. Actuators B: Chem. 177, 841–847. Available from: https://doi.org/10.1016/j.snb.2012.11.111.

El-Nashar, R.M., Abdel Ghani, N.T., Hassan, S.M., 2012. Construction and performance characteristics of new ion selective electrodes based on carbon nanotubes for determination of meclofenoxate hydrochloride. Anal. Chim. Acta 730, 99–111. Available from: https://doi.org/10.1016/j.aca.2011.10.069.

Ensafi, A.A., et al., 2011. Highly sensitive voltammetric sensor based on catechol-derivative-multiwall carbon nanotubes for the catalytic determination of captopril in patient human urine samples. Colloids Surf. B: Biointerfaces 87 (2), 480–488. Available from: https://doi.org/10.1016/j.colsurfb.2011.06.013.

Farquharson, S., et al., 2005. Detection of 5-fluorouracil in saliva using surface-enhanced Raman spectroscopy. Vib. Spectrosc. 38, 79–84.

Feng, G.-H., Yang, S.-X., 2010. Quartz crystal microbalance biochip with ultrasonic standing wave enhancement. Sens. Actuators B: Chem. 150 (2), 601–608. Available from: http://linkinghub.elsevier.com/retrieve/pii/S0925400510006945.

Gao, Z.-N., et al., 2006. Electrochemical oxidation of isoniazid catalyzed by (FcM)TMA at the platinum electrode and its practical analytical application. Anal. Bioanal. Chem. 385 (7), 1324–1329.

Gopal, M.S., et al., 2016. Development of a 2-methoxyphenyl-4,5-dicarbmethoxy-cycl [3.2.2]azine fluorescen sensor for Co^{2+}. J. Lumin. 179, 361–365. Available from: http://linkinghub.elsevier.com/retrieve/pii/S0022231316300205.

Goyal, R.N., et al., 2008. Voltammetric determination of bisoprolol fumarate in pharmaceutical formulations and urine using single-wall carbon nanotubes modified glassy carbon electrode. Electrochim. Acta 53 (6), 2802–2808. Available from: http://linkinghub.elsevier.com/retrieve/pii/S0013468607013448.

Goyal, R.N., Bishnoi, S., 2009. Simultaneous voltammetric determination of prednisone and prednisolone in human body fluids. Talanta 79 (3), 768–774. Available from: http://linkinghub.elsevier.com/retrieve/pii/S0039914009003725.

Goyal, R.N., Gupta, V.K., Chatterjee, S., 2010. Voltammetric biosensors for the determination of paracetamol at carbon nanotube modified pyrolytic graphite electrode. Sens. Actuators B: Chem. 149 (1), 252–258. Available from: https://doi.org/10.1016/j.snb.2010.05.019.

Gupta, V.K., Jain, A.K., Shoora, S.K., 2013. Multiwall carbon nanotube modified glassy carbon electrode as voltammetric sensor for the simultaneous determination of ascorbic acid and caffeine. Electrochim. Acta 93, 248–253. Available from: https://doi.org/10.1016/j.electacta.2013.01.065.

Hassan, S., et al., 1998. Membrane sensors for the selective determination of fluorouracil. Anal. Chim. Acta 363 (1), 81–87.

Hofmann, M.C., et al., 2010. Development of a four electrode sensor array for impedance spectroscopy in high content screenings of fermentation processes. Sens. Actuators B: Chem. 147 (1), 93–99. Available from: http://linkinghub.elsevier.com/retrieve/pii/S0925400510002534.

Hua, X., et al., 2013. Electrochemical behavior of 5-fluorouracil on a glassy carbon electrode modified with bromothymol blue and multi-walled carbon nanotubes. Anal. Methods 5 (10), 2470.

Jain, K.K., 2009. The role of nanobiotechnology in drug discovery. Adv. Exp. Med. Biol. 655, 37–43. Available from: http://link.springer.com/10.1007/978-1-4419-1132-2_4.

Jain, R., Dhanjai, 2013. Nano graphene based sensor for antiarrhythmic agent quinidine in solubilized system. Colloids Surf. B: Biointerfaces 105, 278–283. Available from: https://doi.org/10.1016/j.colsurfb.2013.01.015.

Jain, R., Rather, J.A., 2011. Voltammetric determination of antibacterial drug gemifloxacin in solubilized systems at multi-walled carbon nanotubes modified glassy carbon electrode. Colloids Surf. B: Biointerfaces 83 (2), 340–346. Available from: https://doi.org/10.1016/j.colsurfb.2010.12.003.

Jain, R., Sharma, S., 2012. Glassy carbon electrode modified with multi-walled carbon nanotubes sensor for the quantification of antihistamine drug pheniramine in solubilized systems. J. Pharm. Anal. 2 (1), 56–61. Available from: https://doi.org/10.1016/j.jpha.2011.09.013.

Kalambate, P.K., et al., 2015. Simultaneous voltammetric determination of paracetamol and domperidone based on a graphene/platinum nanoparticles/nafion composite modified glassy carbon electrode. Sens. Actuators B: Chem. 213, 285–294. Available from: https://doi.org/10.1016/j.snb.2015.02.090.

Kalambate, P.K., Rawool, C.R., Srivastava, A.K., 2016. Voltammetric determination of pyrazinamide at graphene-zinc oxide nanocomposite modified carbon paste electrode employing differential pulse voltammetry. Sens. Actuators B: Chem. 237, 196–205. Available from: http://linkinghub.elsevier.com/retrieve/pii/S0925400516308711.

Karadas, N., Ozkan, S.A., 2014. Electrochemical preparation of sodium dodecylsulfate doped over-oxidized polypyrrole/multi-walled carbon nanotube composite on glassy carbon electrode and its application on sensitive and selective determination of anticancer drug: Pemetrexed. Talanta 119, 248–254. Available from: https://doi.org/10.1016/j.talanta.2013.10.065.

Karadas, N., et al., 2013. Analytical application of polymethylene blue-multiwalled carbon nanotubes modified glassy carbon electrode on anticancer drug irinotecan and determination of its ionization constant value. Talanta 115, 911–919. Available from: https://doi.org/10.1016/j.talanta.2013.07.006.

Keyvanfard, M., Alizad, K., 2016. Determination of isoproterenol in pharmaceutical and biological samples using a pyrogallol red multiwalled carbon nanotube paste electrode as a sensor. Chin. J. Catal. 37 (4), 579–583. Available from: http://linkinghub.elsevier.com/retrieve/pii/S1872206715610361.

Khaleghi, F., Arab, Z., et al., 2016a. Fabrication of novel electrochemical sensor for determination of vitamin C in the presence of vitamin B9 in food and pharmaceutical samples. J. Mol. Liq. 221, 666–672. Available from: http://www.sciencedirect.com/science/article/pii/S0167732216313290.

Khaleghi, F., Irai, A.E., et al., 2016b. Highly sensitive nanostructure voltammetric sensor employing Pt/CNTs and 1-butyl-3-methylimidazolium hexafluoro phosphate for determination of tryptophan in food and pharmaceutical samples. J. Mol. Liq. 223, 431–435. Available from: http://linkinghub.elsevier.com/retrieve/pii/S0167732216315446.

Kinouchi, H., et al., 2016. Surface plasmon resonance assay of inhibition by pharmaceuticals for thyroxine hormone binging to transport proteins. Anal. Biochem. 492, 43–48. Available from: http://linkinghub.elsevier.com/retrieve/pii/S0003269715004339.

Kubo, T., Otsuka, K., 2016. Recent progress for the selective pharmaceutical analyses using molecularly imprinted adsorbents and their related techniques: A review. J.

Pharm. Biomed. Anal. 130, 68–80. Available from: http://linkinghub.elsevier.com/retrieve/pii/S0731708516302837.

Kumar, N., Rosy, Goyal, R.N., 2016. Nanopalladium grained polymer nanocomposite based sensor for the sensitive determination of Melatonin. Electrochim. Acta 211, 18–26. Available from: http://linkinghub.elsevier.com/retrieve/pii/S0013468616313408.

Kutluay, A., Aslanoglu, M., 2014. An electrochemical sensor prepared by sonochemical one-pot synthesis of multi-walled carbon nanotube-supported cobalt nanoparticles for the simultaneous determination of paracetamol and dopamine. Anal. Chim. Acta 839, 59–66. Available from: https://doi.org/10.1016/j.aca.2014.05.018.

Lee, S.H., et al., 2016. Determination of acetaminophen using functional paper-based electrochemical devices. Sens. Actuators B: Chem. 232, 514–522. Available from: https://doi.org/10.1016/j.snb.2016.03.169.

Leite, F.R.F., et al., 2012. Development of a sensor for L-Dopa based on Co(DMG)2ClPy/multi-walled carbon nanotubes composite immobilized on basal plane pyrolytic graphite electrode. Bioelectrochemistry 86, 22–29. Available from: https://doi.org/10.1016/j.bioelechem.2012.01.001.

Lu, T.-L., Tsai, Y.-C., 2010. Electrocatalytic oxidation of acetylsalicylic acid at multi-walled carbon nanotube–alumina-coated silica nanocomposite modified glassy carbon electrodes. Sens. Actuators B: Chem. 148 (2), 590–594. Available from: https://doi.org/10.1016/j.snb.2010.05.037.

Lu, T.-L., Tsai, Y.-C., 2011. Sensitive electrochemical determination of acetaminophen in pharmaceutical formulations at multiwalled carbon nanotube–alumina-coated silica nanocomposite modified electrode. Sens. Actuators B: Chem. 153 (2), 439–444. Available from: https://doi.org/10.1016/j.snb.2010.11.013.

Mahmoud, A.M., et al., 2015. Carbon nanotubes versus polyaniline nanoparticles; which transducer offers more opportunities for designing a stable solid contact ion-selective electrode. J. Electroanal. Chem. 755, 122–126. Available from: https://doi.org/10.1016/j.jelechem.2015.07.045.

Majidi, M.R., Jouyban, A., Asadpour-Zeynali, K., 2006. Voltammetric behavior and determination of isoniazid in pharmaceuticals by using overoxidized polypyrrole glassy carbon modified electrode. J. Electroanal. Chem. 589 (1), 32–37.

Malode, S.J., Shetti, N.P., Nandibewoor, S.T., 2012. Voltammetric behavior of theophylline and its determination at multi-wall carbon nanotube paste electrode. Colloids Surf. B: Biointerfaces 97, 1–6. Available from: https://doi.org/10.1016/j.colsurfb.2012.04.010.

Mirčeski, V., et al., 2000. Square-wave voltammetry of 5-fluorouracil. J. Electroanal. Chem. 490 (1–2), 37–47.

Mohammadi-Behzad, L., et al., 2016. Highly sensitive voltammetric sensor based on immobilization of bisphosphoramidate-derivative and quantum dots onto multi-walled carbon nanotubes modified gold electrode for the electrocatalytic determination of olanzapine. Mater. Sci. Eng. C 60, 67–77.

Ndamanisha, J.C., Guo, L., 2012. Ordered mesoporous carbon for electrochemical sensing: a review. Anal. Chim. Acta 747, 19–28.

Nemutlu, E., et al., 2007. Validation of a rapid micellar electrokinetic capillary chromatographic method for the simultaneous determination of isoniazid and pyridoxine hydrochloride in pharmaceutical formulation. J. Chromatogr. B 854 (1–2), 35–42.

Nguyen, N., et al., 2015. Toxicity detection using lysosomal enzymes, glycoamylase and thioredoxin fused with fluorescent protein in Saccharomyces cerevisiae. J. Biotechnol. 214, 141–146. Available from: http://linkinghub.elsevier.com/retrieve/pii/S0168165615301310.

Nigović, B., Sadiković, M., Sertić, M., 2014. Multi-walled carbon nanotubes/Nafion composite film modified electrode as a sensor for simultaneous determination of ondansetron and morphine. Talanta 122, 187–194. Available from: http://linkinghub.elsevier.com/retrieve/pii/S003991401400037X.

Patil, R.H., Hegde, R.N., Nandibewoor, S.T., 2011. Electro-oxidation and determination of antihistamine drug, cetirizine dihydrochloride at glassy carbon electrode modified with multi-walled carbon nanotubes. Colloids Surf. B: Biointerfaces 83 (1), 133–138. Available from: https://doi.org/10.1016/j.colsurfb.2010.11.008.

Pisoschi, A.M., et al., 2014. Electrochemical methods for ascorbic acid determination. Electrochim. Acta 121, 443–460. Available from: https://doi.org/10.1016/j.electacta.2013.12.127.

Prakash, S., et al., 2011. Polymeric nanohybrids and functionalized carbon nanotubes as drug delivery carriers for cancer therapy. Adv. Drug Deliv. Rev. 63 (14–15), 1340–1351. Available from: https://doi.org/10.1016/j.addr.2011.06.013.

Prasad, B.B., et al., 2012. Nonhydrolytic sol–gel derived imprinted polymer–multiwalled carbon nanotubes composite fiber sensors for electrochemical sensing of uracil and 5-fluorouracil. Electrochim. Acta 71, 106–115.

Qi, H., et al., 2011. Electrogenerated chemiluminescence sensor for the determination of propranolol hydrochloride. Anal. Methods 3 (2), 446. Available from: https://doi.org/10.1039/C0AY00591F.

Raoof, J.B., Baghayeri, M., Ojani, R., 2012. A high sensitive voltammetric sensor for qualitative and quantitative determination of phenobarbital as an antiepileptic drug in presence of acetaminophen. Colloids Surf. B: Biointerfaces 95, 121–128. Available from: https://doi.org/10.1016/j.colsurfb.2012.02.038.

Rastogi, P.K., Ganesan, V., Azad, U.P., 2016. Electrochemical determination of nanomolar levels of isoniazid in pharmaceutical formulation using silver nanoparticles decorated copolymer. Electrochim. Acta 188, 818–824. Available from: https://doi.org/10.1016/j.electacta.2015.12.058.

Rattanarat, P., et al., 2016. Graphene–polyaniline modified electrochemical droplet-based microfluidic sensor for high-throughput determination of 4-aminophenol. Anal. Chim. Acta 925, 51–60. Available from: http://linkinghub.elsevier.com/retrieve/pii/S0003267016303270.

Rosy, Raj, M., Goyal, R.N., 2016. A facile method to anchor reduced graphene oxide polymer nanocomposite on the glassy carbon surface and its application in the voltammetric estimation of tryptophan in presence of 5-hydroxytryptamine. Sens. Actuators B: Chem. 233, 445–453. Available from: http://linkinghub.elsevier.com/retrieve/pii/S0925400516305792.

Sadiković, M., et al., 2014. Voltammetric determination of ropinirole in the presence of levodopa at the surface of a carbon nanotubes based electrochemical sensor in pharmaceuticals and human serum. J. Electroanal. Chem. 733, 60–68. Available from: http://linkinghub.elsevier.com/retrieve/pii/S1572665714004111.

Safavi, A., Bagheri, M., 2008. Design of an optical sensor for indirect determination of isoniazid. Spectrochim. Acta A: Mol. Biomol. Spectrosc. 70 (4), 735–739.

Safavi, A., Karimi, M., Hormozi Nezhad, M., 2004. Flow-injection determination of isoniazid using sodium dichloroisocyanurate- and trichloroisocyanuric acid-luminol chemiluminescence systems. IL Farmaco 59 (6), 481−486.

Santos Pereira, L.N., et al., 2016. Fast quantification of α-lipoic acid in biological samples and dietary supplements using batch injection analysis with amperometric detection. Talanta 154, 249−254. Available from: http://linkinghub.elsevier.com/retrieve/pii/S0039914016301734.

Satija, J., Karunakaran, B., Mukherji, S., 2014. A dendrimer matrix for performance enhancement of evanescent wave absorption-based fiber-optic biosensors. RSC Adv. 4 (31), 15841. Available from: http://xlink.rsc.org/?DOI = C4RA00198B.

Satyanarayana, M., Reddy, K.K., Gobi, K.V., 2014. Multiwall carbon nanotube ensembled biopolymer electrode for selective determination of isoniazid in vitro. Anal. Methods 6 (11), 3772. Available from: http://xlink.rsc.org/?DOI = c4ay00154k (accessed 16.11.14).

Satyanarayana, M., et al., 2015. Biopolymer stabilized nanogold particles on carbon nanotube support as sensing platform for electrochemical detection of 5-fluorouracil in vitro. Electrochimica Acta 178, 608−616. Available from: http://linkinghub.elsevier.com/retrieve/pii/S0013468615302863.

Shahrokhian, S., Amiri, M., 2007. Multi-walled carbon nanotube paste electrode for selective voltammetric detection of isoniazid. Microchim. Acta 157 (3−4), 149−158.

Shahrokhian, S., Asadian, E., 2010. Simultaneous voltammetric determination of ascorbic acid, acetaminophen and isoniazid using thionine immobilized multi-walled carbon nanotube modified carbon paste electrode. Electrochim. Acta 55 (3), 666−672. Available from: http://linkinghub.elsevier.com/retrieve/pii/S0013468609010925.

Silva, T.A., et al., 2015. Electrochemical determination of rosuvastatin calcium in pharmaceutical and human body fluid samples using a composite of vertically aligned carbon nanotubes and graphene oxide as the electrode material. Sens. Actuators B: Chem. 218, 51−59. Available from: https://doi.org/10.1016/j.snb.2015.04.085.

Song, S., et al., 2010. Functional nanoprobes for ultrasensitive detection of biomolecules. Chem. Soc. Rev. 39 (11), 4234. Available from: http://xlink.rsc.org/?DOI = c000682n.

Strehlitz, B., et al., 2012. Aptamers for pharmaceuticals and their application in environmental analytics. Bioanal. Rev. 4 (1), 1−30. Available from: http://link.springer.com/10.1007/s12566-011-0026-1.

Taei, M., et al., 2016. Simultaneous detection of morphine and codeine in urine samples of heroin addicts using multi-walled carbon nanotubes modified $SnO_2-Zn_2SnO_4$ nanocomposites paste electrode. Appl. Surf. Sci. 363, 490−498. Available from: https://doi.org/10.1016/j.apsusc.2015.12.074.

Thapliyal, N., et al., 2015. A categorical review on electroanalytical determination of non-narcotic over-the-counter abused antitussive drugs. Talanta 142, 157−163. Available from: https://doi.org/10.1016/j.talanta.2015.04.061.

Tothill, I.E., 2009. Biosensors for cancer markers diagnosis. Semin. Cell Devel. Biol. 20 (1), 55−62. Available from: http://linkinghub.elsevier.com/retrieve/pii/S1084952109000196.

Uslu, B., Ozkan, S., 2007. Electroanalytical application of carbon based electrodes to the pharmaceuticals. Anal. Lett. 40 (5), 817−853. Available from: http://www.tandfonline.com/doi/abs/10.1080/00032710701242121.

Vinoth, V., et al., 2015. Simultaneous detection of dopamine and ascorbic acid using silicate network interlinked gold nanoparticles and multi-walled carbon nanotubes. Sens. Actuators B: Chem. 210, 731–741. Available from: http://linkinghub.elsevier.com/retrieve/pii/S0925400515000568.

Walcarius, A., 2013. Mesoporous materials and electrochemistry. Chem. Soc. Rev. 42 (9), 4098–4140.

Wang, H., et al., 2014. Low-frequency shift dispersive Raman spectroscopy for the analysis of respirable dosage forms. Int. J. Pharm. 469 (1), 197–205. Available from: http://linkinghub.elsevier.com/retrieve/pii/S0378517314002944.

Wang, S., Fu, S., Ding, H., 2012. Determination of 5-fluorouracil using disposable gold nanoparticles modified screen-printed electrode. Sens. Lett. 10 (3), 974–978.

Wei, Y., et al., 2016. An ultrasensitive electrochemiluminescence sensor based on reduced graphene oxide–copper sulfide composite coupled with capillary electrophoresis for determination of amlodipine besylate in mice plasma. Biosens. Bioelectron. 86, 714–719. Available from: https://www.scopus.com/inward/record.uri?eid = 2-s2.0-84982182083&partnerID = 40&md5 = 74483111e0f5fbafcb8b69cbff63ecc5.

Yan, X., Bo, X., Guo, L., 2011. Electrochemical behaviors and determination of isoniazid at ordered mesoporous carbon modified electrode. Sens. Actuators B: Chem. 155 (2), 837–842.

Yang, L., et al., 2011. Aptamer-conjugated nanomaterials and their applications. Adv. Drug Deliv. Rev. 63 (14–15), 1361–1370. Available from: http://linkinghub.elsevier.com/retrieve/pii/S0169409X11002699.

Yu, J.-G., et al., 2014. Aqueous adsorption and removal of organic contaminants by carbon nanotubes. Sci. Tot. Environ. 482–483 (1), 241–251. Available from: https://doi.org/10.1016/j.scitotenv.2014.02.129.

Zhan, T., et al., 2011. Electrochemical behavior of 5-fluoro-1H-pyrimidine-2 on an ionic liquid modified carbon paste electrode. Anal. Methods 3 (11), 2651.

Zhao, D., Wan, Y., Zhou, W., 2013. In: Zhao, D., Wan, Y., Zhou, W. (Eds.), Ordered mesoporous materials. Wiley-VCH Verlag GmbH & Co. KGaA, Weinheim.

Zhou, Z., et al., 2009. Study of cloud point extraction and high-performance liquid chromatographic determination of isoniazid based on the formation of isonicotinylhydrazone. J. Chromatogr. A 1216 (1), 30–35.

CHAPTER

8 The evolution of dendrimers to composite dendrimers: a review of the state of the art

Maha Nasr, Enas Elmowafy and Mahmoud Eid Soliman

Department of Pharmaceutics and Industrial Pharmacy, Faculty of Pharmacy, Ain Shams University, Cairo, Egypt

CHAPTER OUTLINE

8.1 INTRODUCTION

Dendrimers, originating from the Greek word *dendron* (Medina and El-Sayed, 2009), which means "tree," belong to a class of macromolecules with highly branched monodisperse structure, which provides them with several unique traits. Being globular with voids in their structure, they can efficiently accommodate many valuable compounds in their interior (Zhou et al., 2009). Moreover, the

Nanoparticles in Pharmacotherapy. DOI: https://doi.org/10.1016/B978-0-12-816504-1.00004-1

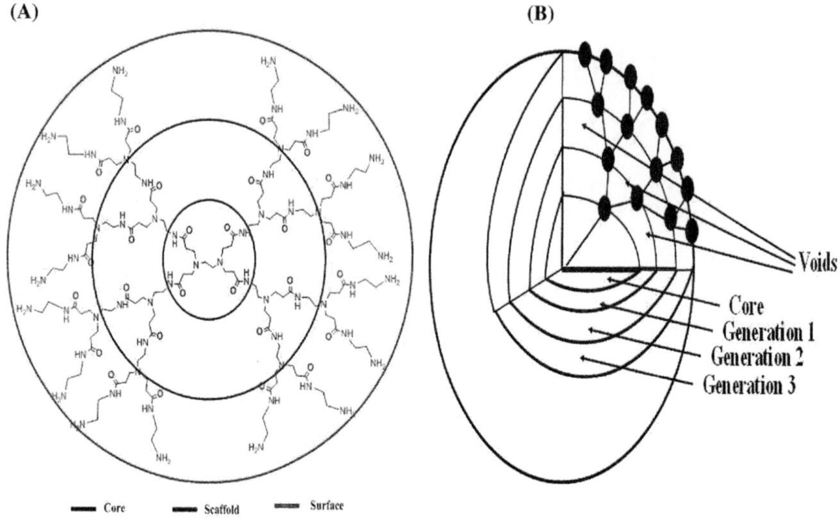

FIGURE 8.1

General structure of PAMAM dendrimer showing different parts forming the macromolecule (A), voids and generations of dendrimers (B).

peculiar geometrical architecture of these molecules leads to unique physical and chemical properties.

The branched-tripropylamine macromolecule was first synthesized by Vogtle et al. (1978), starting from a primary amine by repeated Michael addition and reduction processes to develop regular shaped hyperbranched product called POPAM. However, *dendrimer* as a word for this specific kind of macromolecules was first coined by Tomalia in 1985 who synthesized branched polyamidoamines (PAMAM) (Tomalia et al., 1985). Dendrimers are tree-like macromolecules with regular, repetitive branching, fixed molecular structure, and specific molecular weight. They consist of three parts, namely core, scaffold, and surface (Fig. 8.1). The core is centered within the macromolecule connected with a number of branches comprised of branching units (the scaffold) with terminal outer surface groups, dictating the dendrimer's generation. The structure of surface groups has a strong effect on the physicochemical properties of the dendrimer.

8.2 DENDRIMER SYNTHESIS STRATEGIES

Dendrimers are assembled in a stepwise manner using repeating synthetic steps (Abbasi et al., 2014). Each cycle creates a higher generation, however, the degree of branching is dependent on the used building blocks and their valencies.

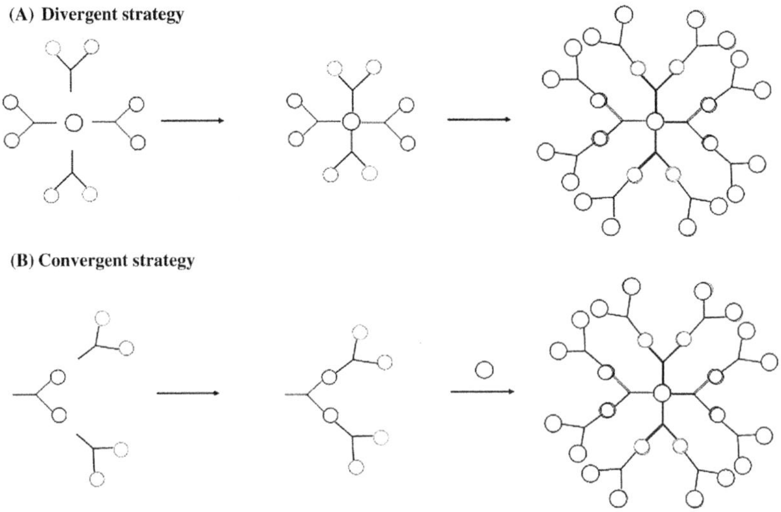

FIGURE 8.2

Techniques utilized for dendrimer synthesis: divergent strategy (A) and convergent strategy (B).

Dendrimers can be synthesized by either the divergent or convergent strategies (Fig. 8.2) (Fischer, Vogtle, 1999).

Using the divergent method, dendrimers are built inside out (from the central core out to the periphery). Repetitive reaction cycles allow the interaction of the reactive groups with equivalent monomer units to increase the dendrimer's generation, and the coupling reactions increase with each successive generation. For the convergent strategy, dendrimers are created from the outside towards the inside. The synthesis starts first by building dendronic monomer units then the monomer units are coupled to a single reactive group of a multivalent core to produce dendrimers (Aguilar, 2013).

As divergent and convergent approaches are tedious and time consuming, other synthesis methods were developed to decorate dendrimers with various functional groups in the outer surface, for example, double-stage convergent strategy, double exponential dendrimer growth (DEDG), comb-burst, and hyper branching strategies (Inoue, 2000).

8.3 PAMAM DENDRIMERS

PAMAM dendrimers are among the most commonly used dendrimers, as they are easily synthesized by two-step Michael reaction of ethylenediamine and methylacrylate using divergent strategy (Boas et al., 2006). The end group of each chain has free amine groups that interact with two methylacrylate monomers and two ethylenediamine molecules, to form new generation possessing cationic surfaces of amine groups (Klajnert and Bryszewska, 2001). Half generations for PAMAM dendrimers can be prepared to have anionic carboxylate groups on their surfaces (Malik et al., 2000). Due to the steric repulsion between terminal functionalities, PAMAM dendrimers at generations higher than 4 have defects in the surface groups (Esfand and Tomalia, 2001). Moreover, the terminal groups can be tailored with different valuable moieties, for example, alkyl chains, oligomers, polymers, enzymes, proteins, and drugs (Inoue, 2000).

The molecular mass and number of terminal groups of PAMAM dendrimers were found to exhibit exponential increase as the dendrimer generation increases (Maiti et al., 2004); however, their size was found to increase by about 1 nm for each generation (Esfand and Tomalia, 2001). There was also an accompanying nonlinear increase of the density of dendrimer surface groups with the generation (Biricova and Laznickova, 2009). The crowdedness of surface groups forces the dendrimer to acquire a spherical or globular shape with rather hollow interior. Hence, PAMAM dendrimers can accommodate ions and molecules in their interior (endoreception) or by interaction with surface groups (exoreception) (Fig. 8.3). Generation 4 PAMAM dendrimers are large and possess a hydrophilic interior and surface groups; hence, they dissolve in polar vehicles such as water, alcohols, and some other polar organic solvents (Liu et al., 2011).

The voids inside dendrimers can selectively accommodate ions and molecules based on the void size and the chemical nature of its groups (Ravoo, 2008). The accommodated species are kept inside the dendrimers by different forces including electrostatic and hydrophobic forces, formation of complexes, and van der Waals and hydrogen bonds (Jain and Tekade, 2013).

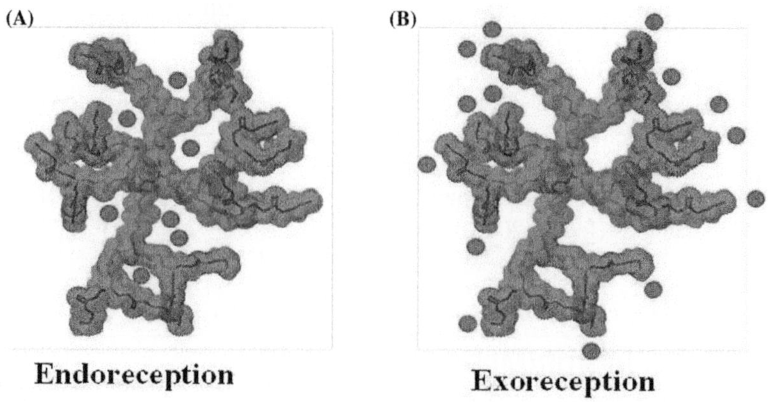

FIGURE 8.3

Accommodation of ions and molecules in PAMAM dendrimers by endoreception (A), exoreception (B).

Many applications involving drug delivery and diagnostic markers encapsulation depend on the hosting of small molecules in dendrimers. Hosts of medicinal applications were incorporated within dendrimers, for example, docetaxel and oxaliplatin for cancer therapy using EPR effect (Madaan et al., 2014), beclomethasone for asthma treatment (Nasr et al., 2014), erythromycin for sustainment of antibacterial action (Bosnjakovic et al., 2011), ketoprofen and diflunisal for improved skin permeation (Cheng et al., 2007), and brimonidine and timolol maleate for enhanced corneal uptake (Holden et al., 2012). Moreover, dendrimers can be used as gene delivery vectors (Jin et al., 2014) by forming condensed complexes that can be easily uptaken by cells (Márquez-Miranda et al., 2016). Commercially available transfection reagents, for example, Superfect and Priofect that could be used to form stable dendrimer/nucleic acid polyplexes (Jin et al., 2014) were developed, which can buffer endosomes and escape endolysosomal damage of nucleic acid by proton sponge effect (Freeman et al., 2013). The transfection and cytotoxic abilities of PAMAM dendrimers depend on their generation, that is, low generations 1–3 have lower transfection efficiency and cytotoxicity while high generations 4–8 have higher transfection efficiency and cytotoxicity (Kesharwani et al., 2015).

Diagnostic probes, for example, Rose Bengal were also efficiently incorporated in dendrimers (Madaan et al., 2014). In addition, Gadolinium (III) loaded PAMAM dendrimers were successfully used to visualize the tumor blood and lymphatic circulation (Swanson et al., 2008). Moreover, Ag, Au, Cu, Ni, Pd, Pt, and Ru were successfully bound to dendrimers (Mazzitelli and Brodbelt, 2006) for catalytic application, for example, reduction, electrocatalysis, and hydrogenation catalysis (Crooks et al., 2001).

8.4 OTHER DENDRIMER TYPES

Poly(propyleneimine) (PPI) dendrimers and poly(ethyleneimine) (PEI) dendrimers were also introduced as amine-terminated dendrimer types (Newkome and Shreiner, 2008; Yemul and Imae, 2008; Shao et al., 2011). Other monomers have been used to synthesize biocompatible dendritic and dendrigraft polymers are amino acids forming polypeptide dendrimers. Dendrimers based on L-lysine, L-ornithine, D- and L-glutamate, and L-serine were also efficiently synthesized (Mezö et al., 2000), and found to be less cytotoxic than traditional dendrimers (PAMAM and PEI) (Lee et al., 2005; Madaan et al., 2014). Glyco-, melamine, and triazine, biodegradable and silicon based dendrimers were also prepared and utilized in a number of biologically relevant applications such as sensors, as novel drug carriers, and for active targeting owing to their safety (Madaan et al., 2014; Undre et al., 2016).

8.5 DENDRIMER PROPERTIES AND CHARACTERIZATION

Dendrimers can be classified based on the flexibility of their backbones into flexible and rigid dendrimers (Fréchet, 2002). Due to their aqueous solubility and biocompatibility, flexible dendrimers, for example, PAMAM dendrimers were used in drug and gene delivery into cells (e.g., DNA and siRNA delivery) (Liu et al., 2011). However, rigid dendrimers have constant shape, with well-defined topology that allows precise arrangements of functional groups, for example, polyphenylene dendrimers (Hammer et al., 2015).

Due to their unique molecular shape, dendrimers have different physicochemical properties compared with the linear analogs (Hawker and Piotti, 2000). Linear polymers manifest themselves as flexible coils in their solution; however, dendrimers are arranged in the form of highly packed spheres. This significantly affects the rheology of their solutions (Uppuluri et al., 1998), in which linear polymers have higher viscosity than dendritic analogs (Ren et al., 2016). Dendrimers' solubility depends mainly on the nature of terminal groups on their surface, that is, dendrimers with hydrophilic groups on their surface are soluble in polar solvents, while those having hydrophobic surface groups dissolve in nonpolar solvents. Being globular in shape and with large internal cavities, dendrimers can efficiently encapsulate guest molecules, for example, drugs and diagnostic agents (Madaan et al., 2014).

Because of their complex structure and high degree of branching, it is very difficult to fully characterize dendrimers (Inoue, 2000; Lederer and Burchard, 2015). Although several characterization methodologies can be used to study dendrimer structure (such as NMR, FTIR, elemental analysis, CZE, and LALLS), none of these methods can fully describe the exact structure of the macromolecule.

The average molecular mass of dendrimers could be determined using chromatographic methods (e.g., GPC and SEC) by comparing dendrimer molecules with reference macromolecules of known molecular mass, chemical structure, and conformation. However, all of the previously mentioned methods cannot accurately determine the molecular mass of dendrimers and their capabilities to separate complex mixtures are low.

Other techniques such as electrospray (ES) mass spectrometry and MALDI are capable of measuring the absolute mass of small dendrimers (with molar masses less than 10,000) with low error and good accuracy (Van Dongen et al., 2013). Although field desorption mass spectrometry can be used for determination of dendrimers' molecular mass slightly above 10,000, plasma desorption mass spectrometry can be effectively used for dendrimer mass characterization up to the value of 50,000 (Zeng and Zimmerman, 1997). Other characterization techniques include circular dichroism for assessment of the optical activity of dendrimers, dielectric spectroscopy for assessing their molecular dynamics, differential scanning calorimetry for the assessment of dendrimers' solid state, and their morphology assessment using AFM and TEM microscopic techniques (Madaan et al., 2014).

8.6 MECHANISMS OF DENDRIMER–LIGAND INTERACTIONS

The unique dendrimer architecture offers various opportunities for interaction with guest molecules. Such interactions involve covalent coupling of drugs to dendrimers' terminal group or physical incorporation inside internal cavities induced by hydrogen bonding, hydrophobic attachment, and electrostatic interactions (Hughes, 2005; Howell and Fan, 2010). A direct relationship was proven to exist between the number of generations and the tendency for encapsulation based on previous studies. The higher the dendrimer generation, the larger the number of functional groups and hence, more space for drug loading (Singh et al., 2016).

Covalent conjugation of the guest molecules could occur directly using surface functional groups that facilitate the formation of dendrimer–drug conjugates, or indirectly through spacers and biodegradable ester or amide linkages (Gillies and Fréchet, 2005; Patri et al., 2005). The utilization of prodrug approach could lead to improved drug stability, solubility, controlled release, and targeting (D'Emanuele et al., 2004). A variety of therapeutic agents including anticancer drugs showed promising attributes upon complexation with dendrimers, for example, venlafaxine (Yang and Lopina, 2005), naproxen (Namazi et al., 2011), propranolol (D'Emanuele et al., 2004), cisplatin (Nguyen et al., 2015), doxorubicin (Papagiannaros et al., 2005), epirubicin (Madaan et al., 2014), methotrexate (Jiang et al., 2010), paclitaxel (Khandare et al., 2006), and chlorambucil (Assadi et al., 2016).

Concerning physical incorporation of drugs, inherent architecture and the presence of hydrophobic internal voids could result in enhanced loading of drugs inside dendrimers via hydrophobic interactions or hydrogen bonding (Liu and Fréchet, 1999; Medina and El-Sayed, 2009; Kesharwani et al., 2011). Unimolecular micelle characteristics of dendrimers have been reported previously, resulting in promising solubilization and targeting of different drug molecules (Fu et al., 2000). Additionally, complexation via electrostatic interactions might happen between ionizable groups of various drugs and oppositely charged multifunctional surface of dendrimers (Milhem et al., 2000; Madaan et al., 2014). Such approach occurs more rapidly without adversely affecting drug pharmacological activity, when contrasted to chemical conjugation. However, it might result in low stability, batch-to-batch variation in solubilized drug concentration, and low-drug encapsulation (Caminade and Turrin, 2014).

Molecular modeling was used to study the interaction of drugs and dendrimers in many biomedical applications (Martinho et al., 2014). It can be used to predict structural and conformational changes in case of drug–dendrimer interaction by molecular simulations (Tanis and Karatasos, 2009). Docking tools and molecular dynamics simulations could be used to understand the mechanism of drug–dendrimer complex formations as well as their binding modes (Martinho et al., 2014). Maingi et al. analyzed the release pattern of four drugs encapsulated inside the PAMAM dendrimer using molecular dynamics simulations (Maingi et al., 2012). The release for these drugs from dendrimer was found to be in the order alanine > salicylic acid and primidone > phenylbutazone, which was in accordance with the calculated force values.

Furthermore, several experiments were done for modeling dendrimer–protein interactions including dendrimer interaction with A2A adenosine receptor agonist nucleoside (CGS21680) (Ivanov and Jacobson, 2008), HAS (Giri et al., 2011), TLR4–MD2 complex (Barata et al., 2011), and human L-chain ferritin (Camarada et al., 2015).

To study the cellular uptake of dendrimers as well as their interaction with endolysosomal and nuclear membranes, molecular dynamics simulations were used to test the dendrimer–model membrane interactions (Mu et al., 2014). Dimyristoylglycero-3-phosphocholine (DMPC) supported lipid bilayers were used to simulate the interaction of PAMAM dendrimers carrying different surface charges with as model membrane (Kelly et al., 2008). Simulation studies managed to predict the possible structural deformation of charged dendrimers when approaching model membrane compared with dendrimers of neutral surface. However, this type of study has the drawbacks of failing to simulate pore formations in the tested bilayered structure with high generation dendrimers (Hong et al., 2004) and lack of solvent/counterion effects on dendrimer–bilayer interactions (Kelly et al., 2009).

8.7 EVOLUTION OF COMPOSITE DENDRIMERS FROM CONVENTIONAL DENDRIMERS

A substantial amount of research work was conducted on dendrimers, to test their applicability in biomedicine, material science, electronics, chemistry, and engineering, in which several review articles comprehensively detailed such applications (Cheng et al., 2008; Abbasi et al., 2014; Noriega-Luna et al., 2014; Undre et al., 2016). All these applications made use of the unique properties of the dendrimers; namely their highly branched multivalent nature, solubility enhancement, and toxicity reduction of drugs. According to Crooks et al. (2001), dendrimers present a very favorable platform for creation of composites with nanoparticles owing to their monodisperse reproducible structure, tendency for nanoparticles stabilization, and the tailorability of the terminal groups for controlling the solubility of the generated composite and binding to other polymers and moieties. Therefore, dendrimer nanocomposites serve as unique hybrid molecules with organic and inorganic nanoparticles with molecular mixing of minute nanoscopic domains and the polymeric dendrimers, mostly leading to more intensified desirable properties compared with the individual components, owing to the substantial interfacial interactions between them (Tang et al., 2011). Therefore, in the current chapter, a highlight would be shed on these applications in the form of a succinct overview, reporting both the biomedical and nonbiomedical applications of composite dendrimers.

8.8 COMPOSITE DENDRIMERS FOR BIOMEDICAL APPLICATIONS

8.8.1 DENDRIMER–CARBON NANOTUBES COMPOSITES

Carbon nanotubes are cylindrical nanoparticles based on graphite first discovered by Iijima in 1991, which have been successfully used in the biomedical field owing to their strong mechanical properties, nanometer size, and their surface modification potential (Iijima, 1991; Bianco et al., 2005; He et al., 2013). Owing to their unique properties, they were able to be conjugated to other delivery systems in a composite form (Spitalsky et al., 2010).

In the work reported by Vasumathi et al. (2013), the researchers reported the preparation of composite dendrimers constructed as a complex of single walled carbon nanotubes with protonated/nonprotonated PAMAM dendrimers of generations 2, 3, and 4, as well as with polypropyl-ether imine PETIM dendrimers. Their microscopic picture generated using all atom molecular dynamics simulations is illustrated in Fig. 8.4. The complexation process was found to depend on the type of dendrimers, in which PAMAM dendrimers showed much stronger noncovalent van der Waals charge transfer interaction with carbon nanotubes than PETIM dendrimers; what they referred to as "conformational wrapping" of the dendrimers on the nanotubes. The efficiency of complex formation was also reported to be dependent on the dendrimers' generation, in which the larger the dendrimers' generation, the greater was the complexation potential, in addition to the dendrimers' protonation, being stronger with nonprotonated dendrimers. Furthermore, PAMAM dendrimer–nanotube complexes were proven to have better stability compared with their PETIM complex counterparts. Possible applications of such a composite system are the enhancement of organic solvent solubility and dispersion of carbon nanotubes from their bundle geometry (Pramanik and Maiti, 2016), hence, overcoming the drawbacks of the conventional solubility enhancement methods of carbon nanotubes, and achieving better delivery efficiency for drugs and DNA and siRNA with lesser toxicity. Apart from biomedical applications, they can also be used as efficient photoactive surfaces in photovoltaics.

Another interesting composite system based on multiwalled carbon nanotubes, dendrimers, and hydroxyapatite was prepared by Murugan and Arumgam (2014) for bone tissue engineering purposes. They can either be employed in the prevention or treatment of bone diseases as bone cancer. This work was actually a

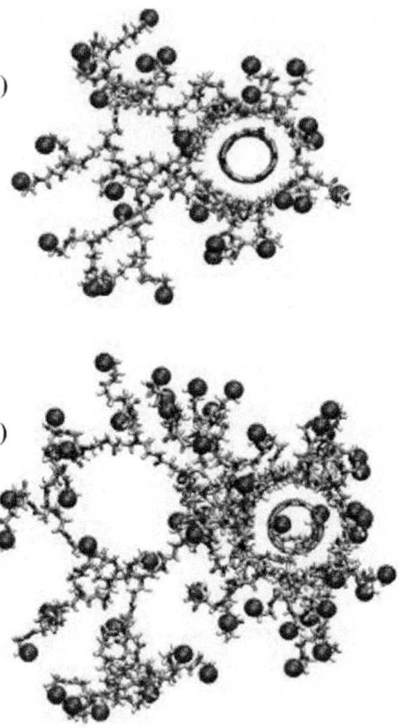

FIGURE 8.4

Representation of nanotube–dendrimer complex formation encountered with protonated dendrimers of generation 3 (A) and generation 4 (B).

Adapted from Vasumathi, V., Pramanik, D., Sood, A.K., Maiti, P.K., 2013. Structure of a carbon nanotube–dendrimer composite. Soft matter 9, 1372–1380 with permission of the Royal Society of Chemistry. https://doi.org/10.1039/C2SM26804C.

continuation of the research reported by Donners et al. (2003), who synthesized hydroxyapatite/polypropylene dendrimer composites, aiming at exhibiting plastic deformation within bones while maintaining considerable stiffness. Briefly, hydroxyapatite nanocomposites were synthesized by simultaneous controlled speed addition of calcium carbonate and potassium dihydrogen phosphate to a dispersion of dendrimer and surfactant template. The prepared biomaterials were characterized for their formation, morphology, and mechanical properties. FTIR study confirmed the formation of the hydroxyapatite in the nanocomposites. TEM images showed the presence of micrometer-sized small clusters. Following sonication for better visualization, needle-like crystals of high charge density were apparent. Promising mechanical properties were also displayed, suggesting their outstanding stiffness and ultimate strength. Meanwhile, in the work reported by Murugan and Arumgam (2014), the role of hydroxyapatite in the composite system was to confer biocompatibility by mimicking the extracellular matrix of the bone and stimulate osteoblast proliferation, leading to promotion of new bone formation. The role of carbon nanotubes was to confer rigidity and mechanical strength to the composite, and the presence of dendrimers would allow functionalization of the carbon nanotubes' surface, and hence enhance their water solubility. Two types of nanocomposites, generation 2 and 3 multiwalled carbon nanotubes/PPI dendrimer/hydroxyapatite scaffolds, were prepared. The composite systems were characterized for their composition using FTIR and energy dispersive spectroscopy, X-ray diffraction, and photoelectron spectroscopy and field emission scanning electron microscopy. Upon examining the anticancer activity of the nanocomposites on human osteosarcoma cell line MG-63 using the MTT and optical microscopic studies, it was concluded that both nanocomposites exhibited anticancer potential on the aforementioned cell line, with that prepared using generation 3 dendrimer being superior to that prepared using generation 2 dendrimers, which was ascribed by the authors to the higher amount of hydroxyapatite in the former composite compared with the latter.

A double composite system was reported by Herrero et al. (2010) in which dendrimer-encapsulated gold nanoparticles were attached to multiwalled carbon nanotubes while preserving the structure of both systems, which interestingly opens a wide range of possibilities for introducing new functionalities to carbon nanotubes to explore several medical applications such as drug and gene delivery, as well as image contrast.

A review article by Sun et al. (2011) describes possible methods of functionalization of carbon nanotubes with dendrimers, taking advantages of the high solubility of the dendrimers and the low viscosity of the produced nanohybrids composed of the carbon nanotubes and the dendrimers, leading to ease of manipulation and processing. Possible means of functionalization can either be classified into covalent or noncovalent modifications. Covalent modifications occur by either Π–Π interactions or other physical associations such as physical entanglements or van der Waals forces, while noncovalent modifications occur by grafting techniques.

8.8.2 DENDRIMER ALGINATE AND DENDRIMER POLYLACTIDE-*CO*-GLYCOLIDE COMPOSITES

Alginate and polylactide-*co*-glycolide (PLGA) are very commonly used polymers in the biomedical field (Nasr et al., 2011a,b, 2013; Elmowafy et al, 2014; Said-Elbahr et al., 2016). Both polymers were reported to be favorably engaged in the formation of composites owing to their biodegradability, biocompatibility, and surface modification potential (Venkatasubbu et al., 2011; Lee and Mooney, 2012; Jain and Datta, 2014; Deng et al., 2016).

Shahbazi et al. (2014) reported the preparation of various generations of PAMAM and alginate–PAMAM dendrimer nanocomposites for antimicrobial actions. The antibacterial potential of the prepared nanocomposites was assessed by calculating the minimum inhibitory/bacterial concentrations. Their mutagenic properties were also investigated using Ames assay. The results confirmed the broad antibacterial activity of different dilutions of PAMAM dendrimers

(G2–G5) and alginate-based nanocomposites on five bacterial species. Ames test revealed that the reverted colonies' numbers in case of the alginate PAMAM nanocomposite G2 were lower than encountered with the parent dendrimer. A direct relationship between the minimum inhibitory concentration and minimum bacterial concentration on one side and the reverted colonies on the other side with generation number was found to exist, delineating the nanocomposite alginate G2 PAMAM to have better and safer antibacterial properties in comparison with other PAMAM generations.

In the work done by Sherwani et al. (2015), the researchers hypothesized that the combination of two delivery systems (PAMAM dendrimers and PLGA nanoparticles) in one nanocomposite system will allow some sort of cooperative inhibition of cancer, by coloading an anticancer drug (cisplatin) and siRNA. Therefore, they prepared a novel antibody decorated dendrimer–PLGA nanocomposite (what they referred to as immunonanocomposite), and tested its efficacy in treatment of hepatocellular carcinoma. Each component of the nanocomposite system had a specific function; the antibody conjugation was responsible for targeting the composite system to the cancer cells, the PLGA nanoparticles would serve as carrier for cisplatin and a scaffold for the dendrimers allowing them to retain their three-dimensional shape, while the dendrimers themselves would serve as carrier for the siRNA allowing its subtle sensitization of the cancerous cells to the toxic action of the coadministered cisplatin. A schematic representation of how this composite system is synthesized is illustrated in Fig. 8.5. The composite system was characterized for its drug entrapment efficiency, morphology using SEM, and assessed for dendrimer content using ninhydrin assay. Upon testing the efficacy of this system in hepatocellular carcinoma bearing mice, it was found that the aforementioned system successfully knocked down the target therapeutic gene with higher apoptotic induction compared with the other control groups leading to significant recovery, maintenance of the liver architecture, and enhanced survival. Furthermore, the system didn't show any systemic, renal, or hepatic toxicity. The novelty of the system lies in making use of the biocompatibility of PLGA, which ameliorates the toxicity of the cationic dendrimers, while preserving the proton sponge effect of the regular PAMAM dendrimers attributed to their cationic nature (Freeman et al., 2013; Kesharwani et al., 2015), consequently leading to the intracellular cystolic delivery of the loaded siRNA cargo while escaping endosomes. PLGA nanoparticles further serve as a carrier of chemotherapeutic agents, which opens vast possibilities for this polyfunctional nanocomposite system in several delivery purposes that require coencapsulation and exhibition of similar pharmacokinetics for different cargos (Fig. 8.6).

Another report on the use of composite dendrimers for nonviral gene delivery was made by Zhang et al. (2007), in which they conjugated dendrimers to PLGA microparticles that were encapsulated with plasmid DNA, and contrasted the properties of the composite system to that of the conventional PLGA microparticles. The composite systems were characterized for their particle size and surface charge using zetasizer and for morphology using SEM. The composite system

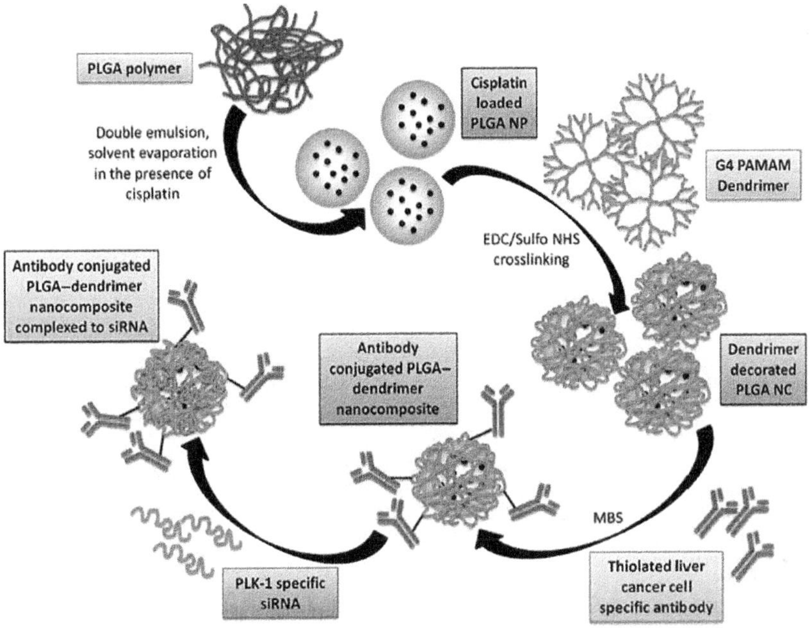

FIGURE 8.5

A schematic diagram for the formulation of immunodendrimer PLGA nanocomposites.

Adapted from Sherwani, M.S., Tufail, S., Khan, A.A., Owais, M., 2015. Dendrimer–PLGA based multifunctional immuno-nanocomposite mediated synchronous and tumor selective delivery of siRNA and cisplatin: potential in treatment of hepatocellular carcinoma. RSC Adv. 5, 39512–39531 with permission of the Royal Society of Chemistry. https://doi.org/10.1039/C5RA03651H.

was proven to significantly enhance the efficiency of transfection on human embryonic kidney cells (HEK293) and monkey African green kidney fibroblast like cell line (COS7) compared with the unconjugated PLGA microparticles, in addition to demonstrating about sevenfold higher uptake by the former cell line. Interestingly, the composite system was able to abolish the toxicity of dendrimers, in which the authors reported that the cytotoxicity of the composite system was comparable to the conventional PLGA microparticles. This work represents another successful attempt for conjugating dendrimers with a PLGA based formulation, imparting interesting properties to the whole system.

Another interesting composite system was highlighted by Ramalingam et al. (2016), in which the researchers immobilized a complex of pDNA codifying bone morphogenetic protein BMP-2 with generation 5 PAMAM dendrimer within a composite PLGA electrospun fiber made by layer-by-layer deposition of chitosan and alginate nanolayers. The composite system was intended to serve as a biodegradable and biocompatible scaffold for enhancement of cellular differentiation towards osteoblastic lineage. The composite system was tested for its in vitro

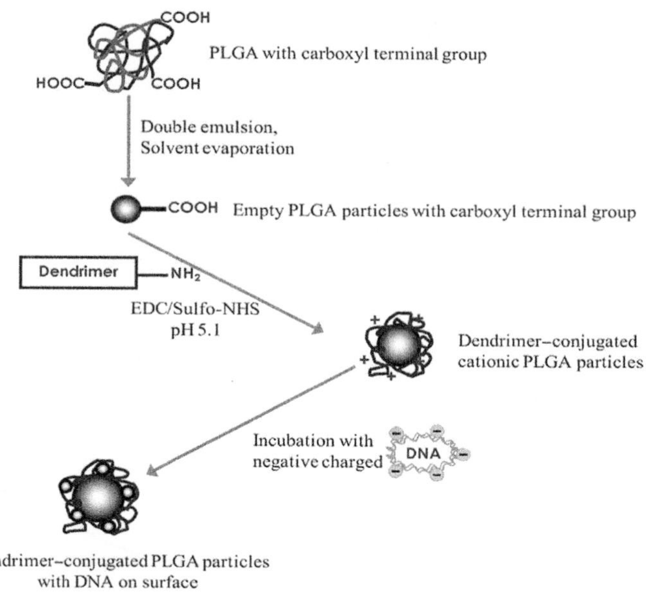

FIGURE 8.6

Synthesis of PAMAM dendrimer conjugated PLGA microparticles composites.

Adapted with permission from Zhang, X.Q., Intra, J., Salem, A.K., 2007. Conjugation of polyamidoamine dendrimers on biodegradable microparticles for nonviral gene delivery. Bioconjug. Chem. 18, 2068–2076. Copyright (2007) American Chemical Society.

release of the pDNA/dendrimer complex, and for its in vitro gene transfection efficiency. The composite system displayed a sustained release behavior coupled with high transfection efficiency, providing a very promising system for tissue engineering purposes to facilitate the attachment and proliferation of human mesenchymal stem cells.

In another work by Galan et al. (2016), PLGA nanoparticles were functionalized with anionic carbosilane dendrons, which are dendrimers with a chemically accessible core, for treatment of HIV infection. The composite system was characterized using thermogravimetric analysis, dynamic light scattering, morphology using TEM, surface charge measurement, and FTIR. The efficacy of the composite system was assessed in peripheral mononuclear blood cells infected with HIV-1, which was shown to exhibit superior antiviral efficacy compared with mere dendrons or mere nanoparticles, which was ascribed by the authors to the enhanced interaction of the composite system with either virus receptors or nanoparticles receptors. Furthermore, the system was proven biocompatible in PBMC and HEC-1A cells as displayed by MTT assay. The described system presents an interesting delivery system to be loaded with antiviral medications to exhibit a combined action.

8.8.3 DENDRIMER HYDROGEL COMPOSITES

Owing to their cell and drug carrying potential, and their ability to confer desirable properties to composite systems, hydrogels were successfully utilized in the biomedical field (Hoffman, 2002; Thoniyot et al., 2015). In a work described by Yang et al. (2012), they first converted PAMAM dendrimers into a hydrogel form by photocuring using polyethylene glycol PEG moieties. This hydrogel served as a platform for inclusion of PLGA nanoparticles, creating a safe composite system for the ocular coadministration of the two drugs brimonidine and timolol maleate for management of glaucoma. A schematic diagram of the proposed system is provided in Fig. 8.7. The idea behind developing such a system came from a previous report by Vandamme and Brobeck (2005) that regular PAMAM dendrimers possessed bioadhesive properties leading to strong interaction with the corneal surface. The described composite system in the work by Yang et al. (2012) was able to exhibit sustained presence in both aqueous humor and cornea up to one week as evident from ocular tissue extraction samples, with 2.2- to 2.6-folds

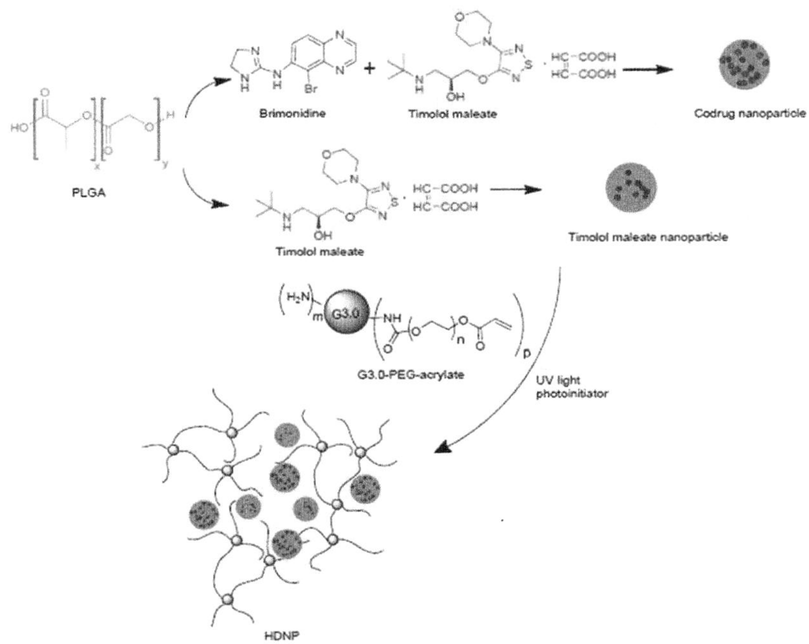

FIGURE 8.7

PAMAM dendrimers hydrogel–PLGA nanoparticles composite.

Adapted with permission from Yang, H., Tyagi, P., Kadam, R.S., Holden, C.A., Kompella, U.B., 2012. Hybrid dendrimer hydrogel/PLGA nanoparticle platform sustains drug delivery for one week and antiglaucoma effects for four days following one time topical administration. ACS Nano 6, 7595–7606. Copyright (2012) American Chemical Society.

higher uptake of drugs from the composite system compared with drugs loaded in mere nanoparticles, in addition to reduction of intraocular pressure up to 4 days compared with 2 days encountered with mere hydrogel and mere nanoparticles. This proved the superiority of the composite system over the dendrimer hydrogel or the PLGA nanoparticles loaded with the drugs administered in noncomposite form. Moreover, the system displayed negligible toxicity to human corneal epithelial cells. This application opens an opportunity for enhancement of patient compliance in glaucoma therapy, and can be applied in other chronic ocular diseases as well owing to its amphiphilic nature, which can allow the encapsulation of hydrophobic drugs within the dendrimers' interior and hydrophilic drugs within the hydrophilic PEG network, with better control of their release properties and better cellular uptake should the drugs be entrapped compositely within PLGA nanoparticles first.

PAMAM dendrimer composites with hydrogel domains were also found promising for bone tissue engineering when combined with 3,4-dihydroxy-L-phenylalanine leading to the formation of 3D bioinspired porous hydrogel with good mechanical strength and biocompatibility, as reported by Wang et al. (2015a), which were successfully used for cranial defect repair. Another composite for treatment of dentin hypersensitivity was prepared by Wang et al. (2015b), in which they treated dentin discs using generation 3 PAMAM dendrimers to induce in situ mineralization inside dentinal tubules. PAMAM dendrimers were able to attract phosphates, leading to effective mineralization and occlusion of dentinal tubules, serving as promising treatment modality for dentin hypersensitivity.

Another work involving the formation of dendrimer–hydrogel constructs was reported by Kabanov et al. (2005). Cationic PPI dendrimers of different generations were sorbed by anionic hydrogels of poly(sodium acrylate) and poly (sodium-2-acrylamido-2-methylpropane sulfonate) via a diffusional coupling reaction by ion pairing of the cationic dendrimer with the anionic hydrogel units. The complex formed was referred to as a "core–shell" construct composed of an outer weakly swollen complex shell and an inner strongly swollen hydrogel core. The final structure of the construct formed was found to be influenced by pH and sorption extent. Though the construct was not developed for drug delivery purposes, it would be a very promising carrier for drugs.

Another similar work was reported by Kojima et al. (2011), in which dendrimers were linked to collagen, creating what they referred to as "collagen mimetic dendrimers." In this work, dendrimers served as a scaffold for the collagen peptides owing to their multifunctionality and their ability to enhance the triple helix formation at their surface. The composite system was produced by acetylation of the peptides followed by condensation with PAMAM dendrimer of generation four, to demonstrate a collagen-like triple helix formation. The obtained system was characterized using circular dichroism, differential scanning calorimetry, transmittance, and TEM imaging. A hydrogel was successfully produced from this composite system exhibiting thermoreversibility, which may successfully serve as an injectable delivery system or a functional biomaterial. A schematic

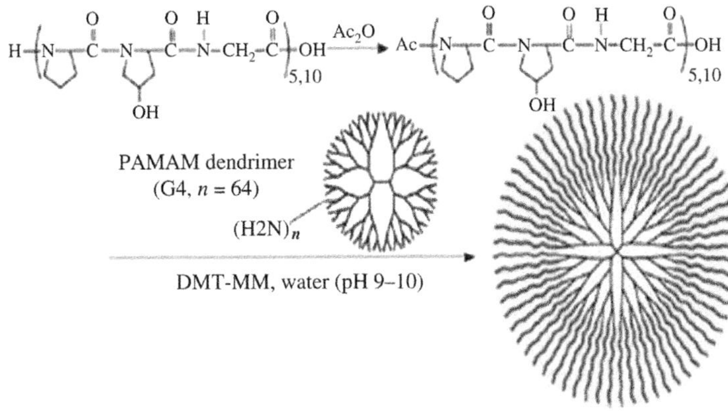

FIGURE 8.8

Fabrication of collagen–mimetic dendrimers.

Adapted from Kojima, C., Suchiro, T., Tada, T., Sakamoto, Y., Waku, T., Tanaka, N., 2011. Preparation of heat-induced artificial collagen gels based on collagen–mimetic dendrimers. Soft Matter. 7, 8991–8997 with permission of the Royal Society of Chemistry. https://doi.org/10.1039/C1SM06157G.

representation of how collagen–mimetic dendrimers are formed is displayed in Fig. 8.8.

Dendritic scaffolds were also utilized as crosslinkers for preparation of PEG-based hydrogels by Hed et al. (2013), to be employed for bone adhesion in fracture incidents. This opens perspectives for implantation in bone fixation applications in fractured or osteoporotic patients. Dendrimers were also employed as an initiator core for synthesizing star-shaped polypeptide based hydrogels by Shen et al. (2015). The authors confirmed the formation of hydrogels of these star-shaped polypeptides in water at low concentration instead of micelles. Shear thinning and rapid recovery properties of the hydrogels were revealed, suggesting their injectability and applicability for controlled peptide release systems. In addition, the hydrogel networks also exhibited entangled and branched fibrils in TEM and AFM images, displaying their three-dimensional hydrogel networks.

8.8.4 DENDRIMER GOLD, SILVER, AND MAGNETIC NANOPARTICLES COMPOSITES

Metallic nanoparticles namely gold, silver, and magnetic nanoparticles have gained interest in the biomedical field over the previous years, owing to their huge functionalization potential with different antibodies, ligands, and drugs, hence, leading to effective drug targeting and diagnostic imaging applications (Mody et al., 2010; Conde et al., 2012).

In the work published by Pan et al. (2007), magnetic nanoparticles were conjugated to PAMAM dendrimers of different generations, and loaded with antisense surviving oligodeoxynucleotide to be tested for application in cancer therapeutics. The composite system was characterized for size using high resolution TEM, and its structure was verified using FTIR and TGA analyses. Furthermore, the composite system was tested for its transfection efficiency, cellular uptake, and biological effect on breast and liver cancer cell lines, and visualization was performed using confocal laser microscopy. Results revealed that the presence of dendrimers allowed the gene to be surface attached to the magnetic nanoparticles, consequently leading to enhancement of gene delivery efficiency as especially encountered with generation 5 dendrimer composites, in addition to increasing the solubility of magnetic nanoparticles. This nominates this composite system as a transcellular membrane gene and drug delivery carrier. Significant downregulation of the survivin gene/protein, and inhibition of cell growth were also reported.

A system involving dendrimers and gold nanoparticles in the form of composite film was reported by Park et al. (2011). The purpose of such system was to create a platform that allows both localization and controlled release of drugs or dyes. The authors created composite films of PAMAM dendrimers and gold nanoparticles using dithiocarbamate chemistry for tunable drug/dye release, as illustrated in Fig. 8.9. The moieties tested for their release behaviors were fluorescent

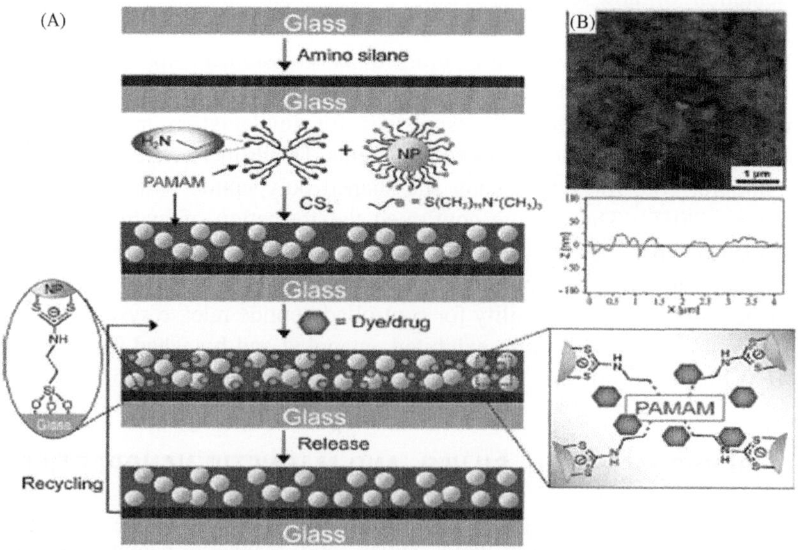

FIGURE 8.9

Fabrication of dendrimer–gold nanoparticle composite film using dithiocarbamate cross-linking chemistry (A) and its validation by atomic force microscopy (B).

Adapted with permission from Park, M.H., Agasti, S.S., Creran, B., Kim, C., Rotello, V.M., 2011. Controlled and sustained release of drugs from dendrimer–nanoparticle composite films. Adv. Mater. 23, 2839–2842.

dyes namely rhodamine, fluorescein, calcein, and Nile red, and the model drug tested was doxorubicin. The composite system was able to modulate the release profile of the dyes and the drug up to 12 or 24 h, with the release profile found to depend on the dendrimers' generation as well as the size of the gold nanoparticles. The proposed scaffold system was reported by the authors to be nontoxic, reusable, and biocompatible, which can be loaded by a wide variety of guest molecules.

Another composite system based on dendrimers and gold nanoparticles was reported by Serafim et al. (2013), in which acid-terminated PAMAM dendrimers were used to form a self-assembled crystalline composite with gold nanoparticles to serve as nanostructured coatings. Characterization of the nanocomposite system was done by UV-spectroscopy and assessment of morphology using both SEM and TEM. The authors reported modulation of the nanocomposite properties such as morphology and assembling propensity by varying the dendrimers' generation and concentration. Although this system was not developed for drug delivery purposes, it still seems promising for pharmaceutical purposes.

Another conjugation attempt between polyphenylene dendrimers and gold nanoparticles was reported by Taubert et al. (2003), leading to the formation of porous nanocomposites, which also seem promising for drug delivery applications. Validation of nanocomposite formation was performed using powder X-ray diffraction, high resolution TEM, and UV–visible spectroscopy.

A composite film formed by self-assembling of sulfur containing dendron-grafted phenylene ethynylenes with gold nanoparticles was also reported by Tozawa (2004), exhibiting blue photoluminescence, leading to the creation of dendrimer–nanoparticles composite with photoelectronic properties. Another composite film prepared by layer-by-layer self-assembly of gold nanoparticles with PAMAM and PPI dendrimers was also reported for sensing applications was also reported by Krasteva et al. (2002) and Joseph et al. (2004). Similarly, a three-dimensional network nanocomposite made of first generation PAMAM dendrimers and gold nanoparticles was made by Luo et al. (2011) for biosensing as well.

Gold nanorods were also utilized in fabrication of composite systems with dendrimers, in which Kim et al. (2016) described the synthesis of gold nanorod dendrimer composite that was sensitive to NIR light, leading to pulstatile release of loaded cargo via localized heat generation, which stimulates their release. The idea was based on a previous work done by the same group in the paper published by Park et al. (2008). The gold nanorod/dendrimer composite was fabricated in the form of a porous membrane fixed to an amine functionalized aluminum oxide membrane as seen in Fig. 8.10. The aforementioned system would be very useful as an implantable device, in which controlled release of drugs can be easily achieved leading to improvement of therapeutic efficacy and reduction of side effects.

Additionally, Kasturirangan et al. (2013) designed multifunctional composite nanodevices for cancer imaging and therapy, which was based on poly

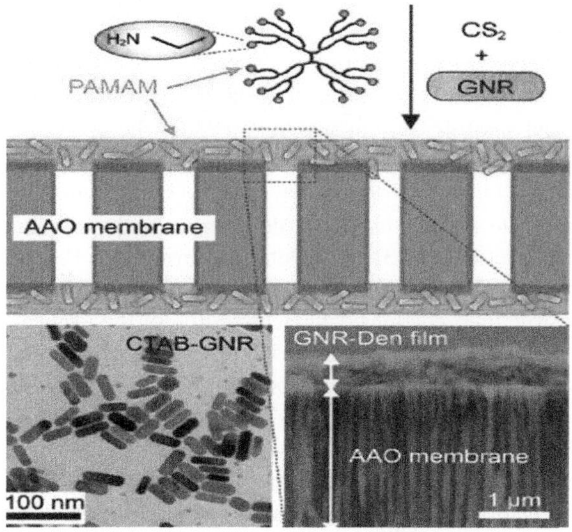

FIGURE 8.10

Fabrication of gold nanorods/dendrimer composite membrane.

Abstracted from Kim, K., Jo, M.-C., Jeong, S., Palanikumar, L., Rotello, V.M., Ryu, J.-H., Park, M.-W., 2016. Externally controlled drug release using a gold nanorod contained composite membrane. Nanoscale 8, 11949–11955 with permission of the Royal Society of Chemistry. https://doi.org/10.1039/C6NR00362A.

(amidoamine) dendrimer as organic component and gold as inorganic component. In this study, the biological safety, organ toxicity, and compatibility were assessed on the nanodevices of different surface charges in mice. Results showed that 5-nm dendrimer nanodevices were produced and significant changes in the zeta potential of dendrimer were displayed upon gold incorporation, diminishing the toxicity of the latter. More interestingly, the biosafety of the dendrimer nanodevices was confirmed with no signs of organ injury or decrease in functionality. Thus, incorporation of inorganic gold to dendrimers significantly changed the surface charge, thus conferring in vivo detoxification.

Another example of a composite system made of dendrimers with metal nanoparticles was reported by Namazkar et al. (2014), in which PAMAM dendrimers of different generations were combined with silver nanoparticles by chemical reaction, and they were examined for their antimicrobial purposes via their ability to release silver ions within bacterial cells. The composite system was formed by dendrimers acting as ligands for the metal ions, leading to preorganization of the silver ions and their homogenous distribution within the dendrimer, with the reduction of the ions into nanoparticulate form. The formation of dendrimer-stabilized nanoparticles was verified using UV–visible spectroscopy, and was found to be dependent on the structure of the dendrimers. Upon testing their efficacy on four bacterial strains, the dendrimer–silver nanoparticles composite exhibited the highest inhibition of bacterial growth at 600 ppm, and these were also found to be stable for 1 year's time. It was hypothesized by the authors that the whole composite system was adsorbed to the bacterial cell wall, after which it penetrates it via the association of the positive charges on the dendrimers with the negative charges of the bacterial constituents, followed by release of silver nanoparticles and ions within the cells causing inhibition of replication and cellular death. This presents an interesting opportunity for utilization of dendrimers as antiseptic material for biomedical applications.

A possible shape evolution of the silver nanoparticles from spherical to cubic shaped within the dendrimer composites as evidenced by TEM imaging was achieved with high generation dendrimers of peripheral hydroxyl sites and stabilizing core, as reported by Vassilieff et al. (2008).

Based on previously conducted studies delineating the promising anticancer effect of drug-loaded PAMAM dendrimers, the idea of exploring the anticancer effect of composite dendrimers in conjunction with silver nanoparticles emerged. When silver nanoparticles were stabilized with PAMAM dendrimers of generation 5, promising synergistic anticancer activity was obtained against lung and breast cancer cells. In the work reported by Matai et al. (2015), generation 5 PAMAM dendrimer-stabilized silver nanoparticles were prepared in situ and coloaded with fluorouracil as anticancer drug, to test the efficacy of the aforementioned composite system as drug carrier. The composite system was characterized for its fluorouracil entrapment efficiency, morphology/size distribution using TEM, in vitro drug release, zeta potential using the zetasizer device and ^{1}HNMR spectroscopy. Moreover, elemental mapping was performed using field emission scanning electron microscopy. Upon testing the cancer cell viability using MTT assay, it was concluded that the composite system displayed superior cytotoxic potential on the cells compared with either drug alone or drug loaded in dendrimers or drug loaded in silver nanoparticles, suggesting their exhibition of synergistic rather than additive effect attributed to their apoptotic potential caused by caspase signaling and ROS generation. This study has further proven the versatility of dendrimers as a dual functional delivery system; being able to act as a stabilizer for another nanoparticulate system (the silver nanoparticles), and at the same time to act as carrier for anticancer cargos, with enhancement of therapeutic efficacy.

Dendrimers in a composite form with silver nanoparticles were also reported to amplify the detection signal for proteins compared with mere silver particles, as reported by Pei et al. (2013), leading to increased sensitivity of the electrochemical immunoassay, and opening new opportunities for biological sensing of low abundance proteins and biomolecules. Similarly, when dendrimers were conjugated in composite form with gold nanoparticles by Dharuman et al. (2012) they were able to effectively anchor DNA molecules for development of affinity biosensors.

If neutralization of the surface positive charges of the dendrimer/gold or silver nanoparticles was needed for further functionalization for biomedical purposes such as targeted cancer therapeutics, acetylation reaction could be conducted, as reported by Shi et al. (2008). This allows the composite to avoid nonspecific bindings to negatively charged cell membranes and avoid toxicities by enhancing biocompatibility. Another composite based on dendrimers and gold nanoparticles was fabricated by Shi et al. (2009) to target and image cancer cells. Generation 5 PAMAM dendrimers were first conjugated with folic acid, then loaded with fluorescein FITC, then complexed with gold ions, leading to spontaneous formation of gold nanoparticles. Introduction of acetyl groups to the amine groups was conducted for the above-mentioned purposes. Characterization of the complex was conducted using UV–visible spectrometry, ^{1}HNMR, and TEM. Results revealed that this composite system was a successful platform for targeting cancer cells, which opens vast possibilities in the area of cancer diagnosis and therapy.

Acetylated dendrimer-stabilized silver nanoparticles were prepared by Liu et al. (2010) to serve as contrast agent for X-ray imaging applications, and characterized using ^{1}HNMR and UV–visible spectroscopy, surface charge, and morphology using TEM. The obtained X-ray attenuation efficiency of the prepared composite systems was similar to the iodine based contrast agents while exhibiting good stability in several media of different pH and temperatures and much slower clearance from the blood owing to their nanometer size, thus allowing longer imaging times. A reported targeted dendrimer metal nanoparticles composite was prepared by the same group, in which Liu et al. (2013) used generation five PAMAM dendrimers as stabilizing agents for reduction of gold and silver salts into nanoparticles; what they referred to in their paper as "dendrimer stabilized gold–silver nanowires" or "nanoprobes." The PAMAM dendrimers were folic acid terminated to achieve cancer cell targeting for computed tomography imaging purposes. Introducing acetyl groups instead of the terminal amine groups of the dendrimers was also performed to neutralize the surface, as demonstrated in the schematic representation in Fig. 8.11. The system was characterized using

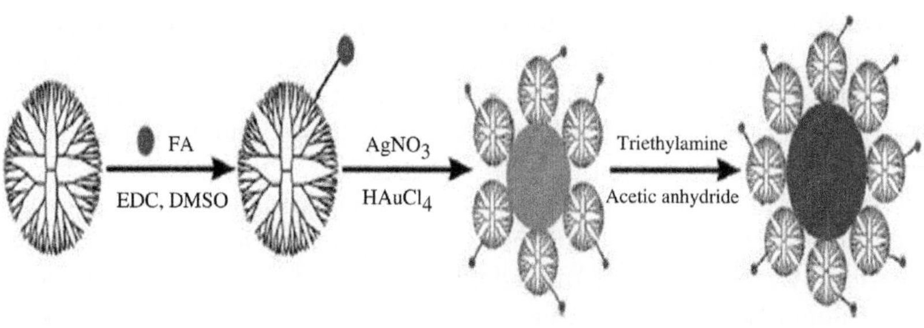

FIGURE 8.11

Synthesis of targeted dendrimer–gold/silver nanoprobes.

Adapted from Liu, H., Shen, M., Zhao, J., Zhu, J., Xiao, T., Cao, X., Zhang, G., Shi, X., 2013. Facile formation of folic acid-modified dendrimer-stabilized gold–silver alloy nanoparticles for potential cellular computed tomography imaging applications. Analyst 138, 1979–1987 with permission of the Royal Society of Chemistry. https://doi.org/10.1039/C3AN36649A.

^{1}HNMR spectroscopy, TEM imaging, zeta potential, and stability using UV-spectroscopy. Their cytocompatibility was verified on KB cells, and their targeting potential was proven on high folate receptor containing KB cells by assessment of the extent of cellular uptake.

A mixed dendrimer/gold/silver nanocomposite was described by Liu et al. (2012), in which dendrimer entrapped or stabilized gold/silver alloy nanoparticles were prepared for biomedical applications as contrast agents, taking advantage of the special optical and electronic properties of metallic nanoparticles. The composite system was synthesized using sodium borohydride reduction reaction, followed by acetylation of amine terminals, and was characterized using ^{1}HNMR spectroscopy, surface charge, and morphology by TEM. Its stability was also assessed by UV−vis spectroscopy, and its cytotoxicity was evaluated using MTT assay. The X-ray attenuation properties were verified using CT scan imaging. Results revealed that the produced composite was stable in different pH conditions. It also displayed satisfactory X-ray attenuation intensity, which was augmented by its gold content and its acetylated nature. Furthermore, acetylation served to improve the cytocompatibility of the composite, hence, delineating it as a subtle system for CT imaging purposes.

8.8.5 DENDRIMER SILICA NANOPARTICLES COMPOSITES

Silica nanoparticles were also extensively utilized in the biomedical field, due to their porosity, the wide possibility for functionalization, and their ease of production (Bitar et al., 2012; Argyo et al., 2014).

A multicomposite system involving dendrimers and silica particles was reported by Hameau et al. (2013), in which phosphorus containing dendrimers were grafted onto silica particles by covalent bonding, further functionalized with PEG chains, then loaded with silver and silver oxide nanoparticles, creating a novel functional nanocomposite system that was preliminary assessed for its antibacterial activity. A stepwise synthetic approach of this system is shown in Fig. 8.12. The aforementioned system was found to be effective against Gram positive and negative bacteria at minute concentrations 50−500 ppm, which opens exciting perspectives for applications with other colloidal delivery systems and nanomaterials.

8.8.6 MISCELLANEOUS DENDRIMER COMPOSITES

Another reported nanocomposite system was fabricated from generation 4 PAMAM dendrimer and activated carboxymethyl starch for parenteral administration by Saboktakin et al. (2008), to be explored for controlled drug delivery purposes. The nanocomposite system was loaded with a model drug 5-aminosalicylic acid, and was characterized for morphology using SEM, and for structure using ^{1}HNMR and FTIR, and finally for release kinetics, providing a promising nanocomposite for controlled drug release at inflammation sites.

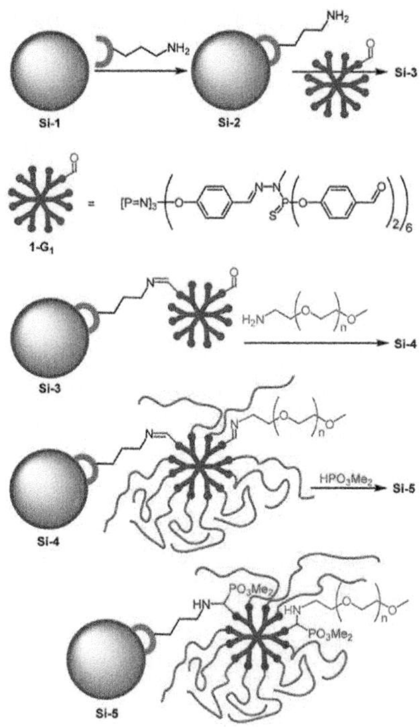

FIGURE 8.12

Step-by-step synthesis of dendrimers grafted on silica nanoparticles, to be loaded with silver based nanoparticles.

Abstracted from Hameau, A., Colliere, V., Grimoud, J., Fau, P., Roques, C., Caminade, A.-M., Turrin, C.-O., 2013. PPH dendrimers grafted on silica nanoparticles: surface chemistry, characterization, silver colloids hosting and antibacterial activity. RSC Adv. 3, 19015–19026 with permission of the Royal Society of Chemistry. https://doi.org/10.1039/C3RA43348J.

Elucidation of potential of dendrimers as nonviral gene delivery vectors has been studied by Bayele et al. (2004), in a nontraditional composite with lipidic amino acids. Reasons behind utilization of dendrimers as nonviral vectors stemmed from their structural versatility and possibility of changing properties. Bayele et al. (2004) synthesized bifunctional dendrimers with lipidic amino acid cores, in which their bifunctionality was attributed to the cationic lysine available for association with nucleic acids and the neutral lipidic molecules. The authors investigated the relationship between their structure and their functions by performing mobility shift assay, cytotoxicity, oligonucleotide delivery, and in vitro transcription and RNA transfection. They suggested that these vectors possessed unexpected pleitropy of functions such as plasmid and oligonucleotide delivery. Findings also showed a modulated luciferase activity by RNA interference and its suppression of luciferase expression. As a macromolecular carrier, this type of denrimer–lipidic composite could offer a change of loading different oligonucleotides cargo.

8.9 COMPOSITE DENDRIMERS FOR NONBIOMEDICAL APPLICATIONS

Nanocomposite dendrimers were also utilized for energy production through fuel cells. Alvarez et al. (2014) described the successful synthesis and characterization of composite Nafion and PAMAM (G4OH) dendrimer membrane, achieved by incorporating PAMAM dendrimers into the polymeric Nafion matrix, which acted as barrier against methanol access, allowing this composite membrane to be used in fuel cells field allowing good proton conductivity.

Considering catalysis applicability, Crooks et al. (2001) reported the synthesis and characterization of hybrid dendrimer-encapsulated metal nanoparticles for application in catalytic reactions. Another dendrimer composite application was reported by Krasteva et al. (2002), who prepared films containing gold nanoparticles and different types of dendrimers to serve as vapor-sensitive thin-film resistors. Gold nanoparticles provided the electricity conducting film material, whereas dendrimers made the cross-linking with the nanoparticles and provided areas for selective sorption of analytes. The dendrimers were also proven to control the film characteristics and it was found that the lower the hydrophobicity of the dendrimers, the lower their sensitivity to toluene vapor and the higher their response to propanol and water vapor. Similarly, Joseph et al. (2004) presented and characterized sensitive vapor-sensing assembled films containing gold (Au) nanoparticle and dodecylamine-stabilized gold nanoparticles with PPI dendrimers (G1–G5) or hexadecanedithiol (HDT) as linker compounds (AuPPI films and the AuHDT film). The results showed that the size of the particles within the film material was similar to those of the dodecylamine-stabilized nanoparticles as shown by TEM and field emission scanning electron microscopy images.

In another work, Taubert et al. (2003) synthesized porous gold nanoparticle–dendrimer composites. Owing to the wide size distribution of the nanoparticles and the large number of thiomethyl groups available for cross-linking, they can be used in the field of sensing due to their easy synthesis and processing, and their strong plasmon band at 560 nm.

Yi et al. (2004) reported the fabrication of polystyrene lattices based on PAMAM dendrimers using emulsion polymerization technique. Lattices were evaluated by their size and morphology using TEM, which displayed the formation of mostly single particles. These lattices were reported by the authors to be beneficial for the coating, rubber, lubricant, food, and pharmaceutical industries.

8.10 CONCLUSIONS AND PERSPECTIVES

As observed from the previous studies, the formation of dendrimer composites with organic and nonorganic molecules was possible, owing to the advancement in synthesis techniques. The applications were numerous, in both the biomedical field (drug delivery, tissue engineering, diagnosis), and the nonbiomedical field. Owing to the success of what we would refer to as "first generation dendrimer composites" described in the current chapter, a second generation would soon follow, which might involve the incorporation several targeting moieties with simultaneous loading of drugs or drugs and genes, as started by Sherwani et al. (2015), which would definitely lead to further evolutions.

REFERENCES

Abbasi, E., et al., 2014. Dendrimers: synthesis, applications, and properties. Nanoscale Res. Lett. 9, 247.

Aguilar, Z.P., 2013. Types of nanomaterials and corresponding methods of synthesis. In: Aguilar, Z.P. (Ed.), Nanomaterials for Medical Applications. Elsevier, pp. 33–82.

Alvarez, A., et al., 2014. Composite membranes based on Nafion and PAMAM dendrimers for PEMFC applications. Int. J. Hydrogen Energy 39, 16686–16693.

Argyo, C., et al., 2014. Multifunctional mesoporous silica nanoparticles as a universal platform for drug delivery. Chem. Mater. 26, 435–451.

Assadi, A., et al., 2016. Novel chlorambucil-conjugated anionic linear-globular PEG-based second-generation dendrimer: in vitro/in vivo improved anticancer activity. OncoTargets Ther. 9, 5531–5543.

Barata, T.S., et al., 2011. Partially glycosylated dendrimers block MD-2 and prevent TLR4–MD–2-LPS complex mediated cytokine responses. PLoS Comput. Biol. 7, e1002095.

Bayele, H.K., et al., 2004. Verstaile peptide dendrimers for nucleic acid delivery. J. Pharm. Sci. 94, 446–457.

Bianco, A., Kostarelos, K., Prato, M., 2005. Applications of carbon nanotubes in drug delivery. Curr. Opin. Chem. Biol. 9, 674–679.

Biricova, V., Laznickova, A., 2009. Dendrimers: analytical characterization and applications. Bioorg. Chem. 37, 185–192.

Bitar, A., 2012. Silica-based nanoparticles for biomedical applications. Drug Discov. Today 17, 1147–1154.

Boas, U., Christensen, J.B., Heegaard, P.M.H., 2006. Dendrimers: design, synthesis and chemical properties. In: Boas, U., Christensen, J.B., Heegaard, P.M.H. (Eds.), Dendrimers in Medicine and Biotechnology: New Molecular Tools. RSC, pp. 3785–3798.

Bosnjakovic, A., et al., 2011. Poly(amidoamine) dendrimer–erythromycin conjugates for drug delivery to macrophages involved in periprosthetic inflammation. Nanomedicine 7, 284–294.

Camarada, M., et al., 2015. PAMAM G4 dendrimers as inhibitors of the iron storage properties of human L-chain ferritin. Phys. Chem. Chem. Phys. 17, 19001–19011.

Caminade, A.-M., Turrin, C.-O., 2014. Dendrimers for drug delivery. J. Mater. Chem. B 2, 4055–4066.

Cheng, Y., Man, N., Xu, T., 2007. Transdermal delivery of nonsteroidal anti-inflammatory drugs mediated by polyamidoamine (PAMAM) dendrimers. J. Pharm. Sci. 96, 595–602.

Cheng, Y., et al., 2008. Dendrimers as drug carriers: applications in different routes of drug administration. J. Pharm. Sci. 97, 123–143.

Conde, J., Doria, G., Baptista, P., 2012. Noble metal nanoparticles applications in cancer. J. Drug Deliv. 2012, 1–12.

Crooks, R.M., et al., 2001. Dendrimer-encapsulated metal nanoparticles: synthesis, characterization, and applications to catalysis. Acc. Chem. Res. 34, 181–190.

D'Emanuele, A., et al., 2004. The use of a dendrimer–propranolol prodrug to bypass efflux transporters and enhance oral bioavailability. J. Control. Release 95, 447–453.

Deng, W., et al., 2016. PLGA nanocomposites loaded with verteporfin and gold nanoparticles for enhanced photodynamic therapy of cancer cells. RSC Adv. 6, 112393–112402.

Dharuman, V., et al., 2012. Graphene core low generation PAMAM dendrimers and gold nano particle composites for the label free electrochemical DNA sensors. J. Biosens. Bioelectron. 2, 137.

Donners, J.J.J.M., Nolte, R.J.M., Sommerdijk, N.A.J.M., 2003. Dendrimer-based hydroxyapatite composite with remarkable materials properties. Adv. Mater. 15, 313–316.

Elmowafy, E., et al., 2014. Nanocomplexes of an insulinotropic drug: optimization, microparticle formation and antidiabetic activity in rats. Int. J. Nanomed. 9, 4449–4465.

Esfand, R., Tomalia, D.A., 2001. Poly(amidoamine) (PAMAM) dendrimers: from biomimicry to drug delivery and biomedical applications. Drug Discov. Today 6, 427–436.

Fischer, M., Vögtle, F., 1999. Dendrimers: from design to application—a progress report. Angew. Chem. Int. Ed. 38, 884–905.

Fréchet, J.M.J., 2002. Dendrimers and supramolecular chemistry. Proc. Natl. Acad. Sci. USA 99, 4782–4787.

Freeman, E.C., Weiland, L.M., Meng, W.S., 2013. Modeling the proton sponge hypothesis: examining proton sponge effectiveness for enhancing intracellular gene delivery through multiscale modeling. J. Biomater. Sci. Polym. Ed. 24, 398–416.

Fu, K., Kitaygorodskiy, A., Sun, Y.-P., 2000. Fullerene-centered macromolecules as unimolecular micellar structures. Chem. Mater. 12, 2073–2075.

Galan, M., et al., 2016. Dendronized PLGA nanoparticles with anionic carbosilane dendrons as antiviral agents against HIV infection. RSC Adv. 6, 73817–73826.

Gillies, E.R., Fréchet, J.M.J., 2005. Dendrimers and dendritic polymers in drug delivery. Drug Discov. Today 10, 35–43.

Giri, J., et al., 2011. Interactions of poly(amidoamine) dendrimers with human serum albumin: binding constants and mechanisms. ACS Nano 5, 3456–3468.

Hameau, A., et al., 2013. PPH dendrimers grafted on silica nanoparticles: surface chemistry, characterization, silver colloids hosting and antibacterial activity. RSC Adv. 3, 19015–19026.

Hammer, B.A.G., et al., 2015. The polar side of polyphenylene dendrimers. Chem. Soc. Rev. 44, 4072–4090.

Hawker, C.J., Piotti, M., 2000. Dendritic macromolecules: hype or unique specialty materials. In: Havelka, K.O., McCormik, C.L. (Eds.), Specialty Monomers and Polymers. American Chemical Society, pp. 107–118.

He, H., et al., 2013. Carbon nanotubes: applications in pharmacy and medicine. Biomed. Res. Int. 2013, 1–12.

Hed, Y., et al., 2013. Multipurpose heterofunctional dendritic scaffolds as crosslinkers towards functional soft hydrogels and implant adhesives in bone fracture applications. J. Mater. Chem. 8, 6015–6019.

Herrero, M.A., et al., 2010. Gold dendrimer encapsulated nanoparticles as labeling agents for multiwalled carbon nanotubes. ACS Nano 4, 905–912.

Hoffman, A.S., 2002. Hydrogels for biomedical applications. Adv. Drug Deliv. Rev. 54, 3–12.

Holden, C.A., et al., 2012. Polyamidoamine dendrimer hydrogel for enhanced delivery of antiglaucoma drugs. Nanomedicine 8, 776–783.

Hong, S., et al., 2004. Interaction of poly(amidoamine) dendrimers with supported lipid bilayers and cells: hole formation and the relation to transport. Bioconjug. Chem. 15, 774–782.

Howell, B.A., Fan, D., 2010. Poly(amidoamine) dendrimer-supported organoplatinum antitumour agents. Proc. Roy. Soc. A: Math. Phys. Eng. Sci. 466, 1515–1526.

Hughes, G.A., 2005. Nanostructure-mediated drug delivery. Nanomed. Nanotechnol. Biol. Med. 1, 22–30.

Iijima, S., 1991. Helical microtubules of graphitic carbon. Nature 354, 56–58.

Inoue, K., 2000. Functional dendrimers, hyperbranched and star polymers. Prog. Polym. Sci. 25, 453–571.

Ivanov, A.A., Jacobson, K.A., 2008. Molecular modeling of a PAMAM–CGS21680 dendrimer bound to an A(2A) adenosine receptor homodimer. Bioorg. Med. Chem. Lett. 18, 4312–4315.

Jain, N.K., Tekade, R.K., 2013. Dendrimers for enhanced drug solubilization. In: Douroumis, D., Fahr, A. (Eds.), Drug Delivery Strategies for Poorly Water-Soluble Drugs. John Wiley & Sons Ltd, Oxford, UK, pp. 373–410.

Jain, S., Datta, M., 2014. Montmorillonite–PLGA nanocomposites as an oral extended drug delivery vehicle for venlafaxine hydrochloride. Appl. Clay Sci. 99, 42–47.

Jiang, Y.-Y., et al., 2010. PEGylated PAMAM dendrimers as a potential drug delivery carrier: in vitro and in vivo comparative evaluation of covalently conjugated drug and noncovalent drug inclusion complex. J. Drug Target. 18, 389–403.

Jin, L., et al., 2014. Current progress in gene delivery technology based on chemical methods and nano-carriers. Theranostics 4, 240–255.

Joseph, Y., et al., 2004. Gold–nanoparticle/organic linker films: self-assembly, electronic and structural characterisation, composition and vapour sensitivity. Faraday Discuss. 125, 77–97.

Kabanov, V.A., et al., 2005. Self-organization of cationic dendrimers in polyanionic hydrogels. Faraday Discuss. 128, 341–354.

Kasturirangan, V., Nair, B.M., Kariapper, M.T., 2013. In vivo toxicity evaluation of gold–dendrimer composite nanodevices with different surface charges. Nanotoxicology 74, 441–451.

Kelly, et al., 2008. Poly(amidoamine) dendrimers on lipid bilayers I: free energy and conformation of binding. J. Phys. Chem. B 112, 9337–9345.

Kelly, et al., 2009. Stoichiometry and structure of poly(amidoamine) dendrimer–lipid complexes. ACS Nano 3, 1886–1896.

Kesharwani, P., et al., 2011. Cancer targeting potential of some ligand-anchored poly(propylene imine) dendrimers: a comparison. Nanomed. Nanotechnol. Biol. Med. 7, 295–304.

Kesharwani, P., et al., 2015. PAMAM dendrimers as promising nanocarriers for RNAi therapeutics. Mater. Today 18, 565–572.

Khandare, J.J., et al., 2006. Dendrimer versus linear conjugate: influence of polymeric architecture on the delivery and anticancer effect of paclitaxel.. Bioconjug. Chem. 17, 1464–1472.

Kim, K., et al., 2016. Externally controlled drug release using a gold nanorod contained composite membrane. Nanoscale 8, 11949–11955.

Klajnert, B., Bryszewska, M., 2001. Dendrimers: properties and applications. Acta Biochim. Polonica 48, 199–208.

Kojima, C., et al., 2011. Preparation of heat-induced artificial collagen gels based on collagen–mimetic dendrimers. Soft Matter. 7, 8991–8997.

Krasteva, N., et al., 2002. Self-assembled gold nanoparticle/dendrimer composite films for vapor sensing applications. Nano Lett. 2, 551–555.

Lederer, A., Burchard, W., 2015. Separation and determination of molar mass. In: Tang, B.-Z. (Ed.), Hyperbranched Polymers: Macromolecules in Between Deterministic Linear Chains and Dendrimer Structures. RCS, pp. 69–87.

Lee, K.Y., Mooney, D.J., 2012. Alginate: properties and biomedical applications. Prog. Polym. Sci. 37, 106–126.

Lee, C.C., et al., 2005. Designing dendrimers for biological applications. Nat. Biotech. 23, 1517–1526.

Liu, M., Fréchet, J.M.J., 1999. Designing dendrimers for drug delivery. Pharm. Sci. Technol. Today 2, 393–401.

Liu, H., et al., 2010. Size-controlled synthesis of dendrimer-stabilized silver nanoparticles for X-ray computed tomography imaging applications. Polym. Chem. 1, 1677–1683.

Liu, X., et al., 2011. Structurally flexible triethanolamine core PAMAM dendrimers are effective nanovectors for DNA transfection in vitro and in vivo to the mouse thymus. Bioconjug. Chem. 22, 2461–2473.

Liu, H., et al., 2012. Tunable synthesis and acetylation of dendrimer-entrapped or dendrimer-stabilized gold–silver alloy nanoparticles. Colloids Surf. B: Biointerfaces 94, 58–67.

Liu, H., et al., 2013. Facile formation of folic acid-modified dendrimer-stabilized gold–silver alloy nanoparticles for potential cellular computed tomography imaging applications. Analyst 138, 1979–1987.

Luo, J., et al., 2011. Three-dimensional network polyamidoamine dendrimer–Au nanocomposite for the construction of a mediator-free horseradish peroxidase biosensor. Analyst 136, 4500–4506.

Madaan, K., et al., 2014. Dendrimers in drug delivery and targeting: drug-dendrimer interactions and toxicity issues. J. Pharm. Bioallied Sci. 6, 139–150.

Maingi, V., Kumar, M.V.S., Maiti, P.K., 2012. PAMAM dendrimer–drug interactions: effect of pH on the binding and release pattern. J. Phys. Chem. B116, 4370–4376.

Maiti, P.K., et al., 2004. Structure of PAMAM dendrimers: generations 1 through 11. Macromolecules 37, 6236–6254.

Malik, N., et al., 2000. Dendrimers: relationship between structure and biocompatibility in vitro, and preliminary studies on the biodistribution of 125I-labelled polyamidoamine dendrimers in vivo. J. Control. Release 65, 133–148.

Márquez-Miranda, V., et al., 2016. Effect of terminal groups of dendrimers in the complexation with antisense oligonucleotides and cell uptake. Nanoscale Res. Lett. 11, 66.
Martinho, N., et al., 2014. Molecular modeling to study dendrimers for biomedical applications. Molecules 19, 20424.
Matai, I., Sachdev, A., Gopinath, P., 2015. Multicomponent 5-fluorouracil loaded PAMAM stabilized-silver nanocomposites synergistically induce apoptosis in human cancer cells. Biomater. Sci. 3, 457–468.
Mazzitelli, C.L., Brodbelt, J.S., 2006. Investigation of silver binding to polyamidoamine (PAMAM) dendrimers by ESI tandem mass spectrometry. J. Am. Soc. Mass Spectrom. 17, 676–684.
Medina, S.H., El-Sayed, M.E.H., 2009. Dendrimers as carriers for delivery of chemotherapeutic agents. Chem. Rev. 109, 3141–3157.
Mezö, G., et al., 2000. Synthesis and conformational studies of poly(L-lysine) based branched polypeptides with Ser and Glu/Leu in the side chains. J. Control. Release 63, 81–95.
Milhem, O.M., et al., 2000. Polyamidoamine starburst® dendrimers as solubility enhancers. Int. J. Pharm. 197, 239–241.
Mody, V.V., et al., 2010. Introduction of metallic nanoparticles. J. Pharm. Bioallied Sci. 2, 282–289.
Mu, Q., et al., 2014. Chemical basis of interactions between engineered nanoparticles and biological systems. Chem. Rev. 114, 7740–7781.
Murugan, E., Arumugam, S., 2014. New dendrimer functionalized multi-walled carbon nanotube hybrids for bone tissue engineering. RSC Adv. 4, 35428–35441.
Namazi, H., Motamedi, S., Namvari, M., 2011. Synthesis of new functionalized citric acid-based dendrimers as nanocarrier agents for drug delivery. BioImpacts 1, 63–69.
Namazkar, S., Endud, S., Asadollahi, Z., 2014. Synthesis of poly(amidoamine)–dendrimer–silver nanoparticles composite for application as bactericides. Biointerf. Res. Appl. Chem. 4, 704–711.
Nasr, M., et al., 2011a. Different modalities of NaCl osmogen in biodegradable microspheres for bone deposition of risedronate sodium. Eur. J. Pharm. Biopharm. 79, 601–611.
Nasr, M., et al., 2011b. A reliable predictive factorial model for entrapment optimization of a sodium bisphosphonates into biodegradable microspheres. J. Pharm. Sci. 100, 612–621.
Nasr, M., et al., 2013. Hydrophilic versus hydrophobic porogens for engineering of poly (lactide-co-glycolide) microparticles containing risedronate sodium. Pharm. Dev. Technol. 18, 1078–1088.
Nasr, M., et al., 2014. PAMAM dendrimers as aerosol drug nanocarriers for pulmonary delivery via nebulization. Int. J. Pharm. 461, 242–250.
Newkome, G.R., Shreiner, C.D., 2008. Poly(amidoamine), polypropylenimine, and related dendrimers and dendrons possessing different 1→2 branching motifs: an overview of the divergent procedures. Polymer 49, 1–173.
Nguyen, H., et al., 2015. Improved method for preparing cisplatin–dendrimer nanocomplex and its behavior against NCI-H460 lung cancer cell. J. Nanosci. Nanotechnol. 15, 4106–4110.
Noriega-Luna, B., et al., 2014. Applications of dendrimers in drug delivery agents, diagnosis, therapy, and detection. J. Nanomater. 2014, 1–19.

Pan, B., et al., 2007. Dendrimer-modified magnetic nanoparticles enhance efficiency of gene delivery system. Cancer Res. 67, 8156–8163.

Papagiannaros, A., et al., 2005. Doxorubicin–PAMAM dendrimer complex attached to liposomes: cytotoxic studies against human cancer cell lines. Int. J. Pharm. 302, 29–38.

Park, M.-H., et al., 2008. Nanoparticle immobilization on surfaces via activatable heterobifunctional dithiocarbamate bond formation. Adv. Mater. 20, 4185–4188.

Park, M.H., et al., 2011. Controlled and sustained release of drugs from dendrimer–nanoparticle composite films. Adv. Mater. 23, 2839–2842.

Patri, A.K., Kukowska-Latallo, J.F., Baker, J.R., 2005. Targeted drug delivery with dendrimers: comparison of the release kinetics of covalently conjugated drug and noncovalent drug inclusion complex. Adv. Drug Deliv. Rev. 57, 2203–2214.

Pei, X., et al., 2013. Electroactive dendrimer-encapsulated silver nanoparticles for sensing low-abundance proteins with signal amplification. Anal. Methods 5, 3235–3241.

Pramanik, D., Maiti, P.K., 2016. Dendrimer assisted dispersion of carbon nanotubes: a molecular dynamics study. Soft Matter. 12, 8512–8520.

Ramalingam, K., et al., 2016. Gene delivery using dendrimer/pDNA complexes immobilized in electrospun fibers using the layer-by-layer technique. RSC Adv. 6, 97116–97128.

Ravoo, B.J., 2008. Nanofabrication with metal containing dendrimers. Dalton Trans. 12, 1533–1537.

Ren, Y., et al., 2016. Relationships between architectures and properties of highly branched polymers: the cases of amorphous poly(trimethylene carbonate) and crystalline poly (ε-caprolactone). J. Phys. Chem. B 120, 4078–4090.

Saboktakin, M.R., Maharramov, A., Ramazanov, M.A., 2008. Poly(amidoamine) PAMAM/CMS dendritic nanocomposite for controlled drug delivery. J. Am. Sci. 4, 48–52.

Said-Elbahr, R., et al., 2016. Nebulizable colloidal nanoparticles co-encapsulating a COX-2 inhibitor and a herbal compound for treatment of lung cancer. Eur. J. Pharm. Biopharm. 103, 1–12.

Serafim, A., et al., 2013. Self-assembled gold–dendrimer composite nanoparticles as surface nanostructuring features. Dig. J. Nanomater. Biostruct. 8, 613–620.

Shahbazi, B., et al., 2014. Mutagenicity survey and antibacterial activity of various generations of poly(amid amine) (PAMAM) dendrimers and algonac acid poly(amid amine) (PAMAM) dendrimer nano composite G2 on some enteric pathogenic bacteria. Int. J. Enteric Pathog. 2, e200012.

Shao, N., et al., 2011. Comparison of generation 3 polyamidoamine dendrimer and generation 4 polypropylenimine dendrimer on drug loading, complex structure, release behavior, and cytotoxicity. Int. J. Nanomed. 6, 3361–3372.

Shen, Y., et al., 2015. Hydrogels assembled from star-shaped polypeptides with a dendrimer as the core. Soft Matter. 11, 2945–2951.

Sherwani, M.S., et al., 2015. Dendrimer–PLGA based multifunctional immunonanocomposite mediated synchronous and tumor selective delivery of siRNA and cisplatin: potential in treatment of hepatocellular carcinoma. RSC Adv. 5, 39512–39531.

Shi, X., Lee, I., Baker, J.R., 2008. Acetylation of dendrimer-entrapped gold and silver nanoparticles. J. Mater. Chem. 18, 586–593.

Shi, X., et al., 2009. Targeting and detecting cancer cells using spontaneously formed multifunctional dendrimer-stabilized gold nanoparticles. Analyst 134, 1373–1379.

Singh, J., et al., 2016. Dendrimers in anticancer drug delivery: mechanism of interaction of drug and dendrimers. Artif. Cells Nanomed. Biotechnol. 44, 1626–1634.

Spitalsky, Z., et al., 2010. Carbon nanotube–polymer composites: chemistry, processing, mechanical and electrical properties. Prog. Polym. Sci. 35, 357–401.

Sun, J.-T., Hong, C.-Y., Pan, C.-Y., 2011. Surface modification of carbon nanotubes with dendrimers or hyperbranched polymers. Polym. Chem. 2, 998–1007.

Swanson, S.D., et al., 2008. Targeted gadolinium-loaded dendrimer nanoparticles for tumor-specific magnetic resonance contrast enhancement. Int. J. Nanomed. 3, 201–210.

Tang, et al., 2011. Metallodendrimers and dendrimer nanocomposites. Curr. Pharm. Des. 17, 2308–2330.

Tanis, I., Karatasos, K., 2009. Association of a weakly acidic anti-inflammatory drug (ibuprofen) with a poly(amidoamine) dendrimer as studied by molecular dynamics simulations. J. Phys. Chem. B 113, 10984–10993.

Taubert, A., Wiesler, U.-M., Mullen, K., 2003. Dendrimer-controlled one-pot synthesis of gold nanoparticles with a bimodal size distribution and their self-assembly in the solid state. J. Mater. Chem. 13, 1090–1093.

Thoniyot, P., et al., 2015. Nanoparticle–hydrogel composites: concept, design and applications of these promising, multifunctional materials. Adv. Sci. 2, 1–13.

Tomalia, D.A., et al., 1985. A new class of polymers: starburst–dendritic macromolecules. Polym. J. 17, 117–132.

Tozawa, T., 2004. Dendron-grafted sulfur-terminated phenyleneethynylene molecular rods and blue luminescence self-assembly with Au nanoparticles. Chem. Commun. (Camb) 17, 1904–1905.

Undre, S.B., et al., 2016. Dendrimers as smart materials for developing the various applications in the field of biomedical sciences. Adv. Mater. Lett. 7, 502–516.

Uppuluri, S., et al., 1998. Rheology of dendrimers. I. Newtonian flow behavior of medium and highly concentrated solutions of polyamidoamine (PAMAM) dendrimers in ethylenediamine (EDA) solvent. Macromolecules 31, 4498–4510.

Vandamme, T.F., Brobeck, L., 2005. Poly(amidoamine) dendrimers as ophthalmic vehicles for ocular delivery of pilocarpine nitrate and tropicamide. J. Control. Release 102, 23–38.

Van Dongen, M.A., et al., 2013. Quantitative analysis of generation and branch defects in G5 poly(amidoamine) dendrimer. Polymer 54, 4126–4133.

Vassilieff, T., Sutton, A., Kakkar, A.K., 2008. Shape control in silver metal nanoparticle construction using dumb-bell dendrimers. J. Mater. Chem. 18, 4031–4033.

Vasumathi, V., et al., 2013. Structure of a carbon nanotube–dendrimer composite. Soft Matter. 9, 1372–1380.

Venkatasubbu, G.D., et al., 2011. Hydroxyapatite–alginate nanocomposite as drug delivery matrix for sustained release of ciprofloxacin. J. Biomed. Nanotechnol. 7, 759–767.

Vogtle, F., Buhleier, E.W., Wehner, W., 1978. Cascade and nonskid-chain-like syntheses of molecular cavity topologies. Synthesis 2, 155–158.

Wang, Y., et al., 2015a. A high stiffness bio-inspired hydrogel from the combination of a poly(amido amine) dendrimer with DOPA. Chem. Commun. 51, 16786–16789.

Wang, T., et al., 2015b. Use of poly(amidoamine) dendrimer for dentinal tubule occlusion: a preliminary study. PLoS ONE 10, e0124735.

Yang, H., Lopina, S.T., 2005. Extended release of a novel antidepressant, venlafaxine, based on anionic polyamidoamine dendrimers and poly(ethylene glycol)-containing semi-interpenetrating networks. J. Biomed. Mater. Res. 72A, 107–114.

Yang, H., et al., 2012. Hybrid dendrimer hydrogel/PLGA nanoparticle platform sustains drug delivery for one week and antiglaucoma effects for four days following one time topical administration. ACS Nano 6, 7595–7606.

Yemul, O., Imae, T., 2008. Synthesis and characterization of poly(ethyleneimine) dendrimers. Colloid Polym. Sci. 286, 747–752.

Yi, C., Xu, Z., Ford, W.T., 2004. Nano-sized dendrimer PAMAM/polystyrene composite polymer emulsion. Colloid Polym. Sci. 282, 1054–1058.

Zeng, F., Zimmerman, S.C., 1997. Dendrimers in supramolecular chemistry: from molecular recognition to self-assembly. Chem. Rev. 97, 1681–1712.

Zhang, X.Q., Intra, J., Salem, A.K., 2007. Comjugation of polyamidoamine dendrimers on biodegradable microparticles for nonviral gene delivery. Bioconjug. Chem. 18, 2068–2076.

Zhou, X.-H., et al., 2009. Supramolecular self-assembled dendritic nonlinear optical chromophores: fine-tuning of arene–perfluoroarene interactions for ultralarge electro-optic activity and enhanced thermal stability. Adv. Mater. 21, 1976–1981.

CHAPTER 9

Physicochemical characterization of PAMAM dendrimer as a multifunctional nanocarriers

Barbara Jachimska

Jerzy Haber Institute of Catalysis and Surface Chemistry, Polish Academy of Sciences, Kraków, Poland

CHAPTER OUTLINE

9.1 INTRODUCTION

Considering the enormous therapeutic benefits associated with the use of drug carriers of active substances, many researchers are currently focused on the development of targeted nanocarriers such as inorganic particles, synthetic polymers, or natural biopolymers. In recent years, enormous progress has been made through the development of new materials that can act as carriers of controllable properties such as size, shape, surface charge, or mechanical strength.

Dendrimers belong to branched polymers with a spherical molecular shape. The structure of the molecule consists of a core and an outer shell containing functional groups in its structure; the number of which is precisely defined. The number of layers in the outer shell is strictly related to the generation of the

Nanoparticles in Pharmacotherapy. DOI: https://doi.org/10.1016/B978-0-12-816504-1.00003-X

dendrimer. The higher the generation of the dendrimer, the more layers in the outer shell surrounding the core of the molecule. Dendrimers differ both in the type of core used and functional groups located within the outer layer (Zhao et al., 2009).

Previous experimental studies show that dendrimers have great potential use as carriers in drug delivery systems, gene delivery, molecular imaging, nanomaterial template creation, the design of high-performance catalysts, and environmental protection (Hu et al., 2009a). Such broad possibilities of using dendrimers resulted in an increased interest in these nanoparticles. As experimental research shows, the specific, well-defined structure of the dendrimer molecule enables two types of active agent encapsulation: in the internal pockets of the molecular structure or on the peripheries of the structure in the outer layer (Bugno et al., 2015; Cheng et al., 2015a,b; Kesharwani et al., 2014; Akesson et al., 2012; Yang et al., 2011; Ruggeri et al., 2013).

The physicochemical properties of the dendrimers are determined mainly by the functional groups present in the structure. The specific structure of the dendrimer enables the encapsulation of both polar and nonpolar molecules. Nonpolar molecules can be enclosed in the hydrophobic interior of the molecule, while polar molecules bind via the electrostatic interaction to the surface of the dendrimer structure. The use of dendrimers as drug carriers is particularly crucial because of its synergistic action resulting in increased solubility, stability, and bioavailability of some drugs (Srinivasan et al., 2015; Cheng et al., 2015a,b; Zhao et al., 2009).

On the basis of many studies, it was noted that (1) higher-generation dendrimers are more efficient in encapsulating the active agent inside the molecule than lower-generation, (2) lower-generation dendrimers are used for electrostatic attachment of the active agent on the outer shell, and (3) electrostatic interactions decide to increase solubility of poorly soluble drugs and are more effective than hydrophobic or hydrogen interactions (Hu et al., 2009b; Kesharwani et al., 2014; Garcia-Fernandez et al., 2015). The process of encapsulation of the active agent depends on its structure and hydrophobic properties. Predicting the location of the active agent in the structure of the dendrimer molecule is still difficult, despite the knowledge of the physicochemical properties of a number of dendrimer types (Bugno et al., 2015; Yang et al., 2011).

Physicochemical properties of dendrimers are determined on the basis of experimental and theoretical studies. The application of molecular dynamics (MD) allowed for the verification of many properties of the dendrimer molecule structure at the atomic scale, which are difficult to obtain experimentally (Welch and Muthukumar, 1998; Welch, 2013; Maiti and Goddard, 2006; Maiti and Messina, 2008; Pande and Crooks, 2011; Nisato et al., 2000). The branched structure of the dendrimer molecule forms a porous structure typical of branched polymer systems. The density of the external structure varies depending on the degree of protonation of functional groups present in the structure. As a result of the increasing repulsive interactions of protonated functional groups located in the outer layer, dendrimers swell and form gel-like structures with a high degree of hydration (Lin et al., 2006; Liu et al., 2009; Jachimska et al., 2013). High hydration of dendrimer systems is partly because water also builds into the internal structure of the molecule. As a result of swelling, the hydrodynamic radius of the molecule is larger compared with in dry conditions.

Particular attention should be paid to the fact that dendrimers are characterized by specific physicochemical properties. Therefore they can have different functions compared with polyelectrolytes with a linear structure (Hu et al., 2009a).

9.2 CHARACTERIZATION OF PAMAM DENDRIMERS IN BULK SOLUTION

Dendrimers are synthetic polyelectrolytes that belong to ionic polymers. The presence of ionic groups in the structure of the dendrimer molecule causes them to undergo a dissociation process in polar solvents. Interactions within the polyelectrolyte chain (short-range interactions), as well as interactions between charged chains (long-range interactions), affect specific thermodynamic and dynamic properties of polymer solutions, especially when the charged chains are not sufficiently shielded (Mandel, 1993). The short- and long-range interactions affect the local rigidity of the chain (the length constant in the worm-like polyelectrolyte chain model), on the excluded volume effect in what is referred to as good solvents and on the intermolecular interactions (e.g., the second viral coefficient—Mandel, 1993).

When considering the nature of polyelectrolytes in the bulk solution, one should take into account the condensation (adsorption) of counterions and the effects of shielding of the polymer molecule charge (Manning, 1969). Counterions are attracted to polyelectrolyte molecules by long-range electrostatic forces, but typically their physical association leads to the formation of a cloud of ions loosely bound to the molecule. The cloud is polarized and diffuses along with the polymer chain, thus affecting the bulk properties, such as the osmotic pressure and the cloud shift entropy is responsible for good solubility of polyelectrolytes in water. Polyelectrolytes can assume various shapes in the water. Solvent properties and polymer molecule conformations are closely related (Mandel, 1993; Gennes, 1972; Gennes et al., 1976).

Dynamic properties of polyelectrolyte solutions differ from those of the neutral polymer solutions. The differences pertain primarily to such properties as the diffusion coefficient, electrophoretic mobility, and dynamic viscosity. Polymer molecules present in the solution are not only in a translational motion but also in a rotational motion around the molecules axes of symmetry. For this reason, the translational diffusion coefficient D and the rotational diffusion coefficient D_r can be distinguished (Adamczyk et al., 2004a).

$$D = \frac{kT}{\eta}\mathbf{M}^{-1} = \frac{kT}{\eta}\begin{vmatrix} 1/K_{\parallel} & & 0 \\ & 1/K_{\perp} & \\ 0 & & 1/K_{\perp} \end{vmatrix}, \quad D_r = \frac{kT}{\eta}\mathbf{M_r}^{-1} = \frac{kT}{\eta}\begin{vmatrix} 1/K_{\parallel} & & 0 \\ & 1/K_{r_\perp} & \\ 0 & & 1/K_{r_\perp} \end{vmatrix} \tag{9.1}$$

where **M** and $\mathbf{M_r}$ are the translational and rotational mobility matrices, are the hydrodynamic resistance coefficients for translational motion relative to the axis of symmetry, and are hydrodynamic resistance coefficients for rotational motion.

Dendrimers belong to the group of spherical molecules. In the case of spherical molecules, assuming that these are rigid bodies, the translational and rotational diffusion coefficients are equal to each other, as they have an infinite number of symmetry axes. Thus Eq. (9.1) can be simplified to:

$$\langle D \rangle = \frac{kT}{3\eta}\left(\frac{1}{K_{\parallel}} + \frac{2}{K_{\perp}}\right) = \frac{kT}{6\pi\eta R_H} \tag{9.2}$$

While progress in the field of polymer structural characterization has been rapid, there are only limited experimental methods that can be capably used to determine these parameters in dilute aqueous solutions of a colloidal system. Noninvasive methods such as dynamic light scattering (DLS), electrophoretic mobility, and dynamic viscosity are the best suited for characterization of polymer systems.

Physicochemical characteristics of dendrimer systems were presented in the example of G6 PAMAM dendrimers. The critical physicochemical properties are listed in Table 9.1. The measurements of the diffusion coefficient as a function of concentration, ionic strength, and pH, performed using DLS, show a strong dependence on these properties. The diffusion coefficients of G6 PAMAAM dendrimers were measured in the range from 500 to 4000 ppm. The effect of pH on the diffusion coefficient was determined for the pH range of 3.5–9.0 at the ionic strength $I = 0.15$ M. The diffusion coefficient is 5.7×10^{-7} cm^2 s^{-1} and is mostly unrelated to the concentration. A higher value of the diffusion coefficient, 7.6×10^{-7} cm^2 s^{-1}, was noticed only at low ionic strengths (Jachimska et al., 2013).

Table 9.1 List of Physicochemical Properties for Six Generations of PAMAM Dendrimers

Characteristic	Value, unit	Remarks
Molar mass, M_w	58.0 kDa	Specified by Dendritech
Primary amine groups	256	Calculated
Tertiary amine groups	254	Calculated
Hydrodynamic radius, R_H	3.75 ± 0.05 nm	DLS
Diffusion coefficient, D	5.7×10^{-7} cm^2 s^{-1}	DLS
Isoelectric point (i.e.p.)	10.0 ± 0.1	Electrophoretic mobility
Bare charge, N_m	510	Calculated from structural formula
Max. effective charge, N_c	28	Calculated from electrophoretic mobility

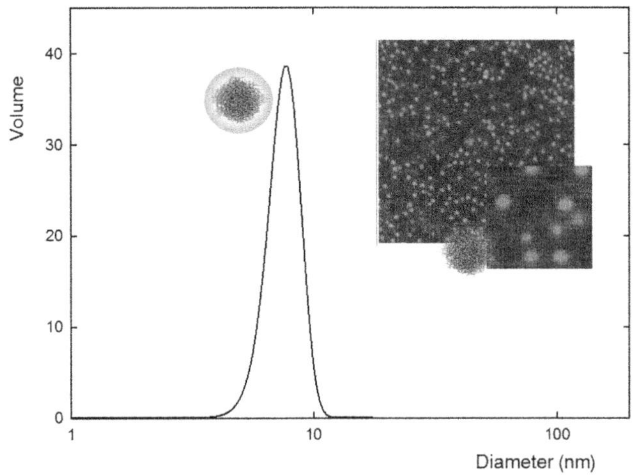

FIGURE 9.1

Hydrodynamic diameter [D (nm)] of G6 PAMAM dendrimers obtained by DLS measurement. SEM image shows the G6 PAMAM dendrimers adsorbed on the surface.

The hydrodynamic radius of a dendrimer molecule, determined from the diffusion coefficient using the equation $R_H = (kT/6\pi\eta D)$, is 3.7 nm for the ionic strength range $I = 0.01-0.15$ M and pH 9.5. The dendrimer molecule radius determined for the same ionic strength and pH using the cryo-TEM technique is 3.4 nm (Jackson et al., 1998). The higher value obtained using the DLS method is justified as in this case it is the hydrodynamic radius that is measured (Fig. 9.1).

Simulation of the G8 PAMAM dendrimers (of the 8th generation) performed using the MD method showed a dependence of the dendrimer molecule radius on the solution's pH. The change in pH value from 12 to 4 results in an increase in the inertia radius from 3.78 to 4.31 nm (Maiti et al., 2005). Similarly, Monte Carlo simulations confirmed that the profile of the intermolecular dendrimer density in the solution could be controlled by changing the pH or the ionic strength of the solution (Welch and Muthukumar, 1998). Theoretical calculations were supported by measurements using the SAXS and SANS methods.

The size or monodispersity of dendrimer systems can be verified using scanning electron microscopy (SEM) or atomic force microscopy (AFM) (Li et al., 2000). Images from SEM show, as expected, that the dendrimer systems are spherical and monodisperse, whereas the AFM method reveals other properties of dendrimer molecules: they are not only spherical regarding their structure but also belong to what is known as soft molecules. The structure's elasticity and its susceptibility to deformation manifest themselves especially during adsorption on a solid surface. As shown in the literature, the degree of flattening depends on both the dendrimer generation and the type of adsorption surface (hydrophobicity level, surface charge density). Flattening phenomenon is confirmed by AFM topographic measurements for individual positively charged PAMAM dendrimer molecules adsorbed on the surface of mica, which is characterized by high negative surface charge (at pH 6.5 and $I = 1 \times 10^{-3}$ M). The diameter of individual dendrimers varies from 5 to 10 nm. The measured height of adsorbed molecules shows that dendrimers undergo partial deformation during adsorption (Jackson et al., 1998; Li et al., 2000). The height of an individual dendrimer adsorbed on the surface strongly depends on the interaction between the dendrimer molecule and the surface (Betley et al., 2001). The degree of deformation is generally smaller in the case of high generation dendrimers in which high density of functional groups in the shell of the structure increases the rigidity of the whole molecule.

9.2.1 ELECTROPHORETIC MOBILITY AND DENDRIMER EFFECTIVE CHARGE

As a result of the specific polyelectrolyte structure, Coulomb's interactions are of great importance. Long-range impacts will determine the conformation of the polyelectrolyte chain and consequentially hydrodynamic properties such as ionic mobility, conductivity, electrophoretic mobility, and dynamic viscosity. Experimental research shows that these quantities are proportional to the square root of the concentration of polyelectrolyte (Barrat and Joanny, 1996). The observed specific properties of polyelectrolyte solutions are a consequence of the relaxation effect and the phenomenon of electrophoresis (Evers et al., 1986). The electrophoresis effect is particularly significant in the case of an external electric field. The forced movement of polyelectrolyte molecules in the electric field affects the surrounding ion cloud. As a result, the solvent modifies the speed of movement of the charged molecule in solution (Barrat and Joanny, 1996).

At a distance of κ^{-1}, the adjacent volume of the solvent moves with the charged molecule. The friction coefficient of the medium is in the order of $\eta_s \kappa^{-1}$, where η_s corresponds to the viscosity of the solvent. In the dynamics of polymers, an electrophoretic effect can be described as a process of shielding hydrodynamic interactions when the ion movement occurs under the influence of an applied electric field (Barrat and Joanny, 1996). Hydrodynamic interactions can be described by the Oseen tensor, which defines the velocity field of the studied system. The velocity field describes the resultant force created as a result of the electric field E acting on the charge q. The polarized cloud formed around the charged particle is determined by the approximation of Debye–Hückel.

$$V = \exp(-\kappa R) H(r) q E \tag{9.3}$$

where $H(r)$ tensor Oseen.

The velocity field is shielded like electrostatic interactions and decays exponentially from the $\exp(-\kappa R)/r$ distance. It is assumed that hydrodynamic interactions between the molecules themselves are shielded at a distance greater than $\kappa-1$ (Barrat and Joanny, 1996).

Electrophoretic mobility of a polyelectrolyte molecule in an ideal dilute solution containing an electrolyte can be described by the formula (Barrat and Joanny, 1996):

$$\nu_e = \mu_e E \tag{9.4}$$

where μ_e is the molecule's electrophoretic mobility, and E is the electric field applied.

The measurement of the electrophoretic mobility μ_e makes it possible to determine the electrokinetic charge q of a dendrimer molecule from the Lorentz–Henry equation:

$$q = \frac{\langle U \rangle}{ME} = \frac{kT}{D}\mu_e = 6\pi\eta R_H \mu_e \tag{9.5}$$

where $\langle U \rangle$ is the average migration velocity of a dendrimer molecule in a uniform electric field E, and M is the hydrodynamic mobility (Jachimska et al., 2013).

The latter equation above can be used for calculating the average number of elementary charges per molecule N_c. Taking into account that $|e| = 1602 \times 10^{-19}$ C, N_c is given by:

$$N_C = \frac{6\pi\eta R_H}{f(\kappa R_H)}(1 + \kappa R_H)\mu_e / e \tag{9.6}$$

This equation can be used for molecules of any shape, if the condition $2\kappa R < 1$ is satisfied, and where R is the molecule radius and κ^{-1} is the thickness of the electric double layer.

If we know the nominal charge per molecule, N_m, and the effective charge of the molecule, N_c, we can find the effective ionization degree α_e of a dendrimer molecule from the equation:

$$\alpha_e = \frac{N_c}{N_m} \tag{9.7}$$

The measurement of the electrophoretic mobility μ_e allows us not only to evaluate the effective ionization level of a dendrimer molecule but also to determine the isoelectric point of the system studied (Jachimska et al., 2013, 2016; Jachimska and Tokarczyk, 2016). Changes in the electrophoretic mobility, the zeta potential, and the effective charge of a G6 PAMAM dendrimer molecule are given in Table 9.2.

To optimize the dendrimer adsorption process, it becomes crucial to determine the effective charge of the dendrimer molecule dependence on pH and ionic strength. The structure of G6 PAMAM dendrimers contains primary amines (in sum 256) and tertiary amines (in sum 254). These groups determine the mechanism of PAMAM molecule protonation. At low pH, all primary and tertiary amines are protonated (pH $<$ 4). At intermediary or neutral pH, no protonation is observed. This behavior is confirmed by the position of the

Table 9.2 Physicochemical Properties of G6 PAMAM Dendrimers in Different pH of the Solution: Electrophoretic Mobility (μ_D), Zeta Potential (ζ_D), Number of Uncompensated Charges (N_D), Effective ionization degree (α_e) and Charge Density (σ_D)

G6 PAMAM Dendrimer, $I = 1 \times 10^{-2}$ M					
pH	μ_D (μm cm V^{-1} s^{-1})	ζ_D (mV)	N_D (e)	α_e (%)	σ_D (e nm^{-2})
4.0	5.6	106.3	43.7	17.2	0.254
6.0	5.0	96.0	39.3	15.4	0.228
8.0	2.8	53.0	21.5	8.5	0.124
10.0	− 0.1	− 2.0	− 1.2	0	− 0.006

Electrophoretic mobility calculated from the equation $\zeta_D = (3\eta/2\varepsilon F(\kappa a))\mu_e$, the uncompensated charge calculated from Eg. 3.6 charge density calculated from the equation $\sigma_D = (N_D/4\pi R_H^2)$.

dendrimer isoelectric point, which is at pH 10. The explanation of the mechanism of protonation of the nonbranched structure is complicated, especially that their behavior cannot be directly compared with protonation of linear chains. Dendrimer molecules contain two types of amine groups, which are located in different chemical surroundings. Consequently, primary groups are much more alkaline than the tertiary ones. The mechanism of dendrimer protonation can be described on the basis of potentiometric titration curves (Cakara et al., 2003). All primary amines are protonated in the alkaline pH range. Due to weak interaction between these two types of functional groups, tertiary amines undergo protonation almost independently. All of them are protonated in the more acidic environment. From this, we conclude that at high pH values the shell of the dendrimer molecule is protonated, whereas its core undergoes protonation only at lower pH. The pH at which 50% of the G6 PAMAM dendrimer groups are protonated is about 8.0.

The nominal level of dendrimer ionization, as determined from Eq. (9.7), is given in Table 9.2. The nominal ionization level strongly depends on pH and is much below one, which means that the nominal charge of PAMAM molecules is compensated mainly by counterions. This phenomenon can be explained by specific adsorption of counterions (in this case Cl^-). According to the Manning–Ossawa theory, the condensation of counterions takes place when the Manning parameter $q_0 = l_B/l_c$ is higher than one (Manning, 1969; Muthukumar, 2004). The Bjerrum length (l_B) is given by $l_B = (e^2/4\pi\varepsilon kT)$ (l_B corresponds to the distance for which the Coulomb interaction between the two elementary charges e is equal to the thermal energy kT).

It should be noted that the maximum protonation degree of the G6 PAMAM dendrimer molecule at low ionic strengths, when the screening of the charge by counterions is weak, accounts for only 20% of the nominal charge.

9.2.2 DYNAMIC VISCOSITY AND THE CORE/SHELL STRUCTURE OF DENDRIMERS

Measurements of dynamic viscosity for dilute solutions of colloidal systems allow obtaining information about not only the shape of the molecule in the suspension but also the stability of the system and the tendency to form aggregates. Dendrimers are a unique model system because they characterize not only a well-defined spherical shape and monodisperse size distribution but also a well-controlled surface charge density (Jachimska et al., 2013).

The relative viscosity is defined as the slope of the η/η_s dependence (where η, η_s are the viscosities of the suspension and the solvent, respectively) on the volume fraction of the solid part of the suspension Φ_v (Brenner, 1974; Harding, 1995).

For rigid uncharged molecules, the Einstein model provides this dependence:

$$\frac{\eta}{\eta_s} = 1 + 2.5\Phi_v \tag{9.8}$$

For nonrigid, nonspherical, and uncharged molecules, the Einstein equation can be put as:

$$\frac{\eta}{\eta_s} = 1 + \nu\Phi_v \tag{9.9}$$

where ν is the increase in viscosity (Simha coefficient), which is related to the solution intrinsic viscosity $[\eta]$ by:

$$[\eta]_{\Phi v} = \lim_{\Phi \to 0}\left(\frac{\eta - \eta_s}{\eta_s \Phi_v}\right) = \nu \tag{9.10}$$

The viscosity increment, ν, referred to as "universal shape function," is directly associated with both the shape of the molecule and its volume in the solution (Jachimska and Pajor, 2012; Jachimska et al., 2012).

If a molecule has a surface charge, it can also contribute to the increase in the solution viscosity. Three electroviscous components can be distinguished in polyelectrolyte solutions: (1) the primary component, resulting from the diffusion resistance of the double layer surrounding the molecules, (2) the secondary component, resulting from the repulsion between double layers of molecules, and (3) the tertiary component, resulting from intermolecular interactions, which affect the shape of macromolecules (Jachimska, 2010; Jachimska et al., 2010).

The experimental value of the Simha coefficient v for dendrimers (Fig. 9.2) is three times greater than might result from the Einstein formula (Eq. (9.8)), which is fulfilled in hard uncharged molecules.

In the case of charged molecules of polyelectrolytes, the deviation from the Einstein formula can be interpreted as the primary electroviscous effect, defined by (Adamczyk et al., 2004b)

$$\frac{\eta}{\eta_s} = 1 + 2.5(1 + p)\Phi_v \tag{9.11}$$

where p is the primary electroviscous function. Assuming that ion mobilities do not differ considerably from each other, p can be expressed as

$$p = \frac{2\varepsilon\xi^2(1+\kappa a)^2 f(\kappa a)}{3\pi\eta D_i} \tag{9.12}$$

where ε is the dielectric constant of water (relative permittivity), D_i is the ion diffusion coefficient, and $F(\kappa a)$ is a function of the dimensionless κa parameter.

From Eq. (9.12) we can determine the electroviscous contribution, which, considering the value of the zeta potential and the dendrimer size, is of the order of $p = 1 \times 10^{-3}$. This value is much lower than what results from experimental data (experimental $p = 2$). This is why the electroviscous effect cannot explain the increase in dendrimer solution viscosity at low concentrations. This behavior can probably be explained by the structure of the dendrimer molecule, which is similar to the core/shell type structure (Jachimska and Adamczyk, 2007).

Assuming that the shell of the dendrimer molecule has a well-hydrated gel-like structure as a result of contact with the solution, it should lead to gradual structural changes inside the molecule (Li et al., 2011). The shell is characterized by a much lower density than the central part of the molecule (Welch and Muthukumar, 1998; Maiti and Goddard, 2006; Maiti and Messina, 2008). Consequently, the effective hydrodynamic volume of the molecule in an aqueous solution is higher than that of a dry substance. This assumption is corroborated by MD simulations, which show that the dendrimer core zone has a higher density than its shell, the latter being relatively small. The density of the dendrimer structure is unusually low for low values of ionic strength and at acidic environment when the dendrimers possess a high degree of protonation. Such behavior of dendrimers can be interpreted as swelling caused by the interaction between solvent molecules and tertiary amine groups (Jachimska et al., 2013) (Fig. 9.2).

Assuming that the adopted model of dendrimer molecule structure is correct, the effective volume of the molecule in the solution must be taken into account. The effective volume fraction $\Phi_v{}^*$ of dendrimer molecules can be expressed as:

$$\Phi_v{}^* = (1 + p^*)\Phi_v \tag{9.13}$$

where $p^* = tg\ \alpha/2.5-1$ is the apparent electroviscous effect and $tg\ \alpha$ is the slope of η/η_s versus Φ_v.

In the relevant literature, no model has described the dependence of the dynamic viscosity on the bulk concentration for systems with the core/shell structure. As changes in dendrimer viscosity are inseparably associated with the molecule's internal structure, they can be compared with the behavior of porous silica molecules (Adamczyk et al., 2004a). Experimental data shows that the relative viscosity of diluted silica solutions is higher than that resulting from the Einstein model. Such anomalous behavior is well described by the adopted core/shell

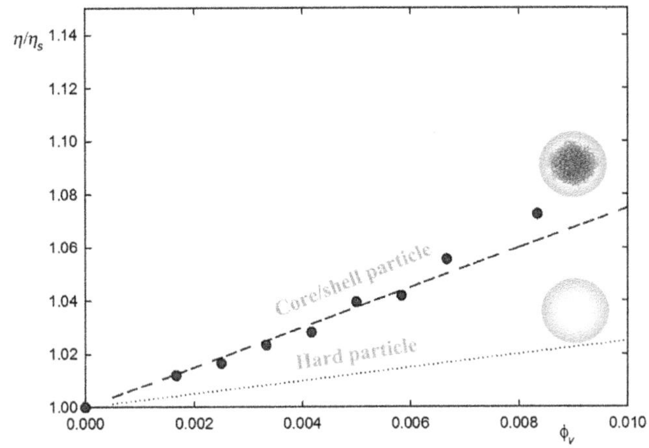

FIGURE 9.2

The relative viscosity η/η_s of the G6 PAMAM dendrimers versus volume fraction Φ_v at $I = 0.15$ M and pH 9.5 (*points*—experimental data, *solid line*—the theoretical results calculated from the Einstein formula $\eta_s/\eta = 1 + 2.5\Phi_v$ for hard spherical particles, *short-dash line*—the linear regression for core/shell particles).

model, which assumes a dense structure of the molecule core and a porous shell (Jachimska and Adamczyk, 2007).

9.3 DENDRIMER ADSORPTION AT INTERFACES

Adsorption of polyelectrolytes on the surface is determined by long-range electrostatic interactions, in contrast to neutral polymers, where short-range interactions dominate. Theoretical description of polymer adsorption is involved as the entropy of particle conformations (soft molecules) must be taken into consideration. Most models of polyelectrolytes were based on the assumptions made for neutral polymers and they "wrestled" with the problem of describing a system in which there are both long-range electrostatic interactions and short-range interactions of chemical nature, as well as the effects associated with the change in conformation. The first theoretical description of polyelectrolyte adsorption was given by Hesselink (1977) by the polymer adsorption model developed by Hoeve (1996). The best-known multilayer model, developed by Stern (Van der Schee and Lyklema, 1984), takes into account Roes' network theory (Roe, 1974), the Scheutjens–Fleer (SF) theory (Scheutjens and Fleer, 1985) and it postulates the existence of a network in which the centers can be occupied by monomers, solvent molecules, or small ions. In this approximation, the electrostatic potential is determined consistently (the average-field theory) with profiles of polymers and small ions. In other models, the electrostatic potential and polyelectrolyte concentration are considered as continuous functions.

Essential parameters describing polyelectrolyte adsorption on a surface are the surface coverage, the surface charge density, the electrolyte ionic strength and the solution's pH Electrostatic interactions between identical charges (functional groups) are always repulsive and lead to effectively stiffening of the structure of the polyelectrolyte molecule. Consequently, no densely packed adsorption layer can be formed on the surface. The addition of salt will increase this effect through the screening of repulsive interactions between molecules and increase the attractive interactions between the surface and the electrolyte (Jachimska and Pajor, 2012; Jachimska et al., 2012, 2013, 2016).

9.3.1 SELF-ASSEMBLING DENDRIMER ON A SOLID SUPPORT

To control the properties of dendrimer systems, one should understand their behavior not only in an electrolyte solution but also in the adsorption state (Jachimska, 2010; Jachimska et al., 2010). Of particular importance are the techniques allowing for the controlled creation of functional layers with dedicated properties. To characterize nanostructure layers it is indispensable to use methods that will allow their analysis at the molecular level. Ellipsometry, AFM in force-spectroscopy mode and quartz crystal microbalance (QCM-D) are frequently used. To study the conformation state of adsorbed molecules, Fourier transformed infrared spectroscopy (FTIR) in ATR mode and fluorescence recovery after photobleaching (FRAP) are also used. The significant advantages of these techniques include the possibility to follow interface processes directly in situ and to measure both the deposition of organic and inorganic materials.

Systems that build on the surface of hierarchical structures ordered with appropriate functionality require the use of advanced optical methods to study, allowing for high-resolution images and the capacity for 3D observation. Atomic force microscopy is one method that performs surface imaging with high resolution. The majority of AFM microscopes work in several measurement modes allowing the determination of characteristic features of the scanned surface such as topography, flexibility, friction forces, adhesion, and distribution of magnetic or electrical properties of the tested sample. The appropriate mode of operation of the microscope helps to determine the imaging of biological samples depending on environmental parameters or temperature. 16-Bit resolution in all three axes gives the opportunity to get an atomic-scale image across the entire scanning range. Quartz crystal microbalance with energy dissipation (QCM-D) is a method using the piezoelectric effect. The central element of the system is a quartz crystal, whose resonant frequency variation allows the determination of the mass adsorbed on the surface of the oscillator. Additionally, the system measures changes in the dissipation of the energy of the system under examination, which allows for the identification of the viscoelastic properties of the layers deposited

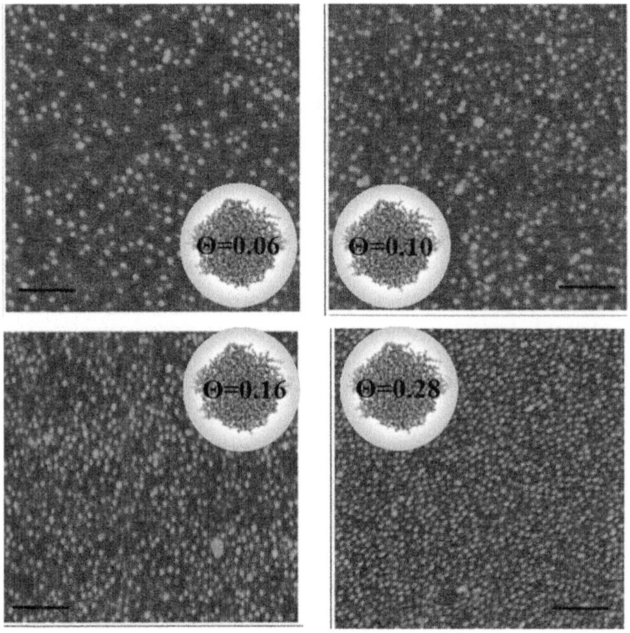

FIGURE 9.3

Images of PAMAM dendrimer after adsorption on Au surface obtained using AFM method (the scales bars represent 100 nm) for concentration $c = 5$ ppm at the ionic strength $I = 1 \times 10^{-2}$ M NaCl.

on the quartz crystal. The QCM-D method is beneficial for the characterization of layers, as well as polymeric and biopolymer or hybrid multilayers.

A relatively new experimental method is surface scanning using a confocal microscope with a high degree of resolution. This method allows performing 3D analysis of a tested system and measurements in real time, which is very useful for the observation of the kinetics of polymer adsorption. This method also makes it possible to determine experimentally 3D fluctuation phenomena and radial distribution functions of adsorbed particles (Fig. 9.3).

Metals are characterized by a high density of free (unbound) electrons that originate from the valence shells of metal atoms. Surface plasmons are the resulting surface collective oscillations of free electrons induced by the oscillations of electromagnetic fields. Surface plasmons form surface electromagnetic waves that propagate parallel to the area of mutual interactions at the metal–dielectric interface; they are susceptible to any changes in this boundary, such as the adsorption of molecules through the metal surface. Surface plasmon resonance (SPR) uses the phenomenon of total internal reflection of light from a surface covered with a conductive metal layer. The angle of incidence for polarized light *p*, at which light reflection is minimal, is defined as the resonant angle. The minimum change in the refractive index of the medium near the metal surface results in a change in the resonant angle. Registration of changes in the resonant angle allows for monitoring with high sensitivity both the adsorption of molecules on the surface of the sensor and the effectiveness of intermolecular interactions.

The SPR technique has been used for real-time and label-free detection of many types of biomolecular recognition phenomena such as ligand–receptor interaction, or polymer–protein interaction. SPR and QCM are robust methods that enable highly sensitive, qualitative, real-time, label-free, and noninvasive detection of adsorbed dendrimers (Jachimska et al., 2013, 2016).

The SPR method determines the surface excess based on the changes of the dielectric constant at the liquid/metal interface. In the QCM method, the surface excess is determined based on the frequency variations of the piezoelectric sensor. Therefore, in the QCM-D method, the surface excess is a summary of the system response resulting both from the adsorption of particles on the surface of the sensor and the water molecules present in the adsorption layer.

The combination of these two techniques gives insight into both the adsorption mechanism of dendrimers on a metal surface, and the structural changes of the system. Also, it allows the determination of the degree of reversibility of the process and also changes in the hydration degree of the system depending on environmental parameters such as pH, ionic strength, type of electrolyte, and temperature.

The quartz crystal microbalance method with dissipation energy (QCM-D) allows the simultaneous measurement of two parameters of the studied system: changes of the resonant frequency (Δf) and energy dissipation (D). When a rigid flat layer is formed on the surface, it is assumed that changes in the resonance frequency are in proportion to the adsorbed mass (Γ_{QCM}) on the QCM-D. This dependence is described by the Sauerbrey equation below:

$$\Gamma_{QCM} = -C\frac{\Delta f}{n} \tag{9.14}$$

where C is the constant of the crystal, which is equal to 17.7 ng/cm^2 Hz based on the physical properties of quartz crystal, and n is the overtone number.

Along with the frequency, changes are also monitored in the dissipation energy D_{dis}, which determines viscoelastic properties of the adsorbed layer:

$$D_{dis} = \frac{E_{dis}}{2\pi E_{stor}} \tag{9.15}$$

where E_{dis} is the energy lost during one oscillation and E_{stor} is the energy stored in the oscillating circuit. The system's dissipation energy increases as the elasticity of the layers increases.

Also, QCM-D measurements are used to determine the thickness of the films that are formed. To do this, the effective thickness of the film formed on the surface must be known.

$$\rho_{\text{film}} = \frac{\Phi_{VD}\rho_D + \Phi_{VS}\rho_s}{\Phi_{VD} + \Phi_{VS}} \tag{9.16}$$

where ρ_s is the density of the solvent, ρ_D is the density of PAMAM dendrimers, Φ_{VD} is the volume fraction of the dendrimer in the solution, and Φ_{VS} is the volume fraction of solvent in the film. Eq. (9.16) gives only the approximate value of ρ_{film} under the assumption of ideal mixing and a homogeneous structure of the film (Jachimska et al., 2013).

The effectiveness of PAMAM dendrimer adsorption on the surface of gold as a function of pH determined using the QCM-D method is shown in Fig. 9.4A. Dendrimer adsorption is irreversible, which is indicated by the low level of desorption. The effectiveness of dendrimer adsorption consistently increases with increasing pH. At pH 10, the adsorbed mass is six times that adsorbed at pH 4. It should be remembered that when using the QCM-D method, the adsorbed mass is the sum of the mass of dendrimers adsorbed on the QCM sensor and the mass of water contained in the dendrimer film (Jachimska and Tokarczyk, 2016) (Fig. 9.4).

A method complementary to QCM-D is the SPR, which belongs to optical methods. In this method, the surface coverage can be calculated according to the equation:

$$\Gamma_{\text{SPR}} = \frac{\Delta\Theta_{\text{SPR}} k d_{\text{G6}}}{(dn/dc)} \tag{9.17}$$

where $\Delta\Theta_{\text{SPR}}$ is the change in the MP-SPR angle, k is an MP-SPR instrumental constant obtained after calibration, d is the thickness of the adsorbed layer, and dn/dc is the refractive increment. The results obtained using the MP-SPR method confirm that the effectiveness of dendrimer adsorption is strongly affected by pH Fig. 9.4B. The highest adsorption of dendrimers occurs at high pH values when the degree of molecule ionization is at a minimum. The MP-SPR measurements make it possible to determine the structure of the formed layer and thus to determine the surface coverage Θ; mainly that the size, shape, and charge of dendrimer molecules are known. The surface excess of the dendrimer adsorbed per unit geometric surface area expressed in ng/cm^2 can be determined by the formula:

$$\Gamma_{\text{Ad}} = \frac{M_w}{S_{\text{G6}} A_V}\Theta = \frac{M_w}{\pi R_H^2 A_v}\Theta \tag{9.18}$$

where M_w is the molecular weight; S_{G6} is the cross-section area, which for the near-spherical molecule is $S_{\text{G6}} = \pi {R_H}^2 = 43\ \text{nm}^2$ (in the case of G6 dendrimer $R_H = 3.7$ nm as determined by DLS measurements); Av is the Avogadro number; and Θ is the surface coverage (Jachimska et al., 2013).

The surface coverage degree can be correlated with the surface charge of dendrimer molecules (Fig. 9.5). A decrease in the zeta potential of a dendrimer molecule is accompanied by a decrease in mutual interaction between molecules, which consequently initiates the formation of a packed layer. The decrease from

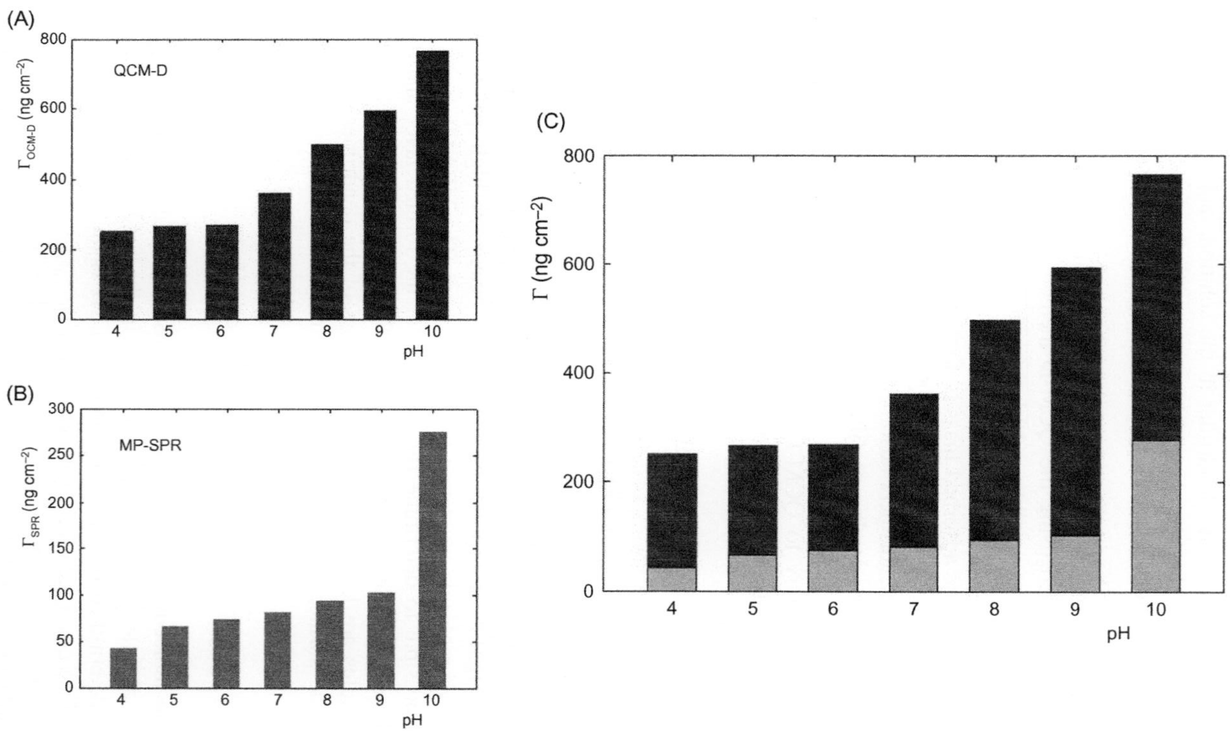

FIGURE 9.4

The surface excess (Γ) of G6 PAMAM ($c = 5$ ppm, $I = 1 \times 10^{-2}$ M NaCl) on the Au surface at different pH solution obtained by (A) QCM-D and (B) MP-SPR method. (C) Comparison data from QCM-D and MP-SPR method.

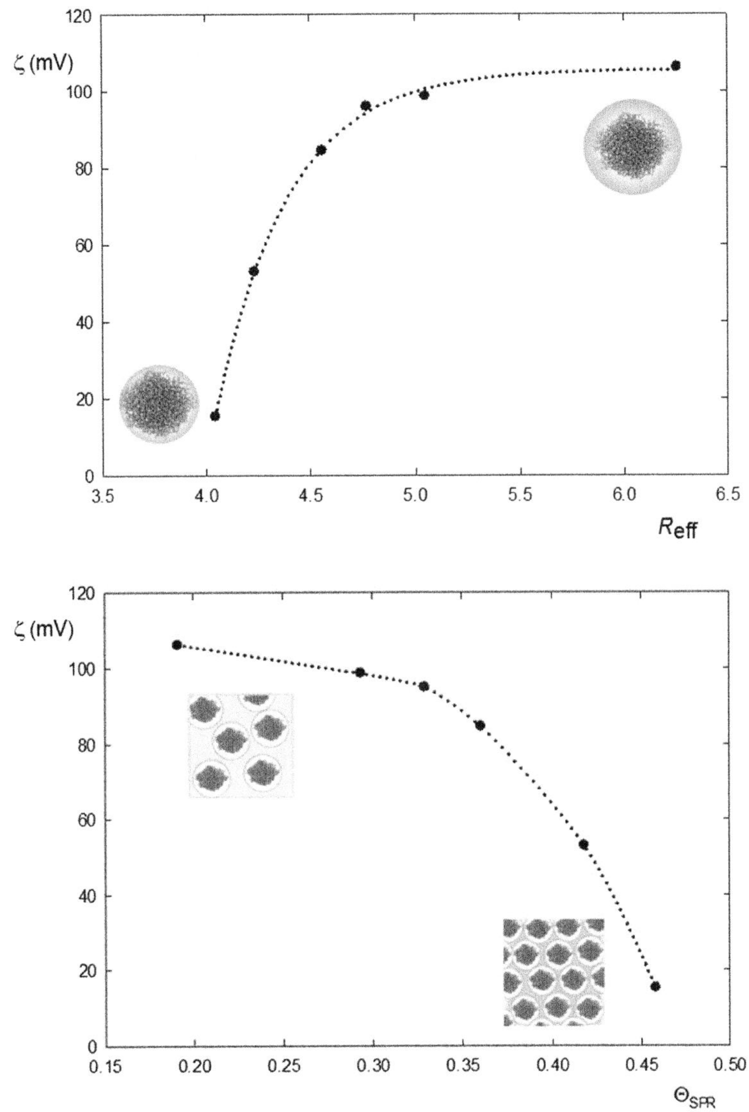

FIGURE 9.5

Zeta potential (ζ) of G6 PAMAM dendrimer versus effective radius R_{eff} and the surface coverage Θ_{SPR} on the Au surface obtained by MP-SPR measurements (*points*—experimental data, *dashed line*—interpolations of experimental data) (Jachimska and Tokarczyk, 2016).

100 mV (pH 4.0) to 15 mV (pH 9.0) at an ionic strength of $I = 1 \times 10^{-2}$ M NaCl results in an increase in the surface coverage from 0.19 to as much as 0.46.

It should be emphasized that the random sequential adsorption (RSA) theory for spherical monodisperse molecules (hard particles) suggests that the maximum surface coverage should be 0.547. The maximum surface coverage found for the planned system was much lower than that suggested by the RSA theory. This phenomenon can be attributed to repulsive electrostatic forces between positively charged dendrimer molecules. The distance between molecules in the adsorption layer characterizes the range of interaction between the overlapping double layers. The surface coverage found in the MP-SPR experiments can be used to determine the effective radius of dendrimer molecules adsorbed on the surface at various pH values (Fig. 9.5). The effective radius of dendrimer molecules decreases with the decrease in pH from 6.26 nm at pH 4.0 to 4.04 nm at pH 9.0 (Jachimska and Tokarczyk, 2016; Tokarczyk and Jachimska, 2017). The concept of the effective molecule radius can be used to describe the effect of blocking the adsorption layer due to the strength of interaction between the adsorbed molecules, and even to determine the charge of a molecule on the adsorption layer. The possibility of precisely controlling the surface structures of nanocolloid systems has a high application potential. One of the areas is the production of biomaterials with controlled structure and functionality (Jachimska and Tokarczyk, 2016; Jachimska et al., 2016).

9.3.2 THE DEGREE OF PAMAM DENDRIMER PROTONATION AND THE REVERSIBLE SWELLING PROCESS OF DENDRIMER SYSTEMS

Simultaneous measurements using MP-SPR and QCM-D methods make it possible to determine parameters that govern the immobilization of PAMAM dendrimers on the Au surface. Changes in pH have a considerable effect on the effectiveness of dendrimer bounding to the surface and on their self-organization. The observed tendency for effective binding and saturating the surface is associated with the degree of protonation of adsorbed molecules.

In both measurement methods, the greatest adsorption value is obtained for pH 10.0 (near G6 PAMAM i.e.p.) and the lowest at pH 4.0. In all measurements, the adsorbed mass measured using the QCM method is six times greater than that obtained using the optical MP-SPR method. Comparing surface excess obtained from both methods and taking into account that the surface excess obtained by QCM-D method contains water, the degree of hydration of the surface layer can be expressed by the equation:

$$\Gamma_{H_2O}[\%] = \frac{\Gamma_{QCM} - \Gamma_{SPR}}{\Gamma_{QCM}} \times 100 \tag{9.19}$$

where Γ_{SPR} is the adsorbed mass obtained from the MP-SPR method, Γ_{QCM} is the adsorbed mass from the QCM-D method, Γ_{H_2O} [%] is the mass fraction of water in the dendrimer film (Fig. 9.4C). A comparison made it possible to determine

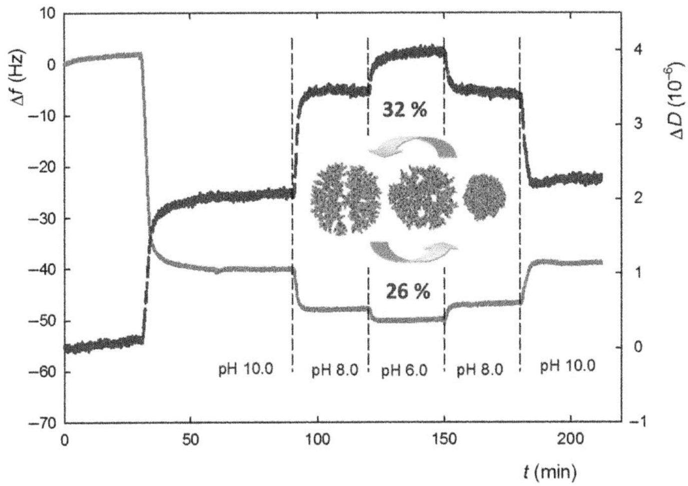

FIGURE 9.6

The frequency shift (Δf) and the dissipation energy (ΔD) profile after adsorption of $c = 5$ ppm G6 PAMAM at $I = 1 \times 10^{-2}$ M and various pH ranging from 10.0 to 6.0 on the Au surface using QCM-D method (Jachimska et al., 2013).

the quantity of water bound in the dendrimer film and to determine how the hydration level changes as a function of pH. These data clearly show how the dendrimer structure can affect properties of the dendrimer film formed on the surface of gold. Experimental data show that PAMAM films on the surface of gold contain as much as 80% water (Jachimska and Tokarczyk, 2016; Jachimska et al., 2016). This is particularly considerable as compared with the percentage of water bound with dendrimer molecules in the process of their swelling. The reversible swelling of PAMAM-type dendrimers can be monitored using the QCM-D method (Fig. 9.6). Studies on changes in dendrimer structure hydration in the pH range from 4.0 to 10.0 show that the hydration level changes by 25%–30% (Jachimska and Tokarczyk, 2016). High degree of hydration is an additional feature of dendrimer systems. The protonation degree of PAMAM dendrimer molecules strongly depends on pH. Protonation of amines, in particular, tertiary amines located inside the dendrimer structure, causes additional solvent molecules to penetrate the inner part of the dendrimer molecule, which leads to an increase in the molecule volume and manifests itself as swelling of the structure.

The inner part of the G6 dendrimer structure is not very packed or rigid, and has voids that can be occupied by additional molecules of water or by ions. Based on simulations, Maiti found that at low pH values the dendrimer structure is available to solvent (Maiti et al., 2005). Under such conditions, solvent molecules were observed to penetrate the dendrimer molecule. Based on MD simulations, it was found that at high pH values, the number of water molecules per a tertiary amine group is equal to 3 and this value increases to 6 at low pH values. The process of reversible swelling of dendrimers has practical advantages because there are indications that dendrimers are a good candidate for intelligent, controlled drug carriers whose conformation is induced by a change in pH or ionic strength.

9.4 CONCLUSION

Experimental studies so far confirm that dendrimers due to their specific physicochemical properties resulting mainly from their structure have high application potential. It seems particularly attractive to use them as carriers of drugs for molecular targeted therapy.

ACKNOWLEDGMENT

This work was partially supported by the statutory research fund of ICSC PAS and grant NCN OPUS 2016/23/B/ST5/02788.

LIST OF SYMBOLS

A_v	Avogadro number
C	constant of the QCM-D crystal
c	concentration
$\boldsymbol{D}$	translation diffusion coefficient matrix
$\boldsymbol{D}_r$	rotation diffusion coefficient matrix
D	diffusion coefficient
ΔD	dissipation shift
d	thickness
$\boldsymbol{E}$	electric field
E_{dis}	energy lost
E_{stor}	energy storage
e	elementary charge
$\boldsymbol{F}$	force vector
Δf	frequency shift
$\boldsymbol{H}$	Oseen tensor
I	ionic strength
$K_{\parallel}, K_{\perp}$	hydrodynamic resistance coefficients for the translation motion
$K_{r_{\parallel}}, K_{r_{\perp}}$	hydrodynamic resistance coefficients for the rotary motion
k	Boltzmann constant
l_B	Bjerrum length
l_c	distance between charged groups
$\boldsymbol{M}$	translation mobility matrix

M_r	rotation mobility matrix
M_w	molecular mass
N	number of monomers
N_c	number of elementary charges per molecule
N_D	uncompensated charge per dendrimer
N_m	nominal number of charges per molecule
n	overtone number
p	primary electroviscous function
q	electric charge
q_0	Manning parameter
R_{eff}	effective radius
R_H	hydrodynamic radius
S_{G6}	cross-section area
T	absolute temperature
t	time
U	migration velocity
V	characteristic velocity
V_{sus}	suspension volume
ν	Simha viscosity
v_e	mobility

Greek

α	ionization degree
Γ	adsorbed mass per unit surface
ε	dielectric constant
κ	specific conductivity of the solution
κ^{-1}	electric double layer thickness
Θ	surface coverage
$\Delta\Theta$	angle shift
μ_e	electrophoretic mobility
ρ	density
ρ_p	specific density
η	viscosity of suspension
η_s	viscosity of solvent
$[\eta]$	intrinsic viscosity
Φ_v	volume fraction
σ_D	charge density per dendrimer
ζ	zeta potential

REFERENCES

Adamczyk, Z., Zembala, M., Warszyński, P., Jachimska, B., 2004a. Characterization of polyelectrolyte multilayers by the streaming potential method. Langmuir 20, 10517–10525.

Adamczyk, Z., Jachimska, B., Kolasińska, M., 2004b. Structure of colloid silica determined by viscosity measurements. J. Colloid Interface Sci. 273, 668–674.

Akesson, A., Cárdenas, M., Elia, G., Monopoli, M.P., Dowson, K.A., 2012. The protein corona of dendrimers: PAMAM binds and activates complement proteins in human plasma in generation dependent manner. RCS Adv. 2, 11245–11248.

Barrat, J.L., Joanny, J.F., 1996. Theory of polyelectrolyte solutions. Advances in Chemical Physics. Wiley, New York.

Betley, T.A., Banaszak, M.M., Orr, B.G., Swanson, D.R., Tomalia, D.A., Baker, J.R., 2001. Tapping mode atomic force microscopy investigation of poly(amidoamine) dendrimers: effects of substrate and pH on dendrimer deformation. Langmuir 17, 2768–2773.

Brenner, H., 1974. Rheology of a dilute suspension of axisymmetric Brownian particles. J. Multiphase Flow 1, 195–341.

Bugno, J., Hsu, H., Hong, S., 2015. Recent advances in targeted drug delivery approaches using dendritic polymers. Biomater. Sci. 3, 1025–1034.

Cakara, D., Kleimann, J., Borkovec, M., 2003. Microscopic protonation equilibria of poly (amidoamine) dendrimers from macroscopic titrations. Macromolecules 36, 4201–4207.

Cheng, L., Hu, Q., Cheng, L., Hu, W., Xu, M., Zhu, Y., 2015a. Construction and evaluation of PAMAM–DOX conjugates with superior tumor recognition and intracellular acid-triggered drug release properties. Colloids Surf. B 136, 37–45.

Cheng, W., Kumar, J.N., Zang, Y., Liu, Y., 2015b. pH- and redox-responsive self-assembly of amphiphilic hyperbranched poly(amidoamine) for controlled doxorubicin delivery. Biomater. Sci. 3, 597–607.

Evers, O.A., Fleer, G.J., Scheutjens, J.M.H.M., Lyklema, J., 1986. Adsorption of weak polyelectrolytes from aqueous solution. J. Colloid Interface Sci. 111, 446–454.

Garcia-Fernandez, E., Paulo, P.M., Costa, S.M.B., 2015. Evaluation of electrostatic binding PAMAM dendrimers and charged phthalocyanines by fluorescence correlation spectroscopy. Phys. Chem. 17, 4319–4327.

Gennes, P.G., 1972. Exponents for the excluded volume problem as derived by the Wilson method. Phys. Lett. A 38, 339–340.

Gennes, P.G., Pincus, P., Velasco, R.M., Brochard, F., 1976. Remarks on polyelectrolyte conformation. J. Phys. 37, 1461–1473.

Harding, S.E., 1995. On the hydrodynamic analysis of macromolecular conformation. Biophys. Chem. 55, 69–93.

Hesselink, T.T., 1977. On the theory of polyelectrolyte adsorption: the effect on adsorption behavior of the electrostatic contribution to the adsorption free energy. J. Colloid Interface Sci. 60, 448–466.

Hoeve, C.A.J., 1996. Adsorption isotherms for polymer chains adsorbed from solvents. J. Chem. Phys. 44, 1505–1510.

Hu, J., Cheng, Y., Wu, Q., Zhao, L., Xu, T., 2009a. Host–guest chemistry of dendrimer–drug complexes. 2. Effects of molecular properties of guests and surface functionalities of dendrimers. J. Phys. Chem. B 113 (31), 10650–10659.

Hu, J., Cheng, Y., Ma, Y., Wu, Q., Xu, T., 2009b. Host–guest chemistry and physicochemical properties of the dendrimer–mycophenolic acid complex. J. Phys. Chem. B 113 (1), 64–74.

Jachimska, B., 2010. Physicochemical characterization of anisotropic molecules (polyelectrolytes and proteins) in bulk solutions. In: Fanun, M. (Ed.), Colloids in Biotechnology, Series: Surfactant Sciences, vol. 152. CRC Press (Chapter 4).

Jachimska, B., Adamczyk, Z., 2007. Characterization of rheological properties of colloid zirconia. J. Eur. Ceram. Soc 27, 2209–2215.

Jachimska, B., Pajor, A., 2012. Physico-chemical characterization of bovine serum albumin in solution and as deposited on surfaces. Bioelectrochemistry 87, 136–146.

Jachimska, B., Tokarczyk, K., 2016. Combining surface plasmon resonance and quartz crystal microbalance to determine hydration of dendrimer monolayers. J. Phys. Chem. C 120, 19678–19685.

Jachimska, B., Jasiński, T., Warszyński, P., Adamczyk, Z., 2010. Conformations of poly (allylamine hydrochloride) in electrolyte solutions: experimental measurements and theoretical modeling. Colloid Surf. A 355, 7–15.

Jachimska, B., Kozlowska, A., Pajor-Swierzy, A., 2012. Protonation of lysozymes and its consequences for the adsorption onto a mica surface. Langmuir 28 (2012), 11502–11510.

Jachimska, B., Łapczyńska, M., Zapotoczny, S., 2013. Reversible swelling process of 6th generation poly(amidoamine) (PAMAM) dendrimers molecule as determined by quartz crystal microbalance (QCM) technique. J. Phys. Chem. C 137, 1136–1145.

Jachimska, B., Tokarczyk, K., Łapczyńska, M., Paciul-Malinowska, A., Zapotoczny, S., 2016. Structure of bovine serum albumin adsorbed on silica investigated by quartz crystal microbalance. Colloids Surf. A 489, 163–172.

Jackson, C.L., Chanzy, H.D., Booy, F.P., Drake, B.J., Tomalia, D.A., Bauer, B.J., et al., 1998. Visualization of dendrimer molecules by transmission electron microscopy (TEM): staining methods and cryo-TEM of vitrified solutions. Macromolecules 31, 6259–6265.

Kesharwani, P., Jain, K., Jain, N., 2014. Dendrimers as nanocarriers for drug delivery. Progr. Polym. Sci. 39, 268–307.

Li, J., Piehler, L.T., Qin, D., Baker, J.R., Tomalia, D.A., 2000. Visualization and characterization of poly(amidoamine) dendrimers by atomic force microscopy. Langmuir 16, 5613–5616.

Li, X., Zamponi, M., Hong, K., Porcar, L., Shew, C., Jenkins, T., et al., 2011. pH responsiveness of polyelectrolyte dendrimers: a dynamical perspective. Soft Matter. 7, 618–622.

Lin, S., Maiti, P.K., Goddard, W.A., 2006. Dynamics and thermodynamics of water in PAMAM dendrimers at subnanosecond time scales. J. Phys. Chem. B 109, 8663–8672.

Liu, Y., Bryantsev, V.S., Diallo, M.S., Goddard, W.A., 2009. PAMAM dendrimers undergo pH-responsive conformational changes without swelling. J. Am. Chem. Soc. 131, 2796–2799.

Maiti, P.K., Goddard, W.A., 2006. Solvent quality changes the structure of G8 PAMAM dendrimer, a disagreement with some experimental interpretations. J. Phys. Chem. B. 110, 25628–25632.

Maiti, P.K., Messina, R., 2008. Counterion distribution and ζ-potential in PAMAM dendrimer. Macromolecules 41, 5002–5006.

Maiti, P.K., Cagin, T., Wang, G., Goddard, W.A., 2005. Effect of solvent and pH on the structure of PAMAM dendrimers. Macromolecules 38, 979–991.

Mandel, M., 1993. Some properties of polyelectrolyte solution and the scaling approach. In: Hara, M. (Ed.), Polyelectrolytes: Science and Technology. Dekker M., Inc, New York.

Manning, G.S., 1969. Limiting laws and counterion condensation in polyelectrolyte solutions I. Colligative properties. J. Chem. Phys. 51, 924–934.

Muthukumar, M., 2004. Theory of counter-ion condensation on flexible polyelectrolytes: adsorption mechanism. J. Phys. Chem. 120, 9343–9350.

Nisato, G., Ivkov, R., Amis, E.J., 2000. Size invariance of polyelectrolyte dendrimers. Macromolecules 33, 4172–4176.

Pande, S., Crooks, R.M., 2011. Analysis of poly(amidoamine) dendrimer structure by UV–vis spectroscopy. Langmuir 27, 9609–9613.

Roe, R.J., 1974. Multilayer theory of adsorption from a polymer solution. J. Chem. Phys. 60, 4192–4208.

Ruggeri, F., Akesson, A., Chapuis, P., Nielsen, C.A.S., Monopoli, M.P., Dawson, K.A., et al., 2013. The dendrimer impact on vesicles can be tuned based on the lipid bilayer charge and the presence of albumin. Soft Matter. 9, 8862–8870.

Scheutjens, J.M.H.M., Fleer, G.J., 1985. Interaction between two adsorbed polymer layers. Macromolecules 18, 1882–1900.

Srinivasan, M., Rajabi, M., Mousa, S., 2015. Multifunctional nanomaterials and their applications in drug delivery and cancer therapy. Nanomaterials 5, 1690–1703.

Tokarczyk, K., Jachimska, B., 2017. Quantitative interpretation of PAMAM dendrimers adsorption on silica surface. J. Colloids and Surface 503, 86–94.

Van der Schee, H.A., Lyklema, J., 1984. A lattice theory of polyelectrolyte adsorption. J. Phys. Chem. 88, 6661–6667.

Welch, P.M., 2013. The escape of a charged dendrimer from and oppositely charged planar surface. J. Chem. Phys. 139, 164906-1–164906-8.

Welch, P.M., Muthukumar, M., 1998. Tuning the density profile of dendritic polyelectrolytes. Macromolecules 31, 5892–5897.

Yang, K., Weng, L., Cheng, Y., Zhang, J., Wu, Q., Xu, T., 2011. Host–guest chemistry of dendrimer–drug complexes. 6. Fully acetylated dendrimers as biocompatible drug vehicles using dexamethasone 21-phosphate as a model drug. J. Phys. Chem. B 115, 2185–2195.

Zhao, L., Cheng, Y., Hu, J., Wu, Q., Xu, T., 2009. Host–guest chemistry of dendrimer–drug complexes. 3. Competitive binding of multiple drugs by a single dendrimer for combination therapy. J. Phys. Chem. B 113 (43), 14172–14179.

CHAPTER

Nanostructure lipid carriers 10

Jan Sobczyński and Gabriela Bielecka

Department of Applied Pharmacy, Medical University of Lublin, Lublin, Poland

CHAPTER OUTLINE

10.1 INTRODUCTION

10.1.1 NANOCARRIERS IN DRUG DELIVERY

New drugs introduced to the pharmaceutical market are in 40% of cases sparingly water-soluble, leading to problems regarding formulation, bioavailability, and stability (Merisko-Liversidge and Liversidge, 2008, 2011). It is believed that the use of nanocarriers in medicine and pharmacy has begun to revolutionize the market due to their unique properties. Drugs in nanocarriers are protected from the impairing environmental factors by absorbing them on the material that nanoparticle is made of, closing them in the particle matrix, or enabling the drug

Nanoparticles in Pharmacotherapy. DOI: https://doi.org/10.1016/B978-0-12-816504-1.00006-5

dissolution within the matrix constituents (Bamrungsap et al., 2012). Moreover, use of such drug delivery system helps to circumvent adverse reactions occurring in case of conventional dosage forms. Interestingly, nanoparticles show enhanced permeability and retention effect (EPR) that is based on their preferential accumulation in tumor tissue. Partitioning of nanocarriers into other tissues with enhanced blood flow (inflammatory, infectious) gives a chance to use them in a targeted therapy. Moreover, lipid nanoparticles possess a significant potential to overcome the first-pass metabolism. Because of those features, nanoparticles enhance therapeutic efficacy by improving pharmacokinetics and biodistribution, and protect tissues from a drug's toxic effects (Bamrungsap et al., 2012; Maeda, 2012; Venishetty et al., 2012).

10.1.2 SOLID–LIPID NANOPARTICLES

In the 1990s solid–lipid nanoparticles (SLNs), the first generation of lipid nanocarriers comprising a solid lipid matrix, were begun to be studied as one of the attempts to solve these problems (Müller et al., 2000). SLNs are made of lipids that remain solid also at body temperature. The production does not require the use of organic solvents, and there is less toxicological risk in as compared with other nanosized drug–carrier system, for example, polymeric carriers made of nonbiodegradable polymers or biodegradable macromolecular materials (Beloqui et al., 2016). Lipid carriers also surpass polymeric nanocarriers in regard to their biodegradability and possibility to release substances in a controlled manner. However, it turned out that there are some limitations regarding the use of SLNs. Because of the crystalline "brick wall" structure, a polymorphic transition may take place during storage. It can result in drug displacement from the carrier, which reduces drug loading over time (Müller et al., 2002b). Those drawbacks of SLNs led to development of a new generation of lipid nanocarriers—*nanostructured lipid carriers*, which are made of solid matrix fused with liquid lipids.

10.1.3 NANOSTRUCTURED LIPID CARRIERS

Nanostructured lipid carriers (NLCs) are a delivery system that showed very short time between their invention by Muller in 1999/2000 and their introduction to the market. The very first product was Nanorepair Q10 Cream (Dr. Rimpler, Wedemark, Germany) cosmetic, marketed since 2005. This short development term shows its significant potential. As previously said, NLCs were introduced to overcome the limitations of SLNs. To improve a carrier's properties an ordered structure of SLNs was replaced by a unstructured matrix made of a mixture of solid and liquid lipids. As a hybrid particle core immobilizes the drug, it prevents its outflow from the carrier. Moreover, the liquidity of oil droplets enhances the drug loading capacity in comparison to SLNs. It has also been suggested that the presence of a liquid lipid may facilitate emulsification as the energy of heat is more evenly distributed throughout the melted lipid (Gokce et al., 2012). Because

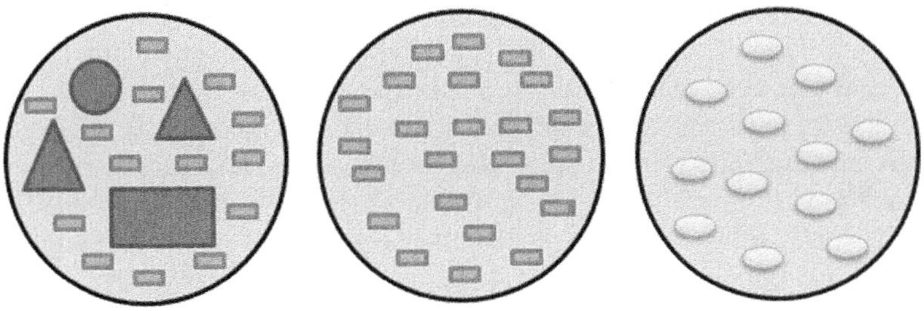

FIGURE 10.1

Schematic diagrams of NLC. (A) NLC type I. (B) NLC type II. (C) NLC type III.

of all of those advantages, NLCs have been investigated and developed on a large scale in recent years (Müller et al., 2000, 2002a,b, 2011).

Depending on the manner that drugs are enclosed into NLCs, three types of NLCs might be distinguished. An imperfect crystal model (NLC type I) is composed of a mixture of lipids that are in different shapes and an ordered structure is not possible to obtain. A structure full of imperfection allows to encapsulate active agents. It is formed after mixing a small amount of oils with solid lipids.

An amorphous model (NLC type II) is obtained after combining special types of lipids like hydroxyoctacosanyl hydroxystearate and isopropyl myristate. These lipids do not recrystallize, thus the drug expulsion is avoided. Lipids stay solid, because they are transferred into the polymorphic α form. A multiple model (NLC type III) is formulated to increase loading capacity of agents of better solubility in liquid lipids than in solid ones. In the formulation process a little compartment of oil in a matrix composed of solid lipids is created. Due to the fact that the oil contains more drug and a presence of a barrier in the form of solid lipids enclosing oil droplets the release process is controlled (Müller et al., 2002a). Fig. 10.1 shows the schemes of NLC types.

10.2 NLCS FABRICATION AND CHARACTERIZATION

10.2.1 NLC EXCIPIENTS

NLCs consist of a few fundamental ingredients: mixed solid and liquid lipid for building unstructured core (blended usually in 70:30 to even 99.9:0.1 ratio), water, and emulsifiers (1.5%–5% content) for aqueous phase (Pardeike et al., 2009).

Typically used solid and liquid lipids can be subdivided into groups based on their saturation, chain length, and degree of substitution. The aliphatic chains of short-chain fatty acids are shorter than 6 carbon atoms, medium-chain possess

6–12 carbon atoms, long-chain 13–21 carbon atoms, and the C number for very-long-chain exceeds 22. Saturated fats are typically used as a solid phase in NLCs. Stearic acid and saturated palm oil, a mixture of fatty acids C8–C18 (Softisan), are the most commonly used solid lipids. Other excipients include diglycerides (glycerol and two fatty acid chains connected by ester bond)—for example, glyceryl dibehenate (Compritol 888 ATO), glyceryl palmitostearate (Precirol ATO 5). Another popular group comprises triglycerides–glycerol esters of three identical fatty acids, for example, tripalmitin (Dynasan 116) and tristearin (Dynasan 118). It is also possible to use waxes, for example, cetyl palmitate (Crodamol CP, Cutina CP) as well as aromatic and aliphatic alcohols. They include sterols, for example, cholesterol and cetyl alcohol (Lanette 16, Crodacol C90). Other lipids include ethoxylated long-chain fatty alcohols, for example, ceteth-20 (polyethylene glycol ether of cetyl alcohol) or steareth-20 (polyethylene glycol ether of stearyl alcohol) or their mixtures (e.g., Emulcire 6). Table 10.1 shows more information regarding the lipid excipients. All these lipids are solid at room temperature and meltable, so that one can obtain nanocarriers in laboratory conditions. (http://www.lipidlibrary.aocs.org/; Strickley, 2004; Nanjwade et al., 2011; Jenning et al., 2000a,b). The processing temperature is usually set at 10°C above the melting point of the solid lipid. The solid and liquid lipid excipients described above (also those in Table 10.1) are "generally recognized as safe" (GRAS), which means they are approved in the United States and Europe for clinical use. According to the US FDA, the general recognition of safety through scientific procedures is based upon the application of generally available and accepted scientific data, information, or methods, which ordinarily are published, as well as the application of scientific principles, and may be corroborated by the application of unpublished scientific data, information, or methods (http://www.accessdata.fda.gov/scripts/fdcc/?set = SCOGS).

Liquid lipids commonly utilized in NLCs manufacturing are biocompatible, mostly naturally occurring obtained from plants, animals, or petroleum refining process. The medium chain triglycerides (MCT), known under brand name Miglyol 812, are the most commonly employed liquid lipid (Jenning et al., 2000a,b). It is also possible to use oleic acid, liquid paraffin, a refined oil from petroleum or squalene, a hydrocarbon obtained from shark liver, or olives (Fang et al., 2013). Furthermore, long chain fatty alcohol (e.g., 2-octyl dodecanol), esters or diesters like isopropyl myristate (Crodamol IPM) and propylene glycol dicaprylocaprate (Labrafac) may be used. Some excipients serve dual role as components of the lipidic core and dermal permeation enhancers, for example, oleic acid, linoleic acid, and decanoic acid. There are examples of using an oil obtained from Siberian pine seed or Siberian fish oil as a liquid lipid phase in NLCs (Averina et al., 2010, 2011). Another alternative is the use of tocol derivatives, for example, tocopherol acetate that can additionally protect the drug from oxidation. Tsai et al. (2012) encapsulated baicalein in NLCs consisting of vitamin E as a liquid lipid showing improved baicalein stability during storage and enhanced penetration into the central nervous system. Emami et al. (2012) prepared a tumor

Table 10.1 Excipients Used for NLC—Examples of Formulations Available and not Available on the Market

		Name	Chemical Group(s)	Brand Name	Example of Appliance
Lipids	Solid	Stearic acid	Saturated C-17 chain fatty acid		Flurbiprofen (transdermal application)
		Saturated palm oil	Mixture of C8–C18 fatty acids	Softisan	
		Glyceryl dibehenate	Diglycerides	Compritol 888 ATO	Benzocaine; cyproterone (dermal formulations)
		Glyceryl palmitostearate	Diglycerides	Precirol ATO 5	Actitretin (dermal formulations) Buprenorphine (transdermal formulation)
		Trimyristin	Triglycerides	Dynasan 114	Flurbiprofen (transdermal application)
		Tripalmitin	Triglycerides	Dynasan 116	Simvastatin + Olanzapine (transdermal formulation)
		Tristearin	Triglycerides	Dynasan 118	
		Cetyl palmitate	Wax	Crodamol CP, Cutina CP	Azelex cream (Allergan)
		Cholesterol	Aromatic alcohol		Minoxidil (dermal application)
		Cetyl alcohol	Aliphatic alcohol	Lanette 16, Crodacol C90	
		Cetearyl alcohol		Lanette O	Lornoxicam (transdermal application)
		Ceteth-20	Polyethylene glycol ether of cetyl alcohol	In mixture with Steareth-20–Emulcire 61	
		Steareth-20	Polyethylene glycol ether of stearyl alcohol	In mixture with Ceteth-20–Emulcire 61	
		Vegetable oil	Mono- and diglycerides	Imwitor 900	
		A mixture of monoglycerides, mainly monostearoylglycerol, with di- and triacylglycerides	Mono-, di-, triglycerides	Geleol	

(Continued)

Table 10.1 Excipients Used for NLC—Examples of Formulations Available and not Available on the Market *Continued*

	Name	Chemical Group(s)	Brand Name	Example of Appliance
Liquid	MCT (medium chain triglycerides)	Captylic/capric tryglycerides	Miglyol; Labrafac Lipophile WL 1349	Avodart soft gelatin capsules (GlaxoSmithKline)
	Liquid paraffin			
	Squalane			Diphenciprone (dermal formulation)
	Vitamin E	Tocols		Paclitaxel formulation
				NeoralSoft Gelatin Capsules, Neoral Oral Solution (Novartis)
	Diethylene glycol monoethyl ether	Ether/alcohol	Transcutol HP	All-trans retinoic acids (dermal formulation)
	Mixture of propylene glycol mono- and diesters of lauric acid	Esters	Lauroglycol FCC	
	Propylene glycol monocaprylate	Esters	Capryol 90	
	Lauroyl polyoxylglycerides		Gelucire 44/14	Lipofen capsules
	Glyceryl monostearate	Esters	Myverol 18-04K	
	2-Octyl dodecanol	Long chain fatty alcohol		
	Isopropyl myristate	Esters	Crodamol IPM	
	Propylene glycol dicaprylocaprate	Esters	Labrafac	
	Oleic acid	Long-chain fatty acids		Econazole (dermal formulation)
				Simvastatin + Olanzapine (transdermal formulation)
	Linoleic acid			
	Decanoic (Capric) acid	Medium-chain fatty acids		
	Siberian pine seed oil			Siberian pine seed oil used as an active agent (dermal application)
	Siberian fish oil			

Emulsifiers	Hydrophilic	Poloxamers	Copolymers of poly (propylene oxide) and poly(ethylene oxide)	Pluronic L44; F68; F108; F127	Diphenciprone (dermal formulation)
		Polysorbates	Polyoxyethylene (20, 40, 80) sorbitan monolaurate-polymer	Tween 20, 40, 80	All-trans retinoic acids Epidermal growth factor Finasteride (dermal formulations) Targretin (Valeant)
		Polyvinyl alcohol	Alcohol		
		Macrogol/polyethylene glycol (PEG)	Polymer		
		Polyoxyethylated 12-hydroxystearic acid	Polymer	SolutolHS 15/ Kolliphor HS 15	Oxidize (Beta S.A.)
		Sodium deoxycholate	Salt of bile acid		
	Lipophilic	Mixture of glyceryl monostearate and glyceryl monopalmitate	Monoglycerides	Myverol18-04K	
		Sorbitan laurate		Span20	Coenzyme Q10 (dermal application)
		Sorbitan monopalmitate		Span40	
		Sorbitan stearate		Span60	
	Amphiphilic	Egg/soy lecithin	Mixture of phospholipides	Epikuron200	Fluconazole (dermal formulation) Baycip (Bayer)
		Phosphatidylocholines	Phospholides with incorporated choline group		
		Stearoyl macrogol-32 glycerides	Polymer	Gelucire 50/13	

Beloqui et al. (2016), Averina (2010, 2011).

targeted NLCs composed of cholesterol and vitamin E, loaded with paclitaxel. The formulation exhibited antineoplastic efficiency comparable to the approved paclitaxel preparation against colon cancer cells.

To stabilize the emulsion such as NLC dispersed in water, there is a need to use emulsifiers. They may be divided into hydrophilic, lipophilic, and amphiphilic. The most popular emulsifiers are thought to be hydrophilic ones, that is, ethylene-oxide derivatives like Pluronics, polysorbates (Tween), and polyvinyl alcohol (Fang et al., 2013). While the European Pharmacopoeia lists five Pluronics (L44, F68, F88, F108, and F127), the US Food and Drug Administration (FDA) has approved several Pluronics as pharmaceutical excipients (http://www.accessdata.fda.gov/scripts/fdcc/?set = SCOGS). The hydrophilic surface composed of a low molecular mass, biocompatible polyethylene glycol (PEG), or polyethylene oxide block (PEO) blocks seems advantageous, due to an increase in circulation time of the drug administered intravenously, as a consequence hindering protein absorption on the NLCs surface (Moghimi et al., 2001). It also prevents particle aggregation as well as opsonization in the bloodstream and subsequent clearance by the mononuclear phagocytes system (MPS). It should be noted that in case of NLC, a low surfactant concentration is required to obtain a stable emulsion, reducing the toxicity concerns. Furthermore, there is a range of emulsifiers suitable for use in NLC preparation, contrary to other lipid-based delivery systems (lipid solutions, emulsions, self-emulsifying drug delivery system, self-microemulsifying drug delivery system, micellar systems) (Khan et al., 2015). Emulsifiers typically used in NLCs are listed in Table 10.1.

10.2.2 NLC PREPARATION

Although there are a variety of preparation methods, high pressure homogenization (HPH) at high temperature is the most popular and preferred. As there is no need to use any solvents during the process, the technique is more environment-friendly and less costly. It is also simple—an aqueous solution in double distilled water of hydrophilic surfactants is added to a drug and lipophilic emulsifier(s) dissolved in a mixture of melted lipids (about 10°C above melting point). The resulting mixture is homogenized in a high-shear homogenizer afterwards, resulting in a hot nanoemulsion. At the end, solidification takes place by cooling the mixture, and the final product is obtained. It has been observed that the HPH technique yields smaller particles, as compared with other techniques, especially when the proportion of liquid to solid lipid is sufficiently high (Vitorino et al., 2013). Shear stress and cavitation forces will cause the disruption of larger particles and subsequent formation of nanosized vesicles. On the other hand an increase of the homogenization time yields only small improvement in particle size with various emulsifier concentration. Despite the mentioned advantages, the method is not applicable for heat-sensitive drugs, but trials to overcome the problem have been performed. By reducing the temperature of the process, Hung et al. (2011) obtained better stability of thermolabile vitamin E and β-karoten.

Cold high-pressure homogenization is another method, which begins with melting lipids. Further, the mixture is solidified in liquid nitrogen or dry ice and ground by a powder mill. The obtained microparticles are added to a cool solution of emulsifiers. Dispersed particles are homogenized directly to form NLCs at room or lower temperature. The process has the advantage of avoiding drug decomposition under high temperature, but not completely as the drug must be added to a hot lipid mixture to be dissolved or suspended. The disadvantage of the method is the fact that the formulation produced is not always nanosized. The latter is attributed to the shortage of hot treatment (Müller et al., 2000; Joshi and Patravale, 2006; Das and Chaudhury, 2011).

Microemulsion technique is another alternative, also suitable for industrial scale. The hot lipid and aqueous phase are blended together to give transparent or semitransparent mixture. Next, the obtained microemulsion is added to a cold water medium (2°C–3°C) where a precipitation of particles takes place. In case of diluted samples, additional concentration process may be performed (Joshi and Patravale, 2006). Emulsification–sonication starts with melting lipids and adding the drug into mixture. Further, aqueous solution of hydrophilic substances (heated to same temperature as liquid mixture) is added to melted lipids and homogenized by high-shear homogenizer till emulsion is obtained. Next, the emulsion is subjected to sonication, and in the end, during cooling at room temperature, NLCs are obtained (Das and Chaudhury, 2011). To obtain a narrower size distribution of the particles, the process may be performed in a probe-type sonicator.

Solvent emulsification–evaporation is suitable for thermolabile drugs, however an important drawback is the use of organic solvents (e.g., cyclohexane and chloroform) required to dissolve lipids. The lipid mixture is added to aqueous phase with emulsifiers. Constant stirring must take place to form an emulsion and enhance solvent evaporation. During this process lipids precipitate and form nanoparticles. A modification of this method is called double emulsion, although it is used in NLCs containing hydrophilic drugs. Solvent diffusion is a method that employs partly miscible organic solvents like benzyl alcohol. In the process a diluted mixture is produced that needs to be further concentrated by a lyophilization or ultrafiltration, as similar to the microemulsion method (Das and Chaudhury, 2011). Solvent injection technique uses a water-miscible solvent (e.g., acetone, methanol, isopropanol, or their mixture) to dissolve lipid phase. Lipid phase is then quickly injected into aqueous solution containing hydrophilic surfactants. The method is easy and fast (Schubert and Müller-Goymann, 2003).

As mentioned in previous paragraphs, numerous production and composition variables can be selected to optimize the NLC formulation. Often a computational approach is used to find the right composition and set most advantageous fabrication conditions. Negi et al. used the Plackett–Burman design to optimize preparation of NLCs containing antineoplastic agent irinotecan. The authors selected 11 production and composition variables and tested their influence on NLC size. Four parameters, that is, temperature, needle size, stirring, and injection speed were found not to influence average particle diameter. Out of the remaining factors, total lipid content and concentration of three surfactants (Pluronic F68, Tween 80 and sodium deoxycholate) were found to influence the size. The last three parameters, namely the phase ratio, drug to lipid ratio, and sonication time influenced the entrapment efficiency (EE). These three variables were varied using Box–Behnken design response surface method to optimize the EE. The optimal NLC were characterized by the following properties: 143.52 ± 1.2 nm diameter, zeta potential of −32.6 ± 0.54 mV, and 98.22 ± 2.06% EE. Overall, 38 trials were needed to select the optimal conditions.

10.2.3 NLC CHARACTERIZATION

To optimize properties of the complex drug delivery such as NLCs, it is crucial to evaluate their physicochemical features. Many features, for example, particle size and zeta potential, influence the stability and quality of the formulation and subsequently determine the NLCs' therapeutic efficacy. Carriers dimensions depend on the excipients used in the formulation process. A careful selection of NLC emulsifier usually yields more well defined, smaller particles (Jenning et al., 2000a,b). Particle size is typically assessed by dynamic light scattering (DLS), also called photon correlation spectroscopy (PCS). DLS measures particles ranging from <1 nm up to few micrometers. The Brownian motion of particles or molecules results in scattering of laser light at different intensities. DLS works by measuring fluctuations of the intensity of dispersed light due to particle movements (Mehnert and Mäder, 2001). The larger particles will move more slowly, resulting in fewer intensity fluctuations. Analysis of these intensity fluctuations enables to determine the particle size using the Stokes–Einstein relationship. While the knowledge of the viscosity of the liquid is required, the technique requires no calibration or sample preparation steps. DLS is also useful in specifying polydispersity indices of the colloidal particles. Another technique, laser diffraction (LD), allows to measure particles of larger size, that is, hundreds of nanometers up to several millimeters by assessing the diffraction angle of the particle radius. As the laser beam passes through the sample, small particles scatter light at large angles. On the other hand, large particles dispersed within the sample scatter light at small angles relative to the laser beam. The angular scattering intensity data is used to calculate the size of the particles using the Mie theory of light scattering. As particles are treated as spheres, their size is reported as a volume equivalent to sphere diameter (Fig. 10.2).

Due to the fact that the analyzed particles' shape often deviates from spherical, it is advisable to conduct a study using two size measuring methods. The most widely reported supplementary techniques are based on electron microscopy, namely scanning electron microscopy (SEM) and transmission electron microscopy (TEM). While SEM is based on electron transmission from the surface of the tested formulation, conducting TEM analysis engages electrons passed throughout the sample (Iqbal et al., 2012). The techniques allow measuring particle radius and size distribution, as in PCS. Contrary to PCS, electron microscopy

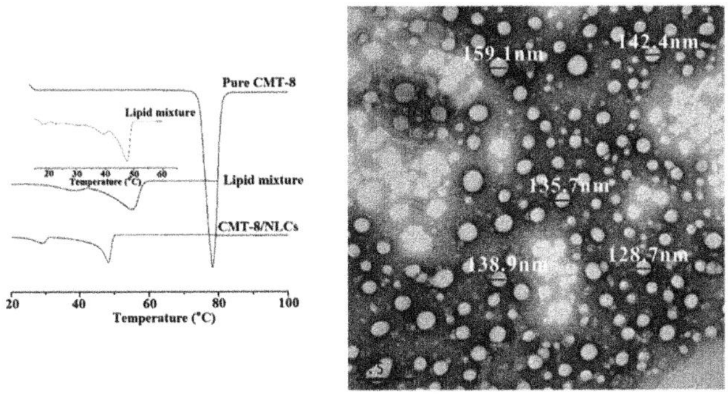

FIGURE 10.2

Sample graph showing results of DSC (left) and TEM (right) measurements.

Reprinted with permission from Zhao, S., Yang, X., Garamus, V.M., Handge, U.A., Bérengère, L., Zhao, L., et al., 2014. Mixture of nonionic/ionic surfactants for the formulation of nanostructured lipid carriers: effects on physical properties. Langmuir 30(23), 6920–6928. Copyright 2014 American Chemical Society.

reveals the nanocarrier morphology showing particle shape, roughness, and irregularities (Hwang et al., 2009). Furthermore, TEM and SEM are helpful in visualizing the carriers in two dimensions. Atomic force microscopy (AFM) is also useful in characterizing NLCs. It allows a wide range of information about nanocarrier surface of particles reaching even 10^{-10} m, and also includes 3D illustration of a formulation (ZurMühlen et al., 1996).

Another parameter that accounts for the stability of colloidal dispersions is the zeta (ζ) potential. The zeta potential value exceeding +30 mV or lower than −30 mV for negatively charged particles ensures repulsion between the charged particles. Consequently the tendency to aggregate is diminished and NLCs are not likely to undergo flocculation (Müller et al., 2000; Freitas and Müller, 1998). Interestingly, to accomplish satisfactory penetration to the central neural system, particles have to be charged positively. This is due to the negatively charged sites on the blood–brain barrier (Parveen and Sahoo, 2008).

Other important features of NLCs are the drug encapsulation efficiency and loading capacity. A high solubility of the drug in the melted lipids is a prerequisite for efficient drug solubilization within the lipid matrix. It should be noted that the NLC fabrication process often includes high processing temperature. After the mixture of NLC components is cooled down, the solubility may decrease and subsequently the loading capacity is reduced (Müller et al., 2000). The entrapment efficiency specifies the content of the drug in the nanoparticles and provides insight into the drug content successfully incorporated into the carriers. It is calculated indirectly from the difference between amount of drug added initially and amount of drug remaining in the nonlipid phase after preparation. The latter is then divided by the whole amount of drug added and multiplied by 100 to obtain a result as a percentage (Joshi and Patravale, 2008).

Highly valuable technique that was employed by, for example, Esposito et al. (2013) is the field flow fractionation. The technique separates particles based on their size using various forces. While the asymmetric field flow fractionation (AF4) employs gravitation, a centrifugal fractionator separates particles using centrifugal force. The particles are separated in dedicated channels as similar to HPLC columns. The unfiltered sample is introduced into the channel and is concentrated in the small area by manipulating flow forces. This ensures good resolution for low-concentration samples. Further sample is separated and can be directed to one or a few detectors. While, the multiangle light scattering detector can be used to determine the gyration radius of particles, the DLS device enables to determine the hydrodynamic particle radius. Typical concentration detectors, for example, UV or diode array detector (DAD) can be used to assess the content of drug in different particle populations. Aside from its analytical potential, the technique may also be used to collect particle fractions for further analysis.

Another useful technique seems to be surface tension measurements. Wilhelmy plate method is utilized in measurement to find proportion of emulsifiers to be used as NLC stabilizers. Two surfactants are mixed in different proportions and the critical micelle concentration of mixtures is measured, as for optimizing mixed micellar formulations. Briefly, a low critical micelle concentration of surfactant pair denotes their synergistic interactions. Such surfactant combination is expected to perform better when mixed with NLC lipids (Forny et al., 2009).

The crystallinity of the lipids in NLCs is associated with the composition of the lipid nanoparticle. The more liquid lipid content, the more imperfections can be found in the core structure. The greater extent of imperfections consequently yields greater encapsulation capacity for drug particles and less crystalline structure of the particles. To analyze the lipids in the particles differential scanning calorimetry (DSC) and X-ray diffraction may be employed. DSC operates on the principle of measuring the enthalpy and melting points of the lipids (Hu et al., 2006) whilst the other technique is based on detecting the intensity of scattered X-ray beam diffraction. Moreover by means of X-ray diffraction the results obtained with DSC can be confirmed (Teeranachaideekul et al., 2007) (Fig. 10.3).

Small-angle X-ray scattering (SAXS) and small-angle neutron scattering (SANS) are used to characterize particles of nanometer size, including NLC. With this method it is possible to elucidate the 3D structure of the particles (Zhao et al., 2014). SAXS is thought to be a complicated and laborious method (Craievich, 2002). SANS is a method employing elastic neutron scattering and providing a wide range of data: shape, size, surface of the investigated material (Baccile et al., 2010). Several other methods are in use to evaluate the status of the particle. A fluorometric spectroscopy is in use to assess the polarity and molecular environment of NLCs (Jores et al., 2005). If the solubilized drug exhibits fluorescence, it will show shifts of the fluorescence emission maximum, depending on the polarity of microenvironment, where the drug is located

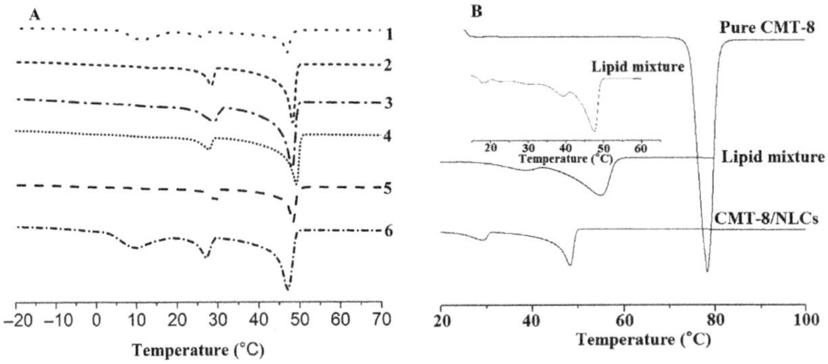

FIGURE 10.3

Sample graph showing results of DSC measurements.

Reprinted with permission from Zhao, S., Yang, X., Garamus, V.M., Handge, U.A., Bérengère, L., Zhao, L., et al., 2014. Mixture of nonionic/ionic surfactants for the formulation of nanostructured lipid carriers: effects on physical properties. Langmuir 30(23), 6920–6928. Copyright 2014 American Chemical Society.

(Schäfer-Korting et al., 2007; Borgia et al., 2005). Thus the location of the drug further influencing the release properties can be evaluated.

Table 10.2 sums up the analytical techniques applied in NLC research.

10.3 SITE-SPECIFIC DRUG DELIVERY

10.3.1 PERORAL NLC FORMULATIONS

10.3.1.1 Lipids in gastrointestinal system

As the NLCs consist mostly of digestible and biocompatible lipids, they will possess a distinct absorption mechanism in the gastrointestinal tract as compared with other nanoparticulate delivery systems. The mechanism will correspond to dietary lipid uptake found for natural lipidic substances. The first step of lipid digestion takes place in the stomach, where solid lipids are melted at body temperature. Following peristaltic movements of the stomach muscle, lipids are mixed with water and an emulsion is made. While triglycerides constitute 90% of dietary fat, the rest comprises cholesterol, cholesteryl esters, phospholipids, and nonesterified fatty acids. The decomposition of triglycerides begins with a gastric lipase action and about 30% of the triglycerides undergo lipolysis by gastric lipase. Although the process starts slowly, it accelerates in time, when the level of free fatty acids increases. It happens because free fatty acids facilitate formation of the emulsion (Dominiczak and Broom, 2014). Lipid molecules on the surface of emulsified droplets are easily accessible to the lipase, as compared with molecules dispersed in aqueous phase. The smaller the droplets are, the faster the lipid degradation takes place, as smaller droplets exhibit more beneficial surface

Table 10.2 Methods of Evaluating Physicochemical Features

Method Name	Basis of Action	Relevance for NLC
Scanning electron microscopy (SEM)	Scanning a sample with a beam of electrons and detecting signals from the sample	Measuring: particle radius and size distribution; observing shape and morphology
Transmission electron microscopy (TEM)	Detecting signals after transmission of a beam of electrons through a sample	Measuring: particle radius and size distribution; observing shape and morphology; visualization of particles
Atomic Force Microscopy (AFM)	Measuring variety of forces between the sharp-tipped probe and a sample	Measuring morphology and surface of extremely small particles
Wilhelmy plate method	Measuring the force exerted on a plate placed perpendicular to the interface between liquid and air	Finding proportion of emulsifiers used as NLC stabilizers
Differential scanning calorimetry (DSC)	Measuring the amount of heat absorbed during melting and the amount of heat released during crystallization	Determination the state of lipids, fusion and crystallization; suggests in which lipid (solid or liquid) drug has a better solubility
X-ray diffraction	Detecting the intensity of X-ray beam diffraction, resulting from interaction of this radiation with the electron clouds of atoms forming a sample	Determination the status of lipids
Field flow fractionation	Separating particles based on their mobility in a channel filled with solution or suspension on which various forces are exerted	Separating different particles populations and possibility of estimating drug content in each population (with using detector)
Small-angle X-ray scattering (SAXS)	A sample is exposed to X-ray beam, and a scattered radiation is registered closely to the primary beam	Elucidating three-dimensional structure of particles, shape, size, distribution
Small-angle neutron scattering (SANS)	A sample is exposed to a beam of neutrons, then a scattered neutrons are detected	
Fluorometric spectroscopy	Registering shifts of fluorescence emission maximum of compounds exhibiting fluorescence	Evaluation of location of the drug in the delivery system
Laser diffraction (LD)	Passing beam of the laser light through the sample and measuring angle of scattered light	Evaluating size of NLC
Dynamic light scattering (DLS)	Measuring fluctuations of the intensity of dispersed light	
Zeta potential measurements	Measuring electrophoretic mobility of nanoparticles	Evaluation of particle surface charge

Esposito et al. (2013), Baccile (2010), Zhao (2014), Fang et al. (2013), Sahana (2010), http://biosaxs.com/technique.html (accessed 20.01.2017).

area-to-volume ratio. Since the reaction speeds up due to locally increased substrate concentration in a lipid droplet, the size of digested lipid droplets is the crucial variable controlling the digestion of fats.

After the alimentary products pass to the duodenum the pancreatic lipase digests the remaining triglycerides yielding 2-monoacylglycerols (2MAGs), monoglycerides (to a lesser extent) and free long-chain fatty acids. The degradation products associate with bile acids and form micelles that absorb into the bloodstream through the small intestine wall. Most fatty acids and 2-MAGs are absorbed through the enterocyte membrane by diffusion. Inside the enterocyte, the triglycerides are reassembled and transferred to chylomicrons, which subsequently are transported to lymph. Contrary to long-chain fatty acids, the absorption of medium- and short-chain fatty acids and their respective triglycerides follows a simpler transport mechanism. They essentially do not require digestion, forming micelles, and any transformation inside the enterocyte. Instead they relocate directly to hepatic portal system (Dominiczak and Broom, 2014; Feinle-Bisset and Azpiroz, 2013). The very same mechanism accounts for increased oral bioavailability of the drug encapsulated in lipid-based formulations.

It has been found that the higher the lipid content is in the meal, the longer it stays in the stomach. Consequently, the longer residence time results in more efficient lipid emulsification. Moreover, high-fat food stimulates bile flow yielding an improved fat solubilization in the small intestine. Due to these factors, absorption of hydrophobic drugs loaded into NLC is enhanced. The drug is incorporated into micelles made of hydrolyzed triglycerides and bile and further transferred with lipids into enterocytes. The following drug biodistribution depends on the excipients and drug lipophilicity. Most APIs are transferred into the portal venous system, although lipids accompanying lipophilic drugs and macromolecules enhance lymph transportation (Li et al., 2001; Martinez et al., 2002; Das and Chaudhury, 2011). Interestingly, nanosized particles may be transferred more easily through lymphatic capillaries than blood capillaries (Sanjula et al., 2009). It is observed that long-chain triglycerides are more efficient lymphatic absorption promoters than medium-chain triglycerides and that C14 to C18 free fatty acids improve the absorption (Khoo et al., 2003; Porter and Charman, 2001). According to these factors and to the fact that drugs entering lymph instead of blood from the intestine avoid first-pass metabolism, it is thought that lipid-based formulations, including NLCs, improve bioavailability (Trevaskis et al., 2008). Liquid dosage forms are easily swallowed and are thus acceptable by patients. However, liquid preparations may require taste masking as they tend to interact with the taste buds due to their unrestricted mobility.

10.3.1.2 Examples of peroral NLC delivery

Aditya et al. (2013) prepared NLC-based nutraceuticals containing genistein and curcumin intended for oral delivery. Although the two agents exhibit a range of biological activities, they are not used as drugs. Their flaws include high lipophilicity and resulting low aqueous solubility as well as low permeation across biological membranes. They are easily degraded by oxidation, light, and heat. In the human body they undergo significant first-pass metabolism. The agents were incorporated into NLCs composed of glyceryl monostearate, oleic acid, lecithin, and Tween 80 and stabilized by the addition of polyvinyl alcohol. Nanoemulsion was obtained using high speed homogenizer followed by ultrasonication. NLC formulation was characterized with distinctive zeta potential, that is, −46 mV, −49 mV, and −47 mV for blank, curcumin- or genistein-loaded, and genistein + curcumin coloaded formulations, respectively. The NLC size was 125 ± 6 nm, 108 ± 7, and 122 ± 6 nm for blank, curcumin- or genistein-loaded, and genistein + curcumin co-loaded formulations, respectively. At the formulation optimization step a decrease in particle size was observed upon increasing lecithin content. The polydispersity index (PDI) was around 0.3 irrespective of NLC loading with any of the active ingredients. The entrapment efficiency was increased to 93% upon coloading NLCs with curcumin and genistein, as compared with 78% loading obtained for NLCs loaded with a single agent. While the formulations were stable in simulated gastric fluid (SGM) during 2 hours, the simulated intestinal fluids (SIM) contained enzymes that promoted NLC degradation and aggregation. The hydrolysis of superficial particle layer by lipase exposed the particle core to hydrolysis by digestive enzymes. To study the stability of active agents, NLCs were mixed with SGM and SIM. After 6 hours, 8% of curcumin and 15% of genistein was degraded. The NLC formulation showed a prolonged release of active agents, as around 61% of genistein and 55% of curcumin was released within 8 hours. Further antineoplastic activity of various formulations was tested on human prostate cancer cell line PC3 employing a WST-1 viability assay in vitro. Coloaded formulation showed higher activity as compared with NLCs loaded with single agent, genistein and curcumin solutions, or blank NLCs.

10.3.2 NLC AS DRUG DELIVERY CARRIERS FOR USE IN SUPERFICIAL INFECTIONS

One of the attempts to use a new form of drug was by incorporating NLCs into the form of cream or gel. The beneficial features of nanocarriers, that is, uniformity, safety, reproducibility can be combined with the advantages of more traditional formulations, that is, semisolid dosage forms, that can easily be administered in the desired location. Moreover, a cream or gel with incorporated 5% of lipid nanoparticles is favorable because of achieving satisfactory occlusive effect needed in dermal absorption. The skin air pores become smaller following application of a semisolid preparation, thus resulting in reduced water evaporation. Hydrogels provide an inert matrix for NLC formulation intended for use on mucosal lining. Hydrogels swell upon hydration subsequently forming hydrogen bonds that lead to binding to mucosal surfaces. Although hydrogel use may be associated with significant burst drug release, there are ways to address the issue. The dissolution-delaying excipients may be added to avoid rapid drug clearance from the targeted site (Kjøniksen et al., 2014, 2015). To obtain a gel, it is only required to add a NLCs suspension of high concentration to a mixed components of the gel and then conduct a gelation process. Gelling agent should be uncharged to avoid aggregation of nanocarriers. Following gelation the particles are immobilized in the gel network rendering them stable (Jenning et al., 2000a,b; Müller et al., 2005).

A cream is another popular dermal formulation. NLC-containing cream is prepared in a slightly different way than regular, that is, NLCs are typically combined with an existing cream. A volume of water added during preparation of the plain cream is reduced. Subsequently the remaining water is added later as a diluent phase of highly concentrated (~50%) aqueous NLCs suspension. To inhibit melting of lipid particles and preserve the desired particle structure, the process of mixing NLCs in the cream must be performed at room temperature. Second, lipid nanoparticles that are loaded in a cream melt and produce a compact layer when applied on skin (Müller et al., 2005). It is has been shown that the occlusive factor is higher for nanoparticles compared with microparticles and higher for NLCs compared with SLNs (De Vringer and Yamanouchi Europe, 1999; Souto et al., 2004). Occlusion is beneficial because it results in noticeable improvement in skin hydration by preventing transepidermal water loss (TEWL). The latter is not only a positive cosmetic effect, but also accounts for enhanced skin penetration by the drug of interest (Wissing and Müller, 2003; Cevc and Vierl, 2010). Occlusion strongly depends on the particles size—the smaller they are, the more densely packed on the skin surface they are and therefore occlusion is more effective (Guimarães and Ré, 2011). The effect leads to an elevation of water content in a stratum corneum from 10% to 20% (healthy skin) even to 50%. It was found that this hydration effect might lead to expressing effect similar to surfactants—impairing lipid barrier of stratum corneum and therefore enhancing penetration of applied substances, especially lipophilic (Zhai and Maibach, 2002).

Interestingly, the delivery system might be used not only as a carrier of irritating sunscreens like titanium dioxide, but as it turned out, the ability of NLC to scatter the UV light results in a sun-protective effect (Hagedorn-Leweke and Lippold, 1998).

Mendes et al. (2013) prepared NLC containing the antifungal agent miconazole using ultrasound and dispersed the miconazole NLC in a hydrogel. Miconazole, clotrimazole, ketoconazole, and bifonazole are broad-spectrum antifungal agents of the imidazole group, widely used for the treatment of candidiasis. They are characterized by low aqueous solubility. The excipient screening phase of the study included numerous lipids. Nine solid lipids were Precirol ATO 5, Suppocire A, Compritol 888 ATO, Gelucire 39/01, Gelucire 43/01, Gelucire 50/13, Suppocire CM, Imwitor 900 and stearic acid. The 12 liquid lipids included Labrafil M 2125 CS, Miglyol 812, Lauroglycol 90, Lauroglycol FCC, Labrafac PG, Capryol 90, Capryol PGMC, Labrafac Lipophile WL 1349, Labrafil M 1944 CS, Labrasol, oleic acid, and sesame oil. The authors chose Gelucire 43/01 (glycerol esters of saturated C12–C18 fatty acids) as a solid carrier lipid and Miglyol

812 (triglycerides of capric/caprylic acids) as a liquid lipid to dissolve miconazole. NLC dispersions were prepared by high-speed stirring followed by ultrasonication. NLC (213 nm diameter; PDI = 0.23) remained physically stable during storage, as assessed with an optical analyzer. Samples presented values of backscattered light that remained stable throughout the entire analysis period both on production day and 3 months later. The melting peak of miconazole was not recorded on DSC thermogram, suggesting the solubilization of the drug in the amorphous lipid matrix. Upon incorporating NLCs into the hydrogel matrix, its viscosity was reduced and no changes in viscosity were detected during the storage period. Miconazole amount corresponding to 8% of encapsulated dose was released from the hydrogel in 30 minutes followed by a slow release profile (22% after 48 hours). The release of miconazole from NLC containing hydrogel corresponded to the Korsmeyer–Peppas model. The predominant mechanism release from the formulation was assumed to be anomalous non-Fickian type, which combines mechanisms of diffusion and erosion. Reference formulation, the commercially available miconazole oral gel, exhibited similar antifungal activity, as compared with the developed hydrogel containing 17 times lower concentration of the active agent.

Esposito et al. (2013) produced two gels containing clotrimazole encapsulated as either NLC or in monoolein aqueous dispersion (MAD). MAD is a heterogeneous system built of various lyotropic liquid crystalline nanostructures, for example, lamelles, micelles, and cubic phases obtained via monoolein self-assembly in water. NLCs were prepared by stirring and ultrasonication method by combining mixture of melted lipids, that is, tristearin and Miglyol with an aqueous Pluronic F68 solution. The emulsion was subjected to ultrasonication and cooled down to room temperature. Plain MAD and NLC exhibited particle diameter of 198 and 196 nm, respectively. Stability studies showed that CLO–MAD could maintain 90% of CLO stability for 8 months. The macroscopic aspect of NLC dispersions didn't change by that time. Authors determined the deposition of clotrimazole in MAD and NLC using sedimentation field flow fractionation. No clotrimazole was found outside the nanocarriers. While for MAD, clotrimazole was evenly distributed in various particle fractions, in NLC, clotrimazole was mostly found in scarce fraction of big 200–300 nm particles. NLC formulation was further incorporated into thermoreversible gel made of Pluronic F127. Above gelling temperature, the elastic modulus was found to increase maintaining a large gap with respect to G''. The authors suggested that the gel structure was stabilized by the presence of NLCs. Further drug release rate from clotrimazole loaded NLCs, clotrimazole loaded gel, and clotrimazole loaded NLCs incorporated into gel. Clotrimazole diffused more slowly from NLC than from Pluronic F127 gel, probably because Pluronic F127 micelles improved clotrimazole solubility and consequently its diffusion. In the case of clotrimazole–NLC Pluronic F127 gel, NLC trapped by Pluronic F127 gel matrix decreases drug diffusion with respect to the plain gel. It can then be concluded that the diffusion of the drug is dependent both on the NLC matrix and on the network of the Pluronic F127 gel, as shown previously by the rheological study. Finally, clotrimazole nanoformulations showed higher activity against planktonic fungal cultures of *Candida albicans* (*C. albicans*), as compared with plain clotrimazole in dimethylsulfoxide.

Das et al. (2012) used clotrimazole as a model drug to compare SLN and NLC in regard to physicochemical formulation properties. SLN and NLC formulations were prepared using emulsification–ultrasonication method. In this study the authors meticulously studied the influence of formulation type and composition on numerous variables, that is, particle size, PDI, zeta potential, and encapsulation efficiency. The authors employed Tween 80, Tween 20, Pluronic F68, and Cremophor EL [polyethylene glycol (35) castor oil] as emulsifiers. The lipids tested in the study included Compritol 888 ATO (glycerol dibehenate/behenate), Precirol ATO 5 (glyceryl distearate), Geleol (Glyceryl monostearate), Suppocire NC (semisynthetic triglycerides of C10 to C18 saturated fatty acids), Labrafac CC (Caprylic/Capric Triglyceride), Dynasan 114 (glycerol trimyristate), Dynasan 118 (glycerol tristearate), and Imwitor 900K (glyceryl stearate). Emulsification–ultrasonication technique was used to prepare SLNs and NLCs. For both formulation types particle size increased upon increase in lipid content and drug content. Particle size was reduced upon prolonging sonication, increasing surfactant content, and using Cremophor as emulsifier instead of Pluronic F68. Zeta potential was around −20 mV for all formulations, and it decreased with increasing drug content. PDI decreased with increasing sonication time and increasing drug content. SEM imaging revealed that the surface morphology of both SLNs and NLCs was smooth and no visible difference between them was detected. Encapsulation efficiency increased upon increasing surfactant content. Clotrimazole peak on DSC curve disappeared in both NLC and SLN, but some slight differences were observed in DSC thermograms of the two evaluated formulation types. Similarly, the crystalline peaks of clotrimazole were absent in clotrimazole-loaded SLNs and NLCs as shown by XRD. Moreover, the XRD study clearly indicated the change in crystalline behavior of the NLC lipid matrix following encapsulation of clotrimazole. NLC exhibited faster drug release and better stability when loaded with high clotrimazole content. While SLN released clotrimazole following zero order kinetics, NLC showed better fit with Higuchi model. While a decreasing trend of zeta potential, drug encapsulation efficiency, and drug loading was observed, an increasing trend of particle size and polydispersity index was observed with storage time at 2°C–8°C. The results also indicate relatively better stability of NLCs than SLNs, especially at 25°C with high drug loading.

Esposito et al. (2013) obtained an antifungal vaginal gel containing clotrimazole encapsulated in NLC. The NLCs composed of tristearin, tricaprin, and Pluronic F68 were prepared by stirring and ultrasonication method. Further NLCs were loaded into Pluronic F127 thermoresponsive gel. While the gelling temperature was influenced by the addition of simulated vaginal fluid, it did not change considerably upon incorporating clotrimazole NLC. The micro-DSC thermogram showed a reduction of critical micelle temperature of Pluronic F127 caused by addition of NLC. It was suggested that at body temperature the network of NLC-gel would be able to efficiently perturb and subsequently release the active agent. The clotrimazole penetration was evaluated using porcine vaginal lining in Franz diffusion cells. No clotrimazole permeated through the vaginal pig tissue and only 0.3% and 0.03% of the loaded content was retained within the vaginal mucosa following application of NLC and NLC-loaded gel, respectively. Clotrimazole toxicity was tested in HeLa human cervical cancer cell line, where clotrimazole is reported to cause cell death by dissociating hexokinases from mitochondria. Following 24 hours incubation 1 mg mL^{-1} clotrimazole-loaded NLC gel exhibited significantly lower toxicity as compared with clotrimazole–NLC or clotrimazole dissolved in ethanol. Clotrimazole-loaded NLC gel showed similar minimal inhibitory concentration (MIC) value, as compared with clotrimazole dissolved in ethanol. It was suggested that the Pluronic F127 gel provides a protective role for vaginal mucosa and at the same time it does not hinder the antifungal activity of clotrimazole.

10.3.3 TRANSDERMAL NLCS

10.3.3.1 Transport across the skin

Skin is the biggest human organ, and separates the inside of the body and the external environment. It allows to maintain proper body temperature and other stable conditions, and protects the organism from UV radiation, pathogens, chemical substances, excessive water loss, etc. Skin bears receptors allowing perception of pain, heat, and touch. Skin is composed of epidermis and dermis, located over the subcutaneous tissues. The outermost layer, epidermis, consists of a few types of lipids, like cholesterol, phospholipids, triglycerides, and also cells—keratinocytes are the most common. The most external layer of the epidermis, the stratum corneum, is formed by nonviable cells, corneocytes filled with keratin arranged very closely to each other. On the top of this layer there is a lipid coating enhancing skin's protective properties. Dermis is a viable layer of skin containing blood vessels, and thus immune system cells. Other structures present in dermis include sweat glands, hair follicles, sebaceous glands, and other skin appendices. The transport through the stratum corneum is possible in three ways. The intracellular route is based on dissolving the substance in the extracellular lipids, and subsequent transportation into deeper tissues. While the transcellular route leads across the corneocytes, the appendageal route passes through skin appendages (sweat glands or hair follicles). The latter transport mechanism is considered minor, due to the fact the appendices occupy approximately 1‰ of the skin surface. However, the appendageal route will be important for transport of hydrophilic macromolecules (Shim, 2007). When the substance penetrates the epidermis, it might be carried further into deeper tissues or to blood vessels and then to the systemic circulation (Contri et al., 2011; Shim, 2007).

Substances applied to the skin may act on its surface (sunscreens, cosmetics), in appendages in dermis (hair restorers, antiacne), in local tissues (antihistamines, antiinflammatories), or may be transferred to systemic circulation (contraceptives, nicotine, opioid analgesics). Due to the complex skin structure the penetration of some substances may not be sufficient enough to obtain a therapeutic effect. Therefore, new transdermal drug delivery carriers need to be developed. One of the advantages of NLCs' use in dermal formulations is a premise of obtaining stabilized forms of some drugs vulnerable to temperature or oxidation, for example, coenzyme Q10.

10.3.3.2 Examples of transdermal NLC delivery

Gokce et al. (2012) compared resveratrol-loaded NLCs and SLN intended for dermal application. Resveratrol is a natural phytoalexin found in many fruits, possessing antiinflammatory and antiproliferative properties. As it is capable of inactivating reactive oxygen species (ROS), it exhibits anticancer action. It was also found to modulate prostaglandin production. Both carrier types were prepared by high shear homogenization. The NLC formulations comprised Compritol 888 ATO, MCT (Miglyol), Pluronic F68, and Tween 80. Particle size was found to decrease with increasing surfactant content (for SLN) and increasing liquid to solid lipid proportion (for NLC). Overall the mean NLC diameter ranged from 106.1 ± 1.6 to 136.3 ± 1.4 nm and from 90.6 ± 1.7 to 148.7 ± 1.5 nm for blank and resveratrol-loaded formulations, respectively. NLC formulations yielded smaller particles as compared with SLN. The latter was attributed to a better emulsification of Compritol 888 ATO. Melted Compritol may return to solid state during mixing, as it requires much energy to become liquefied. In the presence of Miglyol the heat is more equally distributed throughout the lipid mixture contributing to better emulsification, which subsequently yields smaller particles. Resveratrol encapsulation did not influence particle size or PDI (0.21–0.29). However a negative zeta potential was decreased upon addition of the active agent. The DSC thermograms showed lack of resveratrol peak indicating its solubilization within the particle core. TEM imaging showed that NLCs were well defined, and had smoother edges and more regular surface as compared with SLN. This was attributed to resveratrol recrystallization on the outermost layer of SLNs in contrast to efficient incorporation of the active agent into the NLC matrix. The cell viability was measured by WST-1 test in normal human dermal fibroblasts cell culture. Further intracellular ROS accumulation in fibroblasts was monitored using fluorescent probe, 2,7-dichlorofluoresceine acetate, DCFH-DA. Both cell viability and the reduction of ROS accumulation were found to be independent of carrier type used. Ex vivo diffusion studies were performed using rat abdominal skin in Franz cells. While dermis accumulation of resveratrol was more significant for NLC, the epidermal accumulation of the active agent was observed for SLN.

Chen-yu et al. (2012) developed a NLC system for the delivery of quercetin (QT), an efficient free radical scavenger that could alleviate oxidative stress-induced skin damage. NLCs were produced using emulsion–solidification technique at low temperature and further loaded with QT yielding entrapment efficiency (EE) of 89.95 ± 0.16% and drug loading of 3.05 ± 0.001%. Based on single factor experiment the following variables were optimized: the amount of the drug, surfactant and cosurfactant concentration, the liquid lipid content, and emulsifying temperature. The optimized formulation comprised QT (35 mg), D-α-Tocopherol polyethylene glycol 1000 succinate (TPGS; 4%), glyceryl monostearate 200 mg, stearic acid 200 mg, MCT 20%. The emulsification was performed at 70°C for 4 hours and yielded large nanoparticles of 215.2 nm with zeta potential of −20.10 ± 1.22 mV. The formulation was more efficient in scavenging free radicals in pyrogallol autoxidation test, as compared with free QT. This was attributed to the antioxidative properties of tocopherol-derivative emulsifier. Skin penetration in vitro of QT administered in the form of NLC followed zero order kinetics. The amount of permeated QT was 1.52 and 3.03 times higher for epidermis and dermis, respectively, as compared with plain QT solution in propylene glycol. Owing to their small size NLCs could adhere to skin improving the hydration of stratum corneum and subsequently enhance QT penetration. The occlusive action of NLCs was confirmed by visual observation of excised skin sections recovered from mice treated with NLC. They appeared swollen and thicker. Further NLCs were suspended in Carbopol gels and applied to murine ears in xylene-induced edema test to determine the antiinflammatory properties of the NLCs. The inhibition of ear edema exerted by NLCs was higher as compared with QT administered as intraperitoneal suspension in saline.

Vitorino et al. (2013) coencapsulated simvastatin (SV) and olanzapine (OL) in NLCs and incorporated them into a transdermal gel. OL is a 2nd generation antipsychotic drug, commonly used for treating schizophrenia and bipolar disease. However, long-term use of OL is reported to elevate cholesterol level in blood, leading to an increased cardiovascular risk. Therefore, a preparation comprising OL and a lipid-lowering medication SV was proposed. To ensure optimal solubility of both SV and OL, oleic acid was selected as liquid lipid constituent as well as transdermal permeation enhancer. The optimal excipient and manufacturing conditions were selected using the two-level, three-variable, 2^k full factorial planning. The optimized formulation was further enriched with various permeation enhancers. Interaction of limonene, cineole, and menthol with lipidic bilayer was studied by means of molecular dynamics simulations. It was shown that 5% limonene addition was responsible for significant increase in the dipalmitoylphosphatidylcholine (DPPC) lateral diffusion, as compared with other penetrants. The water content in the first hydration shell of DPPC was increased upon limonene addition. The permeation of SV and OL was measured in excised newborn porcine skin. Two mechanisms contributed to the enhanced permeability of tested drugs. While the NLC constituted a drug reservoir, the limonene promoted drug diffusion and partitioning to deeper skin layers. As a result SV and OL permeation rate was enhanced 8–10 times. Further the optimized formulation was incorporated into Carbopol Ultrez hydrogel. Although, the resulting NLC-gel was homogenous, the nanoparticles were entrapped in the entanglements of the gelling agent forming grains. Cytotoxicity of NLC-gel was studied in human adult fibroblast cell line Df and spontaneously immortalized human keratinocyte cell line HaCaT. The NLC-gel exhibited a similar cytotoxicity level as compared with plain drugs. Confocal laser microscopy revealed that during the incubation NLCs were efficiently internalized by the studied cells, which shows formulation biocompatibility. The stability studies showed that the NLC size and drug entrapment efficiency were not affected by 6-month storage.

Gupta and Vyas (2012) aimed to improve transdermal delivery of fluconazole by using SLNs and NLCs. While lipid phase was composed of Compritol ATO 888 and oleic acid, Pluronic F68 and phosphatidylcholine were used as emulsifiers. Solvent diffusion in aqueous medium yielded small round particles of 134.3 ± 5.2 and 178.9 ± 3.8 nm for NLCs and SLNs, respectively. The carriers were characterized by a similar negative charge of (-29 and -25 mV for NLCs and SLNs, respectively), as well as entrapment efficiency ($81.4 \pm 3.89\%$ and $75.7 \pm 4.94\%$ for NLCs and SLNs, respectively). Ex vivo permeation studies performed using hairless rat skin in Franz diffusion cells showed that fluconazole exhibited delayed and enhanced permeation form nanovesicles, as compared with plain drug solution. The use of NLCs resulted in drug accumulation in the stratum corneum and viable skin as compared with SLNs. NLCs formulation also showed the highest amount of total permeated drug. The amount of fluconazole retained in the skin in vivo was evaluated 12 hours after application. The retention was 1.7-fold and 1.5-fold higher than for plain drug solution in case of NLC and SLN formulations, respectively. Interestingly, the intradermal retention of fluconazole was lower as compared with ex vivo results. The latter was attributed to lipase activity and nanoparticle interactions with viable skin components that was not significant in previous ex vivo tests. No irritation was found in Draize patch test performed on rabbits for nanoparticulate fluconazole formulations. The antifungal activity was evaluated in vivo in male albino rats immunosuppressed with cyclophosphamide and infected with a clinical isolate of *C. albicans* (10^7 colony forming units; cfu mL^{-1}). The infection sites on rats' backs were further covered with an occlusive dressing and treated for 3 consecutive days with plain drug solution or a nanoparticulate formulation. While plain fluconazole solution only reduced the fungal load by 1 log cfu, the SLN and NLC yielded a 3 log reduction lasting for 5 days after administering the formulation. The better therapeutic efficacy was achieved due to higher accumulation in different skin layers and subsequent formation of drug depot.

10.3.4 OCULAR NLC DELIVERY

The bioavailability of different ocular formulations is limited due to numerous protection mechanisms that shield eye outer tissues from external factors. They include closing eyelids, blinking, and flushing the eye surface with tears, which comprise lysozyme and immunoglobulins possessing antiinfectious properties.

Moreover, the lacrimal liquid is transferred into the throat through the nasolacrimal routes and subsequently removed by expectoration or swallowed. On account of this mechanism, the residence time of ocular preparation on the eye surface is often not long enough to safeguard the required contact with the targeted tissues and consequently deliver therapeutic substances to treat eye diseases (Furrer et al., 2008). The permeation into deeper ocular tissues is also impeded. The outermost layer of eye cornea is divided into three layers, which differ in respect to the drug permeability. The first one, the corneal epithelium, strongly limits the transfer of the hydrophilic drugs. The next layer, the stroma, hinders permeation of lipophilic drugs. Those two parts of the cornea are thought to be major causes of poor ocular permeability (Gaudana et al., 2010). To enhance the bioavailability of ophthalmic drugs it is necessary to develop new, better formulations. The NLCs exhibit several interesting features making them a suitable nanocarrier for ocular delivery. They include biocompatibility of its lipidic components, bioavailability enhancement of lipophilic drugs as well as stabilization of easily disintegrating substances. When introduced onto the cornea, NLCs are expected to undergo degradation process by lipases. In the eye, retinal pigment epithelium is responsible for lysosomal lipid metabolism. Upon the NLC degradation, the active ingredient is released and may interact with ocular tissues. Therefore this drug delivery system started to be investigated in recent years, yielding some successful new formulations. Among others, triamcinolone acetonide, flurbiprofen, and ibuprofen were incorporated into nanoparticles and efficiently delivered through the cornea (Gonzalez-Mira et al., 2012; Araújo et al., 2011; Li et al., 2008; Luo et al., 2011).

Üstündağ-Okur et al. (2015) aimed to obtain an ocular insert containing a second-generation fluoroquinolone, ofloxacin loaded into NLCs. NLCs comprising the drug (0.3%) Compritol 888 ATO (0.73%), oleic acid (1.47%), Tween 80 (0.73%) in water were produced using high shear homogenization method. The formulation was characterized by hydrodynamic diameter of 153.5 ± 2.3 nm, 0.188 PDI, and slightly negative zeta potential of -4.63 ± 0.26 mV. NLC-based inserts composed of chitosan oligosaccharide lactate (COL) and glycerol or PEG 400 as plasticizers were fabricated with a novel solvent casting evaporation technique. The inserts were characterized in respect to thickness, moisture, hygroscopicity, sterility, stability, bioadhesiveness, roughness, and homogeneity. The results were satisfactory. DSC thermograms revealed lack of ofloxacin peak pointing out its solubilization within the NLC lipidic matrix. The peak of COL glass transition was also gone suggesting a favorable interaction between NLC excipients and the COL matrix. The ofloxacin release was analyzed for 48 hours in pH 7.4 buffer under constant stirring. Within 24 hours the analyzed inserts exhibited 70%–86% drug release following first-order kinetics. The release mechanism was suggested to base on a diffusion and swelling of COL matrix resulting in a controlled release of the active agent. Ex vivo permeation studies in excised leporine corneas in semiautomated Franz cells showed poor permeation, that is, 5% after 48 hours. However the conditions did not mimic any natural tear flow, which was expected to cause swelling of inserts and subsequently promote drug release. Further dissolution tests were performed with inserts dissolved in aqueous humor acquired from rabbit eyes. Correspondingly to permeability studies, the desired ofloxacin release could not be achieved in aqueous humor. This was attributed to the lack of tear flow and the absence of lipases in the medium. Finally the efficiency of inserts was tested in vivo in rabbits infected with a Gram-positive pathogenic bacterium *Staphylococcus aureus*. The group treated with NLC-containing insert recovered with no significant sign of conjunctival redness or corneal opacity. While the untreated group became blind, rabbits treated with commercial ofloxacin formulation scored higher in modified Draize test, meaning more conjunctival redness and corneal opacity, as compared with NLC-based insert.

10.3.5 PULMONARY NLC DELIVERY

Many nanosized drug delivery systems have been introduced into the market as a novel way to overcome some conventional drugs' limitations. Owing to their nanometric size and bioadhesive properties, NLC may be used to administer lipophilic pulmonary medications, for example, antineoplastic drugs locally to lungs. While these drugs are poorly absorbed following an oral administration, loading them into progressive formulation such as NLCs provides a better bioavailability as well as longer drug residence in the pulmonary area (Taratula et al., 2013). First of all, the direct administration to the pulmonary epithelium restricts the drug distribution to the nontargeted healthy tissues, as opposed to parenteral administration. The latter is of critical importance in case of cancer treatment. Moreover, drugs administered directly into lungs avoid the first-pass effect and therefore the drug metabolism prior to reaching the intended site of action is diminished (Jaques and Kim, 2000). As the effective concentration is increased, the intermission between doses could be prolonged with maintaining the drug action. Aside from cancer treatment, direct administration of drug into lungs or the epithelium of the lower respiratory tract proves useful when treating, for example, inflammatory diseases, various infections, as well as hereditary diseases, for example, cystic fibrosis (Ungaro and Vanbever, 2014; Andrade et al., 2013).

Pardeike et al. (2011) designed an itraconazole-loaded NLCs composed of Precirol ATO 5 and oleic acid. The results indicated there was no change in particle size and drug entrapment before and after nebulization. Drug-loading was maintained after 6-month storage demonstrating the formulation's suitability in inhalational administration. Kardara et al. (2013) proved protective properties of NLCs in the acute lung injury model (ALI). Before treating mice with 0.1 N hydrochloric acid administered intratracheally, NLCs or saline were given in intravenous injection. No harm was made to the lung structure pretreated with NLCs. These findings give hope to use the drug delivery system in treating pulmonary diseases not only in inhalational administration, but also as injections. Taratula et al. (2013) obtained NLC combined with a ligand of pulmonary cancer cell receptors. The formulation was loaded with an antineoplastic agent paclitaxel or siRNA, with aim to suppress drug resistance in lung tumor cell lines. The group conducted research on mice lacking thymus with implemented lung cancer. Results demonstrated that NLC administered inhalational accumulates in the amount of three times larger (83% of formulation stays in lungs) than given intravenously, which implies minimum drug migration to other tissues after inhalation.

10.3.6 PARENTERAL NLC DELIVERY

Nanocarriers as parenteral drugs were first introduced into the market in form of a submicron microemulsion containing propofol (Diazemuls, Actavis) and diazepam (AstraZeneca, London, UK). Then, liposomes loaded with number of agents, for example, doxorubicin, amphotericin B were produced. Main obstacles to obtain full success were the substantial cost of manufacturing and problems with obtaining formulation with sufficient stability (Gill et al., 1995). NLCs, as a lipid carriers formulated of biodegradable excipients, are perfect for parenteral route. What is more, the great advantage of NLCs is a possibility of using it as a controlled release drug administered in injection (Marcato and Durán, 2008), making it very useful in neurodegenerative and other neurological disorders. The NLCs loaded with bromocriptine, medicine for Parkinson's disease, were evaluated by Esposito et al. (2012). It was demonstrated that the formulation containing Pluronic F68 favors brain targeting of the bromocriptine. NLCs were shown to prolong the therapeutic effect in rats using 6-hydroxydopamine lesion (6-OHDA) Parkinson model, as compared with plain drug solution. The reason why NLCs reach the brain easier than the free drug is probably the same for NLCs' predecessor, the SLN. Chattopadhyay et al. (2008) evaluated SLN formulation of atazanavir, an anti-HIV agent, and came to the conclusion that the good brain targeting occurs due to improved intracellular transport. Because of opening the connections between the cell membranes called occluding junctions, endocytosis of the particular carriers is more efficient. Another drug used to manage Parkinson's disease that was investigated is apomorphine, a medicine that administered orally performs poorly, due to its insufficient bioavailability. NLCs containing the agent managed to circumvent the BBB following intravenous (IV) administration and reach brain vessels, as shown in rats (Subramony, 2006; Hsu et al., 2010).

Zhao et al. (2014) studied the influence of NLCs composition on their physicochemical and in vitro biological properties. They chose modified 4-dedimethylaminosancycline, CMT-8, a tetracycline analog with high antibiotic and antineoplastic activity. CMT-8 exhibits low permeability and low aqueous solubility. Moreover it causes some serious adverse effects, for example, nausea, vomiting, and liver dysfunction following iv injection. CMT-8 was loaded into NLCs obtained using high-pressure homogenization at 600 bars within five cycles. Based on previous preformulation experiments, stearic acid and monoglycerides were chosen as solid lipids and oleic acid and caprylic/capric triglycerides were chosen as liquid lipids. The formulation was further lyophilized with 3%

trehalose as cryoprotectant. The authors evaluated two surfactants, a nonionic Pluronic F68 and octenylsuccinic modified gum arabic, GA-OSA. While Pluronic F68 provides a sterical stabilization the GA-OSA is responsible for electrostatic repulsion contributing to the overall good physical stability of NLC aqueous dispersions. The particle size of six different formulations varied between 130.5 ± 6.3 and 205 ± 5.5 nm. The PDI ranged from 0.11 to 0.32 and the zeta potential was between -31.04 and -61.2 mV. All formulations showed entrapment efficiency exceeding 60%. TEM revealed a spherical and oval shape of NLCs. Rheological analysis indicated that the storage modulus of NLC formulation lacking GA-OSA decreased with decreasing frequency to a larger extent, as compared with other evaluated NLCs. This proved that GA-OSA at the interface of NLC particle led to weaker flocculated network or disappearance of the network structure and subsequently enhanced dispersion stability. Formulations were further examined by means of small angle X-ray scattering (SAXS). In SAXS experiments, fluctuations of the electron density give rise to the characteristic patterns for the scattering of incident X-rays. Formulations emulsified by Pluronic F68 showed uniformly dense core and perfectly smooth surface. The analysis of DSC thermograms showed that blending behavior of NLCs was significantly influenced by addition of GA-OSA. The crystallinity of GA-OSA containing formulations was increased, as shown by the sharpening of peaks at around 30°C and 50°C, attributed to the unstable α and stable β polymorphic forms, respectively. No peak of pure CMT-8 was detected in either formulation's thermograms. The release rate of CMT-8 from NLC formulations was highly dependent on GA-OSA content. All formulations exhibited a constant drug release rate resulting from diffusion or bulk erosion. The addition of GA-OSA yielded comparatively slower release. A higher fraction of emulsifier led to nonsustained release. In vitro cell viability MTT assay was performed on HeLa cell line. Cell viability was around 75%–80% for all evaluated formulations, showing little influence of NLC composition. Further the authors performed a phagocytosis assay using promonocytic leukemia cell line U937. In the test cells were coincubated with NLC formulations for 24 hours. The amount of CMT-8 was determined in phagocytized cells by means of high performance liquid chromatography following homogenization of cell debris. Increasing content of GA-OSA decreased the ratio of phagocytized nanoparticles due to disturbed interaction of NLCs with leukemic cells. The macrophages failed to recognize the particles, as a result of their high hydrophilicity attributed to the interfacial localization of GA-OSA.

Another study involving encapsulation of a chemically modified tetracyline, 4-dedimethylamino sancycline (CMT-3), was reported by Yang et al. (2013). The group used HPH homogenization, as similar to the previously described study. Stearic acid, oleic acid, MCT, Pluronic F68, and Cremophor EL were used as NLC components. While the increase of total surfactant concentration had only slight influence on particle size, it decreased the stability of NLC dispersion. The latter was attributed to extensive attraction between PEO chains of Pluronic F68 hydrophilic block by hydrogen bonding. As a result, the mean size of NLCs increased substantially during 10 days storage time, even for formulations with low surfactant content. The entrapment efficiency and drug loading values were found to be in the range 90%–96% and 4.2%–10.6%, respectively. As CMT-3 is a hydrophobic molecule, the increase in liquid lipid ratio yielded better solubilization and higher entrapment efficiency. SEM microscopy showed that while the shape of the blank NLC was almost spherical, the loaded NLCs tended to be more elongated. Small angle neutron scattering (SANS) data seemed to contradict the DLS size measurements, as lower particle diameters were detected. However, the SANS data indicated only the lower limit of the maximal aggregate size. Moreover DLS measures the hydrodynamic radius of nanoparticles, while SANS reports the "dry size." In this regard, it was concluded that that the evaluated NLCs were characterized by a dense core and a rough surface. The formulations were further studied by SAXS and wide angle X-ray powder diffraction (XRD). The addition of CMT-3 to blank NLCs was responsible for increasing the nanoparticle crystallinity. The intensities of several diffraction peaks attributed to CMT-3 were reduced in freeze-dried CMT-3 loaded NLCs. Moreover the diffraction intensity of CMT-3 NLC was clearly enhanced as compared with blank NLC. The in vitro drug release was examined by using the dynamic dialysis method. While the release of CMT-3 was prolonged by acidic medium, the active ingredient exhibited a burst release in both pH 5.6 and 7.4 Tween-enriched buffers. The encapsulation of CMT-3 in NLCs yielded a sustained release formulation. The authors suggested that in the first stage the CMT-3 loaded in the shell was released. Further an extended release was obtained as the drug molecules bound to core-forming solid lipids were liberated. The cubic structure of CMT-3 loaded NLCs protected its loading from leaking from the carrier. The in vitro cytotoxicity assay was performed in HeLa cell line. The cell viability was strongly dependent on the incubation time. While at $t = 0$ the cell viability of the CMT-3 as well as blank and loaded NLCs was similar, after 20 hours the least cells survived in dishes treated with CMT-3 loaded NLCs. The latter was in line with previous observations revealing the prolonged release of the active ingredient. Further, cell morphology of HeLa cells treated with various formulations was analyzed under a microscope. Cells treated with CMT-3 loaded NLCs shrunk; cell condensation and fragmentation was also observed. HeLa cells were also incubated with Rhodamine B-loaded NLCs to corroborate the previous observations.

Mussi et al. (2014) took advantage of NLCs' capability to solubilize multiple drugs in one carrier. The use of two different treatment modalities may reduce the drug doses and exploit a synergistic mechanism of action. Here, the authors encapsulated doxorubicin, an anthracycline broad-spectrum antineoplastic agent with docosahexaenoic acid (DHA). DHA is able to sensitize tumors to doxorubicin due to increased level of oxidative stress. The formulation containing doxorubicin and DHA comprised of oleic acid, Compritol 888 ATO, Tween 80, and triethanolamine as excipients. Hot melting homogenization method was used employing emulsification–ultrasound. The obtained particles were in the range of

76–86 nm with PDI < 0.22, depending on the formulation. The zeta potential ranged between −23 and −36 mV and the entrapment efficiency reached nearly 100% with a drug loading 31 mg g^{-1}. The drug release was performed using Dulbecco's modified Eagle medium as a release fluid. While formulations with high DHA content promoted faster doxorubicin release, the NLCs containing 0.4% DHA released 30% of their load during 4 hours, followed by a plateau for up to 48 hours. The formulation was stable in the presence of fetal bovine serum for 24 hours. In vitro cytotoxicity studies performed using human breast adenocarcinoma MCF-7 cell line and its doxorubicin-resistant version, MCF-7/Adr, proved that the combination of DHA and doxorubicin in NLCs was more potent than doxorubicin alone or concomitant treatment with DHA and doxorubicin. Further study on cancer cell spheroids confirmed the results from cancer cell monolayers. Spheroids are a more relevant in vitro model as they mimic tumor architecture, pH gradient, pO_2, and vascularization. Three-dimensional microscopic examination using confocal laser scanning microscopy as well as uptake studies in cancer cell monolayers proved better penetration of doxorubicin and DHA coloaded NLCs. The formulation penetration exceeded that of free drug as well as commercially available preparation, the liposomal doxorubicin, Lipodox.

10.4 CONCLUSIONS AND FUTURE PERSPECTIVES

In this chapter we presented various examples of using nanostructured lipid carriers as drug delivery systems. NLCs provide good drug loading, size uniformity, and possibility of size and surface charge control. It is worth noting that many excipients used in NLC systems are registered by FDA, widely used in other drug products and well known for their overall safety. Subsequently they exhibit good storage stability and biocompatibility, which is often difficult to achieve with systems such as polymeric micelles or nanoemulsions. NLCs are also very flexible carriers, capable of delivering active agents across very different biological barriers. These nanoparticulate drug delivery systems are also capable to encapsulate a wide range of drugs and encapsulate two active agents with good encapsulation efficiency. We thus anticipate that owing to their versatility and multifunctionality many nanoparticle-based NLC products will be developed in the next 10 years.

REFERENCES

Aditya, N.P., Shim, M., Lee, I., Lee, Y., Im, M.H., Ko, S., 2013. Curcumin and genistein coloaded nanostructured lipid carriers: in vitro digestion and antiprostate cancer activity. J. Agric. Food Chem. 61 (8), 1878–1883.

Andrade, F., Rafael, D., Videira, M., Ferreira, D., Sosnik, A., Sarmento, B., 2013. Nanotechnology and pulmonary delivery to overcome resistance in infectious diseases. Adv. Drug Deliv. Rev. 65 (13), 1816–1827.

Araújo, J., et al., 2011. Nanostructured lipid carriers for triamcinolone acetonide delivery to the posterior segment of the eye. Colloids Surf. B: Biointerfaces 88 (1), 150–157.

Averina, E.S., et al., 2010. Nanostructured lipid carriers (NLC) on the basis of Siberian pine (*Pinus sibirica*) seed oil. Die Pharmazie: Int. J. Pharm. Sci. 65 (1), 25–31.

Averina, E.S., Müller, R.H., Popov, D.V., Radnaeva, L.D., 2011. Physical and chemical stability of nanostructured lipid drug carriers (NLC) based on natural lipids from Baikal region (Siberia, Russia). Die Pharmazie: Int. J. Pharm. Sci. 66 (5), 348–356.

Baccile, N., Nassif, N., Malfatti, L., Van Bogaert, I.N., Soetaert, W., Pehau-Arnaudet, G., et al., 2010. Sophorolipids: a yeast-derived glycolipid as greener structure directing agents for self-assembled nanomaterials. Green Chem. 12 (9), 1564–1567.

Bamrungsap, S., Zhao, Z., Chen, T., Wang, L., Li, C., Fu, T., et al., 2012. Nanotechnology in therapeutics: a focus on nanoparticles as a drug delivery system. Nanomedicine 7 (8), 1253–1271.

Beloqui, A., Solinís, M.Á., Rodríguez-Gascón, A., Almeida, A.J., Préat, V., 2016. Nanostructured lipid carriers: promising drug delivery systems for future clinics. Nanomed. Nanotechnol. Biol. Med. 12 (1), 143–161.

Borgia, S.L., Regehly, M., Sivaramakrishnan, R., Mehnert, W., Korting, H.C., Danker, K., et al., 2005. Lipid nanoparticles for skin penetration enhancement—correlation to drug localization within the particle matrix as determined by fluorescence and parelectric spectroscopy. J. Control. Release 110 (1), 151–163.

Cevc, G., Vierl, U., 2010. Nanotechnology and the transdermal route: a state of the art review and critical appraisal. J. Control. Release 141 (3), 277–299.

Chattopadhyay, N., Zastre, J., Wong, H.L., Wu, X.Y., Bendayan, R., 2008. Solid lipid nanoparticles enhance the delivery of the HIV protease inhibitor, atazanavir, by a human brain endothelial cell line. Pharm. Res. 25 (10), 2262–2271.

Chen-yu, G., Chun-fen, Y., Qi-lu, L., Qi, T., Yan-wei, X., Wei-na, L., et al., 2012. Development of a quercetin-loaded nanostructured lipid carrier formulation for topical delivery. Int. J. Pharm. 430 (1), 292–298.

Contri, R.V., Fiel, L.A., Pohlmann, A.R., Guterres, S.S., Beck, R., 2011. Transport of substances and nanoparticles across the skin and in vitro models to evaluate skin permeation and/or penetration. In: Beck, R., Guterres, S., Pohlmann, A. (Eds.), Nanocosmetics and Nanomedicines. New Approaches for Skin Care. Springer-Verlag, Berlin Heidelberg, pp. 3–5.

Craievich, A.F., 2002. Synchrotron SAXS studies of nanostructured materials and colloidal solutions: a review. Mater. Res. 5 (1), 1–11.

Das, S., Chaudhury, A., 2011. Recent advances in lipid nanoparticle formulations with solid matrix for oral drug delivery. AAPS PharmSciTech 12 (1), 62–76.

Das, S., Ng, W.K., Tan, R.B., 2012. Are nanostructured lipid carriers (NLCs) better than solid lipid nanoparticles (SLNs): development, characterizations and comparative evaluations of clotrimazole-loaded SLNs and NLCs? Eur. J. Pharm. Sci. 47 (1), 139–151.

Dominiczak, M.H., Broom, J.I., 2014. Digestion and absorption of nutrients: the gastrointestinal tract. In: Baynes, J.W., Dominiczak, M.H. (Eds.), Medical Biochemistry. Elsevier Health Sciences, pp. 111–126.

Emami, J., Rezazadeh, M., Varshosaz, J., 2012. Formulation of LDL targeted nanostructured lipid carriers loaded with paclitaxel: a detailed study of preparation, freeze drying condition, and in vitro cytotoxicity. J. Nanomater. 2012, 3.

Esposito, E., Mariani, P., Ravani, L., Contado, C., Volta, M., Bido, S., et al., 2012. Nanoparticulate lipid dispersions for bromocriptine delivery: characterization and in vivo study. Eur. J. Pharm. Biopharm. 80 (2), 306–314.

Esposito, E., Ravani, L., Contado, C., Costenaro, A., Drechsler, M., Rossi, D., et al., 2013. Clotrimazole nanoparticle gel for mucosal administration. Mater. Sci. Eng. C 33 (1), 411–418.

Fang, C.L., Al-Suwayeh, S.A., Fang, J.Y., 2013. Nanostructured lipid carriers (NLCs) for drug delivery and targeting. Recent Pat. Nanotechnol. 7 (1), 41–55.

Feinle-Bisset, C., Azpiroz, F., 2013. Dietary lipids and functional gastrointestinal disorders. Am. J. Gastroenterol. 108 (5), 737–747.

Forny, L., Saleh, K., Denoyel, R., Pezron, I., 2009. Contact angle assessment of hydrophobic silica nanoparticles related to the mechanisms of dry water formation. Langmuir 26 (4), 2333–2338.

Freitas, C., Müller, R.H., 1998. Effect of light and temperature on zeta potential and physical stability in solid lipid nanoparticle (SLN) dispersions. Int. J. Pharm. 168 (2), 221–229.

Furrer, P., Delie, F., Plazonnet, B., 2008. Ophthalmic drug delivery. In: Rathbone, M.J., Hadgraft, J., Roberts, M.S., Lane, M.E. (Eds.), Modified-Release Drug Delivery Technology. CRC Press, London, pp. 59–84.

Gill, P.S., Espina, B.M., Muggia, F., Cabriales, S., Tulpule, A., Esplin, J.A., et al., 1995. Phase I/II clinical and pharmacokinetic evaluation of liposomal daunorubicin. J. Clin. Oncol. 13 (4), 996–1003.

Gokce, E.H., Korkmaz, E., Dellera, E., Sandri, G., Bonferoni, M.C., Ozer, O., 2012. Resveratrol-loaded solid lipid nanoparticles versus nanostructured lipid carriers: evaluation of antioxidant potential for dermal applications. Int. J. Nanomed. 7 (1), 1841–1850.

Gaudana, R., Ananthula, H.K., Parenky, A., Mitra, A.K., 2010. Ocular drug delivery. AAPS J. 12 (3), 348–360.

Gonzalez-Mira, E., Nikolić, S., Calpena, A.C., Egea, M.A., Souto, E.B., García, M.L., 2012. Improved and safe transcorneal delivery of flurbiprofen by NLC and NLC-based hydrogels. J. Pharm. Sci. 101 (2), 707–725.

Gupta, M., Vyas, S.P., 2012. Development, characterization and in vivo assessment of effective lipidic nanoparticles for dermal delivery of fluconazole against cutaneous candidiasis. Chem. Phys. Lipids 165 (4), 454–461.

Guimarães, K.L., Ré, M.I., 2011. Lipid nanoparticles as carriers for cosmetic ingredients: the first (SLN) and the second generation (NLC). In: Beck, R., Guterres, S., Pohlmann, A. (Eds.), Nanocosmetics and Nanomedicines. New Approaches for Skin Care. Springer-Verlag, Berlin Heidelberg, pp. 117–119.

Hagedorn-Leweke, U., Lippold, B.C., 1998. Accumulation of sunscreens and other compounds in keratinous substrates. Eur. J. Pharm. Biopharm. 46 (2), 215–221.

Hu, F.Q., Jiang, S.P., Du, Y.Z., Yuan, H., Ye, Y.Q., Zeng, S., 2006. Preparation and characteristics of monostearin nanostructured lipid carriers. Int. J. Pharm. 314 (1), 83–89.

Hung, L.C., Basri, M., Tejo, B.A., Ismail, R., Nang, H.L.L., Hassan, H.A., et al., 2011. An improved method for the preparations of nanostructured lipid carriers containing heat-sensitive bioactives. Colloids Surf. B: Biointerfaces 87 (1), 180–186.

Hsu, S.H., Wen, C.J., Al-Suwayeh, S.A., Chang, H.W., Yen, T.C., Fang, J.Y., 2010. Physicochemical characterization and in vivo bioluminescence imaging of nanostructured lipid carriers for targeting the brain: apomorphine as a model drug. Nanotechnology 21 (40), 405101.

Hwang, T.L., Lin, Y.K., Chi, C.H., Huang, T.H., Fang, J.Y., 2009. Development and evaluation of perfluorocarbon nanobubbles for apomorphine delivery. J. Pharm. Sci. 98 (10), 3735–3747.

Iqbal, M.A., Md, S., Sahni, J.K., Baboota, S., Dang, S., Ali, J., 2012. Nanostructured lipid carriers system: recent advances in drug delivery. J. Drug Target. 20 (10), 813–830.

Jaques, P.A., Kim, C.S., 2000. Measurement of total lung deposition of inhaled ultrafine particles in healthy men and women. Inhalation Toxicol. 12 (8), 715–731.

Jenning, V., Schäfer-Korting, M., Gohla, S., 2000a. Vitamin A-loaded solid lipid nanoparticles for topical use: drug release properties. J. Control. Release 66 (2), 115–126.

Jenning, V., Thünemann, A.F., Gohla, S.H., 2000b. Characterisation of a novel solid lipid nanoparticle carrier system based on binary mixtures of liquid and solid lipids. Int. J. Pharm. 199 (2), 167–177.

Jores, K., Haberland, A., Wartewig, S., Mäder, K., Mehnert, W., 2005. Solid lipid nanoparticles (SLN) and oil-loaded SLN studied by spectrofluorometry and Raman spectroscopy. Pharm. Res. 22 (11), 1887–1897.

Joshi, M., Patravale, V., 2006. Formulation and evaluation of nanostructured lipid carrier (NLC)–based gel of Valdecoxib. Drug Dev. Ind. Pharm. 32 (8), 911–918.

Joshi, M., Patravale, V., 2008. Nanostructured lipid carrier (NLC) based gel of celecoxib. Int. J. Pharm. 346 (1), 124–132.

Kardara, M., Hatziantoniou, S., Sfika, A., Vassiliou, A.G., Mourelatou, E., Magkou, C., et al., 2013. Caveolar uptake and endothelial-protective effects of nanostructured lipid carriers in acid aspiration murine acute lung injury. Pharm. Res. 30 (7), 1836–1847.

Khan, S., Baboota, S., Ali, J., Khan, S., Narang, R.S., Narang, J.K., 2015. Nanostructured lipid carriers: an emerging platform for improving oral bioavailability of lipophilic drugs. Int. J. Pharm. Investig. 5 (4), 182.

Khoo, S.M., Shackleford, D.M., Porter, C.J., Edwards, G.A., Charman, W.N., 2003. Intestinal lymphatic transport of halofantrine occurs after oral administration of a unit-dose lipid-based formulation to fasted dogs. Pharm. Res. 20 (9), 1460–1465.

Kjøniksen, A.L., Calejo, M.T., Zhu, K., Nyström, B., Sande, S.A., 2014. Stabilization of pluronic gels in the presence of different polysaccharides. J. Appl. Polym. Sci. 131 (13).

Kjøniksen, A.L., Calejo, M.T., Zhu, K., Nyström, B., Sande, S.A., 2015. Stabilization of pluronic gels by hydrophobically modified hydroxyethylcellulose. Int. J. Polym. Mater. Polym. Biomater. 64 (2), 76–83.

Li, C., Fleisher, D., Li, L., Schwier, J.R., Sweetana, S.A., Vasudevan, V., et al., 2001. Regional-dependent intestinal absorption and meal composition effects on systemic availability of LY303366, a lipopeptide antifungal agent, in dogs. J. Pharm. Sci. 90 (1), 47–57.

Li, X., Nie, S.F., Kong, J., Li, N., Ju, C.Y., 2008. A controlled-release ocular delivery system for ibuprofen based on nanostructured lipid carriers. Int. J. Pharm. 363 (1), 177–182.

Luo, Q., Zhao, J., Zhang, X., Pan, W., 2011. Nanostructured lipid carrier (NLC) coated with chitosan oligosaccharides and its potential use in ocular drug delivery system. Int. J. Pharm. 403 (1), 185–191.

Maeda, H., 2012. Macromolecular therapeutics in cancer treatment: the EPR effect and beyond. J. Control. Release 164 (2), 138–144.

Marcato, P.D., Durán, N., 2008. New aspects of nanopharmaceutical delivery systems. J. Nanosci. Nanotechnol. 8 (5), 2216–2229.

Martinez, M., Amidon, G., Clarke, L., Jones, W.W., Mitra, A., Riviere, J., 2002. Applying the biopharmaceutics classification system to veterinary pharmaceutical products: Part II. Physiological considerations. Adv. Drug Deliv. Rev. 54 (6), 825–850.

Mehnert, W., Mäder, K., 2001. Solid lipid nanoparticles: production, characterization and applications. Adv. Drug Deliv. Rev. 47 (2), 165–196.

Mendes, A.I., Silva, A.C., Catita, J.A.M., Cerqueira, F., Gabriel, C., Lopes, C.M., 2013. Miconazole-loaded nanostructured lipid carriers (NLC) for local delivery to the oral mucosa: improving antifungal activity. Colloids Surf. B: Biointerfaces 111, 755–763.

Merisko-Liversidge, E.M., Liversidge, G.G., 2008. Drug nanoparticles: formulating poorly water-soluble compounds. Toxicol. Pathol. 36 (1), 43–48.

Merisko-Liversidge, E., Liversidge, G.G., 2011. Nanosizing for oral and parenteral drug delivery: a perspective on formulating poorly-water soluble compounds using wet media milling technology. Adv. Drug Deliv. Rev. 63 (6), 427–440.

Moghimi, S.M., Hunter, A.C., Murray, J.C., 2001. Long-circulating and target-specific nanoparticles: theory to practice. Pharmacol. Rev. 53 (2), 283–318.

Müller, R.H., Mäder, Gohla, S., 2000. Solid lipid nanoparticles (SLN) for controlled drug delivery—a review of the state of the art. Eur. J. Pharm. Biopharm. 50 (1), 161–177.

Müller, R.H., Radtke, M., Wissing, S.A., 2002a. Solid lipid nanoparticles (SLN) and nanostructured lipid carriers (NLC) in cosmetic and dermatological preparations. Adv. Drug Deliv. Rev. 54, S131–S155.

Müller, R.H., Radtke, M., Wissing, S.A., 2002b. Nanostructured lipid matrices for improved microencapsulation of drugs. Int. J. Pharm. 242 (1), 121–128.

Müller, R.H., Mehnert, W., Souto, E.B., 2005. Solid lipid nanoparticles (SLN) and nanostructured lipid carriers (NLC) for dermal delivery. In: Robert, B.L., Maibach, H.I. (Eds.), Percutaneous Absorption: Mechanisms–Methodology–Drug Delivery, vol. 155. Taylor & Francis Group, Boca Raton, pp. 719–734.

Müller, R.H., Shegokar, R., Keck, C.M., 2011. 20 years of lipid nanoparticles (SLN & NLC): present state of development & industrial applications. Curr. Drug Discov. Technol. 8 (3), 207–227.

Mussi, S.V., Sawant, R., Perche, F., Oliveira, M.C., Azevedo, R.B., Ferreira, L.A., et al., 2014. Novel nanostructured lipid carrier co-loaded with doxorubicin and docosahexaenoic acid demonstrates enhanced in vitro activity and overcomes drug resistance in MCF-7/Adr cells. Pharm. Res. 31 (8), 1882–1892.

Nanjwade, B.K., Patel, D.J., Udhani, R.A., Manvi, F.V., 2011. Functions of lipids for enhancement of oral bioavailability of poorly water-soluble drugs. Sci. Pharm. 79 (4), 705–728.

Pardeike, J., Hommoss, A., Müller, R.H., 2009. Lipid nanoparticles (SLN, NLC) in cosmetic and pharmaceutical dermal products. Int. J. Pharm. 366 (1), 170–184.

Pardeike, J., Weber, S., Haber, T., Wagner, J., Zarfl, H.P., Plank, H., et al., 2011. Development of an itraconazole-loaded nanostructured lipid carrier (NLC) formulation for pulmonary application. Int. J. Pharm. 419 (1), 329–338.

Parveen, S., Sahoo, S.K., 2008. Polymeric nanoparticles for cancer therapy. J. Drug Target. 16 (2), 108–123.

Porter, C.J., Charman, W.N., 2001. Intestinal lymphatic drug transport: an update. Adv. Drug Deliv. Rev. 50 (1), 61–80.

Sanjula, B., Shah, F.M., Javed, A., Alka, A., 2009. Effect of poloxamer 188 on lymphatic uptake of carvedilol-loaded solid lipid nanoparticles for bioavailability enhancement. J. Drug Target. 17 (3), 249–256.

Schäfer-Korting, M., Mehnert, W., Korting, H.C., 2007. Lipid nanoparticles for improved topical application of drugs for skin diseases. Adv. Drug Deliv. Rev. 59 (6), 427–443.

Schubert, M.A., Müller-Goymann, C.C., 2003. Solvent injection as a new approach for manufacturing lipid nanoparticles—evaluation of the method and process parameters. Eur. J. Pharm. Biopharm. 55 (1), 125–131.

Shim, J., 2007. Transdermal applications of nanoparticulates. In: Thassu, D., Deleers, M., Pathak, Y.V. (Eds.), Nanoparticulate Drug Delivery Systems, vol. 166. Informa Healthcare, New York, pp. 330–331.

Subramony, J.A., 2006. Apomorphine in dopaminergic therapy. Mol. Pharm. 3 (4), 380–385.

Souto, E.B., Wissing, S.A., Barbosa, C.M., Müller, R.H., 2004. Development of a controlled release formulation based on SLN and NLC for topical clotrimazole delivery. Int. J. Pharm. 278 (1), 71–77.

Strickley, R.G., 2004. Solubilizing excipients in oral and injectable formulations. Pharm. Res. 21 (2), 201–230.

Taratula, O., Kuzmov, A., Shah, M., Garbuzenko, O.B., Minko, T., 2013. Nanostructured lipid carriers as multifunctional nanomedicine platform for pulmonary co-delivery of anticancer drugs and siRNA. J. Control. Release 171 (3), 349–357.

Teeranachaideekul, V., Müller, R.H., Junyaprasert, V.B., 2007. Encapsulation of ascorbyl palmitate in nanostructured lipid carriers (NLC)—effects of formulation parameters on physicochemical stability. Int. J. Pharm. 340 (1), 198–206.

Trevaskis, N.L., Charman, W.N., Porter, C.J., 2008. Lipid-based delivery systems and intestinal lymphatic drug transport: a mechanistic update. Adv. Drug Deliv. Rev. 60 (6), 702–716.

Tsai, M.J., Wu, P.C., Huang, Y.B., Chang, J.S., Lin, C.L., Tsai, Y.H., et al., 2012. Baicalein loaded in tocol nanostructured lipid carriers (tocol NLCs) for enhanced stability and brain targeting. Int. J. Pharm. 423 (2), 461–470.

Ungaro, F., Vanbever, R., 2014. Improving the efficacy of inhaled drugs for severe lung diseases: emerging pulmonary delivery strategies. Adv. Drug Deliv. Rev. 75, 1–2.

Üstündağ-Okur, N., Gökçe, E.H., Bozbıyık, D.İ., Eğrilmez, S., Ertan, G., Özer, Ö., 2015. Novel nanostructured lipid carrier-based inserts for controlled ocular drug delivery: evaluation of corneal bioavailability and treatment efficacy in bacterial keratitis. Expert Opin. Drug Deliv. 12 (11), 1791–1807.

Venishetty, V.K., Chede, R., Komuravelli, R., Adepu, L., Sistla, R., Diwan, P.V., 2012. Design and evaluation of polymer coated carvedilol loaded solid lipid nanoparticles to improve the oral bioavailability: a novel strategy to avoid intraduodenal administration. Colloids Surf. B: Biointerfaces 95, 1–9.

De Vringer, T., Yamanouchi Europe, B.V., 1999. *Topical preparation containing a suspension of solid lipid particles*. U.S. Patent 5,904,932.

Vitorino, C., Almeida, J., Gonçalves, L.M., Almeida, A.J., Sousa, J.J., Pais, A.A.C.C., 2013. Co-encapsulating nanostructured lipid carriers for transdermal application: from experimental design to the molecular detail. J. Control. Release 167 (3), 301–314.

Wissing, S.A., Müller, R.H., 2003. The influence of solid lipid nanoparticles on skin hydration and viscoelasticity—in vivo study. Eur. J. Pharm. Biopharm. 56 (1), 67–72.

Yang, X., Zhao, L., Almasy, L., Garamus, V.M., Zou, A., Willumeit, R., et al., 2013. Preparation and characterization of 4-dedimethylamino sancycline (CMT-3) loaded nanostructured lipid carrier (CMT-3/NLC) formulations. Int. J. Pharm. 450 (1), 225–234.

Zhai, Hongbo, Maibach, Howard I., 2002. Occlusion vs. skin barrier function. Skin Res. Technol. 8 (1), 1–6.

Zhao, S., Yang, X., Garamus, V.M., Handge, U.A., Bérengeère, L., Zhao, L., et al., 2014. Mixture of nonionic/ionic surfactants for the formulation of nanostructured lipid carriers: effects on physical properties. Langmuir 30 (23), 6920–6928.

Zur Mühlen, A., Zur Mühlen, E., Niehus, H., Mehnert, W., 1996. Atomic force microscopy studies of solid lipid nanoparticles. Pharm. Res. 13 (9), 1411–1416.

CHAPTER

Lipid nanocarriers as skin drug delivery systems 11

María Guadalupe Nava-Arzaluz[1], Elizabeth Piñón-Segundo[2] and Adriana Ganem-Rondero[1]

[1]División de Estudios de Posgrado (Tecnología Farmacéutica), Facultad de Estudios Superiores Cuautitlán, Universidad Nacional Autónoma de México, Cuautitlán Izcalli, Mexico [2]Laboratorio de Sistemas Farmacéuticos de Liberación Modificada, Facultad de Estudios Superiores Cuautitlán, Universidad Nacional Autónoma de México, Cuautitlán Izcalli, Mexico

CHAPTER OUTLINE

11.1 INTRODUCTION

One goal of pharmaceutical research is to design new molecules with pharmacological activity that respond to unresolved diseases such as cancer, HIV infection, Alzheimer's disease, among others. Most of these new molecules have poor

Nanoparticles in Pharmacotherapy. DOI: https://doi.org/10.1016/B978-0-12-816504-1.00007-7

solubility or permeability. On the other hand, there are many molecules with proven pharmacological action for various conditions that present problems of stability, solubility, permeability across the membranes, very short plasma half-lives, severe adverse effects, etc. Therefore, simultaneous research is focused on solving these problems, developing new release systems, and applying novel routes of administration. In this sense, an alternative to the conventional routes of administration is the skin drug delivery.

The skin is the largest organ of the body, composed of three main layers: the epidermis, dermis, and hypodermis, all of them giving a thickness less than 2 mm. The outermost layer of epidermis, the stratum corneum (SC), composed of lipids (ceramides, triglycerides, fatty acids, cholesterol, etc.), represents a very effective barrier to most molecules, especially the hydrophilic ones. This permeability barrier is not absolute and it is possible that certain substances penetrate the underlying tissue or even reach the general circulation. This has given rise to numerous studies regarding the penetration of all kinds of substances through the skin (Ahad et al., 2013). Depending on the desired effect, it is possible to have a local effect, within the skin (dermal delivery) or a systemic effect (transdermal delivery). In both cases, for local and systemic delivery, the SC has to be overcome (Akomeah, 2010; Ganem Rondero, 2011). Transdermal delivery offers advantages like avoiding the first pass effect, control over rate of drug release, better patient compliance, the option of immediate termination of the therapy, etc. In addition, this route of administration is noninvasive and can bypass variables influencing oral absorption such as pH, gastrointestinal motility, and food intake (Ahad et al., 2013; Neubert, 2011).

The low permeability of most substances through the skin has guided studies towards the search for strategies to improve the transport of substances through it. One of the first strategies is the use of chemical absorption promoters such as Azone, glycols, ethanol, terpenes, etc., which facilitate the transport of drugs by fluidization or partial extraction of lipids from the skin or by increasing drug partitioning (Williams and Barry 2012; Lane 2013). A second strategy is the use of physical absorption promoters such as sonophoresis, electroporation, magnetophoresis, microneedles, thermal ablation, microdermabrasion, iontophoresis, and microwaves, either alone or in combination. These approaches avoid the SC and release the drug directly into target site of the skin (Azagury et al., 2014; Han and Das, 2015; Hao, 2014; Larrañeta et al., 2016; Mitragotri, 2013; Nawaz and Wong, 2017; Shahzad et al., 2015; Wang et al., 2005; Wong, 2014).

Both chemical and physical promoters have been shown to be successful in releasing various types of drugs. However, most physical promoters are painful or have the risk of causing skin damage, are expensive, and lack patient compliance while chemical permeation promoters can cause irritation and permanent damage to the skin. A third approach is the use of drug–nanocarrying systems. These systems can increase the transport of drugs through the skin by increasing the solubility of the drug in the formulation, improving the distribution of the drug in the skin, and modifying the structure of the skin. Lipid-based systems have shown a great potential for both dermal and transdermal drug delivery. These systems are composed of biocompatible and biodegradable lipids that can be used to achieve controlled release, targeted delivery, and drug protection. In 1988, the first lipid-based product containing econazole was commercialized. Since then, many reports have been published indicating the success of these drug delivery systems (Bilia et al., 2016; Cevc and Vierl 2010; Elsayed et al., 2007; Garcês et al., 2018; Jain et al., 2017; Sala et al., 2018; Singh et al., 2017). The lipid-based skin delivery systems can be classified as vesicular systems and lipid particles. The vesicular systems comprise liposomes, ethosomes, ultradeformable liposomes (UDLs), among others. The lipid particles include solid lipid nanoparticles (SLNs) and nanostructured lipid carriers (NLCs) (Jain et al., 2017).

It became evident that polymeric and lipid nanoparticles, as well as conventional liposomes, are not able to deeply penetrate skin, remaining confined to the upper layers of the SC or forming a depot in the hair follicles. That is why incorporation of the drug within flexible liposomal vesicles has become one of the most popular and well-studied approaches for delivering drugs to deep layers of the skin or even to systemic circulation. Flexible/deformable/elastic liposomal systems such as invasomes, transfersomes, ethosomes, niosomes, etc., have demonstrated encouraging results in delivering small molecules and large proteins to the skin (Ashtikar et al., 2016).

Nanocarriers can act by different mechanisms: (1) being adsorbed, fusing with the skin, transferring the vesicle-entrapped drug molecules into the skin; and (2) acting as penetration enhancers, where carriers enter the SC, alter the fluidity of the SC lipids, change the intercellular lipid lamellae, and therefore, facilitate the penetration of drug molecules into and through the SC (Al-Mahallawi et al., 2015; Barua and Mitragotri, 2014). Ashtikar et al. (2016) and Jain et al. (2017) summarize the results of some recent studies performed with different vesicular and particulate carriers.

Once applied on the skin, nanocarrier structure can suffer changes. Some of the conditions that impact a formulation when it is applied on the skin surface are the following: (1) skin surface temperature, which can vary between 25°C and 33°C depending on the weather conditions; (2) application of formulation on an open skin surface causes changes in composition because of drying of formulation, which can provoke rearrangement, giving rise to other structures, which in some cases are larger than the originals; (3) in the case of nanoparticles, exposition to dryness may provoke coalescence, shedding their stabilizing surfactant into the surrounding environment. This coalescence can be positive, since in some cases, the goal is to form a (semi)occlusive film.

The remarkable progress of transdermal drug delivery during the last few years has positioned it as a major alternative to other administration routes. The use of nanocarriers offers different advantages such as (1) easy access into the skin due to their size, providing deep skin penetration; (2) inclusion of biocompatible components approved for pharmaceutical and cosmetic use; (3) sustained release; (4) inclusion of molecules with a wide range of solubility; (5) improved permeation through the skin by combination with physical and mechanical enhancers; and (6) in the case of local activity, nanocarriers offer targeted delivery and decreased systemic side effects. Administration of drugs into and through the skin is both a challenge and an opportunity to treat not only local skin disorders, but also as a site of entry of drugs intended to act systemically.

Special attention has been given to the development of transdermal systems as an alternative for the treatment of diseases that have become world health problems. This is the case of diabetes mellitus, which is in fact one of the most studied diseases. Ahad et al. (2015) published a review about the transdermal options to administer antidiabetic drugs. Ethosomes, proniosomes, and NLCs are proposed as suitable delivery systems.

In the case of skin diseases, nanocarriers represent a new alternative to treat disorders such as atopic dermatitis (the most prevalent disease affecting predominantly children). A plethora of carriers can be found in literature for the treatment of this skin disease, such as SLNs, nanoemulsions, liposomes, transfersomes, ethosomes, and cubosomes, loaded with drugs such as cetirizine, levocetirizine, tacrolimus, cyclosporine A, adenosylcobalamin, and KIOM-MA-128 (a novel herbal medicine). Excellent reviews on nanocarrier systems for the treatment of atopic dermatitis are presented by Akhtar et al. (2017) and Shao et al. (2016). Regarding psoriasis, Pradhan et al. (2013) published a review about the delivery of antipsoriatic drugs through novel colloidal carriers.

Treatment of inflammatory pathologies using nanotherapeutics has attracted great attention. Colloidal nanocarriers for analgesic drugs to provide a more efficient pain management have included liposomes, ethosomes, cubosomes, niosomes, and PEVs (penetration enhancer-containing vesicles (Di Marzio et al., 2016; Maestrelli et al., 2016)).

Studies have focused on vaccine delivery through the skin (transcutaneous immunization) including the use of nanocarriers to overcome the challenges imposed by the skin barrier (Kaurav et al., 2016).

This chapter provides an overview of different lipid nanocarriers destined to dermal and transdermal delivery. Rigid carriers such as nanoparticles and liposomes are described, and their characteristics and advantages are compared with those of elastic vesicles. Table 11.1 shows the general composition of these systems.

11.2 VESICULAR CARRIERS FOR SKIN DRUG DELIVERY

Colloidal carriers have received considerable interest for dermal and transdermal drug delivery, being the main characteristic of these systems their submicron size.

Vesicular carriers are constituted by amphiphilic molecules. These molecules can form vesicular structures after the addition of an excess of water (>45%),

Table 11.1 General Composition of Lipid-Based Nanocarriers

Vesicle	Composition
Bilosomes	Nonionic surfactants and bile salts
Cerasomes	Liposome structure with a layer of polyorganosiloxane network on the surface
Cryptosomes	Phospholipids and poloxamer
Cubosomes	Bicontinuous lipid bilayer (e.g., glyceryl monooleate) and interpenetrating independent aqueous nanochannels
Ethosomes	Phospholipids and ethanol
Glycerosomes	Phospholipids and glycerol
Hyalurosomes	Phospholipids and hyaluronan
Invasomes	Phospholipids, terpenes, and ethanol
Liposomes	Phospholipids and cholesterol
Nanostructured Lipid Carriers	Solid and liquid lipids and stabilizers
Niosomes	Nonionic surfactants, cholesterol and molecules like fatty alcohols
Novasomes	Polyoxyethylene fatty acids monoesters, cholesterol and free fatty acids
Pharmacosomes	Amphiphilic compounds from drugs esterified with a lipid
Photosomes	Light sensitive lipids
Phytosomes	Phospholipids and a plant extract
Solid lipid nanoparticles	Solid lipids and stabilizers
Transfersomes	Phospholipids, cholesterol and surfactant
Vesosomes	Large vesicles with many small vesicles inside them

which is the result of its amphiphilic character due to the presence of a polar or hydrophilic region (head) and a nonpolar or lipophilic region (tail). The hydrophilic heads are oriented towards the aqueous phase and the lipid tails are oriented towards one another in the presence of water (Chen, 2011). Vesicles can be prepared from a wide variety of lipids and surfactants; among the most common are phospholipids and nonionic surfactants. Vesicles prepared with phospholipids are known as liposomes, and those prepared with nonionic surfactants, as niosomes (Honeywell-Nguyen and Bouwstra, 2005; Pierre and dos Santos, 2011).

Liposomes are classified, according to the presence of surfactant or cosolvent in their structure, as conventional (classic) liposomes, deformable (elastic) liposomes, ethosomes, invasomes, and propylene glycol-containing liposomes, among others. These vesicles are biocompatible, biodegradable, and structurally similar to the skin constituents (Prasanthi and Lakshmi, 2012; Vanić, 2015).

11.2.1 LIPOSOMES

Liposomes are composed of one or more phospholipid bilayers surrounding inner aqueous compartment(s). Structure of liposomes is quite like the intercellular lipids of biological membranes. Liposomes are prepared with a single phospholipid or a mixture of them, cholesterol, and aqueous medium (water or buffer solution). The bilayers can be composed of neutral phospholipids or combination of neutral and surface charged phospholipids. Cholesterol is added in most formulations as a structure stabilizer, which acts by providing rigidity to the bilayer membrane and then reducing the permeability of hydrosoluble molecules. The liposomes can incorporate into their structure hydrophilic, hydrophobic, and amphiphilic molecules. Hydrophilic drugs are encapsulated into the aqueous compartment(s), lipophilic inside the bilayer, while amphiphilic partition between these two regions (Liu and Hu, 2007).

The mimetic character of liposomes in relation to the structural organization of the SC lipids provides a strategy to carry drugs into the skin. Liposomes were introduced in 1965 by Bangham et al., and were initially used as models for studying the cell membrane. It was not until 1980 that Mezei and Gulasekhram reported increased deposition in the epidermis and dermis of triamcinolone acetonide formulated in liposomes, and in 1988, the first topical liposomal product, an econazole preparation for topical dermatomycosis therapy, was approved (Egbaria and Weiner, 1990; El Maghraby et al., 2008). Since then, liposomes have been extensively investigated as drug delivery systems, both for topical and transdermal routes. Studies focusing on transdermal delivery of drugs incorporated in liposomes have yielded contradictory results. Some works have shown that liposomes have little or no value as carriers for the transdermal delivery of drugs, since they cannot penetrate deep layers of the skin and remain confined in the upper layer of the SC, suggesting that these are useful only as topical delivery systems (Touitou et al., 2000). Others, however, have suggested that the inclusion of drugs in liposomes may lead to therapeutic drug concentrations in the systemic circulation and that therefore, they are suitable for transdermal delivery (El Maghraby et al., 2006; El Maghraby et al., 2008). These inconsistent results may be explained in part by the fact that the studies were carried out with liposomes of different composition and preparation methods, and therefore with different physicochemical characteristics. The physical and chemical characteristics of liposomes determine their stability and drug delivery abilities. Some of these characteristics are lipid composition, membrane rigidity/elasticity, particle size, surface charge, number of lamellae, and drug loading capacity (Honeywell-Nguyen and Bouwstra, 2005; Pierre and dos Santos, 2011).

Generally, large liposomes with sizes ≥ 600 nm are not able to release their content into deep skin layers. These vesicles tend to remain in or on the SC and form a film after drying. Vesicles with sizes ≤ 300 nm can release their content to some degree in deep skin layers, while vesicles with sizes ≤ 100 nm show maximal deposition in both the viable epidermis and the dermis (Boakye et al., 2015; Verma et al., 2003). However, it is still subject to discussion if the lipid vesicles can pass intact through the skin.

The surface charge of vesicles plays a crucial role on its ability to remain within the SC or to penetrate into deeper skin layers. It has been proposed that positively charged vesicles are absorbed to a greater extent across the SC layers than negatively charged liposomes. All this because of strong electrostatic interaction between the positively charged vesicles and the negatively charged skin surface, which favors permeation (Boakye et al., 2015). This was demonstrated by Hasanovic et al. (2010) who showed that the addition of cationic polymers, such as chitosan, to liposomes (coated liposomes) increased the skin permeation compared to uncoated liposomes. The positive charged liposomes seemed to show higher absorption through the SC than negative charged liposomes. However, according to Gillet et al. (2011) the negative charged liposomes displayed higher absorption through the viable epidermis and dermis.

Another important factor is the composition of liposomes. In this sense, there are studies that demonstrate that phospholipid composition (hydrogenated or not hydrogenated) and the amount of cholesterol have a significant effect on the penetration behavior of liposomes. Composition correlates with membrane fluidity: highest fluidity means major penetration. The thermodynamic state of liposome bilayers (gel-phase or liquid-crystalline) has an important role on the transport of liposomes through the skin. Confocal laser scanning microscopy images demonstrated that gel-state liposomes did not penetrate as deep into the skin as liquid-state liposomes (Bouwstra et al., 2003). In a recent study, Sakdiset and coworkers (2018) prepared liposomes varying the type of phospholipids with different number of carbon atoms (12–18) and double bonds in the alkyl chain (five phosphatidylcholines and five phosphatidylglycerols) and evaluated their effect on the permeation of a hydrophilic model drug (caffeine). Liposomes prepared with 1,2-dipalmitoyl-*sn*-glycero-3-phosphoglycerol, sodium salt (DPPG), 1,2-dilauroyl-*sn*-glycero-3-phosphocholine (DLPC), and 1,2-dimyristoyl-*sn*-glycero-3-phosphocholine (DMPC) promoted caffeine permeation through the skin compared to the control (caffeine solution in PBS), while liposomes prepared with 1,2-dipalmitoyl-*sn*-glycero-3-phosphocholine (DPPC) and 1,2-dioleoyl-*sn*-glycero-3-phosphoglycerol, sodium salt (DOPG) significantly decreased caffeine permeation. No change in caffeine permeation was observed for liposomes prepared with the rest of the phospholipids. Permeation enhancement could be attributed to (1) disruption of the SC lipids by phospholipids, (2) disruption of the SC lipids by liposomes, and (3) carrying the drug associated with the liposomes into the skin. The delayed effect was explained by (1) formation of an extra lipid barrier on the surface of the skin due to the rigid structure of some liposomes, which depends on the type of phospholipid; and (2) interaction between the drug and the phospholipids. The promoter or retarder effect depends on liposomal composition, and more studies should be carried out to understand the molecular mechanisms involved.

Five general mechanisms by which liposomes modulate the permeability of drugs into the skin have been proposed:

1. Liposomes can penetrate intact through the different layers of the skin, acting as drug carriers to release the encapsulated drug into or through the skin.
2. Lipid vesicles act as penetration enhancers by loosening the lipid structure of SC and promoting impaired barrier function with subsequent increase in the skin partitioning of drugs.
3. Liposomes adhere to the surface of the skin and subsequently are destabilized (fusing or mixing with the skin lipid matrix), releasing their content.
4. Liposomes promote transdermal delivery through the appendages of the skin (e.g., hair follicles and sweat ducts).
5. The drug exits from the liposomes and then permeates the skin independently (El Maghraby et al., 2008; Elsayed et al., 2007; Fahr and Chen, 2010; Gillet et al., 2011; Prasanthi and Lakshmi, 2012; Vanić, 2015).

Liposomes can be obtained by various techniques, which determine their characteristics. For detailed discussion about these techniques and their challenges, the readers can refer to the recently published review articles (Akbarzadeh et al., 2013; Çağdaş et al., 2014).

Liposomes have shown to be useful in the dermal application of drugs, which is advantageous for many drugs, since it allows high and sustained drug concentrations to be achieved at the site of action, decreasing the frequency of administration, and reducing adverse effects due to the low systemic absorption, mainly for lipophilic drugs. This has been demonstrated for corticosteroids and anesthetics (Gesztes and Mezei, 1988). Hence, liposomes are suitable drug carriers for skin diseases, able to form deposits of drug in the skin and to release the drug for a long period (Liu and Hu, 2007). A recent review (Vanić, 2015) analyzed the potential use of drug-loaded liposomes in dermal drug delivery for the treatment of some skin disorders like skin fungal infections, acne, psoriasis, rosacea, and the treatment of xerosis cutis with "empty" liposomes.

Liposomes applied topically to the skin tend to remain confined in the upper layers of SC and are not able to penetrate the layers of the viable epidermis, functioning more like a reservoir, providing a localized action (Honeywell-Nguyen and Bouwstra, 2005). Conventional liposomes increase the effectiveness of dermal drug delivery in intact skin compared to the free drug, in addition to providing prolonged release. It has been shown that liposomes cannot diffuse easily through healthy SC, but they can penetrate through damaged skin, for example, precancerous or cancerous skin lesions and burns (Caddeo et al., 2016). Therefore, liposomes are proposed as delivery systems for the wound healing and for the treatment of cancerous lesions of the skin.

The use of multiple therapies with the aim of achieving a synergistic effect in the treatment of diseases is a strategy that has gained interest in recent years. Coencapsulation and corelease of drugs using liposomes could simplify therapy, since the drugs, incorporated in the same carrier, would be released simultaneously at the site of action. These liposomes are beneficial in terms of overcoming the solubility and stability issues of the drugs.

Another approach to improve the efficiency of liposomes as drug delivery system is modifying their structure, for example, the addition of cell penetrating peptides (CPP) like Histidine tagged TAT peptide (TAT peptide), that significantly enhance the permeation of liposomes. TAT peptide is a permeation enhancer that interacts with the skin lipids, temporarily disrupting the lipid arrangement, creating aqueous pores and channels that facilitate liposomes permeation (Boakye et al., 2015). Table 11.2 shows some dermal applications of liposomes.

In conclusion, the use of liposomes as transdermal drug delivery systems is limited since they cannot penetrate deep skin layers. In addition, phospholipid costs, difficulty in scaling of preparation methods, and physical and chemical instability are challenges that need to be solved.

11.2.2 ULTRADEFORMABLE LIPOSOMES

As already mentioned, liposomes have little utility as transdermal delivery systems. Therefore, it is sought to design new vesicular systems that have the advantages of liposomes (high biocompatibility and low toxicity), but are able to promote transdermal drug permeation. Thus in 1992, Cevc and Blume introduced deformable liposomes called transfersomes. These deformable liposomes contain an edge activator (generally a surfactant) to modify the elasticity of the bilayer, increasing its deformability. Due to these properties, they appear as useful carriers for transdermal drug delivery (Elsayed et al., 2007).

The term *transfersome* means "carrying body" and is derived from the Latin word *transferre*, which means "to carry through" and the Greek word *soma* meaning "body." In this chapter, we will refer to transfersomes as UDLs. In functional terms, UDLs can be described as lipid droplets whose deformability allows them to easily pass through pores much smaller than their own size. They are highly adaptable complexes that respond to stress by adapting their shape. This flexibility minimizes the risk of vesicle rupture (Priyanka and Shipra, 2012).

UDLs are modified liposomes, formed by amphipathic natural compounds, suspended in an aqueous solution. Like liposomes, they contain a lipid bilayer that surrounds an aqueous core. Due to the incorporation of a surface-active agent in their membrane, UDLs are highly flexible, a property that facilitates a rapid penetration through the intercellular lipid spaces of the SC (which are considerably smaller than the vesicles diameter). A surface activator is often a single chain surfactant that destabilizes the lipid bilayer of the vesicle and increases the deformability of the bilayer by lowering the interfacial tension (Honeywell-Nguyen and Bouwstra, 2005). Among these activators are sodium cholate, sodium deoxycholate, Spans, Tweens, and dipotassium glycyrrhizinate (Cevc et al., 1996; El Maghraby et al., 1999, 2000; Trotta et al., 2004).

Table 11.2 Examples of Dermal Applications of Liposomes

Active Substance	*Liposome Type*	*Results*	*Reference*
Emodin	Dipalmitoyl phosphatidyl glycerol:palmitoyl oleoyl phosphatidyl glycerol-liposomes	Emodin-loaded liposomes, incorporated in layer by layer films of poly(ethylenimine) and poly(vinyl sulfonate), retard the release of drug with respect to films with free drug. These systems are a feasible alternative for transdermal applications.	Campos et al. (2018)
Lornoxicam	Conventional liposomes	Optimized formulation of liposomes loaded with lornoxicam was obtained using Central Composite Design; the liposomes obtained showed particle size around 200 nm, high entrapment efficiency ($>95\%$) and provided sustained release for 8 h. The permeability flux obtained from the ex vivo skin permeation through goat skin was 23.29 $\mu g\ cm^{-2}$ h, indicating their potential for topical application.	Joshny et al. (2018)
Quercetin	*N*-succinylchitosan and chitooligosaccharide-multilayered liposomes	Multilayered liposomes, prepared with *N*-succinyl-chitosan and chitooligosaccharide showed a better stability to surfactants (Triton X-100) regarding conventional liposomes. The quercetin was slowly release by diffusion from multilayered liposomes. The release was dependent on the number of layers and pH. In the same form, the skin permeation of drug was slightly higher from multilayered liposomes than from uncoated liposomes.	Seong et al. (2018)
Chloramphenicol (CAM) and benzoyl peroxide (BPO) (acne treatment)	Polyethylene glycol (PEG)-liposomes	Coencapsulation of CAM and BPO in PEG-liposomes led to the sustained release of CAM, while BPO seemed retained within the liposomes. Liposomal formulations were less cytotoxic than the corresponding drug solutions used as reference evaluated by MTT assay on HaCaT cells.	Ingebrigtsen et al. (2017)

Madecassoside (wound healing and psoriasis)	PEG-liposomes	The transdermal property and wound healing effect of madecassoside liposomes prepared by a double-emulsion method were superior to those prepared by film dispersion method. Both liposomes were endowed with excellent performance by PEG modification. PEG modification improves encapsulation efficiency, stability, and enhanced wound healing activity. These liposomes complete healing at day 12 of treatment in the wound induced rat model.	Li et al. (2016)
Allantoin	Argan oil–liposomes	The addition of argan oil to the liposomes improved the vesicle ability to modify SC and its hydration favoring the allantoin skin delivery. The combination of argan oil and phospholipids may facilitate the distribution of drug into the liposomes, and avoiding drug leakage. Argan oil improves allantoin accumulation in the skin and in the dermis compared with water dispersion and commercial gel.	Manca et al. (2016b)
Usnic acid (wound healing)	Gelatin-based membrane containing usnic acid-loaded liposome (UAL)	Transdermal absorption studies of UAL showed that usnic acid can penetrate through the skin, since amounts of usnic acid were detected in different skin layers (epidermis, dermis) and in the receptor fluid, although, a larger amount of usnic acid was quantified in the dermis. These experiments were realized in skin without SC to simulate first degree injury. The UAL system promoted the development and maturation of the granulation reaction over the scar repair, and collagen deposition.	Nunes et al. (2016)

(*Continued*)

Table 11.2 Examples of Dermal Applications of Liposomes *Continued*

Active Substance	*Liposome Type*	*Results*	*Reference*
Quercetin (antioxidant)	Hyaluronate–chitosan multilayered liposomes	The increase of the number of hyaluronate–chitosan layers on liposome surface result in a sustained delivery of quercetin. The deposition of polysaccharides contributes to an improvement of the drug permeability.	Jeon et al. (2015)
Retinyl palmitate (carcinogenesis prevention)	Multilamellar liposomes	Significant higher skin retention than solid lipid nanoparticles.	Clares et al. (2014)
Meglutamine antimoniate (treatment of cutaneous leishmaniasis)	Liposomes	Mice treated with drug-loaded liposomes showed a significantly smaller lesion size than the group treated with empty liposomes. However, when the treatment was stopped, the lesion grew up again, but to a somewhat lesser degree than the control groups.	Moosavian Kalat et al. (2014)
Mupirocin (wound healing)	Liposomes	Prolonged drug release in both in vitro and ex vivo studies. The liposomes and associated drug remained on the skin, indicating their potential as wound healing.	Hurler et al. (2012)
Calcipotriol (treatment of psoriasis)	Gel-state liposomes and liquid-state liposomes	A higher penetration of lipid into barrier-impaired skin was observed from liquid-state liposomes compared to gel-state liposomes. Liposomal delivery to diseased skin is affected by the fluidity of the liposomal membrane.	Knudsen et al. (2011)
Basic fibroblast growth factor (bFGF) (wound healing)	Conventional liposomes	The bFGF-liposomes could effectively protect the biological activity of bFGF, promoting the wound-healing process.	Xiang et al. (2011)

The average size of UDLs is generally smaller than those that do not contain the edge activator using the same amount of phospholipids. This is due to the addition of the edge activator, which induces perturbation of the packed acyl chains of phospholipid bilayer and modifies the fluidity of the lipid membrane, increases membrane hydration, and facilitates the formation of particles smaller than rigid liposomes. This effect is proportional to the amount of edge activator added. However, a superfluous amount of edge activator may decrease the vesicles' stability by excessively disturbing the fluidity of the membrane, thus leading to leakage of the cargo drug from the vesicle, resulting in a lower entrapment efficiency. It is therefore necessary to optimize the type and amount of surface activator for each formulation (Zhang et al., 2015a).

Issues concerning preparation methods and optimization, as well as characterization of UDLs, are addressed in recent reviews (Merin et al., 2016; Ranju and Manoj, 2015; Sarmah et al., 2013; Singh, 2013).

According to Cevc et al. (1998), UDLs meet all requirements to penetrate permeability barriers efficiently, including intact mammalian skin, with an efficiency close to 100% and a release efficiency greater than 50%. These authors suggest that this depends on the transdermal osmotic pressure difference. In addition, the passage of UDLs through the skin is a function of their flexibility, hydrophilicity, and ability to retain their integrity while undergoing drastic changes in shape. UDLs can act as drug-carrying systems, whereby intact vesicles enter the SC, transporting drug molecules into the skin. The driving force for the vesicles to penetrate the skin is called xerophobia (tendency to avoid dry surroundings). The main difference between UDLs and traditional liposomes is the high stress-dependent adaptability of these deformable vesicles, which make it possible for them to be squeezed between the cells of the SC, despite their large size. Thus, they can enter intact and spontaneously in the skin, under the influence of the transcutaneous hydration gradient that occurs naturally in vivo, without disintegration of the vesicles (Fig. 11.1). Occlusion could eliminate this gradient and thus have a detrimental effect on the action of UDLs (Cevc and Blume, 1992; Honeywell-Nguyen and Bouwstra, 2005). UDLs can penetrate even if the pores are five times smaller than their own diameter (Cevc et al., 2002; Hofer et al., 2004). The UDLs reach regions in the deep layers of the skin, with high water content. Subsequently, the vesicles are distributed between the cells. Because they are too large to penetrate cutaneous blood vessels, the UDLs deviate from them and reach the subcutaneous tissue. Finally, the vesicles reach the systemic circulation via the fenestrated lymphatic system, which has openings with sufficient amplitude (Cevc et al., 1995). For the vesicles to remain swollen, they should follow the local hydration gradient and penetrate the deeper, more hydrated layers of the skin (viable epidermis and dermis) (Dhamecha et al., 2009).

Cevc et al. (2002) found that the chemical composition of the elastic vesicles does not change significantly during their passage through the narrow pores. Based on the information obtained to date, it is believed that the elastic vesicles

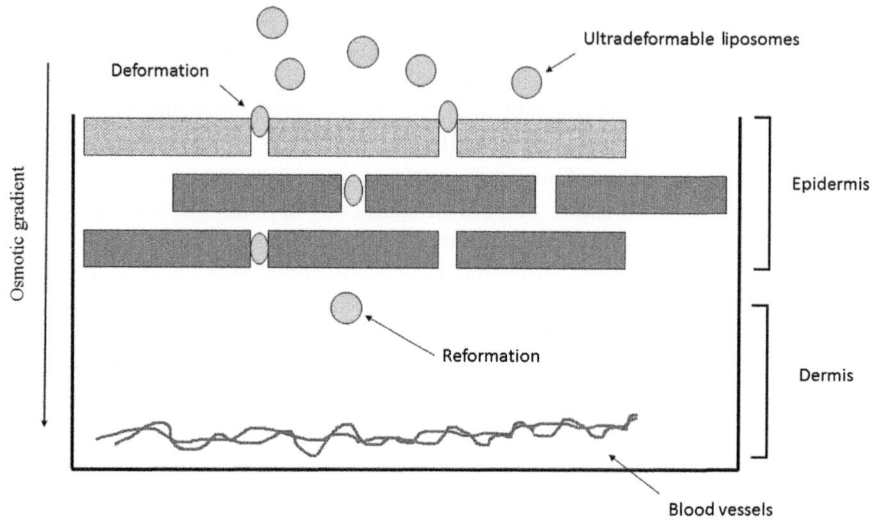

FIGURE 11.1

Mechanism of penetration for UDLs through the skin.

retain their physical stability, being transported through the skin as intact dynamic entities, being able to cross the SC, independently of the concentration of the drug.

The passage of UDLs through the skin and epithelial barriers is influenced by the flexibility of their membrane, which can be achieved by employing suitable proportions of edge activators. The flexibility of the membrane of these liposomes also reduces the possibility of rupture in the skin (Kumar et al., 2012).

Another mechanism that has been proposed is that UDLs may act as permeation promoters. Upon entering the SC, vesicles modify lamellae of intercellular lipids, facilitating the permeation of free drug molecules (Dhamecha et al., 2009).

The effectiveness of UDLs as dermal and transdermal drug delivery systems has been successfully demonstrated using molecules with different physicochemical characteristics such as molecular size and lipophilicity, for example lidocaine, tetracaine, cyclosporine, hydrocortisone, dexamethasone, triamcinolone acetonide, diclofenac, ibuprofen, tamoxifen, testosterone, etc. (Cevc, 1997).

Several studies have reported that deformable liposomes could enhance in vitro drug release through the skin (El Maghraby et al., 1999, 2001; Trotta et al., 2002, 2004), and penetrate intact skin in vivo, transferring therapeutic amounts of drug (Cevc and Blume, 2001, 2004), with efficacy comparable to that of subcutaneous administration (Cevc et al., 1995, 1998; Cevc, 2003; Paul et al., 1995).

Morilla and Romero (2015) and Hussain et al. (2017) performed extensively reviews about transdermal applications of UDLs, so Table 11.3 only shows some recent studies. More recently, the combined use of UDLs with some physical

Table 11.3 Recent Applications of UDLs

Active Substance	*Treatment*	*Edge Activators*	*Results*	*Reference*
Baicalin	Inflammatory skin disorders	Tween 80	Baicalin-loaded gellan–cholesterol–UDLs inhibited the inflammation markers (edema, myeloperoxidase activity, tumor necrosis factor-α and interleukin-1β) and favored the skin repair in mice with induced cutaneous inflammation. These effects were higher than those obtained using a commercial betamethasone cream.	Manconi et al. (2018)
Raloxifene HCl	Hormone replacement therapy	Span 80	The loaded-UDLs enhanced the ex vivo transdermal drug flux compared to conventional liposomes. Confocal laser scanning microcopy studies showed that UDLs were distributed through the SC, viable epidermis, and dermis.	Mahmood et al. (2018)
Naringin	Antiinflammatory skin therapy	Tween 80	Cell viability studies showed low toxicity of UDLs loaded with naringin. The transdermal drug delivery assay indicated a low flavonoid bioavailability. The cumulative amount of naringin in the receptor compartment was negligible and only ~10% of drug remained in the epidermis. However, in vivo studies, showed that administration of naringin-loaded UDLs were more effective in reducing skin edema and myeloperoxidase activity, compared to betamethasone cream and to a drug dispersion. Thus, this vesicular system is a promising formulation for treatment of skin inflammation.	Pleguezuelos-Villa et al. (2018)
Minoxidil and caffeine	Androgenic alopecia treatment	Tween 20, Tween 80	UDLs are appropriate systems for codelivery of minoxidil and caffeine through the skin to promote hair growth. Hair length and weight was studied in rats over a period of 1 month to evaluate the therapeutic efficacy of UDLs loaded with both drugs. The rats treated with UDLs showed faster growth in terms of length compared to commercially hydroalcoholic and aqueous solutions. Similar results were also obtained in terms of hair weight.	Ramezani et al. (2018)

(Continued)

Table 11.3 Recent Applications of UDLs *Continued*

Active Substance	*Treatment*	*Edge Activators*	*Results*	*Reference*
Sertaconazole nitrate	Antifungal	Sodium deoxycholate (SDC)	The authors evaluated different vesicular systems: liposomes, UDLs, glycerosomes, and ethosomes. They found that UDLs showed highest flux through rat skin compared to other vesicular formulations. The UDL containing Sertaconazole nitrate was selected to incorporate in 1% hydroxypropyl methylcellulose gel. The UDLs gel showed higher permeation compared to commercial cream (Dermofix) of Sertaconazole nitrate. The antifungal activity of UDLs gel was significantly higher in therapeutic efficacy than commercial cream in dermal fungal infection animal model with *Candida albicans*.	Abdellatif et al. (2017)
Eprosartan mesylate	Antihypertensive	Span 80 (SP), SDC	The transdermal flux of drug through rat skin was dependent on the type of edge activator. The flux increased linearly with an increment in the concentration of SDC from 10.46 $\mu g\ cm^{-2}$ h (5%) to 27.22 $\mu g/cm^{-2}$ h (25%). An increase in SP concentration results in a higher flux up to 15%, after that a lower flux was observed. Confocal laser microscopy showed that the UDLs permeated into deeper layers of rat skin.	Ahad et al. (2017)
Procyanidins	Antioxidants	Tween 80, SDC	Transdermal The penetration rate of UDLs was 6.9-fold greater than that of drug solution and 3.7-fold greater than that of conventional liposomes. Also, UDLs increase the procyanidins deposition both in the SC and epidermis compared to drug solution and conventional liposomes. In addition, UDLs improve the stability of procyanidins.	Chen et al. (2017a)

Papain	Hypertrophic scar treatment	Tween 80	UDLs were more effective in enhancing transdermal release than papain solution (PS), using abdominal rat skin. The cumulative amount of papain in receptor was ~5 times higher with UDLs than with PS. Also, skin deposition showed higher papain deposition from UDLs as compared with PS. After topical application of UDLs with papain, the hypertrophic scar in rabbit ears was inhibited, showing decreased scar elevation index, decreased microvascular density, and decreased collagen fiber with regular arrangement.	Chen et al. (2017b)
Growth factors (GFs). Epidermal growth factor (EGF), insulin-like growth factor-I (IGF-I), and platelet-derived growth factor-A (PDGF-A).	Diabetic wound healing	Tween 80	The incorporation of low-molecular-weight protamine (LMWP) fused GFs and hyaluronic acid (HA) in cationic elastic liposomes accelerated the wound healing process in the diabetic mice model compared with mice treated with control vehicle and cationic elastic liposomes loaded with native GFs and HA. This was due to the enhanced and prolonged skin permeation of the LMWP-fused GFs loaded in cationic elastic liposomes.	Choi et al. (2017)
Risperidone	Antipsychotic treatment of schizophrenia and bipolar disorder management	SDC, Tween 80	The ex vivo transdermal permeation of risperidone from UDLs gel was evaluated using porcine ear skin. The flux through skin increased with decreasing soya lecithin to cholesterol ratio and increased with SDC to Tween 80 ratio. The skin permeation of risperidone from UDLs gel was sustained over 24 h. The iontophoresis application (0.5 mA cm^{-2}) improved skin permeation of risperidone from UDLs gel.	Das et al. (2017)
Zinc phthalocyanine (ZnPc) and nitrosyl ruthenium complex $[Ru(NH{\cdot}NHg)(tpy)NO]^{3+}$ (RuNO)	Photodynamic therapy of skin melanoma	Tween 20	The coencapsulation of ZnPc and RuNO in Dimyristoyl and dioleoyl-phosphatidylcholine UDLs enhanced ex vivo transdermal permeation of ZnPC through pig ear skin. In the absence of irradiation, the viability cell (B16-F10 melanoma cells) was unaffected by empty UDLs and UDLs with RuNO, while UDLs containing both drugs showed some toxicity towards the cells, even in the absence of irradiation. With irradiation (635 nm) the empty UDLs didn't affect cell viability, while cell viability decreased 90% with UDLs containing ZnPc, in the presence or absence of RuNO.	de Lima et al. (2017)

(*Continued*)

Table 11.3 Recent Applications of UDLs *Continued*

Active Substance	*Treatment*	*Edge Activators*	*Results*	*Reference*
Psoralen and resveratrol	Treatment of vitiligo	Sodium deoxycholate	The psoralen and resveratrol coloaded UDLs showed slow release of both drugs compared to drug solutions. UVA therapy with psoralen-UDLs enhanced stimulation of melanin levels in a dose dependent compared to psoralen solution, while resveratrol-UDLs had no effect on melanin content in mouse melanoma cells. Psoralen and resveratrol coloaded UDLs also showed enhanced stimulation of melanin levels. It was also observed that coloaded UDLs increased the tyrosinase activity, this effect attributable to psoralen. Additionally, coloaded UDLs also exhibited potential free radical scavenging activity attributed mainly to resveratrol. Thus, psoralen and resveratrol coloaded UDLs have a potential use in the treatment of vitiligo.	Doppalapudi et al. (2017)
Atorvastatin calcium	Treatment of hyperlipidemia	PEG fatty acid esters: Labrasol, Cremophor EL, Gelucire 44/14, and Tween 80	The ex vivo permeation studies in rat skin showed that all liposomal formulations enhanced drug permeation through skin compared with drug solution. The UDLs prepared with Gelucire 44/14 as edge activator showed a greater promoter effect attributable to smaller vesicle size. The formulation UDLs-Gelucire 44/14 gel was more efficient in delivering the drug to systemic circulation compared with both the traditional gel and the oral suspension. The $AUC_{0-\alpha}$ for the UDLs gel was three times greater than the $AUC_{0-\alpha}$ for the oral suspension. Furthermore, the in vivo study using poloxamer 407-induced hyperlipidemic rats demonstrated the effectiveness of drug-UDLs-Gelucire 44/14 gel in the treatment of hyperlipidemia, decreasing the atherogenic lipids, being biocomparable to an oral suspension of the drug. No side effects in the liver were found.	Mahmoud et al. (2017)

Methotrexate	Antirheumatic	Tween 80	The in vitro skin permeation assay through rat skin showed that the flux increased in the following order: drug-UDLs gel > drug-conventional liposomes gel > drug plain gel. The in vivo antirheumatic effect of gel with methotrexate loaded in UDLs was evaluated in adjuvant-induced arthritis rat model. The drug-UDLs gel reduced the severity of edema volume and suppressed the expression of pro-inflammatory cytokines in paw tissues to a greater extent than drug-conventional liposomes gel and drug plain gel.	Zeb et al. (2017)
Pentoxifylline	Treatment of intermittent claudication and other forms of chronic occlusive arterial diseases of the limbs	Sodium cholate, Tween 21, Tween 20, Span 80, Span 20	The type and amount of edge activator used affected the UDLs properties like elasticity, drug loading, and drug release; these properties were optimal when it was used sodium cholate. Too, cholate-UDLs enhance pentoxifylline permeation through the excised rat skin by ninefold in comparison with the control. The in vivo pharmacokinetics study showed that cholate-UDLs increased drug absorption and prolonged its half-life comparable to oral tablets.	AL-Shuwaili et al. (2016)
1,1-Bis(3-indolyl)-1-(p-chlorophenyl) methane (DIM-D)	Cancer chemoprevention	Tween 80	The ultraflexible liposomes (cationic) were more effective in overcoming the skin barrier function to enhance topical release of DIM-D compared with PEG solution. This led to inhibition of skin carcinogenesis. Also, reduced the number of tumors induced by UV radiation.	Boakye et al. (2016)
BRAF-siRNA (v-Raf murine sarcoma viral oncogene homolog (ii)	Anticancer, melanoma therapy	Sodium cholate	Cationic-UDLs or lipoplexes could overcome the SC barrier and permeate the skin and deliver siRNA principally in the basal epidermis and upper dermis. The lipoplexes could internalize into melanoma cells and knockdown the expression of target proteins. In addition, lipoplexes reduce the viability of the cultured melanoma cells.	Dorrani et al. (2016)

(Continued)

Table 11.3 Recent Applications of UDLs *Continued*

Active Substance	*Treatment*	*Edge Activators*	*Results*	*Reference*
Niacinamide	Antimelanogenesis (skin whitening)	Dipotassium glycyrrhizate (DPG)	DPG-liposomes increase the cumulative ratio of drug penetration by 85.5% in comparison with drug solution, and increased 25.4% in comparison with conventional liposomes. DPG-liposomes also enhance skin whitening in volunteers with melasma.	Lee et al. (2016)
Oxaprozin	Analgesic and antipyretic	Sodium cholate	UDLs were twofold more effective than nanostructured lipid carriers in promoting drug permeation through excised human skin. Also found was a synergistic effect of the combined use of cyclodextrin and nanocarriers in improving the drug permeation, flux, and apparent permeability.	Mennini et al. (2016)
Baicalin	Antiinflammatory	Tween 80	UDLs liposomes provided the greatest penetration rate of drug compared with that obtained with the drug saturated solution through excised human epidermis. And, the antiinflammatory activity was greater than dexamethasone.	Mir-Palomo et al. (2016)
Timolol maleate	Antihypertensive (β-adrenergic blocker)	Span 80, Tween 80	Span-liposomes had better permeation profile than Tween-liposomes. Transdermal UDLs prolong plasma profile up to 72 h compared with oral solution.	Morsi et al. (2016)
Amphotericin B (AmB)	Antifungal and antileishmanial	Sodium cholate (NaChol), Tween 80 (T80)	Even though AmB-T80 liposomes exhibited analogous antifungal and antileishmanial activities with loaded AmB unilamellar liposomal formulation, only AmB-T80 liposomes provided a considerable increased AmB deposition in SC and viable epidermis. The total AmB accumulation in skin was 40 times higher when applied AmB-T80 liposomes than AmB unilamellar liposomal formulation.	Perez et al. (2016)
Cytarabine	Treatment of leukemia	SDC	SDC-liposomes showed higher flux and drug deposition in ex vivo rat skin compared with conventional liposomes and drug solution. In addition, the SDC-liposomes exhibit a lower lag time. In the in vivo study, it was found that transdermally administered UDLs deliver an appreciable amount of the drug and it was detected in plasma after 1 h of application while with conventional liposomes no detectable amount was found at the same time.	Raj et al. (2016)

Resveratrol (RSV) and 5-fluorouracil (5-FU)	Treatment of cancer	Sodium cholate (SC)	Coencapsulation of 5-FU and RSV in UDLs improves their percutaneous permeation with respect to the free drugs. Coencapsulation of 5-FU and RSV in UDLs favored their synergistic antitumor action on skin cancer cells.	Cosco et al. (2015)
Antigen of *Plasmodium falciparum*, PfMSP-119	Transdermal vaccination	Span 80	UDLs can overcome the skin permeability barrier and deliver antigen loaded to inmmunologically active cells of skin and draining lymph nodes. The UDLs showed minimum lag time (0.9 h) compared with conventional liposomes that show a 2.1 h lag time.	Tyagi et al. (2015)
Isooctyl p-methoxycinnamate	Chemical sunscreen	SDC	SDC-liposomes loaded with isooctyl p-methoycinnamate achieved superior cutaneous absorption of the drug and exhibit greater skin tissue antioxidative capacity than the control.	Zhang et al. (2015a)
Itraconazole-hydroxypropyl-β-cyclodextrin	Antifungal	Tween 20 (T20), Tween 80 (T80), SDC	UDLs formulation enhance itraconazole disposition into skin in the following order: SDC-liposomes > T80-liposomes > T20-liposomes > conventional liposomes. UDLs -cyclodextrin improve the amount of drug in SC compared with ones containing drug alone. Also, the antifungal activity was maintained.	Alomrani et al. (2014)
Tretinoin	Acne	Tween 80	After 24 h in the in vitro skin permeation study, no drug was detected in the receptor phase. The drug was retained in SC, which served the therapeutic purpose of the drug. In addition, tretinoin transdermal delivery is not desirable due its teratogenic effect.	Ascenso et al. (2014)
Lornoxicam	Analgesic	Tween 80	The in vitro release is dependent on content of Tween 80, the lowest content shows a slow release due to higher rigidity of vesicles. UDLs enhanced drug permeation through ex vivo rabbit skin over traditional liposomes.	Taha (2014)

Note: *All the permeation experiments were performed under nonocclusive conditions.*

promoters such as ultrasound and iontophoresis has been evaluated (Kasetvatin et al., 2015; Malinovskaja-Gomez et al., 2017).

UDLs are potential carriers for dermal and transdermal administration of drugs, being able to encapsulate a wide variety of drugs as a wide range of small molecules, peptides, proteins, and DNA. Studies to date have shown that UDLs are more effective in penetrating through the skin than conventional liposomes, thanks to their high elasticity. However, there is a lack of in vivo studies that demonstrate the effectiveness (not only penetration through the skin but also therapeutic effectiveness), since the current evidence is the high in vitro penetration through the skin of various animals (rat, rabbit, pig) of these specialized vesicles. In addition, several problems such as cost, long-term stability, and scaling of the manufacturing process have to be solved.

11.2.3 ETHOSOMES

Ethosomes (ethanolic liposomes) are vesicles composed of soy-phosphatidylcholine (SPC) and a high ethanol content (up to 50%). Although the mechanisms by which they enhance drug penetration are not yet fully understood, some reports (Touitou et al., 2000) suggest the following: as ethanol lowers the transition temperature of phospholipids, it confers fluidity, elasticity and high membrane permeability, being able to penetrate deep into the skin. In general, ethosomal formulations have shown higher permeation and longer action as compared to conventional liposomes or hydroalcoholic solutions. A synergistic mechanism was suggested between ethanol and phospholipids, as both of them are capable of fluidizing the SC lipids. In addition, ethanol is able to permeate across the skin, carrying drugs into the tissue, (as ethanol traverses); and finally, ethanol can extract some of the lipids from within the SC. Presence of ethanol also provides a negative surface charge (negative Z potential) that prevents aggregation of vesicles due to electrostatic repulsion (Jain et al., 2017). In fact, the average size and size distribution of ethosomes remains constant for at least two years at room temperature (Romero and Morilla, 2013). Ethosomes are generally characterized by a narrow size distribution (Touitou and Godin, 2007). Ethosomes are composed of generally recognized as safe (GRAS) constituents, however, it is important to consider that although ethanol is a permeation enhancer, it can also lead to skin irritation. In general, an ethanol concentration up to 40% is considered acceptable for skin delivery (Jain et al., 2017). In fact, the safety of ethosomal systems has been investigated in numerous in vitro and in vivo studies on skin cultured cells, on animals, and humans. The results of these works revealed that ethosomes are not irritants. Ethosomes offer the unique property of controlling their size as a function of ethanol content with no need for ad hoc equipment, that is, as the ethanol concentration increases, the size can be reduced (Romero and Morilla, 2013; Touitou and Godin, 2007).

Elsayed et al. (2006) found that ethosomes were not able to improve skin delivery of nonentrapped drug (they used ketotifen as model drug), but showed superior skin deposition when ethosomes had drug both inside and outside the vesicles. Unlike transfersomes, ethosomes can provide effective delivery under both occlusive and nonocclusive conditions (Jain et al., 2017).

Some studies (e.g., Dayan and Touitou, 2000; Fang et al., 2008) have compared the efficacy of conventional liposomes with ethosomes and, in general, found that in contrast to liposomes, ethosomes had smaller particle sizes, exhibited higher entrapment capacity and deeper penetration into the skin, and higher transdermal delivery.

Positive results have been obtained for a great variety of drugs when they are included in ethosomes, for example, erythromycin, bacitracin, melatonin, lamivudine, indinavir, methotrexate, testosterone, cannabidiol, clindamycin phosphate, salicylic acid, acyclovir, ammonium glycyrrhizinate, and trihexyphenidyl HCl (Ainbinder and Touitou, 2005; Ashtikar et al., 2016; Dayan and Touitou, 2000; Dubey et al., 2007a, 2007b, 2010; Godin and Touitou, 2004, 2005; Lodzki et al., 2003; Paolino et al., 2005; Shumilov and Touitou, 2010; Touitou et al., 2008). Examples of the studies performed during the last five years are presented in Table 11.4.

In conclusion, the penetration enhancing ability of ethosomes has been related (1) to its elasticity, which allows an easy penetration through the intercellular spaces of the SC; and (2) to the interaction of ethanol and phospholipids with lipid molecules, resulting in a reduction in the T_m of the SC lipids, increasing their fluidity.

11.2.4 INVASOMES

Invasomes are bilayer structures, composed of SPC (invasomal bilayer matrix), lysophosphatidylcholine (flexibility agent), terpenes, and ethanol (Ashtikar et al., 2016). The combination of all of these components (i.e., unsaturated phospholipids, terpenes, and ethanol) synergistically increases the amount of drug in deeper layers of the skin than with conventional liposomes or a solution of the drug. This was noted by Verma et al. (2004) when studying the deposition of invasomal cyclosporine-A, with promising results for the therapy of alopecia areata; Dragicevic-Curic et al. (2008) with temoporfin used for photodynamic therapy in the treatment of skin cancer; or Chen et al. (2010) with ferulic acid. The physicochemical properties of a drug and of the components of the vesicles have influence on the penetration degree, depending on the drug–vesicle interactions, which resulted in different partitioning between vesicles and biological membranes (Shah et al., 2015). Some pharmacokinetic studies have proven that topically applied drug-loaded invasomal formulation increased bioavailability and allowed a longer duration of action, compared to conventional administration, for example, the oral route (Kamran et al., 2016). A synergistic effect between terpenes and ethanol results in higher penetration rate through the skin for invasomes, compared to liposomes and ethosomes (Lakshmi et al., 2013).

Table 11.4 Examples of Studies Performed with Ethosomes, Invasomes, and Niosomes

Ethosomes				
Active Substance	***Formulation***	***Model***	***Results***	***Reference***
Curcumine	Ethosomes with varying proportions of lecithin, ethanol, and cholesterol.	In vitro permeation through human cadaver skin. In vivo: the antiinflammatory activity was evaluated using carrageenan-induced hind paw edema method in male Wistar rats.	Maximum antiinflammatory effects for the formulation containing 10 mg lecithin, 4.5% ethanol and 10 mg cholesterol. A significant increase in percent inhibition (58.8% for 7 h) for this formulation as compared with oral administration.	Pathan et al. (2018)
Ranolazine	Ethosomes with varying proportions of lecithin and ethanol.	In vitro penetration through rat skin.	Higher penetration and transdermal flux of ethosomes as compared with hydroethanolic drug solution.	Bisht et al. (2017)
Ovalbumin and saponin	Liposomes, transfersomes, and ethosomes.	In vivo transdermal immunization in BALB/c mice.	Ethosomes showed the best stability and the highest titer of serum antibody.	Zhang et al. (2017)
Terbinafine HCl	Ethosomes and vesicles with labrasol, Transcutol, and terpenes (cineole and limonene) in a chitosan gel.	Rat induced superficial *Candida* infection. Randomized double-blind clinical study in patients.	Highest deposition in the skin and a cure rate of 86% in 7 days with limonene (only 20% for market product) (Lamisil cream).	AbdelSamie et al. (2016)
Glimepiride	Ethosomes varying ratios of phospholipids, cholesterol, ethanol, and PG. Ethosomes were incorporated in polymeric films with PG and DMSO as enhancers.	Ex vivo skin permeation studies using Wistar rat skin. In vivo pharmacokinetics with human volunteers.	Permeation was superior for ethosomes than for pure drug transdermal films. Extended drug release and lower maximum drug plasma level for ethosomes/film.	Ahmed et al. (2016)
Terbinafine HCl	Ethosomes prepared by an unsonicated and a sonicated method. Ethosomes were included in a Carbopol 934 gel.	Ex vivo permeation studies through male albino rats.	Sonication yielded a better formulation. Higher drug diffusion for ethosomes than from a marketed cream.	Iizhar et al. (2016)

Griseofulvin	Ethosomes with different amounts of drug.	In vitro permeation studies using newborn pig skin. Cytotoxicity in HaCaT cells. Antifungal activity versus *Trichophyton rubrum*.	No cytotoxicity to HaCaT was observed. Antifungal activity with 0.1% drug-loaded ethosomes. Ethosomes pass easily between corneocytes.	Marto et al. (2016)
1-(1-Naphthyl) piperazine	Transethosomes and vesicles containing DMSO.	In vitro skin permeation through pig ear skin. In vivo photoprotective effect in hairless mice. Myeloperoxidase activity in mice skin.	DMSO vesicles showed good permeation in vitro, and in vivo prevented UVB-induced inflammation.	Menezes et al. (2016)
Osthole	Ethosomes, transfersomes, liposomes.	In vitro skin permeation through porcine ears. Plasma protein binding in rat plasma. In vivo pharmacokinetic study.	Ethosomes showed enhanced transdermal flux, decreased lag time and increased AUC and C_{max} compared with the other formulations.	Meng et al. (2016)
Sinomenine HCl	Ethosomes obtained by the injection method.	In vitro permeation studies through rat skin. Skin irritation study on New Zealand white rabbits. In vivo antiinflammatory study in mice.	Good skin tolerability, with significant inhibition of ear edema.	Yan et al. (2016)
Polyphenolic phytocomplex from *Fraxinus angustifolia*	Ethosomes and vesicles with Transcutol or ethylene glycol.	Cell viability, cell uptake, and antioxidant activity in human keratinocytes. In vivo edema and myeloperoxidase assay in CD-1 mice.	Ethosomes and vesicles with ethylene glycol were small, unilamellar, and monodispersed, while those with Transcutol were larger, multilamellar, and less homogeneous. Extracts were not effective in free form, but significantly inhibited H_2O_2 stress damages when included in vesicles, and showed high antioxidant and antiinflammatory effects, favoring wound healing.	Moulaoui et al. (2015)

(*Continued*)

Table 11.4 Examples of Studies Performed with Ethosomes, Invasomes, and Niosomes *Continued*

Ethosomes				
Active Substance	***Formulation***	***Model***	***Results***	***Reference***
Mitoxantrone	Ethosomes obtained by thin film dispersion and incorporated in a hydroxypropyl methylcellulose gel.	In vitro permeation study through rat skin. Pharmacodynamic study on nude mice with melanoma.	Ethosomes showed higher permeability and higher antimelanoma effect than drug aqueous solutions.	Yu et al. (2015)
Meloxicam	Ethosomes prepared by thin layer evaporation.	In vitro permeation studies through Wistar albino rats.	Ethosomes showed lesser sizes, better EE%, and improved transdermal flux than rigid liposomes. Ethosomes showed higher inhibition of rat paw edema than an oral administration.	Ahad et al. (2014)
Apigenin	Ethosomes, liposomes, and deformable liposomes.	In vitro permeation studies through rat skin. In vivo skin deposition in rats. UVB-induced skin inflammation in mice.	Increasing the amount of phospholipids in ethosomes, increased EE%, skin deposition, and transdermal flux. Ethosomes showed the strongest effect on reduction skin inflammation.	Shen et al. (2014)
Psoralen	Ethosomes and liposomes	In vitro permeation study through rat skin. Biocompatibility with glabrous rat skin.	Higher permeability for ethosomes than for liposomes. Ethosomes showed better biocompatibility with human skin fibroblasts.	Zhang et al. (2014a)
Psoralen	Ethosomes and liposomes.	In vitro skin permeation study and in vivo deposition using rat skin. Evaluation of skin viability. Cellular uptake with HaCaT and CCC-ESF-1 cell lines.	Skin deposition from ethosomes and liposomes first increased and then reached a plateau, indicating a saturation of intracellular drug delivery. This agreed with a reduced deposition of ethosome and liposome in skin with reduced viability.	Zhang et al. (2014b)

Valsartan	Ethosomes prepared by thin layer evaporation with different amounts of ethanol.	In vitro skin permeation studies and pharmacokinetic studies with albino Wistar rats.	Ethosomes showed better permeation, penetrating into deep skin layers, compared with rigid liposomes. Increased bioavailability for ethosomes compared with an oral suspension.	Ahad et al. (2013)
Tretinoin	Liposomes, ethosomes, SLNs, NLCs.	In vitro skin permeation studies through shaved Laca mice skin. Skin compliance study with Laca mice. Antipsoriatic activity using mouse tail model.	SLNs and NLCs offered enhanced photostability, skin transport, and antipsoriatic activity versus ethosomes, liposomes, and a marketed product.	Raza et al. (2013)
Resveratrol	Transfersomes and ethosomes.	Reactive oxygen species and malondialdehyde assay in HaCaT cells. Ex vivo permeation studies with pig ear skin.	Only blank nanocarriers with Tween 80 or ethanol were cytotoxic. Only ethosomes were able to promote resveratrol permeation.	Scognamiglio et al. (2013)
Terbinafine HCl	Ethosomes and liposomes in Carbopol, chitosan, and poloxamer gels.	Ex vivo permeation studies through human cadaver nails.	The liposome poloxamer gel formulation showed the best accumulation and easiest application.	Tanriderdi and Özer (2013)
Curcumin	PG liposomes, ethosomes, and traditional liposomes.	In vitro permeation studies through rat skin. Antiinflammatory studies on Sprague–Dawley rats.	The particle size, EE%, and inhibition of paw edema followed this order: PG liposomes > ethosomes > traditional liposomes.	Zhao et al. (2013)
Tacrolimus	Ethosomes with ethanol, PG, or ethanol–PG and traditional liposomes.	Skin samples from male Wistar rats. BALB/c mice induced with atopic dermatitis.	Higher amount in epidermis with ethosomes than with commercial ointment (Protopic). Higher penetration with ethanol than with propylene glycol. Lower ear swelling in mice for ethosomes than for liposomes and commercial ointment.	Li et al. (2012)

(Continued)

Table 11.4 Examples of Studies Performed with Ethosomes, Invasomes, and Niosomes *Continued*

Ethosomes				
Active Substance	***Formulation***	***Model***	***Results***	***Reference***
Vitamin E acetate	Ethosome-like cationic vesicles.	Physicochemical characterization of vesicles.	Enhanced stability, rigidity, and encapsulation efficiency with the addition of cholesterol. Most stable vesicles were found with 20% ethanol.	Liu et al. (2012)
Clotrimazole	Ethosomes (with different concentrations of ethanol) and ultradeformable liposomes with Span 80.	Permeation studies with male Sprague–Dawley rats. Standard disk diffusion method for in vitro anticandidal testing.	Enhanced transdermal flux, decreased lag time, and better aptness in anticandidal testing for ethosomes in comparison with liposomes.	Maheshwari et al. (2012)
Paclitaxel	Ethosomes with different ethanol concentrations.	Human skin obtained from abdominal reduction surgery.	Improved permeation of paclitaxel and increased antiproliferative activity as compared with free drug.	Paolino et al. (2012)
Voriconazole	Transethosomes (containing ethanol and surfactants or oleic acid as permeation enhancer).	In vitro: hairless mouse skin. In vivo: 6-week-old hairless male mice.	Transethosomes enhanced skin permeation compared with control and other vesicles (ethosomes, deformable liposomes, conventional liposomes). Permeation enhancer increased elasticity compared with vesicles with surfactants.	Song et al. (2012)
Econazole nitrate	Ethosomes with different ratios phospholipid:ethanol.	Albino rats (Wistar strain). Cup plate method with strains of C. albicans.	Controlled drug release into deep skin layers. Percentage drug diffused from ethosomes twofold higher than liposomes and hydroethanolic gels. Better antifungal activity for ethosomes.	Verma and Pathak (2012)
5-Fluorouracil	Ethosomes	Human hypertrophic scar (HS) and skin obtained from female plastic surgery patients.	Delivery more efficient in HS than in skin in vitro. Delivery of ethosomes greater than control solutions in HS.	Zhang et al. (2012)

Invasomes				
Active Substance	*Formulation*	*Model*	*Results*	*Reference*
–	A mixture of d9-DLPC, soy-LPC, and citral:cineol:D-limonene.	SC of the volar surface of the left forearm of a male (30 years old).	Intact invasomes were visualized deep in the SC by tip-enhanced Raman scattering.	Ashtikar et al. (2017)
Olmesartan medoxomil	Nanoinvasomes with β-citronellene and different amounts of ethanol, incorporated in a gel.	Ex vivo permeation studies using rat skin. In vivo pharmacokinetic study on albino Wistar rats.	Optimized formulation with phospholipid: 11.11 mg, ethanol: 4.35%, and β-citronellene: 1.09%. Entrapment efficiency of about 65%. Deep permeation into the skin according to confocal laser microscopy. Transdermal administration showed 1.15 times improvement in bioavailability versus a marketed tablet formulation.	Kamran et al. (2016)
Idebenone and azelaic acid	DDAB and CTAB LeciPlex, invasomes, and liposomes incorporated in a gel.	Ex vivo penetration: human skin from abdominal surgery. In vitro cytotoxicity: B16F10 melanoma cell lines for idebenone. Antimicrobial study: *Propionibacterium acnes* for azelaic acid. In vivo antiacne efficacy: Wistar rats.	*Idebenone*. Highest penetration in LeciPlex. Cytotoxicity activity: LeciPlex > invasomes > liposomes. *Azelaic acid*. Highest penetration in invasomes. Antimicrobial activity: DDAB LeciPlex > invasomes ≈ CTAB LeciPlex > liposomes. Antiacne efficacy: Invasomes > liposomes > LeciPlex.	Shah et al. (2015)

(*Continued*)

Table 11.4 Examples of Studies Performed with Ethosomes, Invasomes, and Niosomes *Continued*

Niosomes				
Active Substance	***Formulation***	***Model***	***Results***	***Reference***
Tetracycline	Niosomes using a cationic gemini lysine-based surfactant.	Rabbit ear skin.	Niosomes behave as drug percutaneous permeation enhancers. The innate antibacterial properties of the surfactant represent an additional advantage for this material.	Muzzalupo et al. (2017)
Diacerein	Niosomes with different amount of Span 60 and Cholesterol.	In vitro penetration: albino rats.	Optimum formulation: Span 60 (90 mg), cholesterol (10 mg). EE% about 83. Niosomes favor penetration of drug in the epidermal and dermal layers.	Moghddam et al. (2016)
Moxifloxacin	Niosomes obtained by three methods (passive loading, active loading, and lyophilization). Three formulations: 1. Drug/niosomes 2. Drug/chitosan gels 3. Drug/niosomes/chitosan gels.	Antibacterial assays: Agar diffusion with *Pseudomona aeruginosa* and *Staphylococcus aureus*.	More sustained drug release of niosomes/gel than niosomes. Niosomes were the most efficient against *P. aeruginosa.* Gel was the most efficient against *S. aureus*.	Sohrabi et al. (2016)
Calcein	Temperature sensitive MLVs niosomes with Pluronic L64 or L64ox and cholesterol.	In vitro release studies.	Spontaneous thermosensitive properties with L64-based niosomes. Drug release was more pronounced at 42°C.	Tavano et al. (2016)
Methotrexate	Niosomes with different ratios of dicetyl phosphate, Span 60, and cholesterol.	In vivo skin deposition and histopathological study in Wistar rats.	Optimal formulation with an EE% of about 78. Skin deposition higher with niosomes (22.45%) than for a drug solution (13.87%). Histopathological studies confirmed the safety of niosomes.	Abdelbary and AbouGhaly (2015)

Itraconazole	Niosomes with different surfactants (Tweens, Spans, and Brij).	In vitro deposition in Sprague–Dawley rat skin. Antifungal activity versus *C. albicans*.	Tri-alkyl chain surfactants showed the smallest sizes and higher EE% compared with mono-alkyl chain. Tri-alkyl chains showed more accumulation in SC and stripped skin. Tween 80 and 60 yielded the highest amount of drug in the receptor fluid. Antifungal activity was not compromised when drug was loaded in niosomes.	Alomrani et al. (2015)
Resveratrol	Niosomes obtained by thin film hydration–sonication and by ethanol injection, containing Gelot 64 and 2 penetration enhancer: oleic acid and linoleic acid.	Ex vivo penetration and permeation studies with skin of newborn pig.	The ethanol injection method produced smaller niosomes with a higher EE% and better penetration (for both enhancers) versus the film hydration–sonication method.	Pando et al. (2015)
Salidroside	Niosomes with different levels of cholesterol and Span 40 and sodium dodecyl sulfate as stabilizer.	In vitro skin permeation and skin irritation with male Sprague–Dawley rats. Biocompatibility and internalization with HaCaT and CCC-ESF cell lines.	Highest transdermal flux with niosomes with Span 40:cholesterol (4:3) (higher than with a drug solution). Good biocompatibility with HaCaT and CCC-ESF. Internalization of niosomes in HaCaT cells by pinocytotic vesicles and micropinocytosis, and in CCC-ESF cells by pinocytotic and lysosomes.	Zhang et al. (2015b)
Tretinoin	MLVs containing Phospholipon 50, Plurol oleique and Labrasol.	In vitro penetration and permeation studies with human skin obtained from abdominal surgery. Internalization through human keratinocytes.	High vesicle affinity for skin appendages with Labrasol. All samples internalized into human keratinocytes. All tested formulations promoted drug accumulation in the skin and did not allow drug permeation into the receptor compartment.	Manca et al. (2014b)

(*Continued*)

Table 11.4 Examples of Studies Performed with Ethosomes, Invasomes, and Niosomes *Continued*

Niosomes				
Active Substance	***Formulation***	***Model***	***Results***	***Reference***
Sodium sulfadiazine	Niosomes with alcohols (ethanol, propylene glycol or glycerol) from 5% to 40%.	Ex vivo permeation studies with rabbit ear skin.	Niosome size increased as follow: ethanol > propylene glycol > glycerol, as well as with the amount of alcohol. Permeation increased with alcohol concentration up to 20%, decreasing with a higher concentration. The best results were obtained with glycerol. Permeation was better than with a drug solution and with traditional niosomes.	Muzzalupo et al. (2014)
Hydroxychloroquine	Niosomes prepared by different techniques (ether injection, sonication, hand shaking, reverse phase evaporation). Different surfactants were used (Spans, Tween, and Brij), varying the surfactant: cholesterol ratio. Niosomes were included in gels from different polymers.	Double-blind clinical study in patients with oral lichen planus.	Optimum formulation was obtained by reverse phase evaporation using Brij 98:cholesterol (1:1.5) in a Pluronic F-127 gel. Patients receiving this formulation showed a 64.28% reduction in the size of the lesions and in the score of pain (from 4 to 1).	Bendas et al. (2013)
Ellagic acid	Niosomes with Span 60, Tween 60 cholesterol and Solulan C24 obtained by reverse phase evaporation.	In vitro permeation studies with human skin from abdominoplasty surgery, using dimethylsulfoxide or N-methyl-2-pyrrolidone as enhancers.	Highest amount of drug in epidermis with DMSO, whereas highest amount in receptor medium with NMP. DMSO niosomes can be used for epidermis delivery, while NMP niosomes for dermis delivery.	Junyaprasert et al. (2013)

Papain	Niosomes prepared by the thin hydration method. Sodium cholate was added as edge activator (elastic niosomes) and PLGA nanospheres, both in a Carbopol 980 gel.	In vitro permeation studies using Sprague–Dawley rats. Skin irritation test in New Zealand White rabbits. Hypertrophic scar model in rabbit ears.	All formulations gave no irritation on rabbit skin. Elastic liposomes < gel gave: Higher reduction of hypertrophic scars than gel containing free papain or nonelastic niosomes; a decreased number of collagen fibers and height of scars; a greater amount of papain in the skin and in the receptor solution.	Manosroi et al. (2013)
Sodium sulfadiazine, propranolol HCl, tyrosol	Niosomes and novel liquid crystal-niosomes using Pluronic L64 and Aerosol OT as surfactants.	In vitro permeation studies through rabbit ear skin.	Aerosol OT and Pluronic L64 act as percutaneous permeation enhancers for all the drugs, obtaining higher permeated amounts than with drug aqueous solutions. It is not possible to establish a trend, since several variables (vehicle composition, drug solubility and structure) could be considered.	Tavano et al. (2013a)
Ellagic acid	Niosomes obtained by reverse phase evaporation using different ratios Span 60:Tween 60 and solubilizers (PEG 400, PG or methanol).	In vitro skin permeation studies with human skin from abdominoplasty surgery.	PEG 400 niosomes gave the highest EE %. The highest entrapment was found with 2:1 Span 60:Tween 60. Niosomes were more efficient in delivering the drug through the skin than a drug aqueous solution.	Junyaprasert et al. (2012)
Semipurified fraction 3 of *Oryza sativa*	Cationic niosomes prepared with various cationic lipids, by the supercritical CO_2 fluid technique-sonication and incorporated in a Carbopol 980 gel.	Cytotoxicity in human skin fibroblast. Transfollicular delivery in newborn skin pig.	Cetyl trimethyl ammonium bromide cationic niosomes in gel was shown to be the best formulation for androgenic alopecia.	Manosroi et al. (2012)

(*Continued*)

Table 11.4 Examples of Studies Performed with Ethosomes, Invasomes, and Niosomes *Continued*

Niosomes				
Active Substance	***Formulation***	***Model***	***Results***	***Reference***
Ammonium glycyrrhizinate	Niosomes with Tween 85, Span 20 and cholesterol at various ratios by the thin layer evaporation method.	In vitro skin permeation studies with human skin from abdominoplasty surgery. Cytotoxicity in human keratinocytes (NCTC 2544). In vivo skin tolerability and antierythema activity in human volunteers. Antiinflammatory activity in CD-1 mice.	Vesicles showed no toxicity, good skin tolerability and improved antiinflammatory activity in mice and in humans.	Marianecci et al. (2012)

MLVs, multilamellar vesicles; NMP, N-methyl-2-pyrrolidone; PEG, polyethylene glycol; PG, propylene glycol; DMSO, dimethyl sulfoxide; EE%, entrapment efficiency; NLCs, nanostructured lipid carriers; PEVs, penetration enhancer-containing vesicles; SLNs, solid lipid nanoparticles.; CTAB, cetyltrimethylammonium bromide; DDAB, didodecyldimethyl ammonium bromide; d9-DLPC, 1,2-dilinoleoyl-3-phosphocholine-N,N,N-trimethyl-d9; soy-LPC, soy lysophosphatidylcholine.

11.2.5 NIOSOMES

Reported for the first time in the 1970s, niosomes were first introduced as a feature of the cosmetic industry by L'Oreal. These are vesicles composed of nonionic surfactants (e.g., alkyl ethers, alkyl esters, alkyl amides, fatty acid and amino acid compounds), cholesterol, and molecules like fatty alcohols. Charge inducers such as dicetyl phosphate, stearylamine and diacylglycerol, provide charge to the vesicles, increase entrapment efficiency, and contribute as well to electrostatic stabilization (Ashtikar et al., 2016; Gangwar et al., 2012). Because of their composition, they are nontoxic and nonimmunogenic. Transdermal enhancement is attributed to the disrupting activity of niosomal surfactants on the intercellular lipids of the SC. Even if a sustained release can be desirable, in some cases, the high content of cholesterol or the presence of surfactants with long alkyl-chains, resulted in slow drug release, with long lag times (Ashtikar et al., 2016; El-Laithy et al., 2011). Importantly, the hydrophilic–lipophilic balance (HLB) of a surfactant has an influence on drug entrapment: a surfactant with an HLB value between 14 and 17 is not suitable to form niosomes, whereas one with an HLB of 8.6 gives niosomes with the highest entrapment efficiency. Entrapment efficiency decreases for lower HLB values. Entrapment efficiency is affected as well by the phase transition temperature of the surfactant. In this sense, Span 60, with a high transition temperature, exhibits the highest entrapment efficiency (Kumar and Rajeshwarrao, 2011). Ritwiset et al. (2016) studied the dynamical properties of Span 60 niosomes with and without cholesterol. The authors found that the presence of cholesterol enhances the niosome stability but also increase the fluidity. In the presence of cholesterol, the Span 60 molecules present a less ordered structure, which will play an important role in niosomal drug delivery. Further details about the factors affecting niosome formulation, characterization, and applications can be found in several reviews published during the last few decades by Mahale et al. (2012), Marianecci et al. (2014, 2016), Moghassemi and Hadjizadeh (2014), Rajera et al. (2011), Uchegbu and Florence (1995), Uchegbu and Vyas (1998).

Compared with liposomes, niosomes share unique advantages: (1) they are quite stable (physically and chemically), without needing special storage conditions such as low temperature or inert atmosphere; (2) low cost materials are used in their manufacture; (3) niosomal dispersions in the aqueous phase can be emulsified in a nonaqueous phase, modulating the delivery rate of drug; and (4) niosomal surfactants are usually biodegradable, biocompatible, and nonimmunogenic (Gangwar et al., 2012).

As liposomes, niosomes can be divided according to the size as small unilamellar vesicles (SUVs) (10–100 nm), large unilamellar vesicles (LUVs) (100–3000 nm), and multilamellar vesicles (MLVs) where more than one bilayer is present (the term *nanoniosome* refers to SUVs) (Moghassemi and Hadjizadeh, 2014).

Niosomes have been thoroughly studied for the transdermal delivery of different drugs such as nonsteroidal antiinflammatory drugs (e.g., aceclofenac, diclofenac, vinpocetine, meloxicam, rofecoxib, and flurbiprofen) (Das and Palei, 2011; El-Badry et al., 2014; El-Laithy et al., 2011; Manosroi et al., 2008; Solanki et al., 2010; Zidan and Mokhtar, 2011); baclofen, a skeletal muscle relaxant (Keservani et al., 2010); and papain, a protease enzyme (Manosroi et al., 2013). Patel et al. (2012) compared the transdermal permeation of lopinavir, used for the treatment of HIV, formulated in niosomes and ethosomes, observing a higher ex vivo flux with the niosomal formulation. In vivo studies reported a higher C_{max} and area under the curve (AUC) for the transdermal application of lopinavir compared to oral administration. Muzzalupo et al. (2011) prepared niosomal formulations using nonionic surfactants belonging to the class of Pluronics and the class of sucrose esters, for the percutaneous delivery of sulfadiazine sodium salt, and found an enhancing effect on the ex vivo percutaneous penetration only for niosomes, but not when the skin was pretreated with a submicellar solution of surfactants or with empty niosomes. Tavano et al. (2011) prepared niosomes and microemulsions containing capsaicin, using mixtures of Tween 80 and Span 80 and found that in both cases, the systems with HLB 12 gave the highest permeated amount of capsaicin. The authors indicate that even if both carriers are effective vehicles for the topical delivery of capsaicin, niosomes could better promote its transdermal delivery. α-Tocopherol availability and stability were improved when included in niosomes. The molar ratio of Span 60:Tween 60 and the presence of cholesterol play an important role on the stability, encapsulation efficiency, and release rate of dicetyl phosphate-based niosomes (Basiri et al., 2017a, 2017b). Ethnobotanical literature reported that *Nerium oleander*, commonly known as oleander, is effective in the treatment of asthma, cardiac illnesses, eye and skin diseases, abortifacient, ringworm, leprosy, indigestion, corns, and warts, as well as epilepsy. Gunes et al. (2017) formulated niosomes containing oleander, obtaining vesicles with a good physical stability, antioxidant activity and without toxic effects on HeLa and A549 cells. Mahdiun et al. (2017) found enhanced antibacterial activity and reduced resistance to antibiotic after encapsulation of tobramycin in combination with bismuth-ethanedithiol in niosomes, with a reduction of the minimal inhibitory concentration of tobramycin, effectively inhibiting the biofilm formation. Muzzalupo et al. (2017) prepared cationic niosomes, using a cationic gemini lysine-based surfactant. The advantages of this surfactant over traditional synthetic cationic surfactants are related to their biodegradability, low cytotoxic potential, and low critical micelle concentration. The authors performed ex vivo permeation studies with tetracycline-loaded niosomes finding that these vesicles behave as drug percutaneous permeation enhancers. A high accumulation of the carriers within the SC may be explained by their ability to interact with the negative charge of the skin lipids. Furthermore, the innate antibacterial properties of the surfactant represent an additional advantage for this material.

Interestingly, niosomes have been used to stabilize gold nanoparticles (De et al., 2012). To achieve a control over the shape and size distribution of nanoparticles, the authors synthesized gold nanoparticles in situ in Span-based niosomes. They found that the use of surfactants with low HLB values led to smaller vesicles, restricting the size of nanoparticles.

Many types of specialized niosomes can be found in the literature (Kumar and Rajeshwarrao, 2011; Moghassemi and Hadjizadeh, 2014; Rajera et al., 2011; Tavano et al., 2013a, 2013b; Yuksel et al., 2016):

- *Proniosomes*. Dry niosomes, which are hydrated immediately before use.
- *Elastic niosomes*. Composed of nonionic surfactants, ethanol and water.
- *Discosomes*. Large discoid structures that exist under certain conditions of the phase diagram of nonionic surfactants. The structural nature of surfactant, surfactant concentration, and even the temperature play a role in the obtention of these structures and others such as tubules.
- *Aspasomes*. Contain a combination of ascorbyl palmitate, cholesterol, and highly charged lipid diacetyl phosphate.
- *Magnetoniosomes*. The aqueous core is loaded with a magnetic material (e.g., magnetite).

11.2.6 OTHER SYSTEMS

11.2.6.1 Bilosomes

Bilosomes are vesicles formed by a bilayer membrane of nonionic surfactants and bile salts (bile salts containing niosomes). These vesicular carriers were first described by Conacher et al. (2001) with the purpose of administering oral vaccines, based on their ability to resist bile salts and enzymes in the gastrointestinal tract. Al-Mahallawi et al. (2015) proposed the use of these carriers systems for the transdermal delivery of tenoxicam. The authors found a better ex vivo permeation and an in vivo penetration of tenoxicam included in bilosomes than a tenoxicam solution. The efficacy of these vesicles is related to the enhancing ability of surfactants and bile salts.

Bilosomes can be prepared by the thin film hydration technique, forming a dry film of surfactant and cholesterol and hydrating the film with an aqueous solution of bile salt (Al-Mahallawi et al., 2015).

11.2.6.2 Cerasomes

Cerasomes (partially ceramic- or silica-coated liposomes), have a liposomal structure, with a layer of polyorganosiloxane networks on its surface. These systems were developed with the aim of avoiding the drawbacks of liposomes (e.g., low stability) and silica nanoparticles (e.g., nonbiodegradability, high rigidity, and low biocompatibility). In this sense, the siloxane surface provides stability in a slightly alkaline medium or in a medium with high salt content. Furthermore, the silanol groups on the surface can be functionalized with hydrophilic polymers, therapeutic moieties, and targeting ligands. On the other hand, the liposomal structure yields a less rigid system. Cerasomes can encapsulate not only both hydrophobic and hydrophilic molecules, but also amphiphilic molecules; in addition, they show long circulation residence times (Dai 2016; Katagiri and Koumoto, 2013; Yue and Dai, 2014). Cao et al. (2012) demonstrated that it is possible to modulate the release rate of paclitaxel, by preparing composite cerasomes obtained from mixtures of cerasome-forming lipid (*N,N*-dihexadecyl-*N*-(3-triethoxysilyl)propylsuccinamide) and 1,2-distearoyl-*sn*-glycero-3-phosphocholine. A wide range of the release profiles were obtained by altering the ratio of cerasome-forming lipid and phospholipid.

11.2.6.3 Cubosomes

Cubosomes are self-assembled lipid bilayer structures, forming a three dimensional structure with a bicontinuous lipid bilayer and interpenetrating independent aqueous nanochannels. This particular structure increases significantly the loading capacity, provides protection to labile molecules, increases the contact time with the absorption/action site, and allows sustained release. Among the materials reported to produce cubosomes, monoolein (glyceryl monooleate) and phytantriol (3,7,11,15-tetramethyl-1,2,3-hexadecanetriol) are the most commonly used. These materials form spontaneously bicontinuous cubic phases upon dispersion in water (Akhlaghi and Loh, 2017; Mulet et al., 2013). An advantage of phytantriol in relation to monoolein is that it has greater chemical stability due to the absence of the ester group. In this sense, Akhlaghi and Loh (2017) prepared phytantriol cubosomes loaded with two palmitoyl peptides, namely palmitoyl-GHK and palmitoyl-GQPR. The authors reported a sustained release and a greater stability when the peptides were incorporated into the cubosomes. Other lipids such as monopalmitolein, monovaccenin, and 1-(7Z-tetradecenoyl)-rac-glycerol have also been used to form cubosomes (Meikle et al., 2017). These authors included three antimicrobial peptides (gramicidin A, alamethicin, and melittin) with different solubility characteristics, in cubosomes formed by the lipids mentioned above in addition to monoolein and phytantriol. The systems were characterized in terms of loading capacity, particle size, morphology, and bilayer structure. The findings may be useful in the development of drug delivery systems.

Boge et al. (2017) proposed the postloading of antimicrobial peptides into cubosomes by allowing water-soluble peptides (with a positive net charge) to adsorb/associate onto the surface of preformed cubosomes (with a slightly negative charge) by means of electrostatic and/or hydrophobic interactions. The idea was to locate the peptides on the cubosome surface. The authors reported a lower release when the ionic strength of the medium was adjusted to the physiological conditions, and in general, the preservation of antimicrobial activity.

Peng et al. (2015) compared the permeation of capsaicin loaded in phytantriol- and glycerol monooleate-based cubosomes against a cream. Although cubosomes showed a reduction in skin absorption, a higher skin retention was observed with both cubosomes than with the cream, obtaining the highest retention with phytantriol-based cubosomes.

In a very interesting work, Rattanapak et al. (2013) studied the in vitro permeation of a peptide loaded in phytantriol-based cubosomes combined with microneedles. The results indicated that once the skin was treated with the microneedles, only the peptide in free form (in solution) was able to permeate across the skin, while peptide-loaded cubosomes remained in the microchannels formed by microneedles. This led to a higher amount of peptide retained in skin when included in cubosomes. A review about cubosomes and their potential use as a drug delivery system was published by Karami and Hamidi (2016).

11.2.6.4 Phytosomes

Phytosomes are complexes of a phospholipid (e.g., phosphatidyl choline) and a plant extract. The goal is to improve the bioavailability of plant extracts, to increase solubility and stability, to retain biological actions, to decrease toxicity, and to modulate drug release. Indena, an Italian pharmaceutical and nutraceutical company, developed and patented this technology. In addition, as the active substance represents an integral part of the phospholipid bilayer, the loading capacity is higher in comparison with liposomes. Most of the phytosomal formulations are directed to the oral route. One formulation with *Ginkgo biloba* with a cardioprotective and antioxidant activity is destined to the subcutaneous route (Akhtar and Khan, 2016). Djekic et al. (2016) incorporated 18β-glycyrrhetinic acid phytosome, well known by its antiinflammatory, antiirritant, and soothing effects, in carbomer gels, to have a suitable vehicle for application on the skin. Although no differences regarding the organoleptic, morphological characteristics, and spreadability of the hydrogels were found, the authors obtained better results in terms of stability with Carbopol Ultrez 10 than with Carbopol 980.

In addition to phytosomes, other systems including plant extracts are, for example, liposomes and deformable liposomes with phenolic compounds, quercetin, verbascoside, resveratrol, and different essential oils; niosomes with ammonium glycyrrhizinate and resveratrol; and PEVs with curcumine, quercetine, and resveratrol. Information about these kinds of systems can be found in Alexander et al. (2016) and Bilia et al. (2016).

11.2.6.5 Glycerosomes

These are elastic vesicles obtained from different phospholipids and high concentrations of glycerol (up to 50%) (Vitonyte et al., 2017). They are prepared by hydrating phospholipid–cholesterol films with glycerol aqueous solutions. Thanks to glycerol, glycerosomes are in a more fluid state than conventional liposomes; furthermore, glycerol is able to reduce the skin barrier, enhancing the hydration of the SC (Manca et al., 2013). In this sense, Manca et al. (2014a) compared the delivery to the skin of quercetin formulated in liposomes and glycerosomes, finding that glycerosomes were more efficient (over 20% of the applied dose) compared to liposomes (10%–20%). In recent works, Manca et al.

(2016a, 2017) reported that glycerosomes containing 20%–30% of glycerol, and containing either hydrogenated SPC or 1,2-dimyristoyl-*sn*-glycero-3-phosphatidy-choline, favor the accumulation and permeation of diclofenac into and through the skin. In general, the skin delivery performances of glycerosomes can be improved when combined with poloxamer, which increases the viscosity of the dispersion, avoiding leakage and favoring the contact with the skin (Caddeo et al., 2017).

11.2.6.6 Pharmacosomes

Pharmacosomes are defined as colloidal dispersions of drugs esterified with a lipid, obtaining an amphiphilic compound. As other vesicles, they are able of showing mesomorphic behavior. Pharmacosomes avoid problems of other vesicles such as drug incorporation (vesicles are formed by drug conjugated with lipids); even entrapment efficiency can be predetermined and there is no need for removing the unentrapped drug. However, although there is no leakage from the carrier, loss may occur by hydrolysis. The membrane fluidity of pharmacosomes depends on the phase transition temperature of the drug–lipid complex. Drug is released from pharmacosome by hydrolysis (Gangwar et al., 2012). Pharmacosomes have been used as carriers for a number of drugs including proteins (Akhtar and Khan, 2016).

11.2.6.7 Photosomes

These carriers include a light sensitive lipid, absorbing light between 300 and 400 nm. Release from lipid membrane occurs in response to an irradiation at predetermined wavelength (Jain et al., 2017). Photosomes are liposomes including photolyases (bacterial enzymes able to repair ultraviolet B-induced cyclobutane pyrimidine dimers (CPD) in eukaryotic cells). CPD may cause mutation when the DNA is replicated, and these genetic alterations may contribute to trigger sunlight-induced carcinogenesis. Decome et al. (2005) evaluated the repair activity of this enzyme after a single dose of UVB (312 nm, 0.06 J cm^{-2}) in human keratinocytes. The importance of these systems is their capacity to reduce the risks of development of skin cancer; therefore, authors point out the importance of including photosomes as an additional component of sunscreens.

11.2.6.8 Vesosomes

Vesosomes, also known as fusion liposomes, are multicompartment structures with inner compartments. They can be seen as larger vesicles that encapsulate many smaller vesicles inside them. Release from vesosomes can occur from hours to days (Kaurav et al., 2016). For the administration of vaccines, the vesosomes are able to produce a better immune response when applied topically. The vesosomes are capable of incorporating colloidal particles and even macromolecules.

11.2.6.9 Novasomes

These modified liposomes are composed of a mixture of polyoxyethylene fatty acids monoesters, cholesterol, and free fatty acids, arranged in several bilayers. Novasomes can encapsulate both hydrophilic and lipophilic drugs (Akhtar and Khan, 2016).

11.2.6.10 Cryptosomes

Cryptosomes contain phospholipids and poloxamer molecules. They are reported as very stable entities of choice for hydrophobic drugs (Akhtar and Khan, 2016).

11.2.6.11 Hyalurosomes

These vesicles are obtained by hydrating phospholipids with an aqueous dispersion of sodium hyaluronate. Hyaluronan plays an important role in skin wound healing. Manca et al. (2015) proposed a one-step, environmentally friendly preparation method to obtain hyalurosomes. This fact, as well as its biocompatibility, its protection against oxidative stress damages, and enhancement of tissue remodeling, makes hyalurosomes potential nanocarriers for cosmetic and pharmaceutical substances. Manconi et al. (2016) prepared glycerol containing hyalurosomes and found that they were able to promote the viability of both keratinocytes and fibroblast, preventing oxidative stress.

11.2.6.12 Penetration enhancer-containing vesicles (PEVs)

This is a new family of liposomes containing an enhancer agent, for example, diethylene glycol monoethyl ether (Transcutol) or propylene glycol. The permeation enhancer has a dual effect, improving the vesicular bilayer fluidity and reducing the SC barrier. Therefore, the presence of the penetration enhancer agent facilitates the transdermal diffusion as intact structures, most of them at least (Manconi et al., 2012). It is important to mention that PEVs are a rather confusing denomination, since ethosomes, invasomes and glycerosomes are also PEVs (Romero and Morilla, 2013). In this sense, Vitonyte et al. (2017) prepared liposomes, glycerosomes, and PEVs including, simultaneously, resveratrol and gallic acid. The vesicles promoted the accumulation of drugs in the skin (specially gallic acid). Coloading of the two phenols showed good antimicrobial activity and protection against oxidative damage, being an interesting option to treat skin infections.

11.2.6.13 Hybrid systems

Hybrid systems have been proposed to overcome some limitations associated with entrapment capacity, stability, and restricted percutaneous delivery. Desai et al. (2013) developed biodegradable lipid-polymer hybrid nanoparticles containing cationic amphiphiles with cyclic pyrrolidinium head group (structurally similar to Azone, a well-known permeation enhancer) and poly(D,L-lactic-*co*-glycolic acid) (PLGA). The authors include capsaicin and siRNA against TNF-α

(tumor necrosis factor-α) toward treating chronic skin inflammation. The results showed that coating the system with the cationic lipid (with cyclic pyrrolidinium head group) enhanced skin permeation of nanoparticles, carrying efficiently the siRNA and capsaicin into deeper skin layers.

11.3 SOLID LIPID NANOPARTICLES AND NANOSTRUCTURED LIPID CARRIERS

11.3.1 HISTORICAL BACKGROUND AND DEFINITIONS

Since the beginning of the 1990s, several research groups focused on the development of any alternative carrier system to polymeric nanoparticles and liposomes (Montenegro et al., 2016; Müller et al., 2000; Prow et al., 2011; Roberts et al., 2016). The history of lipid nanoparticles begins in 1990 with the first experiments performed in parallel by Maria Rosa Gasco (Italy), and Rainer H. Müller, and J. Stefan Lucks (Germany) (Montenegro et al., 2016; Müller et al., 2016).

These carriers showed a structure similar to nanoemulsions, nevertheless, they were constituted by a solid lipid matrix. For this reason, the inventors decided to call them SLNs (Beloqui et al., 2016). When the first publications and patent applications related to these colloidal systems came to light, the attention of scientists all over the world in the field immediately increased. Subsequently, SLNs have been investigated as delivery system for hydrophobic drugs administered by different routes (e.g., oral, parenteral, dermal, transdermal, ocular, and pulmonary), with respect to their production, optimization, characterization, and applications (e.g., Jensen et al., 2011; Kathe et al., 2014; Lauterbach and Müller-Goymann, 2015; Lombardi Borgia et al., 2005; Müller et al., 2000; Prow et al., 2011; Roberts et al., 2016; Schäfer-Korting et al., 2007; Souto et al., 2013; Tapeinos et al., 2017; Teixeira et al., 2017; Zhai and Zhai, 2014).

SLNs represented an alternative carrier system to traditional colloidal carriers, such as emulsions, liposomes, and polymeric micro- and nanoparticles. When SLNs came on the scene they were considered as the most promising lipid-based colloidal carriers because they had the capability of combining advantages of other nanometric carriers while minimizing the problems associated with these systems (Kathe et al., 2014; Montenegro et al., 2016; Zhai and Zhai, 2014). SLNs showed many advantages such as good biocompatibility, low toxicity, feasibility to scale up and sterilize, organic solvent-free preparation processes, high in vivo stability, and wide application spectrum (Beloqui et al., 2016; Montenegro et al., 2016; Müller et al., 2000; Roberts et al., 2016).

SLNs are colloidal drug delivery systems composed of physiological and biodegradable lipids, stabilized by a suitable surfactant, and ranging in sizes between 50 and 1000 nm. These lipids form a solid lipophilic matrix at both room and body temperature (or at least at skin temperature) in which hydrophilic or lipophilic drug molecules can be incorporated (Doktorovová et al., 2016; Guimarães and Ré, 2011; Jain et al., 2017; Murthy and Shivakumar, 2010). The SLNs are produced only from solid lipids, either triglycerides, partial glycerides, fatty acids, steroids, or waxes, or mixtures thereof. In general, only one solid lipid is used (Montenegro et al., 2016; Müller et al., 2016). Typical SLNs formulations include 0.1%–30% w/w of solid lipid content and 0.5%–5% w/w of the surfactant stabilizers (Jain et al., 2017). Structurally, SLNs have spherical shape with a mean particle size ranging between 50 and 1000 nm, showing a narrow particle size distribution. SLNs with solid particle matrix are derived from o/w emulsions for parenteral nutrition by simply replacing the liquid lipid (oil) of the emulsion droplets by a solid lipid (Müller et al., 2000). Recently, Chantaburanan and coworkers (2017) analyzed the effect of the binary lipid matrix of cetyl palmitate (CP) and Softisan 378 (S378) on the physicochemical properties of ibuprofen-loaded SLN dispersions for dermal delivery. SLNs were prepared at different ratios of CP and S378, that is, 1:0, 3:1, 1:1, 1:3, and 0:1, by hot high pressure homogenization (HPH). On the whole, in vitro drug release studies across a mix cellulose ester membrane showed that the addition of S378 into CP-matrix led to slow ibuprofen release owing to the drug–lipid interaction when increasing S378 loading. The results from proton nuclear magnetic suggested the formation of tiny liquid S378 nanocompartments within CP-matrix similar to the multiple systems of NLCs. Therefore, the authors concluded that the localization of ibuprofen in the S378 nanocompartments and the interaction between ibuprofen and S378 had an impact on the release profiles of ibuprofen-loaded SLNs depending on the ratios of CP and S378. Due to the structural similarity between the lipid matrix of SLNs and the epidermal lipids in skin, SLNs have been extensively investigated as transdermal drug carriers. Compared to lipid-based vesicular carriers, SLNs provide flexibility in modulating drug release, higher drug loading of lipophilic compounds, and enhancing drug or active stability from chemical degradation, oxidation, light degradation, and moisture; furthermore, the final formulations have a higher physical stability (Jain et al., 2017; Müller et al., 2016). Obviously, SLNs show some limitations: poor drug loading capacity, particle growth, polymorphic transitions, and unpredictable gelation tendency during the cooling of dispersion, as well as component degradation and possible drug release during long-term storage (Jain et al., 2017; Musicanti and Gasco, 2016; Zhai and Zhai, 2014).

The SC is the principal barrier for the exchange of substances between the body and the environment and implies the hardest challenge for active agent delivery into and through the skin. The small particle size and high surface area of SLNs allow a close contact with superficial junction of corneocyte clusters and channels of SC, leading to film formation on the skin (Jain et al., 2017). SLNs produce adhesive films onto the skin, like liposomes. Increased adhesiveness to surfaces is a general property of ultrafine materials; the adhesiveness increases with decreasing particle size. The effects of SLNs on skin barrier properties have been well established (Müller et al., 2007, 2002). It has been reported that SLNs form an invisible, occlusive film with affinity for the SC, which leads to an increased skin hydration due to a reduced transepidermal water loss (TEWL); despite this, SLNs may enhance the penetration of drugs through SC and a drug release for a long period is possible. Furthermore, SLNs enhance the penetration and transport of active substances, particularly lipophilic agents, into the skin (Jain et al., 2017; López-García and Ganem-Rondero, 2015; Souto and Müller, 2008).

The major drawback of SLNs is the low drug loading and the drug expulsion during storage. SLNs are composed of solid lipids assembled in a high-energy crystallized form (α and β'), which eventually convert into low-energy form (β) resulting in increased orderliness of the system; this can lead to the expulsion of the drug from the now more perfect lipid particle matrix (like blocks in a wall). This expulsion effect occurs directly during or immediately after the cooling process when SLNs are prepared and is particularly true when high purity lipids are used (Abla et al., 2016; Müller et al., 2016). The pure lipids can crystalize in a perfect crystalline lattice, which leaves no amorphous domains for the incorporation of drugs or active substances (Guimarães and Ré, 2011). This limitation gave rise to the second-generation lipid carriers, the NLCs proposed by Müller et al. (2002). In contrast to SLNs, NLCs are produced by controlled mixing of solid lipids with liquid lipids (oils), leading to special nanostructures with improved drug incorporation and release properties (Fig. 11.2). The blend of solid and liquid lipids induces a melting point depression, which does not happen with the pure solid lipid, however, this melting point still needs to be above body/skin temperature. When lipid molecules of different molecular size and conformational shape (e.g., solid lipids with longer chain fatty acids and oils with shorter chain fatty acids) are blended, less perfect crystalline matrixes are formed. The high number of imperfections in the matrix provides more space to accommodate the drug, subsequently, the NLCs system minimizes drug expulsion on storage and

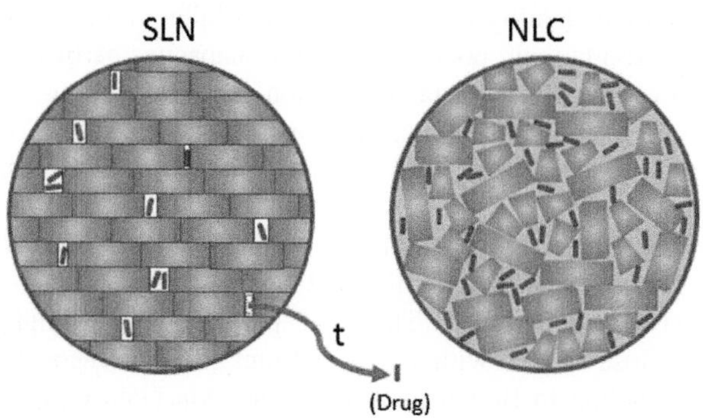

FIGURE 11.2

Structure of SLNs composed of a pure solid lipid (left) and NLCs containing a blend of solid and liquid lipids (right). The less ordered lipid matrix of NLCs facilitates a higher drug loading and avoids drug expulsion during storage.

offers higher drug incorporation capacity (Montenegro et al., 2016; Müller et al., 2007, 2016). Commonly, to obtain NLCs, solid lipids are mixed with liquid lipids in a ratio of 70:30 up to a ratio of 99.9:0.1, whereas the surfactant content ranges from 1.5% to 5% w/v (Beloqui et al., 2016). NLCs have been extensively used for medical and cosmetic products with good drug targeting, improved drug stability, and easy scale-up.

11.3.2 PREPARATION METHODS

The methods to produce SLNs and NLCs are identical, just the excipients are different. The solid lipid of SLNs is replaced by a lipid blend of solid lipid and oil for NLCs. Different combinations of lipids and/or surfactants have been used to incorporate several actives and drugs in SLNs and NLCs for dermal and transdermal administration. Several reviews list the ingredients most used to obtain them (e.g., Beloqui et al., 2016; Geszke-Moritz and Moritz, 2016; Kaur et al., 2014; Montenegro et al., 2016; Souto et al., 2013). The lipid screening is very important and in the case of requiring a blend of lipids, the mixture of solid lipids (SLNs) or solid and liquid lipids must be uniform (one phase) in the molten and in the solid state (Müller et al., 2016).

Many different techniques to produce lipid nanoparticles have been described in the literature: HPH, microemulsion technique, solvent emulsification–evaporation (or diffusion), solvent injection, ultrasonication, membrane contractor technique, coflowing microchannels, among others. Advantages and limitations of these preparation methods have been discussed elsewhere (Geszke-Moritz and Moritz, 2016; Guimarães and Ré, 2011; Mishra et al., 2012; Souto et al., 2013; Yoon et al., 2013). In general, for any preparation method it is necessary to obtain a liquid phase where lipid(s) and water can be emulsified before lipid nanoparticle formation, thus lipids in the solid state at room temperature are melted or they are dissolved into solvents (Musicanti and Gasco, 2016). Nowadays, the most used methods are HPH and the microemulsion technique, so these will be described in the next sections (Müller et al., 2016).

11.3.2.1 High pressure homogenization

This is a well-established technique on the large scale and it is already available in the pharmaceutical industry. A high-pressure homogenizer pushes a liquid with high pressure (100–2000 bar) through a micron size gap, then the liquid accelerates to high velocity (over 1000 km h^{-1}), and the resulting shear stress and cavitation forces break down the accelerated particles to submicron size. This method includes both hot and cold techniques. Cold homogenization is recommended to prepare lipid nanoparticles containing highly thermosensitive or thermolabile drugs. It is also necessary to use this technique when formulating hydrophilic drugs because they would partition between the melted lipid and the water phase during the hot homogenization process. For both techniques, the drug is dissolved or solubilized in the solid lipid or lipid blend, being melted at approximately 5°C–10°C above the melting point of the mixture.

For the hot homogenization technique, the melt lipid blend containing the active compound is emulsified in a hot surfactant solution (equal temperature) using a high speed or high shear stirring. Then, the resultant o/w emulsion is passed through a high-pressure homogenizer, applying repeated cycles at a specified pressure condition. The high temperature is kept during the whole process and the system is cooled down to room temperature only after finishing the homogenization cycles, to promote the solidification/recrystallization of the SLNs. Recrystallization can also be induced by lyophilization.

In contrast, in the cold homogenization technique, the melt lipid blend containing the active compound is cooled down, just prior to passing through the high-pressure homogenizer, resulting in a mass that must be crushed and ground to obtain lipid microparticles. Then, the micronized lipid particles are dispersed in a cold surfactant solution yielding a cold presuspension, which is passed through a high-pressure homogenizer applying repeated cycles. The number of cycles and working pressure typically selected for the hot homogenization method vary between 2–3 bar and 500–800 bar, respectively. For the cold homogenization, 5–10 cycles and 1500 bar are usually applied (Guimarães and Ré, 2011; Mishra et al., 2012; Montenegro et al., 2016; Müller et al., 2000; Souto et al., 2013; Musicanti and Gasco, 2016).

11.3.2.2 Microemulsion technique

Microemulsions are dispersions of two immiscible liquids (oil and water) stabilized by surfactants and optionally by cosurfactants. Warm microemulsions obtained with melted lipids (fatty acids and/or glycerides) have been used as precursor for the preparation of SLN: after their preparation at temperature ranging from 55°C to 85°C depending on melted lipid used, they are then dispersed in cold aqueous medium (2°C–10°C) under mild mechanical mixing, and subsequently melted lipid drops solidify into SLN. Multiple w/o/w emulsion can also be used to prepare based lipid nanoparticles with hydrophilic drugs. Excessive use of surfactants is the first drawback of this method (Mishra et al., 2012; Montenegro et al., 2016; Müller et al., 2000; Musicanti and Gasco, 2016; Souto et al., 2013; Yoon et al., 2013).

11.3.3 DERMAL AND TRANSDERMAL APPLICATIONS

SLNs and NLCs are very attractive colloidal lipidic nanocarriers for dermal and transdermal drug delivery. Some of the desirable effects include the ability to enhance the chemical stability of compounds sensitive to light, oxidation, and hydrolysis; occlusive properties; bioavailability enhancement; physical stability; and the possibility to modulate drug release (Montenegro et al., 2016; Müller et al., 2007; Roberts et al., 2016; Wang et al., 2014; Yoon et al., 2013).

Incorporation of drugs into SLNs and NLCs has been widely analyzed for topical and for transdermal administration (e.g., Costa and Santos, 2017; Lauterbach and Müller-Goymann, 2015; Müller et al., 2000; Prow et al., 2011; Roberts et al., 2016; Schäfer-Korting et al., 2007). Nevertheless, at this time, these systems have been marketed only in the cosmetic field, where they have found a great success. In fact, it is too difficult to estimate the real number of available products in the worldwide market. The first commercial cosmetic product with NLCs was a formulation of coenzyme Q10 for an antiaging therapy (Abla et al., 2016; Müller et al., 2007). In general, the formulation of cosmetic products (e.g., creams and lotions) is identical for both SLNs and NLCs. Novel products can be obtained by addition of lipid nanoparticle concentrates to already existing products, addition of viscosity enhancers to the aqueous phase of SLNs/NLCs to obtain a gel, or the direct production of a final product containing only nanoparticles in a single step process (Fireman et al., 2011; Müller et al., 2016).

Topical application of aqueous SLNs and NLCs onto the skin dispersions creates a mono-layered lipid film onto the skin, which prevents TEWL, thus increasing moisture and hydration (Beloqui et al., 2016; Zhai and Zhai, 2014). To determine differences between occlusive effects of SLNs and NLCs, López-García and Ganem-Rondero (2015) compared their occlusive effect and their penetration enhancement, through skin in vitro. SLNs (composed of glyceryl dibehenate) and NLCs (including glyceryl dibehenate/caprylic capric triglycerides 90:10) were prepared by high shear homogenization. Briefly, a hot 2.5% poloxamer 188 solution (90°C) was added to the melted lipid phase (solid lipid or lipids blend). This mixture was emulsified at 20,000 rpm for 5 min at 90°C. The resultant emulsion was cooled down in a cold water bath (7°C) to room temperature to obtain nanoparticles. Occlusive effect was assessed by an in vitro test and by measuring TEWL using pig skin. Results showed that both SLNs and NLCs produced an occlusive effect of similar magnitude (36%–39% for both systems), while a reduction in TEWL of 34.3% $\pm$ 14.8% and 26.2% $\pm$ 6.5% was seen after treatment with SLNs and NLCs, respectively. In contrast, NLCs allowed the penetration of a greater amount of the marker Nile red (4.7 $\pm$ 1.3 μg) than SLNs (1.7 $\pm$ 0.4 μg). Another study demonstrated that hydrogel formulations containing SLNs or NLCs increased the skin hydration more than the hydrogel formulation without lipid nanoparticles; nevertheless the skin hydration increase provided by SLNs and NLCs had no significant differences (Üner et al., 2005).

The occlusive properties of SLNs and NLCs assure an increased skin hydration effect and normalize the living conditions for the cells underneath (Montenegro et al., 2016). Furthermore, as consequence of this occlusion, the corneocytes become less compact and the intercorneocyte gaps become wider, increasing the permeability of the SC and enhance drug penetration into deeper skin layers (Schäfer-Korting et al., 2007). These occlusive properties cause that SLNs and NLCs are suitable for use on damaged or inflamed skin, besides their composition based on nonirritant and nontoxic lipids (Yoon et al., 2013). Hung and coworkers (2015) published an interesting paper where the impact of ultraviolet (UV) irradiation on the cutaneous penetration of NLCs and polymer nanoparticles containing retinoic acid was analyzed. SLNs exhibited a capability to intrude into epidermis and dermis and fuse with the lipid matrix of the SC in photodamaged skin.

Despite the early and successful entry of SLNs and NCLs in the cosmetics field, their expected entry into the pharmaceutical market has not happened to date (Doktorovová et al., 2016). Nevertheless, there is abundant evidence of the incorporation of drugs into SLNs and NLCs and their efficiency and quality.

As already mentioned for the cosmetic products, the development of pharmaceutical dosage forms containing SLNs or NLCs usually involves (1) their incorporation in preformed topical products (e.g., lotions, gels, or creams); (2) addition of viscosity enhancers to obtain a gel; or (3) the direct production of a final product containing only nanoparticles in a one-step process using high lipid concentrations (Beloqui et al., 2016). Due to the rigid nature of SLNs, cutaneous uptake should be closer to the cutaneous penetration of drugs from gel-state liposomes. Nevertheless, drug distribution within the particle can vary considerably and each system must be studied separately to determine its influence on percutaneous absorption (Kaur et al., 2014; Montenegro et al., 2016; Schäfer-Korting et al., 2007). It is important to consider that skin penetration depends not only on particle formulation, but also on the vehicle in which SLNs or NLCs are included.

There are numerous examples of the successful incorporation of different substances into SLNs and NLCs for a topical delivery. Examples of these actives or drugs include (1) nonsteroidal antiinflammatory drugs (NSAIDs) (indomethacin, ketorolac, ketoprofen, naproxen, flufenamic acid, meloxicam, dexflurbiprofen, etc.); (2) antifungal agents (econazole nitrate, ketoconazole, terbinafine hydrochloride, griseofulvin, amphotericin B, etc.); (3) antioxidants (curcumin, hydroquinine, idebedone, isotretinoin, lutein, coenzyme Q10, resveratrol, etc.); (4) drugs for skin diseases (adapalene, psoralen, curcuminoids, etc.); (5) local anesthetics (benzocaine, lidocaine, dibucaine, etc.); (6) glucocorticoids; (7) drugs for treatment of chronic wounds; and (8) skin whitening agents (ascorbyl palmitate, resveratrol, etc.), photosensitizers, among others. The are several reviews related to dermal and transdermal applications of SLNs and NLCs (e.g., Garcês et al., 2018; Ephrem et al., 2017; Jain et al., 2017; Lauterbach and Müller-Goymann, 2015; Md et al., 2017; Montenegro et al., 2016; Müller et al., 2016, 2000; Prow et al., 2011; Roberts et al., 2016; Schäfer-Korting et al., 2007; Sala et al., 2018; Souto et al., 2013; Yoon et al., 2013; Zhai and Zhai, 2014). Table 11.5 shows some examples of the studies performed in the last years to evaluate these kind of lipid nanocarriers.

Incorporation of two drugs in NLCs can be a good option for a simultaneous release. To cite an example, Lin et al. (2013) proposed the production of dual-drug NLCs as topical formulations. They combined tretinoin and tetracycline included in NLCs and nanoemulsions, for the treatment of acne vulgaris. NLCs may be superior carriers to nanoemulsions because of the higher

Table 11.5 Examples of Dermal and Transdermal Applications With SLNs and NLCs

Active substance	*Composition*	*Models*	*Results*	*Reference*
Auraptene, 7-geranyloxycoumarin (AUR)	Glycerol monostearate (GMS), glyceryl palmitostearate (Precirol ATO 5), poloxamer 188, Span 80 (SLNs dispersions).	In vitro: For the drug release study, male BALB/c mice were anesthetized, and the abdominal skin was excited after shaving with electric and hand razors. The full thickness of the skin was faced to the donor compartment of the Franz diffusion cell, and the dermis was faced to the receiver compartment. In vivo: Croton oil-induced dermatitis in mouse ears was carried out and histological evaluation of samples of treated and untreated ears was performed. Skin sensitization study using Guinea pigs (0.45–0.56 kg).	There was a significant drug diffusion difference between the AUR-conventional cream and the AUR-loaded SLNs over the 24-h period. The drug accumulations in the skin were 44.64 $\mu g\ cm^{-2}$ and 133.77 $\mu g\ cm^{-2}$ after 24 h, when AUR-cream and AUR-loaded SLNs were applied, respectively. The antiinflammatory and histopathological studies exhibited no significant differences between AUR-loaded SLNs and indomethacin solution (positive control). AUR-loaded SLNs did not induce skin sensitization in Guinea pigs. The results suggest that SLNs are promising carriers for the topical application of AUR as a natural antiinflammatory agent.	Daneshmand et al. (2018)
Aceclofenac (ACE)	GMS, stearic acid (SA), cetyl alcohol, Transcutol, Labrafac, Labrasol, Phospolipon S 100, Pluronic F-68, Tween 80 (NLCs/gel).	Ex vivo skin permeation studies on porcine ear pinna skin, in dark. In vitro cell uptake on the human HaCaT cell lines (hyperproliferative keratinocyte). Carrageenan-induced paw edema in mice. Skin compliance study by modified Draize test with female LACA mice.	The optimized NLCs-based gel formulation showed better skin permeation efficiency and improved dermatokinetic profile, biocompatibility, and cell uptake in HaCaT cells. Furthermore, better antiinflammatory activity in carrageenan-induced paw edema was observed. The safety profile of the developed formulation was confirmed by skin compliance assessment and compared with a marketed gel formulation.	Garg et al. (2017)

(*Continued*)

Table 11.5 Examples of Dermal and Transdermal Applications With SLNs and NLCs *Continued*

Active substance	*Composition*	*Models*	*Results*	*Reference*
Progesterone (PRG)	Stearic triglyceride (tristearin), caprylic/capric triglycerides (Miglyol 812N), poloxamer 188, (SLNs or NLCs in a viscous gel).	In vitro: PRG release studies by the dialysis tube method. Permeation studies using stratum corneum/epidermis (SCE) membranes of adult human skin obtained from breast reduction operations. In vivo: Tape-stripping involving both sex humans (25–55 years old).	PRG is more slowly released from PRG-loaded NLCs, compared with PRG-loaded SLNs and PRG suspension. PRG content in the receptor phase was below the detection limit of the analytical method employed in skin permeability experiments. The analysis of the PRG content in SCE, demonstrated that after 6 h similar amounts of PRG were present in gels with PGR loaded in SLNs or NLCs. After 24 h, the presence of PRG in SCE was statistically higher in the case of PRG-loaded SLNs/Gel with respect to PRG-loaded NLCs/gel (1.5-fold higher). NLCs appear as vehicle able to better retain PRG within lipid matrix with respect to SLNs. Tape-stripping experiments indicated that PRG penetrated into stratum corneum and deeper skin layers. PRG penetration was slightly higher in the case of PRG-loaded SLNs/gel, possibly due to a more pronounced occlusive effect induced by the SLNs.	Esposito et al. (2017)
Aluminum chloride phthalocyanine (ClAlPc)	Compritol 888 CG ATO, stearic acid, sorbitan isostearate polyoxyethylene-40 hydrogenated castor oil (SLNs dispersions).	In vitro: Biocompatibility test using NIH-3T3 cells (murine fibroblasts). Dark cytotoxicity test on the cancerous cell line (B16-F10, pigmented melanoma). Photodynamic therapy (PDT) efficiency using B16-F10 cell line. MTT, (4,5-dimethylthiazol-2-yl)2,5-dyphenyl-tetrazolium bromide assay for NIH-3T3 and B16-F10 cell viability.	The fibroblast culture exposed to dispersion of unloaded SLNs (0.1 and 0.2% v/v) showed a cell viability of over 70%. Toxicity studies revealed suitable biocompatibility of ClAlPc-SLNs with NIH-3T3 cells and with B16-F10 cells under dark conditions. PDT efficiency was directly related to the energy applied to the B16-F10 cells and were more effective when these cells were treated with ClAlPc-loaded SLNs. Therefore, all of the results confirmed that SLNs are potential lipid nanocarriers for the encapsulation of hydrophobic photosensitizer drugs.	Goto et al. (2017)

Vancomycin free base (VM-FB)	An acid cleavable solid lipid (SA-3M), Poloxamer 188 (dispersions of pH-responsives SLNs).	In vitro: Cytotoxicity study for the solid lipid on MCF-7, Hep G2 and A549 cell lines using 3-(4,5-dimethylthiazol-2-yl)-2,5-diphenyltetrazolium bromide (MTT) assay. Drug release studies using dialysis tubing. Minimum inhibitory concentration (MIC) values against methicillin-sensitive *S. aureus* (MSSA) and methicillin-resistant *S. aureus* (MRSA) at pH 6.5 and 7.4. MTT assay to determine the cell viability of MSSA and MRSA cells after treatment with VM-FB-loaded SA-3M SLNs In vivo: Antibacterial activity of loaded and unloaded SLNs against MRSA using BALB/c mice.	Percentage cell viability for SA-3M at all concentrations using MTT assay was > 75% across all tested cell lines. A pH-dependent drug release was observed for VM-FB-loaded SA-3M SLNs. At pH 6.5, the fast VM_FB release was observed, and almost 100% of the VM-FB was released at the end of 4 h. In vitro antibacterial activity against. MSSA and MRSA revealed that unloaded SLNs had enhanced activity at pH 6.5 than pH 7.4. In contrast, VM-FB-loaded SA-3M SLNs had lower MICs at pH 6.5 than at pH 7.4. In vivo study showed that the amount of MRSA remaining in the skin of VM-FB-loaded SA-3M_SLNs treated mice was approximately 22-fold lower than VM-FB and the blank SA-3M SLNs treated mice. Histological investigations revealed that signs of inflammation in the skin treated with VM-FB-loaded_SA-3M_SLNs were minimal, whereas significantly greater number of lymphocytes and leukocytes were observed in skin treated with unloaded SLNs and with VM-FB. In conclusion, this study confirmed that SA-3M can form pH-responsive SLNs capable of releasing antibiotic specifically at acidic infection sites.	Kalhapure et al. (2017)

(*Continued*)

Table 11.5 Examples of Dermal and Transdermal Applications With SLNs and NLCs *Continued*

Active substance	*Composition*	*Models*	*Results*	*Reference*
Capsaicin	Tripalmitin (Dynasan116), Caprylic/Capric/Myristic/ Stearic triglyceride (Softisan 378), Poloxamer 188, Tween 80. Capsaicin-loaded SLNs were prepared with Dynasan 116 or Softisan 378. (SLNs dispersions).	In vitro: Skin permeation study using an excised abdominoplasty surgery skin from female patients (25–60 years old).	At the first 8 h, Dynasan 116 SLNs showed the highest amount of capsaicin in receptor medium, followed by Miglyol 812 nanoemulsion and Softisan 378 SLNs, respectively. After 4 h, a film formation on the donor phase of Franz diffusion cells was observed for Dynasan 116 SLNs. Therefore, the film formation could promote the penetration of the drug through epidermis. In addition, the amount of capsaicin permeated through epidermis is related to crystallinity index (CI%) of the lipid matrix. The high degree of crystallinity of Capsaicin-loaded Dynasan 116 SLNs (CI = 91.8%) provided the high occlusion factor. The CI for Capsaicin-loaded Softisan 378 SLNs was 27.3%. At 8 h, the occlusion factor of Dynasan 116 SLNs and Softisan 378 SLNs was ~55% and ~19%, respectively.	Tapeinos et al. (2017)
Flufenamic acid and fluconazole (as model drugs with different polarity)	Monoacyl-phosphatidylcholine (MAPL), medium chain triglycerides (Lipoid MCT), Precirol-ATO5 SLNs: Ratio liquid to solid lipid 2:7; three phospholipid mixtures of variable MAPL content were evaluated, namely LPC20, LPC65 and LPC80, containing 20%, 65% and 80% w/w of MAPL, respectively. Olive oil was selected to replace medium chain triglycerides and produce NLCs (SLNs and NLCs dispersions).	In vitro: Skin permeation studies using porcine abdominal skin treated at 60°C with a scalding and dehairing machine.	In vitro studies were carried out to analyze the impact of both MAPL and the total lipid content on skin permeation of two model drugs. A low surfactant content and a low total lipid content led to increased skin permeation of flufenamic acid. In case of fluconazole, higher surfactant content in combination with different total lipid content yielded the highest skin permeation.	Wolf et al. (2017)

LL37, a human antimicrobial peptide	Precirol ATO 5, Miglyol 812N, Poloxamer 188, Tween 80 (freeze-dried NLCs).	In vitro: Cell viability studies using HFF cells; inhibition of macrophages activation induced by lipopolysaccharides (LPS); antimicrobial assay (*E. coli*). In vivo: wound healing study with 16 eight-week-old male db/db mice; histological analysis; immunohistochemistry.	The cytotoxicity assay performed in human foreskin fibroblasts demonstrated that the NLC-LL37 did not affect cell viability. The encapsulated LL37 maintained its bioactivity, as it was evidenced by the immunomodulatory and antimicrobial in vitro assays.The in vivo full thickness wound healing assay demonstrated that the administration of 6 μg of LL37 topically improved wound healing compared to the administration of the free LL37. Overall, these findings suggest a promising potential of the NLC-LL37 formulation for chronic wound healing.	Garcia-Orue et al. (2016)
Olanzapine	Glyceryl dibehenate/behenate (Compritol ATO 888), Glyceryl palmitostearate (Precirol ATO 5), Oleic acid, Stearic acid, Polysorbate 80 (Tween 80). Different liquid-to-solid lipid ratios (SLNs and NLCs dispersions).	Newborn pig (3 weeks) epidermis.	Particles with a higher proportion of solid lipid presented the fastest flux, in particular the SLNs. Oleic acid is thought to disrupt packing within the stratum corneum and act as a chemical penetration enhancer. The highest amount permeated per day was below transdermal target values.	Iqbal et al. (2016)
Lansoprazole (LPZ)	GMS, Stearylamine (SA), Pluronic F68, SDS, Isopropyl myristate (IPM), Menthol. Different GMS/SA and SDS/F68 ratios. (NLCs/hydrogel with different type and concentration of enhancers).	In vitro skin penetration studies using the skins of hairless genius pigs or Wistar rats. In vivo pharmacokinetic study in Wistar rats (intravenous injection and transdermal applications).	SDS facilitated skin penetration of LPZ NLCs. IPM and menthol successfully acted as the transdermal enhancers to enhance penetration flux of lansoprazole NLCs hydrogels that was concentration dependent. The transdermal application of NLCs hydrogel not only sustained drug release but also maintained drug concentration in blood for longer period as compared to intravenous solution. It implied that the drug was accumulated in the skin followed by penetration through the skin continuously.	Lin and Duh (2016)

(Continued)

Table 11.5 Examples of Dermal and Transdermal Applications With SLNs and NLCs *Continued*

Active substance	*Composition*	*Models*	*Results*	*Reference*
			The steady state flux of LPZ NLCs hydrogel containing 3.75% IPM was the highest, which was enhanced by 2.7-fold as compared to enhancer-free NLCs hydrogel. In vivo pharmacokinetics of LPZ following transdermal delivery of NLCs hydrogel showed that the elimination of drug was significantly reduced, and the mean residence time of drug was prominently prolonged as compared to intravenous drug solution.	
Piroxicam (Pir)	Glycerol monostearate, Lecithin, Tween 80, Oleic acid (OA).	In vitro: Permeation in excised rat skin. Fluorescence studies using coumarin 6 in frozen rat skin sections. In vivo: Antiinflammation studies using Sprague–Dawley rats. Carrageenan-induced paw model in rats. Paw edema inhibition rate. Histopathological examination. ELISA test of the PGE2 level in the rat paw.	The combined use of SLNs and OA significantly promoted the skin penetration of piroxicam by 38-fold. Skin appendages were one of the major pathways for the permeation of the Pir loaded in the SLNs. Piroxicam-loaded SLNs demonstrated as a promising therapeutic for the treatment of paw edema in relieving the paw edema, decreasing the inflammatory responses in the edema site and reducing the secretion of inflammatory cytokines.	Peng et al. (2017)
Flurbiprofen (FP)	Glyceryl dibehenate/behenate (Compritol ATO 888), Miglyol, Lecithin, Poloxamer 188, SDC, and Tween 80. Different liquid-to-solid lipid ratios (SLNs and NLCs dispersions).	Excised skin from dorsal surface of male Wistar rats. In vitro occlusion test.	The liquid-to-solid lipid ratio influenced the particle size of SLNs and NLCs, as well as percutaneous permeation profile. Both SLNs and NLCs improved the permeated amounts of FP, but only NLCs were stable after long-term storage. All formulations effectively reduced the water loss.	Song et al. (2016)

All-trans retinoic acid (ATRA)	Soybean oil (S), linoleic acids (L), oleic acids (O), medium chain triglycerides (M), cetyl palmitate (CP), lecithin (Epikuron 200), Polysorbate 80 (Tween 80) (different amounts of ATRA and S, M, O, and L) (SLNs and NLCs dispersions).	Shed skin of the snake, *Naja kaouthia*.	The amount of the drug and solid lipids and the types of oil significantly affected the skin penetration of the ATRA-loaded SLNs. The SLNs and NLCs formulations containing oleic acids demonstrated high skin permeation rates of ATRA. Smaller SLNs particles led to a greater occlusive effect and greater flux of ATRA.	Charoenputtakhun et al. (2014)
Methotrexate (MTX)	Witepsol S51, Oleic acid, Polysorbates 60 and 80 (Tween 60, Tween 80) (NLCs).	In vitro: skin permeation using pig ear skin.	MTX-loaded NLCs with Tween 60 had higher skin penetration when compared to free MTX, providing drug fluxes of 0.88 mg cm^{-2} h^{-1}, higher than with the free drug (control, 0.59 mg cm^{-2} h^{-1}). The choice of surfactant affected the flux rate of MTX.	Pinto et al. (2014)
Meloxicam (MLX)	Glyceryl monostearate 40–55 (Geleol), Glyceryl behenate (CompritolATO), Glyceryl palmitostearate (Precirol ATO5), Miglyol 812, Poloxamer 188 (Pluronic F68) (SLNs or NLCs/ hydrogels).	In vivo: UV-induced erythema test using adult male Wistar rats. The antiinflammatory activity was determined by erythema visual scoring and histopathological analysis.	MLX-loaded SLNs and NLCs hydrogels compared to blank SLNs and NLCs gels. The order of the antiinflammatory activity of the investigated gels is: MLX-NLCs gel > MLX-SLNs > commercial gel > Blank NLCs gel > Blank SLNs gel.	Khalil et al. (2014)
Tretinoin and tetracycline HCl	Squalene/Precirol ATO 5 (a mixture of 8%–22% monoacyl, 40%–60% diacyl, and 25%–35% triacyl glyceryl palmitate and stearate) and Myverol 18-04K (palmitinic acid monoglyceride); Pluronic F68, Forestall (soyaethyl morpholinium ethosulfate) (NLCs dispersion).	Skin from the dorsal region of female nude mice and female hairy mice. Antibacterial activities by an agar diffusion assay (against S. aureus, P. aeruginosa, and Propionibacterium acnes). Bicinchoninic acid assay for total protein.	NLCs enhanced tetracycline flux by about 2-times over the control solution in vitro. Tretinoin permeation was generally unaffected after nanoparticulate encapsulation. There was no significant difference in tretinoin delivery before or after tetracycline incorporation, while tetracycline permeation significantly decreased by twofold in the dual-drug system.The antibacterial activity of tetracycline was preserved after nanoparticulate loading. NLCs may be superior carriers to nanoemulsions.	Lin et al. (2013)

(*Continued*)

Table 11.5 Examples of Dermal and Transdermal Applications With SLNs and NLCs *Continued*

Active substance	*Composition*	*Models*	*Results*	*Reference*
Corticosteroid betamethasone-17-valerate (BMV)	Glycerol distearate (Precirol ATO 5) and cetylpalmitate, glycerol tripalmitate (Dyanasan 116); Polysorbate 80 (SLNs dispersion).	Intact and barrier-impaired porcine skin in vitro.	A higher amount of BMV remained in the skin, intact as well as barrier impaired, when SLNs were used as a vehicle. The reservoir effect was more evident for intact skin. The drug partitioning into the skin layers was dependent on the lipid properties of the SLNs. A high level of BMV in the skin when drug was administered into SLNs, as compared to the BMV ointment.	Jensen et al. (2011)
Flurbiprofen (FP)	Compritol ATO 888, Miglyol 812, Lecithin, Poloxamer 188, SDC, Tween 80 (NLCs dispersion).	Abdominal excised skin of male Wistar rats.	The skin permeation rate at steady-state of FP from NLCs was 5.04 times higher than FP solution.	Han et al. (2008)
Econazole nitrate (ECN)	Precirol ATO (glycerol palmitostearate); Tween 80 (SLNs/hydrogels).	Ex vivo permeation studies with porcine SC. In vivo: 5 healthy Caucasian females.	Controlled drug release through the SC. Release rate depends upon the lipid content. SLNs with lower lipid concentration improve the ECN penetration rate into upper skin after a short period.	Sanna et al. (2007)

tetracycline permeation, more significant tretinoin uptake by follicles, and conservation of bacterial growth inhibition. In a more recent work Andrade et al. (2017) proposed a coencapsulation of clobetasol (CLO) and tracolimus in uncoated NLCs and NLCs coated with chitosan oligosaccharide. CLO was released faster from the NLCs, and in a greater amount than TAC, which showed affinity for the solid lipid in the matrix. A fivefold increase in TAC penetration across full thickness porcine ear skin was observed from TAC + CLO–NLCs in comparison to TAC–NLCs; coencapsulation of TAC and CLO favored TAC skin penetration, especially in deeper skin layers. Chitosan coating also increased TAC and CLO permeation to deeper skin layers (1.8-fold and 1.6-fold, respectively).

The combined use of SLNs and NLCs with physical promoters has been a matter of interest in recent years. As an example, the effect of low frequency sonophoresis (20 kHz) on the permeation of sodium fluorescein (NaFI)-loaded lipid nanocarriers (such as liposomes, niosomes and SLNs) through porcine skin has been studied (Rangsimawong et al., 2017). Surprisingly, the authors found that SLNs alone had a significantly higher NaFI flux than the SLNs with sonophoresis. The sonication energy could break the solid core of the lipid nanoparticles, although the temperature of the SLNs dispersion was lower than the melting point of the solid core; thus, the skin surface was covered by a solid lipid fraction, leading to a reduced NaFI transport through the skin. The combined effects of cathodal iontophoresis and lipid particles (SLNs and NCLs) on drug delivery across human epidermal membrane were investigated by Charoenputtakun et al. (2015). The behavior of some lipophilic and hydrophilic model drugs included in lipid nanoparticles were analyzed in passive and iontophoresis experiments (0.5 mA cm^{-2}). The results suggest that the SLNs and NLCs could be a promising drug delivery method to improve topical and transdermal drug delivery.

The application of marketing authorization for medicinal products requires detailed data on efficacy, quality, and safety, and, obviously is more complex than a cosmetic authorization. SLNs and NLCs for dermal and transdermal are mostly composed of GRAS excipients approved for use in human medicines and several studies have been carried out to probe their efficiency and quality. Recently, an interesting review about the preclinical safety of these nanocarriers was published by Doktorovová et al. (2016).

In conclusion, several studies have evidenced the ability of SLNs and NLCs to improve skin permeation and/or to achieve cutaneous targeting. SLNs existed almost three decades ago, and have been constantly improved. NLCs are the most tangible signal of these changes. Undoubtedly, this development will go further in the coming years. In the near future, SLNs and NLCs will reach the pharmaceutical market.

11.4 PROPOSED MECHANISMS FOR THE PENETRATION OF LIPID-BASED NANOCARRIERS INTO AND THROUGH THE SKIN

Different mechanisms by which nanocarriers promote drug penetration into and through the skin have been proposed, however, there is no one predominant mechanism; in fact, penetration occurs most of the time, by a combination of some of them. The extent of contribution of each mechanism depends on the composition and the size of the system, the properties of the drug, the mode of application and the model used. In general, higher penetration has been observed for nanocarriers 120 nm and smaller (Verma et al., 2003) and elastic carriers penetrate more easily than the rigid ones. Some of the proposed mechanisms are described below and are summarized in Fig. 11.3.

1. *Adsorption and/or fusion with the SC lipids*. Campani et al. (2016) found by attenuated-total reflectance/Fourier transform infrared spectroscopy (ATR–FTIR) that vesicles fused with upper layers of the SC. In general, lipid nanocarriers can blend with SC lipids releasing their content (Hofland et al., 1995). The interaction between SLNs and lipids of the SC, facilitating

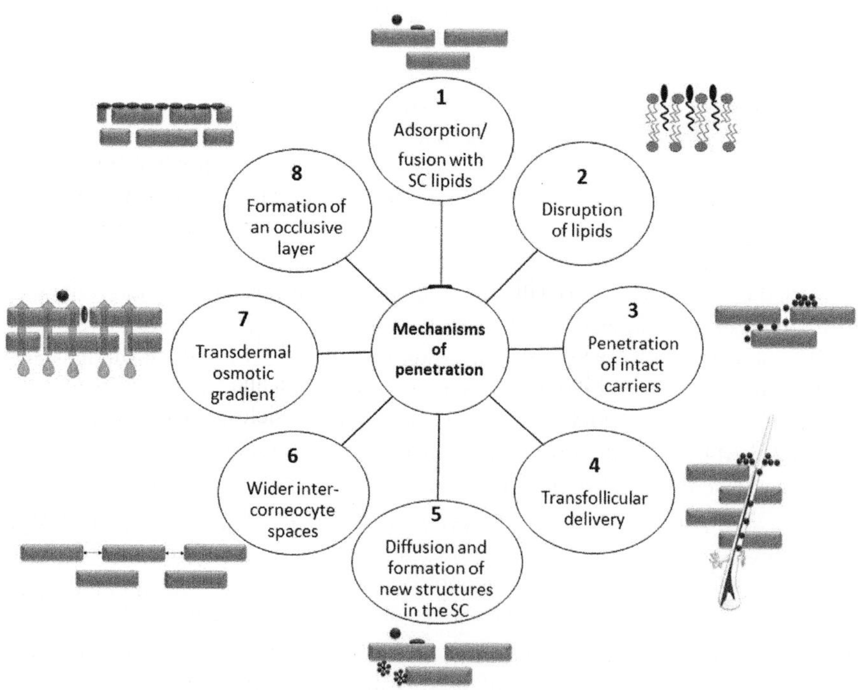

FIGURE 11.3

Proposed penetration mechanisms for lipid-based nanocarriers.

permeation of lipid soluble compounds, has been demonstrated by scanning electron microscopy, differential scanning calorimetry, and FTIR (Lauterbach and Müller-Goymann, 2015; Roberts et al., 2016).

2. *Disruption of the intercorneocyte lipid organization.* Although many authors agree that insertion of vesicle components between SC lipids results in fluidization, this is controversial. On one hand, some authors like Hasanovic et al. (2011) found by ATR–FTIR that treatment of the skin with dipalmitoyl phosphatidylcholine (DPPC) liposomes results in a decrease of the orthorhombic organization in the SC lipids; on the other hand, Campani et al. (2016) found that treatment of the skin with SPC liposomes, transfersomes, and ethosomes did not change the lipid organization. Van den Bergh et al. (1999b) reported the presence of lamellar stacks in the intercellular space, forming separated lamellar domains when elastic vesicles are applied on the skin, while rigid vesicles do not induce the formation of lamellar stacks.
3. *Intact vesicles carrying drug cargo can penetrate into or through the SC.* Intact vesicular formulations seem to be able to penetrate the top 2–3 layers of the SC, being detected in the intercellular space (Van den Bergh et al., 1999a, 1999b). Phagocytosis of intact liposomes was observed in cultured human keratinocytes (Korting et al., 1993; Schaller et al., 1996), although no evidence of phagocytosis was observed in reconstructed skin.
4. *Transfollicular delivery.* This route is particularly relevant for nanoparticles, but also for rigid liposomes and elastic vesicles, where they form a depot from which prolonged release can occur. Occlusion as well as massage after application of formulation may promote penetration into hair follicles (Lieb et al., 1992; Pardeike et al., 2009; Trauer et al., 2014).
5. Hofland et al. (1991) have proposed that surfactants and/or lipids diffuse in the SC, where they mix with SC lipids, forming vesicles or new structures within the SC; however, no evidence for this mechanism has been provided by other authors.
6. Appearance of wider spaces between corneocytes, especially in the top few layers of the SC (Chantassart et al., 2007).
7. Transdermal osmotic gradient. The hydration force drives vesicles through the SC into the viable epidermis (Merin et al., 2016).
8. Formation of an occlusive layer on the skin is one of the mechanisms of drug penetration enhancement when lipid nanoparticles are used as carriers (Pardeike et al., 2009; Roberts et al., 2016).

Furthermore, there are two possible pathways when nanocarriers penetrate through the SC: (1) interconeocyte pathway and (2) intercluster pathway (through wrinkles). Cells are organized in clusters separated by intercluster regions (canyons). These canyons start at the skin surface and go deep in the epidermis. Therefore, molecules can penetrate through the intercluster pathway, reaching even the stratum basale (Carrer et al., 2008; Cevc and Vierl, 2010).

11.5 CONCLUSIONS

The skin has become an important drug delivery route over the past decades, not just to treat or alleviate skin diseases, but for achieving systemic effects. However, the natural barrier imposed by the SC (the outermost layer of the skin) represents a major challenge. Although different chemical and physical penetration enhancers have been proposed for circumventing the SC, the inclusion of active substances in nanocarriers has received considerable attention, given the advantages that these systems provide, especially those composed of lipids. Biocompatibility and biodegradability are among the attractive properties of these materials. Rigid carriers such as liposomes, SLNs, and NLCs, as well as elastic vesicles, including in their structure edge activators or surfactants, solvents, and permeation enhancers, offer different features that can be exploited for a variety of purposes, such as to improve the chemical and physical stability of a molecule, to modulate the release rate, to seek a targeting release, or to enhance dermal and transdermal delivery. Lipid nanocarriers are the focus of a large number of research groups, from the pharmaceutical and cosmetic sectors, and this is reflected in the growing number of publications. Research on lipid-based nanocarriers is moving towards the delivery of proteins and other macromolecules, making use of physical or mechanical enhancers to facilitate their penetration deeply into the skin.

ACKNOWLEDGMENT

The authors acknowledge the financial support provided by PAPIIT: IN 226119, and IN 221717 from DGAPA-UNAM and PIAPI1803 (FESC-UNAM).

REFERENCES

AL-Shuwaili, A.H., Rasool, B.K.A., Abdulrasool, A.A., 2016. Optimization of elastic transfersomes formulations for transdermal delivery of pentoxifylline. Eur. J. Pharm. Biopharm. 102, 101–114.

AbdelSamie, S.M., Kamel, A.O., Sammour, O.A., Ibrahim, S.M., 2016. Terbinafine hydrochloride nanovesicular gel: in vitro characterization, ex vivo permeation and clinical investigation. Eur. J. Pharm. Sci. 88, 91–100.

Abdelbary, A.A., AbouGhaly, H.H., 2015. Design and optimization of topical methotrexate loaded niosomes for enhanced management of psoriasis: application of Box–Behnken design, in vitro evaluation and in vivo skin deposition study. Int. J. Pharm. 485, 235–243.

Abdellatif, M.M., Khalil, I.A., Khalil, M.A.F., 2017. Sertaconazole nitrate loaded nanovesicular systems for targeting skin fungal infection: in vitro, ex vivo and in vivo evaluation. Int. J. Pharm. 527, 1–11.

Abla, M.J., Singh, N.D., Banga, A.K., 2016. Role of nanotechnology in skin delivery of drugs. In: Dragicevic, N., Maibach, H.I. (Eds.), Percutaneous Penetration Enhancers Chemical Methods in Penetration Enhancement: Nanocarriers. Springer-Verlag, Berlin Heidelberg, pp. 1–13.

Ahad, A., Aqil, M., Kohli, K., Sultana, Y., Mujeeb, M., 2013. Enhanced transdermal delivery of an anti-hypertensive agent via nanoethosomes: statistical optimization, characterization and pharmacokinetic assessment. Int. J. Pharm. 443, 26–38.

Ahad, A., Raish, M., Al-Mohizea, A.M., 2014. Enhanced anti-inflammatory activity of carbopol loaded meloxicam nanoethosomes gel. Int. J. Biol. Macromol. 67, 99–104.

Ahad, A., Al-Saleh, A.A., Akhtar, N., Al-Mohizea, A.M., Al-Jenoobi, F.I., 2015. Transdermal delivery of antidiabetic drugs: formulation and delivery strategies. Drug Discov. Today 20, 1217–1227.

Ahad, A., Al-Saleh, A.A., Al-Mohizea, A.M., Al-Jenoobi, F.I., Raish, M., Yassin, A.E.B., et al., 2017. Formulation and characterization of novel soft nanovesicles for enhanced transdermal delivery of eprosartan mesylate. Saudi Pharm. J. 25, 1040–1046.

Ahmed, T.A., El-Say, K.M., Aljaeid, B.M., Fahmy, U.A., Abd-Allah, F.I., 2016. Transdermal glimepiride delivery system based on optimized ethosomal nano-vesicles: preparation, characterization, in vitro, ex vivo and clinical evaluation. Int. J. Pharm. 500, 245–254.

Ainbinder, D., Touitou, E., 2005. Testosterone ethosomes for enhanced transdermal delivery. Drug Deliv. 12, 297–303.

Akbarzadeh, A., Rezaei-Sadabady, R., Davaran, S., Joo, S.W., Zarghami, N., Hanifehpour, Y., et al., 2013. Liposome: classification, preparation, and applications. Nanoscale Res. Lett. 8, 102–110.

Akhlaghi, S.P., Loh, W., 2017. Interactions and release of two palmitoyl peptides from phytantriol cubosomes. Eur. J. Pharm. Biopharm. 117, 60–67.

Akhtar, N., Khan, R.A., 2016. Liposomal systems as viable drug delivery technology for skin cancer sites with an outlook on lipid-based delivery vehicles and diagnostic imaging inputs for skin conditions. Prog. Lipid Res. 64, 192–230.

Akhtar, N., Verma, A., Pathak, K., 2017. Exploring preclinical and clinical effectiveness of nanoformulations in the treatment of atopic dermatitis: safety aspects and patent reviews. Bull. Fac. Pharm. Cairo Univ. 55, 1–10.

Akomeah, F.K., 2010. Topical dermatological drug delivery: quo vadis? Curr. Drug Deliv. 7, 283–296.

Alexander, A., Ajazuddin, Patel, R.J., Saraf, S., Saraf, S., 2016. Recent expansion of pharmaceutical nanotechnologies and targeting strategies in the field of phytopharmaceuticals for the delivery of herbal extracts and bioactives. J. Control. Release 241, 110–124.

Al-Mahallawi, A.M., Abdelbary, A.A., Aburahma, M.H., 2015. Investigating the potential of employing bilosomes as a novel vesicular carrier for transdermal delivery of tenoxicam. Int. J. Pharm. 485, 329–340.

Alomrani, A.H., Shazly, G.A., Amara, A.A., Badran, M.M., 2014. Itraconazole–hydroxypropyl-β-cyclodextrin loaded deformable liposomes: in vitro skin penetration studies and antifungal efficacy using candida albicans as model. Colloids Surf. B 121, 74–81.

Alomrani, A.H., Al-Agamy, M.H., Badran, M.M., 2015. In vitro skin penetration and antimycotic activity of itraconazole loaded niosomes: various non-ionic surfactants. J. Drug Deliv. Sci. Technol. 28, 37–45.

Andrade, L.M., Silva, L.A.D., Krawczyk-Santos, A.P., Amorim, I.C.S.M., Rocha, P.B.R. D., Lima, E.M., et al., 2017. Improved tacrolimus skin permeation by co-encapsulation with clobetasol in lipid nanoparticles: study of drug effects in lipid matrix by electron paramagnetic resonance. Eur. J. Pharm. Biopharm. 119, 142–149.

Ascenso, A., Salgado, A., Euletério, C., Garcia Praça, F., Lopes Badra Bentley, M.V., Marques, H.C., et al., 2014. In vitro and in vivo topical delivery studies of tretinoin-loaded ultradeformable vesicles. Eur. J. Pharm. Biopharm. 88, 48–55.

Ashtikar, M., Nagarsekar, K., Fahr, A., 2016. Transdermal delivery from liposomal formulations—evolution of the technology over the last three decades. J. Control. Release 242, 126–140.

Ashtikar, M., Langelüddeckeb, L., Fahr, A., Deckert, V., 2017. Tip-enhanced Raman scattering for tracking of invasomes in the stratum corneum. BBA—Gen. Subjects 1861, 2630–2639.

Azagury, A., Khoury, L., Enden, G., Kost, J., 2014. Ultrasound mediated transdermal drug delivery. Adv. Drug. Deliv. Rev. 72, 127–143.

Barua, S., Mitragotri, S., 2014. Challenges associated with penetration of nanoparticles across cell and tissue barriers: a review of current status and future prospects. Nano Today 9, 223–243.

Basiri, L., Rajabzadeh, G., Bostan, A., 2017a. Physicochemical properties and release behavior of Span 60/Tween 60 niosomes as vehicle for a-Tocopherol delivery. LWT—Food Sci. Technol. 84, 471–478.

Basiri, L., Rajabzadeh, G., Bostan, A., 2017b. a-Tocopherol-loaded niosome prepared by heating method and its release behavior. Food Chem. 221, 620–628.

Beloqui, A., Solinís, M.Á., Rodríguez-Gascón, A., Almeida, A.J., Préat, V., 2016. Nanostructured lipid carriers: promising drug delivery systems for future clinics. Nanomedicine 12, 143–161.

Bendas, E.R., Abdullah, H., El-Komy, M.H.M., Kassem, M.A.A., 2013. Hydroxychloroquine niosomes: a new trend in topical management of oral lichen planus. Int. J. Pharm. 458, 287–295.

Bilia, A.R., Bergonzi, M.C., Guccione, C., Manconi, M., Fadda, A.M., Sinico, C., 2016. Vesicles and micelles: two versatile vectors for the delivery of natural products. J. Drug Deliv. Sci. Technol. 32, 241–255.

Bisht, D., Verma, D., Mirza, M.A., Anwer, Md. K., Iqbal, Z., 2017. Development of ethosomal gel of ranolazine for improved topical delivery: in vitro and ex vivo evaluation. J. Mol. Liq. 225, 475–481.

Boakye, C.H.A., Patel, K., Singh, M., 2015. Doxorubicin liposomes as an investigative model to study the skin permeation of nanocarriers. Int. J. Pharm. 489, 106–116.

Boakye, C.H.A., Patel, K., Doddapaneni, R., Bagde, A., Behl, G., Chowdhury, N., et al., 2016. Ultra-flexible nanocarriers for enhanced topical delivery of a highly lipophilic antioxidative molecule for skin cancer chemoprevention. Colloid Surf. B 143, 156–167.

Boge, L., Umerska, A., Matougui, N., Bysell, H., Ringstad, L., Davoudi, M., et al., 2017. Cubosomes post-loaded with antimicrobial peptides: characterization, bactericidal effect and proteolytic stability. Int. J. Pharm. 526, 400–412.

Bouwstra, J.A., Honeywell-Nguyen, P.L., Gooris, G.S., Ponec, M., 2003. Structure of the skin barrier and its modulation by vesicular formulations. Prog. Lipid Res. 42, 1–36.

Caddeo, C., Nacher, A., Vassallo, A., Armentano, M.F., Pons, R., Fernández-Busquets, X., et al., 2016. Effect of quercetin and resveratrol co-incorporated in liposomes against inflammatory/oxidative response associated with skin cancer. Int. J. Pharm. 513, 153–163.

Caddeo, C., Manca, M.L., Matos, M., Gutierrez, G., Díez-Sales, O., Peris, J.E., et al., 2017. Functional response of novel bioprotective poloxamer-structured vesicles on inflamed skin. Nanomedicine 13, 1127–1136.

Çağdaş, M., Sezer, A.D., Bucak, S., 2014. Liposomes as potential drug carrier systems for drug delivery. In: Sezer, A.D. (Ed.), Application of Nanotechnology in Drug Delivery. InTech, pp. 1–50.

Campani, V., Biondi, M., Mayol, L., Cilurzo, F., Franze, S., Pitaro, M., et al., 2016. Nanocarriers to enhance the accumulation of vitamin K1 into the skin. Pharm. Res. 33, 893–908.

Campos, P.P., Fernandes-Fraceto, L., Ferreira, M., 2018. Layer-by-layer films containing emodin or emodin encapsulated in liposomes for transdermal applications. Colloid Surf. B 162, 69–75.

Cao, Z., Yue, X., Jin, Y., Wu, X., Dai, Z., 2012. Modulation of release of paclitaxel from composite cerasomes. Colloids Surf. B 98, 97–104.

Carrer, D.C., Vermehren, C., Bagatolli, L.A., 2008. Pig skin structure and transdermal delivery of liposomes: a two photon microscopy study. J. Control. Release 132, 12–20.

Cevc, G., 1997. Drug delivery across the skin. Expert. Opin. Inv. Drug. 6, 1887–1937.

Cevc, G., 2003. Transferosomes: innovative transdermal drug carriers. In: Rathbone, M.J., Hadgraft, J., Roberts, M.S. (Eds.), Modified Release Drug Delivery Technology. Marcel Dekker Inc, New York, pp. 533–560.

Cevc, G., Blume, G., 1992. Lipid vesicles penetrate into intact skin owing to the transdermal osmotic gradients and hydration force. Biochim. Biophys. Acta 1104, 226–232.

Cevc, G., Blume, G., 2001. New, highly efficient formulation of diclofenac for the topical, transdermal administration in ultradeformable drug carriers, Transfersomes. Biochim. Biophys. Acta 1514, 191–205.

Cevc, G., Blume, G., 2004. Hydrocortisone and dexamethasone in very deformable drug carriers have increased biological potency, prolonged effect, and reduced therapeutic dosage. Biochim. Biophys. Acta 1663, 61–73.

Cevc, G., Vierl, U., 2010. Nanotechnology and the transdermal route. A state of the art review and critical appraisal. J. Control. Release 141, 277–299.

Cevc, G., Schätzlein, A., Blume, G., 1995. Transdermal drug carriers: basic properties, optimization and transfer efficiency in the case of epicutaneously applied peptides. J. Control. Release 36, 3–16.

Cevc, G., Blume, G., Schtitzlein, A., Gebauer, D., Paul, A., 1996. The skin: a pathway for systemic treatment with patches and lipid-based agent carriers. Adv. Drug. Deliv. Rev. 349–378.

Cevc, G., Gebauer, D., Stieber, J., Schätzlein, A., Blume, G., 1998. Ultraflexible vesicles, Transfersomes, have an extremely low pore penetration resistance and transport therapeutic amounts of insulin across the intact mammalian skin. Biochim. Biophys. Acta 1368, 201–215.

Cevc, G., Schätzlein, A., Richardsen, H., 2002. Ultradeformable lipid vesicles can penetrate the skin and other semi-permeable barriers unfragmented. Evidence from double label CLSM experiments and direct size measurements. Biochim. Biophys. Acta 1564, 21–30.

Chantaburanan, T., Teeranachaideekul, V., Chantasart, D., Jintapattanakit, A., Junyaprasert, V.B., 2017. Effect of binary solid lipid matrix of wax and triglyceride on lipid crystallinity, drug–lipid interaction and drug release of ibuprofen-loaded solid lipid nanoparticles (SLN) for dermal delivery. J. Colloid Interface Sci. 504, 247–256.

Chantassart, D., Sa-Nguandeekul, P., Prakongpan, S., Li, S.K., Higuchi, W.I., 2007. Comparison of the effects of chemical permeation enhancers on the lipoidal pathways of human epidermal membrane and hairless mous skin and the mechanism of enhancer action. J. Pharm. Sci. 96, 2310–2326.

Charoenputtakhun, P., Opanasopit, P., Rojanarata, T., Ngawhirunpat, T., 2014. All-trans retinoic acid-loaded lipid nanoparticles as a transdermal drug delivery carrier. Pharm. Dev. Technol. 19, 164–172.

Charoenputtakun, P., Li, S.K., Ngawhirunpat, T., 2015. Iontophoretic delivery of lipophilic and hydrophilic drugs from lipid nanoparticles across human skin. Int. J. Pharm. 495, 318–328.

Chen, M., 2011. Skin Drug Delivery from Different Lipid Vesicular Systems (Ph.D. thesis). Faculty of Biology and Pharmacy of the Friedrich-Schiller-Universität Jena.

Chen, M., Liu, X., Fahr, A., 2010. Skin delivery of ferulic acid from different vesicular systems. J. Biomed. Nanotechnol. 6, 577–585.

Chen, R., Li, R., Liu, Q., Bai, C., Qin, B., Ma, Y., et al., 2017a. Ultradeformable liposomes: a novel vesicular carrier for enhanced transdermal delivery of procyanidins: effect of surfactants on the formation, stability, and transdermal delivery. AAPS PharmSciTec 18, 1823–1832.

Chen, Y.Y., Lu, Y.H., Ma, C.H., Tao, W.W., Zhu, J.J., Zhang, X., 2017b. A novel elastic liposome for skin delivery of papain and its application on hypertrophic scar. Biomed. Pharmacother. 87, 82–91.

Choi, J.U., Lee, S.W., Pangeni, R., Byun, Y., Yoon, I.S., Park, J.W., 2017. Preparation and in vivo evaluation of cationic elastic liposomes comprising highly skin-permeable growth factors combined with hyaluronic acid for enhanced diabetic wound-healing therapy. Acta Biomater. 57, 197–215.

Clares, B., Calpena, A.C., Parra, A., Abrego, G., Alvarado, H., Fangueiro, J.F., et al., 2014. Nanoemulsions (NEs), liposomes (LPs) and solid lipid nanoparticles (SLNs) for retinyl palmitate: effect on skin permeation. Int. J. Pharm. 473, 591–598.

Conacher, M., Alexander, J., Brewer, J.M., 2001. Oral immunisation with peptide and protein antigens by formulation in lipid vesicles incorporating bile salts (bilosomes). Vaccine 19, 2965–2974.

Cosco, D., Paolino, D., Maiuolo, J., Di Marzio, L., Carafa, M., Ventura, C.A., et al., 2015. Ultradeformable liposomes as multidrug carrier of resveratrol and 5-fluorouracil for their topical delivery. Int. J. Pharm. 489, 1–10.

Costa, R., Santos, L., 2017. Delivery systems for cosmetics—from manufacturing to the skin of natural antioxidants. Powder Technol. 322, 402–416.

Dai, Z., 2016. China nanomedicine abstracts. Nanomed. Nanotechnol. Biol. Med. 12, 449–575.

Daneshmand, S., Jaafari, M.R., Movaffagh, J., Malaekeh-Nikouei, B., Iranshahi, M., Seyedian Moghaddam, A., et al., 2018. Preparation, characterization, and optimization of auraptene-loaded solid lipid nanoparticles as a natural anti-inflammatory agent: in vivo and in vitro evaluations. Colloids Surf. B: Biointerfaces 164, 332–339.

Das, M.K., Palei, N.N., 2011. Sorbitan ester niosomes for topical delivery of rofecoxib. Indian J. Exp. Biol. 49, 438–445.

Das, B., Sen, S.O., Maji, R., Nayak, A.K., Sen, K.K., 2017. Transferosomal gel for transdermal delivery of risperidone: formulation optimization and ex vivo permeation. J. Drug Deliv. Sci. Technol. 38, 59–71.

Dayan, N., Touitou, E., 2000. Carriers for skin delivery of trihexyphenidyl HCl: ethosomes vs. liposomes. Biomaterials 21, 1879–1885.

De, S., Kundu, R., Biswas, A., 2012. Synthesis of gold nanopaticles in niosomes. J. Colloid Int. Sci. 386, 9–15.

Decome, L., De Méo, M., Geffard, A., Doucet, O., Duménil, G., Botta, A., 2005. Evaluation of photolyase (photosome) repair activity in human keratinocytes after a single dose of ultraviolet B irradiation using the comet assay. J. Photochem. Photobiol. B 79, 101–108.

Desai, P.R., Marepally, S., Patel, A.R., Voshavar, C., Chaudhuri, A., Singh, M., 2013. Topical delivery of anti-TNFα siRNA and capsaicin via novel lipid–polymer hybrid nanoparticles efficiently inhibits skin inflammation in vivo. J. Control. Release 170, 51–63.

Dhamecha, D.L., Rathi, A.A., Saifee, M., Lahoti, S.R., Dehghan, M.H.G., 2009. Drug vehicle based approaches of penetration enhancement. Int. J. Pharm. Pharm. Sci. 1, 24–46.

Di Marzio, L., Ventura, C.A., Cosco, D., Paolino, D., Di Stefano, A., Stancanelli, R., et al., 2016. Nanotherapeutics for anti-inflammatory delivery. J. Drug Deliv. Sci. Technol. 32, 174–191.

Djekic, L., Krajisnik, D., Micic, Z., Calija, B., 2016. Formulation and physicochemical characterization of hydrogels with 18b-glycyrrhetinic acid/phospholipid complex phytosomes. J. Drug Deliv. Sci. Technol. 35, 81–90.

Doktorovová, S., Kovačević, A.B., Garcia, M.L., Souto, E.B., 2016. Preclinical safety of solid lipid nanoparticles and nanostructured lipid carriers: current evidence from in vitro and in vivo evaluation. Eur. J. Pharm. Biopharm. 108, 235–252.

Doppalapudi, S., Mahira, S., Khan, W., 2017. Development and in vitro assessment of psoralen and resveratrol co-loaded ultradeformable liposomes for the treatment of vitiligo. J. Photochem. Photobiol. B 174, 44–57.

Dorrani, M., Garbuzenko, O.B., Minko, T., Michniak-Kohn, B., 2016. Development of edge-activated liposomes for siRNA delivery to human basal epidermis for melanoma therapy. J. Control. Release 228, 150–158.

Dragicevic-Curic, N., Scheglmann, D., Albrecht, V., Fahr, A., 2008. Temoporfin-loaded invasomes: development, characterization and in vitro skin penetration studies. J. Control. Release 127, 59–69.

Dubey, V., Mishra, D., Dutta, T., Nahar, M., Saraf, D.K., Jain, N.K., 2007a. Dermal and transdermal delivery of an anti-psoriatic agent via ethanolic liposomes. J. Control. Release 123, 148–154.

Dubey, V., Mishra, D., Jain, N.K., 2007b. Melatonin loaded ethanolic liposomes: physicochemical characterization and enhanced transdermal delivery. Eur. J. Pharm. Biopharm. 67, 398–405.

Dubey, V., Mishra, D., Nahar, M., Jain, V., Jain, N.K., 2010. Enhanced transdermal delivery of an anti-HIV agent via ethanolic liposomes. Nanomedicine 6, 590–596.

Egbaria, K., Weiner, N., 1990. Nanoparticulate systems for drug delivery. Adv. Drug Deliver. Rev. 5, 287–300.

El-Badry, M., Fetih, G., Fathalla, D., Shakeel, F., 2014. Transdermal delivery of meloxicam using niosomal hydrogels: in vitro and pharmacodynamic evaluation. Pharm. Dev. Technol. 9, 1–7.

El-Laithy, H.M., Shoukry, O., Mahran, L.G., 2011. Novel sugar esters proniosomes for transdermal delivery of vinpocetine: preclinical and clinical studies. Eur. J. Pharm. Biopharm. 77, 43–55.

El Maghraby, G.M., Williams, A.C., Barry, B.W., 1999. Skin delivery of oestradiol from deformable and traditional liposomes: mechanistic studies. J. Pharm. Pharmacol. 51, 1123–1134.

El Maghraby, G.M., Williams, A.C., Barry, B.W., 2000. Oestradiol skin delivery from ultradeformable liposomes: refinement of surfactant concentration. Int. J. Pharm. 196, 63–74.

El Maghraby, G.M., Williams, A.C., Barry, B.W., 2001. Skin delivery of 5-fluorouracil from ultradeformable and standard liposomes in vitro. J. Pharm. Pharmacol. 53, 1069–1077.

El Maghraby, G.M., Williams, A.C., Barry, B.W., 2006. Can drug bearing liposomes penetrate intact skin? J. Pharm. Pharmacol. 58, 415–429.

El Maghraby, G.M., Barry, B.W., Williams, A.C., 2008. Liposomes and skin: from drug delivery to model membranes. Eur. J. Pharm. Sci. 34, 203–222.

Elsayed, M.M.A., Abdallah, O.Y., Naggar, V.F., Khalafallah, N.M., 2006. Deformable liposomes and ethosomes: mechanism of enhanced skin delivery. Int. J. Pharm. 322, 60–66.

Elsayed, M.M., Abdallah, O.Y., Naggar, V.F., Khalafallah, N.M., 2007. Lipid vesicles for skin delivery of drugs: reviewing three decades of research. Int. J. Pharm. 332, 1–16.

Ephrem, E., Elaissari, H., Greige-Gerges, H., 2017. Improvement of skin whitening agents efficiency through encapsulation: current state of knowledge. Int. J. Pharm. 526, 50–68.

Esposito, E., Sguizzato, M., Drechsler, M., Mariani, P., Carducci, F., Nastruzzi, C., et al., 2017. Progesterone lipid nanoparticles: scaling up and in vivo human study. Eur. J. Pharm. Biopharm. 119, 437–446.

Fahr, A., Chen, M., 2010. Liposomes as promising tools for the skin delivery and protection of damaged skin. In: Monteiro-Riviere, N.A. (Ed.), Toxicology of the Skin. Informa Healthcare, New York, pp. 289–300.

Fang, Y.P., Tsai, Y.H., Wu, P.C., Huang, Y.B., 2008. Comparison of 5-aminolevulinic acid-encapsulated liposome versus ethosome for skin delivery for photodynamic therapy. Int. J. Pharm. 356, 144–152.

Fireman, S., Toledano, O., Neimann, K., Loboda, N., Dayan, N., 2011. A look at emerging delivery systems for topical drug products. Dermatol. Ther. 24, 477–488.

Ganem Rondero, A., 2011. What do you know about... transdermal drug delivery? Rev. Mex. Cienc. Farma 42, 65–68.

Gangwar, M., Singh, R., Goel, R.K., Nath, G., 2012. Recent advances in various emerging vescicular systems: an overview. Asian Pac. J. Trop. Biomed. S1176–S1188.

Garcês, A., Amaral, M.H., Sousa Lobo, J.M., Silva, A., 2018. Formulations based on solid lipid nanoparticles (SLN) and nanostructured lipid carriers (NLC) for cutaneous use: a review. Eur. J. Pharm. Sci. 112, 159–167.

Garcia-Orue, I., Gainza, G., Girbau, C., Alonso, R., Aguirre, J.J., Pedraz, J.L., et al., 2016. LL37 loaded nanostructured lipid carriers (NLC): a new strategy for the topical treatment of chronic wounds. Eur. J. Pharm. Biopharm. 108, 310–316.

Garg, N.K., Sharma, G., Singh, B., Nirbhavane, P., Tyagi, R.K., Shukla, R., et al., 2017. Quality by design (QbD)-enabled development of aceclofenac loaded-nano structured lipid carriers (NLCs): an improved dermatokinetic profile for inflammatory disorder(s). Int. J. Pharm. 517, 413–431.

Geszke-Moritz, M., Moritz, M., 2016. Solid lipid nanoparticles as attractive drug vehicles: composition, properties and therapeutic strategies. Mater. Sci. Eng. C 68, 982–994.

Gesztes, A., Mezei, M., 1988. Topical anesthesia of the skin by liposome-encapsulated tetracaine. Anesth. Analg. 67, 1079–1081.

Gillet, A., Evrard, B., Piel, G., 2011. Liposomes and parameters affecting their skin penetration behavior. J. Drug Deliv. Sci. Technol. 21, 35–42.

Godin, B., Touitou, E., 2004. Mechanism of bacitracin permeation enhancement through the skin and cellular membranes from an ethosomal carrier. J. Control. Release 94, 365–379.

Godin, B., Touitou, E., 2005. Erythromycin ethosomal systems: physicochemical characterization and enhanced antibacterial activity. Curr. Drug Deliv. 2, 269–275.

Goto, P.L., Siqueira-Moura, M.P., Tedesco, A.C., 2017. Application of aluminum chloride phthalocyanine-loaded solid lipid nanoparticles for photodynamic inactivation of melanoma cells. Int. J. Pharm. 518, 228–241.

Guimarães, K.L., Ré, M.I., 2011. Lipid nanoparticles as carriers for cosmetic ingredients: the first (SLN) and the second generation (NLC). In: Beck, R., Guterres, S., Pohlmann, A. (Eds.), Nanocosmetics and Nanomedicines. Springer-Verlag, Berlin Heidelberg, pp. 101–122.

Gunes, A., Guler, E., Un, R.N., Demir, B., Barlas, F.B., Yavuz, M., et al., 2017. Niosomes of *Nerium oleander* extracts: in vitro assessment of bioactive nanovesicular structures. J. Drug Deliv. Sci. Technol. 37, 158–165.

Han, T., Das, D.B., 2015. Potential of combined ultrasound and microneedles for enhanced transdermal drug permeation: a review. Eur. J. Pharm. Biopharm. 89, 312–328.

Han, F., Li, S., Yin, R., Shi, X., Jia, Q., 2008. Investigation of nanostructured lipid carriers for transdermal delivery of flurbiprofen. Drug Dev. Ind. Pharm. 34, 453–458.

Hao, J., 2014. Topical iontophoresis for local therapeutic effects. J. Drug Deliv. Sci. Technol. 24, 255–258.

Hasanovic, A., Hollick, C., Fischinger, K., Valenta, C., 2010. Improvement in physicochemical parameters of DPPC liposomes and increase in skin permeation of aciclovir and minoxidil by the addition of cationic polymers. Eur. J. Pharm. Biopharm. 75, 148–153.

Hasanovic, A., Winkler, R., Resch, G.P., Valenta, C., 2011. Modification of the conformational skin structure by treatment with liposomal formulations and its correlation to the penetration depth of Aciclovir. Eur. J. Pharm. Biopharm. 79, 76–81.

Hofer, C., Randenborgh, H., Lehmer, A., Hartung, R., Breul, J., 2004. Transcutaneous IL-2 uptake mediated by Transfersomes depends on concentration and fractionated application. Cytokine 25, 141–146.

Hofland, H.E.J., Bouwstra, J.A., Ponec, M., Boddé, H.E., Spies, F., Verhoef, J.C., et al., 1991. Interactions of non-ionic surfactant vesicles with cultured keratinocytes and human skin in vitro: a survey of toxicological aspects and ultrastructural changes in stratum corneum. J. Control. Release 16, 155–167.

Hofland, H.E., Bouwstra, J.A., Bodde, H.E., Spies, F., Junginger, H.E., 1995. Interactions between liposomes and human stratum corneum in vitro: freeze fracture electron microscopical visualization and small angle X-ray scattering studies. Br. J. Dermatol. 132, 853–866.

Honeywell-Nguyen, P.L., Bouwstra, J.A., 2005. Vesicles as a tool for transdermal and dermal delivery. Drug Discov. Today Technol. 2, 67–74.

Hung, C.F., Chen, W.Y., Hsu, C.Y., Aljuffali, I.A., Shih, H.C., Fang, J.Y., 2015. Cutaneous penetration of soft nanoparticles via photodamaged skin: lipid-based and polymer-based nanocarriers for drug delivery. Eur. J. Pharm. Biopharm. 94, 94–105.

Hurler, J., Berg, O.A., Skar, M., Conraid, A.H., Johnsen, P.J., Škalko-Basnet, N., 2012. Improved burns therapy: liposomes-in-hydrogel delivery system for mupirocin. J. Pharm. Sci. 101, 3906–3915.

Hussain, A., Singh, S., Sharma, D., Webster, T.J., Shafaat, K., Faruk, A., 2017. Elastic liposomes as novel carriers: recent advances in drug delivery. Int. J. Nanomed. 12, 5087–5108.

Iizhar, S.A., Syed, I.A., Satar, R., Ansari, S.A., 2016. In vitro assessment of pharmaceutical potential of ethosomes entrapped with terbinafine hydrochloride. J. Adv. Res. 7, 453–461.

Ingebrigtsen, S.G., Škalko-Basnet, N., Cavalcanti Jacobsen, C.A., Holsæter, A.M., 2017. Successful co-encapsulation of benzoyl peroxide and chloramphenicol in liposomes by a novel manufacturing method—dual asymmetric centrifugation. Eur. J. Pharm. Sci. 97, 192–199.

Iqbal, N., Vitorino, C., Taylor, K.M.G., 2016. How can lipid nanocarriers improve transdermal delivery of olanzapine? Pharm. Dev. Technol. 22, 587–596.

Jain, S., Patel, N., Shah, M.K., Khatri, P., Vora, N., 2017. Recent advances in lipid-based vesicles and particulate carriers for topical and transdermal application. J. Pharm. Sci. 106, 423–445.

Jensen, L.B., Petersson, K., Nielsen, H.M., 2011. In vitro penetration properties of solid lipid nanoparticles in intact and barrier-impaired skin. Eur. J. Pharm. Biopharm. 79, 68–75.

Jeon, S., Yoo, C.Y., Park, S.N., 2015. Improved stability and skin permeability of sodium hyaluronate–chitosan multilayered liposomes by layer-by-layer electrostatic deposition for quercetin delivery. Colloid Surf. B 129, 7–14.

Joshny, J., Vedha-Hari, B.N., Ramya-Devi, D., 2018. Experimental optimization of lornoxicam liposomes for sustained topical delivery. Eur. J. Pharm. Sci. 112, 38–51.

Junyaprasert, V.B., Singhsa, P., Suksiriworapong, J., Chantasart, D., 2012. Physicochemical properties and skin permeation of Span 60/Tween 60 niosomes of ellagic acid. Int. J. Pharm. 423, 303–311.

Junyaprasert, V.B., Singhsa, P., Jintapattanakit, A., 2013. Influence of chemical penetration enhancers on skin permeability of ellagic acid-loaded niosomes. Asian J. Pharm. Sci. 8, 110–117.

Kalhapure, R.S., Sikwal, D.R., Rambharose, S., Mocktar, C., Singh, S., Bester, L., et al., 2017. Enhancing targeted antibiotic therapy via pH responsive solid lipid nanoparticles from an acid cleavable lipid. Nanomedicine 13, 2067–2077.

Kamran, M., Ahad, A., Aqil, M., Imam, S.S., Sultana, Y., Ali, A., 2016. Design, formulation and optimization of novel soft nano-carriers for transdermal olmesartan medoxomil delivery: in vitro characterization and in vivo pharmacokinetic assessment. Int. J. Pharm. 505, 147–158.

Karami, Z., Hamidi, M., 2016. Cubosomes: remarkable drug delivery potential. Drug Discov. Today 21, 789–801.

Kasetvatin, C., Rujivipat, S., Tiyaboonchai, W., 2015. Combination of elastic liposomes and low frequency ultrasound for skin permeation enhancement of hyaluronic acid. Colloid Surf. B 135, 458–464.

Katagiri, K., Koumoto, K., 2013. Organic–inorganic hybrid materials prepared through supramolecular assembly. In: Somiya, S. (Ed.), Handbook of Advanced Ceramics. Elsevier Inc, Amsterdam, pp. 1011–1023.

Kathe, N., Henriksen, B., Chauhan, H., 2014. Physicochemical characterization techniques for solid lipid nanoparticles: principles and limitations. Drug. Dev. Ind. Pharm. 40, 1565–1575.

Kaur, I.P., Bhandari, R., Yakhmi, J.V., 2014. Lipids as biological materials for nanoparticulate delivery. In: Bhushan, B., et al., (Eds.), Handbook of Nanomaterials Properties. Springer-Verlag, Berlin Heidelberg, pp. 409–455.

Kaurav, M., Minz, S., Sahu, K., Kumar, M., Madan, J., Pandey, R.S., 2016. Nanoparticulate mediated transcutaneous immunization: myth or reality. Nanomed. Nanotechnol. Biol. Med. 12, 1063–1081.

Keservani, R.K., Sharma, A.K., Ramteke, S., 2010. Novel vesicular approach for topical delivery of baclofen via niosomes. Lat. Am. J. Pharm. 29, 1364–1370.

Khalil, R.M., Abd-Elbary, A., Kassem, M.A., Ghorab, M.M., Basha, M., 2014. Nanostructured lipid carriers (NLCs) versus solid lipid nanoparticles (SLNs) for topical delivery of meloxicam. Pharm. Dev. Technol. 19, 304–314.

Knudsen, N.Ø., Jorgensen, L., Hansen, J., Vermehren, C., Frokjaer, S., Foged, C., 2011. Targeting of liposome-associated calcipotriol to the skin: effect of liposomal membrane fluidity and skin barrier integrity. Int. J. Pharm. 416, 478–485.

Korting, H.C., Schmid, M.H., Hartinger, A., Maierhofer, G., Stolz, W., Braun-Falco, O., 1993. Evidence for the phagocytosis of intact oligolamellar liposomes by human keratinocytes in vitro and consecutive intracellular disintegration. J. Microencapsul. 10, 223–228.

Kumar, G.P., Rajeshwarrao, P., 2011. Nonionic surfactant vesicular systems for effective drug delivery—an overview. Acta Pharm. Sin. B 1, 208–219.

Kumar, A., Pathak, K., Bali, V., 2012. Ultra-adaptable nanovesicular systems: a carrier for systemic delivery of therapeutic agents. Drug Discov. Today 17, 1233–1241.

Lakshmi, P.K., Kalpana, B., Prasanthi, D., 2013. Invasomes-novel vesicular carriers for enhanced skin permeation. Syst. Rev. Pharm. 4, 26–30.

Lane, M.E., 2013. Skin penetration enhancers. Int. J. Pharm. 447, 12–21.

Larrañeta, E., Lutton, R.E.M., Woolfson, A.D., Donnelly, R.F., 2016. Microneedle arrays as transdermal and intradermal drug delivery systems: materials science, manufacture and commercial development. Mater. Sci. Eng. R 104, 1–32.

Lauterbach, A., Müller-Goymann, C.C., 2015. Applications and limitations of lipid nanoparticles in dermal and transdermal drug delivery via the follicular route. Eur. J. Pharm. Biopharm. 97, 152–163.

Lee, M.H., Lee, K.K., Park, M.H., Hyun, S.S., Kahn, S.Y., Joo, K.S., et al., 2016. In vivo anti-melanogenesis activity and in vitro skin permeability of niacinamide-loaded flexible liposomes (Bounsphere™). J. Drug Deliv. Sci. Technol. 31, 147–152.

Li, G., Fan, Y., Fan, C.H., Li, X., Wang, X., Li, M., et al., 2012. Tacrolimus-loaded ethosomes: physicochemical characterization and in vivo evaluation. Eur. J. Pharm. Biopharm. 82, 49–57.

Li, Z., Liu, M., Wang, H., Du, S., 2016. Increased cutaneous wound healing effect of biodegradable liposomes containing madecassoside: preparation optimization, in vitro dermal permeation, and in vivo bioevaluation. Int. J. Nanomed. 11, 2995–3007.

Lieb, L.M., Ramachandran, C., Egbaria, K., Weiner, N., 1992. Topical delivery enhancement with multilamellar liposomes into pilosebaceous units: I. In vitro evaluation using fluorescent techniques with the hamster ear model. J. Investig. Dermatol. 99, 108–113.

de Lima, R.G., Tedesco, A.C., da Silva, R.S., Lawrence, M.J., 2017. Ultradeformable liposome loaded with zinc phthalocyanine and $[Ru(NH{\cdot}NHq)(tpy)NO]^{3+}$ for photodynamic therapy by topical application. Photodiagn. Photodyn. Ther. 19, 184–193.

Lin, C.H., Fang, Y.P., Al-Suwayeh, S.A., Yang, S.Y., Fang, J.Y., 2013. Percutaneous absorption and antibacterial activities of lipid nanocarriers loaded with dual drugs for acne treatment. Biol. Pharm. Bull. 36, 276–286.

Lin, W.J., Duh, Y.S., 2016. Nanostructured lipid carriers for transdermal delivery of acid labile lansoprazole. Eur. J. Pharm. Biopharm. 108, 297–303.

Liu, J., Hu, G., 2007. Advances in studies of phospholipids as carriers in skin topical Application. J. Nanjin Med. Univ. 21, 349–353.

Liu, Y.S., Wen, C.F., Yang, Y.M., 2012. Development of ethosome-like catanionic vesicles for dermal drug delivery. J. Taiwan Inst. Chem. Eng. 43, 830–838.

Lodzki, M., Godin, B., Rakou, L., Mechoulam, R., Gallily, R., Touitou, E., 2003. Cannabidiol transdermal delivery and anti-inflammatory effect in a murine model. J. Control. Release 93, 377–387.

Lombardi Borgia, S., Regehly, M., Sivaramakrishnan, R., Mehnert, W., Korting, H.C., Danker, K., et al., 2005. Lipid nanoparticles for skin penetration enhancement—correlation to drug localization within the particle matrix as determined by fluorescence and parelectric spectroscopy. J Control. Release 110, 151–163.

López-García, R., Ganem-Rondero, A., 2015. Solid lipid nanoparticles (SLN) and nanostructured lipid carriers (NLC): occlusive effect and penetration enhancement ability. J. Cosm. Dermatol. Sci. Appl. 5, 62–72.

Maestrelli, F., Bragagni, M., Mura, P., 2016. Advanced formulations for improving therapies with anti-inflammatory or anaesthetic drugs: a review. J. Drug Deliv. Sci. Technol. 32, 192–205.

Mahale, N.B., Thakkar, P.D., Mali, R.G., Walunj, D.R., Chaudhari, S.R., 2012. Niosomes: novel sustained release nonionic stable vesicular systems—an overview. Adv. Colloid Interface Sci. 183-184, 46–54.

Mahdiun, F., Mansouri, S.H., Khazaeli, P., Mirzaei, R., 2017. The effect of tobramycin incorporated with bismuth–ethanedithiol loaded on niosomes on the quorum sensing and biofilm formation of *Pseudomonas aeruginosa*. Microb. Pathog. 107, 129–135.

Maheshwari, R.G.S., Tekade, R.K., Sharma, P.A., Darwhekar, G., Tyagi, A., Patel, R.P., et al., 2012. Ethosomes and ultradeformable liposomes for transdermal delivery of clortrimazole: a comparative assessment. Saudi Pharm. J. 20, 161–170.

Mahmood, S., Chatterjee, B., Mandal, U.K., 2018. Nano transfersomes vesicles of raloxifene HCl with sorbitan 80: formulation and characterization. Bioequiv. Bioavail. Int. J. 2, 000121.

Mahmoud, M.O., Aboud, H.M., Hassan, A.H., Ali, A.A., Johnston, T.P., 2017. Transdermal delivery of atorvastatin calcium from novel nanovesicular systems using polyethylene glycol fatty acid esters: ameliorated effect without liver toxicity in poloxamer 407-induced hyperlipidemic rats. J. Control. Release 254, 10–22.

Manca, M.L., Zaru, M., Manconi, M., Lai, F., Valenti, D., Sinico, C., et al., 2013. Glycerosomes: a new tool for effective dermal and transdermal drug delivery. Int. J. Pharm. 455, 66–74.

Manca, M.L., Castangia, I., Caddeo, C., Pando, D., Escribano, E., Valenti, D., et al., 2014a. Improvement of quercetin protective effect against oxidative stress skin damages by incorporation in nanovesicles. Colloids Surf. B 123, 566–574.

Manca, M.L., Manconi, M., Nacher, A., Carbone, C., Valenti, D., Maccioni, A.M., et al., 2014b. Development of novel diolein–niosomes for cutaneous delivery of tretinoin: influence of formulation and in vitro assessment. Int. J. Pharm. 477, 176–186.

Manca, M.L., Castangia, I., Zaru, M., Nácher, A., Valenti, D., Fernández-Busquets, X., et al., 2015. Development of curcumin loaded sodium hyaluronate immobilized vesicles (hyalurosomes) and their potential on skin inflammation and wound restoring. Biomaterials 71, 100–109.

Manca, M.L., Cencetti, C., Matricardi, P., Castangia, I., Zaru, M., Diez Sales, O., et al., 2016a. Glycerosomes: use of hydrogenated soy phosphatidylcholine mixture and its effect on vesicle features and diclofenac skin penetration. Int. J. Pharm. 511, 198–204.

Manca, M.L., Matricardi, P., Cencetti, C., Peris, J.E., Melis, V., Carbone, C., et al., 2016b. Combination of argan oil and phospholipids for the development of an effective liposome-like formulation able to improve skin hydration and allantoin dermal delivery. Int. J. Pharm. 505, 204–211.

Manca, M.L., Manconi, M., Zaru, M., Valenti, D., Peris, J.E., Matricardi, P., et al., 2017. Glycerosomes: investigation of role of 1,2-dimyristoyl-sn-glycero-3-phosphatidycholine (DMPC) on the assembling and skin delivery performances. Int. J. Pharm. 532, 401–407.

Manconi, M., Caddeo, C., Sinico, C., Valenti, D., Mostallino, M.C., Lampis, S., et al., 2012. Penetration enhancer-containing vesicles: composition dependence of structural features and skin penetration ability. Eur. J. Pharm. Biopharm. 82, 352–359.

Manconi, M., Manca, M.L., Marongiu, F., Caddeo, C., Castangia, I., Petretto, G.L., et al., 2016. Chemical characterization of Citrus limon var. pompia and incorporation in phospholipid vesicles for skin delivery. Int. J. Pharm. 506, 449–457.

Manconi, M., Manca, M.L., Caddeo, C., Valenti, D., Cencetti, C., Diez-Sales, O., et al., 2018. Nanodesign of new self-assembling core–shell gellan-transfersomes loading baicalin and in vivo evaluation of repair response in skin. Nanomedicine 14, 569–579.

Manosroi, A., Jantrawut, P., Manosroi, J., 2008. Anti-inflammatory activity of gel containing novel elastic niosomes entrapped with diclofenac diethylammonium. Int. J. Pharm. 360, 156–163.

Manosroi, A., Ruksiriwanich, W., Abe, M., Manosroi, W., Manosroi, J., 2012. Transfollicular enhancement of gel containing cationic niosomes loaded with unsaturated fatty acids in rice (*Oryza sativa*) bran semi-purified fraction. Eur. J. Pharm. Biopharm. 81, 303–313.

Manosroi, A., Chankhampan, C., Manosroi, W., Manosroi, J., 2013. Transdermal absorption enhancement of papain loaded in elastic niosomes incorporated in gel for scar treatment. Eur. J. Pharm. Sci. 48, 474–483.

Malinovskaja-Gomeza, K., Espuelas, S., Garrido, M.J., Hirvonena, J., Laaksonen, T., 2017. Comparison of liposomal drug formulations for transdermal iontophoretic drug delivery. Eur. J. Pharm. Sci. 106, 294–301.

Marianecci, C., Rinaldi, F., Mastriota, M., Pieretti, S., Trapasso, E., Paolino, D., et al., 2012. Anti-inflammatory activity of novel ammonium glycyrrhizinate/niosomes delivery system: human and murine models. J. Control. Release 164, 17–25.

Marianecci, C., Di Marzio, L., Rinaldi, F., Celia, C.H., Paolino, D., Alhaique, F., et al., 2014. Niosomes from 80s to present: the estate of the art. Adv. Colloid Interface Sci. 205, 187–206.

Marianecci, C., Petralito, S., Rinaldi, F., Hanieh, P.N., Carafa, M., 2016. Some recent advances on liposomal and niosomal vesicular carriers. J. Drug Deliv. Sci. Technol. 32, 256–269.

Marto, J., Vitor, C., Guerreiro, A., Severino, C., Elautério, C., Ascenso, A., et al., 2016. Ethosomes for enhanced skin delivery of griseofulvin. Colloid Surf. B 146, 616–623.

Md, S., Haque, S., Madheswaran, T., Zeeshan, F., Meka, V.S., Radhakrishnan, A.K., et al., 2017. Lipid based nanocarriers system for topical delivery of photosensitizers. Drug Discov. Today 22, 1274–1283.

Meikle, T.G., Zabarac, A., Waddington, L.J., Separovic, F., Drummond, C.J., Conn, C.H. E., 2017. Incorporation of antimicrobial peptides in nanostructured lipid membrane mimetic bilayer cubosomes. Colloid Surf. B 152, 143–151.

Menezes, A.C., Campos, P.M., Euletério, C., Simoes, S., Garcia Praca, F.S., Lopes Badra Bentley, M.V., et al., 2016. Development and characterization of novel 1-(1-naphthyl) piperazine loaded lipid vesicles for prevention of UV-induced skin inflammation. Eur. J. Pharm. Biopharm. 104, 101–109.

Meng, Sh, Zhang, C.H., Shi, W., Zhang, X.W., Liu, D.H., Wang, P., et al., 2016. Preparation of osthole-loaded nano-vesicles for skin delivery: characterization, in vitro skin permeation and preliminary in vivo pharmacokinetic studies. Eur. J. Pharm. Sci. 92, 49–54.

Mennini, N., Cirri, M., Maestrelli, F., Mura, P., 2016. Comparison of liposomal and NLC (nanostructured lipid carrier) formulations for improving the transdermal delivery of oxaprozin: effect of cyclodextrin complexation. Int. J. Pharm. 515, 684–691.

Merin, P.E., Flowerlet, M., Neethusha, J.M., 2016. Transfersomes—a review. Int. J. Pharm. Pharm. Res. 6, 436–452.

Mir-Palomo, S., Nácher, A., Díez-Sales, O., Vila Busó, M.A.O., Caddeo, C., Manca, M.-L., et al., 2016. Inhibition of skin inflammation by baicalin ultradeformable vesicles. Int. J. Pharm. 511, 23–29.

Mishra, D.K., Dhote, V., Bhatnagar, P., Mishra, P.K., 2012. Engineering solid lipid nanoparticles for improved drug delivery: promises and challenges of translational research. Drug Deliv. Trans. Res. 2, 238–253.

Mitragotri, S., 2013. Devices for overcoming biological barriers: the use of physical forces to disrupt the barriers. Adv. Drug Deliv. Rev. 65, 100–103.

Moghassemi, S., Hadjizadeh, A., 2014. Nano-niosomes as nanoscale drug delivery systems: an illustrated review. J. Control. Release 185, 22–36.

Moghddam, S.R.M., Ahad, A., Aqil, M., Imam, S.S., Sultana, Y., 2016. Formulation and optimization of niosomes for topical diacerein delivery using 3-factor, 3-level Box–Behnken design for the management of psoriasis. Mater. Sci. Eng. C 69, 789–797.

Montenegro, L., Lai, F., Offerta, A., Sarpietro, M.G., Micicchè, L., Maccioni, A.M., et al., 2016. From nanoemulsions to nanostructured lipid carriers: a relevant development in dermal delivery of drugs and cosmetics. J. Drug Deliv. Sci. Technol. 32, 100–112.

Moosavian Kalat, S.A., Khamesipour, A., Bavarsad, N., Fallah, M., Khashayarmanesh, Z., Feizi, E., et al., 2014. Use of topical liposomes containing meglumine antimoniate (Glucantime) for the treatment of L. major lesion in BALB/c mice. Exp. Parasitol. 143, 5–10.

Morilla, M.J., Romero, E.L., 2015. Ultradeformable phospholipid vesicles as a drug delivery system: a review. Res. Rep. Trans. Drug Deliv. 4, 55–69.

Morsi, N.M., Aboelwafa, A.A., Dawoud, Marwa, H.S., 2016. Improved bioavailability of timolol maleate via transdermal transfersomal gel: statistical optimization, characterization, and pharmacokinetic assessment. J. Adv. Res. 7, 691–701.

Moulaoui, K., Caddeo, C., Manca, M.L., Castangia, I., Valenti, D., Escribano, E., et al., 2015. Identification and nanoentrapment of polyphenolic phytocomplex from *Fraxinus angustifolia*: in vitro and in vivo wound healing potential. Eur. J. Med. Chem. 89, 179–188.

Mulet, X., Boyd, B.J., Drummond, C.J., 2013. Advances in drug delivery and medical imaging using colloidal lyotropic liquid crystalline dispersions. J. Colloid Interface Sci. 393, 1–20.

Müller, R.H., Mäder, K., Gohla, S., 2000. Solid lipid nanoparticles (SLN) for controlled drug delivery—a review of the state of the art. Eur. J. Pharm. Biopharm. 50, 161–177.

Müller, R.H., Radtke, M., Wissing, S.A., 2002. Nanostructured lipid matrices for improved microencapsulation of drugs. Int. J. Pharm. 242, 121–128.

Müller, R.H., Petersen, R.D., Hommoss, A., Pardeike, J., 2007. Nanostructured lipid carriers (NLC) in cosmetic dermal products. Adv. Drug Deliv. Rev. 59, 522–530.

Müller, R.H., Alexiev, U., Sinambela, P., Keck, C.M., 2016. Nanostructured lipid carriers (NLC): the second generation of solid lipid nanoparticles. In: Dragicevic, N., Maibach, H.I. (Eds.), Percutaneous Penetration Enhancers Chemical Methods in Penetration Enhancement: Nanocarriers. Springer-Verlag, Berlin Heidelberg, pp. 161–185.

Murthy, S.N., Shivakumar, H.N., 2010. Topical and transdermal drug delivery. In: Kulkarni, V.S. (Ed.), Handbook of Non-Invasive Drug Delivery Systems. Elsevier Inc, USA, pp. 1–36.

Musicanti, C., Gasco, P., 2016. Solid lipid nanoparticles (SLN). In: Bhushan, B. (Ed.), Encyclopedia of Nanotechnology. Springer, Dordrecht, pp. 3761–3780.

Muzzalupo, R., Tavano, L., Cassano, R., Trombino, S., Ferrarelli, T., Picci, N., 2011. A new approach for the evaluation of niosomes as effective transdermal drug delivery systems. Eur. J. Pharm. Biopharm. 79, 28–35.

Muzzalupo, R., Tavano, L., Lai, F., Picci, N., 2014. Niosomes containing hydroxyl additives as percutaneous penetrationenhancers: effect on the transdermal delivery of sulfadiazine sodium salt. Colloid Surf. B 123, 207–212.

Muzzalupo, R., Pérez, L., Pinazo, A., Tavano, L., 2017. Pharmaceutical versatility of cationic niosomes derived from amino acid-based surfactants: skin penetration behavior and controlled drug release. Int. J. Pharm. 529, 245–252.

Nawaz, A., Wong, T.W., 2017. Microwave as skin permeation enhancer for transdermal drug delivery of chitosan-5-fluorouracil nanoparticles. Carbohyd. Polym. 157, 906–919.

Neubert, R.H.H., 2011. Potentials of new nanocarriers for dermal and transdermal drug delivery. Eur. J. Pharm. Biopharm. 77, 1–2.

Nunes, P.S., Silva Rabelo, A., Souza, J.C.C., Vasconcelos Santana, B., Menezes da Silva, T.M., Serafini, M.R., et al., 2016. Gelatin-based membrane containing usnic acid-loaded liposome improves dermal burn healing in a porcine model. Int. J. Pharm. 513, 473–482.

Pando, D., Matos, M., Gutiérrez, G., Pazos, C., 2015. Formulation of resveratrol entrapped niosomes for topical use. Colloid Surf. B 128, 398–404.

Paolino, D., Lucania, G., Mardente, D., Alhaique, F., Fresta, M., 2005. Ethosomes for skin delivery of ammonium glycyrrhizinate: in vitro percutaneous permeation through human skin and in vivo anti-inflammatory activity on human volunteers. J. Control. Release 106, 99–110.

Paolino, D., Celia, C., Trapasso, E., Cilurzo, F., Fresta, M., 2012. Paclitaxel-loaded ethosomes: potential treatment of squamous cell carcinoma, a malignant transformation of actinic keratoses. Eur. J. Pharm. Biopharm. 81, 102–112.

Pardeike, J., Hommoss, A., Müller, R.H., 2009. Lipid nanoparticles (SLN, NLC) in cosmetic and pharmaceutical dermal products. Int. J. Pharm. 366, 170–184.

Patel, K.K., Kumar, P., Thakkar, H.P., 2012. Formulation of niosomal gel for enhanced transdermal lopinavir delivery and its comparative evaluation with ethosomal gel. AAPS Pharm. Sci. Technol. 13, 1502–1510.

Pathan, I.B., Jaware, B.P., Shelke, S., Ambekar, W., 2018. Curcumin loaded ethosomes for transdermal application: formulation, optimization, in vitro and in vivo study. J. Drug Deliv. Sci. Technol. 44, 49–57.

Paul, A., Cevc, G., Bachhawat, B.K., 1995. Transdermal immunization with large proteins by means of ultradeformable drug carriers. Eur. J. Immunol. 25, 3521–3524.

Peng, X., Zhou, Y., Han, K., Qin, L., Li, G., Pan, X., et al., 2015. Characterization of cubosomes as a targeted and sustained transdermal delivery system for capsaicin. Drug Des. Dev. Ther. 9, 4209–4218.

Peng, L.H., Wei, W., Shan, Y.H., Chong, Y.S., Yu, L., Gao, J.Q., 2017. Sustained release of piroxicam from solid lipid nanoparticle as an effective anti-inflammatory therapeutics in vivo. Drug Dev. Ind. Pharm. 43, 55–66.

Perez, A.P., Altube, M.J., Schilrreff, P., Apezteguia, G., Celes, F.S., Zacchino, S., et al., 2016. Topical amphotericin B in ultradeformable liposomes: formulation, skin penetration study, antifungal and antileishmanial activity in vitro. Colloid Surf. B 139, 190–198.

Pierre, M.B.R., dos Santos Miranda Costa, I., 2011. Liposomal systems as drug delivery vehicles for dermal and transdermal applications. Arch. Dermatol. Res. 303, 607–621.

Pinto, M.F., Costa Moura, C., Nunes, C., Segundo, M.A., Costa Lima, S.A., Reis, S., 2014. A new topical formulation for psoriasis: development of methotrexate-loaded nanostructured lipid carriers. Int. J. Pharm. 477, 519–526.

Pleguezuelos-Villa, M., Mir-Palomo, S., Díez-Sales, O., Vila-Buso, M.A.O., Ruiz-Sauri, A., Nácher, A., 2018. A novel ultradeformable liposomes of Naringin for anti-inflammatory therapy. Colloid Surf. B 162, 265–270.

Pradhan, M., Singh, D., Sing, M.R., 2013. Novel colloidal carriers for psoriasis: current issues, mechanistic insight and novel delivery approaches. J. Control. Release 170, 380–395.
Prasanthi, D., Lakshmi, P.K., 2012. Vesicles—mechanism of transdermal permeation: a review. Asian J. Pharm. Clin. Res. 5, 18–25.
Priyanka, R., Shipra, D., 2012. Transferosomes: a novel Carrier for transdermal drug delivery system. Int. J. Pharm. Technol. 4, 1854–1865.
Prow, T.W., Grice, J.E., Lin, L.L., Faye, R., Butler, M., Becker, W., et al., 2011. Nanoparticles and microparticles for skin drug delivery. Adv. Drug Deliv. Rev. 63, 470–491.
Raj, R., Raj, P.M., Ram, A., 2016. Lipid based noninvasive vesicular formulation of cytarabine: nanodeformable liposomes. Eur. J. Pharm. Sci. 88, 83–90.
Rajera, R., Nagpal, K., Singh, S.K., Mishra, D.N., 2011. Niosomes: a controlled and novel drug delivery system. Biol. Pharm. Bull. 34, 945–953.
Ramezani, V., Honarvar, M., Seyedabadi, M., Karimollah, A., Ranjbar, A.M., Hashemi, M., 2018. Formulation and optimization of transfersome containing minoxidil and caffeine. J. Drug Deliv. Sci. Technol. 44, 129–135.
Rangsimawong, W., Opanasopit, P., Rojanarata, T., Panomsuk, S., Ngawhirunpat, T., 2017. Influence of sonophoresis on transdermal drug delivery of hydrophilic compound-loaded lipid nanocarriers. Pharm. Dev. Technol. 22, 597–605.
Ranju, P., Manoj, P., 2015. Transfersomes: a novel carrier for enhanced dermal delivery of drug. World J. Pharm. Pharm. Sci. 4, 1816–1832.
Rattanapak, T., Birchall, J., Young, K., Ishii, M., Meglinski, I., Rades, T., et al., 2013. Transcutaneous immunization using microneedles and cubosomes: mechanistic investigations using optical coherence tomography and two-photon microscopy. J. Control. Release 172, 894–903.
Raza, K., Singh, B., Lohan, S., Sharma, G., Negi, P., Yachha, Y., et al., 2013. Nano-lipoidal carriers of tretinoin with enhanced percutaneous absorption, photostability, biocompatibility and anti-psoriatic activity. Int. J. Pharm. 456, 65–72.
Ritwiset, A., Krongsuka, S., Johns, J.R., 2016. Molecular structure and dynamical properties of niosome bilayers with and without cholesterol incorporation: a molecular dynamics simulation study. Appl. Surf. Sci. 380, 23–31.
Roberts, M.S., Mohammed, Y., Pastore, M., Namjoshi, S., Yousef, S., Alinaghi, A., et al., 2016. Topical and cutaneous delivery using nanosystems. J. Control. Release 10, 247–286.
Romero, E.L., Morilla, M.J., 2013. Highly deformable and highly fluid vesicles as potential drug delivery systems: theroretical and practical considerations. Int. J. Nanomed. 8, 3171–3186.
Sakdiset, P., Okada, A., Todo, H., Sugibayashi, K., 2018. Selection of phospholipids to design liposome preparations with high skin penetration–enhancing effects. J. Drug Deliv. Sci. Technol. 44, 58–64.
Sala, M., Diab, R., Elaissari, A., Fessi, H., 2018. Lipid nanocarriers as skin drug delivery systems: properties, mechanisms of skin interactions and medical applications. Int. J. Pharm. 535, 1–17.
Sanna, V., Gavini, E., Cossu, M., Rassu, G., Giunchedi, P., 2007. Solid lipid nanoparticles (SLN) as carriers for the topical delivery of econazole nitrate: in vitro characterization, ex vivo and in vivo studies. J. Pharm. Pharmacol. 59, 1057–1064.

Sarmah, P.J., Kalita, B., Sharma, A.K., 2013. Transfersomes based transdermal drug delivery: an overview. Int. J. Adv. Pharm. Res. 4, 2555–2563.
Schäfer-Korting, M., Mehnert, W., Korting, H.C., 2007. Lipid nanoparticles for improved topical application of drugs for skin diseases. Adv. Drug Deliv. Rev. 59, 427–443.
Schaller, M., Korting, H.C., Schmid, M.H., 1996. Interaction of cultured human keratinocytes with liposomes encapsulating silver sulphadiazine: proof of the uptake of intact vesicles. Br. J. Dermatol. 134, 445–450.
Scognamiglio, I., De Stefano, D., Campani, V., Mayol, L., Carnuccio, R., Fabbrocini, G., et al., 2013. Nanocarriers for topical administration of resveratrol: a comparative study. Int. J. Pharm. 440, 179–187.
Seong, J.S., Yun, M.E., Park, S.N., 2018. Surfactant-stable and pH-sensitive liposomes coated with N-succinylchitosan and chitooligosaccharide for delivery of quercetin. Carbohydr. Polym. 181, 659–667.
Shah, S.M., Ashtikar, M., Jain, A.S., Makhija, D.T., Nikam, Y., Gude, R.P., et al., 2015. LeciPlex, invasomes, and liposomes: a skin penetration study. Int. J. Pharm. 490, 391–403.
Shahzad, Y., Louw, R., Gerber, M., du Plessis, J., 2015. Breaching the skin barrier through temperature modulations. J. Control. Release 202, 1–13.
Shao, M., Hussain, Z., Thu, H.E., Khan, Sh, Katas, H., Ahmed, T.A., et al., 2016. Drug nanocarrier, the future of atopic diseases: advanced drug delivery systems and smart management of disease. Colloid Surf. B 147, 475–491.
Shen, L.N., Zhang, Y.T., Wang, Q., Xu, L., Feng, N.P., 2014. Enhanced in vitro and in vivo skin deposition of apigenin delivered using ethosomes. Int. J. Pharm. 460, 280–288.
Shumilov, M., Touitou, E., 2010. Buspirone transdermal administration for menopausal syndromes, in vitro and in animal model studies. Int. J. Pharm. 387, 26–33.
Singh, D., 2013. Ultradeformable system: a carrier for transdermal drug delivery. Int. J. Pharm. Sci. Res. 4, 4098–4104.
Singh, R.P., Gangadharappa, H.V., Mruthunjaya, K., 2017. Phospholipids: Unique carriers for drug delivery systems. J. Drug Deliv. Sci. Technol. 39, 166–179.
Sohrabi, S., Haeri, A., Mahboubi, A., Mortazavi, A., Dadashzadeh, S., 2016. Chitosan gel-embedded moxifloxacin niosomes: an efficientantimicrobial hybrid system for burn infection. Int. J. Biol. Macromol. 85, 625–633.
Solanki, A., Parikh, J., Parikh, R., Patel, M., 2010. Evaluation of different compositions of niosomes to optimize aceclofenac transdermal delivery. Asian J. Pharm. Sci. 5, 87–95.
Song, C.H.K., Balakrishnan, P., Shim, C.H.-K., Chung, S.-J., Chong, S., Kim, D.-D., 2012. A novel vesicular carrier, transethosome, for enhanced skin delivery of voriconazole: characterization and in vitro/in vivo evaluation. Colloid Surf. B 92, 299–304.
Song, A., Zhang, X., Li, Y., Mao, X., Han, F., 2016. Effect of liquid-to-solid lipid ratio on characterizations of flurbiprofen-loaded solid lipid nanoparticles (SLNs) and nanostructured lipid carriers (NLCs) for transdermal administration. Drug Dev. Ind. Pharm. 42, 1308–1314.
Souto, E.B., Müller, R.H., 2008. Cosmetic features and applications of lipid nanoparticles (SLN, NLC). Int. J. Cosmet. Sci. 30, 157–165.
Souto, E.B., Fangueiro, J.F., Müller, R.H., 2013. Solid lipid nanoparticles (SLN™). In: Uchegbu, I.F., et al., (Eds.), Fundamentals of Pharmaceutical Nanoscience. Springer, Nueva York, pp. 91–116.

Taha, E.I., 2014. Lipid vesicular systems: formulation optimization and ex vivo comparative study. J. Mol. Liq. 196, 211–216.

Tanriderdi, S.T., Özer, Ö., 2013. Novel topical formulations of Terbinafine–HCl for treatment of onychomycosis. Eur. J. Pharm. Sci. 48, 628–636.

Tapeinos, C., Battaglini, M., Ciofani, G., 2017. Advances in the design of solid lipid nanoparticles and nanostructured lipid carriers for targeting brain diseases. J. Control. Release 264, 306–332.

Tavano, L., Alfano, P., Muzzalupo, R., de Cindio, B., 2011. Niosomes vs microemulsions: new carriers for topical delivery of capsaicin. Colloid Surf. B 87, 333–339.

Tavano, L., Gentile, L., Rossi, C.O., Muzzalupo, R., 2013a. Novel gel-niosomes formulations as multicomponent systems for transdermal drug delivery. Colloid Surf. B 110, 281–288.

Tavano, L., Vivacqua, M., Carito, V., Muzzalupo, R., Caroleo, M.C., Nicoletta, F., 2013b. Doxorubicin loaded magneto-niosomes for targeted drug delivery. Colloid Surf. B 102, 803–807.

Tavano, L., Rossi, C.O., Picci, N., Muzzalupo, R., 2016. Spontaneous temperature-sensitive Pluronic1 based niosomes: triggered drug release using mild hyperthermia. Int. J. Pharm. 511, 703–708.

Teixeira, M.C., Carbone, C., Souto, E.B., 2017. Beyond liposomes: recent advances on lipid based nanostructures for poorly soluble/poorly permeable drug delivery. Prog. Lipid Res. 68, 1–11.

Touitou, E., Dayan, N., Bergelson, L., Godin, B., Eliaz, M., 2000. Ethosomes-novel vesicular carriers for enhanced delivery: characterization and skin penetration properties. J. Control. Release 65, 403–418.

Touitou, E., Godin, B., 2007. Ethosomes for skin delivery. J. Drug Deliv. Sci. Technol. 17, 303–308.

Touitou, E., Godin, B., Shumilov, M., Bishouty, N., Ainbinder, D., Shouval, R., et al., 2008. Efficacy and tolerability of clindamycin phosphate and salicylic acid gel in the treatment of mild to moderate acne vulgaris. J. Eur. Acad. Dermatol. Venereol. 22, 629–631.

Trauer, S., Richter, H., Kuntsche, J., Büttemeyer, R., Liebsch, M., Linscheid, M., et al., 2014. Influence of massage and occlusion on the ex vivo skin penetration of rigid liposomes and invasomes. Eur. J. Pharm. Biopharm. 86, 301–306.

Trotta, M., Peira, E., Debernardi, F., Gallarate, M., 2002. Elastic liposomes for skin delivery of dipotassium glycyrrhizinate. Int. J. Pharm. 241, 319–327.

Trotta, M., Peira, E., Carlotti, M.E., Gallarate, M., 2004. Deformable liposomes for dermal administration of methotrexate. Int. J. Pharm. 270, 119–125.

Tyagi, R.K., Garg, N.K., Jadon, R., Sahu, T., Katare, O.P., Dalai, S.K., et al., 2015. Elastic liposome-mediated transdermal immunization enhanced the immunogenicity of *P. falciparum* surface antigen, MSP-119. Vaccine 33, 4630–4638.

Uchegbu, I.F., Florence, A.T., 1995. Non-ionic surfactant vesicles (niosomes): physical and pharmaceutical chemistry. Adv. Colloid Interface Sci. 58, 1–55.

Uchegbu, I.F., Vyas, S.P., 1998. Non-ionic surfactant based vesicles (niosomes) in drug delivery. Int. J. Pharm. 172, 33–70.

Üner, M., Wissing, S.A., Yener, G., Müller, R.H., 2005. Skin moisturizing effect and skin penetration of ascorbyl palmitate entrapped in solid lipid nanoparticles (SLN) and nanostructured lipid carriers (NLC) incorporated into hydrogel. Pharmazie 60, 751–755.

Van den Bergh, B.A., Vroom, J., Gerritsen, H., Junginger, H.E., Bouwstra, J.A., 1999a. Interactions of elastic and rigid vesicles with human skin in vitro: electron microscopy and two-photon excitation microscopy. Biochim. Biophys. Acta 1461, 155–173.

Van den Bergh, B.A.I., Bouwstra, J.A., Junginger, H.E., Wertz, P.W., 1999b. Elasticity of vesicles affects hairless mouse skin structure and permeability. J. Control. Release 62, 367–379.

Vanić, Ž., 2015. Phospholipid vesicles for enhanced drug delivery in dermatology. J. Drug Discov. Dev. Deliv. 2, 1010–1019.

Verma, D.D., Verma, S., Blume, G., Fahr, A., 2003. Particle size of liposomes influences dermal delivery of substances into skin. Int. J. Pharm. 258, 141–151.

Verma, D.D., Verma, S., McElwee, K.J., Freyschmidt-Paul, P., Hoffman, R., Fahr, A., 2004. Treatment of alopecia areata in the DEBR model using cyclosporin a lipid vesicles. Eur. J. Dermatol. 14, 332–338.

Verma, P., Pathak, K., 2012. Nanosized ethanolic vesicles loaded with econazole nitrate for the treatment of deep fungal infections through topical gel formulation. Nanomed. Nanotechnol. Biol. Med. 8, 489–496.

Vitonyte, J., Manca, M.L., Caddeo, C., Valenti, D., Peris, J.E., Usach, I., et al., 2017. Bifunctional viscous nanovesicles co-loaded with resveratrol and gallic acid for skin protection against microbial and oxidative injuries. Eur. J. Pharm. Biopharm. 114, 278–287.

Wang, Y., Thakur, R., Fan, Q., Michniak, B., 2005. Transdermal iontophoresis: combination strategies to improve transdermal iontophoretic drug delivery. Eur. J. Pharm. Biopharm. 60, 179–191.

Wang, J., Tang, J., Zhou, X., Xia, Q., 2014. Physicochemical characterization, identification and improved photo-stability of alpha-lipoic acid[HYPHEN]loaded nanostructured lipid carrier. Drug Dev. Ind. Pharm 40, 201–210.

Williams, A.C., Barry, B.W., 2012. Penetration enhancers. Adv. Drug Deliv. Rev. 64, 128–137.

Wolf, M., Klang, V., Halper, M., Stix, C., Heuser, T., Kotisch, H., et al., 2017. Monoacylphospatidylcholine nanostructured lipid carriers: influence of lipid and surfactant content on in vitro skin permeation of flufenamic acid and fluconazole. J. Drug Deliv. Sci. Technol. 41, 419–430.

Wong, T.W., 2014. Electrical, magnetic, photomechanical and cavitational waves to overcome skin barrier for transdermal drug delivery. J. Control. Release 193, 257–269.

Xiang, Q., Xiao, J., Zhang, H., Zhang, X., Lu, M., Zhang, H., et al., 2011. Preparation and characterisation of bFGF-encapsulated liposomes and evaluation of wound-healing activities in the rat. Burns 37, 886–895.

Yan, Y., Zhang, H., Sun, J., Pengchong, W., Dong, K., Dong, Y., et al., 2016. Enhanced transdermal delivery of sinomenine hydrochloride by ethosomes for anti-inflammatory treatment. J. Drug Deliv. Sci. Technol. 36, 201–207.

Yoon, G., Park, J.W., Yoon, I.S., 2013. Solid lipid nanoparticles (SLNs) and nanostructured lipid carriers (NLCs): recent advances in drug delivery. J. Pharm. Investig. 43, 353–362.

Yu, X., Du, L., Li, Y., Fu, G., Jin, Y., 2015. Improved anti-melanoma effect of a transdermal mitoxantrone ethosome gel. Biomed. Pharmacother. 73, 6–11.

Yue, X., Dai, Z., 2014. Recent advances in liposomal nanohybrid cerasomes as promising drug nanocarriers. Adv. Colloid Interface Sci. 207, 32–42.

Yuksel, N., Bayindir, Z.S., Aksakal, E., Ozcelikay, A.T., 2016. In situ niosome forming maltodextrin proniosomes of candesartancilexetil: in vitro and in vivo evaluations. Int. J. Biol. Macromol. 82, 453–463.

Zeb, A., Qureshi, O.S., Yu, C.H., Akram, M., Kim, H.S., Kim, M.S., et al., 2017. Enhanced anti-rheumatic activity of methotrexate-entrapped ultradeformable liposomal gel in adjuvant-induced arthritis rat model. Int. J. Pharm. 525, 92–100.

Zhai, Y., Zhai, G., 2014. Advances in lipid-based colloid systems as drug carrier for topic delivery. J. Control. Release 193, 90–99.

Zhang, Z., Wo, Y., Zhang, Y., Wang, D., He, R., Chen, H., et al., 2012. In vitro study of ethosome penetration in human skin and hypertrophic scar tissue. Nanomed. Nanotech. Biol. Med. 8, 1026–1033.

Zhang, Y.T., Shen, L.N., Wu, Z.H., Zhao, J.H., Feng, N.P., 2014b. Evaluation of skin viability effect on ethosome and liposome-mediated psoralen delivery via cell uptake. J. Pharm. Sci. 103, 3120–3126.

Zhang, Y.T., Shen, L.N., Wu, Z.H., Zhao, J.H., Feng, N.P., 2014a. Comparison of ethosomes and liposomes for skin delivery of psoralen for psoriasis therapy. Int. J. Pharm. 471, 449–452.

Zhang, Y., Shen, L., Zhang, K., Guo, T., Zhao, J., Li, N., et al., 2015a. Enhanced antioxidation via encapsulation of isooctyl p-methoxycinnamate with sodium deoxycholate-mediated liposome endocitosis. Int. J. Pharm. 496, 392–400.

Zhang, Y., Zhang, K., Wu, Z., Guo, T., Ye, B., Lu, M., et al., 2015b. Evaluation of transdermal salidroside delivery using niosomes via in vitro cellular uptake. Int. J. Pharm. 478, 138–146.

Zhang, Y., Ng, W., Feng, X., Cao, F., Xu, H., 2017. Lipid vesicular nanocarrier: quick encapsulation efficiency determination and transcutaneous application. Int. J. Pharm. 516, 225–230.

Zhao, Y.Z., Lu, C.T., Zhang, Y., Xiao, J., Zhao, Y.P., Tian, J.L., et al., 2013. Selection of high efficient transdermal lipid vesicle for curcumin skin delivery. Int. J. Pharm. 454, 302–309.

Zidan, A.S., Mokhtar, M., 2011. Multivariate optimization of formulation variables influencing flurbiprofen proniosomes characteristics. J. Pharm. Sci. 100, 2212–2221.

FURTHER READING

Cevc, G., 2004. Lipid vesicles and other colloids as drug carriers on the skin. Adv. Drug Deliv. Rev. 56, 675–711.

Daddeo, C., Diez Sales, O., Valenti, D., Ruiz Saurí, A., Fadda, A.M., Manconi, M., 2013. Inhibition of skin inflammation in mice by diclofenac in vesicular carriers: liposomes, ethosomes and PEVs. Int. J. Pharm. 443, 128–136.

Döge, N., Hönzke, S., Schumacher, F., Balzus, B., Colombo, M., Hadam, S., et al., 2016. Ethyl cellulose nanocarriers and nanocrystals differentially deliver dexamethasone into intact, tape-stripped or sodium lauryl sulfate-exposed ex vivo human skin—assessment by intradermal microdialysis and extraction from the different skin layers. J. Control. Release 242, 25–34.

Kobierski, S., Ofori-Kwakye, K., Muller, R.H., Keck, C.M., 2009. Resveratrol nanosuspensions for dermal application-production, characterization, and physical stability. Pharmazie 64, 741–747.

Pireddu, R., Caddeo, C., Valenti, D., Marongiu, F., Scano, A., Ennas, G., et al., 2016. Diclofenac acid nanocrystals as an effective strategy to reduce in vivo skin inflammation by improving dermal drug bioavailability. Colloid Surf. B 143, 64–70.

Romero, G.B., Arntjen, A., Keck, C.M., Muller, R.H., 2016. Amorphous cyclosporin A nanoparticles for enhanced dermal bioavailability. Int. J. Pharm. 498, 217–224.

Teeranachaideekulm, V., Chantaburanan, T., Junyaprasert, V.B., 2017. Influence of state and crystallinity of lipid matrix on physicochemical properties and permeation of capsaicin-loaded lipid nanoparticles for topical delivery. J. Drug Deliv. Sci. Technol. 39, 300–307.

CHAPTER

Stimuli sensitive systems for camptothecin delivery 12

Pablo Botella and Eva Rivero-Buceta

Instituto de Tecnología Química, Universitat Politècnica de València-Consejo Superior de Investigaciones Científicas, Valencia, Spain

CHAPTER OUTLINE

12.1 INTRODUCTION

Camptothecin (CPT) is a water insoluble, natural pentacyclic alkaloid isolated from the oriental tree *Camptotheca accuminata* by Wall et al. (1966). CPT is well-known for its antitumor activity against a wide spectrum of human cancers (Beidler et al., 1996; Giovanella et al., 1991; Del Bino et al., 1990; Li et al., 1972; Chen and Liu, 1994; Oberlies and Kroll, 2004; Clements et al., 1996). In addition, a variety of biological activities, such as pesticidal (Liu et al., 2011; Zhang et al., 2013), antipsoriatic (Lin and Huang, 1988; Chiao and Li, 1975), antiparasitic (Bodley and Shapiro, 1995), antifungal (Del Poeta et al., 1999), antimicrobial (Alaghaz et al., 2012), and antiviral (Li et al., 2009; Horwitz et al., 1972) have been described for this compound. Certainly, several researchers have

Nanoparticles in Pharmacotherapy. DOI: https://doi.org/10.1016/B978-0-12-816504-1.00012-0

reported that CPT inhibits a cellular enzyme DNA topoisomerase I and induces apoptosis in various cancer cells (Boven et al., 2000; Hsiang et al., 1985).

CPT is a planar pentacyclic quinoline that includes three rings of pyrrolo-(3,4-β)-quinoline part (A, B, and C rings), an unsaturated pyridone moiety in ring D and a α-hydroxy lactone ring (E ring), containing one chiral center with (S)-configuration (Fig. 12.1) (Tobin and Rivory, 2004; Muqeet and Qudsia, 2014). It is worth noting that the isomer 20(R) hydroxyl shows very little activity, while the isomer 20(S) is between 10- and 100-fold more active (Garcia-Carbonero and Supko, 2002). Also, it is known that the E-ring exists in equilibrium between the lactone form (close ring, water insoluble) and the carboxylate form (open ring, water soluble). At acidic pH, the lactone predominates, which, according to different clinical trials and structure–activities studies, is the only therapeutic molecule. However, at physiological pH, the lactone ring is converted into the carboxylate, much less active, which predominates at neutral and alkaline pH (Carrigan et al., 1997; Mi and Burke, 1994).

Unfortunately, CPT presents some major limitations for therapeutic application, like poor water solubility and the rapid lactone ring hydrolysis at physiological pH, which gives rise to the inactive carboxylate form (Basili and Moro, 2009). In addition, CPT is extremely insoluble in organic compounds except for dimethyl sulfoxide, in which it shows moderate solubility (Hatefi and Amsden, 2002). Due to CPT's insolubility in biocompatible solvents, it is very difficult to distribute this compound into the body through the conventional drug administration routes, including oral, intravenous, or intramuscular injection (Zhi et al., 2016). Furthermore, other negative aspects that limit the use of CPT in clinical trials are a pronounced loss of activity due to lactone-ring hydrolysis, reversibility of drug–target interaction and severe toxicity, including hemorrhagic cystitis and myelotoxicity (Zunino and Pratesi, 2004). All these drawbacks have precluded its clinical application, making it necessary to develop alternative compounds and structures for CPT use in humans.

In an effort to improve CPT solubility, stability in biological fluids, and pharmacokinetics, the prodrug strategy modifies chemically the quinoline ring. Two of these structural derivatives currently approved for clinical use are topotecan and irinotecan (Kunimoto et al., 1987; Liu et al., 2015). Unfortunately, CPT

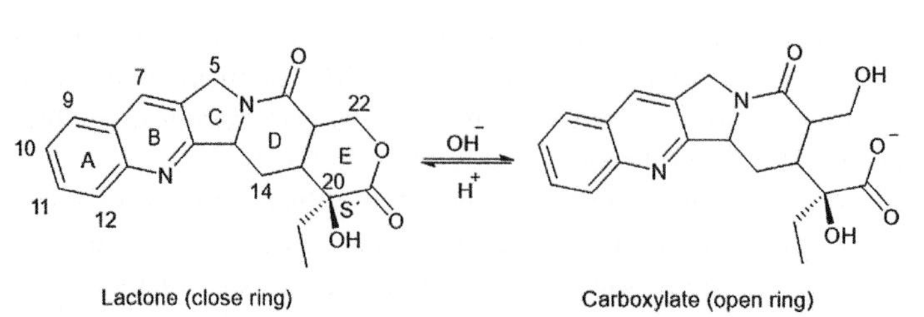

FIGURE 12.1

Camptothecin equilibrium between lactone (active form) and carboxylate (inactive form).

structure modification usually brings about a dramatic drop of cytotoxic activity, which results one order or even lower than pristine drug. Moreover, therapies merely relying on single, small molecules now seem out of step, mostly due to their limited efficacy and unacceptable toxicities. For this reason, huge emphasis is being placed on developing new drug delivery systems (DDSs) that are able to simultaneously deliver CPT and other active agents, as well as incorporating imaging moieties that may ease the diagnosis process. Most specifically, the future seems good for those systems that are able to discharge selectively the drug at the target cells under a specific stimulus, with no previous release. Stimuli-responsive preparations provide a dramatically better spatial and temporal control over the release of drugs, which is crucial for delivery of the very cytotoxic CPT.

In this review, we summarize the most advanced and recent achievements in the design, synthesis, and development of CPT stimuli-responsive DDSs, including a description of the most important formulations currently under clinical use.

12.2 MATERIALS FOR CAMPTOTHECIN DELIVERY SYSTEMS

Many types of nanovehicles have been developed for CPT delivery, and these DDSs may be classified in several ways, according to their chemical nature, structure, composition, morphology, etc. Herein, we have followed a systematic classification of CPT nanocarriers according to their chemical nature in organic, inorganic, and hybrid materials, presenting different properties and behaviors (Botella and Rivero-Buceta, 2017; Ranjit and Baquee, 2013; Gabizon et al., 2014; Re et al., 2012; Doshi and Mitragotri, 2009). These novel nanoplatforms have been designed to solve most of the antitumor drug limitations, improving CPT pharmacokinetics, which is affected by poor water solubility, rapid plasma clearance, high systemic toxicity, and poor selectivity towards cancer cells (Venditto and Simanek, 2010). A variety of DDSs have been developed and allow CPT incorporation. In addition, these DDSs can conjugate the drug in different ways, such as adsorption, encapsulation, or covalent bonding (Mahapatro and Singh, 2011), working as inert supports to transport the drug to the tumor (Shen et al., 2010). In this sense, some major advantages of these DDSs over free CPT are improved solubility, lactone ring stability, half-life extension, biocompatibility, and control drug release rates, which allow significant advances at the clinical stage (Duncan, 1997; Venditto and Szoka, 2013).

The first pharmaceutical nanocarriers used to enhance CPT efficacy had fully organic composition (Torchilin, 2006). To prepare such smart systems, therapeutic moieties must be simultaneously encapsulated or assembled on the surface (López-Dávila, Seifalian and Loizidou, 2012). Here, micelles, liposomes, dendrimers, and polymers (e.g., β-cyclodextrins (β-CDs), poly(L-glutamic) acid, chitosan and polyhyaluronic acid, among others), have been formulated as DDSs for CPT

(Venditto and Simanek, 2010; Matsumura, 2011; Wong et al., 2013; Lim and Simanek, 2012). In general terms, these nanovehicles are highly biocompatible and present good drug loading ratios. However, organic moieties are barely able to accept structural modifications that promote zero release before reaching the target cells, due to their low surface area and chemical stability.

On the other hand, a wide variety of inorganic nanoparticles show high stability in the biological environment, ease of functionalization (allowing multifunctionality), and unique physicochemical properties (optical, magnetic and electronic) that can be tailored by controlling particle structure, composition, size, and shape. Materials such as semiconductor quantum dots (QDs), magnetic nanoparticles, and plasmonic nanoparticles, among others, have been proposed for imaging and therapy (Huang et al., 2011; Giner-Casares et al., 2016). However, despite the interest raised by inorganic nanoparticles as DDSs, most of these systems suffer from biocompatibility issues, mostly related to the toxicity of some structural components and the strong interaction with serum proteins, which results in rapid blood clearance and hepatic removal (Nam et al., 2013). Moreover, occasionally, elimination from organic tissues may be an issue, involving accumulation of highly toxic elements in the body.

Therefore, inorganic nanocarriers are better handled surface modified or combined with organic components, which provide good stability in physiological fluids, low toxicity and immunogenicity, and minimized reactions with serum proteins, to control their fate in biological environment (Nam et al., 2013). These composites consisting of two or more constituents distributed in the nanoscale or molecular range are hybrid materials. Typically, the hybrid nanocontainers are made of inorganic and organic building blocks, presenting unique properties depending on the spatial and size distribution of their components (Kickelbick, 2007; Fahmi et al., 2009). Actually, efforts have been made to design hybrid devices combining different functions or strengths of the individual components in a synergic way, or even to compensate for the weaknesses of one another (Li et al., 2016). Whilst the organic component provides good biocompatibility and better solubility in physiological fluids, the inorganic moiety offers almost unlimited external surface area to bind different functionalities, improving the capabilities of the DDS. On the other hand, the main handicap of these systems is the lack of long-term toxicity studies about possible interactions within the body of the different components. A myriad of compositions and models have been reported with potential use in nanomedicine. Most of these hybrid nanocarriers for CPT delivery carry molecular devices sensitive to specific physicochemical stimuli, which act as triggers for selective release activation at the targeting tissues and cells.

Overall, the incorporation of CPT into a nanocarrier definitely improves its therapeutic efficacy. All vehicles provide a clear increase of drug stability, as the encapsulated or bonded molecule is protected from enzymatic degradation and lactone ring cleavage during its transit through the blood stream and tissues to the tumor location. Moreover, colloidal stability of these formulations can boost drug solubility by several orders, which, together with the rise in stability, improves bioavailability. However, the most significant effect of incorporating CPT into a nanocontainer is the dramatic toxicity reduction and therapeutic window extension, as the transported molecule has very little activity before it is discharged at the tumor location. As a consequence, in vivo antitumor activity increases with regards to the free drug, due to an improved bioavailability and the possibility of increasing the dose with no side effects. In addition, many of these nanomedicines offer the possibility of delivering several drugs in a row, which is very important for the treatment of MDR cancer, and allow for the incorporation of targeting molecules that favor selective therapy. Finally, inorganic and hybrid materials can present unique properties (e.g., optical activity, magnetism, semiconductor activity), that enhance the potential of these systems in therapeutics and diagnostics.

12.3 STIMULI SENSITIVE SYSTEMS

These are currently the most representative DDSs in nanomedicine. Their most important characteristic is the ability to deliver to target cells a therapeutic agent with no previous release. This is due to the accurate control of the release process through a specific intracellular stimulus, which minimizes undesired side effects. Stimuli-responsive nanocarriers can be classified as endogenous or exogenous, depending on the stimulus that activates the drug release mechanism. Examples of endogenous stimuli are pH, redox potential, and enzyme activity. On the other hand, classical exogenous stimuli include magnetic fields, ultrasound, light, and temperature (Ganta et al., 2008; Mura et al., 2013; Fleige et al., 2012; Tong et al., 2014). The different examples described in the present work have been compiled according to the triggering stimulus in Table 12.1.

12.3.1 PH-SENSITIVE SYSTEMS

Several pH-sensitive DDSs have been developed based on the pH difference between normal and tumor tissue. Here, some of these platforms take advantage of the fact that, in some cases, tumor tissue pH profile is slightly lower than in normal tissue. However, the heterogenicity of extracellular pH both within and between tumors poses a significant challenge, as low extracellular pH regions are likely to be less accessible to nanoparticles that do not distribute readily within the tumor interstitium. This factor, together with the lower pH values inside endosomes, suggests that enhancing endosomal escape is the most realistic objective for pH-sensitive nanocarriers in the short term (Kanamala et al., 2016).

Polymeric, pH-responsive micelles have been prepared by a solvent evaporation method. For this purpose, hydrophilic methyl ether poly(ethylene glycol) (MPEG) and biodegradable poly(β-amino ester) (PAE) were copolymerized using a Michael type step polymerization to give MPEG–PAE block copolymer

Table 12.1 Stimuli-Responsive Systems for Intracellular Camptothecin Delivery Described in the Present Work[a]

Stimulus	Material	Release System	Drug Loading	Release Mechanism	Cell Line	References
pH	MPEG–PAE micelles	Ionizable linear polyamines	Gating	Amines protonation and demicellization	MDA-MB231	Min et al. (2010)
	Chitosan–b-poly(p-dioxanone) micelles	Ionizable linear polyamines	Gating	Amines protonation	HeLa L929 fibroblast	Tang et al. (2013)
	MAP – CPT nanoparticles	Polymer-based nanoparticles	Ester bond	Nanoparticle disruption by fat Ester bond hydrolysis	BT-474 HER2	Han and Davis (2013)
	Hollow silica nanoparticles sealed with ZnO QDs	ZnO QDs	Gating	ZnO QDs disintegration	A549	Li et al. (2015)
	Gold nanoclusters–poly-(L-lactide)–GPPS-folate	Poly(L-lactide) derivative	Adsorption	Hydrophobic–hydrophobic interactions Swelling of PLA	HeLa	Chen et al. (2012)
Redox	Uncross-linked and core–cross-linked micelles	Polyester derivative	Disulfide bond	Disulfide reduction and disassembly of micelles	MCF-7	Wang et al. (2013)
	Poly(ethylene glycol)–poly(L-lysine) micelles	PEG–poly(L-lysine) derivative	Disulfide bond	Reductive conditions of the cytosol. Glutathione	AY27	Yen et al. (2014)
	CPT–SS–CPT	2,6-Bis(hydroxymethyl)-aniline derivative	Disulfide bond	Cleavage of disulfide bond and decomposition of dimer	HeLa	Cai et al. (2015)
	Drug amphiphiles	Disulfylbutyrate derivative	Disulfide bond	Disulfide reduction and dissociation of nanostructure	9L F98L	Cheetham et al. (2013)
	Drug amphiphiles	Disulfanyl(ethyl) carbonate derivative	Disulfide bond	Disulfide reduction and dissociation of nanostructure	MCF-7	Cheetham et al. (2014)
	SiO_2@MSN–SS–CPT	Self-immolative ligand	Disulfide bond	Disulfide reduction and intramolecular cyclization	HeLa	Muniesa et al. (2013), Botella et al. (2016)

Enzyme	CPT–PGA–SNPs	Poly(L-glutamic acid) derivative	Ester bond	Ester bond hydrolysis	MCF-7 C57BL/6	Chen et al. (2012)
	PBAsp–CPT	α,β-Poly[(N-carboxybutil)-L-aspartamide]	Ester bond	Ester bond hydrolysis	L929	Fan et al. (2010)
	PGA–CPT	Poly-(L-glutamic acid)-glycine derivative	Amide bond	Hydrolysis of glycine molecule	NCI-H460	Singer et al. (2001)
	HA–CPT	Hyaluronic acid derivative	Ester bond	Ester bond hydrolysis	A2780 HepG2	Xu et al. (2013)
	PAMAM–CPT	Poly(amido amine) dendrimer	Ester bond	Ester bond hydrolysis	HTC-116	Thiagarajan et al. (2010)
	PAMAM-G2	Poly(amido amine) dendrimer	Carbamate bond	Self-immolative reaction sequence triggered by PGA	MOLT-3 JURKAT Hek-293	Gopin et al. (2006)
Enzyme	PEGylated PAMAM	Poly(amido amine) dendrimer	Ester bond	Ester bond hydrolysis	U1242	Zolotarskaya et al. (2015)
	CDP–CPT	β-Cyclodextrin–glycine derivative	Amide bond	Hydrolysis of glycine molecules	LS174T HT29 PC3 A2780	Cheng et al. (2003), Cheng et al. (2004), Schluep et al. (2009), Davis (2009)
	SiO_2 nanoparticles	Carboxyl group modified surface	Ester bond	Ester bond hydrolysis	HeLa U87-MG HCT-116 HT-29	Botella et al. (2011)

(*Continued*)

Table 12.1 Stimuli-Responsive Systems for Intracellular Camptothecin Delivery Described in the Present Work[a] *Continued*

Stimulus	Material	Release System	Drug Loading	Release Mechanism	Cell Line	References
Light	Mesoporous silica nanoparticles	Azobenzene derivative ligand	Gating	Trans–cis photoisomerization of azobenzene	PANC-1 SW480	Lu et al. (2008)
	Nanoimpellers	Two-photon fluorophore	Entrapped	Photoisomerization of azobenzene by two photon-triggered fluorophore	MCF-7	Croissant et al. (2013)
	Mesoporous silica nanoparticles	Spiropyran molecule	Gating	Conformational conversion of spiropyran	EA. hy926 HeLa	Chen et al. (2014a, 2014b)
	DNA–CPT	Self-immolative ligand	Ester bond	Disintegration of core by an irreversible self-immolative process	SK-BR-3	Tan et al. (2015)
	GNC@SiO_2	Gold nanoclusters and organosilica coating	Adsorption	Combination of photothermolysis and mechanical disruption	42-MG-BA	Botella et al. (2012)
	PMO NPs	Fluorescent fluorophores with different emission wavelength	Gating	Variation between the ratio of different fluorophores	Hela	Lu et al. (2012)
Magnetic	MSN@Fe_3O_4	Fe_3O_4	Adsorption	Segregation of Fe_3O_4 nanocaps from surfaces of mesoporous silica	A549	Chen et al. (2011)
	Self-assembly of iron oxide@SiO_2	Fe_3O_4	Gating	Disruption of the nanostructure	A549	Hu et al. (2008)
	Poly(lactic-co-glycolic) acid@Fe_3O_4	Fe_3O_4	Gating	Degradation of polymeric matrices	MT2	Kong et al. (2013)
Temperature	PNIPAM–b-hydroxyl-terminated polybutadiene	Thermo-triggered polymers	Entrapped	Structural changes of the micelles by temperature	MDA-MB231	Luo et al. (2014)
Ultrasound	Acoustically active nanoemulsions	Cavitation-triggered polymers	Encapsulation	Degradation of bubbles	B16-F0 ES-2	Fang et al. (2009)
Glucose	Magnetic mesoporous silica nanoparticles	Phenylboronic acid molecules	Gating	Competition between glucose and dextran for binding phenylboronic acid	HeLa RIN-5F CHO	Sinha et al. (2014)

[a]Abbreviations: *see text.*

(Fig. 12.2). Then, CPT was encapsulated into the micelles (CPT/MPEG–PAE) through hydrophobic interactions between drug and PAE. At pH 6.4–6.8 CPT/MPEG–PAE was demicellized and the encapsulated CPT was rapidly released from these nanocarriers, but under pH 7.4 release was clearly slower. In vitro cytotoxicity assays in MDA-MB231 human breast tumor cells confirmed that these micelles provoked severe cytotoxicity at acidic environment due to rapid CPT release, whereas cytotoxicity decreased at physiological pH (Min et al., 2010).

Encapsulation of CPT in micelles with a hydrophilic shell of chitosan and a hydrophobic core of poly(p-dioxanone) has been also used to control drug release on the basis of environment pH changes (Tang et al., 2013). CPT was incorporated in the core of this system and its release was triggered at pH 7.4, 6.2, and 5.0. In vitro drug release studies in HeLa human cervix cancer cells and L929 fibroblast demonstrated that micelles presented a much faster CPT release at pH 5.0 than at pH 7.4, whereas a control sample without CPT was found to be

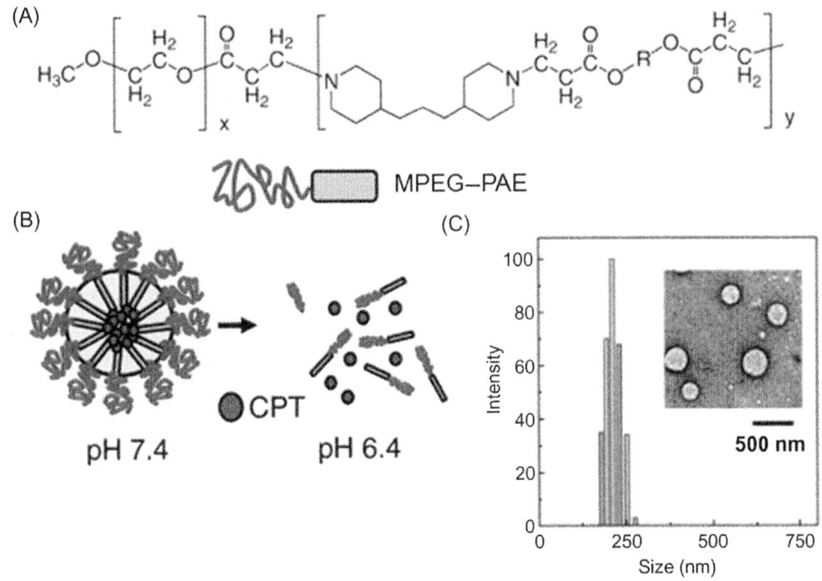

FIGURE 12.2

(A) Chemical structure of methyl ether poly(ethylene glycol)–poly(β-aminoester) block copolymer (MPEG–PAE). (B) Schematic diagram of pH-responsive CPT/MPEG–PAE micelles at weakly pH condition. (C) Size distribution of MPEG–PAE micelles in PBS at pH 7.4.

Reprinted from Min, K.H., Kim, J.H., Bae, S.M., Shin, H., Kim, M.S., Park, S., et al., 2010. Tumoral acidic pH-responsive MPEG–poly(β-amino ester) polymeric micelles for cancer targeting therapy. J. Control. Release, 144 (2), 259–266. Copyright © 2010, with permission from Elsevier.

nontoxic in both cell lines. The good internalization effect of these DDSs was explained by the electrostatic interactions between positively charged micelles and negatively charged cell membranes.

Another pH-sensitive nanoscaled system for CPT delivery has been obtained by assembling a targeted nanoparticle through reversible covalent complexation between boronic acids and diols (Han and Davis, 2013). CPT was conjugated to a biocompatible, hydrophilic copolymer of mucic acid and polyethylene glycol (MAP). When this polymer–drug conjugate was placed in water, it self-assembled into MAP – CPT nanoparticles of c.30 nm. Then, antibody herceptin was attached to a boronic acid via a polyethylene glycol spacer, and this boronic acid-containing targeting moiety was complexed with MAP–CPT to form a targeted nanoparticle. A strong dependence of release rate with pH was observed but, unlike the standard behavior, as pH increased from 6.5 to 7.4, the release half-lives dropped from 338 to 58 hours for MAP – CPT nanoparticles. In vitro testing over BT-474 and HER2 human breast cancer cell lines showed that cellular uptake of nanoparticles was enhanced by 70%, compared with a nontargeted version, by the incorporation of a single herceptin antibody molecule per nanoparticle. Moreover, CPT was released from MAP–CPT nanoparticles by hydrolysis and nanoparticle disruption by fat. The targeted system carried c.60 CPT molecules per nanoparticle and showed prolonged plasma circulation with an elimination half-life of 21.2 hours and an area under the curve (AUC) value of 2766 μg h mL^{-1} at a 10 mg CPT kg^{-1} tail vein injection in mice.

Besides the organic DDSs, inorganic pH-sensitive nanocarriers are able to deliver different chemotherapeutic drugs simultaneously for a better synergetic effect. One example of these nanoplatforms was based on hollow silica nanoparticles sealed with ZnO QDs. Subsequently, CPT was encapsulated in the hollow core through hydrophobic interactions. Then, doxorubicin (DOX) was deposited on the mesoporous shell via electrostatics interactions with carboxylate groups incorporated on the mesoporous surface (Li et al., 2015). This cooperative drug loading largely increased the drug encapsulation efficiency of both CPT and DOX up to 89.28% and 44.98%, respectively. Drug release was compared at pH 7.4 and pH 5.0, concluding that in acid medium the amount of discharged drugs was higher due to ZnO QDs dissolution.

Moreover, specific functionalization can be achieved in inorganic–organic hybrid materials, for example, core–shell systems obtained by protecting a metal nanoparticle with and organic coating. In one representative case, multifunctional nanocarriers of gold nanoclusters (GNCs) as core and folate (FA)-conjugated amphiphilic hyperbranched block copolymer as shell (based on poly(L-lactide) inner arm and FA-conjugated sulfated polysaccharide outer arm) were synthesized for CPT targeted delivery (Chen et al., 2012). These nanocarriers presented a release pattern curve at different pH in two stages: an initial rapid release (1 hour) was followed by a sustained release period (up to 15 hours), and then reached a plateau. Cytotoxicity studies over HeLa human cervix cancer cells showed that, with regards to the free drug, the CPT-loaded nanocarrier provided high anticancer activity, also gaining specificity to cancer cells with overexpressed FA receptor moieties.

12.3.2 REDOX-SENSITIVE SYSTEMS

It is well known the difference in redox potential between the oxidizing extracellular space and reducing intracellular space, which is related to glutathione (GSH) concentration. This fact has found application in the building of specific DDSs that are able to selectively discharge their therapeutic load after cell uptake. Here, drug covalent linking to nanocarriers is mostly carried out by disulfide bridge, which makes it possible to design different sensitive systems. The main concern regarding in vivo application of these systems is the presence of GSH in serum that, despite the fact that its concentration is less than one order below that of cytosol, may provoke some premature release during CPT transit at bloodstream (Balendiran et al., 2004).

Hybrid micelles subject to redox-responsive cleavage have been developed by cross-linking CPT-loaded organic units with disulfide bonds (Wang et al., 2013). These micelles were obtained via coprecipitation of disulfide-containing CPT−poly(tyrosine(alkynyl)-*O*-carboxyanhydride) conjugate and monomethoxy poly(ethylene glycol)−*b*-poly(tyrosine(alkynyl)-*O*-carboxyanhydride), followed by cross-linking of the micellar core via azide−alkyne click chemistry. In this case, CPT was rapidly released due to the high concentration of cytosolic glutathione that disorganizes micelle structure by disulfide bond cleavages, which are present both in the CPT prodrug and the cross-linker. In vitro cytotoxic assays against MCF-7 human breast cancer cells suggested high anticancer activity due to the disassembly of the cross-linked micelle and rapid CPT release.

In addition, CPT was also conjugated in different quantities to 45-nm-diameter micelles of poly(ethylene glycol)−poly(L-lysine) block copolymer by using 3,3′-dithiodipropionic acid as linker (Yen et al., 2014). Then, CPT release was studied in response to cytosol reductive conditions by exploiting the permeabilization of endosomes upon light irradiation induced by a clinically approved photosensitizer agent, Photofrin. Experiments in BALB/c nu/nu mice inoculated subcutaneously with AY27 urothelial cancer cells confirmed that after endosomal permeabilization CPT loaded micelles escaped (from endocytic vesicles) into the cytosol, promoting drug release from micelles in the irradiated tissue.

CPT can be encapsulated in hydrophobic polymers or amphiphilic copolymers for safe delivery. At this point, one interesting approach that allows exceptionally high drug loading is to polymerize the drug itself. In one representative example, a dimeric CPT derivative, CPT−SS−CPT, bearing a trigger responsive domain was used as the core-constructing unit of the nanoparticle (Cai et al., 2015). Here, CPT molecules are stably conjugated via carbonate linkages that are subject to triggered bond cleavage and subsequent drug release by a reducing reagent. Upon triggering, cleavage of the disulfide bond in CPT−SS−CPT would result in decomposition of the drug dimer, releasing CPT in its authentic form.

Unfortunately, CPT–SS–CPT has much lower water solubility (less than 10 ng mL^{-1}) than CPT (3 μg mL^{-1}), presumably due to stronger intermolecular interactions between molecules. However, when CPT–SS–CPT was encapsulated in methoxy poly(ethylene glycol)–poly(lactide) (mPEG–PLA, M_w 3500–5000) empty micelles, drug-loaded nanoparticles were formed, stable in phosphate buffer saline (PBS) and human serum for at least 3 days, with a particle diameter of about 180 nm. The anticancer effect of these CPT–SS–CPT/mPEG–PLA nanoparticles was then evaluated in vitro by MTT assay over HeLa human cervix carcinoma cell line. The IC_{50} of CPT–SS–CPT/mPEG–PLA was 114 nM, compared with 17 nM for free CPT, but the main advantages of this nanoconjugate are the lack of systemic toxicity, as CPT release only takes place through disulfide bond reduction by intracellular thiols, and the stabilization of the lactone ring.

A supramolecular strategy was reported to directly assemble the hydrophobic CPT molecule to a β-sheet-forming peptide sequence derived from the cysteine-containing Tau protein through the reducible disulfylbutyrate (buSS) linker, obtaining discrete, stable, well-defined nanostructures with high and quantitative drug loading (Cheetham et al., 2013). Drug content was precisely controlled using the two amine functionalities of the amino acid lysine to create branching points that allow the attachment of one, two, or four CPT molecules, corresponding to respective drug loadings of 23%, 31%, and 38%. Depending on the number of CPT molecules in the supramolecular design, the resulting nanostructures can be either nanofibers or nanotubes. Such nanostructures provide CPT protection from the external environment. Release profiles showed that glutathione degraded the designed buSS linker, releasing a mixture of intermediates and CPT. Under tumor-relevant conditions, these drug nanostructures released the active form of CPT and showed moderated in vitro efficacy against MCF-7 human breast cancer cells and 9L and F98L rat gliosarcoma cell lines. Moreover, to avoid the formation of intermediates, the same researchers prepared these CPT nanostructures using a (disulfanyl)ethyl carbonate linker, that after disulfide bridge reduction by glutathione followed a self-immolative process to release free CPT, improving cytotoxicity over the MCF-7 cell line (Cheetham et al., 2014).

Very recently, our group implemented a new synthetic strategy by direct coupling of as-synthesized (pyridin-2-yldisulfanyl)alkyl carbonate derivatives of CPT with thiol groups of silica hybrid nanoparticles containing a nonporous core and a mesoporous shell (Muniesa et al., 2013). Upon reaction with thiols in physiological conditions, disulfide bridge cleavage occurs, releasing the naked drug after an intramolecular cyclization mechanism (Fig. 12.3). Additional incorporation of a fluorophore into particles core facilitated imaging at the subcellular level for the monitoring of uptake and delivery. Confocal microscopy experiments in HeLa human cervix cancer cells confirmed that nanoparticles enter the cells by endocytosis but are able to escape from endolysosomes and enter the cytosolic compartment to release their cargo by glutathione reducing activity. Moreover, in a separate study (Botella et al., 2016), it was shown that side chain carbon number (n, Fig. 12.3) determines material hydrophobic properties and, as a consequence,

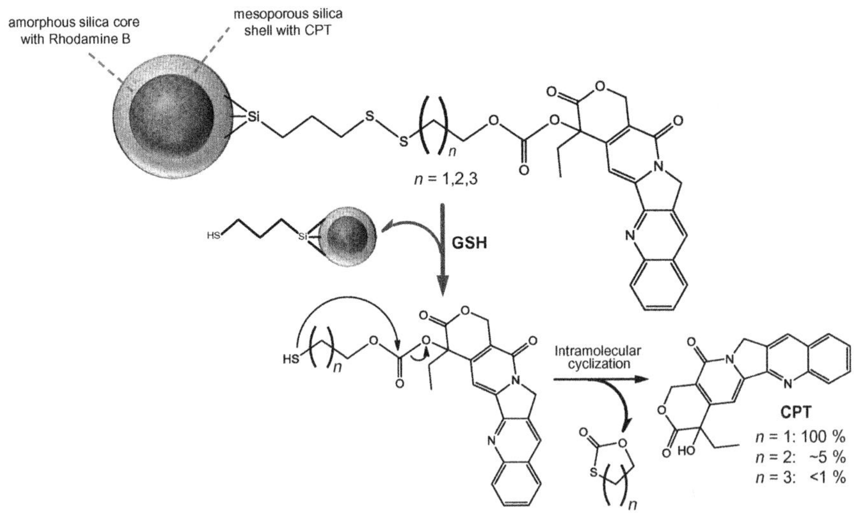

FIGURE 12.3

Schematic preparation of the redox-responsive nanoplatform by direct coupling of (pyridin-2-yldisulfanyl)alkyl carbonate derivative with thiol groups of silica hybrid nanoparticles. Upon reaction with thiols in physiological conditions, disulfide bridge cleavage occurs, releasing the naked drug after an intramolecular cyclization mechanism.

Adapted from Muniesa, C., Vicente, V., Quesada, M., Sáez-Atiénzar, S., Blesa, J.R., Abasolo, I., et al., 2013. Glutathione-sensitive nanoplatform for monitored intracelular delivery and controlled release of camptothecin. RSC Adv. 3 (35), 15121–15131. With permission of The Royal Society of Chemistry.

its stability in aqueous medium and cell uptake kinetics. When the *n* value increases, the negative surface charge decreases dramatically due to a shielding effect provoked by hydrophobic ligands, which promotes particle aggregation and favors cell internalization. Furthermore, the *n* value also determines the type of products released and, subsequently, the cytotoxic activity. Although full disulfide bridge reduction occurs in all cases within the cell, the subsequent intramolecular cyclization releasing CPT, CO_2, and the corresponding thiolactone only happened quantitatively for $n = 1$, whereas higher homologs released low CPT quantities (5% for $n = 2$ and traces for $n = 3$).

12.3.3 ENZYME-SENSITIVE SYSTEMS

In these drug delivery nanoplatforms, the therapeutic molecule is engaged to nanoparticles by a covalent bond that may be cleavaged by enzymatic activity. Enzyme-responsive nanomaterials facilitate the targeting of specific tissue by programming drug release via enzymatic digestion of the nanocarrier or linking moiety (Rica et al., 2012). The biological recognition of the substrate by the enzyme leads to the selective and specific triggering event to release the cargo, which minimizes toxicological issues related to CPT. In this context, due to the high concentration of carboxylesterases in the cytosol, most of these models have been developed over the basis of an ester bond at 20-OH position between CPT and the nanostructured surface. On the other hand, there are two main shortcomings in the use of enzyme-responsive devices for CPT delivery: (1) the presence of carboxylesterases in plasma may motivate some nonspecific release outside the tumor cells, and (2) the binding of CPT to a nanoparticle surface can limit the accessibility to enzyme active site due to steric effect, slowing down the release process, which reduces the cytotoxic activity (Botella et al., 2011).

In a singular approach, CPT molecules were directly grafted to carboxyl groups of poly(L-glutamic acid) (CPT−PGA) through ester bond (Chen et al., 2012). Subsequently, a supramolecular nanoparticle system was obtained by conjugation of CPT−PGA with three different polymeric derivatives: poly(ethylene glycol), poly(ethylenimine), and poly(amido)amine. This system released CPT under physiological conditions by means of esterase-mediated hydrolysis. Cell viability assays in MCF-7 human breast cancer cells indicated nanoparticle delivery system IC_{50} was four times that of the free drug, probably due to the slow release of CPT from polymer. However, a positron emission tomography study with ^{64}Cu-labeled nanoparticles in C57Bl/6 mice bearing Lewis lung carcinoma tumors revealed significant tumor specific uptake of nanoparticles. Moreover, antitumor studies with a 7% CPT containing PGA over sc H322 tumors in nude mice showed a 32-day tumor growth delay after four 40 mg kg^{-1} intravenous (IV) injections were administered every 4 days (Zou et al., 2001).

A nanomicelle system was developed by using α,β-poly[(*N*-carboxybutil)-L-aspartamide] (PBAsp) with different CPT loading (PBAsp−CPT), through direct esterification between PBAsp carboxyl groups and CPT 20-hydroxyl (Fan et al., 2010; Zhang et al., 2010). Subsequently, spherical nanomicelles were produced in aqueous medium. In this case, CPT was sequestered toward the core whereas PBAsp backbone formed the hydrophilic shell. Unfortunately, PBAsp−CPT solubility decreased quickly as the amount of CPT increased. Anyway, in vitro cytotoxicity assays of PBAsp−CPT (2.1% CPT) against L929 mouse muscular cell line showed less toxicity than free CPT (5%−10% activity drop depending on CPT concentration) due to its linear sustained drug release profile.

Conversely, CPT conjugation to nanoparticles by ester bond is usually performed in two steps. First, a prodrug is prepared by esterification with a short linker and then this prodrug is covalently linked to a supramolecular structure. For instance, CPT can be substituted at 20-OH position with glycine through ester bond and, afterwards, this prodrug can be incorporated to the polymeric backbone. In a representative example, CPT−glycine was reacted (amidation) with PGA (Singer et al., 2001), obtaining loadings of nearly 40%. In particular, by working with polymer molecular weight of up to 50 kDa and CPT loading of up to 37%, a tumor growth delay to 50% was reported after a 30 mg kg^{-1} dose in preliminary in vivo studies over athymic NCr (nu/nu) mice bearing NCI-H460 human lung large cell carcinoma tumors.

More recently, CPT has been conjugated with hyaluronic acid (HA–CPT) with variable molecular weight, to overcome the low solubility and poor targeting ability of the naked drug (Xu et al., 2013). Here, CPT was incorporated using two linkage molecules such as succinate and adipic acid dihydrazide. The authors studied the influence of the polymer molecular weight in the solubility, drug loading, and stability. Although solubility assays confirmed that these conjugates increase CPT solubility, the sampler with lower molecular weight showed higher solubility. Furthermore, in vitro efficacy assays demonstrated that the antitumor activity of HA–CPT polymers were higher than the naked drug against HepG2 human liver cancer cell line as the HA receptors are overexpressed.

Dendrimers have received much attention in drug delivery, as they are highly branched, multivalent, and tunable polymers for different architecture, size, shape, and surface properties (Duncan and Gaspar, 2011). They are usually classified by generation (denominated G), which refers to the number of repeated branching cycles that are performed during its synthesis. Due to their positive charge, cationic dendrimers as poly(amido)amine (PAMAM) and triazine derivatives conjugate easily with the phenolic 20-hydroxyl group of CPT, resulting in stable nanomedicines. In a significant study, CPT was incorporated into a generation 4 PAMAM dendrimer (PAMAM-G4) by amidation of CPT–glycine prodrug, to give PAMAM–CPT (Thiagarajan et al., 2010). This material was stable in plasma and the presence of proteolytic enzymes such as cytosolic esterases promoted conjugate cleavage and CPT release. Cell internalization and in vitro assays of this nanomedicine in HTC-116 human colorectal cancer cells indicated a high cytotoxic effect due to nuclear fragmentation and formation of apoptotic bodies.

A relevant PAMAM second generation (PAMAM-G2) model was synthesized with four molecules of CPT per unit and a trigger that allows activation with penicillin-G-amidase (PGA). In this system, the cleavage of the phenylacetamide group by PGA triggered the disassembly of dendron through a known self-immolative reaction sequence (Gopin et al., 2006). This conjugate demonstrated outstanding improvement of CPT cytotoxic activity, as cell-growth inhibition assays indicated that half maximal inhibitory concentration (IC_{50}) value for the dendrimer was between 100- and 1000-fold lower than free CPT against MOLT-3 lymphoblastic leukemia, JURKAT human leukemia T, and HEK-293 human kidney embryonic cell line.

Actually, click chemistry has been widely used to improve coupling reaction efficacy in the synthesis of CPT–dendrimer conjugates (Zolotarskaya et al., 2015). First, CPT was functionalized with a spacer 1-azido-3,6,9,12,15-pentaoxaoctadecan-18-oic acid (APO). On the other hand, propargylamine (PPA) and methoxypoly(ethylene glycol) amine were conjugated to PAMAM dendrimer G4.5 by using an effective coupling agent 4-(4,6-dimethoxy-(1,3,5)triazin-2-yl)-4-methyl-morpholinium chloride (DMTMM). Then, CPT-APO was coupled to the PEGylated PAMAM dendrimer G4.5-PPA via click chemistry. This new dendrimer–CPT was more cytotoxic than free CPT (IC_{50} value decrease of

185-fold) in U1242 human glioma cells. Furthermore, nearly 100% of CPT molecules were covalently conjugated to the dendrimer and the drug was released by ester bond hydrolysis.

One of the best performances to improve CPT stability and solubility in biological conditions has been to covalently conjugate it with a β-CD derivative condensed with difunctionalized PEG comonomers, obtaining linear high molecular weight (over 50 kDa) β-CD–CPT based polymers (Cheng et al., 2003, 2004; Schluep et al., 2009; Davis, 2009). For this purpose, a CPT–glycine ester derivative was prepared, which was subsequently linked to the β-CD derivative by amide bond (Fig. 12.4). This way, the solubility of CPT is increased by more than three orders of magnitude (Cheng et al., 2003). Moreover, this nanomedicine showed higher concentration in tumor and superior efficacy than irinotecan in the treatment of CD-1 female athymic nude mice bearing LS174T human colon carcinoma, HT-29 human colorectal carcinoma, H1299 human nonsmall cell lung cancer (NSCLC) model, or Panc-1 human pancreatic cancer cell tumors (Cheng et al., 2004; Schluep et al., 2009, 2006a, 2006b). As a consequence, a commercial product (CRLX-101, formerly IT-101) has been released, which is currently under clinical trials (Davis, 2009).

As a different model, our group recently prepared a hybrid organic–inorganic silica nanoplatform for CPT delivery based on enzyme-triggered release. For this sake, CPT–glycine ester was covalently linked to nanoparticles through an amide

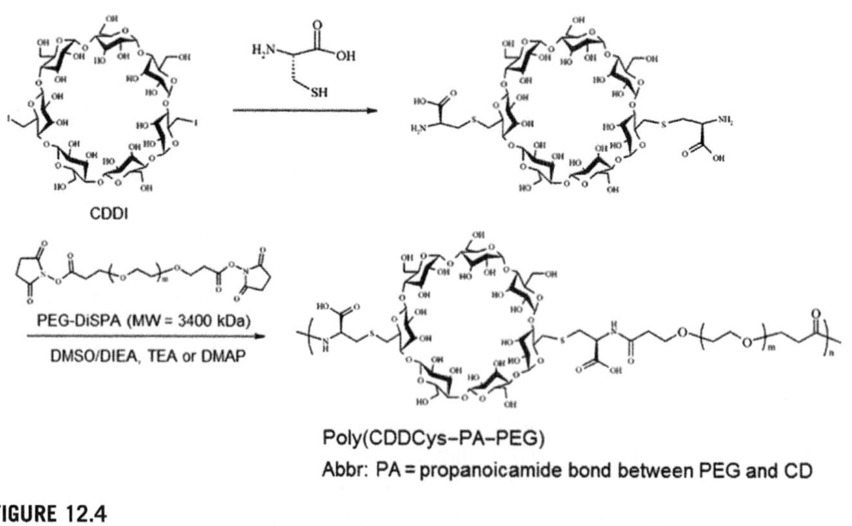

FIGURE 12.4

Schematic illustration of the synthesis steps involved in preparing $6^A,6^D$-Diiodo-$6^A,6^D$-dideoxy-cyclodextrin (CDDI) polymerized derivative.

Reprinted with permission from Cheng, J., Khin, K.T., Jensen, G.S., Liu, A., Davis, W.E., 2003. Synthesis of linear, β-cyclodextrin-based polymers and their camptothecin conjugates. Bioconjug. Chem. 14 (5), 1007–1017. Copyright © 2003 American Chemical Society.

bond with the 20-hydroxy moiety, to stabilize its lactone ring and to avoid unspecific release of the drug (Botella et al., 2011). The obtained material was highly stable in plasma, with low release of the cargo at physiological pH. Cell internalization and in vitro efficacy assays demonstrated that nanoparticles carrying CPT (SNP (silver nanoparticle)–CPT) entered cells via endocytosis and the intracellular release of the cargo induced cell death with half maximal inhibitory concentration (IC_{50}) values and cell cycle distribution profiles similar to those observed for the naked drug. Furthermore, in vivo biodistribution, therapeutic efficacy, and biocompatibility of the SNP–CPT were evaluated in hsd:athymic nude-Foxn1 mice bearing HT-29. Fluc subcutaneous tumors showed that although SNP–CPT tended to accumulate in organs of the reticulo-endothelial system, nanoparticles delayed the growth of subcutaneous tumors while significantly reducing the systemic toxicity associated to CPT administration.

12.3.4 LIGHT-SENSITIVE SYSTEMS

In addition to chemically triggered delivery systems, remote activation of nanoparticles by light for biomedical applications has become an important area of research (Tong and Kohane, 2012). This type of stimuli-responsive nanoplatform presents a component sensitive to ultraviolet (UV), visible (vis), and/or near infrared (NIR) light that is able to activate drug release under irradiation. So far, many different systems have been described, involving photocleavage, photodissociation, photoisomerization, photorelease, photothermal, photoplasmonic heating, and upconversion photoisomerization (Croissant et al., 2013).

In a pioneering work, nanoimpeller-controlled mesostructured nanoparticles were developed to deliver and release CPT into living cells upon light activation (Lu et al., 2008). These light-triggered MSNs were functionalized with azobenzene moieties in the pore interiors, with one end attached to the pore walls and the other end free to undergo photoisomerization. The impeller in the nanopores trapped drug molecules in the dark, and released them in response to UV light (413 nm) excitation. In vitro studies in PANC-1 human pancreatic and SW480 human colon cancer cell lines proved the photoresponsive behavior of the impeller inside of cells. After 3 hours of incubation in the dark, cells were irradiated with $\sim$0.1 W cm^{-2}, 413 nm light, for various excitation times (0–10 minutes). Cell death was induced under photocontrol. In the absence of light excitation, the CPT remained in the particles and cells were not damaged. However, illumination promptly expelled CPT from the particles, causing cancer cell apoptosis, which is demonstrated by nuclear fragmentation and chromatin condensation.

Unfortunately, the application of UV/vis light in biomedical studies is mostly limited to proof-of-concept, because irradiation at this wavelength can damage cells and does not penetrate deep inside the tissues. However, using two-photon excitation (TPE) in the NIR region instead of UV/vis light leads to better tissue penetration, lower scattering loss, and three-dimensional spatial resolution. For this sake, the former nanoimpeller-controlled MSNs device was functionalized with a high absorption cross-section two-photon fluorofore, suitable for Forster resonance energy transfer (FRET) to photoizomerize azobenzene moieties in the NIR region (Croissant et al., 2013). The two-photon absorption properties of the fluorophore were retained in the materials and no decrease of cross-sections $\sigma 2$ was noticed after encapsulation. The nanoimpeller groups pending in the porous framework allow the physical entrapment of CPT, which is then kicked out of the pores by two-photon-triggered photoisomerization. Here, the nanoimpellers configured for two-photon activation were loaded with CPT and then screened under TPE in MCF-7 human breast cancer cells. Cells were incubated with nanoparticles and further irradiated for a very short time with a focused laser beam (input 3 W, output 900 mW cm^{-2}). The MTT assay was performed 2 days after irradiation. Nanoimpellers loaded with CPT and high fluorofore/azobenzene ratio significantly increased cancer cell death under TPE compared with no irradiation.

The surface of MSNs may be modified to deposit a hydrophobic layer able to monitor drug delivery by adjusting the wetting of particle surface. In this context, a light-sensitive release system was developed, based on a hybrid with MSN core and spiropyran and perfluorodecyltriethoxysilane coating (Chen et al., 2014a,b). First, CPT was loaded inside MSN channels and, next, MSN surface was modified with the optimal molecular ratio of spiropyran to fluorinated silane (0.249:1), to prevent particles from being wetted by water, which successfully inhibited drug release. Upon irradiation with UV light (365 nm), the conformational conversion of spiropyran from a "closed" state to an "open" state caused the surface to be wetted, leading to the release of encapsulated CPT from the pores. In vitro experiments in EA.hy926 human somatic cell hybrid and HeLa human cervix carcinoma cells demonstrated an accurate drug controlled release. Here, CPT loaded nanoparticles (10 mg g^{-1}) were incubated in cells for 24 hours and then, upon UV light irradiation, a noticeable loss of cell viability was observed with regards to control of nonirradiated cell cultures or with no nanoparticles.

Another representative light-responsive prototype was developed as a self-deliverable form of nucleic acid–CPT nanostructure (DNA–CPT) that is composed almost entirely of payload molecules (Tan et al., 2015). To tackle this strategy, three CPT molecules were attached to a phenol group connected to a photolabile 2-nitrobenzyl ether moiety. This photolabile group was linked to a DNA strand to give an amphiphile, which self-assembled into a DNA–CPT nanostructure. Under UV irradiation (365 nm), the 2-nitrobenzyl was cleaved, releasing the DNA from the conjugate and leaving a decapped self-immolative drug core that, afterwards, rapidly underwent a spontaneous and irreversible degradation process by light treatment to release free CPT molecules. In vitro efficacy assays in SK-BR-3 human breast cancer cell line confirmed that cytotoxicity of DNA–CPT nanostructure increased significantly by light triggering, achieving IC_{50} values similar to those observed for the naked drug.

Fully inorganic core–shell hybrids can also be used for CPT delivery. In a most usual configuration, the nanocarrier is obtained by coating a metal nanoparticle with a silica outer layer, CPT being encapsulated inside the silica pores. In this sense, our group has recently carried out combined chemo- and photothermal therapy in in vitro testing by means of multifunctional nanoparticles formed by plasmonic GNCs with a protective shell of porous silica that contains CPT (GNC@SiO_2) (Botella et al., 2012, 2009). This therapeutic nanoplatform associated the optical activity of small GNCs (longitudinal surface plasmon resonance peak, LSPR, at 660 nm) with the cytotoxic activity of CPT simultaneously released for the efficient destruction of cancer cells. A method was used for the controlled assembly of 15 nm diameter gold colloids into stable clusters with a tailored absorption cross-section in the vis/NIR spectrum (Fig. 12.5). Clusters were further encapsulated in an ordered homogeneous mesoporous silica coating that incorporates CPT molecules within the pores. After internalization in

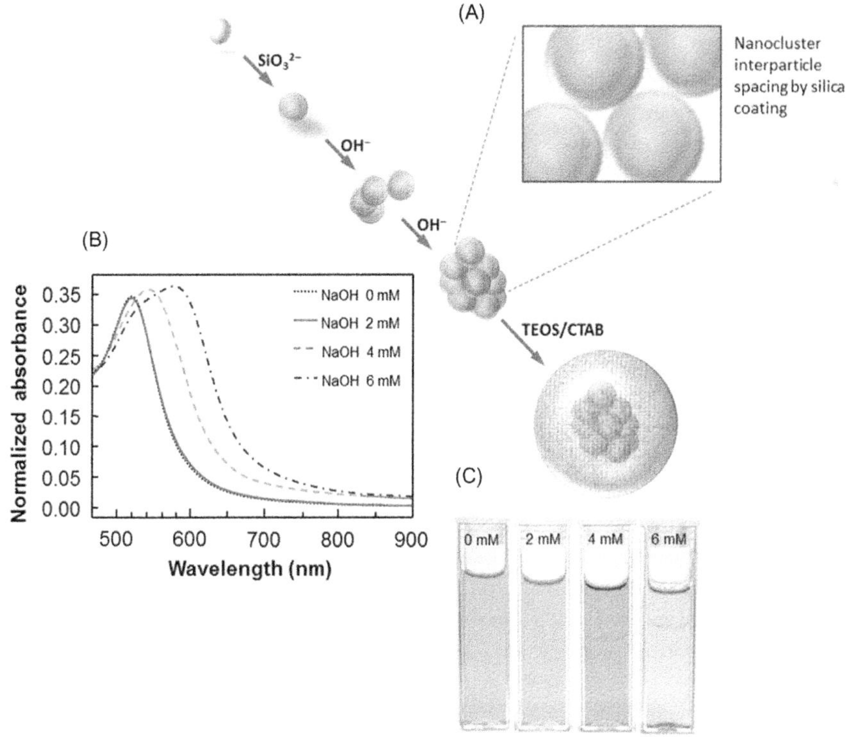

FIGURE 12.5

(A) Schematic of synthesis and controlled aggregation of gold colloids protected with a thin silica layer into gold nanoclusters (GNC@SiO_2). (B) UV–vis/NIR absorbance spectra of gold nanoclusters obtained at different NaOH concentration. (C) Photograph of the different colloidal gold dispersions.

Adapted from Botella, P., Ortega, I., Quesada, M., Madrigal, R.F., Muniesa, C., Fimia, A., et al., 2012. Multifunctional hybrid materials for combined photo and chemotherapy of cancer. Dalton Trans. 41 (31), 9286–9296. With permission of The Royal Society of Chemistry.

42-MG-BA human glioma cells, GNC@SiO_2 was able to produce effective photothermolysis under femtosecond pulse laser irradiation of 790 nm. Moreover, the simultaneous release of CPT during the process increased cell death up to 70%. This therapeutic model could be of interest for application in the treatment and suppression of nonsolid tumors like glioblastoma multiforme.

The integration of functional organic fragments into the framework of mesoporous silica to produce periodic mesoporous organosilica materials (PMOs) was reported in 1999 by three different groups (Asefa et al., 1999; Inagaki et al., 1999; Melde et al., 1999), by using bridged organoalkoxysilane precursors $(R'O)_3Si\text{-}\boldsymbol{R}\text{-}Si(OR')_3$ (***R***: organic group). PMOs feature molecularly organic–inorganic hybrid compositions where the organic ***R*** moiety homogeneously distributes within the whole framework without blocking pore channels. The recent development shift from bulk materials to nanosized PMOs has resulted in mesoporous organosilica nanoparticles (MONs) with biomedical applications (Chen and Shi, 2016). As a consequence of this hybridization strategy, the chemical nature of the silica framework can be significantly altered, offering a number of specific biological effects and functions suitable for biomedicine, as tunable biodegradation and improved biocompatibility. However, with regards to MSNs, the in vivo evaluation of the biological effects and the biosafety of MONs is still in its infancy, and complete studies of biodistribution, excretion, biodegradation, and toxicity are still pending (Chen et al., 2014a,b). In this context, in a singular study, monodisperse MONs codoped with fluorescence resonance energy transfer (FRET) cascades composed of fluorophores anthracene (Anth), naphthalimide (Naph), and biphenyl (Biph) at various ratios were prepared by a modified Stöber method, and CPT was loaded inside the pores (Lu et al., 2012). Next, release experiments in PBS demonstrated that the sample containing the three fluorophores (Biph/Anth/Naph) presented quicker release profiles than the samples with two fluorophores (Biph/Anth) or with only Biph. Actually, this CPT tunable retention ability is governed by the quantity of incorporated fluorophores and their polarities, which follows the order of Naph > Anth ~ Biph. Moreover, although the release rate can be more finely tuned in an asynchronous way (a single drug) by varying the ratio between fluorophores, it is also possible to monitor the synchronous discharge of several drugs. In this sense, it can be said that drug releasing performance of MONs can be well correlated with its color. Studies on cellular uptake efficiency were carried out on HeLa human cervix carcinoma cell line, showing that by varying the ratio between the different fluorophores, a colorful bioimaging system benefited from FRET may be obtained. Also, cell viability evaluated using MTT assay indicated similar growth inhibition potential for MONs than the free drug.

12.3.5 MAGNETIC-SENSITIVE SYSTEMS

Magnetic gradients may be used to target therapeutics to specific tissues, while alternating magnetic fields produce frequency-dependent effects at the nanoparticle level. This type of nanocarrier implicates paramagnetic or superparamagnetic materials that are embedded into organic polymeric moieties, allowing the release of the drug under an external magnetic stimulus. The inorganic component is designed to impart responsivity to external magnetic fields, both static and dynamic, to impose a remote magnetic control of drug cargo release. On the other hand, soft matter architectures provide a very convenient structural scaffold for insertion of therapeutic principles (Bonini et al., 2013).

One of the most illustrative drug delivery models of hybrid magnetic materials is based on MSN pore capping with iron oxide nanoparticles (Giri et al., 2005). This may be preferentially done through chemical bonding. For this sake, MSNs surface was functionalized with 3-aminopropyltrimethoxy silane groups. Afterwards, CPT was filled into the pores and these were covalently capped through amidation of the 3-aminopropyl groups at surface with meso-2,3-dimercaptosuccinic acid functionalized superparamagnetic iron oxide nanoparticles (Chen et al., 2011). The Fe_3O_4 nanocaps formed a dense and uniform layer tightly bound to the MSN surface (MSN@Fe_3O_4). Under magnetic stimulus, cumulative drug release was very quick, and prolonged even after the stimulus was stopped, due to detachment of some Fe_3O_4 nanoparticles (Fig. 12.6). Such a magnetically induced removal of the nanocaps implies a cleavage of the chemical bond between the Fe_3O_4 nanoparticles and silanized MSNs. The energy induced from the magnetic stimulus over various timespans is sufficiently large to cleave the chemical bond, resulting in an effective removal of the nanocaps. Actually, the rate of nanocap removal and drug diffusion are a function of the magnetic energy. Then, cytotoxicity studies in A549 human lung adenocarcinoma epithelial cell line by MTT assay indicated that nano-Fe_3O_4-capped MSN effectively constrained the CPT from free elution and were well tolerated, resulting in cell viability of over 80%. However, upon magnetic stimulus, about 42% of cells were killed, and the CPT released from the MSN@Fe_3O_4 was the major cause of cell death.

Magnetic stimulus can be also used to accelerate drug release from core–shell hybrid nanoparticles constructed by self-assembly of iron oxide in the presence of polyvinyl alcohol. Then, CPT was incorporated to the organic component of the core, followed by coating with a thin layer of silica to eliminate undesirable drug leakage (Tung et al., 2011). Here, the presence of an external magnetic stimulus allows for a pulse-type drug release to be readily achieved in a real-time responsive manner without undesirable delays in dosing accuracy. The intracellular release of CPT from nanocarriers in A549 cells was stimulated by high frequency magnetic field (HFMF, 2.5 kA m^{-1}, 50 kHz) for 30–120 seconds. Upon magnetic stimulus, CPT release profile for the nanocarriers in different eluting media indicated that the rate of drug diffusion was considerably increased, which is due to thermally induced disruption of the nanostructure of the particles (Hu et al., 2008). Subsequently, cells were incubated for 18 hours, and the anticancer effect was measured by MTT assay. It was found that CPT IC_{50} value measured on A549 human lung adenocarcinoma epithelial cell line was around five times lower compared with the free drug when CPT was loaded into magnetically sensitive nanocarriers and released intracellularly. Moreover, a similar model was performed by

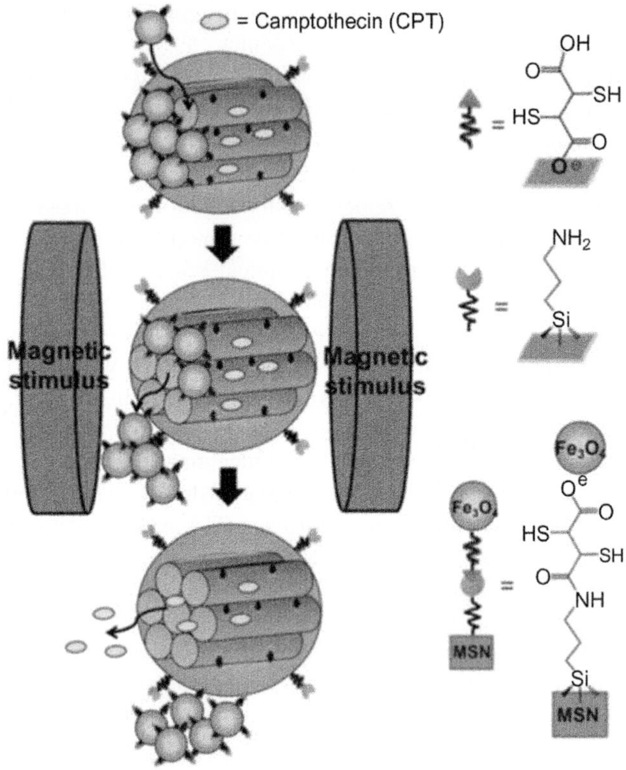

FIGURE 12.6

Schematic illustration of the synthesis and structure of mesoporous silica nanoparticles capped with iron oxide nanoparticles. On the right, a brief illustration of all different ligands used in the synthesis of this material.

Reprinted from Chen, P.J., Hu, S.H., Hsiao, C.S., Chen, Y.Y., Liu, D., Chen, S.Y., 2011. Multifunctional magnetically removable nanogated lids of Fe_3O_4-capped mesoporous silica nanoparticles for intracellular controlled release and MR imaging. J. Mater. Chem. 21 (8), 2535–2543. With permission of The Royal Society of Chemistry.

preparation of embedded Fe_3O_4 nanoparticles in a PLGA core containing CPT, which is protected by a lipid–polymer hybrid shell (Kong et al., 2013). In vitro testing of these drug delivery capsules in MT2 mouse breast cancer cells showed significantly suppressed cancer cell growth upon magnetic field activation.

12.3.6 THERMORESPONSIVE SYSTEMS

Thermoresponsive nanomaterials are a kind of adaptive nanoparticle sensitive to environmental temperature changes, which may help to monitor spatial delivery of a therapeutic payload. Under pathological local high temperature records (for instance, those reported in inflammation and cancer) these systems undergo a phase transition and conformational change that compromise the integrity of the nanostructure and activate payload release. At this point, polymers can be designed in a manner so that this phase transition, which is termed as the lower critical solution temperature (LCST), takes place at around 37°C. Under LCST, the hydrogel nanopreparation is soluble for in vivo injection, but becomes insoluble and aggregated in the tumor regions heated above its LCST, leading to local drug discharge (Jhaveri et al., 2014). In this context, CPT thermosensitive DDSs can be prepared based on spherical core–shell nanoscale micelles. First, two kinds of thermosensitive poly(*N*-isopropylacrylamide) (PNIPAM) four-armed star multiblock copolymers were synthesized by ATRP with hydroxyl-terminated polybutadiene as central blocks. Then, the multiblock copolymers could spontaneously assemble into regular spherical core–shell nanoscale micelles of about 100–120 nm diameter and LCST about 33°C–35°C (Luo et al., 2014). Drug release was thermoresponsive, accompanied by the temperature-induced structural changes of the micelles. At 37°C, CPT release from micelles is clearly quicker than at 25°C (respectively, 74% and 53% at 48 hours) due to temperature-induced structural alterations of multiblock copolymers. This accelerated drug release above the LCST is correlated with the enhanced hydrophobicity of PNIPAM shells, which leads to micelle shrinkage, and causes CPT to quickly diffuse out. In vitro cytotoxicity studies carried out by MTT assay over MDA-MB231 human breast cancer cell line demonstrated slightly lower activity for the CPT-loaded micelles than the free CPT, probably due to the time-consuming CPT release from the micelles.

12.3.7 ULTRASOUND-SENSITIVE SYSTEMS

Ultrasound represents another effective method for attaining spatiotemporal controlled release at the target cells, preventing side effects over healthy tissues. Ultrasound waves can trigger the release of the drug through the thermal and/or mechanical effects generated by cavitation phenomena or radiation forces. Here, physical forces associated with cavitation can induce nanocarrier destabilization, drug release, and transient increase in vessel permeability, leading to the cellular uptake of therapeutic molecules. Acoustically active perfluorocarbon nanoemulsions have been proposed for CPT encapsulation, to circumvent its well-known bioavailability issues as insolubility, instability, and toxicity (Fang et al., 2009). The nanoemulsions were firstly prepared with perfluoropentane, perfluorohexane, and coconut oil as the core of the inner phase, which incorporates CPT molecules. Then, a stabilizing coating made of phospholipids and/or Pluronic F68 was applied, obtaining a mean droplet diameter of 220–420 nm. For the study of ultrasound influence on CPT release, the nanoemulsion was exposed to ultrasound using a 1 MHz probe with a 1.5 W cm^{-2} intensity for 2 hours. However, CPT released from the nanoemulsion was limited, with less than 15% of the drug being discharged over 48 hours. Moreover, it was observed that droplet diameter influenced release rate, as the amount of exposed surface decreased in bigger droplets, which slows down drug liberation. Furthermore, to simulate parenteral administration conditions, plasma was added to the nanoemulsions. Next, sonication with ultrasound energy was used to stimulate the liquid-to-gas transition of perfluorocarbons (Unger et al., 2004). As the droplets became gas bubbles, they began to oscillate or resonate and, eventually, were destroyed resulting in local drug release. Unfortunately, no biological study is reported with this interesting CPT controlled release model.

12.3.8 GLUCOSE-SENSITIVE SYSTEMS

Recently, glucose-responsive materials have attracted research interest due to their potential application as DDSs. Most of these systems have been developed as glycemia control platforms in diabetes mellitus patients. Among them, the most studied model is the one based on phenylboronic acid (Ma and Shi, 2014). In aqueous medium, phenylboronic acid holds an equilibrium between an uncharged form and a charged form. The first one is hydrophobic while the second one is hydrophilic. This last one is able to form a stable phenylborate with cis-diol compounds, like glucose, which shifts the equilibrium in the direction of increasing the hydrophilic form. Usually, this shift is enough to increase the swelling degree of cross-linked nanogels or to disintegrate self-assembled micelles, causing the entrapped payload to be released into solution. This provides a way for designing polymeric carriers for self-regulated delivery not only of insulin but also of other therapeutic agents. For the purpose of CPT delivery, magnetic mesoporous silica was loaded with the drug and further capped with a phenylboronic–dextran conjugate (Sinha et al., 2014). First, hybrid nanoparticles with iron oxide cores and mesoporous silica shell were synthesized and functionalized at the surface with amine groups. Then, primary amine groups were reacted with carboxyphenylboronic acid and, subsequently, succinic anhydride was also conjugated over amine groups, providing nanoparticles with terminal carboxylate moieties. In a further step, CPT was encapsulated in silica mesopores followed by incubation with dextran (M_w 6000 and 100,000), which bonded with phenylborate linker via glucose monomer 1,2-diol groups. Consequently, dextran closed the surface of pores, preventing drug release. CPT release mechanism is based in the competition between soluble glucose and dextran for binding phenylboronic acid. Glucose has one 1,2-cis diol group that strongly binds to phenylborate linker, removing the larger size dextran from the pore surface. In vitro results over HeLa human cervix cancer cells showed a significant reduction of cell viability in glucose containing RPMI cell culture medium, but no effect in the absence of glucose, showing that drug loaded particle offers glucose concentration-dependent cell viability.

To end this section, we can conclude that stimuli-responsive systems are able to deliver CPT to tumor tissue and cells with minimum (or meaningless) premature release, imposing an efficient drug discharge inside cells under a specific stimulus. This is particularly important in the case of very cytotoxic molecules like CPT, as these DDSs increase antitumor activity and minimize side effects. It must be taken into account that for most DDSs, less than ~5% of the injected dose reaches tumor site (Svenson et al., 2011). In this context, the design of nanocarriers that are sensitive to exogenous or endogenous stimuli represents a safe alternative to regular chemotherapy with CPT structural analogs. The main concern with most of these systems, for translation from the bench to the bedside, is related to their chemical complexity, as most of them are hybrid materials with organic and inorganic components, and several-step synthesis protocols. This hampers not only the scaling-up of their synthesis, but also their degradability and biocompatibility. Moreover, the ability of these models to be sensitive to discrete variations of pH, redox potential, or enzyme activity is not always straightforward to achieve, whereas the scope of externally applied stimulus strongly depends on the penetration depth ability. Nevertheless, some of these systems are on the way towards achieving clinical application, as it will be shown below.

12.4 CLINICAL TESTING

One of the most disputable arguments against nanomedicines is that, despite the strong publication statistics in the last decade, very few drugs have made it to the commercial market (Venditto and Szoka, 2013). Although it is accepted that nanocarriers improve solubility, efficacy, and biodistribution, while decreasing the adverse side effects of very cytotoxic drugs as CPT, strict regulations for their use in humans have hampered, so far, a quicker translation of these therapeutic platforms into the clinical stage. At this point, it must be taken into account that nanocarriers are evaluated for each of their components within the formulation, and aspects like colloid stability in biological fluids, low immunogenicity, and long-term toxicity may not always be assessed. Moreover, the necessary refining in the delivery process performance is yet to come (Wilhelm et al., 2016). In this context, only a few organic polymers have been approved by the Food and Drug Administration (FDA) as CPT delivery systems for cancer therapy; below we briefly describe some of the most representative examples as well as their current position in the pipeline. All these cases have been compiled in Table 12.2. We accept that some of these DDSs might not form solid nanoparticles in real in vivo conditions (Svenson, 2014), but in all cases, the methods followed for material preparation, drug administration, biodistribution profile study, and therapeutic validation are fully comparable.

CRLX101 (formerly IT-101) is a pharmaceutical polymeric nanoparticle developed at Calando Pharmaceuticals, Inc. composed of a cyclodextrin–PEG copolymer covalently linked to CPT, and was designed to enhance drug solubility and biodistribution (Davis, 2009). The antitumor drug was released by enzymatic hydrolysis of the ester bond between the cyclodextrin and CPT hydroxyl group at

Table 12.2 Nanomedicines of CPT Under Clinical Investigation[a]

Compound	Formulation	Clinical Trial	Number of Patients	Pathology	Dose Regimen	MTD[b]	Toxicity	References
CRLX-101	Cyclodextrin–PEG copolymer	Phase Ia	10	Advanced Gastric, Gastroesophageal, or Esophageal Squamous or Adenocarcinoma	60 min IV infusions on days 1, 8, and 15 of a 28 day cycle	30 mg m^{-2} $month^{-1}$	Pancytopenia	Svenson et al. (2011)
		Phase Ib/IIa	62	Solid malignances	60 min IV infusion for 3 consecutive weeks every 28 days	15 mg m^{-2} IV every 2 weeks	Myelosuppression, grade 3/4 neutropenia and fatigue.	Weiss et al. (2013)
		Phase II	29	Ovarian cancer Fallopian tube cancer Primary peritoneal cancer	IV infusion once every 14 days. Each cycle is 28 days	15 mg m^{-2} IV every 2 weeks	G3 vasovagal reaction G3 pneumonia G3 pulmonary embolism	Davis (2009)
		Phase II	157	Nonsmall cell lung cancer	15 mg m^{-2} IV every other week	15 mg m^{-2} IV every 2 weeks	Fatigue/asthenia, dypsnea	Davis (2009)
CT-2106	Poly(L-glutamic acid) polymer conjugate	Phase I	24	Adult solid tumors	10 min IV infusion every 3 weeks	75 mg CPT-2106 m^{-2}	Neutropenia and thrombocytopenia, febrile neutropenia, grade 2 hematuria, stomatitis	Springett et al. (2004)
		Phase I	26	Advanced solid tumors (50% melanoma)	10 min IV infusion weekly × 3 every 4 weeks	25 mg CPT-2106 m^{-2} $week^{-1}$	Grade 3 fatigue and grade 3 and 4 thrombocytopenia, grade 2 hematuria	Homsi et al. (2007)

XMT-1001	Polyacetal poly(1-hydroxymethylethylene hydroxymethylformal)	Phase I	24	Small and nonsmall cell lung cancer	1–20.5 mg CPT eq m^{-2} IV infusion once every 3 weeks	113 mg CPT (eq) m^{-2}	Myelosuppresion, diarrhea	Sausville et al. (2007)
		Phase Ib	11	10 with nonsmall cell lung cancer and 1 with gastric cancer	IV infusion once every 3 weeks	initially 113 mg CPT (eq) m^{-2}	—	Yurkovetskiy and Fram (2009)
EZ-246	Pegylated CPT conjugate	Phase I	37	Solid malignances (15 with colorectal cancer)	1 h IV infusion once every 3 weeks	7000 mg m^{-2} (1.67% CPT weight load)	Myelosuppression, bladder toxicity at 4800 mg m^{-2}	Rowinsky et al. (2003)
		Phase II	35	Advanced and metastatic adenocarcinomas of the stomach gastroesophageal junction	1 h IV infusion once every 3 weeks	3240 mg m^{-2}	Neutropenia, thrombocytopenia, anemia	Scott et al. (2004)
MAG–CPT	*N*-(hydroxypropyl) methacrylamide polymer	Phase I	23	Metastatic, refractory solid tumors (6 with colorectal cancer)	30 min IV infusion once every week	100 mg CPT eq^{-1} m^{-2} every 4 weeks	Myelosuppression, grade 2 dysuria/ hematuria	Schoemaker et al. (2002)

[a]*Additional information may be found at www.clinicaltrials.gov.*
[b]MTD, *maximum tolerated dose.*

C20 position. With regards to irinotecan, this compound showed improved cytotoxicity against MDA-MB-231, Panc-1 and HT29 cell lines (Svenson et al., 2011), entering the Phase Ia trial for the treatment of solid tumors over 10 patients in 2006. Also, a combined Phase Ib/IIa trial over 62 patients with advanced solid malignancies demonstrated encouraging safety, pharmacokinetics, and efficacy (Weiss et al., 2013). This preparation was tolerated with no evident side effects. Consequently, a Phase II trial was started in 2008 for therapy of 29 ovarian cancer patients and, later on, in 2011, a Phase II trial was started over 157 patients with NSCLC, to assess CRLX101 activity and safety. In addition, comparative preclinical studies suggested that CRLX101 behavior in animal models could be the same than in humans (Eliasof et al., 2013). Actually, it must be taken into account that, although nanomedicine accumulation in solid tumors by EPR effect is quite rationalized for animal models of human disease, these models poorly represent human tumors. Nevertheless, in a very recent study over nine gastric tumor patients consisting of collecting tumor and nonneoplastic tissue biopsies 24–48 hours after CRLX101 administration, evidence was found of the intact deposition of nanoparticles in tumor tissue in five of these nine patients, whereas no CRLX101 trace was found in nonneoplastic tissue. Moreover, in all cases, sufficient CPT concentration reached the tumor to cause downregulation of topoisomerase I and carbonic anhydrase IX (Clark et al., 2016).

CT-2106 is a CPT–poly-L-glutamate conjugate developed to improve the solubility and prevent CPT lactone ring cleavage. In this supramolecular system, CPT binds the glutamate polymer through esterification at 20-OH, and is released under the activity of cytosolic carboxyl-esterases. In preclinical models, CT-2106 showed antitumor activity against multiple human tumor cell lines, such as lung and colon lines, and retained substantial anti-tumor activity in syngeneic and xenogeneic murine tumor models (Singer et al., 2000). CT-2106 Phase I trial was started by Cell Therapeutics, Inc. in 2003. A dose escalation study was carried out over 24 patients with different advanced solid tumors (Springett et al., 2004). The only serious related toxicity was grade 4 hypersensitivity in a patient with a history of hypersensitivity to irinotecan. Furthermore, Cell Therapeutics, Inc. extended this Phase I trial in 2006. For this sake, 26 patients of advanced solid tumors (50% melanoma), were intravenously weekly administered for three consecutive weeks during a 28-day cycle (Homsi et al., 2007). Dose limiting toxicities were thrombocytopenia and fatigue. The pharmacokinetic profile of conjugated CPT showed rapid urinary excretion, and similar terminal half-life than the free drug. Twenty-five patients were assessable for response, but no objective responses (complete or partial) were noted. Only three patients had a stable disease: one, breast cancer; another one, histiocytoma; and a third one, melanoma, whereas the rest exhibited progression of disease. Overall, this trial provided a strong indication that CT-2106 may have the ability to substantially improve standard CPT toxicity profile.

XMT-1001 is a water soluble macromolecular conjugate (M_w ~70 kDa) synthesized by tethering CPT by hydroxyl group at C20 position to a hydrophilic, biodegradable polyacetal polymer, poly(1-hydroxymethylethylene hydroxymethyl-formal), also called PHF or Fleximer (Yurkovetskiy and Fram, 2009). This compound was designed to improve efficacy and toxicity with regards to CPT, and showed higher antitumor efficacy compared with irinotecan against HT-29 human colon carcinoma xenografts (Walsh et al., 2012). CPT intracellular discharge involves a dual-step release mechanism controlled by low pH and esterases activity: first, XMT-1001 follows an intramolecular transacylation releasing two CPT prodrugs, camptothecin-20-O-(*N*-succinimidoglycinate) (CPT-SI) and camptothecin-20-*O*-(*N*-succinami-doyl-glycinate) (CPT-SA) and, then, these prodrugs hydrolyze to yield CPT. A Phase I study with XMT-1001 was launched in 2007 by Mersana Therapeutics, Inc. to determine the maximum tolerable dose (MTD), and assess the safety and pharmacokinetics of XMT-1001 and its release products. Then, a total of 24 patients with small cell lung cancer (SCLC) and NSCLC were administered XMT-1001 by intravenous infusion every 21 days (Sausville et al., 2007). No drug related serious adverse events were observed during the treatment, while several patients showed promising evidence of clinical activity, including tumor shrinkage and stable disease for several weeks. Also, preliminary pharmacokinetic findings confirm the formation of release products (CPT-SI, CPT-SA, and CPT) in plasma. Subsequently, a Phase Ib extension study of XMT-1001 was initiated in 2011 over 11 patients (10 with NSCLC and 1 with gastric cancer), to fully explore the broad antitumor potential of this formulation, but no further results have been reported yet.

Conversely, EZ-246 (also known as pegamotecan) is an ester conjugate of CPT with poly(ethylene glycol) (M_w ~40 kDa), developed by Enzon Pharmaceuticals. In preclinical in vivo studies, this compound showed higher activity than irinotecan and topotecan against human tumor xenografts (including colon, lung, breast, and pancreatic origin) (Greenwald et al., 2000). Then, in a Phase I trial of this formulation, 37 patients with different solid malignancies (15 with colorectal cancer) were administered by intravenous infusion for 1 hour every 3 weeks. Antineoplastic activity was observed in two patients, but significant toxicity was associated to the treatment, mostly neutropenia and genitourinary toxicity (Rowinsky et al., 2003). Actually, free CPT accumulated slowly in plasma, with a dose proportional pharmacokinetics, which forced to reduce the dose limiting toxicity in six patients. Subsequently, a Phase II trial was started at 2004 over 35 patients with advanced and metastatic adenocarcinomas of the stomach and gastroesophageal junction (Scott et al., 2004, 2009). Partial antitumor response was observed in five subjects, with a median time to progression of about 12 weeks, and median overall survival of 38 weeks. However, severe toxicity, including neutropenia, thrombocytopenia, fatigue, nausea, vomiting, and anorexia, appeared again in some patients, showing the need of further studies, maybe combined with other active agents.

Finally, MAG–CPT (also known as PNU 166148 and mureletecan) comprises CPT linked to a water-soluble polymeric backbone methacryloylglycynamide (M_w ~18 kDa), developed by Pharmacia. CPT derivatives were synthetized by conjugation of a *N*-(2-hydroxypropyl)methacrylamide copolymer precursor with modified CPT at the C20 α-hydroxy with glycine (Duncan, 2009). These conjugates were designed primarily to improve the clinical delivery of CPT, and utilized a hydrolytically labile spacer able to liberate drug intratumorally in a pH and enzyme-dependent manner. Preclinical studies of MAG–CPT showed that this compound was active in colon, stomach, pancreas, ovary, breast, lung, and melanoma xenograft models (Conover et al., 1998). Then, a Phase I study over 23 patients with metastatic, refractory solid tumors (6 with colorectal cancer) was carried out in 1999 (Schoemaker et al., 2002). Unfortunately, no objective antitumor responses were seen, and severe drug-related toxicity effects were reported, such as myelosuppression, neutropenic sepsis, and diarrhea. Also, one patient died after the first treatment cycle at the maximum dose. Based on the results of the Phase I studies and the intratumoral kinetic study, Pharmacia made a strategic decision to discontinue further clinical development of MAG–CPT (Bissett et al., 2004).

12.5 CONCLUSION AND OUTLOOK

CPT is one of the most active molecules for cancer treatment, but its particular limitations such as poor solubility, lack of stability of the lactone ring, and strong toxicity have encouraged the development of different delivery forms to achieve a safety approach. Here, it is a fact that current cancer treatment cannot be limited to the administration of a single therapeutic molecule but, instead, the new target is personalized medicine, which selects the appropriate drug combination and dosing schedules for individual patients.

In this context, despite that irinotecan and topotecan present improved pharmacokinetic parameters, and they are the only commercial CPT derivatives approved by the FDA, therapies merely relying on single, small molecules seem now out of step, mostly due to their limited efficacy and unacceptable toxicities. For this reason, huge emphasis is being placed on developing new DDSs that are able to simultaneously deliver CPT and other active agents, as well as incorporating imaging moieties that may ease the diagnosis process. Most specifically, the future seems good for those systems that are able to selectively discharge the drug at the target cells under a specific stimulus, with no previous release. Stimuli-responsive preparations provide a dramatically better spatial and temporal control over the release of drugs, which is crucial for delivery of the very cytotoxic CPT.

Although no CPT nanomedicine has got to a commercial form, there are several organic polymer compositions currently undergoing different phases of clinical trials. In all cases, there is a significant increase of the MTD with regards to CPT structural derivatives. Among them, cyclodextrin conjugates as CRLX101 look the best positioned to achieve a regular clinical use, according to the promising efficacy and tolerability results, as well as plasma half-life values obtained in Phase II trials. However, most of these systems show limited functionalization ability to covalently bind the drug, which limits the loading capacity, as well as the possibility to combine CPT with other therapeutic molecules and/or imaging agents. In this sense, the next generation of CPT delivery vehicles should merge properties of organic and inorganic materials. Whilst organic components will provide good stability in physiological fluids, low toxicity, and immunogenicity, and minimized reactions with serum proteins, inorganic materials (which alone could present strong limitations for biological use, due to possible compositional toxicity and severe interaction with serum proteins), may offer almost unlimited external surface area to bind different active molecules and stimuli-sensitive linkers and devices. Obviously, from the regulatory point of view, it is much more challenging to develop hybrid nanocarriers, which requires the evaluation of every single component and, possibly, their clinical translation will not be so smooth as organic polymers, but the potential of these systems aims to solve the strong interpatient variability that currently hampers the widespread use of CPT nanopreparations. In this chapter, we have tried to reflect some of the most interesting candidates to achieve clinical use soon.

ACKNOWLEDGMENTS

Financial support of the Spanish Ministry of Economy and Competitiveness (projects TEC2016-80976-R and SEV-2016-0683) is gratefully acknowledged. Prof. Eduardo Fernández and Dr. Ibane Abasolo are acknowledged for the useful discussion.

REFERENCES

Alaghaz, A.N.M.A., El-Sayad, B.A., Albohy, S.A.H., 2012. Synthesis, spectroscopic, and antimicrobial activity studies of novel 10-substituted camptothecin phosphorothioate analogs. Phosphorus Sulfur Silicon Relat. Elem. 187 (7), 799–807.

Asefa, T., MacLachlan, M.J., Coombs, N., Ozin, G.A., 1999. Periodic mesoporous organosilicas with organic groups inside the channel walls. Nature 402, 867–871.

Balendiran, G.K., Dabur, R., Fraser, D., 2004. The role of glutathione in cancer. Cell Biochem. Funct. 22 (6), 343–352.

Basili, S., Moro, S., 2009. Novel camptothecin derivatives as topoisomerase I inhibitors. Expert Opin. Ther. Pat. 19 (5), 555–574.

Beidler, D.R., Chang, J.Y., Zhou, B.S., Cheng, Y.C., 1996. Camptothecin resistance involving steps subsequent to the formation of protein-linked DNA breaks in human camptothecin-resistant KB cell lines. Cancer Res. 56 (2), 345–353.

Bissett, D., Cassidy, J., Bono, J.S., Muirhead, F., Main, M., Robson, L., et al., 2004. Phase I and pharmacokinetic (PK) study of MAG–CPT (PNU 166148): a polymeric derivative of camptothecin (CPT). Br. J. Cancer 91 (1), 50–55.

Bodley, A.L., Shapiro, T.A., 1995. Molecular and cytotoxic effects of camptothecin, a topoisomerase I inhibitor, on trypanosomes and Leishmania. Proc. Nat. Acad. Sci. USA 92 (9), 3726–3730.

Bonini, M., Berti, D., Baglioni, P., 2013. Nanostructures for magnetically triggered release of drugs and biomolecules. Curr. Opin. Colloid Interface Sci. 18 (5), 459–467.

Botella, P., Rivero-Buceta, E., 2017. Safe approaches for camptothecin delivery: structural analogues and nanomedicines. J. Control. Release 247, 28–54.

Botella, P., Corma, A., Navarro, M.T., Quesada, M., 2009. Design of optically active nanoclusters of gold particles with mesostructured silica coating. J. Mater. Chem. 19 (20), 3168–3175.

Botella, P., Abasolo, I., Fernández, Y., Muniesa, C., Miranda, S., Quesada, M., et al., 2011. Surface-modified silica nanoparticles for tumor-targeted delivery of camptothecin and its biological evaluation. J. Control. Release 156 (2), 246–257.

Botella, P., Ortega, I., Quesada, M., Madrigal, R.F., Muniesa, C., Fimia, A., et al., 2012. Multifunctional hybrid materials for combined photo and chemotherapy of cancer. Dalton Trans. 41 (31), 9286–9296.

Botella, P., Muniesa, C., Vicente, V., Cabrera-García, A., 2016. Effect of drug precursor in cell uptake and cytotoxicity of redox-responsive camptothecin nanomedicines. Mater. Sci. Eng. C 58, 692–699.

Boven, E., Van Hattum, A.H., Hoogsteen, I., Schlüper, H.M.M., Pinedo, H.M., 2000. New analogues of camptothecins: activity and resistance. Ann. N.Y. Acad. Sci. 922, 175–177.

Cai, K., He, X., Song, Z., Yin, Q., Zhang, Y., Uckun, F.M., et al., 2015. Dimeric drug polymeric nanoparticles with exceptionally high drug loading and quantitative loading efficiency. J. Am. Chem. Soc. 137 (10), 3458–3461.

Carrigan, S.W., Fox, P.C., Wall, M.E., Wani, M.C., Bowen, J.P., 1997. Comparative molecular field analysis and molecular modeling studies of 20-(S)-camptothecin analogs as inhibitors of DNA topoisomerase I and anticancer/antitumor agents. J. Comput. Aided Mol. Des. 11 (1), 71–78.

Cheetham, A.G., Zhang, P., Lin, Y.A., Lock, L.L., Cui, H., 2013. Supramolecular nanostructures formed by anticancer drug assembly. J. Am. Chem. Soc. 135, 2907–2910.

Cheetham, A.G., Ou, Y.C., Zhang, P., Cui, H., 2014. Linker-determined drug release mechanism of free camptothecin from self-assembling drug amphiphiles. Chem. Commun. 50 (45), 6039–6042.

Chen, A.Y., Liu, L.F., 1994. DNA topoisomerases: essential enzymes and lethal targets. Annu. Rev. Pharmacol. Toxicol. 34, 191–218.

Chen, Y., Shi, J., 2016. Chemistry of mesoporous organosilica in nanotechnology: molecularly organic–inorganic hybridization into frameworks. Adv. Mater. 28 (17), 3235–3272.

Chen, P.J., Hu, S.H., Hsiao, C.S., Chen, Y.Y., Liu, D., Chen, S.Y., 2011. Multifunctional magnetically removable nanogated lids of Fe_3O_4-capped mesoporous silica nanoparticles for intracellular controlled release and MR imaging. J. Mater. Chem. 21 (8), 2535–2543.

Chen, T., Xu, S., Zhao, T., Zhu, L., Wei, D., Li, Y., et al., 2012. Gold nanocluster-conjugated amphiphilic block copolymer for tumor-targeted drug delivery. ACS Appl. Mater. Interfaces 4 (11), 5766–5774.

Chen, L., Wang, W., Su, B., Wen, Y., Li, C., Zhou, Y., et al., 2014a. A light-responsive release platform by controlling the wetting behavior of hydrophobic surface. ACS Nano 8 (1), 744–751.

Chen, Y., Meng, Q., Wu, M., Wang, S., Xu, P., Chen, H., et al., 2014b. Hollow mesoporous organosilica nanoparticles: a generic intelligent framework-hybridization approach for biomedicine. J. Am. Chem. Soc. 136 (46), 16326–16334.

Cheng, J., Khin, K.T., Jensen, G.S., Liu, A., Davis, W.E., 2003. Synthesis of linear, β-cyclodextrin-based polymers and their camptothecin conjugates. Bioconjug. Chem. 14 (5), 1007–1017.

Cheng, J., Khin, K.T., Davis, M.E., 2004. Antitumor activity of β-cyclodextrin polymer–camptothecin conjugates. Mol. Pharm. 1 (3), 183–193.

Chiao, C., Li, H., 1975. Effect of topical use of camptothecin–dimethyl sulfoxide solution in psoriasis. Chin. Med. J. (Engl.) 1, 355–360.

Clark, A.J., Wiley, D.T., Zuckerman, J.E., Webster, P., Chao, J., Lin, J., et al., 2016. CRLX101 nanoparticles localize in human tumors and not in adjacent, nonneoplastic tissue after intravenous dosing. Proc. Nat. Acad. Sci. USA 113 (14), 3850–3854.

Clements, M.K., Wasi, S., Daoud, S.S., 1996. Camptothecin exhibits selective cytotoxicity towards human breast carcinoma as compared with normal bovine endothelial cells in vitro. Anticancer Drugs 7 (8), 851–857.

Conover, C.D., Greenwald, R.B., Pendri, A., Gilbert, C.W., Shum, K.L., 1998. Camptothecin delivery systems: enhanced efficacy and tumor accumulation of camptothecin following its conjugation to polyethylene glycol via a glycine linker. Cancer Chemother. Pharmcol. 42 (5), 407–414.

Croissant, J., Maynadier, M., Gallud, A., N'Dongo, H.P., Nyalosaso, J.L., Derrien, G., et al., 2013. Two-photon-triggered drug delivery in cancer cells using nanoimpellers. Angew. Chem. Int. Ed. 52 (51), 13813–13817.

Davis, M.E., 2009. Design and development of IT-101, a cyclodextrin-containing polymer conjugate of camptothecin. Adv. Drug Deliv. Rev. 61 (13), 1189–1192.

Del Bino, G., Skierski, J.S., Darzynkiewicz, Z., 1990. Diverse effects of camptothecin, an inhibitor of topoisomerase I, on the cell cycle of lymphocytic (L1210, MOLT-4) and myelogenous (HL-60, KG1) leukemic cells. Cancer Res. 50 (18), 5746–5750.

Del Poeta, M., Chen, S.F., Von Hoff, D., Dykstra, C.C., Wani, M.C., Manikumar, G., et al., 1999. Comparison of in vitro activities of camptothecin and nitidine derivatives against fungal and cancer cells. Antimicrob. Agents Chemother. 43 (12), 2862–2868.

Doshi, N., Mitragotri, S., 2009. Designer biomaterials for nanomedicine. Adv. Funct. Mater. 19 (24), 3843–3854.

Duncan, R., 1997. Polymer therapeutics for tumor specific delivery. Chem. Ind. London 7, 262–264.

Duncan, R., 2009. Development of HPMA copolymer–anticancer conjugates: clinical experience and lessons learnt. Adv. Drug Deliv. Rev. 61 (13), 1131–1148.

Duncan, R., Gaspar, R., 2011. Nanomedicine(s) under the microscope. Mol. Pharm. 8 (6), 2101–2141.

Eliasof, S., Lazarus, D., Peters, C.G., Case, R.I., Cole, R.O., Hwang, J., et al., 2013. Correlating preclinical animal studies and human clinical trials of a multifunctional, polymeric nanoparticles. Proc. Nat. Acad. Sci. USA 110 (13), 15127–15132.

Fahmi, A., Pietsch, T., Mendoza, C., Cheval, N., 2009. Functional hybrid materials. Mater. Today 12 (5), 44–50.

Fan, N., Duan, K., Wang, C., Liu, S., Luo, S., Yu, J., et al., 2010. Fabrication of nanomicelle with enhanced solubility and stability of camptothecin based on α,β-poly[(N-carboxybutyl)-L-aspartamide]-camptothecin conjugate. Colloids Surf. B, Biointerfaces 75 (2), 543–549.

Fang, J.Y., Hung, C.F., Hua, S.C., Hwang, T.L., 2009. Acoustically active perfluorocarbon nanoemulsions as drug delivery carriers for camptothecin: drug release and cytotoxicity against cancer cells. Ultrasonics 49 (1), 39–46.

Fleige, E., Quiadir, M.A., Haag, R., 2012. Stimuli-responsive polymeric nanocarriers for the controlled transport of active compounds: concepts and applications. Adv. Drug Deliv. Rev. 64 (9), 866–884.

Gabizon, A., Bradbury, M., Prabhakar, U., Zamboni, W., Libutti, S., Grodzinski, P., 2014. Cancer nanomedicines: closing the translational gap. Lancet 384 (9961), 2175–2176.

Ganta, S., Devalapally, H., Shahiwala, A., Amiji, M., 2008. A review of stimuli-responsive nanocarriers for drug and gene delivery. J. Control. Release 126 (3), 187–204.

Garcia-Carbonero, R., Supko, J.G., 2002. Current perspectives on the clinical experience, pharmacology, and continued development of the camptothecins. Clin. Cancer Res. 8 (3), 641–661.

Giner-Casares, J.J., Henriksen-Lacey, M., Coronado-Puchau, M., Liz-Marzán, L.M., 2016. Inorganic nanoparticles for biomedicine: where materials scientists meet medical research. Mater. Today 19 (1), 19–28.

Giovanella, B.C., Hinz, H.R., Kozielski, A.J., Stehlin Jr., J.S., Silber, R., Potmesil, M., 1991. Complete growth inhibition of human cancer xenografts in nude mice by treatment with 20-(S)-camptothecin. Cancer Res. 51 (11), 3052–3055.

Giri, S., Trewyn, B.G., Stellmaker, M.P., Lin, V.S., 2005. Stimuli-responsive controlled-release delivery system based on mesoporous silica nanorods capped with magnetic nanoparticles. Angew. Chem. Int. Ed. 44 (32), 5038–5044.

Gopin, A., Ebner, S., Attali, B., Shabat, D., 2006. Enzymatic activation of second-generation dendritic produgs: conjugation of self-immolative dendrimers with poly(ethylene glycol) via click chemistry. Bioconjug. Chem. 17 (6), 1432–1440.

Greenwald, R.B., Conover, C.D., Choe, Y.H., 2000. Poly(ethylene glycol) conjugated drugs and prodrugs: a comprehensive review. Crit. Rev. Ther. Drug Carrier Syst. 17 (2), 101–161.

Han, H., Davis, M.E., 2013. Targeted nanoparticles assembled via complexation of boronic-acid-containing targeting moieties to diol-containing polymers. Bioconjug. Chem. 24 (4), 669–677.

Hatefi, A., Amsden, B., 2002. Camptothecin delivery methods. Pharm. Res. 19 (10), 1389–1399.

Homsi, J., Simon, G.R., Garrett, C.R., Springett, G., De Conti, R., Chiappori, A., et al., 2007. Phase I trial of poly-L-glutamate camptothecin (CT-2106) administered weekly in patients with advanced solid malignancies. Clin. Cancer Res. 13 (19), 5855–5861.

Horwitz, S.B., Chang, C.K., Grollman, A.P., 1972. Antiviral action of camptothecin. Antimicrob. Agents Chemother. 2 (5), 395–401.

Hsiang, Y.H., Hertzberg, R., Hecht, S., Liu, L.F., 1985. Camptothecin induces protein-linked DNA breaks via mammalian DNA topoisomerase I. J. Biol. Chem. 260 (25), 14873–14878.

Hu, S.H., Liu, T.Y., Huang, H.Y., Liu, D.M., Chen, S.Y., 2008. Magnetic-sensitive silica nanospheres for controlled drug release. Langmuir 24 (1), 239–244.

Huang, H.C., Barua, S., Sharma, G., Dey, S.K., Rege, K., 2011. Inorganic nanoparticles for cancer imaging and therapy. J. Control. Release 155 (3), 344–357.

Inagaki, S., Guan, S., Fukushima, Y., Ohsuna, T., Terasaki, O., 1999. Novel mesoporous materials with a uniform distribution of organic groups and inorganic oxide in their frameworks. J. Am. Chem. Soc. 121 (41), 9611–9614.

Jhaveri, A., Deshpande, P., Torchilin, V., 2014. Stimuli-sensitive nanopreparations for combination cancer therapy. J. Control. Release 190, 352–370.

Kanamala, M., Wilson, W.R., Yang, M., Palmer, B.D., Wu, Z., 2016. Mechanisms and biomaterials in pH-responsive tumour targeted drug delivery: a review. Biomaterials 85, 152–167.

Kickelbick, G., 2007. In: Kickelbick, G. (Ed.), Hybrid Materials: Synthesis, Characterization, and Applications. Wiley-VCH.

Kong, S.D., Sartor, M., Hu, C.M.J., Zhang, W., Zhang, L., Jin, S., 2013. Magnetic field activated lipid–polymer hybrid nanoparticles for stimuli-responsive drug release. Acta Biomater. 9 (3), 5447–5452.

Kunimoto, T., Nitta, K., Tanaka, T., Uehara, N., Baba, H., Takeuchi, M., et al., 1987. Antitumor activity of 7-ethyl-10-[4-(1-piperidino)-1-piperidino]carbonyloxycamptothecin, a novel water soluble derivative of camptothecin, against murine tumors. Cancer Res. 47 (22), 5944–5948.

Li, L.H., Fraser, T.J., Olin, E.J., Bhuyan, B.K., 1972. Action of camptothecin on mammalian cells in culture. Cancer Res. 32 (12), 2643–2650.

Li, Y.Y., Chen, S.W., Yang, L.M., Wang, R.R., Pang, W., Zheng, Y.T., 2009. The anti-HIV actions of 7- and 10-substituted camptothecins. Molecules 15 (1), 138–148.

Li, Z., Li, H., Liu, L., You, X., Zhang, C., Wang, Y., 2015. A pH-sensitive nanocarrier for co-delivery of doxorubicin and camptothecin to enhance chemotherapeutic efficacy and overcome multidrug resistance in vitro. RSC Adv. 5, 77097–77105.

Li, Z., Ye, E., David, Lakshminarayanan, R., Loh, X.J., 2016. Recent advances of using hybrid nanocarriers in remotely controlled therapeutic delivery. Small 12 (35), 4782–4806.

Lim, J., Simanek, E.E., 2012. Triazine dendrimers as drug delivery systems: from synthesis to therapy. Adv. Drug Deliv. Rev. 64 (9), 826–835.

Lin, X.R., Huang, T., 1988. Topical camptothecin in treatment of psoriasis. Int. J. Dermatol. 27 (7), 475–477.

Liu, Y.Q., Dai, W., Tian, J., Yang, L., Feng, G., Zhou, X.W., et al., 2011. Synthesis and insecticidal activities of novel spin-labeled derivatives of camptothecin. Heteroatom Chem. 22 (6), 687–691.

Liu, Y.Q., Li, W.Q., Morris-Natschke, S.L., Qian, K., Yang, L., Zhu, G.X., et al., 2015. Perspectives on biologically active camptothecin derivatives. Med. Res. Rev. 35 (4), 753–789.

López-Dávila, V., Seifalian, A.M., Loizidou, M., 2012. Organic nanocarriers for cancer drug delivery. Curr. Opin. Pharmacol. 12 (4), 414–419.

Lu, J., Choi, E., Tamanoi, F., Zink, J.I., 2008. Light-activated nanoimpeller-controlled drug release in cancer cells. Small 4 (4), 421–426.

Lu, D., Juying, L., Wang, L., Zhang, J., 2012. Multifluorescent traceable nanoparticle by a single-wavelength excitation with color-related drug release performance. J. Am. Chem. Soc. 134 (21), 8746–8749.

Luo, Y.L., Yang, X.L., Xu, F., Chen, Y.S., Zhang, B., 2014. Thermosensitive PNIPAM–b-HTPB block copolymer micelles: molecular architectures and camptothecin drug release. Colloids Surf. B: Biointerfaces 114, 150–157.

Ma, R., Shi, L., 2014. Phenylboronic acid-based glucose-responsive polymeric nanoparticles: synthesis and applications in drug delivery. Polym. Chem. 5 (5), 1503–1518.

Mahapatro, A., Singh, D.K., 2011. Biodegradable nanoparticles are excellent vehicle for site directed in-vivo delivery of drugs and vaccines. J. Nanobiotechnology 9, 55.

Matsumura, Y., 2011. Preclinical and clinical studies of NK012, an SN-38-incorporating polymeric micelles, which is designed based on EPR effect. Adv. Drug Deliv. Rev. 63 (3), 184–192.

Melde, B.J., Holland, B.T., Blanford, C.F., Stein, A., 1999. Mesoporous sieves with unified hybrid inorganic/organic frameworks. Chem. Mater. 11 (11), 3302–3308.

Mi, Z., Burke, T.G., 1994. Differential interactions of camptothecin lactone and carboxylate forms with human blood components. Biochemistry 33 (34), 10325–10336.

Min, K.H., Kim, J.H., Bae, S.M., Shin, H., Kim, M.S., Park, S., et al., 2010. Tumoral acidic pH-responsive MPEG–poly(β-amino ester) polymeric micelles for cancer targeting therapy. J. Control. Release 144 (2), 259–266.

Muniesa, C., Vicente, V., Quesada, M., Sáez-Atiénzar, S., Blesa, J.R., Abasolo, I., et al., 2013. Glutathione-sensitive nanoplatform for monitored intracelular delivery and controlled release of camptothecin. RSC Adv. 3 (35), 15121–15131.

Muqeet, W., Qudsia, B., 2014. Camptothecin and its analogs antitumor activity by poisoning topoisomerase I, their structure activity relationship and clinical development perspective of analogs. J. Appl. Pharm. 6 (3), 286–295.

Mura, S., Nicolas, J., Couvreur, P., 2013. Stimuli-responsive nanocarriers for drug delivery. Nat. Mater. 12, 991–1003.

Nam, J., Won, N., Bang, J., Jin, H., Park, J., Jung, S., et al., 2013. Surface engineering of inorganic nanoparticles for imaging and therapy. Adv. Drug Deliv. Rev. 65 (5), 622–648.

Oberlies, N.H., Kroll, D.J., 2004. Camptothecin and taxol: historic achievements in natural products research. J. Nat. Prod. 67 (2), 129–135.

Ranjit, K., Baquee, A.A., 2013. Nanoparticle: an overview of preparation, characterization and application. Int. Res. J. Pharm. 4 (4), 47–57.

Re, F., Moresco, R., Masserini, M., 2012. Nanoparticles for neuroimaging. J. Phys. D: Appl. Phys. 45 (7), 073001.

Rica, R., Aili, D., Stevens, M.M., 2012. Enzyme-responsive nanoparticles for drug release and diagnostics. Adv. Drug Deliv. Rev. 64 (11), 967–978.

Rowinsky, E.K., Rizzo, J., Ochoa, L., Takimoto, C.H., Forouzesh, B., Schwartz, G., et al., 2003. A phase I and pharmacokinetic study of pegylated camptothecin as a 1-hour infusion every 3 weeks in patients with advanced solid malignancies. J. Clin. Oncol. 21 (1), 148–157.

Sausville, E., Anthony, E., Garbo, L., Shkolny, D., Yurkovetskiy, A., Bethune, C., et al., 2007. A phase I study of the safety, tolerability, and pharmacokinetics of intravenous XMT-1001 in patients with advanced solid tumors. Mol. Cancer Ther. 6 (11), A146.

Schluep, T., Hwang, J., Cheng, J., Heidel, J.D., Bartlett, D.W., Hollister, B., et al., 2006a. Preclinical efficacy of the camptothecin–polymer conjugate IT-101 in multiple cancer models. Clin. Cancer Res. 12 (5), 1606–1614.

Schluep, T., Cheng, J., Khin, K.T., Davis, M.E., 2006b. Pharmacokinetics and biodistribution of the camptothecin–polymer conjugate IT-101 in rats and tumor-bearing mice. Cancer Chemother. Pharmacol. 57 (5), 654–662.

Schluep, T., Hwang, J., Hildebrandt, I.J., Czernin, J., Choi, C.H.J., Alabi, C.A., et al., 2009. Pharmacokinetics and tumor dynamics of the nanoparticle IT-101 from PET imaging and tumor histological measurements. Proc. Nat. Acad. Sci. USA 106 (27), 11394–11399.

Schoemaker, N.E., van Kesteren, C., Rosing, H., Jansen, S., Swart, M., Lieverst, J., et al., 2002. A phase I and pharmacokinetic study of MAG–CPT, a water-soluble polymer conjugate of camptothecin. Br. J. Cancer 87 (6), 608–614.

Scott, L.C., Evans, T., Yao, J.C., Benson, A.I., Mulcahy, M., Thomas, A., et al., 2004. Pegamotecan (EZ-246), a novel PEGylated camptothecin conjugate, for treatment of adenocarcinomas of the stomach and gastroesophageal (GE) junction: preliminary results of a single-agent phase 2 study. J. Clin. Oncol. 22 (14_suppl.), 4030.

Scott, L.C., Yao, J.C., Benson 3rd, A.B., Thomas, A.L., Falk, S., Mena, R.R., et al., 2009. A phase II study of pegylated-camptothecin (pegamotecan) in the treatment of locally advanced and metastatic gastric and gastro-oesophageal junction adenocarcinoma. Cancer Chemother. Pharmacol. 63 (2), 363–370.

Shen, Y., Jin, E., Zhang, B., Murphy, C.J., Sui, M., Zhao, J., et al., 2010. Prodrugs forming high drug loading multifunctional nanocapsules for intracellular cancer drug delivery. J. Am. Chem. Soc. 132 (12), 4259–4265.

Singer, J.W., De Vries, P., Bhatt, R., Tulinsky, J., Klein, P., Li, C., et al., 2000. Conjugation of camptothecins to poly-(l-glutamic acid). Ann. N.Y. Acad. Sci. 902, 136–150.

Singer, J.W., Bhatt, R., Tulinsky, J., Buhler, K.R., Heasley, E., Klein, P., et al., 2001. Water-soluble poly-(L-glutamic acid)-gly-camptothecin conjugates enhance camptothecin stability and efficacy in vivo. J. Control. Release 74 (1–3), 243–247.

Sinha, A., Chakraborty, A., Jana, N.R., 2014. Dextran-gated, multifunctional mesoporous nanoparticle for glucose-responsive and targeted drug delivery. ACS Appl. Mater. Interfaces 6 (24), 22183–22191.

Springett, G.M., Takimoto, C., McNamara, M., Doroshow, J.H., Syed, S., Eastham, E., et al., 2004. Phase I study of CT-2106 (polyglutamate camptothecin) in patients with advanced malignancie. J. Clin. Oncol. 22 (14_suppl.), 3127.

Svenson, S., 2014. What nanomedicine in the clinic right now really forms nanoparticles? Wiley Interdiscip. Rev. Nanomed. Nanobiotechnol. 6 (2), 125–135.

Svenson, S., Wolfgang, M., Hwang, J., Ryan, J., Eliasof, S., 2011. Preclinical to clinical development of the novel camptothecin nanopharmaceutical CRLX101. J. Control. Release 153 (1), 49–55.

Tan, X., Li, B.B., Lu, X., Jia, F., Santori, C., Menon, P., et al., 2015. Light-triggered, self-immolative nucleic acid–drug nanostructures. J. Am. Chem. Soc. 137 (19), 6112–6115.

Tang, D.L., Song, F., Chen, C., Wang, X.L., Wang, Y.Z., 2013. A pH-responsive chitosan–b-poly(p-dioxanone) nanocarrier: formation and efficient antitumor drug delivery. Nanotechnology 24 (14), 145101.

Thiagarajan, G., Ray, A., Malugin, A., Ghandehari, H., 2010. PAMAM–camptothecin conjugate inhibits proliferation and induces nuclear fragmentation in colorectal carcinoma cell. Pharm. Res. 27 (11), 2307–2316.

Tobin, P.J., Rivory, L.P., 2004. Camptothecins and key lessons in drug design. Drug Des. Rev. Online 1 (4), 341–346.

Tong, R., Kohane, D.S., 2012. Shedding light on nanomedicine. Wiley Interdiscip. Rev. Nanomed. Nanobiotechnol. 4 (6), 638–662.

Tong, R., Tang, L., Ma, L., Tu, C., Baumgartner, R., Cheng, J., 2014. Smart chemistry in polymeric nanomedicine. Chem. Soc. Rev. 43 (20), 6982–7012.

Torchilin, V.P., 2006. Multifunctional nanocarriers. Adv. Drug Deliv. Rev. 58 (14), 1532–1555.

Tung, W.L., Hu, S.H., Liu, D.M., 2011. Synthesis of nanocarriers with remote magnetic drug release control and enhanced drug delivery for intracellular targeting of cancer cells. Acta Biomater. 7 (7), 2873–2882.

Unger, E.C., Porter, T., Culp, W., Labell, R., Matsunaga, T., Zutshi, R., 2004. Therapeutic applications of lipid-coated microbubbles. Adv. Drug Deliv. Rev. 56 (9), 1291–1314.

Venditto, V.J., Simanek, E.E., 2010. Cancer therapies utilizing the camptothecins: a review of the in vivo literature. Mol. Pharm. 7 (2), 307–349.

Venditto, V.J., Szoka Jr, F.C., 2013. Cancer nanomedicines: so many papers and so few drugs!. Adv. Drug Deliv. Rev. 65 (1), 80–88.

Wall, M.E., Wani, M.C., Cook, C.E., Palmer, K.H., McPhail, A.T., Sim, G.A., 1966. Plant antitumor agents. I. The isolation and structure of camptothecin, a novel alkaloidal leukemia and tumor inhibitor from *Camptotheca accuminata*. J. Am. Chem. Soc. 88 (16), 3888–3890.

Walsh, M.D., Hanna, S.K., Sen, J., Rawal, S., Cabral, C.B., Yurkovetskiy, A.V., et al., 2012. Pharmacokinetics and antitumor efficacy of XMT-1001, a novel, polymeric topoisomerase I inhibitor, in mice bearing HT-29 human colon carcinoma xenografts. Clin. Cancer Res. 18 (9), 2591–2602.

Wang, H., Tang, L., Tu, C., Song, Z., Yin, Q., Yin, L., et al., 2013. Redox responsive, core-cross-linked micelles capable of on-demand, concurrent drug release and structure disassembly. Biomacromolecules 14 (10), 3706–3712.

Weiss, G.J., Chao, J., Neidhart, J.D., Ramanathan, R.K., Bassett, D., Neidhart, J.A., et al., 2013. First-in-human phase 1/2a trial of CRLX101, a cyclodextrin-containing polymer–camptothecin nanopharmaceutical in patients with advanced solid tumor malignancies. Invest. New Drugs 31 (4), 986–1000.

Wilhelm, S., Tavares, A.J., Dai, Q., Ohta, S., Audet, J., Dvorak, H.F., et al., 2016. Analysis of nanoparticle delivery to tumours. Nat. Rev. Mater. 1, 16014.

Wong, B.S., Yoong, S.L., Jagusiak, A., Panczyk, T., Ho, H.K., Ang, W.H., et al., 2013. Carbon nanotubes for delivery of small molecule drugs. Adv. Drug Deliv. Rev. 65 (15), 1964–2015.

Xu, Z., Zheng, W., Yin, Z., 2013. Synthesis and optimization of a bifunctional hyaluronan-based camptothecin prodrug. Arch. Pharm. (Weinheim) 347 (4), 240–246.

Yen, H.C., Cabral, H., Mi, P., Toh, K., Matsumoto, Y., Liu, X., et al., 2014. Light-induced cytosolic activation of reduction-sensitive camptothecin-loaded polymeric micelles for spatiotemporally controlled in vivo chemotherapy. ACS Nano 8 (11), 11591–11602.

Yurkovetskiy, A., Fram, R.J., 2009. XMT-1001, a novel polymeric camptothecin pro-drug in clinical development for patients with advanced cancer. Adv. Drug Deliv. Rev. 61 (13), 1193–1202.

Zhang, W., Huang, J., Fan, N., Yu, J., Liu, Y., Liu, S., et al., 2010. Nanomicelle with long-term circulation and enhanced stability of camptothecin based on mPEGylated α,β-poly (L-aspartic acid)–camptothecin conjugate. Colloids Surf. B: Biointerfaces 81 (1), 297–303.

Zhang, L., Ma, D., Zhang, Y., He, W., Yang, J., Li, C., et al., 2013. Characterization of DNA topoisomerase-1 in *Spodoptera exigua* for toxicity evaluation of camptothecin and hydroxycamptothecin. PLoS ONE 8 (2), e56458.

Zhi, J., Liu, Q., Li, T., Ren, B., 2016. Measurement and correlation of the solubility for camptothecine in different organic solvents. J. Chem. Eng. Data 61 (6), 2052–2061.

Zolotarskaya, O.Y., Xu, L., Valerie, K., Yang, H., 2015. Click synthesis of a polyamidoamine dendrimer-based camptothecin prodrug. RSC Adv. 5 (72), 58600–58608.

Zou, Y., Wu, Q.P., Tansey, W., Chow, D., Hung, M.C., Charnsangavej, C., et al., 2001. Effectiveness of water soluble poly(L-glutamic acid)camptothecin conjugate against resistant human lung cancer xenografted in nude mice. Int. J. Oncol. 18 (2), 331–336.

Zunino, F., Pratesi, G., 2004. Camptothecins in clinical development. Expert Opin. Investig. Drugs 13 (3), 269–284.

CHAPTER

Functionalized graphene-based nanomaterials for drug delivery and biomedical applications in cancer chemotherapy

13

Mahfoozur Rahman[1], Imran Kazmi[2], Sarwar Beg[3], Abdul Hafeez[2], Muhammad Afzal[4], Vikas Kumar[1], Firoz Anwar[5] and Farhan J Ahmad[3]

[1]Department of Pharmaceutical Sciences, Shalom Institute of Health and Allied Sciences, Sam Higginbottom University of Agriculture, Technology & Sciences (SHUATS), Allahabad, India [2]Glocal School of Pharmacy, Glocal University, Saharanpur, India [3]Department of Pharmaceutics, Nanomedicine Research Lab, School of Pharmaceutical Education and Research, Jamia Hamdard (Hamdard University), New Delhi, India [4]Department of Pharmacology, College of Pharmacy, Al Jouf University, Sakaka, Saudi Arabia [5]Department of Biochemistry, Faculty of Science, King Abdulaziz University, Jeddah, Saudi Arabia

CHAPTER OUTLINE

Nanoparticles in Pharmacotherapy. DOI: https://doi.org/10.1016/B978-0-12-816504-1.00011-9

13.1 INTRODUCTION

Cancer is considered as one of the major growing causes of death worldwide due to multiple problems associated with its treatment. Thus, it is found to be responsible for higher morbidity or mortality with more than 10 million new cases every year (Beg et al., 2017; Ahmad et al., 2015). Several approaches have been investigated for identifying an effective treatment strategy for early stage diagnosis, mitigation, and cure (Aneja et al., 2014; Ahmad et al., 2010). In the past few decades, nanomedicines have emerged as an evolving area of research for effective treatment against cancer (Rahman et al., 2013). The cancer nanomedicines have yielded higher fruition as a boon for early stage detection, diagnosis, and delivery of therapeutic molecules for this purpose (Akhter et al., 2011; Rahman et al., 2015a,b). Of late, a vast range of nanocarriers have been employed in chemotherapeutic research like, polymeric, magnetic, and metallic nanoparticles; carbon nanotubes (CNTs); dendrimers; quantum dots; etc. Although such carriers have been investigated extensively in cancer treatment, allotropic nanostructures of carbon (like CNTs and graphenes) have attracted pharmaceutical scientists with a great deal of interest in the past few years (Rahman et al., 2016, 2017).

Graphenes, the carbon-based nanomaterials, exist as the three-dimensional allotropes of carbon with flat-planar structure. These were first isolated by Novoselov and Geim in 2010, as nanomaterials composed of carbon (sp^2 bound) in honeycomb lattice structure. The carbon atoms in graphenes are arranged in a regular hexagonal pattern, which provides high aspect ratio, tensile strength, innumerable surface area, and typical physiochemical characteristics (Hong et al., 2017). The outstanding electrical, mechanical, and optical properties of graphenes have made them an attractive nanocarrier for the vast majority of applications in cancer therapeutics. Practically graphene is insoluble in water, which restricts its use for in vivo applications. However, the modified form of graphene and their oxides show water solubility and can be easily functionalized with diagnostic agents and drug molecules for applications in drug delivery (Wang et al., 2016).

Over the past decade, the research in the field of oncology has undergone a paradigm shift from the conventional treatment strategies towards the nanotechnology-based approaches for targeted delivery of therapeutic molecules. Undoubtedly, such treatment approaches have shown multifold augmentation in the success rate for treatment of malignant diseases. Graphenes, in this regard, have benchmarked their place in cancer nanomedicines owing to their innumerable merits like functionalization ability with variety of drug molecules, diagnostic agents, proteins, peptides, and cancer cell specific genes such as siRNA, mRNA, and DNA (Wu et al., 2013). Besides the drug molecules, the ultratunable surface properties of graphenes allows them to anchor several nanocomposites like quantum dots, silver, gold, and silica nanoparticles onto their surface for enhanced imaging and therapeutic applications.

The most promising facets of graphenes have been extensively studied in the past few years including their applications in biomedical imaging as a tool for early diagnosis of cancers of varied kinds. Also, such approach helps in monitoring the biodistribution of cellular devices, drug delivery systems, and in biophotoluminescence. Graphenes can be anchored with diverse variety of ligands, antibodies, imaging agents for targeted delivery into the cells for imaging and chemotherapeutic action with high affinity. This approach helps in overcoming the drug resistance during chemotherapeutic treatment. Further, as compared with the 1D carbon-based materials, 2D graphene sheets possess an ultralarge and continuous sensing/interfacing area with easy surface modification. Thus, graphene provides a stable interface for attachment of biological and synthetic molecules. Owing to these innumerable applications in cancer, graphene-based nanomedicines have now been relatively the most promising area of research. This current review article provides insight into the current advances in the field of graphene-based nanosystems for their specific applications in drug delivery and biomedical applications in cancer chemotherapy (Xu et al., 2015).

13.2 GRAPHENE VERSUS C-NANOTUBES

The dramatic development of nanoscience has offered numerous opportunities and innovative nanoparticles in oncology. Among the different types of novel materials discovered at the nanoscale, carbon-based nanomaterials are a superfamily that includes fullerenes, CNTs, carbon nanohorns, diamonds, and graphene (Iijima, 1991). Graphene is a sp^2-hybridized nanomaterial. It includes few-layer graphene (1–5 layers), graphene oxide (GO), reduced graphene oxide (RGO), and graphite. CNTs are carbon-based nanostructures, which are first atomically described by Iijima in 1991 (Kostarelos et al., 2009). It can be classified into single, double, and multiwalled CNTs (SWCNTs, DWCNTs, and MWCNTs). Both CNT and graphene materials have outstanding electronic, mechanical, electrical, optical properties and a chemically tunable surface that has made them an attractive candidate for a broad range of applications, spanning from composites and electronics to nanomedicine (Yang et al., 2009). Despite their common carbon-based elemental consistency, CNTs and graphene are two very distinct nanomaterial entities. Differences on the basis of their shape, tubular versus planer and dimensions 1D (fibrous in shape) versus 2D (nonfibrous in shape), are the main structural differences. Furthermore, a difference in another way, the dispersibility of graphene sheets in various solvents requires a surfactant to facilitate their dispersion as compared with CNT. CNTs at the nanolevel form bundles or entangled aggregates, since graphene and their derivatives have tendency to form stacks in a few layers or cause less aggregation. Purified graphene nanomaterials are usually without metallic impurities, but in the case of CNT, during fabrication metal-based impurities are found (Banks et al., 2006). The main differences on the basis of physiochemical characteristics and cellular uptake are highlighted in Table 13.1 and Fig. 13.1 for CNTs and graphene. As compared with CNT, GO is

Table 13.1 Physiochemical Characteristics Differences Between Carbon Nanotubes (CNTs) and Graphene

Physiochemical Characteristics	Carbon Nanotubes	Graphene	Difference
Shape	Cylindrical, it can be SWCNT and MWCNT	Planar, which are monolayer or multilayer or sheet, ribbon, triangle, and hexagon	Entirely different
Surface area	With SWCNT $>1000\ m^2\ g^{-1}$ and MWCNT contain $100–500\ m^2\ g^{-1}$	Up to $2675\ m^2\ g^{-1}$ and decrease with increasing layer	Different
Colloidal stability	Oxidized CNT dispersible in water and causes aggregation via bundling and tangling	Graphene oxide dispersible in water and aggregation by stacking	Very different
Durability	Enzymatic degradation by unzipping and decrease of length and diameter	Enzymatic degradation by forming of defect in plan	Different
Impurities	Impurities vary with different manufacturing processes, various metal impurity residues such as Fe, Co, Ni, Cr, Zn, and Cu	No metal impurities found, but some other impurities other than metal may sometimes be found, depending on types of manufacturing process	Very different

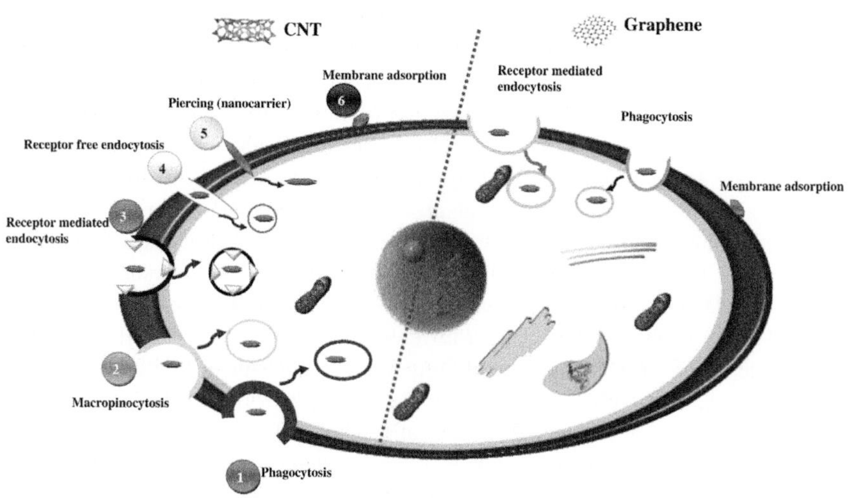

FIGURE 13.1

Basic difference between CNTs and graphene nanoparticles on cellular uptake mechanism.

formed easily at low cost; furthermore it conjugates with functional moieties and provide a large amount of surface area for potential multimodal imaging and therapeutic application in cancer. Targeting studies of CPT-loaded CNT and GO-based nanocarriers reported the higher loading capacity found with PVA–GO, such as that 1 g of PVA–CNT was able to load about 0.1 g of CPT, whereas 1 g of PVA–GO loaded 0.12 g of CPT, verifying the higher loading capacity of GO for CPT. Further it also confirms that PVA–GO entered into the cells through endocytosis, whereas PVA–CNT entered into the cells likely through both nanopenetration and endocytosis mechanism (Zhang et al., 2010). Graphene nanomaterials are younger in development, surrounded by great enthusiasm and expectations, as well as a variety of unanswered questions regarding their safety considerations (Bussy et al., 2013).

13.3 GRAPHENE-BASED NANOCOMPOSITES

Gelatin–graphene nanosheets (GNS) are composed of gelatin and graphene. Gelatin is a heterogeneous linear polypeptide with triple helical structure by linking of 18 amino acids (Jing et al., 2013). It is generally prepared by the partial hydrolysis of collagen, bones and white connective tissues. Gelatin has excellent membrane forming ability, good adhesion, biocompatibility, biodegradability, nontoxicity, and nonimmunogenicity. Therefore, it has been widely using in pharmaceutical applications such as plasma expanders, dispersants for many colloids, and controlled drug delivery (Bourlinos et al., 2003). Chemically, it is an amino acid and easily oxidized into nitrite, thus used as a reductant. This can be immobilized on the surface of graphene via hydrophobic interactions. As poor solubility and stability in physiological solutions hindered the application of graphene in drug delivery system, the novel approach of functionalizing nanographene with gelatin provides novel biocompatible, high solubility, better stability, high drug loading capacity, and availability of large specific surface area for drug delivery (Liu et al., 2008). More recently, methotrexate loaded GNS, formed by interaction with the aromatic portion of MTX with gelatin–graphene nanosheets by p–p stacking and hydrogen bonding, were developed. Furthermore MTX–gelatin–GNS interacts with cell membrane of A549 cell and then releases MTX gradually from this complex (Chang et al., 2011). Moreover, afterwards the MTX diffuses into the tumor cells via controlled sustained release effect, achieving the desired anticancer effect. Furthermore, the release behavior of MTX from functionalized complex is dependent on pH; under acidic conditions it exhibits higher sustained release than under neutral conditions (Zhou et al., 2011). This novel approach with unique advantages provides a promising nanoplatform for the application of graphene in oncology. Surface chemistry of nanographene is very important to improve biocompatibility and their behaviors in the biological system. Use of different techniques such as surface coating strategies including covalent and noncovalent approaches and other inorganic nanostructures attached to make highly enriched functionalized graphene (Yang et al., 2012a). In the early 1950s, Hummers and Offeman developed graphene oxide (GO) with oxidative exfoliation via reaction of graphite with a mixture of potassium permanganate ($KMnO_4$) and concentrated sulfuric acid (H_2SO_4). This is the easiest method to obtain huge quantities of graphene derivatives (Hummers and Offeman, 1958). This GO contains a large number of residual epoxides, hydroxides, and carboxyl groups. Furthermore, the sp^2-hybridized network is destroyed and ultimately its electrical conductivity is completely lost. To regain the electrical conductivity of GO it must be treated with reducing agents to form RGO with high conductivity and good optical absorbance, but its solubility in aqueous solution is lost due to the absence of hydroxyl groups on RGO (Robinson et al., 2011). Many researchers have made a tremendous effort to make functionalized graphene via methods such as chemical vapor deposition (CVD), solvothermal and organic synthesis, or up/down approaches including mechanical, physical, and chemical exfoliation methods (Wang et al., 2009). The presence of many reactive oxygen groups such as carboxylic acid, hydroxyl groups, etc. on GO, these group use for binding to other groups with the help of covalent functionalization method. Among other groups, one is polyethylene glycol (PEG). It is a hydrophilic polymer, which is widely applied to functionalized nanographene and their derivatives to improve the biocompatibility (Zhang et al., 2011a,b). Furthermore, it reduces the nanospecific targeting and also improves their in vivo pharmacokinetics for better tumor targeting. Conjugated PEGylated graphene is classified into (1) ultrasmall nGO–PEG with nearly 23 nm, (2) ultrasmall nRGO–PEG with 27 nm, and (3) larger RGO–PEG in 65 nm with C18PMH–PEG functional group and is radiolabeled by I^{125} for in vivo tracking (Wen et al., 2012). The functionalization of GO by the use of amino terminated branched PEG via the conjugation of amino group with PEG to carboxylic group on GO results in formation of 10–50 nm ultrasmall size PEGylated nano-GO complex, showing excellent stability. Amine modified dextran (DEX) use for covalent functionalization of GO results in significantly improving the stability of GO (Dia et al., 2008). Conjugation of GO with sulfonic acid improves the biocompatibility and colloidal stability of GO in biological solutions (cell medium and serum). Functionalization is made on GO using polyacrylic acid (PAA) to retain biocompatibility (Wang et al., 2011a). Conjugation of chitosan on GO results in better biocompatibility as well as drug and gene delivery. Recent studies report the encapsulation of doxorubicin onto GO with the help of charged folate conjugated chitosan, obtaining DOX–GO–chitosan folate nanocarriers, resulting in development of a pH responsive drug release (Depan et al., 2011). Other than covalent conjugation, noncovalent functionalization between GO and other molecules is made by polymers or biomolecules via weak interactions such as hydrophobic, p–p interaction, van der Waals, and electrostatic interaction (Chen et al., 2011). Hydrophobic interaction based noncovalent functionalization between the graphene surface and amphiphilic substances such as biocompatible polymers and surfactants. For biomedical application, biocompatible polymers are usually preferred over small surfactant molecules (Wang et al., 2009). Noncovalent adhesion of RGO with tween is demonstrated by Park et al. Functionalization of graphene by the use of Pluronic F127 (PF127) is used to obtain a nanohybrid graphene (Park et al., 2010). Noncovalent conjugation by PEGylated phospholipid on RGO to obtain ultrasmall RGO. Recently, functionalization of RGO with PEG grafted poly(maleic anhydride-alt-1-octadecene) (C18PMH–PEG) obtain RGO–PEG, provides better physiological stability and increased ultrablood circulation half-life, which is useful for photothermal treatment of cancer (Yang et al., 2010). Hydrophobic interaction of polyethyleneimine (PEI) with GO, results in improved physiological stability as compared with GO and high gene transfection efficiency. Recently Dong et al. demonstrated functionalization of graphene nanoribbon (NGR) with PEI to obtain PEI grafted GNR for in vitro transfection and in situ detection of micro-RNA (Dong et al., 2011). This approach may be applied to drug delivery. Moreover, natural biomolecules such as DNA and protein also can be functionalized on GO. Graphene and GO have highly localized π-electrons on their surface, various aromatic molecules are able to bind on GO via $\pi-\pi$ stacking such as stacking of DNA. Aromatic molecules have large aromatic plane that is anchored easily on reduced graphene surface without disturbing its electronic conjugation and further provides higher stability for reduced GO and opens the door for other moieties for conjugation (Wu et al., 2013). The representations of various functionalized nanographene are illustrated in Fig. 13.2. Decoration of various inorganic nanostructures such as metal nanocomposites including Au, Ag,

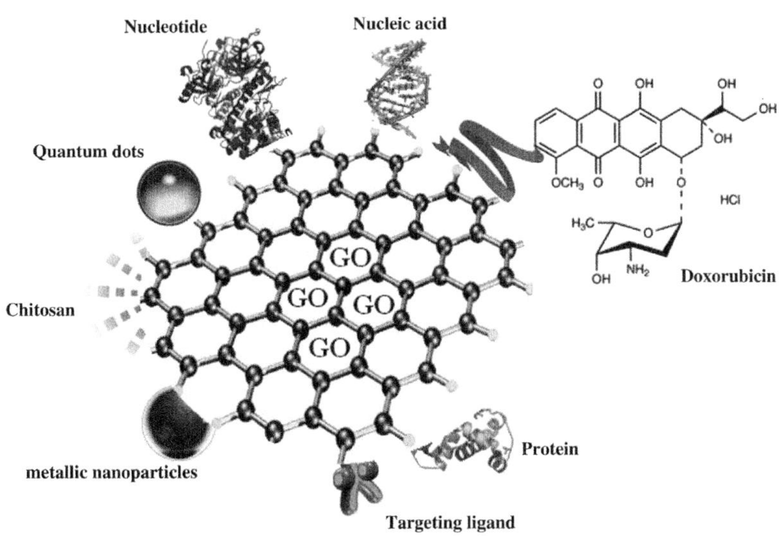

FIGURE 13.2

Diagram showing functionalized graphene nanoparticles.

Pd, Ni, TiO_2, MnO_2, ZnO, and Fe_3O_4 have been incorporated on graphene and applied in various applications such as fluorescent cell imaging, photothermal therapy (PTT), drug delivery, and photodynamic killing of cancer cell (Peng et al., 2010; Fig. 13.2). So, these all conjugate nanographene via covalent, noncovalent, $\pi-\pi$ stacking, and inorganic attachment to make nanographene a theragnostic nanoparticle for cancer treatment in the near future (Yang et al., 2013).

13.4 GRAPHENE NANOCOMPOSITES IN ONCOLOGY

A diverse variety of nanoparticulate systems have been employed in cancer chemotherapy for drug delivery and biomedical imaging applications. Despite this, the quest for identifying the most effective nanosystems with multifunctional characteristics has become the prime focus of research (Zhang et al., 2011a,b). The earlier success of CNTs in cancer chemotherapy has paved the way for researchers to explore other carbon-based nanomaterials for this purpose. Lately, graphenes are identified as the most promising nanomaterials for such applications (Rahman et al., 2016). Several drug molecules have been successfully delivered employing graphenes for treatment of multiple cancers of human existence.

Looking into the mechanism of the functionalization ability of graphenes, it has been proposed that the delocalized π-electrons on the graphene plane are primarily responsible for it (Wu et al., 2013). Doxorubicin and camptothecin are the instances of drug molecules employed for surface functionalization on graphene via $\pi-\pi$ stacking mechanism, while agents like SN38 (camptothecin analog), ellagic acid, β-lapachone, and carmustine are attached via physical adsorption or covalent conjugation mechanisms (Nasongkla et al., 2004). Camptothecin, a cytotoxic quinoline alkaloid used for the treatment of various tumors exhibits poor water solubility and low oral bioavailability, which restricts its clinical applications (Li et al., 2006). It has been observed that functionalization of it with graphene produces enhanced cellular uptake by enhanced interaction with the tumor cell membrane (Li et al., 2013). The chemical modification of native graphenes into their oxides for chemotherapeutic applications has also been published in the literature.

Among the diverse chemical functionalization strategies available for augmenting the stealth effect of the nanomaterials, PEGylation is known to be one of the easily applicable and common approaches for producing modified graphenes for chemotherapeutic applications (Yang et al., 2011a,b). It provides higher solubility, physiological stability, and longer circulation time in vivo in the blood. Moreover, such approach employed for producing PEGylated graphenes showed enhanced delivery of SN38 (a potent chlorpromazine analog) with significant improvement in the anticancer activity in HCT-116 human colon carcinoma cells compared with the water insoluble free drug (Zhang et al., 2011a,b). Further, studies in this area demonstrated that conjugation of chlorpromazine with PVA functionalized graphene oxide is considered to be safer as compared with the native graphene oxides. Also, the studies have reported that implications of these complexes for augmenting the anticancer activity of doxorubicin in MDA-MB-231 human breast cancer cells and A-5RT$_3$ skin cancer cells vis-à-vis the use of free drug alone (Rahman et al., 2016). Depan et al. (2011) reported the loading of DOX to graphene oxide surface using folic acid conjugated chitosan via noncovalent $\pi-\pi$ stacking. This system provides sustained release of DOX owing to strong charge dependent interactions with GO to liberate the drug molecules in the medium. Also, the DOX–GO–CHI–FA complexes exhibited targeted delivery and controlled drug releases (35%) at pH 5.3 as compared with that at pH 7.4 (11%), due to less interaction of hydrogen bonding at pH 5.3 and more prominent interaction of hydrogen bonding reported between DOX–GO at neutral pH (Liu et al., 2010).

Formation of sulfonic acid conjugated CPT–GO–FA nanocarrier for delivering CPT to the targeted folate receptors overexpressed on cancer cells, thus providing dose dependent cell death by synergetic effect. These nanocarriers produce better antitumor action than PTT or chemotherapy alone (Zhang et al., 2014).

Another similar study reported the loading with CPT, where superior in vitro anticancer activity on A-5RT3 cells was observed with respect to the CPT itself (Markovic et al., 2011). More recently, the covalent conjugation of GO with hydrophilic and biocompatible pluronic F38 (PF38), Tween 80, and maltodextrin have used for loading of ellagic acid (EA), resulting to exhibit better cytotoxic activity than free EA dissolved in DMSO (Kakran et al., 2011). Hu et al. reported loading of different concentration of DOX with same concentration of Pluronic F127 and gold nanoparticle clusters via noncovalent conjugation (Wang et al., 2011b). It showed higher loading efficiency and in vitro cytotoxicity activity on MCF-7 human breast cancer cells, much higher than DOX-free PF127/GNS (Gollavelli and Ling, 2012). Wang et al. reported the enhanced anticancer effect by gold nanoparticles clusters (GNC) derived from the noncovalent physisorption of RGO. The loading of DOX to produce DOX–GNC–RGO nanocarriers has further shown to inhibit HepG2 cell growth more strongly as compared with the DOX and GNC–RGO alone (Wang et al., 2008). In another study, Yang et al. reported GO–IONP as DOX carrier as pH-controlled release and dual targeting delivery system through the magnetic property of IONP (Yang et al., 2013). Furthermore, in vitro studies showed that DOX-loaded GO–IONP nanohybrid exhibited much improved cytotoxicity against SK3 human breast cancer cells as compared with the free DOX alone. However, the same concentration of GO–IONP–FA–DOX exhibited much higher cytotoxicity as compared with the GO–DOX–IONP. Recently reports have suggested the attachment of GO–IONP with PEG by hydrothermal conjugation provides higher physiological stability, biocompatibility, magnetically targeted drug delivery, and PTT for cancer treatment (Liao et al., 2011).

Currently, graphenes and their derivatives are blooming in nanotechnology with wide potential in gene delivery to treat gene related disorders including cancer. Availability of enormous surface area with abundant carboxyl groups on its surface, utilized by conjugation of PEI, chitosan, or 1-pyrenemethylamine via an amide bond, electrostatic interaction, or $\pi-\pi$ stacking interactions are effective in DNA and small interfering RNA (siRNA) delivery (Li et al., 2013). Moreover, development of negatively charged GO conjugated with cationic PEI via simple electrostatic interactions to obtain GO–PEI complexes is found to be less toxic as compared with PEI (Zhang et al., 2011a,b). Graphene nanocarriers as anticancer therapeutics are summarized in Table 13.2. A better and smarter design of surface chemistry of graphene nanocomposites is needed to attain multifunctionality for the development of more sophisticated carrier systems that could serve as a gene/drug delivery vector and as a bioimaging tool in the near future.

13.5 GRAPHENE NANOCOMPOSITES: TARGETING OVERVIEW AND RECENT ADVANCEMENT IN ONCOLOGY

First, after administration of anticancer loaded graphene nanoparticles, they reach to desired tumor tissues through penetration of barrier with minimum loss of their activity in the blood circulation (Hong et al., 2012). Second, after reaching the tumor tissues, therapeutic molecules kill the tumor tissue without disturbing the healthy tissues with a controlled release mechanism of the active molecules (Zhang et al., 2010). These two approaches achieve the intracellular concentration of drugs and reduce dose limiting toxicities simultaneously. Currently passive and active tumor targeting strategies are used to deliver nanoparticles to the tumor sites for effective action (Fig. 13.3).

Mostly tumor cells have incomplete vasculature resulting in leaky blood vessels. This phenomenon is induced by vascular permeability factors such as bradykinin, nitric oxide, peroxynitrite, prostaglandins, vascular endothelial growth factor (VEGF), matrix metalloproteinase (MMPs), kallikrein, and tumor necrosis factor (TNF) (Hong et al., 2012). High concentration of these mediators around the tumor tissues induces the formation of new blood vessels, called angiogenesis, for the purpose of supplying nutrients and oxygen to tumor cells. A characteristic feature of angiogenesis is defect in basement membranes of new blood vessels. This results in hyperpermeable leaky blood vessels and formation of defective lymphatic drainage system. Ultimately graphene nanoparticles get accumulated or trapped in the tumor vasculature and release active molecules (Dinauer et al., 2005). This phenomenon is known as enhanced permeability and retention effect (EPR) at tumor sites; these are critical principles for passive targeting for nanographene (Fig. 13.3). Furthermore, after accumulation at the tumor tissues, the drug is released into the extracellular matrix, where it diffuses into intracellular sites. In passive targeting biodistribution of graphene nanoparticles is mediated by their physiological conditions of tumor cells such as pH, temperature, and specific enzymes like alkaline phosphatase and plasmins present in higher amount in the

Table 13.2 Examples of Different Graphene Nanovectors Showing Their Multifunctionality as Therapeutic Carriers for Active Targeting and Self-Theranostic in Oncology

Anticancer Drugs	Surface Functionalization/Drug Loading	Remarks	References
Photosensitizer (PS) Chlorin e6 (Ce6)	Loading of Ce6 on graphene oxide via folic acid conjugation (GO–FA–Ce6)	GO–Ce6 combination conjugated to FA provide targeted delivery by interacts with FA-receptor and facilitates receptor-meditated endocytosis on MGC803 tumor cell and achieves higher photothermal action	Huang et al. (2011)
Photosensitizer: poly(*N*-vinyl pyrrolidone)	Loading of poly(*N*-vinyl pyrrolidone) on GO (less than 50 nm) via folic acid conjugation	Achieves targeted delivery and improves photodynamic action against cancer	Huang et al. (2011)
DOX	Loading of doxorubicin on GO–PEG	Achieves targeted delivery and cellular uptake in β cell lymphoma	Yang et al. (2013)
Camptothecin analog (SN38)	Loading of SN38 with GO–PEG by $\pi-\pi$ stacking/hydrophobic interactions	In comparison to water-soluble prodrugs CPT11, this combination provides targeted action and exhibits 1000 times higher cancer cell killing activity in human colon cancer (HCT-116) cells	Sahoo (2011)
DOX and CPT	Loading of both drugs on GO–FA conjugation	Achieves targeted action and enhances the cellular uptake into MCF breast cancer, and A549 human lung carcinoma cells	Zhang et al. (2010)
Plasmid DNA	Conjugation of plasmid DNA on GO–PEI	Enhanced cellular uptake of plasmid DNA in HeLa cervical carcinoma and greater tumor cell death	Zhang et al. (2011a,b)
DOX	Loading of DOX on GO–pluronic F127	Achieved tumor targeted action in MCF-7 breast cancer	Hu et al. (2012)
Photosensitizer molecules (Chlorin e6)	GO–PEG (covalent functionalization)	Achieved higher cellular uptake by targeted action in kB nasopharyngeal carcinoma	Tian et al. (2011)
DOX	Conjugation of DOX with GO–PEG via $\pi-\pi$ stacking–hydrophobic interaction + NIR	These grafted nanoparticles provided 82.1% tumor inhibition on EMT6 cell, by virtue of targeted action of DOX and higher photothermal action, after exposed with NIR. In vivo study found complete destruction of tumor after 30 days. They served as cancer targeted theranostic agent.	Park (2009)

(Continued)

Table 13.2 Examples of Different Graphene Nanovectors Showing Their Multifunctionality as Therapeutic Carriers for Active Targeting and Self-Theranostic in Oncology *Continued*

Anticancer Drugs	Surface Functionalization/Drug Loading	Remarks	References
DOX	Conjugated with FA–GO–PVP	High drug loading (near 100%) and achieved targeted action into tumor (HeLa cells) via receptor-mediated endocytosis, and produced 70.9% tumor inhibition with reduced drug resistance	Qin et al. (2013)
TRC105, a human/murine chimeric IgG1 monoclonal antibody	TRC105 loaded on GO via PEG conjugation	Achieved tumor targeted action on CD105 and significantly improved photothermal action	Hong et al. (2012)
DOX	Loading of DOX on gold nanoclusters and graphene nanocomposite and further conjugated with FA (DOX–Au–GO–FA)	Achieves targeted action and higher uptake of DOX resulting in greater killing of HepG2 cells	Wang et al. (2011a,b)
1,3-Bis(2-chloroethyl)-1-nitrosourea (BCNU)	Loading of BCNU on GO–PAA	This conjugated nanocarrier system provided long circulation and higher thermal stability as compared with free drug. Furthermore they improved cytotoxicity and there was 77% decrease in the IC50 versus BCNU alone at the same dose toward GL261cells.	Lu (2012)
DOX	GO–Fe_3O_4 (superparamagnetic iron oxide nanoparticles) conjugated with FA	Passive targeted action produced by iron oxide and the specific targeted action achieved by FA conjugation	Wang et al. (2013a,b)
DOX	Amino modified GO–Fe_3O_4	Higher drug loading and greater cytotoxicity to SK3 human breast cancer cells as compared with DOX alone	Yang (2011)
DOX	GO functionalized with FA–CHI and loaded with DOX (triple functionalized dual targeted drug nanocarrier)	High loading efficiency (0.98 mg/mg), targeted action, and higher drug release (35%) at acidic pH (5.3) and lower release 11% at neutral pH (7.4)	Wang et al. (2013a,b)

DOX	FA–NGO–PVP/DOX + NIR (2 μg mL^{-1} of DOX)	Excellent therapeutic potential and served as single agent for combination therapies	Qin et al. (2013)
DOX	FA–NGO–PVP/DOX + NIR (20 μg mL^{-1} of DOX)	At 20 μg mL^{-1} of DOX, this complex provided 90% tumor inhibition as compared with 2 μg mL^{-1} of DOX	
DOX	DOX grafted on graphene conjugated chitosan nanogel	Provided five times (18.81%) higher drug release and 49% tumor cells (LLC1) were killed, after exposure under NIR irradiation	Wang et al. (2013a,b)
DOX	DOX grafted on graphene conjugated chitosan nanogel + NIR light (808 nm)	Achieved higher drug release and cytotoxicity, when exposed under NIR. They served as theranostic agent in combination therapy.	
Hypocrellins, (natural perylenequinonoid pigments)	Grafted to GO and exposed to 470 nm LED irradiation	Higher drug loading (2 mg mg^{-1}) was found and enhanced active uptake and higher killing of Human cervical carcinoma (HeLa), and human hepatocellular carcinoma (SMMC-7721) tumor cells, after exposed under LED irradiation	Zhou et al. (2011)
DOX	DOX grafted on GO	High loading of DOX (186.8 w/w) reported at pH 6.5 and better drug release found	Gao (2013)
TRC105, (monoclonal antibody)	64 Cu–NOTA–RGO–TRC105	Sterically stabilized nanoparticles exhibited selective targeting to CD105 and photothermal action on $4T_1$ murine breast cancer	Hong et al. (2012)
Trastuzumab (anti-HER2 antibody)	Grafted on NGO, radiolabelled with ^{111}In-benzyl-diethylene-triaminepentaacetic acid (BnDTPA)	These radiolabeled grafted nanoparticles exhibited selective targeting action on HER2-overexpressing murine models of human breast cancer with 7.8-fold higher uptake and superior pharmacokinetic action as compared with radiolabeled trastuzumab without NGO	Cornelissen et al. (2013)

(Continued)

Table 13.2 Examples of Different Graphene Nanovectors Showing Their Multifunctionality as Therapeutic Carriers for Active Targeting and Self-Theranostic in Oncology *Continued*

Anticancer Drugs	Surface Functionalization/Drug Loading	Remarks	References
Photosensitizer (PS) Chlorin e6	Ce6 grafted on GO–HA + NIR	Higher enhanced intracellular uptake and 33% tumor cell death of A549 human lung cancer cells, when irradiated with NIR at 810 nm. Whereas no tumor cells are killed under PTT, when exposed to alone Ce6 and HA-Ce6.	Cho et al. (2013)
DOX	DOX grafted on graphene–HQDs (silica coated QDS)–Trf (transferrin)	High drug loading (1.4 mg mg^{-1}) and higher killing of tumor (HeLa) cells achieved by specific targeting on Trf, facilitated enhanced cellular uptake via receptor-mediated endocytosis	
Paclitaxel (Tx)	Tx grafted on single-walled carbon nanotube (SWNT) and graphene oxide (GO)	This combination provides significant enhanced cell death as compared with additive effect of individual agents on A549 and NCI-H460. Further achieved higher efficacies with lower side effects at minimum dose.	Arya (2013)
AgNP	AgNP grafted on RGO–FA	Provided higher uptake into HeLa cells, due to FR mediated endocytosis and had a great potential as a Raman probe in cancer diagnosis	Hu et al. (2013)

DOX, *Doxorubicin;* DOX–PEG, *doxorubicin–polyethylene glycol;* GO–Fe_3O_4, *graphene oxide–iron oxide nanoparticles;* PAA, *polyacrylic acid;* FA, *folic acid;* PVP, *polyvinyl pyrrolidone;* PEI, *polyethylenimine;* CPT, *camptothecin;* CHI, *chitosan;* NIR, *near infrared;* NOTA, *1,4,7-triazacyclononane-1,4,7-triacetic acid;* RGO, *reduced graphene oxide;* Cu, *copper;* HA, *hyaluronic acid;* Trf, *transferrin;* HQDs, *silica coated quantum dots;* Tx, *paclitaxel;* SWNT, *single-walled carbon nanotube;* AgNP, *silver nanoparticle;* FR, *folate receptor;* CPT11, *camptothecin-11;* IC50, *inhibitory concentration 50;* LED, *light-emitting diode.*

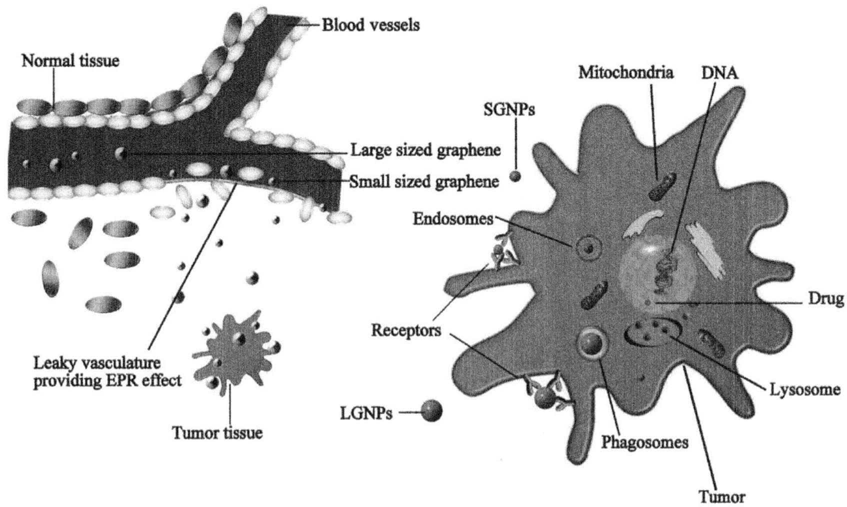

FIGURE 13.3

Schematic representation of enhanced permeability effect.

tumor cells (Cho et al., 2013). When the volume of the tumor reaches 2 mm angiogenesis occurs and diffusion becomes limited. EPR effect results in achieving 10- to 15-fold higher concentration of graphene nanoparticles at the tumor sites in comparison to normal tissues. For efficient passive accumulation of nanocarriers, size is an important factor and it must be less than 400 nm. High passive targeting further depends on degree of tumor vascularization and angiogenesis, and there should be lower interstitial fluid pressure at the center of tumors (Cornelissen et al., 2013).

Two major limitations block the use of graphene nanoparticles as drug carrier in cancer chemotherapy (Rahman et al., 2012a,b). One is lacking in the ability to carry the drug to specific tumor sites, whereas the other is nonspecific toxicity to the normal cells (Akhter et al., 2011). Internalization of nanoparticles by passive targeting to the targeted cell is not assured; therefore, nanoparticles are modified with various conjugates for active targeting (Daniels et al., 2006). Active targeting involves targeting of nanocarrier moieties and corresponding receptors to facilitate the targeting of a carrier to a specific malignant cell (Cornelissen et al., 2013). Many reports in the literature have shown that tumor cells express or overexpress specific receptors, proteins, antibodies, etc.

Folic acid is an essential nutrient for all living cells in the biosynthesis of nucleotide bases (Huang et al., 2011). It belongs to the vitamin B family. Folic acid has high affinity towards folate receptors (FR) including FR-a and FR-b; both are membrane bound proteins. Folate receptors are mostly expressed or overexpressed on tumor cells such as ovarian and endometrial cancers. Literature reported that folic acid binds to tumor cells 20 times greater than normal epithelial cells (Wang et al., 2013a,b). In the last several years, many researchers reported that folic acid can be easily conjugated to various nanoparticles for tumor imaging and targeted drug delivery. Recently, a novel folic acid conjugated graphene oxide develops by use of second-generation photosensitizer (PS) Chlorin e6 (Ce6) (Qin et al., 2013). The targeted delivery is achieved by interaction with FR on MGC803 tumor cell, resulting in receptor-mediated endocytosis of Ce6 and becoming trapped in the lysosome at acidic pH and further releasing from the lysosome to various cell organs to achieve higher accumulation of Ce6. Furthermore, the remarkable photodynamic efficacy on MGC803 significantly increased upon NIR irradiation on conjugated GO nanoparticles. In addition, the controlled size of conjugated GO and surface modification may prevent fluorescence quenching of PS with GO and also improve the photodynamic effect (Maltzahn et al., 2009). Currently, development of conjugated FA nanographene oxide (nGO) less than 50 nm in size by use of PS, called poly(*N*-vinyl pyrrolidone), results in better in vivo imaging and targeting PDT. Furthermore, the PTT action is due to strong optical absorbance of GO in the near infrared region (Pei and Cheng, 2012). These results suggest that folic acid conjugated GO loaded nanoparticles had great potential as an effective targeting drug delivery system and in the near future it could bring novel opportunities of combined cancer treatment. The epidermal growth factor receptor (EGFR) or human epidermal growth factor receptor (HER1 or ErbB1) is a 170-kDa tumor marker and it is a member of a tyrosine kinase family receptor. EGFR is expressed or overexpressed on many types of epithelial cancer; its mutation causes autonomous cell growth and metastasis (Hu et al., 2013). Other members are HER2 (ErbB2), HER3 (ErbB3), and HER4 (ErbB4). Various monoclonal antibody-based inhibitors target EGFR or HER1 in cancer as anticancer therapeutics such as cetuximab, zalutumumab, nimotuzumab, panitumumab, and matuzumab. Cetuximab is a chimeric monoclonal antibody, clinically it is effective against head, neck, and metastatic colorectal cancer. HER2 is overexpressed in 20%−35% of breast carcinoma cells and is also reported in ovarian cancer. Dimerization is required for activation of signal cascade of transmembrane receptors of the tyrosine kinase family. Herceptin, trastuzumab, and pertuzumab are FDA approved monoclonal antibodies that block the activation of HER2 and signaling pathway in breast and ovarian cancer. The use of monoclonal antibodies for molecular imaging has been restricted by their own large size and slow elimination rate from the body, which can result in low tumor to blood ratio and poor image quality (Gollavelli and Ling, 2012). Recently a nanographene conjugated anti-HER2 antibody (trastuzumab) radiolabeled with indium-111 (^{111}In) and metal ion chelator, 2-(4-aminobenzyl)−diethylene−triaminepentaacetic acid via p−p stacking was developed. Trastuzumab directly binds specifically to the HER2, providing superior pharmacodynamic action for imaging of HER2-overexpressing tumors in comparison to radiolabeled trastuzumab without NGO. NGO-based immunoconstructs provide excellent target specificity and affinity of antibodies with rapid pharmacokinetics (rapid clearance from the body). They also provide large payload carrying capacity of NGO and are easily adapted for therapeutic applications, due to availability of large surface area of NGO (Zhu et al., 2011). These meritorious advantages may be very effective for multifunctional nanographene as a carrier of anticancer drugs, photosensitizer for PDT–PTT, and other therapeutic entities like radionuclides targeted radiotherapy in the near future. New blood capillaries formation is essential for carrying nutrients and oxygen. In the latter stage uncontrolled development and disproportionate manner develops leaky and tortuous blood vessels, which leads to inflammation. Many angiogenic cytokines and growth factors are secreted by tumors cells such as vascular endothelial cell growth factor (VEGF-A, VEGF-B, VEGF-C, VEGF-D, VEGF-E, VEGF-F), placenta growth factor, angiopoietin, and integrins (Ferrari, 2005). VEGF is a more important player in angiogenesis and vascular permeability in tumor progression. Recently bevacizumab (anti-VEGF, Avastin) and ranibizumab monoclonal antibody have been used clinically for cancer treatment. More recently anti-VEGF targeting monoclonal antibodies such as IMC-1C11, IMC-1121B, and CDP791 have taken entry under clinical development. Recently, an ideal vascular marker for tumor angiogenesis CD105 (endoglin) is reported (Feng and Liu, 2011). Normally it is not detected in resting endothelial cells or normal organs. Targeting of these mediators or antiangiogenesis has emerged as a promising approach in limiting tumor cell progression by molecular imaging and therapy of cancer. Recently, TRC105, a human/murine chimeric IgG1 monoclonal antibody that specifically binds to both human and murine CD105 has been developed. Right now, it is under phase II clinical trial in patients with various solid tumors. Currently the loaded TRC105 on functionalized GO (10–50 nm in size) by use of six arm branched PEG has been investigated, which has many amino groups on the surface of GO for conjugation of various targeting ligands, ultimately delivering TRC105 to target CD105 and significantly improving tumor uptake for cancer targeted drug delivery and PTT to enhance therapeutic efficacy (Choi et al., 2010). Furthermore, pharmacokinetics and tumor targeting efficacy are studied for conjugated GO labeled with Cu^{64} isotopes (using positron emission tomography) and use of ideal chelator 1,4,7-triazacyclononane-1,4,7-triacetic acid (NOTA) for Cu in tumor bearing mice exhibits excellent stability, targeting is vascular specific, and there is good imaging action (Geim and Novoselov, 2007). Moreover, the therapeutic efficacy of conjugated nanographene in targeting of vasculature tumor markers is significantly improved. In the near future it may be used for targeted/imaging/PTT and to enable cancer theragnostics in cancer nanomedicine.

13.6 GRAPHENE AS THERAGNOSTIC NANOMEDICINE IN CANCER

Theragnostic nanomedicine is a combined nanotherapeutic system that is used for simultaneous diagnosis and treatment by means of imaging contrasting agents, imaging-guided therapeutics/PTT, and targeted drug delivery at the same time (Yang et al., 2013). The term *theragnostic* was first used by Warner. The emergence of theragnostic nanomedicine in oncology is based on understanding of molecular mechanism, diagnostic strategies, therapeutic efficacy, toxicity, and method of preparation for the dual purpose of diagnosis and therapy (Kyriazi et al., 2017). In addition, it decreases the dosage of drugs and may result in lower side effects. The most beneficial potential achieved by multifunctional or combined therapy is to overcome or reduces multidrug resistance of the tumor cells. In the past decade many eminent researchers developed theragnostic nanoparticles and have enormously increased their production to date. Recently, a novel triple functionalized drug delivery system by use of superparamagnetic graphene oxide and doxorubicin with folic acid and functionalized with chitosan has emerged (Yang et al., 2012b). Moreover, chitosan is used as a linker to combine folic acid with functional GO and also provide encapsulation, better stability, biocompatibility, and controlled release of active molecules (Chauhan et al., 2017). This multifunctional complex exhibits high drug loading efficiency and prolonged release rate with higher release rate of DOX. Efficient targeted drug delivery is achieved with specific ligands that recognize specific cell receptors, and this multifunctional complex acts as a dual targeted drug nanocarrier. Passive targeting associated with the use of Fe_3O_4 and active targeting are achieved by the use of folic acid. Furthermore, with the aid of magnetic field drug delivery efficiency is improved via driving the specific drug at the specific tumor tissues. Furthermore, recent research has developed GO–FA based combined therapy in which two anticancer drugs are loaded simultaneously, that is, DOX and CPT, which exhibit better targeting efficiency and higher cancer killing ability than GO with either DOX or CPT (Lee et al., 2015). PEI is a cationic polymer widely used in conjugation with graphene for DNA, RNA, or gene delivery (Teimouri et al., 2016). Recently, Zhang et al. developed conjugated PEI–GO nanocarriers for simultaneous delivery of small interfering RNA (siRNA) and DOX to treat HeLa cells in vitro. A synergistic cancer killing effect is achieved with better therapeutic effectiveness. Interaction of cationic polymer with nanographene for the drug–gene codelivery is still at its infancy stage. In this particular area more in-depth research is required to pave a new way in cancer theragnostics in the near future. Possessing dual magnetic and biological targeting properties with this theragnostic could be a promising element for diagnosis and effective targeting of active molecules in the treatment of cancer (Delpassand, 2016). Compared with single chemotherapy or phototherapy, the combined chemophototherapy finds much higher therapeutic efficacy in animal models. In cancer nanomedicines, phototherapy is increasingly recognized as an alternative to treat various cancers in clinical practice. It uses photoabsorbing agents to generate heat from optical energy. Disadvantages associated with major therapeutic approaches (chemotherapy and radiotherapy) are limited specificity to cancer cells and undesired side effects to normal tissues and organs (Ma et al., 2012). Phototherapies include PTT and photodynamic therapy (PDT), which are able to irradiate specific light on cancer cells. Recently, phototherapies irradiate with the help of phototherapeutic nanoagents, specifically targeting the cancerous cell and killing via either passive or active tumor targeting without significant damage to normal organs (Hu et al., 2012). Therefore, phototherapies with the aid of nanotechnology exhibit remarkable advantages like minimal invasive nature, enhancing cancer killing specificity, reducing side effect as compared with conventional cancer therapy, and the ability to treat patients with repeated doses without initiating resistance or exceeding total dose limitations (as associated with radiotherapy) (Dong et al., 2010). Currently, GO has been reported as ability to absorb energy at the near infrared (NIR) spectrum (700–1100 nm) and subsequently emit heat, it leading to thermal ablation of tissues (Robinson et al., 2011). Furthermore, it provides higher efficacy at lower power radiation and saves healthy tissues from excessive heating. Photothermal ablations of tissues by the nanographene occur via generation of reactive oxygen species (ROS), caspase activation, the mitochondrial membrane depolarization, and DNA fragmentation (Tian et al., 2011). Fig. 13.2 illustrates the multifunctional nature of graphene nanoparticles and different molecules/bioactives that can effectively deliver to the target tissue and cells. Recently, unique behavior found with nanographene such as optical and chemical properties have aroused increasing interest. PTT is an optical absorbing agent to produce heat under light irradiation resulting in local high temperature to kill cancerous cells (Nel et al., 2009). Recently, PTT treatment of cancer was also reported by PEGylated GO and RGO with high NIR absorbance. First, PEGylated nano-GO, after i.v. injection by a fluorescent and imaging method, reported high passive tumor accumulation via enhanced permeability and retention (EPR) effect of cancerous cells (Liu et al., 2017a). In recent years, intravenous administration of nGO–PEG by irradiation with 808 nm laser at a power density of 2 W cm^{-2} for 5 minutes was done on balb/c mice grown on murine $4T_1$ breast cancer; the photothermal agent removed 100% tumors after treatment (Xiong et al., 2014). Ultimately, strong NIR absorption and high tumor passive accumulation of nRGO–PEG achieve excellent beneficial effect at low power. It prevents unwanted effects on the normal tissues and improves treatment efficacy on large tumors. However, these characteristics are not found with the other nanoparticles such as gold nanoparticles, which exhibit photothermal properties for in vivo lesser tumor ablation by inducing cellular toxicity (Yang et al., 2012b). Recently, Dia et al. reported the ultrasmall reduced nano-GO (nRGO) coated with PEG and PEGylated phospholipid has much higher NIR absorbance as compared with NIR absorbance of nGO–PEG with arginine glycine aspartic acid peptide conjugation. So, nRGO–PEG phospholipid acts as a highly efficient targeted PTT for in vitro cancer cell ablation. Further it has also been reported the nGO–PEG and nRGO–PEG with ultrasmall size after i.v. injection exhibits much lower accumulation in RES organs such as liver and spleen as compared with RGO–PEG (Rangsimawong et al., 2016), since it has long blood circulation half-life, higher tumor passive uptake, and low-RES retention, that is, lower toxicity with high optical absorbance for in vivo governed by nRGO–PEG (Chen et al., 2014).

Under suitable light irradiation photosensitizers generate ROS to kill cancer cells. Graphene-based PDT treatment was first reported by Shi et al. (2014). Zinc phthalocyanine (znPc), widely used as PS molecules, was conjugated on the surface of nGO–PEG, and under the Xe light irradiation it exhibits significant efficacy against cancer cells. In another study it was reported that the GO could load chlorine e_6 (Ce_6) for targeting delivery to FR-positive cancer with higher photodynamic action as compared with Ce_6 alone, owing to the intracellular trafficking by loading on the nanocarrier. Table 13.2 summarizes the self-theragnostic application of graphene nanocarriers in cancer treatment.

13.7 TOXICITY WITH GRAPHENE NANOCOMPOSITES

Two-dimensional graphene nanomaterials are unique in chemical and physical determinants for their cellular interaction in comparison to spherical nanoparticles or one-dimensional nanotubes or nanorods (Sasidharan et al., 2011). Graphene nanomaterials interact with cells to produce direct or indirect generation of ROS, leading to oxidative stress in various target cells. Currently, it is the main mechanism proposed for graphene toxicity. A balance of ROS (production and elimination) is maintained in the body by antioxidant enzymes such as superoxide dismutase, catalase, or glutathione peroxidase (He et al., 2017). Overproduction or slow elimination causes oxidative stress, which may lead to alteration of macromolecules like polyunsaturated fatty acids in the membrane lipids, protein denaturation, and finally DNA destruction. Exposure of graphene nanomaterials to various cells generates ROS in a concentration- and time-dependent manner; moreover the presence of higher surface area of graphene interacts with membrane lipids, causing direct or indirect toxicity (Yang et al., 2015). Recently, Chang et al. demonstrated the in vitro toxicity of graphene oxide on the morphology, viability, mortality, and membrane integrity of human lung epithelial cells, with dose dependent oxidative stress at the dose of $50\ \mu g\ mL^{-1}$. Furthermore, they observed that human alveolar basal epithelial cells (A549) show concentration dependent toxicity (Chang et al., 2011). The various in vitro toxicity studies have been reported by use of cell culture lines such as epithelial cells, fibroblasts, alveolar basal epithelial cells, pheochromocytoma, fetal osteoblasts, cervical cells, and red blood cells (RBCs). There are only some in vivo studies that have been reported related to toxicity (Wen et al., 2015). Recently Wang et al. reported the intravenous administration of graphene oxides at three different doses, 0.1, 0.25, and 0.4 mg (low dose, middle, and high dose), in three different groups. At low and middle dose they do not cause toxicity. At high dose four out of nine mice died due to granuloma development in kidney, lungs, liver, and spleen (Wang et al., 2013a,b). Furthermore, they also cause accumulation and are not filtered by the kidney. Recently Zanni et al. demonstrated the in vivo toxicity of graphene nanoplatelets in *Caenorhabditis elegans*; no chronic toxicity of graphene

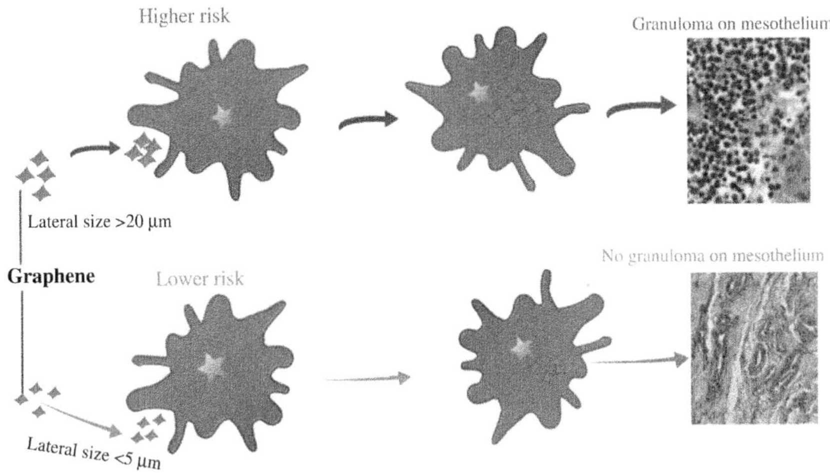

FIGURE 13.4

Potential risk associated with different dimensions of graphene nanoparticles.

nanomaterials was observed (Zanni et al., 2012). In another study, administration of pristine graphene into the lungs of C57BL mice was reported. The authors reported severe with persistent lung injury and inflammatory responses via ROS generation (Shvedova et al., 2008). Schinwald et al. reported large nonfunctionalized graphene with lateral size of 25 μm caused local inflammation followed by formation of mesothelial granuloma, and further no granuloma was reported with lateral in size lower than 5 μm (Schinwald et al., 2012; Fig. 13.4).

13.8 CURRENT IN VITRO AND IN VIVO SCALE RESEARCH: NEED FOR SAFETY REGULATION

Current literature reports that the functionalization of graphene nanomaterials has resulted in enhancement of their biocompatibility and solubility. Unintended exposure to graphene nanomaterials may result in either elimination from the body via glomerular and GI tract or accumulation within tissues (Yin et al., 2015). In both processed, either elimination or tissue accumulation, they interact with host cells that may potentially induce local histopathological responses (Merritt et al., 1992). In recent, Yang et al. reported that the PEG–nanographene oxide causes accumulation in the liver and spleen in mice followed by gradual clearance, and further no toxicity was reported at the end of the study at the dose of 20 mg kg^{-1} (Yang et al., 2009). Wojtoniszak et al. reported the biocompatibility of functionalized GO–PEG, GO–PEG–poly-propylene–GO glycol–PEG (Pluronic P123)–GO, and deoxycholate (DOC)–GO on L929 in fibroblast cells of mice (Wojtoniszak et al., 2012). Better biocompatibility was reported with PEG functionalized graphene as compared with other Pluronic 123 and DOC. In another study, nanographene covalently conjugated with DEX formed a reduced sheet size, increased thickness, and higher solubility. This complex was further demonstrated on human cervical cancer HeLa cells, showing that reduced cellular toxicity and higher clearance were reported (Zhang et al., 2011c). Recently, Sasidharan et al. reported the functionalization of graphene nanoparticles with carboxylic acid and in vivo studies on monkey renal cells without causing any toxicity (Sasidharan et al., 2011). Whereas nonfunctionalized graphene nanomaterials interact with the cell membrane, producing ROS and leading to apoptosis and further causing toxicity. Gollavelli and Ling reported the in vitro studies of multifunctional magnetic graphene functionalized by using of PAA and FMA to human cervical cancer cells, which did not develop any toxicity or ROS generation (Gollavelli and Ling, 2012). Robinson et al. demonstrated that RGOs (20 nm in lateral size dimension) functionalized with PEGylated polymer improved stability, PTT, and in vitro studies on human breast cancer cells reported nonsignificant toxicity or little toxicity (Robinson et al., 2011). Safety consideration is an important factor for graphene nanoparticles exposure on human health. There are three different levels: (1) cellular level, (2) tissue level, and (3) whole-body interaction, which are discussed here. First, at the cellular level, experimental evidence suggests the interaction of graphene nanomaterials with membranes is the first point of contact during their contact with the cells, with the functionalization of graphene surface thought crucial in such molecular interaction with biological membranes and lipid bilayer (Duan et al., 2017). Furthermore, surface charge also plays a critical role in such interactions, with cationic charged materials being more favorable due to the electrostatic attraction forces for membrane interaction. Moreover, colloidal dispersion and structural parameters such as diameter or graphene surface defect are thought to play in these interactions (Miao et al., 2017). Yu et al. reported the interaction of two different lateral dimension such as 350 nm and 2 μm with peritoneal macrophages. On the basis of transmission electron microscopy (TEM) of cells, initially graphene with 350 nm was wrapped by the active filopodia of the macrophages, while 2 μm GO was seen perpendicular to the plasma membrane (Yu et al., 2017). Recently Mu et al. reported on the interaction of protein coated GO with mouse mesenchymal progenitor cells, resulting in cell membrane interaction (Mu et al., 2012). This biological interaction with graphene has now been well-characterized to elucidate the effect of dimension and surface properties interaction on the biological membranes. After membrane interaction, cell internalization may occur by various processes (Bramini et al., 2016). In Fig. 13.1, the mechanisms of internalization of graphene are described. Recently cellular internalization of size 0.5 μm BSA-coated GO uptaken by clathrin mediated endocytosis in C2C12 cells was reported. Whereas those larger in size, around 1 μm, were uptaken by clathrin mediated endocytosis and phagocytosis. Furthermore, the uptake of lateral dimension of GO in phagocytic cells and nonphagocytic cells has been reported; higher uptake was observed in phagocytic cells. So, size is an important factor for internalization that can be chemically modified by different chemical functionalization strategies that enhance dispersibility (Dudek et al., 2016). Fate after cellular internalization also imparts a great importance in safety considerations. Further accumulation of graphene-based nanomaterials will occur in various organs or cytoplasmic organs like kidney, liver, etc. or endoplasmic reticulum, Golgi apparatus, lysosome, mitochondria, and nucleus. All these cellular responses eventually lead to cytotoxicity such as genotoxic, inflammatory, and cytotoxic damages (Urbas et al., 2016). Mostly the functionalized graphene surface has predominantly negative charge like hydroxyl and carboxyl groups; their interaction in comparison to CNT may be expectedly different because the CNT surface mainly contains cationic charge (He et al., 2016). So, more in-depth research is needed to address safety regulations on graphene nanomaterials. In graphene derivatives interacting with human RBCs, hemolytic activity has been reported. It is further removed by the development of chitosan GO (Kim et al., 2016). Kim et al. reported at the concentration of 10 $\mu g\ mL^{-1}$ GO no changes were observed in the shape of erythrocytes. But at 80 $\mu g\ mL^{-1}$ of GO, RBCs were ruptured. And to further extend the research, in intravenous administration of single layered GO sheets of 10–800 nm in lateral size at low dosing 1 $mg\ kg^{-1}$ using mice, no pathological alteration was observed after 14 days postinjection. But at 10 $mg\ kg^{-1}$, it results in lung accumulation with slow clearance caused by granulomatous lesions, pulmonary edema, inflammation, as well as fibrosis (Kim et al., 2016). Yang et al. reported high tumor accumulation of PEGylated GO in tumor bearing mice. The same group reported the long term biodistribution of I^{125} labeled PEGylated GO of 10–30 nm in size causes accumulation in the liver and spleen and low excretion rate by both renal and fecal route (Yang et al., 2017). Furthermore, in a different study reported for the intravenous administration of GO in mice at the dose of 0.1–0.4 mg, no sign of toxicity was reported at lower dose. But at highest dose chronic severe toxicity was found such as dose dependent inflammatory response. From the safety point of view biodegradation is an important factor related to tissue accumulation and clearance of graphene (Wu et al., 2016). For biodegraded materials the degraded product must be toxin-free. Nonbiodegradable graphene nanomaterials are only safe when body clearance and excretion of the nanomaterials are fast (Seabra et al., 2013). For both biodegradation and nonbiodegradable nanomaterials, surface chemistry and size of the materials is a determining factor. In vitro studies reported the enzymatic oxidation of GO. But RGO is not oxidized by a horseradish peroxidase enzyme (HRP). There are very limited numbers of in vivo studies available (Liu et al., 2017b). So, a common distribution profile has not been established for graphene nanomaterials. Inadequate individualization of like structural features such as size, shape, surface charge, functionalization, and lattice defects of graphene and their derivatives will influence their in vivo degree of individualization and will tend to form stacks of graphene sheets. Exposure to graphene at the whole-body level results in immunoreactivity and inflammogenicity (Ali et al., 2016). Graphene fibers that remain in the lungs and pleural

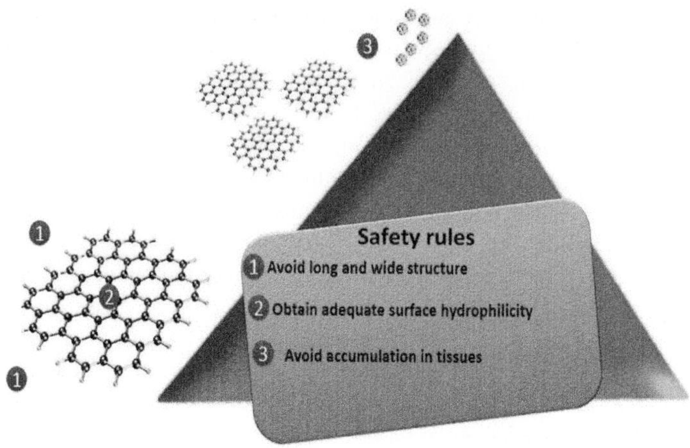

FIGURE 13.5

Safety consideration for the development and biomedical application of graphene nanoparticle.

cavities are likely to induce inflammation (Ishimatsu et al., 2003). Very recent studies reported the production of well-dispersed purified graphene oxide sheets in biological media with no evidence of in vitro cytotoxicity and in vivo pathogenicity (Ali-Boucetta et al., 2013). Finally, regarding the safety profile, colloidal properties, hydrophilicity, biodegradation, and nonbiodegradation parameters are very crucial for pharmacokinetics of the graphene nanomaterials in vitro and in vivo. This is depicted in Fig. 13.5.

13.9 CONCLUSION

Regular synergistic efforts from researchers in various areas including materials science, chemistry, biology, and clinical medicine have explored graphene-based materials in cancer theragnostics. The functionalization of graphene by various techniques has been discussed in detail in the current review. In covalent modification, nucleophilic substitution of an amine-terminated organic modifier is the easiest way to produce functionalized graphene. In this case, large-scale production of functionalized graphene from GO is also possible. However, the electrical conductivity of the functionalized graphene is greatly reduced with this type of modification. Noncovalent modification of a small organic polymer on a prereduced graphene surface can also yield better electrical conductivity. Apart from covalent and noncovalent techniques, preparation of processable graphene directly from graphite has attracted significant research interest. Moreover, the separation of excess organic modifier from the functionalized graphene is a major concern.

Therefore, further research on the functionalization of graphene is required to improve its electrical conductivity, dispersibility, and address the unquestionable need for a facile synthetic route. Therefore, research efforts should focus on achieving large-scale production of graphene to drive modern science and technology in terms of mobility, environmental safety and security, health, and energy production. Surface modification of graphene nanovectors (graphene oxide and RGO, etc.) with some polymeric substrates such as FBS, PEG, DOC, and DEX leads to nonsignificant toxicity reported at normal dose compared with graphene alone. Recently developed combined DOX and NGO–PEG chemophotothermal treatment resulted in complete destruction of the tumors and no recurrence of tumors as compared with use of DOX or NGO–PEG alone. Furthermore, the systematic toxicity study show combined treatment reduced drug side effects in comparison to DOX alone. Conjugates with a tumor specific antibody can access targeted delivery with excellent behavior. There is much need of systematic studies for understanding the interaction of nano-graphenes with biological membranes, and establishing correlation with key physicochemical and material properties to overcome the toxicity. Development of theragnostic nanographene and its interaction with various living biosystems is still a great challenge for in vivo biomedical application.

REFERENCES

Ahmad, M.Z., Akhter, S., Jain, G.K., Rahman, M., Pathan, S.A., Ahmad, F.J., et al., 2010. Metallic nanoparticles: technology overview & drug delivery applications in oncology. Expert. Opin. Drug Deliv. 7 (8), 927–942.

Ahmad, J., Akhter, S., Rizwanullah, M., Amin, S., Rahman, M., Ahmad, M.Z., et al., 2015. Nanotechnology-based inhalation treatments for lung cancer: state of the art. Nanotechnol. Sci. Appl. 8, 55–66.

Akhter, S., Ahmad, Z., Singh, A., Ahmad, I., Rahman, M., Anwar, M., et al., 2011. Cancer targeted metallic nanoparticle: targeting overview, recent advancement and toxicity concern. Curr. Pharm. Des. 17, 1834–1850.

Ali, M.A., Mondal, K., Jiao, Y., Oren, S., Xu, Z., Sharma, A., et al., 2016. Microfluidic immuno-biochip for detection of breast cancer biomarkers using hierarchical composite of porous graphene and titanium dioxide nanofibers. ACS Appl. Mater. Interfaces 8 (32), 20570–20582.

Ali-Boucetta, H., Bitounis, D., Raveendran-Nair, R., Servant, A., Van den Bossche, J., Kostarelos, K., 2013. Purified graphene oxide dispersions lack in vitro cytotoxicity and in vivo pathogenicity. Adv. Healthc. Mater. 2 (3), 433–441.

Aneja, P., Rahman, M., Beg, S., Aneja, S., Dhingra, V., Chugh, R., 2014. Cancer targeted magic bullets for effective treatment of cancer. Recent Pat. Antiinfect. Drug Discov. 9 (2), 121–135.

Banks, C.E., Crossley, A., Salter, C., Wilkins, S.J., Compton, R.G., 2006. Carbon nanotubes contain metal impurities which are responsible for the "Electrocatalysis" seen at some nanotube-modified electrodes. Angew. Chem. Int. Ed. 45, 2533–2537.

Beg, S., Rahman, M., Jain, A., et al., 2017. Nanoporous metal organic frameworks as hybrid polymer–metal composites for drug delivery and biomedical applications. Drug. Discov. Today 22 (4), 625–637.

Bourlinos, A.B., Gournis, D., Petridis, D., Szabó, T., Szeri, A., Dékány, I., 2003. Graphite oxide: chemical reduction to graphite and surface modification with primary aliphatic amines and amino acids. Langmuir 19, 6050–6055.

Bramini, M., Sacchetti, S., Armirotti, A., Rocchi, A., Vázquez, E., León Castellanos, V., et al., 2016. Graphene oxide nanosheets disrupt lipid composition, Ca(2 +) homeostasis, and synaptic transmission in primary cortical neurons. ACS Nano 10 (7), 7154–7171.

Bussy, S., Ali-boucetta, H., kostarelos, K., 2013. Safety considerations for graphene: lessons learnt from carbon nanotubes. Acc. Chem. Res. 46, 692–701.

Chang, Y., Yang, S.T., Liu, J.H., Dong, E., Wang, Y., Cao, A., et al., 2011. In vitro toxicity evaluation of graphene oxide on A549 cells. Toxicol. Lett. 200 (3), 201–210.

Chauhan, G., Chopra, V., Tyagi, A., Rath, G., Sharma, R.K., Goyal, A.K., 2017. "Gold nanoparticles composite–folic acid conjugated graphene oxide nanohybrids" for targeted chemo-thermal cancer ablation: in vitro screening and in vivo studies. Eur. J. Pharm. Sci. 96, 351–361.

Chen, M.L., Liu, J.W., Hu, B., Chen, M.L., Wang, J.H., 2011. Conjugation of quantum dots with graphene for fluorescence imaging of live cells. Analyst 136, 4277–4283.

Chen, J., Wang, X., Chen, T., 2014. Facile and green reduction of covalently PEGylated nanographene oxide via a 'water-only' route for high-efficiency photothermal therapy. Nanoscale Res. Lett. 9 (1), 86.

Cho, Y., Kim, H., Choi, Y., 2013. A graphene oxide–photosensitizer complex as an enzyme-activatable theranostic agent. Chem. Commun. (Camb) 49, 1202–1204.

Choi, W., Lahiri, I., Seelaboyina, R., Kang, Y.S., et al., 2010. Synthesis of graphene and its applications: a review. Crit. Rev. Solid State Mater. Sci. 35, 52–71.

Cornelissen, B., Able, S., Kersemans, V., 2013. Nanographene oxide-based radioimmunoconstructs for in vivo targeting and SPECT imaging of HER2-positive tumors. Biomaterials 34, 1146–1154.

Daniels, T.R., Delgado, T., Helguera, G., Penichet, M.L., 2006. The transferrin receptor part II: targeted delivery of therapeutic agents into cancer cells. Clin. Immunol. 121, 159–176.

Delpassand, E., 2016. TH-AB-206-02: nuclear medicine theronostics: wave of the future; pre-clinical and clinical opportunities. Med. Phys. 43 (6), 3862–3863.

Depan, D., Shah, J., Misra, R.D.K., 2011. Controlled release of drug from folate-decorated and graphene mediated drug delivery system: synthesis, loading efficiency, and drug release response. Mater. Sci. Eng. C 31, 1305–1312.

Dia, H., Liu, Z., Robinson, J.T., Sun, X., Dai, H., 2008. PEGylated nanographene oxide for delivery of water-insoluble cancer drugs. J. Am. Chem. Soc. 130, 10876–10877.

Dinauer, N., Balthasar, S., Weber, C., Kreuter, J., Langer, K., von Briesen, H., 2005. Selective targeting of antibody-conjugated nanoparticles to leukemic cells and primary T-lymphocytes. Biomaterials 26, 5898–5906.

Dong, H., Zhao, Z., Wen, H., 2010. Poly(ethylene glycol) conjugated nano-graphene oxide for photodynamic therapy. Sci. China Chem. 53, 2265–2271.

Dong, H., Ding, L., Yan, F., Ji, H., Ju, H., 2011. The use of polyethylenimine-grafted graphene nanoribbon for cellular delivery of locked nucleic acid modified molecular beacon for recognition of microRNA. Biomaterials 32, 3875–3882.

Duan, G., Zhang, Y., Luan, B., Weber, J.K., Zhou, R.W., Yang, Z., et al., 2017. Graphene-induced pore formation on cell membranes. Sci. Rep. 7, 42767.

Dudek, I., Skoda, M., Jarosz, A., Szukiewicz, D., 2016. The molecular influence of graphene and graphene oxide on the immune system under in vitro and in vivo conditions. Arch. Immunol. Ther. Exp. (Warsz) 64 (3), 195–215.

Feng, L.Z., Liu, Z., 2011. Graphene in biomedicine: opportunities and challenges. Nanomedicine 6, 317–324.

Ferrari, M., 2005. Cancer nanotechnology: opportunities and challenges. Nat. Rev. Cancer 5, 161–171.

Geim, A.K., Novoselov, K.S., 2007. The rise of graphene. Nat. Mater. 6, 183–191.

Gollavelli, G., Ling, Y.C., 2012. Multi-functional graphene as an in vitro and in vivo imaging probe. Biomaterials 33, 2532–2545.

He, Z., Zhang, S., Song, Q., Li, W., Liu, D., Li, H., et al., 2016. The structural development of primary cultured hippocampal neurons on a graphene substrate. Colloids Surf. B Biointerfaces 146, 442–451.

He, Y., Del Valle, A., Qian, Y., Huang, Y.F., 2017. Near infrared light-mediated enhancement of reactive oxygen species generation through electron transfer from graphene oxide to iron hydroxide/oxide. Nanoscale 9 (4), 1559–1566.

Hong, H., Yang, K., Zhang, Y., Engle, J.W., Feng, L., Yang, Y., et al., 2012. In vivo targeting and imaging of tumor vasculature with radiolabelled, antibody-conjugated nanographene. ACS Nano 6, 2361–2370.

Hong, H.K., Jo, J., Hwang, D., Lee, J., Kim, N.Y., Son, S., et al., 2017. Atomic scale study on growth and heteroepitaxy of ZnO monolayer on graphene. Nano Lett. 17 (1), 120–127.

Hu, S.H., Chen, Y.W., Hung, W.T., Chen, I.W., Chen, S.Y., 2012. Quantum-dot-tagged reduced graphene oxide nanocomposites for bright fluorescence bioimaging and photothermal therapy monitored in situ. Adv. Mater. 24, 1748–1754.

Hu, C., Liu, Y., Qin, J., 2013. Fabrication of reduced graphene oxide and silver nanoparticle hybrids for Raman detection of absorbed folic acid: a potential cancer diagnostic probe. ACS Appl. Mater. Interfaces 5, 4760–4768.

Huang, P., Xu, C., Lin, J., Wang, C., Wang, X., Zhang, C., et al., 2011. Folic acid conjugated graphene oxide loaded with photosensitizers for targeting photodynamic therapy. Theranostics 1, 240–250.

Hummers, W.S., Offeman, R.E., 1958. Preparation of graphitic oxide. J. Am. Chem. Soc. 80, 1339.

Iijima, S., 1991. Helical micro tubules of graphitic carbon. Nature 354, 56–58.

Ishimatsu, S., Hori, H., Oyabu, T., Tanaka, I., 2003. Biopersistence of graphite whiskers deposited in rat lungs by 4-week inhalation. Inhal. Toxicol. 15 (1), 53–66.

Jing, A., Gou, Y., Yang, C., Hu, F., Wang, C., 2013. Synthesis of a biocompatible gelatin functionalized graphene nanosheets and its application for drug delivery. Mater. Sci. Eng. C: Mater. Biol. Appl. 33, 2827–2837.

Kakran, M., Sahoo, N.G., Bao, H., Pan, Y., Li, L., 2011. Functionalized graphene oxide as nanocarrier for loading and delivery of ellagic Acid. Curr. Med. Chem. 18, 4503–4512.

Kim, J., Nafiujjaman, M., Nurunnabi, M., Lee, Y.K., Park, H.K., 2016. Hemorheological characteristics of red blood cells exposed to surface functionalized graphene quantum dots. Food Chem. Toxicol. 97, 346–353.

Kostarelos, K., Bianco, A., Prato, M., 2009. Promises, facts and challenges for carbon nanotubes in imaging and therapeutics. Nat. Nanotechnol. 4, 627–633.

Kyriazi, M.E., Heuer-Jungemann, A., Kanaras, A.G., 2017. How can nano-delivery systems selectively kill cancerous cells? Ther. Deliv. 8 (4), 171–173.

Lee, J.H., Sahu, A., Jang, C., Tae, G., 2015. The effect of ligand density on in vivo tumor targeting of nanographene oxide. J. Control. Release 209, 219–228.

Li, Q.Y., Zu, Y.G., Shi, R.Z., Yao, L.P., 2006. Review camptothecin: current perspectives. Curr. Med. Chem. 13, 2021–2039.

Li, X.L., Shan, S., Xiong, M., Xia, X.H., Xu, J.J., Chen, H.Y., 2013. On-chip selective capture of cancer cells and ultrasensitive fluorescence detection of survivin mRNA in a single living cell. Lab. Chip 13 (19), 3868–3875.

Liao, K.H., Lin, Y.S., Macosko, C.W., Haynes, C.L., 2011. Cytotoxicity of graphene oxide and graphene in human erythrocytes and skin fibroblasts. ACS Appl. Mater. Interfaces 3, 2607–2615.

Liu, Z., Robinson, J.T., Sun, X., Dai, H., 2008. PEGylated nanographene oxide for delivery of water-insoluble cancer drugs. J. Am. Chem. Soc. 130, 10876–10877.

Liu, J., Tao, L., Yang, W., Li, D., Boyer, C., Wuhrer, R., et al., 2010. Synthesis, characterization, and multilayer assembly of pH sensitive graphene–polymer nanocomposites. Langmuir 26, 10068–10075.

Liu, Y., Liu, X., Guo, Z., Hu, Z., Xue, Z., Lu, X., 2017a. Horseradish peroxidase supported on porous graphene as a novel sensing platform for detection of hydrogen peroxide in living cells sensitively. Biosens. Bioelectron. 87, 101–107.

Liu, X., Zhang, X., Zhu, M., Lin, G., Liu, J., Zhou, Z., et al., 2017b. PEGylated Au@Pt nanodendrites as novel theranostic agents for computed tomography imaging and photothermal/radiation synergistic therapy. ACS Appl. Mater. Interfaces 9 (1), 279–285.

Ma, X.X., Tao, H.Q., Hang, K., Feng, L., Cheng, L., Shi, X., et al., 2012. A functionalized graphene oxide–iron oxide nanocomposite for magnetically targeted drug delivery, photothermal therapy, and magnetic resonance imaging. Nano Res. 5, 199–212.

Maltzahn, G.V., Park, J.H., Agrawal, A., Bandaru, N.K., Das, S.K., Sailor, M.J., et al., 2009. Computationally guided photothermal tumor therapy using long-circulating gold nanorod antennas. Cancer Res. 69, 3892–3900.

Markovic, Z.M., Harhaji-Trajkovic, L.M., Todorovic-Markovic, B.M., Kepić, D.P., Arsikin, K.M., Jovanović, S.P., et al., 2011. In vitro comparison of the photothermal anticancer activity of graphene nanoparticles and carbon nanotubes. Biomaterials 32, 1121–1129.

Merritt, K., Margevicius, R.W., Brown, S.A., 1992. Storage and elimination of titanium, aluminum, and vanadium salts, in vivo. J. Biomed. Mater. Res. 26 (11), 1503–1515.

Miao, R., Wang, L., Zhu, M., Deng, D., Li, S., Wang, J., et al., 2017. Effect of hydration forces on protein fouling of ultrafiltration membranes: the role of protein charge, hydrated ion species, and membrane hydrophilicity. Environ. Sci. Technol. 51 (1), 167–174.

Mu, Q., Su, G., Li, L., Gilbertson, B.O., Yu, L.H., Zhang, Q., et al., 2012. Size dependent cell up take of protein coated graphene oxide nanosheets. ACS Appl. Mater. Interfaces 4, 2259–2266.

Nasongkla, N., Shuai, X., Ai, H., Weinberg, B.D., Pink, J., Boothman, D.A., et al., 2004. cRGD-functionalized polymer micelles for targeted doxorubicin delivery. Angew. Chem. 116, 6483–6487.

Nel, A.E., Madler, L., Velegol, D., Xia, T., Hoek, E.M., Somasundaran, P., et al., 2009. Understanding biophysicochemical interactions at the nano-bio interface. Nat. Mater. 8, 543–557.

Park, S., Mohanty, N., Suk, J.W., Nagaraja, A., An, J., Piner, R.D., et al., 2010. Biocompatible, robust free-standing paper composed of a TWEEN/graphene composite. Adv. Mater. 22, 1736–1740.

Pei, S., Cheng, H.M., 2012. The reduction of graphene oxide. Carbon 50, 3210–3228.

Peng, C., Hu, W., Zhou, Y., Fan, C., Huang, Q., 2010. Intracellular imaging with a graphene-based fluorescent probe. Small 6, 1686–1692.

Qin, X.C., Guo, Z.Y., Liu, Z.M., Zhang, W., Wan, M.M., Yang, B.W., 2013. Folic acid-conjugated graphene oxide for cancer targeted chemo-photothermal therapy. J. Photochem. Photobiol. B 120, 156–162.

Rahman, M., Ahmad, M.Z., Kazmi, I., et al., 2012a. Emergence of nanomedicine as cancer targeted magic bullets: recent development and need to address the toxicity apprehension. Curr. Drug Discov. Technol. 9 (4), 319–329.

Rahman, M., Ahmad, M.Z., Kazmi, I., Akhter, S., Afzal, M., Gupta, G., et al., 2012b. Advancement in multifunctional nanoparticles for the effective treatment of cancer. Expert Opin. Drug Deliv. 9, 367–381.

Rahman, M., Beg, S., Ahmed, A., Swain, S., 2013. Emergence of functionalized nanomedicines in cancer chemotherapy: recent advancements, current challenges and toxicity considerations. Recent Pat. Nanomed. 2, 128–139.

Rahman, M., Ahmad, M.Z., Ahmad, J., Firdous, J., Ahmad, F.J., Mushtaq, G., et al., 2015a. Role of graphene nano-composites in cancer therapy: theranostic applications, metabolic fate and toxicity issues. Curr. Drug. Metab. 16 (5), 397–409.

Rahman, M., Akhter, S., Ahmad, M.Z., Ahmad, J., Addo, R.T., Ahmad, F.J., et al., 2015b. Emerging advances in cancer nanotheranostics with graphene nanocomposites: opportunities and challenges. Nanomed. (Lond.) 10 (15), 2405–2422.

Rahman, M., Kumar, V., Beg, S., Sharma, G., Katare, O.P., Anwar, F., 2016. Emergence of liposome as targeted magic bullet for inflammatory disorders: current state of the art. Artif. Cells Nanomed. Biotechnol. 44 (7), 1597–1608.

Rahman, M., Beg, S., Verma, A., Kazmi, I., Patel, D.K., Anwar, F., et al., 2017. Therapeutic applications of liposomal based drug delivery and drug targeting for immune linked inflammatory maladies: a contemporary view point. Curr. Drug Targets 18 (13), 1558–1571.

Rangsimawong, W., Opanasopit, P., Rojanarata, T., Duangjit, S., Ngawhirunpat, T., 2016. Skin transport of hydrophilic compound-loaded PEGylated lipid nanocarriers: comparative study of liposomes, niosomes, and solid lipid nanoparticles. Biol. Pharm. Bull. 39 (8), 1254–1262.

Robinson, J.T., Tabakman, S.M., Liang, Y., Wang, H., Casalongue, H.S., Vinh, D., et al., 2011. Ultra small reduced graphene oxide with high near-infrared absorbance for photothermal therapy. J. Am. Chem. Soc. 133, 6825–6831.

Sasidharan, A., Panchakarla, L.S., Chandran, P., Menon, D., Nair, S., Rao, C.N., et al., 2011. Differential nano-bio interactions and toxicity effects of pristine versus functionalized graphene. Nanoscale 3, 2461–2464.

Schinwald, A., Murphy, F.A., Jones, A., MacNee, W., Donaldson, K., 2012. Graphene-based nanoplatelets: a new risk to the respiratory system as a consequence of their unusual aerodynamic properties. ACS Nano 6, 736–746.

Seabra, A.B., Paula, A.J., Durán, N., 2013. Redox–enzymes, cells and micro-organisms acting on carbon nanostructures transformation: a mini-review. Biotechnol. Prog. 29 (1), 1–10.

Shi, S., Chen, F., Ehlerding, E.B., Cai, W., 2014. Surface engineering of graphene-based nanomaterials for biomedical applications. Bioconjug. Chem. 25 (9), 1609–1619.

Shvedova, A.A., Kisin, E., Murray, A.R., Johnson, V.J., Gorelik, O., Arepalli, S., et al., 2008. Inhalation vs. aspiration of single-walled carbon nanotubes in C57BL/6 mice: inflammation, fibrosis, oxidative stress, and mutagenesis. Am. J. Physiol. Lung Cell. Mol. Physiol. 295 (4), L552–L565.

Teimouri, M., Nia, A.H., Abnous, K., Eshghi, H., Ramezani, M., 2016. Graphene oxide–cationic polymer conjugates: synthesis and application as gene delivery vectors. Plasmid 84–85, 51–60.

Tian, B., Wang, C., Zhang, S., 2011. Photothermally enhanced photodynamic therapy delivered by nano-graphene oxide. ACS Nano 5, 7000–7009.

Urbas, K., Jedrzejczak-Silicka, M., Rakoczy, R., Zaborski, D., Mijowska, E., 2016. Effect of GO–Fe_3O_4 and rotating magnetic field on cellular metabolic activity of mammalian cells. J. Biomater. Appl. 30 (9), 1392–1406.

Wang, X., Yang, L., Chen, Z.G., Shin, D.M., 2008. Application of nanotechnology in cancer therapy and imaging. CA Cancer J. Clin. 58, 97–110.

Wang, G., Wang, B., Park, J., Yang, J., Shen, X., Yao, J., 2009. Synthesis of enhanced hydrophilic and hydrophobic graphene oxide nanosheets by a solvothermal method. Carbon 47, 68–72.

Wang, C.S., Li, J.Y., Amatore, C., Chen, Y., Jiang, H., Wang, X.M., 2011a. Gold nanoclusters and graphene nanocomposites for drug delivery and imaging of cancer cells. Angew. Chem. Int. Ed. 50, 11644–11648.

Wang, Y., Li, Z., Wang, J., Li, J., Lin, Y., 2011b. Graphene and graphene oxide: biofunctionalization and applications in biotechnology. Trends Biotechnol. 29, 205–212.

Wang, Z., Zhou, C., Xia, J., 2013a. Fabrication and characterization of a triple functionalization of graphene oxide with Fe_3O_4, folic acid and doxorubicin as dual-targeted drug nanocarrier. Colloids Surf. B Biointerfaces 106, 60–65.

Wang, C., Mallela, J., Garapati, U.S., Ravi, S., Chinnasamy, V., Girard, Y., et al., 2013b. A chitosan-modified graphene nanogel for noninvasive controlled drug release. Nanomedicine 9, 903–911.

Wang, Q., Wang, C., Zhang, M., Jian, M., Zhang, Y., 2016. Feeding single-walled carbon nanotubes or graphene to silkworms for reinforced silk fibers. Nano Lett. 16 (10), 6695–6700.

Wen, H., Dong, C., Dong, H., Shen, A., Xia, W., Cai, X., et al., 2012. Engineered redox-responsive PEG detachment mechanism in PEGylated nano-graphene oxide for intracellular drug delivery. Small 8, 760–769.

Wen, K.P., Chen, Y.C., Chuang, C.H., Chang, H.Y., Lee, C.Y., Tai, N.H., 2015. Accumulation and toxicity of intravenously-injected functionalized graphene oxide in mice. J. Appl. Toxicol. 35 (10), 1211–1218.

Wojtoniszak, M., Chen, X., Kalenczuk, R.J., Wajda, A., Łapczuk, J., Kurzewski, M., et al., 2012. Synthesis, dispersion, and cytocompatibility of graphene oxide and reduced graphene oxide. Colloids Surf. B 89, 79–85.

Wu, W., Guo, W., Zeng, X.C., 2013. Intrinsic electronic and transport properties of graphyne sheets and nanoribbons. Nanoscale 5, 9264–9276.

Wu, W., Yan, L., Wu, Q., Li, Y., Li, Q., Chen, S., et al., 2016. Evaluation of the toxicity of graphene oxide exposure to the eye. Nanotoxicology 10 (9), 1329–1340.

Xiong, H., Guo, Z., Zhang, W., Zhong, H., Liu, S., Ji, Y., 2014. Redox-responsive biodegradable PEGylated nanographene oxide for efficiently chemo-photothermal therapy: a comparative study with non-biodegradable PEGylated nanographene oxide. J. Photochem. Photobiol. B 138, 191–201.

Xu, Z., Zhu, S., Wang, M., Li, Y., Shi, P., Huang, X., 2015. Delivery of paclitaxel using, graphene oxide as a nanocarrier. ACS Appl. Mater. Interfaces 7 (2), 1355–1356.

Yang, X.Y., Zhang, X.Y., Ma, Y., Huang, Y., Wang, Y., Chen, Y., 2009. Superparamagnetic graphene oxide–Fe_3O_4 nanoparticles hybrid for controlled targeted drug carriers. J. Mater. Chem. 19, 2710–2714.

Yang, K., Zhang, S., Zhang, G., Sun, X., Lee, S.T., Liu, Z., 2010. Graphene in mice: ultra-high in vivo tumor uptake and efficient photothermal therapy. Nano Lett. 10, 3318–3323.

Yang, K., Wan, J., Zhang, S., Zhang, Y., Lee, S.T., Liu, Z., 2011a. In vivo pharmacokinetics, long-term biodistribution, and toxicology of PEGylated graphene in mice. ACS Nano 5, 516–522.

Yang, X.Y., Wang, Y.S., Huang, X., Ma, Y., Huang, Y., Yang, R., et al., 2011b. Multi-functionalized graphene oxide based anticancer drug-carrier with dual-targeting function and pH-sensitivity. J. Mater. Chem. 21, 3448–3454.

Yang, K., Hu, L., Ma, X., Ye, S., Cheng, L., Shi, X., et al., 2012a. Multimodal imaging guided photothermal therapy using functionalized graphene nanosheets anchored with magnetic nanoparticles. Adv. Mater. 24, 1868–1872.

Yang, K., Wan, J., Zhang, S., Tian, B., Zhang, Y., Liu, Z., 2012b. The influence of surface chemistry and size of nanoscale graphene oxide on photothermal therapy of cancer using ultra-low laser power. Biomaterials 33, 2206–2214.

Yang, K., Feng, L., Shi, X., Liu, Z., 2013. Nano-graphene in biomedicine: theranostic applications. Chem. Soc. Rev. 42, 530–547.

Yang, C., Denno, M.E., Pyakurel, P., Venton, B.J., 2015. Recent trends in carbon nanomaterial-based electrochemical sensors for biomolecules: a review. Anal. Chim. Acta 887, 17–37.

Yang, M.M., Wilson, W.R., Wu, Z., 2017. pH-sensitive PEGylated liposomes for delivery of an acidic dinitrobenzamide mustard prodrug: Pathways of internalization, cellular trafficking and cytotoxicity to cancer cells. Int. J. Pharm. 516 (1–2), 323–333.

Yin, J.X., Wei, Z., Xu, J.J., Sun, Z.Q., 2015. In vivo pharmacokinetic and tissue distribution investigation of sustained-release cisplatin implants in the normal esophageal submucosa of 12 beagle dogs. Cancer Chemother. Pharmacol. 76 (3), 525–536.

Yu, L., Li, P., Ding, X., Zhang, Q., 2017. Graphene oxide and carboxylated graphene oxide: viable two-dimensional nanolabels for lateral flow immunoassays. Talanta 165, 167–175.

Zanni, E., De Bellis, G., Bracciale, M.P., Bracciale, M.P., Broggi, A., Santarelli, M.L., et al., 2012. Graphite nanoplatelets and *Caenorhabditis elegans*: insights from an in vivo model. Nano Lett. 12, 2740–2744.

Zhang, L., Xia, J., Zhao, Q., Liu, L., Zhang, Z., 2010. Functional graphene oxide as a nanocarrier for controlled loading and targeted delivery of mixed anticancer drugs. Small 6, 537–544.

Zhang, W., Guo, Z.Y., Huang, D.Q., Liu, Z., Guo, X., Zhong, H., 2011a. Synergistic effect of chemo-photothermal therapy using PEGylated graphene oxide. Biomaterials 32, 8555–8561.

Zhang, L., Lu, Z., Zhao, Q., Huang, J., Shen, H., Zhang, Z., 2011b. Enhanced chemotherapy efficacy by sequential delivery of siRNA and anticancer drugs using PEI-grafted graphene oxide. Small 7, 460–464.

Zhang, S., Yang, K., Feng, L., Liu, Z., 2011c. In vitro and in vivo behaviors of dextran functionalized graphene. Carbon 49, 4040–4049.

Zhang, Y., Feng, Y., Zhu, B., Zhao, J., Jiang, T., 2014. Graphene based tunable metamaterial absorber and polarization modulation in terahertz frequency. Opt. Express 22 (19), 22743–22752.

Zhou, L., Wang, W., Tang, J., Zhou, J.H., Jiang, H.J., Shen, J., 2011. Graphene oxide noncovalent photosensitizer and its anticancer activity in vitro. Chem. Eur. J. 17, 12084–12091.

Zhu, S., Zhang, J., Qiao, C., Tang, S., Li, Y., Yuan, W., et al., 2011. Strongly green-photoluminescent graphene quantum dots for bioimaging applications. Chem. Commun. 47, 6858–6860.

FURTHER READING

Falagan-Lotsch, P., Grzincic, E.M., Murphy, C.J., 2017. New advances in nanotechnology-based diagnosis and therapeutics for breast cancer: an assessment of active-targeting inorganic nanoplatforms. Bioconjug. Chem. 28 (1), 135–152.

Koo, H.Y., Lee, H.J., Go, H.A., Lee, Y.B., Bae, T.S., Kim, J.K., et al., 2011. Graphene-based multifunctional iron oxide nanosheets with tunable properties. Chemistry 17 (4), 1214–1219.

Koyama, S., Kim, Y.A., Hayashi, T., Takeuchi, K., Fujii, C., Kuroiwa, N., et al., 2009. In vivo immunological toxicity in mice of carbon nanotubes with impurities. Carbon 47, 1365–1372.

Liu, Z., Cai, W.B., He, L., Nakayama, N., Chen, K., Sun, X., et al., 2007. In vivo biodistribution and highly efficient tumor targeting of carbon nanotubes in mice. Nat. Nanotechnol. 2, 47–52.

Luo, Y., Cai, X., Li, H., Lin, Y., Du, D., 2016. Hyaluronic acid-modified multifunctional Q-graphene for targeted killing of drug-resistant lung cancer cells. ACS Appl. Mater. Interfaces 8 (6), 4048–4055.

Ren, T., Li, L., Cai, X., Dong, H., Liu, S., Li, Y., 2012. Engineered polyethylenimine/graphene oxide nanocomposite for nuclear localized gene delivery. Polym. Chem. 3, 2561–2569.

Udukala, D.N., Wang, H., Wendel, S.O., Malalasekera, A.P., Samarakoon, T.N., Yapa, A. S., et al., 2016. Early breast cancer screening using iron/iron oxide-based nanoplatforms with sub-femtomolar limits of detection. Beilstein J. Nanotechnol. 7, 364–373.

Wu, C., Wang, C., Zheng, J., Luo, C., Li, Y., Guo, S., et al., 2015. Vacuolization in cytoplasm and cell membrane permeability enhancement triggered by micrometer-sized graphene oxide. ACS Nano 9 (8), 7913–7924.

Zeng, X., Yuan, Y., Wang, T., Wang, H., Hu, X., Fu, Z., et al., 2017. Targeted imaging and induction of apoptosis of drug-resistant hepatoma cells by miR-122-loaded graphene–InP nano compounds. J. Nanobiotechnol. 15 (1), 9.

CHAPTER

Antimicrobial activity of silver nanoparticles

14

Suresh Kumar Kailasa[1], Tae-Jung Park[2], Jigneshkumar V. Rohit[3] and Janardhan Reddy Koduru[4]

[1]Department of Applied Chemistry, S.V. National Institute of Technology, Surat, India
[2]Department of Chemistry, Institute of Interdisciplinary Convergence Research, Research Institute of Halal Industrialization Technology, Chung-Ang University, Seoul, Republic of Korea
[3]Department of Chemistry, National Institute of Technology Srinagar, Hazratbal Kashmir, India
[4]Department of Environmental Engineering, Kwangwoon University, Seoul, Korea

CHAPTER OUTLINE

14.1 INTRODUCTION

Over the past few years, nanostructured materials (NMs) have been significantly improved by the growth of research activities of various fields of science and technology (Seil and Webster, 2012). Due to the size dependent (10^{-9} m) properties, NMs have been considered as promising materials in the development of nanomaterials-based biomedical technologies for controlling or curing of various disorders. Furthermore, the small difference in the size of bacteria (10^{-6} m) and the size of NPs (10^{-3} m) allow them to act as potential antimicrobial agents to a wide variety of microorganisms.

NMs exhibit novel and specific physicochemical and biological properties, allowing them to be used as promising materials in nanomedicine, which facilitates to develop effective and novel biomedical modalities for various applications

Nanoparticles in Pharmacotherapy. DOI: https://doi.org/10.1016/B978-0-12-816504-1.00009-0

in pharmacology and medicine (Sirelkhatim et al., 2015). Due to the unique physicochemical properties, that is, high surface area, small size, and chemical reactivity, NMs can easily enter into biological molecules to be incorporated with such systems. Generally, any materials or chemical compounds that are intended to locally kill bacteria or slow down their growth with the least toxic effects to the surrounding tissue are collectively referred to as having antimicrobial activity. As such, NMs can serve as effective and efficient vehicular carriers for various biomedical applications such as drug delivery, antimicrobial agents, vaccine delivery, and immunomodulation, respectively (Salata, 2004).

For a successful integration of nanotechnology into pharmaceuticals, the potent roles of various inorganic NPs (metals—Cu, Au, Ag, and Pt; and metal oxides—Ag_2O, SiO_2, TiO_2, CuO, CaO, ZnO, Fe_2O_3, and MgO) have been explored for their antimicrobial activities against various microorganisms (Rai and Bai, 2011; Dizaj et al., 2014). These studies revealed that the NPs act as potential microbial agents, and have intensively inhibited the growth of microbial species via various biopathways including reactive oxygen species (ROS) generation, inhibition of the electron transport chain, and depletion of antioxidant molecules (through the alteration of enzymatic production of ROS) (Ahn et al., 2014; Maurer and Meyer, 2016).

Several research groups have well described the mechanistic action mode of NPs (Johnston et al., 2010; Prabhu and Poulose, 2012; Koduru et al., 2018; Lemire et al., 2013). Furthermore, several review papers have already summarized the significance of the synthetic routes, characterization, and bioapplication of nanoparticles. Briefly, the key factors on antimicrobial behavior of Ag NPs have been thoroughly illustrated in the literature (Ouay and Stellacci, 2015). Similarly, the advances and possible risk factors of using NPs as microbial agents have also been extensively reviewed in the literature (Hajipour et al., 2012). Lemire's group (Lemire et al., 2013) illustrated the chemical behavior and toxicological effects of metal atoms on various microbial species. In another review, the significance of metals and metal oxide NPs' antibacterial properties and their mode of actions have been well described (Dizaj et al., 2014). Recent developments on the synthesis routes have also been discussed by focusing on antimicrobial properties of Ag NPs against various microbial species (bacteria, fungi, and viruses) (Tran et al., 2013). Moritz and Geszke-Moritz (2013) described the progress of synthesis routes for the fabrication of functional nanomaterials and their antimicrobial properties on various species.

In this chapter, we summarize the several physicochemical aspects that govern the action mode of metal NPs. To this end, we provided a comprehensive overview of how chemistry of metal NPs affects bacterial uptake of essential and nonessential metal ions. The antibacterial, antifungal, and antiviral mechanisms of Ag NPs have been described. The antimicrobial activity of Ag NPs on various microbial species is summarized in the Tables. Finally, we summarize the potential antimicrobial applications of Ag NPs in biomedical and pharmaceutical applications.

14.2 METHODS AVAILABLE FOR EVALUATING ANTIMICROBIAL ACTIVITY OF NANOPARTICLES

To evaluate and to know how Ag NPs show antimicrobial properties on various microbial species, various biotools are available for evaluating antimicrobial activity of nanoparticles. Importantly, the use of different biotools plays a key role for the estimation of antimicrobial behavior of Ag NPs. In view of this, various biological assays including cell counting, spread-plate colony counts, crystal violet staining, Tetrazolium salt reduction, live/dead staining, spread-plate colony counts, and optical density measurement, have been used for the confirmation of antimicrobial properties of various NMs including Ag NPs.

14.3 ANTIMICROBIAL ACTIVITY OF SILVER NANOPARTICLES

In recent years, silver NPs have shown significant advancements in various fields of sciences, medical, analytical sciences, material sciences, and health care fields (Silver et al., 2006). They have shown tremendous and very strong antibacterial properties against various bacterial species (Rai et al., 2009). Furthermore, nanostructured Ag materials have extensively been used as biomedical probes in treatment of wounds, burns, and in controlling various infectious diseases (Elliott, 2010; Aditya et al., 2013). It was noticed that silver acts as an antiseptic with a wide biocidal effect against various microbial species via disruption of enzyme activity and membrane function. Because silver ions might be destabilized and increase permeability of bacterial membranes, this results in the inhibition of functions of essential proteins and respiratory enzymes, which results in breaking up the replication of nucleic acids (RNA and DNA) of bacteria and inhibiting the ion transport process. Briefly, antimicrobial behavior of Ag NPs was evaluated on *Escherichia coli* and bacteriological tests were carried out in Luria–Bertani (LB) medium on solid agar plates using different concentrations of Ag NPs (Sondi and Salopek-Sondi, 2004). Authors noticed that the inhibition rate of *E. coli* is directly proportional to the concentration of Ag NPs, which confirms that concentration of Ag NPs plays an essential role to stop the growth of bacteria. The cells of *E. coli* were significantly damaged and destroyed the protein functions due to the accumulation of Ag NPs. It was also confirmed that the degree of antibacterial activity of Ag NPs is size dependent, since the small size Ag NPs (1–10 nm) have shown high tendency to interact with cell walls of bacteria (Morones et al., 2005). Pal's group (2007) studied the antibacterial activity of different morphology (shapes) of Ag NPs on *E. coli*. Authors noticed that the truncated triangular Ag NPs show higher antibacterial nature than the spherical and rod-shaped Ag NPs. Table 14.1 displays the antibacterial activity of functional Ag NPs on various bacteria.

Interestingly, surface chemistry of Ag NPs plays an essential role on the antibacterial activity of Ag NPs (Kvitek et al., 2008). To estimate the role of surface

Table 14.1 Functionalization of Ag NPs With Various Molecules and Their Antibacterial Properties on Various Bacterial Species

Ag NPs With Capping Agent and Size (nm)	Microorganism	Inhibitory Concentration	Reference
Ag NPs (12.0)	*Escherichia coli*	50–60 μg cm^{-3} and 20 μg cm^{-3}	Sondi and Salopek-Sondi (2004)
Ag NPs (1–100)	*Vibrio cholera*, *E. coli*, *Salmonella typhus* and *Pseudomonas aeruginosa*	75 μg mL^{-1}	Morones et al. (2005)
Different sizes and shapes of Ag NPs	*E. coli*	50–100 μg for spherical NPs, and >100 μg for rod-shape NPs	Pal et al. (2007)
Ag NPs with surfactants/polymers (26)	Bacterial strains isolated form clinical samples	1.69–13.5 μg mL^{-1}	Kvitek et al. (2008)
Ag and Au NPs derived from cell extract of, *Candida albicans* (5 and 30)	*Staphylococcus aureus* and *E. coli*	128–512 μg mL^{-1}	Ahmad et al. (2013)

chemistry on Ag NPs, various organic molecules such as sodium dodecyl sulfate—SDS, polyvinylpyrrolidone—PVP 360, and polyoxyethylenesorbitane monooleate—Tween 80 were functionalized onto Ag NPs surfaces and their antibacterial properties on various bacteria were evaluated. It was noticed that the 1.69–13.5 μg mL^{-1} concentration of Ag NPs exhibits minimum inhibitory concentrations (MICs), and the antibacterial nature of Ag NPs also depends on the bacterial species and surface chemistry of Ag NPs. Similarly, one-pot green synthetic approach was established for the fabrication of Au NPs and Ag NPs using natural surfactant isolated from *Trianthema decandra* (Geethalakshmi and Sarada, 2013). The antibacterial behaviors of both NPs were estimated by Kirby–Bauer method. Ahmad et al. (2013) developed eco-friendly biosynthetic method for the fabrication of highly monodisperse crystalline Au and Ag NPs using fungus cell extracts (*Candida albicans*). The fabricated Ag and Au NPs exhibit maximum absorption at 450 and 540 nm, and calculated their specific surface areas as 184.4 $m^2\ g^{-1}$ and 18.9 $m^2\ g^{-1}$. Authors evaluated their antibacterial behavior on *E. coli* and *Staphylococcus aureus*. Bindhu and Umadevi (2013) also synthesized Ag NPs by using *Hibiscus cannabinus* leaf extract as template. The antimicrobial properties of both NPs have been investigated on various microbial species including *Proteus mirabiles*, *E. coli*, and *Shigella flexneri*. In this work, Ag NPs were fabricated by using various concentration of Ag^+ ion (5 mM, 8 mM and

10^{-2} M) and studied their antimicrobial properties on the above species. It was revealed that Ag NPs synthesized by low concentration of Ag^+ ion (5.0 mM) showed higher degree of antimicrobial activity than that of other concentrations, signifying the formation of small size Ag NPs using low Ag^+ ion concentration, which favor an efficient interaction of small size Ag NPs with cell walls of bacteria.

Apart from these, Ag NPs have been synthesized by eco-friendly approaches that do not contain expensive and toxic chemicals (Ravindran et al., 2013). In these routes, Ag NPs were fabricated by either cell extracts from microbial species or plant extracts as templates, which provides green synthetic approaches that effectively replace the use of expensive/toxic chemicals. Generally, eco-friendly Ag NPs synthetic protocols involve (1) natural resources as template, (2) solvent medium, and (3) stabilization of Ag NPs with nontoxic chemicals (Kim et al., 2009). For example, Ag NPs have synthesized by using polysaccharides extracted from marine macroalgae [*Jania rubins* (Jr), *Pterocladia capillacae* (Pc), *Colpmenia sinusa* (Cs), and *Ulva faciata* (Uf)] as templates (El-Rafie et al., 2013). The fabricated Ag NPs showed antibacterial activity on Gram-negative and positive bacteria, and the cell walls of the bacteria species were significantly damaged by Ag NPs due to the inhibition of cellular protein functions by Ag NPs. Authors noticed that Gram-positive bacteria allowed less quantity of Ag^+ ion whereas Gram-negative bacteria had taken a higher amount of Ag NPs, suggesting the strong interactions between Ag NPs and Gram-negative bacteria. Antibacterial experiments on bacterial communities in activated sludge using 1 mg L^{-1} Ag NPs. The antibacterial property of Ag NPs was evaluated by using 16S rRNA gene based polymerase chain reaction–denaturing gradient gel electrophoresis (PCR–DGGE). These studies revealed that the microbes containing activated sludge showed high sensitivity towards Ag NPs and the number of bacterial cells was estimated by heterotrophic plate counts (HPCs) method.

Similarly, chitosan-4-[(E)-2-(3-hydroxynaphthalen-2-yl)diazen-1-yl] benzoic acid derivative was prepared by dicyclohexyl carbodiimide method (Fig. 14.1A) and then used as encapsulating agent for the encapsulation of Ag NPs (Mathew and Kuriakose, 2013a,b). The functional Ag NPs were characterized by spectroscopic (UV–visible, fluorescence, nuclear magnetic resonance, and Fourier-transform infrared) and electron microscopic (scanning electron microscopy and transmission electron microscopy) techniques. The antibacterial activity of chitosan derivative encapsulated Ag NPs was carried out on bacterial (*E. coli* and *S. aureus*) and fungal (*Aspergillus terreus* and *Aspergillus flavus*) species. In this, *S. aureus* and *E. coli* bacterial species were grown in agar plates by pour plate method. It was clearly observed that the generation of clear zone around the discs is indicative of antibacterial property of functional chitosan encapsulated-Ag NPs (Fig. 14.1B and C). The diameters of the zone of inhibition were found to be 20 ± 0.53 and 20 ± 0.45 nm for *S. aureus* and *E. coli*. The same group encapsulated Ag NPs were with bovine serum albumin (BSA) and studied their antibacterial properties on various bacterial species (*S. aureus*, *Serratia marcescens*,

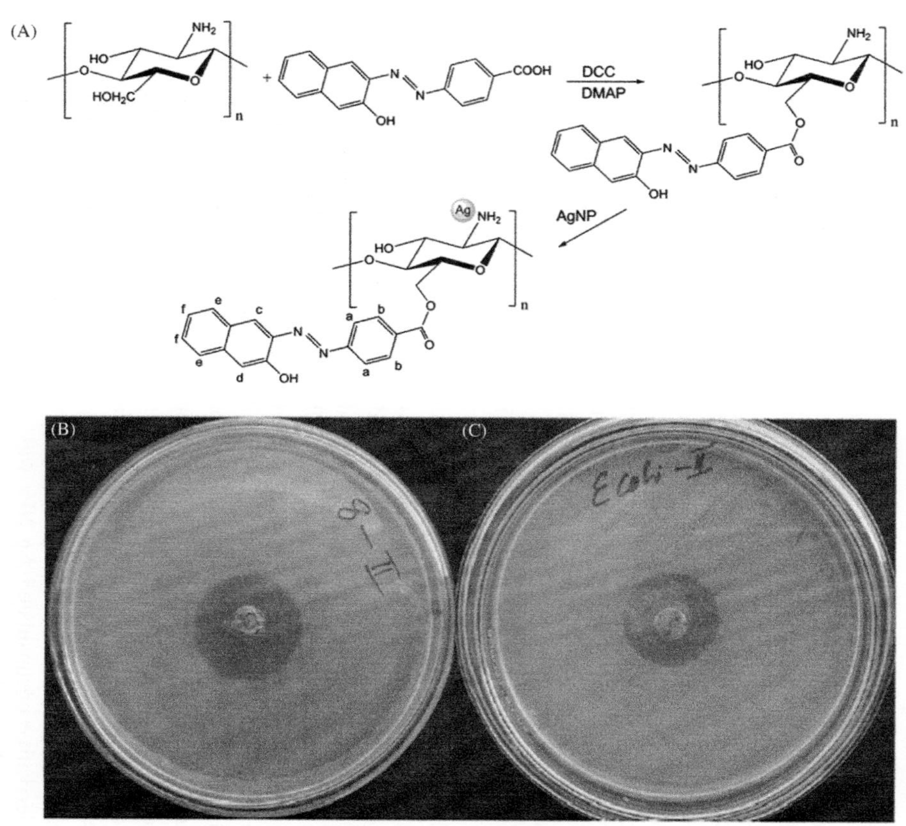

FIGURE 14.1

(A) Functionalization of Ag NPs with chitosan derivative. Study of antibacterial activity of chitosan derivative functionalized Ag NPs carried out on (B) *S. aureus* and (C) *E. coli*.

Reprinted from Mathew, T.V., Kuriakose, S., 2013. Photochemical and antimicrobial properties of silver nanoparticle-encapsulated chitosan functionalized with photoactive groups. Mater. Sci. Eng. C 33, 4409–4415. With permission.

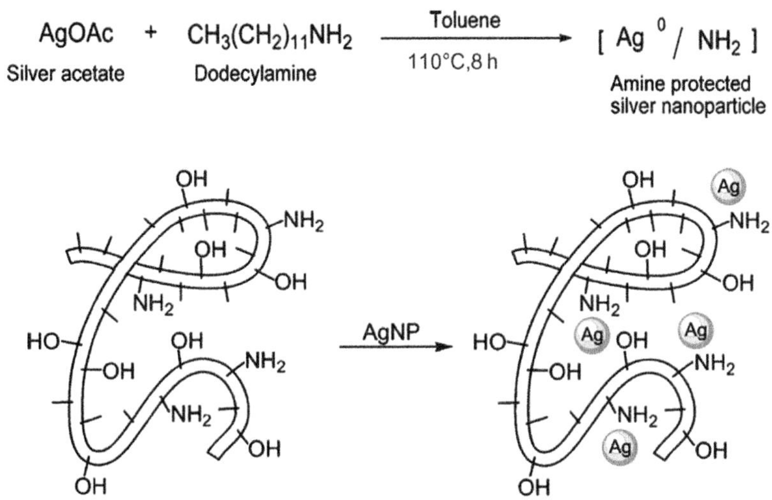

FIGURE 14.2

Fabrication of Ag NPs dodecylamine as a reducing agent and then functionalization of Ag NPs with BSA.

Reprinted from Mathew, T.V., Kuriakose, S., 2013. Studies on the antimicrobial properties of colloidal silver nanoparticles stabilized by bovine serum albumin. Colloids Surf. B 101, 14–18. With permission.

Klebsiella pneumonia, *E. coli*, and *Pseudomonas aeruginosa*) (Mathew and Kuriakose, 2013a,b). In this approach, Ag NPs were fabricated by the reduction of AgOAc using dodecyl amine as a reducing agent. Then, the fabricated Ag NPs were encapsulated with BSA (Fig. 14.2) and their antibacterial property on various bacterial species was studied by agar diffusion method. As shown in Fig. 14.3, clear zones were observed around the discs and the diameters of the inhibition zones were in the range of 24–25 mm for various bacterial species, signifying the antibacterial property of BSA–Ag NPs.

Suman et al. (2013) developed an eco-friendly approach for the fabrication of Ag NPs using root extract of *Morinda citrifolia* as a template. The derived Ag NPs exhibited absorption peak 413 nm, and are in spherical shape with an average diameter of 30–55 nm. The fabricated Ag NPs showed excellent cytotoxicity on HeLa cells. Gupta and coworkers described the use of one-step green synthetic approach for synthesis of Ag NPs using *Psidium guajava* L. as a template (Gupta et al., 2014). The as-prepared Ag NPs exhibited surface plasmon resonance (SPR) band at 487 nm and their mean diameter of ~60 nm. The phytochemicals protected Ag NPs exhibited excellent antimicrobial activity on various bacterial species such as *Pseudomonas diminuta*, *E. coli*, *Mycobacterium smegmatis*, *S. aureus*, and *C. albicans*, respectively. It was observed that bioactive organic

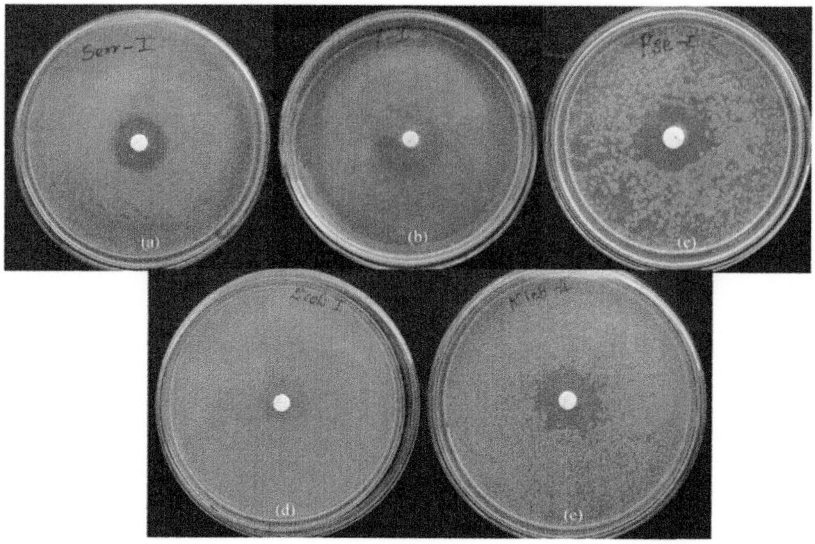

FIGURE 14.3

Examination of antibacterial properties of BSA–Ag NPs on various bacterial species (A) *S. aureus*, (B) *S. marcescens*, (C) *P. aeruginosa*, (D) *E. coli*, and (E) *K. pneumoniae*.

Reprinted from Mathew, T.V., Kuriakose, S., 2013. Studies on the antimicrobial properties of colloidal silver nanoparticles stabilized by bovine serum albumin. Colloids Surf. B 101, 14–18. With permission.

molecules (flavonoids, carbohydrates, terpenoids, and polyphenols) have found to be promising ligands to protect the surfaces of Ag NPs, which enhances their antimicrobial activities with high degree (Burkowska-But et al., 2014; Perni et al., 2014; Shankar et al., 2014). Further, Kumar's group developed a green synthetic approach for the fabrication of Ag NPs aqueous extract of *Alternanthera dentata* (Kumar et al., 2014). The derived Ag NPs showed SPR bank at 430 nm and studied antibacterial activity on various bacterial species such as *K. pneumonia*, *E. coli*, and *Enterococcus faecalis*. Another study explores the use of *P. guajava* leaf extract as promising template for the fabrication of Ag NPs (Parashar et al., 2011). The as-prepared Ag NPs were characterized by spectroscopic and electron microscopic techniques. The antibacterial activity of fabricated Ag NPs was carried out against *E. coli*. The synergetic antibacterial properties of Ag NPs were evaluated on various microbial species including *E. coli*, *S. aureus*, *P. diminuta*, *C. albicans*, *M. smegmatis*, *K. pneumonia*, *B. subtilis*, and *F. oxysporum*, respectively. These studies revealed that the fabricated Ag NPs from various plants extracts exhibited antimicrobial activities against various microbial species.

Table 14.2 summarizes the fabrication of Ag NPs from various plants extracts as templates and their antibacterial activities on various bacterial species. These methods have proven to be eco-friendly and cost-effective platforms for the synthesis of Ag NPs without use of expensive/toxic chemicals. Remarkably, the

Table 14.2 Summary of Ag NPs Derived From Various Plants Extracts and Their Antimicrobial Activities on Various Microbial Species

Synthesis of Ag NPs by Natural Resources	Studied Microbial Species	Reference
Rosmarinus officinalis leaf extract	*Bacillus subtilis*, *S. aureus*, *P. aeroginosa*, and *E. coli*	Ghaedi et al. (2015)
Ocimum tenuiflorum, Solanum tricobatum, Centella asiatica, Syzygium cumini, and *Citrus sinensis*	*P. aeruginosa*, *S. aureus*, *E. coli* and *Klebsiella pneumonia*	Logeswari et al. (2015)
Amphora sp.	*E. coli* and *Streptococcus mutans* and *Bacillus stearothermophilus*	Jena et al. (2015)
Flower broth of *Tagetes erecta*	Bacteria—*Bacills cereus*, *S. aureus*, *P. aeruginosa*, and *E. coli*; Fungi—*Cryptococcae neoformans*, *Candida albicans*, and *Candida glabrata*	Padalia et al. (2015)
Potentilla fulgens	MCF-7 and U-87 cell lines	Mittal et al. (2015)
Banana peel extract	Gram-positive bacteria (*B. subtilis* (local isolate), *S. aureus* (ATCC 6538)) and Gram-negative bacteria (*P. aeruginosa* (ATCC 9027), *P. aeruginosa* (local isolate), *E. coli* (ATCC 8739)) as well as the yeast *C. albicans* (ATCC 120231)	Ibrahim (2015)
Grewia flaviscences	*P. aeruginosa*	Sana et al. (2015)
Eucalyptus camaldulensis	*E. coli*, *P. aeruginosa*, *B. subtilis*, and *S. aureus*	Mohammed (2015)
Carica papaya leaf extract	*B. subtilis* (MCC 2511), *S. aureus* (MCC 2408), *Micrococcus luteus* MCC 2155, *K. pneumonia* (MCC 2451), *E. coli* (MCC 2412)	Banala et al. (2015)
Aeromonas salmonicida, Psychrobacter sp., *Yersinia kristensenii*, and *P. veronii*	*K. pneumoniae* (ATCC 29665), *E. coli* (ATCC 10536), and *S. epidermidis* (ATCC 12128)	Javani et al. (2015)
Mukia maderaspatana leaf extract	*K. pneumoniae* (MTCC 3384), *B. subtilis* (MTCC 1790), *Salmonella typhi* (MTCC 3224), and *S. aureus* (ATCC 25923)	Harshiny et al. (2015)
Leaves of *Moringa oleifera*, petals of *Cucurbita maxima* and rhizome of *Acorus calamus*	*B. subtilis*, *P. aeruginosa*, *V. cholera*, and *E. coli*	Nayak et al. (2015)
Penicillium atramentosum KM	*B. cereus*, *S. aureus*, and *Aeromonas hydrophila*	Sarsar et al. (2015)
Root extracts of *Diospyros paniculata*	Bacteria—*B. subtilis*, *Streptococcus pyogenes*, *Bacillus pumilis*, *S. aureus*, *E. coli*, *Proteus vulgaris*, *K. pneumoniae*, and *P. aeruginosa*. Fungi—*Aspergillus flavus*, *Penicillium notatum*, *Aspergillus niger*, *Saccharomyces cerevisiae*, and *C. albicans*	Rao et al. (2016)

(*Continued*)

Table 14.2 Summary of Ag NPs Derived From Various Plants Extracts and Their Antimicrobial Activities on Various Microbial Species *Continued*

Synthesis of Ag NPs by Natural Resources	Studied Microbial Species	Reference
Ginkgo Biloba leaf extract	*E. coli* and *S. aureus*	Ren et al. (2016)
Mycosynthesis	*B. subtilis* and *S. aureus*	Balakumaran et al. (2016)
Azadirachta indica and *Ficus benghalensis* (Bark extract)	*P. aeruginosa*, *E. coli*, *V. cholera*, *B. subtilis*	Nayak et al. (2016)
Sesbania grandiflora leaf extract	Bacteria—*P.* spp., *E. coli*, *S.* spp., and *B.* spp. Fungi—*A. flavus* spp., *A. niger* spp., and *Penicillium* spp.	Ajitha et al. (2016)
Crude extract of *Bergenia ciliata*	*Micrococcus luteus*, *S. aureus*, *Bordetella bronchiseptica*, and *Enterobactor aerogens*	Phull et al. (2016)
Atrocarpus altilis leaf extract	Bacteria—*P. aeruginosa*, *E. coli*, *S. aureus*; Fungus—*A. vesicolor*	Ravichandran et al. (2016)
Cacao extract	*S. aureus*, *E. coli*, *S. epidermidis*, and *P. aeruginosa*	Chowdhury et al. (2016)
Extract of *Azadirachta indica* (leaves)	*S. aureus* and *E. coli*	Verma and Mehata (2016)
Heterotheca inuloides, *Agave lechuguilla*	*E. coli* and *S. aureus*	Morales-Luckie et al. (2016)
Pod extract of *Cola nitida*	*P. aeruginosa*, *K. granulomatis*, and *E. coli*, *S. aureus*, *A. niger*, *A. flavus*, *P. aeruginosa*, and *A. fumigatus*	Lateef et al. (2016)
Taxus baccata extracts	HeLa cells	Kajani et al. (2016)
Sporosarcina koreensis DC4 strain	*E. coli*, *Salmonella enterica*, *Vibrio parahaemolyticus*, *B. cereus*, *B. anthracis*, and *S.aureus*	Singh et al. (2016)
Prunus x yedoensis tree leaf extract	*B. subtilis* (6 mm) and *P. aeruginosa*	Manikandan et al. (2016)
M. chamomile plant extract	*E. coli*, *S. aureus*, and *C. albicans* fungal species	Ocsoy et al. (2017)
Liquor of *Eucalyptus wood*	Bacteria—*S. aureus*, *E. coli*, and *P. aeruginosa* Fungi—*C. oxysporum*, *C. albicans*, *P. chrysogenum*, and *A. niger*	Shivakumar et al. (2017)
Pelargonium endlicherianum Fenzl. root extracts	*E. coli*, *P. aeruginosa*, *S. epidermidis*	Karatoprak et al. (2017)
Sargassum polycystum	Human colon cancer (HT-29) cells	Palanisamy et al. (2017)
Tree oyster mushroom *Pleurotus ostreatus*	*B. cereus*, *B. subtilis*, *E. coli*, *S. aureus*, and *P. aeruginosa*	Al-Bahrani et al. (2017)

phytochemicals protected Ag NPs exhibited excellent antibacterial properties due to the presence of bioactive molecules on the surfaces of Ag NPs. Importantly, Ag^+ ions are gradually leached from Ag NP and show high ability to bind with either phosphate group of DNA or sulfur-containing bacterial proteins, which results in inhibiting the growth of microbial cells.

14.4 MECHANISM AND ROLES OF AG NPS AND AG^+ ION IN ANTIMICROBIAL ACTIVITY

For better understanding of the antimicrobial property of Ag NPs, it is essential to describe the mode of action and role of Ag NPs and Ag^+ ion in inhibiting the growth of microbial species. In antimicrobial action, two distinctive steps are involved in controlling the growth of microbial species. The first step involves modification stabilization of Ag NPs surfaces according to the microbial environment via various process dissolution, aggregation, photoredox reaction, and releasing of Ag^+ ions from Ag NPs. The second step is the interaction of Ag^+ ion–Ag NPs with the cell walls of bacteria, resulting in inhibition of the functions of bacterial proteins, which stops bacteria growth and leads to bacterial cell death. However, the degree of antimicrobial activity of Ag NPs also depends on various factors including physicochemical properties, size, morphology, and ligand chemistry of Ag NPs, medium and microbial environment and concentration of Ag NPs, respectively (Misra et al., 2012). Even though many studies have well described the synthetic parameters and antibacterial activity of Ag NPs on various bacterial species, unfortunately the exact antibacterial mechanism of Ag NPs is still unexplored. Thus, much efforts need to be focused on understanding and exploring the exact mechanism of the antibacterial nature of Ag NPs in the near future.

It is well known that Ag NPs have shown high tendency to interact with bacteria, allowing the accumulation of Ag NPs onto the surfaces of bacterial membrane, which results in formation of Ag NPs-bacteria aggregates. As a result, the bacterial cell membranes were greatly damaged due to the inhibition of bacterial cell proteins, which leads to bacterial cell death (Li et al., 2010). Furthermore, the antibacterial function of Ag NPs is highly dependent on the interaction of bacterial cell walls and the surfaces of Ag NPs. It was noticed that the different sizes of Ag NPs did not show the same antibacterial action, which is due to the interaction ability of Ag NPs with bacterial cells (Chernousova and Epple, 2013). The degree of cytotoxicity of Ag NPs may also be dependent on the production of ROS, which can be considered as a primary cytotoxic action by Ag NPs (Ouay and Stellacci, 2015). As a result, the bacterial cellular proteins functions were inactivated due to the generation of high oxidative stress by ROS via Ag NPs. These observations revealed that Ag NPs may act as catalyst to generate high-level ROS, generating high oxidative stress, which inhibits the biopathways of bacterial cells.

Generally, Ag^+ ions were gradually released by oxidative dissolution of Ag NPs, suggesting the Ag^+ ion act as a promising antibacterial agent where Ag NPs act as a source. The chemistry of the Ag^+ ion revealed that it has high affinity to bind with bioactive molecules those are containing $-NH_2$, phosphates and mercapto groups, which leads to the formation of quasi-covalent bond between Ag^+ ion and bioactive molecules (Barngrover and Aikens, 2011). Remarkably, Ag^+ ions act as a coordinating agent to bind with several mercapto bioactive molecules, resulting in the formation of large molecular chains/networks, which leads to irreversible aggregation of mercapto bioactive molecules (Parikh et al., 1999). As a result, Ag^+ ions significantly inactivated the biopathways of bacterial cells, thus inhibiting the replication of nucleic acids and mitochondria functions, which results in bacteria death.

Another hypothesis describes that low levels of Ag^+ ions are very sensitive to disturb the cellular function of bacteria, which causes the significant death of bacteria. Further, several experimental investigations revealed that ROS were generated due to the physicochemical action of Ag NPs (Marambio-Jones and Hoek, 2010). It was also noticed that ROS can be formed by photochemical oxidation of H_2O in the presence of catalyst (Campos-Martin et al., 2006). These generated ROS cause significant oxidative stress on the cellular system of bacterial species. Finally, the generated ROS disturbed the functions of bacterial proteins, which suggests that Ag NPs exhibit antibacterial properties.

Another antibacterial study revealed that the nanomolar concentration of Ag NPs significantly caused the death of *E. coli* within minutes because of the rapid dissipation of proton motive force (Sharma et al., 2009). The similar action was also found in another antimicrobial studies of Ag^+ ions (Dibrov et al., 2002). It was observed that trace levels of Ag^+ ions cause a massive leakage of protons via the membrane of *Vibrio cholerae*. The leakage of protons is due to either Ag^+ ion-conjugated bacterial protein or Ag^+ ion-conjugated phospholipids, and this mechanism leads to the death of bacteria (Dibrov et al., 2002). Interestingly, nanomolar Ag NPs show high degree of antibacterial activity whereas micromolar Ag^+ ions exhibit efficient antibacterial activities. The role of Ag^+ ion was also investigated by slight oxidation of Ag NPs (Sharma et al., 2009). The oxygen causes the oxidation of Ag NPs, resulting to generate Ag^+ ion–Ag NPs via chemisorption. The chemisorbed Ag^+ ion–Ag NPs exhibited promising antibacterial activity on *E. coli*, however it may depends on the degree of chemisorbed Ag^+ ion and size of Ag NPs. Overall, the small size Ag NPs have significantly inhibited the bacterial growth via adhesion mechanism by Ag NPs with bacterial cell walls. Then, Ag NPs slowly enter into the bacteria via disrupting cell walls, inhibiting the regular functions of proteins, which stops replication of DNA, and leads to bacteria death.

14.5 ANTIFUNGAL NATURE OF AG NPS

It has been noticed that fungi are considered to be potential pathogens that can cause serious health problems to humans (Enoch et al., 2006). Apart from antibacterial behavior of Ag NPs, several studies have also confirmed the antifungal properties of Ag NPs against various fungal species. For example, Kim's group fabricated nano-Ag particles and studied their antifungal activities on *C. albicans* and *Trichophyton mentagrophytes* (Kim et al., 2009). It was noticed that the nano-size Ag NPs exhibited antifungal activity on *C. albicans* and showed minimum inhibition concentration of 2.0 μg mL^{-1}, which is similar to the antifungal activity of amphotericin B (AmB). Further, nanosize Ag exhibited lower hemolytic activity whereas AmB showed higher hemolytic activity. These results indicate that nanosize Ag can be used as a potential antifungal agent against various fungal diseases. The same group explored the use of spherical Ag NPs for investigation of antifungal activity on *Candida* species and ATCC strains of *T. mentagrophytes* clinical isolates (Kim et al., 2008). The antifungal studies of Ag NPs revealed that the antifungal activity of Ag NPs is equal to amphotericin B, and higher to that of fluconazole. Thus, nanosize Ag NPs act as potential antifungal agent for various clinical applications. Roe et al. (2008) described the simple synthetic route for coating of catheters with Ag NPs using N,N,N′,N′-tetramethylethylenediamine as a reducing agent. The antibacterial and antifungal activities of Ag NPs were carried out on various microbial species including *E. coli*, *S. aureus*, *P. aeruginosa*, and *C. albicans*, respectively. This study revealed that Ag NPs were effectively coated onto plastic catheters and showed antimicrobial activities on various microbial species. Panácek et al. (2009) fabricated Ag NPs by Tollens process and evaluated their antifungal properties on *Candida* spp. via the MIC, and minimum fungicidal concentration (MFC). To evaluate the antifungal property of Ag NPs, authors estimated the growth of inhibition of yeast by measuring the optical density of the samples at different time intervals (0–72 hours) (Fig. 14.4). It was observed that ionic silver exhibited higher growth inhibition rates than Ag NPs. However, the Ag NPs showed MIC 0.21 mg L^{-1} and the Ag NPs inhibitory effect was enhanced by the stabilization of Ag NPs with surfactant (SDS). Authors noticed that antifungal activity of Ag NPs was greatly improved by the modification of Ag NPs with surfactants and polymer. The cell walls of yeasts were significantly disrupted by surfactant activity. Furthermore, authors also investigated that the antifungal activity of Ag NPs with various surface functional molecules (Tween 80, PVP 360, SDS, and Brij) (Fig. 14.5). Among these functional Ag NPs, SDS–Ag NPs exhibited the best antifungal activity against fungal species of *C. albicans I* and *C. albicans II*. These studies revealed that the surface chemistry of Ag NPs also plays a key role in enhancing the antimicrobial properties of Ag NPs. Mallmann et al. (2015) studied the antimicrobial activity of Ag NPs derived from ribose as a reducing agent. The as-synthesized Ag NPs were stabilized with SDS and their antifungal properties on *Candida tropicalis* and *C. albicans* were studied, suggesting the fabricated Ag NPs exhibited antifungal activity on the above fungal species. Narayanan and Park (2014) illustrated green synthetic approach for the fabrication of Ag NPs using the extract of turnip leaf. The fabricated Ag NPs are in spherical shape with a mean diameter of 16.14 nm, and showed zeta potential value at −14.8 mV. The

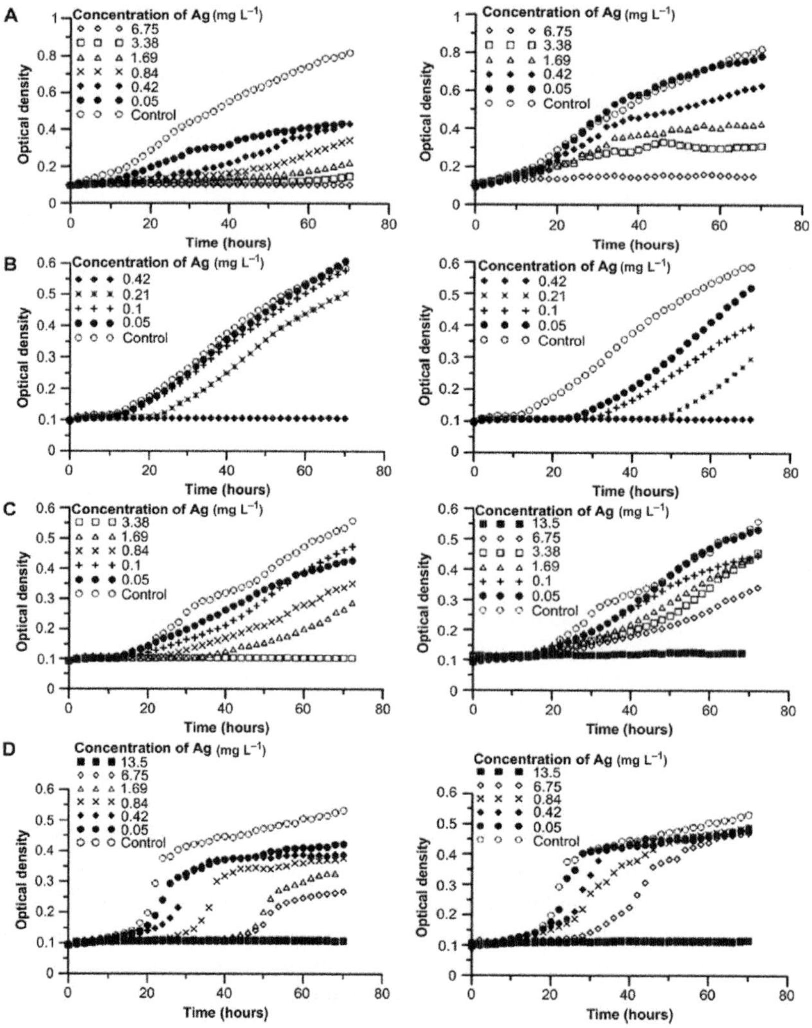

FIGURE 14.4

Evaluation of antifungal activity of Ag NPs through growth curves against (A) *C. albicans I*, (B) *C. albicans II*, (C) *C. parapsilosis* and (D) *C. tropicalis*, respectively.

Reprinted from Panácek, A., Kola, M., Vec, R., Prucek, R., Soukupova, J., Hamal, P., Zbor. R., 2009. Antifungal activity of silver nanoparticles against Candida spp. *Biomaterials 30, 6333–6340. With permission.*

antifungal activity of Ag NPs was carried out against various fungal species (*Gloeophyllum abietinum*, *Chaetomium globosum*, *Gloeophyllum trabeum*, and *Phanerochaete sordid*). These studies suggest that Ag NPs exhibited wide spectrum antifungal properties against wood-degrading fungal species. Monteiro et al. (2013)

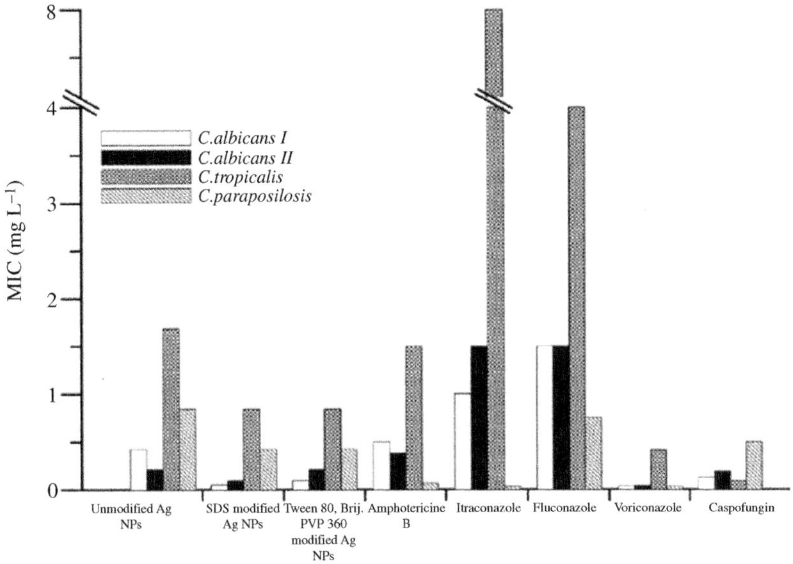

FIGURE 14.5

Evaluation of minimum inhibitory concentration ability of bare and functionalized Ag NPs on isolates of *Candida* spp.

Reprinted from Panáček, A., Kola, M., Vec, R., Prucek, R., Soukupova, J., Hamal, P., Zbor. R., 2009. Antifungal activity of silver nanoparticles against Candida *spp. Biomaterials 30, 6333–6340. With permission.*

studied the antifungal efficacy of Ag NPs with the addition of chlorhexidine digluconate (CHG) and of nystatin (NYT) on *Candida glabrata* and *C. albicans* biofilms. These studies revealed that Ag NPs with CHG and NYT exhibited synergistic antifungal activity and found that this activity is dependent on the fungal species. Thus, Ag NPs with NYT/CHG can be used as promising antifungal agents for clinical implications in the fungal related diseases.

Recently, Elgorban et al. (2016) explored the antifungal application of Ag NPs on *Rhizoctonia solani*. Antifungal activity was carried out using Ag NPs at six different concentration levels (0.0, 0.0002, 0.0005, 0.0007, 0.0009, 0.0014, and 0.0019 mol L^{-1}) on six different *R. solani* anastomosis groups (AGs), which shows serious effects on cotton plants. It was noted that Ag NPs exhibited high tendency to inhibit the growth of R. solani AGs on Czapek Dox agar using various concentration of Ag NPs. Tutaj et al. (2016) fabricated a novel Ag NPs–polymer based hybrid system using AmB as a reducing and stabilizing/capping agent. The as-synthesized AmB–Ag NPs were spherical in shape with an average size of 7.0 nm. The synthesized AmB–Ag NPs antifungal activity was investigated on various fungal species such as *Fusarium culmorum*, *Aspergillus niger*, and *C. albicans*, respectively. Further, authors also evaluated their cytotoxicity studies on cell lines of CCD-841CoTr and THP-1 cell. It was noticed that AmB–Ag NPs act as promising antifungal agent for the above species.

14.6 ANTIVIRAL NATURE OF AG NPS

Apart from the potential antibacterial and antifungal activities, Ag NPs have also been found to have potential antiviral activities on various viral species including hepatitis B virus (HBV), human immunodeficiency virus, respiratory syncytial virus (RSV), herpes simplex virus, and monkey pox virus, respectively (Galdiero et al., 2011; Rai et al., 2016). These studies illustrated that Ag NPs have become novel agents in various antiviral therapies.

In view of this, several research groups have been dedicated their work to evaluate the antiviral properties of Ag NPs on various viruses. Briefly, Sun et al. (2005) synthesized Ag NPs using [4-(2-hydroxyethyl)-1-piperazineethanesulfonic acid] (HEPES) buffer and studied their potent cytoprotective and antiviral activity on HIV-1 toward Hut/CCR5 cells. Similarly, polyvinylpyrrolidone (PVP) was decorated onto the surfaces of Ag NPs and their antiviral properties on HIV-1 were studied (Lara et al., 2010). It was noticed that PVP-Ag NPs exhibited high tendency to stop transmission of HIV-1 infection in human cervical culture, confirming the PVP–Ag NPs act as a promising antiviral agent to prevent the transmission of the HIV-1 virus. Lu et al. (2008) fabricated 10 nm size Ag NPs using HEPES buffer and studied their antiviral activity on HBV. Authors noticed that the fabricated Ag NPs showed high binding affinity towards HBV DNA, inhibiting the production of extracellular virions and HBV RNA, which confirms that Ag NPs act as an antiviral agent. Sun et al. (2008) decorated Ag NPs with various bioactive molecules (PVP, recombinant RSV fusion protein, and BSA) with polylactic acid for the inhibition of RSV infection in cells. Authors noticed that PVP–Ag NPs act as a promising antiviral agent for the inhibition of RSV, which can be used for the treatment of RSV in animal models. Baram-Pinto et al. (2009) described the use of mercaptoethane sulfonate functionalized Ag NPs for antiviral activity against herpes simplex virus type 1 (HSV-1) into cells. Antiviral studies revealed that the mercaptoethane sulfonate functionalized Ag NPs have successfully prevented the viral entry as well as stopped subsequent infection in the cells. Authors concluded that mercaptoethane sulfonate molecule on Ag NPs surfaces plays a key role to bind with HSV-1 virus. Rogers et al. (2008) illustrated the antiviral activity of polysaccharide coated Ag NPs against monkeypox virus.

Further, Speshock et al. (2010) evaluated the novel antiviral activity of Ag NPs on a new type of world arenavirus, that is, Tacaribe virus. It was observed that the Ag NPs effectively blocked the RNA production of virus and stopped the release of progeny, which signifies that Ag NPs act as potential candidates for the inhibition of arenavirus infection in vitro. Elechiguerra et al. (2005) demonstrated the size-dependent interaction of Ag NPs with HIV-1. Authors summarized that Ag NPs have shown high affinity to bind with HIV-1 virus through possible attachment with the gp120 glycoprotein knobs, which leads to stop the binding of virus with host cells. The chitosan was coated on Ag NPs surfaces and their antiviral property on H1N1 influenza A virus was evaluated (Mori et al., 2013). Authors observed that the degree of antiviral activity of chitosan–Ag NPs against H1N1 virus was progressively increased with increasing concentration of Ag NPs. The small size Ag NPs exhibited high antiviral activity than that of large size Ag NPs. Trefry and Wooley (2012) developed a novel colorimetric assay for the rapid assessment of HIV-1 virus. Orlowski et al. (2014) evaluated the antiviral activity of different sizes of tannic acid protected Ag NPs (13, 33 and 46 nm) against herpes simplex virus type 2 (HSV-2) in vitro and in vivo. It was noticed that small size Ag NPs highly induced the generation of cytokines and chemokines, suggesting that Ag NPs act as good candidates for the treatment of HSV-2 infections. Yang et al. (2016) synthesized curcumin capped Ag NPs and studied their antiviral properties on RSV. Authors observed that curcumin capped Ag NPs efficiently inhibited the RSV activity, which can be used as a novel viricide for the treatment of RSV infections. Based on these reports, Ag NPs can be used as promising and potential antiviral agents for the treatment of various viral infections.

14.7 SUMMARY

In this book chapter, we explored the potential antimicrobial (antibacterial, antifungal, and antiviral) applications of Ag NPs on various microbial species. The size, capping agent, and concentration of Ag NPs play essential roles in inhibiting the microbial growth. Based on the observations, the biological activity of Ag NPs is highly dependent on several factors such stabilizing agents, size, and concentration of Ag NPs, respectively. In general, antibacterial mechanism is mainly due to the disruption of bacterial cell walls and inhibition of bacterial proteins functions, which inhibits the growth of bacteria. Regarding the antifungal properties of Ag NPs the most significant antifungal activity was observed on various fungal species including *Candida* species. The functionalized Ag NPs were effectively inhibited the growth of fungal species. Similarly, functionalized Ag NPs have shown high affinity to bind with specific groups of viral species, stopping the production of viral RNA. Overall, functional Ag NPs are promising antimicrobial candidates for the inhibition of a wide spectrum of microbial species in various clinical applications.

ACKNOWLEDGMENT

S.K.K. acknowledges the financial support from Department of Science and Technology, Government of India (EMR/2016/002621/IPC).

REFERENCES

Ahn, J.M., Eom, H.J., Yang, X., Meyer, J.N., Choi, J., 2014. Comparative toxicity of silver nanoparticles on oxidative stress and DNA damage in the nematode, *Caenorhabditis elegans*. Chemosphere 108, 343–352.

Aditya, N.P., Vathsala, P.G., Vieira, V., Murthy, R.S.R., Souto, E.B., 2013. Advances in nanomedicines for malaria treatment. Adv. Colloid Interface Sci. 201–202, 1–17.

Ahmad, T., Wani, I.A., Manzoor, N., Ahmed, J., Asiri, A.M., 2013. Biosynthesis, structural characterization and antimicrobial activity of gold and silver nanoparticles. Colloids Surf. B 107, 227–234.

Ajitha, B., Reddy, Y.A.K., Rajesh, K.M., Reddy, P.S., 2016. *Sesbania grandiflora* leaf extract assisted green synthesis of silver nanoparticles: antimicrobial activity. Mater. Today 3, 1977–1984.

Al-Bahrani, R., Raman, J., Lakshmanan, H., Hassan, A.A., Sabaratnam, V., 2017. Green synthesis of silver nanoparticles using tree oyster mushroom *Pleurotus ostreatus* and its inhibitory activity against pathogenic bacteria. Mater. Lett. 186, 21–25.

Banala, R.R., Nagati, V.B., Karnati, P.R., 2015. Green synthesis and characterization of *Carica papaya* leaf extract coated silver nanoparticles through X-ray diffraction, electron microscopy and evaluation of bactericidal properties. Saudi J. Biol. Sci. 22, 637–644.

Balakumaran, M.D., Ramachandran, R., Balashanmugam, P., Mukeshkumar, D.J., Kalaichelvan, P.T., 2016. Mycosynthesis of silver and gold nanoparticles: optimization, characterization and antimicrobial activity against human pathogens. Microbiol. Res. 182, 8–20.

Baram-Pinto, D., Shukla, S., Perkas, N., Gedanken, A., Sarid, R., 2009. Inhibition of herpes simplex virus type 1 infection by silver nanoparticles capped with mercaptoethane sulfonate. Bioconjug. Chem. 20, 1497–1502.

Barngrover, B.M., Aikens, C.M., 2011. Incremental binding energies of gold(I) and silver (I) thiolate clusters. J. Phys. Chem. A 115, 11818–11823.

Bindhu, M.R., Umadevi, M., 2013. Synthesis of monodispersed silver nanoparticles using *Hibiscus cannabinus* leaf extract and its antimicrobial activity. Spectrochim. Acta A 101, 184–190.

Burkowska-But, A., Sionkowski, G., Walczak, M., 2014. Influence of stabilizers on the antimicrobial properties of silver nanoparticles introduced into natural water. J. Environ. Sci. 26, 542–549.

Campos-Martin, J.M., Blanco-Brieva, G., Fierro, J.L.G., 2006. Hydrogen peroxide synthesis: an outlook beyond the anthraquinone process. Angew. Chem., Int. Ed. 45, 6962–6984.

Chernousova, S., Epple, M., 2013. Silver as antibacterial agent: ion, nanoparticle, and metal. Angew. Chem. Int. Ed. 52, 1636–1653.

Chowdhury, N.R., MacGregor-Ramiasa, M., Zilm, P., Majewski, P., Vasilev, K., 2016. 'Chocolate' silver nanoparticles: synthesis, antibacterial activity and cytotoxicity. J. Colloid Interf. Sci 482, 151–158.

Dibrov, P., Dzioba, J., Gosinl, K.K., Hase, C.C., 2002. Chemiosmotic mechanism of antimicrobial activity of Ag(+) in *Vibrio cholera*. Antimicrob. Agents Chemother. 46, 2668–2670.

Dizaj, S.M., Lotfipour, F., Barzegar-Jalali, M., Zarrintan, Khosro Adibkia, M.H., 2014. Antimicrobial activity of the metals and metal oxide nanoparticles. Mater. Sci. Eng. C 44, 278–284.

Elgorban, A.M., El-Samawaty, Yassin, A.M., Sayed, M.A., Adil, S.R., Elhindi, S.F., et al., 2016. Antifungal silver nanoparticles: synthesis, characterization and biological evaluation. J. Biotechnol. Biotechnol. Equip. 30, 56–62.

Elechiguerra, J.L., Burt, J.L., Morones, J.R., Camacho-Bragado, A., Gao, X., Lara, H.H., et al., 2005. Interaction of silver nanoparticles with HIV-1. J. Nanobiotechnol. 29, 3–6.

Elliott, C., 2010. The effects of silver dressings on hronic and burns wound healing. Br. J. Nur. 19, 32–36.

El-Rafie, H.M., El-Rafie, M.H., Zahran, M.K., 2013. Green synthesis of silver nanoparticles using polysaccharides extracted from marine macro algae. Carbohydr. Polym. 96, 403–410.

Enoch, D.A., Ludlam, H.A., Brown, N.M., 2006. Invasive fungal infections: a review of epidemiology and management options. J. Med. Microbiol. 55, 809–818.

Galdiero, S., Falanga, A., Vitiello, M., Cantisani, M., Marra, V., Galdiero, M., 2011. Silver nanoparticles as potential antiviral agents. Molecules 16, 8894–8918.

Geethalakshmi, R., Sarada, D.V.L., 2013. Characterization and antimicrobial activity of gold and silver nanoparticles synthesized using saponin isolated from *Trianthema decandra* L. Ind. Crop. Prod. 51, 107–115.

Ghaedi, M., Yousefinejad, M., Safarpoor, M., Khafri, H.Z., Purkait, M.K., 2015. *Rosmarinus officinalis* leaf extract mediated green synthesis of silver nanoparticles and investigation of its antimicrobial properties. J. Ind. Eng. Chem. 31, 167–172.

Gupta, K., Hazarika, S.N., Saikia, D., Namsa, N.D., Mandal, M., 2014. One step green synthesis and anti-microbial and anti-biofilm properties of *Psidium guajava* L. leaf extract-mediated silver nanoparticles. Mater. Lett. 125, 67–70.

Hajipour, M.J., Fromm, K.M., Ashkarran, A.A., de Aberasturi, D.J., de Larramendi, I.R., Rojo, T., et al., 2012. Antibacterial properties of nanoparticles. Trends Biotechnol. 30, 499–511.

Harshiny, M., Matheswaran, M., Arthanareeswaran, G., Kumaran, S., Rajasree, S., 2015. Enhancement of antibacterial properties of silver nanoparticles–ceftriaxone conjugate through *Mukia maderaspatana* leaf extract mediated synthesis. Ecotoxicol. Environ. Saf. 121, 135–141.

Ibrahim, H.M.M., 2015. Green synthesis and characterization of silver nanoparticles using banana peel extract and their antimicrobial activity against representative microorganisms. J. Radiat. Res. Appl. Sci. 8, 265–275.

Javani, S., Marín, I., Amils, R., Abad, J.P., 2015. Four psychrophilic bacteria from Antarctica extracellularly biosynthesize at low temperature highly stable silver nanoparticles with outstanding antimicrobial activity. Colloids Surf. A 483, 60–69.

Jena, J., Pradhan, N., Dash, B.P., Panda, P.K., Mishra, B.K., 2015. Pigment mediated biogenic synthesis of silver nanoparticles using diatom *Amphora* sp. and its antimicrobial activity. J. Saudi Chem. Soc. 19, 661–666.

Johnston, H.J., Hutchison, G., Christensen, F.M., Peters, S., Hankin, S., Stone, V., 2010. A review of the in vivo and in vitro toxicity of silver and gold particulates: particle attributes and biological mechanisms responsible for the observed toxicity. Crit. Rev. Toxicol. 40, 328–346.

Kajani, A.A., Zarkesh-Esfahani, S.H., Bordbar, A.–K., Khosropour, A.R., Razmjou, A., Kardi, M., 2016. Anticancer effects of silver nanoparticles encapsulated by *Taxus baccata* extracts. J. Mol. Liq. 223, 549–556.

Karatoprak, G.S., Aydin, G., Altinsoy, B., Altinkaynak, C., Koşar, M., Ocsoy, I., 2017. The effect of *Pelargonium endlicherianum* Fenzl. root extracts on formation of nanoparticles and their antimicrobial activities. Enzyme Microb. Technol. 97, 21–26.

Kim, K.J., Sung, W.S., Moon, S.K., Choi, J.S., Kim, J.G., Lee, D.G., 2008. Antifungal effect of silver nanoparticles on dermatophytes. J. Microbiol. Biotechnol. 18, 1482–1484.

Kim, K.J., Sung, W.S., Suh, B.K., Moon, S.K., Choi, J.S., Kim, J.G., et al., 2009. Antifungal activity and mode of action of silver nano-particles on *Candida albicans*. Biometals 22, 235–242.

Koduru, J.R., Kailasa, S.K., Bhamore, J.R., Kim, K.H., Dutta, T., Vellingiri, K., 2018. Phytochemical-assisted synthetic approaches for silver nanoparticles antimicrobial applications: a review. Adv. Colloid Interface Sci. 256, 326–339.

Kumar, D.A., Palanichamy, V., Roopan, S.M., 2014. Green synthesis of silver nanoparticles using *Alternanthera dentata* leaf extract at room temperature and their antimicrobial activity. Spectrochim. Acta A 127, 168–171.

Kvitek, L., Panacek, A., Soukupova, J., Kolar, M., Vecerova, R., Prucek, et al., 2008. Effect of surfactants and polymers on stability and antibacterial activity of silver nanoparticles (NPs). J. Phys. Chem. C 112, 5825–5834.

Lara, H.H., Ixtepan-Turrent, L., Garza-Treviño, E.N., Rodriguez-Padilla, C., 2010. PVP-coated silver nanoparticles block the transmission of cell-free and cell-associated HIV-1 in human cervical culture. J. Nanobiotechnol. 8, 15–25.

Lateef, A., Azeez, M.A., Asafa, T.B., Yekeen, T.A., Akinboro, A., Oladipo, I.C., et al., 2016. Biogenic synthesis of silver nanoparticles using a pod extract of *Cola nitida*: antibacterial and antioxidant activities and application as a paint additive. J. Taibah Univ. Sci. 10, 551–562.

Lemire, J.A., Harrison, J.J., Turner, R.J., 2013. Antimicrobial activity of metals: mechanisms, molecular targets and applications. Nat. Rev. Microbiol. 11, 371–384.

Li, W.-R., Xie, X.-B., Shi, Q.-S., Zeng, H.-Y., OU-Yang, Y.-S., Chen, Y.-B., 2010. Antibacterial activity and mechanism of silver nanoparticles on *Escherichia coli*. Appl. Microbiol. Biotechnol. 85, 1115–1122.

Logeswari, P., Silambarasan, S., Abraham, J., 2015. Synthesis of silver nanoparticles using plants extract and analysis of their antimicrobial property. J. Saudi Chem. Soc. 19, 311–317.

Lu, L., Sun, R.W., Chen, R., Hui, C.K., Ho, C.M., Luk, J.M., et al., 2008. Silver nanoparticles inhibit hepatitis B virus replication. Antivir. Ther. 13, 253–262.

Mallmann, E.J.J., Cunha, F.A., Castro, N.M.F.B., Maciel, A.M., Menezes, E.A., Fechine, P.B.A., 2015. Antifungal activity of silver nanoparticles obtained from green synthesis. Rev. Inst. Med. Trop. Sao Paulo 57, 165–167.

Manikandan, V., Velmurugan, P., Park, J.-H., Lovanh, N., Seo, S.-K., Jayanthi, P., et al., 2016. Synthesis and antimicrobial activity of palladium nanoparticles from *Prunus* × *yedoensis* leaf extract. Mater. Lett. 185, 335–338.

Maurer, L.L., Meyer, J.N., 2016. A systematic review of evidence for silver nanoparticle-induced mitochondrial toxicity. Environ. Sci. Nano 3, 311–322.

Mathew, T.V., Kuriakose, S., 2013a. Photochemical and antimicrobial properties of silver nanoparticle-encapsulated chitosan functionalized with photoactive groups. Mater. Sci. Eng., C 33, 4409–4415.

Mathew, T.V., Kuriakose, S., 2013b. Studies on the antimicrobial properties of colloidal silver nanoparticles stabilized by bovine serum albumin. Colloids Surf. B 101, 14–18.

Marambio-Jones, C., Hoek, E.M.V., 2010. A review of the antibacterial effects of silver nanomaterials and potential implications for human health and the environment. J. Nanopart. Res. 12, 1531–1551.

Misra, S.K., Dybowskab, A., Berhanub, D., Luomac, S.N., Valsami-Jones, E., 2012. The complexity of nanoparticle dissolution and its importance in nanotoxicological studies. Sci. Total Environ. 438, 225–232.

Mittal, A.K., Tripathy, D., Choudhary, A., Aili, P.K., Chatterjee, A., Singh, I.P., et al., 2015. Bio-synthesis of silver nanoparticles using *Potentilla fulgens* Wall. ex Hook. and its therapeutic evaluation as anticancer and antimicrobial agent. Mater. Sci. Eng. C 53, 120–127.

Mohammed, A.E., 2015. Green synthesis, antimicrobial and cytotoxic effects of silver nanoparticles mediated by *Eucalyptus camaldulensis* leaf extract. Asian Pac. J. Trop. Biomed. 5, 382–386.

Monteiro, D.R., Silva, S., Negri, M., Gorup, L.F., de Camargo, E.R., Oliveira, R., et al., 2013. Antifungal activity of silver nanoparticles in combination with nystatin and chlorhexidine digluconate against *Candida albicans* and *Candida glabrata* biofilms. Mycoses 56, 672–680.

Morales-Luckie, R.A., Lopezfuentes-Ruiz, A., Olea-Mejía, O.F., Liliana, A.-F., Sanchez-Mendieta, V., Brostow, Witold, et al., 2016. Synthesis of silver nanoparticles using aqueous extracts of *Heterotheca inuloides* as reducing agent and natural fibers as templates: *Agave lechuguilla* and silk. Mater. Sci. Eng. C 69, 429–436.

Mori, Y., Ono, T., Miyahira, Y., Nguyen, V.Q., Matsui, T., Ishihara, M., 2013. Antiviral activity of silver nanoparticle/chitosan composites against H1N1 influenza A virus. Nanoscale Res. Lett. 8, 93.

Moritz, M., Geszke-Moritz, M., 2013. The newest achievements in synthesis, immobilization and practical applications of antibacterial nanoparticles. Chem. Eng. J. 228, 596–613.

Morones, J.R., Elechiguerra, J.L., Camacho, A., Holt, K., Kouri, J.B., Ramírez, J.T., et al., 2005. The bactericidal effect of silver nanoparticles. Nanomed. Nanotechnol. 3, 95.

Narayanan, K.B., Park, H.H., 2014. Antifungal activity of silver nanoparticles synthesized using turnip leaf extract (*Brassica rapa* L.) against wood rotting pathogens. Eur. J. Plant Pathol. 140, 185–192.

Nayak, D., Pradhan, S., Ashe, S., Rauta, P.R., Nayak, B., 2015. Biologically synthesised silver nanoparticles from three diverse family of plant extracts and their anticancer activity against epidermoid A431 carcinoma. J. Colloid Interf. Sci. 457, 329–338.

Nayak, D., Ashe, S., Rauta, P.R., Kumari, M., Nayak, B., 2016. Bark extract mediated green synthesis of silver nanoparticles: evaluation of antimicrobial activity and antiproliferative response against osteosarcoma. Mater. Sci. Eng. C 58, 44–52.

Ocsoy, I., Temiz, M., Celik, C., Altinsoy, B., Yilmaz, V., Duman, F., 2017. A green approach for formation of silver nanoparticles on magnetic graphene oxide and highly effective antimicrobial activity and reusability. J. Mol. Liq. 227, 147–152.

Orlowski, P., Tomaszewska, E., Gniadek, M., Baska, P., Nowakowska, J., Sokolowska, J., et al., 2014. Tannic acid modified silver nanoparticles show antiviral activity in Herpes simplex virus type 2 infection. PLoS ONE 9 (8), e104113. Available from: https://doi.org/10.1371/journal.pone.0104113.

Ouay, B.L., Stellacci, F., 2015. Antibacterial activity of silver nanoparticles: a surface science insight. Nano Today 10, 339–354.

Padalia, H., Moteriya, P., Chanda, S., 2015. Green synthesis of silver nanoparticles from marigold flower and its synergistic antimicrobial potential. Arab. J. Chem. 8, 732–741.

Pal, S., Tak, Y.K., Song, J.M., 2007. Does the antibacterial activity of silver nanoparticles depend on the shape of the nanoparticle? A study of the Gram-negative bacterium *Escherichia coli*. Appl. Environ. Microbiol. 73, 1712–1720.

Palanisamy, S., Rajasekar, P., Vijayaprasath, G., Ravi, G., Manikandan, R., Prabhu, N.M., 2017. A green route to synthesis silver nanoparticles using *Sargassum polycystum* and its antioxidant and cytotoxic effects: an in vitro analysis. Mater. Lett. 189, 196–200.

Panácek, A., Kola, M., Vec, R., Prucek, R., Soukupova, J., Hamal, P., et al., 2009. Antifungal activity of silver nanoparticles against *Candida* spp. Biomaterials 30, 6333–6340.

Parashar, U.K., Bera, V.K.T., Saxena, P.S., Nath, G., Srivastava, S.K., Giri, R., et al., 2011. Study of mechanism of enhanced antibacterial activity by green synthesis of silver nanoparticles. Nanotechnology 22, 415104.

Parikh, A.N., Gillmor, S.D., Beers, J.D., Beardmore, K.M., Cutts, R.W., Swanson, B.I., 1999. Characterization of chain molecular assemblies in long-chain, layered silver thiolates: a joint infrared spectroscopy and X-ray diffraction study. J. Phys. Chem. B 103, 2850.

Perni, S., Hakala, V., Prokopovich, P., 2014. Biogenic synthesis of antimicrobial silver nanoparticles capped with l-cysteine. Colloids Surf. A 460, 219–224.

Phull, A.R., Abbas, Q., Ali, A., Raza, H., Kim, S.J., Zia, M., et al., 2016. Antioxidant, cytotoxic and antimicrobial activities of green synthesized silver nanoparticles from crude extract of *Bergenia ciliate*. Fut. J. Pharma. Sci. 2, 31–36.

Prabhu, S., Poulose, E., 2012. Silver nanoparticles: mechanism of antimicrobial action, synthesis, medical applications, and toxicity effects. Int. Nano Lett. 2, 32–42.

Rai, V.R., Bai, A.J., 2011. Nanoparticles and their potential application as antimicrobials. In: Méndez-Vilas, A. (Ed.), Science Against Microbial Pathogens: Communicating Current Research and Technological Advances. Formatex, Microbiology Series no. 3. pp. 197–209.

Rai, M., Yadav, A., Gade, A., 2009. Silver nanoparticles as a new generation of antimicrobials. Biotechnol. Adv. 27, 76–83.

Rai, M., Deshmukh, S.D., Ingle, A.P., Gupta, I.R., Galdiero, M., Galdiero, S., 2016. Metal nanoparticles: the protective nanoshield against virus infection. Crit. Rev. Microbiol. 42, 46–56.

Rao, N.H., Lakshmidevi, N., Pammi, S.V.N., Kollu, P., Ganapaty, S., Lakshmi, P., 2016. Green synthesis of silver nanoparticles using methanolic root extracts of *Diospyros paniculata* and their antimicrobial activities. Mater. Sci. Eng. C 62, 553–557.

Ravindran, A., Chandran, P., Khan, S.S., 2013. Biofunctionalized silver nanoparticles: advances and prospects. Colloids Surf. B 105, 342–352.

Ravichandran, V., Vasanthi, S., Shalini, S., Shah, S.A.A., Harish, R., 2016. Green synthesis of silver nanoparticles using *Atrocarpus altilis* leaf extract and the study of their antimicrobial and antioxidant activity. Mater. Lett. 180, 264–267.

Ren, Y., Yang, H., Wang, T., Wang, C., 2016. Green synthesis and antimicrobial activity of monodisperse silver nanoparticles synthesized using *Ginkgo Biloba* leaf extract. Phys. Lett. A 380, 3773–3777.

Roe, D., Karandikar, B., Bonn-Savage, N., Gibbins, B., Roullet, J.–B., 2008. Antimicrobial surface functionalization of plastic catheters by silver nanoparticles. J. Antimicrob. Chemother. 61, 869–876.

Rogers, J.V., Parkinson, C.V., Choi, Y.W., Speshock, J.L., Hussain, S.M., 2008. A preliminary assessment of silver nanoparticles inhibition of monkeypox virus plaque formation. Nanoscale Res. Lett. 3, 129–133.

Salata, O.V., 2004. Applications of nanoparticles in biology and medicine. J. Nanobiotechnol. 2, 3–8.

Sana, S.S., Badineni, V.R., Arla, S.K., Boya, V.K.N., 2015. Eco-friendly synthesis of silver nanoparticles using leaf extract of *Grewia flaviscences* and study of their antimicrobial activity. Mater. Lett. 145, 347–350.

Sarsar, V., Selwal, M.K., Selwal, K.K., 2015. Biofabrication, characterization and antibacterial efficacy of extracellular silver nanoparticles using novel fungal strain of *Penicillium atramentosum* KM. J. Saudi Chem. Soc. 19, 682–688.

Seil, J.T., Webster, T.J., 2012. Antimicrobial applications of nanotechnology: methods and literature. Int. J. Nanomed. 7, 2767–2781.

Shankar, S., Jaiswal, L., Aparna, R.S.L., Prasad, R.G.S.V., 2014. Synthesis, characterization, in vitro biocompatibility, and antimicrobial activity of gold, silver and gold silver alloy nanoparticles prepared from Lansium domesticum fruit peel extract. Mater. Lett. 137, 75–78.

Sharma, V.K., Yngard, R.A., Lin, Y., 2009. Silver nanoparticles: green synthesis and their antimicrobial activities. Adv. Colloid Interface Sci. 145, 83–96.

Shivakumar, M., Nagashree, K.L., Yallappa, S., Manjappa, S., Manjunath, K.S., Dharmaprakash, M.S., 2017. Biosynthesis of silver nanoparticles using pre-hydrolysis liquor of *Eucalyptus wood* and its effective antimicrobial activity. Enzyme Microb. Technol. 97, 55–62.

Silver, S., Phung, L.T., Silver, G., 2006. Silver as biocides in burn and wound dressings and bacterial resistance to silver compounds. J. Ind. Microbiol. Biotechnol. 33, 627–634.

Singh, P., Singh, H., Kim, Y.J., Mathiyalagan, R., Wang, C., Yang, D.C., 2016. Extracellular synthesis of silver and gold nanoparticles by *Sporosarcina koreensis* DC4 and their biological applications. Enzyme Microb. Technol. 86, 75–83.

Sirelkhatim, A., Mahmud, S., Seeni, A., Kaus, N.H.M., Ann, L.C., Bakhori, S.K.M., Hasan, H., Mohamad, D., 2015. Review on zinc oxide nanoparticles: Antibacterial activity and toxicity mechanism. Nano-Micro Lett. 7, 219–242.

Sondi, I., Salopek-Sondi, B., 2004. Silver nanoparticles as antimicrobial agent: a case study on *E. coli* as a model for Gram-negative bacteria. J. Colloid Interface Sci. 275, 177–182.

Speshock, J.L., Murdock, R.C., Braydich-Stolle, L.K., Schrand, A.M., Hussain, S.M., 2010. Interaction of silver nanoparticles with *Tacaribe* virus. J. Nanobiotechnol. 8, 19–27.

Suman, T.Y., Radhika Rajasree, S.R., Kanchana, A., Elizabeth, S.B., 2013. Biosynthesis, characterization and cytotoxic effect of plant mediated silver nanoparticles using *Morinda citrifolia* root extract. Colloids Surf. B 106, 74–78.

Sun, R.W., Chen, R., Chung, N.P., Ho, C.M., Lin, C.L., Che, C.M., 2005. Silver nanoparticles fabricated in Hepes buffer exhibit cytoprotective activities toward HIV-1 infected cells. Chem. Commun. 40, 5059–5061.

Sun, L., Singh, A.K., Vig, K., Pillai, S., Shreekumar, R., Singh, S.R., 2008. Silver nanoparticles inhibit replication of respiratory sincitial virus. J. Biomed. Biotechnol. 4, 149–158.

Tran, Q.H., Nguyen, V.Q., Le, A., 2013. Silver nanoparticles: synthesis, properties, toxicology, applications and perspectives. Adv. Nat. Sci.: Nanosci. Nanotechnol. 4, 033001–033020.

Trefry, J.C., Wooley, D.P., 2012. Rapid assessment of antiviral activity and cytotoxicity of silver nanoparticles using a novel application of the tetrazolium-based colorimetric assay. J. Virol. Methods 183, 19–24.

Tutaj, K., Szlazak, R., Szalapata, K., Starzyk, J., Luchowski, R., Grudzinski, W., et al., 2016. Amphotericin B-silver hybrid nanoparticles: synthesis, properties and antifungal activity. Nanomedicine 12, 1095–1103.

Verma, A., Mehata, M.S., 2016. Controllable synthesis of silver nanoparticles using Neem leaves and their antimicrobial activity. J. Radiat. Res. Appl. Sci. 9, 109–115.

Yang, X.X., Li, C.M., Huang, C.Z., 2016. Curcumin modified silver nanoparticles for highly efficient inhibition of respiratory syncytial virus infection. Nanoscale 8, 3040–3048.

FURTHER READING

Amna, S., Shahrom, M., Azman, S., Noor, H.M.K., Ling, C.A., Siti, K.M.B., et al., 2015. Review on zinc oxide nanoparticles: antibacterial activity and toxicity mechanism. Nano-Micro Lett. 7, 219–242.

Jeyaraj, M., Varadan, S., Anthony, K.J.P., Murugan, M., Raja, A., Gurunathan, S., 2013. Antimicrobial and anticoagulation activity of silver nanoparticles synthesized from the culture supernatant of *Pseudomonas aeruginosa*. J. Ind. Eng. Chem. 19, 1299–1303.

Park, H.-J., Kim, J.Y., Kim, J., Lee, J.-H., Hahn, J.-S., Gu, M.B., et al., 2009. Silver-ion-mediated reactive oxygen species generation affecting bactericidal activity. Water Res. 43, 1027–1032.

Sun, X., Sheng, Z., Liu, Y., 2013. Effects of silver nanoparticles on microbial community structure in activated sludge. Sci. Total Environ. 443, 828–835.

CHAPTER

Microbially synthesized nanoparticles as next generation antimicrobials: scope and applications

15

Siddhardha Busi and Jobina Rajkumari

Department of Microbiology, School of Life Sciences, Pondicherry University, Puducherry, India

CHAPTER OUTLINE

15.1 INTRODUCTION

The development of multidrug resistant (MDR) infectious microbes is a major challenge to the healthcare system. Today, more than 70% of bacteria responsible for causing infectious disease are resistant to atleast one of the antibiotics

Nanoparticles in Pharmacotherapy. DOI: https://doi.org/10.1016/B978-0-12-816504-1.00008-9

employed in conventional antimicrobial therapy (Samanipour et al., 2016). Moreover, 80% of microbial infections in humans are caused by biofilm-forming pathogenic organisms. With the growing concern of MDR strains and associated biofilm, attention has been devoted to the development of new and effective antimicrobial agents. In the present scenario, the rapidly growing field of nanotechnology is considered to open new windows to combat and prevent microbial disease. Owing to their high surface area to volume ratio and unique chemical properties, the nanomaterials have emerged as a reliable candidate as a potential antiinfective. Among the nanomaterials, metal and metal oxide nanoparticles exhibiting antimicrobial properties serve as a promising tool in controlling MDR strains.

Many chemical and physical methods are employed for the preparation of metal nanoparticles. However, the use of toxic and expensive chemicals greatly limits its biomedical applicability. On the other hand, a broad range of microorganisms have been studied as potential biofactories for eco-friendly and cost-effective production of metallic nanoparticles, such as silver, titanium oxide, cadmium sulfide, gold, etc. Thus, research in microbial nanotechnology provides a reliable way for large-scale synthesis of nanomaterials (Barapatre et al., 2016). The use metallic preparations of gold, copper, silver, iron, and lead, as well as alloys in healthcare, can be traced back to ancient civilizations. Microfine metallic powders were used for both external and internal applications, in a wide variety of diseases. Traditionally, gold has been used as an antipruritic medication to suppress itching palms. It influences the immunological response and controls the anaphylactic release of histamine. Iron compounds act as a rejuvenator and as a restorative agent. They are particularly administered in diseases such as iron-deficiency anemia. They also stimulate the functional activity of all the organs and combat a number of diseases (Galib et al., 2011). Silver in various forms including silver metal, silver nitrate, silver sulfadiazine, silver acetate, and silver protein are well known for their antimicrobial properties (McDonnell and Denver Russell, 1999). Copper exerts broad-spectrum biocidal effects on bacterial cells and spores, fungi, and viruses. It damages the microbial envelope, membrane phospholipids, intracellular proteins, and nucleic acids. It is widely recommended in the treatment of ulcers, skin conditions, syphilis, and tuberculosis (Borkow, 2014). Zinc and its salt forms have been employed as therapeutic agents for centuries in topical as well as oral treatment. They are recommended in a number of dermatological disorders and infections including warts, cutaneous leishmaniasis, leprosy, inflammatory dermatoses, melasma, and basal cell carcinoma. Oral zinc, when given with dapsone (adjuvant), results in significant bacterial clearance in patients with lepromatous leprosy. The use of zinc acetate gel, in vivo, was found to prevent the sexual transmission of HSV-2 and HIV infection. Due to its antibiofilm activity on dental plaque forming microorganisms, zinc citrate or acetate are blended in many oral products, including toothpaste (Gupta et al., 2014).

In the last decade, a strong emphasis has been given on the research and applications of nanotechnology to bring advancement in the area of diagnosis and management of diseases. Nanoparticles of size range 100 nm and less have become an area of extensive research due to their antimicrobial properties attributed to their large surface area. Metal nanoparticles like gold, copper, titanium, silver, and zinc nanoparticles have attracted tremendous attention due to their unique physiochemical characteristics, which include antimicrobial, catalytic, electronic, optical, and magnetic properties. Hence, nanoparticles currently play an important part in diagnostics, drug delivery, gene therapy, and tissue engineering (Hamouda, 2012). With the integration of nanoparticles in the development of new health and pharmaceutical products, it is essential to develop eco-friendly and biocompatible techniques for the production of metal nanoparticles. The use of plant extracts, enzymes, bacteria, fungi, and algae provides an environmentally safe route for the production of nanoparticles. The main focus of nanotechnology is to synthesize monodispersed nanoparticles with predictable shape, size, and polymeric coating for potential biomedical applications (Sharma et al., 2015). Metal nanoparticles with inherent antimicrobial property or when coated with an antimicrobial entity on the surface increase their applicability in biomedical devices, textile industries, water treatment, and food packaging. The nanocomposites constituting of metal nanoparticles and polymers provided enhanced synergistic antimicrobial activity (Gutierrez et al., 2010).

15.2 SYNTHESIS OF METAL NANOPARTICLES USING MICROORGANISMS

Large quantities of metal nanoparticles with defined shapes and sizes are synthesized rapidly using various chemical and physical methods, however, being energy intensive and the use of toxic chemicals as the starting material limits it biomedical applicability. These methods are complicated and result in the release of toxic byproducts that are hazardous to the environment and human health (Li et al., 2011). Therefore, an environmentally benign and nontoxic method for nanoparticle synthesis is required to expand its therapeutic applications. Researchers in the field of nanotechnology and nanomedicine have turned to biological systems. In this regard, the microbial cells are considered as promising biofactories for the synthesis of metal nanoparticles. In the past decade, several research groups have successfully synthesized inorganic nanoparticles such as gold, iron, silver, lead, and calcium, either extracellularly or intracellularly using microorganisms like bacteria, cyanobacteria, fungi, actinomycetes, yeast, and algae. A few of the notable examples are shown in Fig. 15.1.

Due to their abundance in the diverse environment and remarkable adaptability, bacteria are most extensively employed for the synthesis of metal-based nanoparticles. The ease of manipulation and control of physical and chemical growth conditions is also an added advantage for the use of bacteria as nanofactories. *Bacillus subtilis*, *Klebsiella pneumonia*, *Escherichia coli*, *Lactobacillus* sp., and

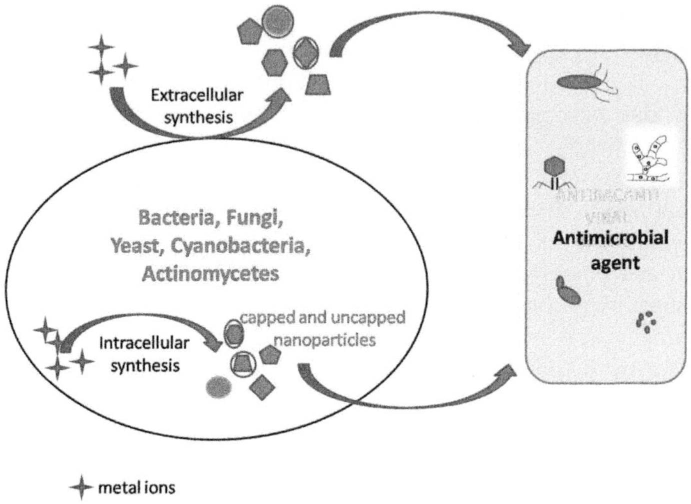

FIGURE 15.1

Schematic representation showing microorganism mediated synthesis of metal-based nanoparticles and its antimicrobial application.

Brevibacterium frigoritolerans have been employed for the synthesis of spherical Ag-NPs, extracellularly (Velmurugan et al., 2014; Kalpana and Lee, 2013; Gurunathan et al., 2009). Manivasagan et al. (2013) reported the synthesis of biofabricated Ag-NPs in size range of 30–90 nm using *Nocardiopsis* sp. MBRC-1. Intracellular synthesis of Ag-NPs was observed when *Ochrobactrum* sp. was used for reduction (Thomas et al., 2014). A similar intracellular synthesis was observed in the synthesis of Au-NPs by *Brevibacterium casei* (Kalishwaralal et al., 2010). A few of the metal-based nanoparticles synthesized using bacteria have been provided in Table 15.1.

A wide range of fungal species have also been reported for the biosynthesis of metal and metal oxide particles (Table 15.2). The large surface area of the fungal mycelium aids in the production of nanoparticles in larger quantity. The advantages of the use of fungi in biosynthesis also lie in the ease of scale-up and downstream processing. Extracellular enzymes and proteins produced by fungi, in some cases, may also act as a reducing and capping agent in the synthesis of nanoparticles.

Algae produce a wide variety of biomolecules and hence are gaining substantial interest in the extracellular synthesis of nanoparticles. A few examples of algal mediated synthesis of nanoparticles are depicted in Table 15.3.

The enzymatic process serves as an easy and inexpensive way to catalyze and synthesize nanoparticles in an aqueous medium at normal temperature, pH, and pressure (Zhang et al., 2011). Numerous microorganisms are reported to produce metallic nanoparticles with properties identical to the ones synthesized chemically

Table 15.1 List of Metal and Metal Oxide Nanoparticles Synthesized Using Bacteria

Nanoparticles	Bacteria	Size (nm)	Location	Duration	Morphology	References
Ag-NPs	*Bacillus subtilis* EWP-46	10–20	Extracellular	12 h	Irregular, capped	Velmurugan et al. (2014)
Ag-NPs	*Klebsiella pneumonia*	15–37	Extracellular	72 h	Spherical, rod, capped	Kalpana and Lee (2013)
Ag-NPs	*Escherichia coli*	10–90	Extracellular	30 min	Spherical, capped	Gurunathan et al. (2009)
Ag-NPs	*Nocardiopsis* sp. MBRC-1	30–90	Extracellular	24 h	Spherical, capped	Manivasagan et al. (2013)
Ag-NPs	*Ochrobactrum* sp.	38–85	Intracellular Extracellular	24 h	Spherical	Thomas et al. (2014)
Ag-NPs	*Lactobacillus casei* subsp. *casei*	25–100	Intracellular	12–13 h	Spherical, capped	Korbekandi et al. (2012)
Ag-NPs	*Brevibacterium frigoritolerans* DC2	50–100	Extracellular	48 h	Spherical	Singh et al. (2015)
Au-NPs	*Rhodopseudomonas capsulata*	10–500	Extracellular, capped	48 h	Triangular nanoplates	He et al. (2008)
Au-NPs	*Deinococcus radiodurans*	50–60	Intracellular Extracellular, capped	8 h	Spherical, triangular	Li et al. (2016)
Au-NPs	*Geobacillus* sp. strain ID17	5–50	Intracellular, capped	16 h	Quasi-hexagonal	Correa-Llanten et al. (2013)
Au-NPs	*Marinobacter pelagius*	2–10	Intracellular, capped	24 h	Spherical, triangles	Sharma et al. (2012)
Au-NPs	*Brevibacterium casei*	10–50	Intracellular, capped	24 h	Spherical	Kalishwaralal et al. (2010)
Au-NPs	*Thermos scotoductus* SA-01	—	Extracellular Intracellular	8 h	Triangular, hexagonal	Erasmus et al. (2014)
Au-NPs	*B. subtilis*	7.6 ± 1.8	Extracellular Intracellular	1 day	—	Reddy et al. (2010)
Au-NPs	*Shewanella oneidensis*	2–50	Extracellular, capped	48 h	Spherical	Suresh et al. (2011)
TiO_2-NPs	*Lactobacillus* sp.	8–35	Extracellular	—	Spherical	Jha et al. (2009)
ZnO-NPs	*Aeromonas hydrophila*	57.72	Extracellular capped	12–48 h	Spherical, oval	Jayaseelan et al. (2012)
Fe_3O_4	*S. oneidensis*	40–50	Extracellular	—	Rectangular, rhombic, hexagonal	Perez-Gonzalez et al. (2010)
Fe_2O_3	*S. oneidensis* MR-1	30–43	Intracellular	—	Pseudohexagonal rhombohedral	Bose et al. (2009)

Table 15.2 List of Metal and Metal Oxide Nanoparticles Synthesized Using Fungi

Nanoparticles	Fungi	Size (nm)	Location	Duration	Morphology	References
Ag-NPs	*Alternaria alternata*	20–60	Extracellular	4 h	Spherical capped	Gajbhiye et al. (2009)
Ag-NPs	*Fusarium semitectum*	10–60	Extracellular	24 h	Spherical capped	Basavaraja et al. (2008)
Ag-NPs	*Humicola* sp.	5–25	Intracellular	96 h	Spherical capped	Syed et al. (2013)
Ag-NPs	*Fusarium oxysporum*	3.4–26.8	Extracellular	5 days	Spherical	Ishida et al. (2014)
Ag-NPs	*Arthroderma fulvum*	13–18	Extracellular	1 h	Spherical	Xue et al. (2016)
Ag-NPs	*Trichoderma viride*	5–40	Extracellular	24 h	Spherical rod-like	Fayaz et al. (2010)
Ag-NPs	*Aspergillus clavatus*	550–650	Extracellular	24 h	Irregular capped	Saravanan and Nanda (2010)
Ag-NPs	*Macrophomina phaseolina*	5–40	Extracellular	24 h	Spherical Capped	Chowdhury et al. (2014)
Ag-NPs	*Coriolus versicolor*	10	Extracellular Intracellular	72 h	Spherical Capped	Sanghi and Verma (2009)
Ag-NPs	*Candida utilis* NCIM 3469	20–80	Extracellular	15 min	Spherical	Waghmare et al. (2015)
Ag-NPs	*Schizophyllum commune*	51–93	Extracellular Intracellular	24 h	Spherical capped	Arun et al. (2014)
Au-NPs	*Neurospora crassa*	28–32	Intracellular Extracellular	24 h	Spherical	Castro-Longoria et al. (2011)
Au-NPs	*A. clavatus*	24.4 ± 11	Extracellular	48–72 h	Triangular, spherical, hexagonal	Verma et al. (2011)
Au-NPs	*F. oxysporum*	8–40	Extracellular	72 h	Spherical, triangular	Mukherjee et al. (2002)
Au-NPs	*Saccharomyces cerevisiae*	15–20	Cell wall	< 24 h	Spherical	Sen et al. (2011)
TiO_2-NPs	*F. oxysporum*	6–13	Extracellular	–	Spherical	Bansal et al. (2005)
ZrO_2-NPs	*F. oxysporum*	3–11	Extracellular	–	Spherical	Bansal et al. (2004)
ZnO-NPs	*Candida albicans*	15–25	Extracellular	15 h	Quasi spherical	Shamsuzzaman et al. (2013)

Table 15.3 List of Metal and Metal Oxide Nanoparticles Synthesized Using Algae and Actinomycetes

Nanoparticles	Algae/Actinomycetes	Size (nm)	Location	Duration	Morphology	References
Ag-NPs	*Spirogyra varians*	~17.6	Extracellular	20 min	Quasi-spheres, capped	Salari et al. (2016)
Ag-NPs	*Cystophora moniliformis*	50–100	Extracellular	30 min	Spherical, capped	Prasad et al. (2013)
Ag-NPs	*Pithophora oedogonia*	25–44	Extracellular	Few min	Cubical, Hexagonal, capped	Sinha et al. (2015)
Ag-NPs	*Chlamydomonas reinhardtii*	5–35	Extracellular	10 h and 13 days	Round, Rectangular	Barwal et al. (2011)
Ag-NPs	*Caulerpa racemosa*	5–25	Extracellular	3 h	Spherical, Triangular	Kathiraven et al. (2015)
Ag-NPs	*Streptomyces* sp. LK3	5 ± 3.9	Extracellular	96 h	Spherical, Disk	Karthik et al. (2014)
Au-NPs	*Streptomyces viridogens* strain HM10	8–20	Intracellular	24 h	Spherical rod	Balagurunathan et al. (2011)
Au-NPs	*Streptomyces* sp.	90	Extracellular	—	Cubical	Vinay Gopal et al. (2013)
	Sargassum muticum	5.42 ± 1	—	30 min	Spherical capped	Namvar et al. (2015)
Au-NPs	*Tetraselmis kochinensis*	5–35	Intracellular	—	Spherical Triangular	Senapati et al. (2012)
Au-NPs	*Stoechospermum marginatum*	18.7–93.7	—	10 min	Spherical, triangle hexagonal	Rajathi et al. (2012)
Au-NPs	*Ecklonia cava*	30 ± 0.25	Extracellular	1 min	Spherical Triangular	Venkatesan et al. (2014)
ZnO-NPs	*Anabaena* strain L31	80	Extracellular	—	Spherical hexagonal	Singh et al. (2014)
CuO-NPs	*Bifurcaria bifurcata*	5–45	Extracellular capped	24 h	Spherical	Abboud et al. (2014)
ZnO-NPs	*S. muticum*	30–57	Extracellular capped	—	Hexagonal	Azizi et al. (2014)

or physically, at ambient temperature, pH, and pressure. However, it is important to study and elucidate the biochemical pathways of the microbes that lead to metal ion reduction and formation of nanoparticles, to exercise a stringent control over size, shape, and constitution of the particles (Gericke and Pinches, 2006). The microorganisms mediated biosynthesis of nanoparticles depends on numerous physical parameters and culture conditions. Hence, standardization of these conditions is necessary for higher yield of nanoparticles with desirable properties. The size of the particle may be controlled by the stringent control of parameters such as temperature, pH, salt content, and time of exposure. For instance, the synthesis medium containing 5 mM of $AgNO_3$, reaction temperature of 60°C, and pH 10 resulted in the synthesis of Ag-NPs, within in 30 minutes using culture supernatant of *E. coli*. Further, by varying the $AgNO_3$ concentration, temperature, and pH the average particle size was controlled between 10 and 90 nm (Li et al., 2011).

15.3 PROPERTIES OF NANOPARTICLES: SIZE AND MORPHOLOGY

Nanoparticles constitute materials with varying dimensions and range in size (in diameter) from a few nanometers to less than 100 nm. Unlike the bulk materials possessing constant physical properties, the nanoparticles exhibit unusual properties different from their bulk counterparts. The surface plasmon resonance and the large surface to volume ratio contribute to the unusual properties of the nanoparticles. The electronic, optical, magnetic, and mechanical properties are greatly influenced by the sizes and shape of the nanoparticles. Tremendous research is directed towards characterization, fabrication, and applications of nanoparticles. The shape and size-dependent chemical and physical properties of the nanoparticles are being extensively studied for various biomedical applications (Li et al., 2011). The nanoparticles are biocompatible to most of the biomolecules and small in size, making them a possible candidate for application as biosensors, drug carriers, and in bioimaging. Another interesting application of nanoparticles in medicine is their use in antimicrobial therapy in combating MDR pathogenic microbes (Li et al., 2011).

The size, shape, and total surface area of the nanoparticles play the central role in the antimicrobial activity. Smaller size particles possessing larger surface to volume ratios are found to have greater antibacterial activity (Duran et al., 2010). The antibacterial property is also influenced by the high particle penetration and total surface area available for interacting with the bacterial cell. Ag-NPs undergo size-dependent interactions with HIV-I and also exhibit shape dependent bactericidal properties against Gram-negative bacteria like *E. coli* (Tak et al., 2015). The truncated triangular silver nanoplates exhibited a comparatively better bactericidal effect as compared with spherical and rod-shaped nanoparticles (Pal et al., 2007).

Different groups of microbes possess the ability to control the shapes and sizes of biologically synthesized nanoparticles. Nanoparticles of various sizes and particle morphologies including spherical, triangular, oval, hexagonal, etc. have been synthesized using microbes. It has been speculated that the particles synthesized intracellularly are usually smaller and monodispersed as compared with those synthesized extracellularly. The intracellular fungal mediated synthesis of gold nanoparticles resulted in nanoparticles of various morphologies and sizes (Gericke and Pinches, 2006). The particles were predominantly spherical in shape and formed in the cytoplasm of the cells. In the case of extracellular synthesis, by tuning the reaction conditions the desired shape and size may be achieved. For example, Gurunathan et al. (2009) reported the synthesis of Ag-NPs and a reduction in particle size by employing optimal a concentration of $AgNO_3$, reaction temperature, and pH. The increase in understanding of shape and size-dependent properties of nanomaterials has motivated new developments on the biosynthesis route. Moreover, better control of size and shape has led to new developments in materials science.

15.4 FABRICATION OF NANOPARTICLES

The biomedical uses of metal NPs are limited by their tendency to agglomerate and form large particles postsynthesis. This unusual characteristic of nanoparticles can be avoided by stabilizing and protecting them from self-aggregation using capping agents. Capping reduces the surface energy and maintains the size of the nanoparticles in the range below 100 nm. The choice of capping agent depends on the characteristics of the nanoparticles and its intended application. Amino acids, polypeptides, and tiopronin have been employed as the capping agent for biological applications. Further, the geometry and solubility of the nanoparticles are also influenced by the properties of this capping agent used. Stabilizing agents such as surfactants, polymers, dendrimers, and biomacromolecules are also used for capping the nanoparticles to enhance their stability and prevent further agglomeration (Perni et al., 2014).

Biosynthesis and fabrication of metal and metal oxide nanoparticles by the aid of intracellular and extracellular functional groups and enzymes from plants, bacteria, fungi, algae, and viruses have recently gained much interest (Siddiqi and Husen, 2016). Bacteria such as *Rhodopseudomonas capsulate*, *Deinococcus radiodurans*, *Geobacillus* sp., *B. subtilis*, and *B. casei* have been documented for the synthesis and capping of Au-NPs (He et al., 2008; Li et al., 2016; Correa-Llanten et al., 2013; Reddy et al., 2010; Kalishwaralal et al., 2010). Perni et al. (2014) documented the synthesis of L-cysteine capped Ag-NPs using cell extract of *E. coli*. The Ag-NPs prepared exhibited remarkable antibactericidal activity against *Staphylococcus aureus* and *E. coli*. Spherical shaped Ag-NPs were prepared using exopolymer isolated from the culture supernatant of *Ochrobactrum rhizosphaerae*. The exopolymer responsible for synthesis and capping of the Ag-NPs was characterized and found to be glycolipoprotein (GLP) in nature. The purified exopolymer was successfully employed to prevent agglomeration and maintain the stability of the Ag-NPs. Further, significant antibacterial activities of the GLP–AgNPs conjugate, more prominent than antibiotic (ciprofloxacin), was observed against *Vibrio cholera*. Owing to their small size and high surface-to-volume ratio, the GLP–AgNPs ruptured the cell wall and caused cell lysis in the test pathogen, *V. cholera* (Gahlawat et al., 2016)

Metal and metal oxide nanoparticles have been also been extracellularly produced and fabricated using algal species such as *Chlorella vulgaris*, *Chlamydomonas reinhardtii*, *Fucus vesiculosus*, *Ecklonia cava*, *Spirulina platensis*, *Oscillatoria willei*, *Sargassum muticum*, *Stoechospermum marginatum*, etc. (Siddiqi and Husen., 2016). Iron oxide nanoparticles of 18 ± 4 nm size were prepared using an aqueous extract of the brown alga, *S. muticum* at 25°C. The reduction in the $FeCl_3$ to Fe_3O_4 nanoparticles was brought about by the polysaccharides present in the algal extract (Mahdavi et al., 2013). In some cases of fungal mediated production of nanoparticles, the biosynthesis of nanoparticles is followed by fabrication of the nanoparticles by stabilizing/capping agents. The capping agents are usually the extracellular proteins from fungal mycelia. The presence of natural capping agents terminates the postproduction steps of capping required for medical applications of the nanoparticles. Silver nanoparticles are well studied and reported for their profound antimicrobial property on several multidrug-resistant bacteria. The extracellular biosynthesis of Ag-NPs using cell-free filtrate of phytopathogenic fungus, *M. phaseolina* (Tassi) Goid was reported by Chowdhury et al. (2014). The synthesized Ag-NPs nanoparticles were found to be naturally coated and stabilized by 85 kDa protein. Arun et al. (2014) reported the extracellular and the intracellular synthesis of biofabricated Ag-NPs by a mushroom fungus, *Schizophyllum commune*, when grown in culture medium supplemented with $AgNO_3$. The Ag-NPs exhibited remarkable antibacterial and anti-dermatophytic activity against *Pseudomonas* sp., *B. subtilis*, *K. pneumonia*, *Trichophyton mentagrophytes*, *E. coli*, *Trichophyton simii*, and *Trichophyton rubrum*. Similarly, Jain et al. (2011) demonstrated a cost-effective and eco-friendly synthesis of biofabricated Ag-NPs using the cell-free extract of *Aspergillus flavus* NJP08. The presence of extracellular capping protein on the Ag-NPs was revealed using Fourier transform infrared (FTIR) spectroscopy. The SDS–PAGE profiles showed the presence of two intense bands of 32 and 35 kDa confirming the presence of extracellular proteins, which might be responsible for the synthesis and stabilization of Ag-NPs. A cost-effective extracellular synthesis of nano bowl-shaped silver nanoparticles using *Trichoderma viride* was documented by Chitra and Annadurai (2013). The presence of two protein bands of 45 and 39 kDa indicated the presence of capping agent on the Ag-NPs. Further, the synthesized Ag-NPs exhibited a considerable bactericidal activity against *Klebsiella planticola* (2277) and *B. subtilis* (3053).

15.5 ANTIMICROBIAL ACTIVITY OF METAL-BASED NANOPARTICLES

The control of infectious diseases has become a major challenge to the healthcare system due to the rise in the number of microorganisms resistant to the conventional antibiotics. In recent times, the advancement in the field of nanotechnology presented the metallic NPs with wide antimicrobial functionality with a profound influence on human health. Nanoparticles (NPs) of a dimension ranging from 10 to 1000 nm exhibit unique physiochemical properties and biological activities. Microbially synthesized metal and metal oxide nanoparticles use multiple mechanisms to kill and/or inhibit the growth of pathogenic microbes, thereby making them promising antimicrobial agent to revolutionize the currently available antimicrobial therapy. Different metallic NPs, including gold (Au), silver (Ag), titanium (Ti), zinc (Zn), copper (Cu), and magnesium (Mg), are known for their efficient antimicrobial properties (Schrofel et al., 2014). These NPs are stable when stored for long duration and can withstand harsh processing conditions including high temperature and pH, without getting inactivated (Ranghar et al., 2014).

The shape and size of nanoparticles greatly influence its antimicrobial activity. Particles with size ranging from 1 to 10 nm have shown to possess higher bactericidal activity compared with its larger counterparts. As a result, small size with improved biocompatibility is widely used in the biomedical field (Allaker and Memarzadeh, 2014). Moreover, NP acts on a wide range of microbial targets to exert their antimicrobial mechanisms. They may directly disrupt the cell membrane or indirectly damage the cellular components by generating free radicals. The free radicals generated in turn cause damage and inhibit the synthesis of DNA, protein, and other cellular components. Nanoantimicrobial agents serve as a cost-effective and nontoxic alternative to the conventional antibiotics in controlling infectious diseases. Metal and metal oxide NPs provides new hope in fighting infectious diseases caused by MDR stains. In addition, nanoparticles loaded with antibiotics when efficiently administered improve the pharmacokinetics and therapeutics of the drug. Some of the metal and metal oxide NPs exhibiting antimicrobial property are described below.

15.5.1 SILVER NANOPARTICLES (AG-NPS)

Since time immemorial, the medicinal properties of silver have been recognized and silver-based formulations have been employed in the treatment of various injuries, burn wounds and infections. Silver salts and their derivatives are well reported for their broad spectrum antagonistic activity against fungi, bacteria, and viruses. According to the recent studies, Ag-NPs have been suggested as a potential medium for delivery of antimicrobial agents, in disinfecting filters and as coating materials (Ravishankar and Jamuna, 2011).

Several mechanisms were proposed to justify the antimicrobial property of Ag-NPs. It is believed that the positive charge of Ag^+ ion of the Ag-NPs causes electrostatic interaction with the negatively charged bacterial cell membrane, thereby contributing to its antimicrobial property (Kim et al., 2007). It is also suggested that Ag-NPs exhibit high affinity for the sulfur-containing protein present inside the cells as well as in the bacterial membrane. Subsequently, the bacterial cell in contact with Ag-NPs takes up the Ag^+ ions, thereby resulting in inhibition of the cellular functions and respiratory enzymes (Kim et al., 2007; Ravishankar and Jamuna, 2011). Ag-NPs affect the permeability of the plasma membrane and cell viability (Chaloupka et al., 2010). It was also suggested that the Ag^+ ion released from Ag-NPs interacts with the phosphorus-containing elements such as DNA, causing a damaging effect on DNA replication and protein synthesis (Ravishankar and Jamuna, 2011; Chaloupka et al., 2010; Gogoi et al., 2006). In fungi, Ag^+ of the Ag-NPs inhibits DNA replication and the respiratory chain as well. At lower concentration of Ag-NPs, the Ag^+ ions have been reported to inhibit respiratory chain enzymes, which resulted in uncoupled respiratory electron transport from oxidative phosphorylation. Due to the high affinity of the Ag^+ for the sulfhydryl (thiol) groups present on the membrane-bound respiratory enzymes, the Ag^+ interferes with the electron transport chain and transfer of energy through the membrane (Bard and Holt, 2005; Allaker and Memarzadeh, 2014). At higher concentration, Ag^+ ions have been shown to cause a detrimental effect on the cytoplasmic components and nucleic acids of the cell (Dibrov et al., 2002; Lara et al., 2010). Like most of the metal NPs, Ag-NPs also have the ability to generate intracellular reactive oxygen species (ROS) in bacterial cells. They show better antimicrobial properties mediated by the production of ROS like H_2O_2, which exert metabolic stress inside the cell, subsequently causing cell death (Ravishankar and Jamuna, 2011).

Furthermore, the size and shape play a crucial part in imparting antimicrobial property to the Ag-NPs. Hence, the antibacterial activity may be regulated by controlling particle size of the Ag-NPs. The large surface–volume ratio facilitates interaction of the Ag-NPs with the cell membrane and allows easy penetration inside the cell, thereby causing complete damage to cell viability (Blecher et al., 2011). Ag-NPs of $\sim$10 nm a diameter readily form pores in the bacterial cell wall. Consequently, the cytoplasmic content gets released out of the cell, ultimately leading to cell death. Ag-NPs of size less than 20 nm (in diameters) increases the permeability of bacterial cells by attaching to sulfur-containing proteins of the bacterial cell membranes (Morones et al., 2005; Gogoi et al., 2006; Ravishankar and Jamuna, 2011). Ag-NPs with triangular and truncated shape readily release Ag^+ and hence exhibit greater antimicrobial activity (Blecher et al., 2011).

Silver nanoparticles, either alone or in combination with other antimicrobial agents, show remarkable synergistic effects against numerous Gram-positive and Gram-negative bacterial pathogens. For instance, Ag-NPs exhibits enhanced antimicrobial effects against *E. coli* and *S. aureus* when administered in combination with amoxicillin, erythromycin, penicillin G, and vancomycin (Fayaz et al., 2010;

Blecher et al., 2011). Silver nanoparticles also exhibit antagonistic effect against drug-resistant bacteria, fungi, and viruses. The broad spectrum antimicrobial activity of Ag-NPs may be attributed to the multiple mechanisms of antimicrobial action exhibited by Ag-NPs. Lara et al. (2010) determined the bactericidal efficacy of Ag-NPs against ampicillin-resistant *E. coli*, multidrug-resistant *Pseudomonas aeruginosa*, and erythromycin-resistant *Streptococcus pyogenes*. It was revealed that the bacteriostatic action of Ag-NPs resulted due to inhibition in the cell wall, protein and nucleic acid synthesis.

A study by Nanda and Saravanan (2009) revealed that *S. pyogenes*, Methicillin-resistant *S. aureus* (MRSA), and *Staphylococcus epidermidis* (MRSE) are significantly susceptible to Ag-NPs. However, it shows a moderate antibacterial activity against Gram-negative pathogenic bacteria, *K. pneumonia* and *Salmonella typhi*. Morones et al. (2005) revealed the antibacterial efficacy of Ag-NPs against Gram-negative bacteria, *P. aeruginosa*, *S. typhi*, *E. coli*, and *V. cholera*. Feng et al. (2000) documented the morphological changes resulting in treatment with Ag-NPs in Gram-positive and Gram-negative bacteria. The cytoplasmic membrane was found to detach from cell walls and the DNA condensed in the middle of the cell. A differential antagonistic activity of Ag^+ observed in Gram-negative and Gram-positive bacteria resulted due to the variation in the thickness of peptidoglycan layer present in the cell wall (Sondi and Salopek-Sondi, 2004). Raheman et al. (2011) reported the extracellular synthesis of Ag-NPs by an endophytic fungus, *Pestalotia* sp., isolated from leaves of *Syzygium cumini*. The synthesized Ag-NPs exhibited antibacterial efficacy against *S. typhi* (ATCC-51812) and *S. aureus* (ATCC-25923).

Various studies have revealed the potential of Ag-NPs as antifungal agent. Ag-NPs were found to effectively inhibit the growth in numerous fungal pathogens such as *Aspergillus fumigatus*, *Candida albicans*, *Cladosporium cladosporioides*, *Chaetomium globosum*, *Penicillium brevicompactum*, *Stachybotrys chartarum*, and *T. rubrum* (Allaker and Memarzadeh, 2014). The antifungal effects of spherical Ag-NPs may be attributed to their damaging effect on the fungal mycelia. Ag-NPs showed significant antifungal efficacy against dermatophytes, *Candida* sp. and *T. mentagrophytes* (Ravishankar and Jamuna, 2011). Ag-NPs have been found to possess antiviral property against hepatitis B virus and human immunodeficiency virus type 1 (HIV 1) (Lara et al., 2010). The above-mentioned reports suggest the possible use of Ag-NPs as a potential antifungal agent in medical devices and healthcare products.

15.5.2 GOLD NANOPARTICLES (AU-NPS)

The use of gold and gold-based agents in the treatment of infectious diseases can be traced back to 5000 years ago in Egypt. Gold is considered as inert and shows comparatively lesser antimicrobial activity to that of silver and copper. However, owing to its high surface-to-volume ratio and remarkable binding efficacy, gold is exceptionally suitable for attaching ligands and enhancing its chemical or photothermal functionality (Allaker and Memarzadeh, 2014). In addition, Au-NPs also possess unique shape and size dependent optical and electronic characteristics (Li et al., 2011).

Au-NPs exert their antimicrobial mechanisms by targeting multiple biological pathways. They generate pores on the cell wall, which in turn leads to leakage of cytoplasmic content and cell death. It also generates photocatalytic ROS, which damage the cellular components of bacteria and viruses. Au-NPs affect the membrane potential and decrease cellular ATP levels by strongly inhibiting ATPase activity. Au-NPs also inhibit binding of tRNA to the ribosome subunit (Schrofel et al., 2014). They prevent unwinding of DNA and inhibit the transcription process (Rudramurthy et al., 2016).

Recent studies have been directed towards conjugation of Au-based nanoparticles with photosensitizers for antimicrobial photodynamic therapy (APDT). On exposure to infrared (NIR) radiation, the Au-based nanomaterials with characteristic NIR absorption destroy the bacterial cell by photothermal heating. Au nanorods fabricated with photosensitizers were found to effectively destroy MRSA via NIR photothermal radiation and photodynamic antimicrobial chemotherapy (Kuo, 2009; Pissuwan et al., 2009). Gold nanorods conjugated with toluidine blue O (a hydrophilic photosensitizer) exhibit dual function against methicillin-resistant *S. aureus* agents and kill the bacteria by hyperthermia and photodynamic inactivation (Gil-Tomas, 2007; Perni, 2009). Light absorbing Au-NPs fabricated with specific antibodies have also been employed to photothermally target *S. aureus* (Zharov, 2006). Khan et al. (2012) determined the possible use of Au-NPs fabricated with methylene blue for prevention of biofilm formation by nosocomial fungus, *C. albicans*.

Au-NPs are doped with biomolecules such as polysaccharides, proteins, antibodies, and oligonucleotides to develop functional moieties for various applications (Rudramurthy et al., 2016). Various studies have revealed the possibility of conjugating gold nanoparticles with antibiotics to enhance their antibacterial efficacy. Rai et al. (2009) reported the synthesis of spherical Au-NPs fabricated with cefaclor, a second generation antibiotic. It was found that the amine group of cefaclor acts as reducing as well as capping agent in the synthesis of Au-NPs. The cefaclor fabricated Au-NPs were found to exhibit better antibacterial activity against both Gram-positive (*S. aureus*) and Gram-negative bacteria (*E. coli*) as compared with gold nanoparticles and cefaclor, taken separately. The synergistic activity was studied and it was revealed that Cefaclor suppresses the synthesis of the peptidoglycan layer of the bacterial cell wall, thereby making the cell wall porous. The Au-NPs also generated holes in the cell wall, causing the leakage of cytoplasmic contents and ultimately leading to cell death. Au-NPs may also inhibit the uncoiling and transcription of DNA by binding to the DNA of bacteria. Coating of aminoglycosidic antibiotics with Au-NPs significantly increases their antibacterial activity against a range of Gram-positive and Gram-negative bacteria (Grace and Pandian, 2007; Saha, 2007). The antibacterial efficacy of vancomycin against vancomycin-resistant enterococci (VRE) was profoundly enhanced when coated with Au-NPs (Gu et al., 2003). Recently, Fayaz et al. (2011) used non-pathogenic fungus, *T. viride*, to synthesize Au-NPs, at room temperature. The synthesized Au-NPs when conjugated with vancomycin by the ionic interaction efficiently suppressed the growth of vancomycin-resistant *S. aureus*. It was found that the Au-NPs penetrate inside the bacterial membrane resulting in the death of the cells. Stable biofabricated Au-NPs coated with antibiotics, rifampicin, vancomycin ciprofloxacin, and gentamycin significantly enhanced the inhibitory effect of the antibiotics against *S. epidermidis* and *Staphylococcus haemolyticus* (Ahangari et al., 2013). Brown et al. (2012) found that Au-NPs conjugated ampicillin (Au-NPs−AMP) exert bactericidal effect against numerous MDR bacteria, including MRSA, *Enterobacter aerogenes*, *P. aeruginosa*, and *E. coli*. It is speculated that the coating of ampicillin molecules on the surface of Au-NPs allows the Au-NPs−AMP to suppress the beta-lactamases expressed by the bacterial pathogens. Further, Au-NPs−AMP also hampers the proper functioning of transmembrane efflux pump, which is responsible for efflux of drug from the bacterial cell. Hence, Au-NPs may be used to coat the surface of a wide variety of medical devices and implants.

15.5.3 COPPER OXIDE NANOPARTICLES (CUO-NPS)

Copper oxide nanoparticles are relatively cheap, photocatalytic, and stable in regard to their chemical and physical properties. The potential application of CuO-NPs as antiinfective agents lies in their extremely high surface area and desirable crystal morphologies (Kwak and Kim, 2005; Stoimenov, 2002). A few mechanisms contributing to the antimicrobial property of the CuO-NPs have been proposed. CuO-NPs may adhere to the bacterial cell walls, owing to their positive charge. CuO-NPs interact with carboxyl and amine groups present on the surfaces of microbial cells. Therefore, bacteria with higher density of these ionic groups on their cell surfaces, such as *B. subtilis*, have higher affinity for and are highly susceptible to CuO-NPs (Blecher et al., 2011; Huh and Kwon, 2011). However, Gram-negative bacteria like *Proteus* spp. and *P. aeruginosa* are less susceptible to the positively charged CuO-NPs (Rupareli et al., 2008).

In living microorganisms, copper is a structural component for numerous enzymes. Hence, a relatively high concentration of Cu^{++} is required to generate toxic effects against microbial pathogens. When administered in high dose, Cu^{++} ions aid in the formation of ROS, which thereby interact with DNA and intercalate nucleic acid strands. The release of Cu^{++} may also disrupt the amino acid synthesis in many microbes (Huh and Kwon, 2011; Rupareli et al., 2008; Esteban et al. 2009). In addition, the generation of ROS may also result in oxidative stress-induced membrane damage in bacteria (Raffi et al., 2010; Pramanik et al., 2012; Weitz et al., 2015).

CuO-NPs exhibit broad range of microbicidal activity against both bacteria (*S. aureus*, *E. coli*, and *Listeria monocytogenes*) and fungi (especially *S. cerevisiae*) (Blecher et al., 2011). The highly ionic CuO nanoparticles effectively target a wide range of bacterial pathogens associated with nosocomial infections. However, a significantly high dose of CuO-NPs is required to generate an antimicrobial effect (Ren et al., 2009). CuO-NPs show profound antibacterial activity against Gram-positive (*S. aureus* and *B. subtilis*) and Gram-negative bacteria (*E. coli*) (Chatterjee et al., 2012). Khashan et al. (2016) documented the antibacterial activity of CuO-NPs against *S. aureus*, *E. coli*, *Proteus vulgaris*, and *P. aeruginosa*. CuO-NPs in combination with fluconazole showed synergistic antifungal effect against pathogenic yeast, *C. albicans* (Raffi et al., 2010).

15.5.4 ALUMINUM OXIDE NANOPARTICLES (AL_2O_3-NPS)

Alumina (also identified as aluminum(III) oxide) is thermodynamically stable and possesses a corundum-like structure, with hexagonal close-packed structure formed by oxygen atoms with two thirds of the octahedral sites in the lattice filled by Al^{3+} ions (Martinez Flores et al., 2003). At near-neutral pH, they carry a positive charge on its surface. The electrostatic interaction together with the hydrophobic interactions and polymer bridging between the particles and negatively charged bacterial cells results in the adhesion of nanoparticles on the cell surfaces (Li and Logan, 2004). The antimicrobial activity of Al_2O_3-NPs resulting due to the release of metal ions has also been studied (Mukherjee et al., 2011). Al_2O_3-NPs generates ROS, which disrupt the bacterial cell wall, causing oxidative damage to cell membrane and eventually leading to cell death (Ravishankar and Jamuna, 2011).

The antimicrobial efficacy of Al_2O_3-NPs is greatly influenced by its shape, surface charge, and particle size. The Al_2O_3-NPs when attached to the bacterial cell surface causes structural deformities to the cell wall. These nanoparticles cross the cell membrane and this eventually results in loss of membrane integrity, as a consequence of intracellular oxidative stress. Al_2O_3-NPs show higher mutagenicity and sensitivity against *Pseudomonas fluorescens* compared with its bulk counterpart (Balasubramanyam et al., 2010). Likewise, nanoalumina also exhibits a better antibacterial effect of towards *E. coli*, *B. subtilis*, and *P. fluorescens* as compared with the bulk materials. The nanoparticles significantly decrease the extracellular protein content of the bacteria. It was found that Al_2O_3-NPs at a concentration of 1000 $\mu g\ mL^{-1}$ show a mild antibacterial effect against *E. coli*. The interaction of the nanoalumina with the bacterial cells results in distortion in cell morphology and agglomeration of particles in the cell wall (Mukherjee et al., 2011). Ansari et al. (2014) investigated the antibacterial efficacy of Al_2O_3-NPs against multidrug-resistant strain of *E. coli* isolated from clinical sample. It was found that Al_2O_3-NPs exert structural modification in the phospholipids layer, subsequently resulting in loss of amphiphilic properties of the cell wall, destruction of the membrane and leakage of cellular components. It is suggested that Al_2O_3-NPs interact with L-α-phosphatidyl-ethanolamine (PE) and lipopolysaccharides (LPS) of the bacterial cell wall through hydrogen bonds and ligand exchange. The binding of NPs to the surface of cell membrane results in the formation of irregular-shaped pores and perforation on the bacterial cell surfaces. It is proposed that the Al_2O_3-NPs might have caused adverse effects by interacting with the intracellular biomolecules and eventually triggering cell death. Ansari et al. (2015) suggested that Al_2O_3-NPs may be exploited as a potential antimicrobial agent to target extended-spectrum, nonextended-spectrum and metallo-β lactamases strains of *P. aeruginosa*.

Al_2O_3-NPs disrupt the cell membranes and exhibits mild inhibitory effect on microbial growth when administered at high concentration. The antibacterial effect of alumina NPs ($10-1000\ \mu g\ mL^{-1}$) on the growth *E. coli* was reported by Sadiq et al. (2009). The inhibitory effect was attributed to the interactions between the nanoparticles and bacterial cell surface due to opposite surface charge. It is revealed that Al_2O_3-NPs can enter inside the cytoplasm of *E. coli* and exert damaging effects. The extracellular protein content was decreased in cells on treatment with $1000\ \mu g\ mL^{-1}$ alumina. Al_2O_3-NPs when conjugated with other metals enhance its antibacterial behavior. According to Bala et al. (2011), Al_2O_3-NPs exhibit enhanced antimicrobial effects against *E. coli* and *S. epidermidis* in conjugation with silver. Similar observations were made in case of TiO_2-Ag and Al_2O_3-Ag nanocomposite. The antimicrobial efficacy of alumina nanoparticles indicates their potential use in the development of effective therapeutic agents against the infections caused by MDR strains.

15.5.5 IRON OXIDE (IO-NPS)

The biological compatibility of the magnetic IO-NPs makes them an attractive candidate for implementation in biomedicine. Like other metal oxide nanoparticles, IO-NPs also exhibit antimicrobial activity by generation of free radicals. The mechanism of antimicrobial action for Fe_3O_4-NPs is believed to be as a result of oxidative stress generated due to the release of ROS such as hydroxyl radicals (OH^-), superoxide radicals (O_2^-), hydrogen peroxide, and singlet oxygen ($^1O_2$) (Behera et al., 2012). These resulting free radicals in turn cause depolymerization of polysaccharides, DNA damage, lipid peroxidation, and/or inactivation of enzymes (Ranghar et al., 2014).

The antibacterial activity of IO-NPs has been determined against various bacteria including *P. vulgaris*, *S. aureus*, *E. coli*, *S. epidermidis*, and *Xanthomonas* sp. (Tran et al., 2010; Prabhu et al., 2015; Lee et al., 2008). According to Tran et al. (2010), IO/PVA generates ROS leading to the growth inhibition in *S. aureus*. The Fe^{2+} reacts with oxygen to produce H_2O_2, which in turn reacts with ferrous irons via Fenton reaction to generate hydroxyl radicals, which damage the cellular constituents. The small size of the nanoparticles also contributes to its antimicrobial property. Nanoparticles ranging in size from 10 to 80 nm readily penetrate inside the cell membrane of *E. coli* and react with intracellular oxygen, thereby resulting in oxidative stress and disruption of the cell membrane. Prabhu et al. (2015) reported that ROS generated by iron oxide nanoparticles significantly inhibited the pathogenic bacteria like *P. vulgaris*, *S. aureus*, *Xanthomonas* sp., and *E. coli*. Chen et al. (2008) demonstrated that immunoglobulin G-bound Fe_3O_4/titania core/magnetic shell NPs successfully suppress the growth of MDR pathogenic bacteria, that is, *S. pyogenes*, MRSA, and *Staphylococcus saprophyticus*.

15.5.6 ZINC OXIDE NANOPARTICLES (ZNO-NPS)

Zinc is an important trace element in the human body and nano ZnO has gained considerable attention due to its biocompatibility and stability under harsh processing conditions (Allaker and Memarzadeh, 2014). Moreover, ZnO-NPs show selective toxicity towards targeted bacteria and exert minimal toxicity to human cells (Ravishankar and Jamuna, 2011). ZnO-NPs possess potent antimicrobial properties and use several mechanisms to combat pathogen, thereby making it unlikely to develop resistance against them. ZnO-NPs are thought to adhere strongly to the cell membranes and damage both the lipids and proteins components of the bacterial membrane, which subsequently leads to increase in permeability of cell membrane and cell death (Hajipour et al., 2012). The generation of ROS and release of Zn^{2+} ions are few of the mechanisms proposed to describe the antimicrobial efficacy of ZnO-NPs. ZnO-NPs induce the release of ROS, such as hydrogen peroxide (H_2O_2) and hydroxyl radicals (OH^-). The generated H_2O_2 on the surface of ZnO-NPs penetrate inside the bacterial cells and damage the cell membrane (Allaker and Memarzadeh, 2014; Padmavathy and Vijayaraghavan, 2008). However, the negatively charged OH^- and superoxides are unable to enter the cell membrane and remain on the surface of the cell. The release of Zn^{2+} ions also damages the cell membrane and the intracellular components (Brayner et al., 2006; Blecher et al., 2011; Huh and Kwon, 2011). It was observed that polyvinyl alcohol (PVA) coated ZnO-NPs increases the permeability of the cell membrane and imposes oxidative stress on entering the cytoplasm (Huh and Kwon, 2011).

The antimicrobial efficacy of ZnO-NPs is significantly governed by their shape, size, crystal structure, and dispensability. ZnO-NPs in different form such as nanorods, nanoplates, flowers, needles, dandelions, helixes, rings, springs, and snowflakes have been synthesized (Yamamoto, 2001). The generation of H_2O_2 increases with the increase in surface area of the ZnO-NPs. It is presumed that with decrease in particle size, the ZnO particles per unit volume increases, resulting in overall increase in surface area and release of hydrogen peroxide (Ravishankar and Jamuna, 2011). According to Vidic et al. (2013), ZnO nanostructures of 100 nm in diameter exhibit significant antibacterial efficacy against *B. subtilis* and *E. coli*. Jin et al. (2009) reported that ZnO nanoparticles of $\sim$12 nm in size inhibited the growth of *E. coli* by disrupting the cell membrane and increasing its permeability. Pati et al. (2014) indicated the potential application of ZnO-NPs as antibacterial agent against *S. aureus*. The destruction of membrane integrity, decrease in cell surface hydrophobicity, and downregulation in the transcription of oxidative stress resistance genes were observed on exposure of the bacterial cell to ZnO-NPs. Liu et al. (2009) investigated the antibacterial efficacy of ZnO-NPs against verocytotoxigenic *E. coli* strain O157:H7. A considerable destruction in the membrane integrity and leakage of intracellular contents was observed without much effect to nucleic acid (Liu et al., 2009). ZnO-NPs also possess significant inhibition efficacy against methicillin resistant strains of *S. aureus* and *Streptococcus agalactiae* (Hajipour et al., 2012).

15.5.7 TITANIUM DIOXIDE-CONTAINING NANOPARTICLES (TIO_2-NPS)

Titanium dioxide (TiO_2) is a well established photocatalytic antimicrobial agent. The anatase form of nano TiO_2 when irradiated with UV light excitation brings about maximum antimicrobial activity of TiO_2-NPs. This photocatalysis process leads to oxidation of the polyunsaturated phospholipid present in the lipid membrane, causing loss of respiratory activity and subsequent cell death (Allaker and Memarzadeh, 2014).

Titanium dioxide-containing nanoparticles employ two antimicrobial mechanisms. In the photocatalysis process, TiO_2-NPs generate ROS when exposed to UV light at wavelength lower than 385 nm. The ROS generated including hydrogen peroxide (H_2O_2) and hydroxyl radicals (*OH) act on the phospholipids present in the bacterial cell resulting in site-specific DNA damage. It also damages the cell membranes of the bacteria, thereby increasing membrane permeability, hampering the process of oxidative phosphorylation, and sometimes resulting in cell death (Rudramurthy et al., 2016; Pelgrift and Friedman, 2013). The antimicrobial effect of TiO_2 photocatalyst against fungi and bacteria in suspension has been reported. The findings have led to the development of photocatalytic methods for combating pathogenic viruses and bacteria using TiO_2-NPs in aqueous media (Ravishankar and Jamuna, 2011). It has been suggested that TiO_2-NPs in combination with UV irradiation may also be employed as an effective way to decrease the disinfection time and eradicate pathogens (Ravishankar and Jamuna, 2011). Researchers have reported the light-induced biocidal property of engineered TiO_2-NPs in controlling *Aspergillus niger* and *E. coli* (Rudramurthy et al., 2016). Hence, the photocatalytic activity of TiO_2-NPs may be exploited in the development of antibiofilm coatings.

However, the major drawback in the using TiO_2-NPs is the need to provide UV light illumination to activate the photocatalyst and initiate the killing of the bacteria and viruses. In recent years, it was found that visible light absorbing photocatalysts of Ag/AgBr/TiO_2 efficiently inhibit *E. coli* and *S. aureus* (Ravishankar and Jamuna, 2011). It has been found that silver coated TiO_2 material exhibits enhanced photocatalytic and antibactericidal property relative to TiO_2 alone (Ranghar et al., 2014). Recently, it is observed that TiO_2-NPs also show antimicrobial efficacy in absence of UV irradiation, suggesting their involvement in other bactericidal mechanism (Pelgrift and Friedman, 2013).

TiO_2-NPs damages viruses, bacterial cell walls, and spores. They show highest antibacterial efficiency for *E. coli*, followed by *P. aeruginosa*, *S. aureus*, *Enterococcus faecium*, and *C. albicans* (Huh and Kwon, 2011).

15.5.8 MAGNESIUM OXIDE NANOPARTICLES (MGO-NPS)

MgO-NPs are usually considered safe for humans and exhibit great stability under severe processing conditions. Unlike TiO_2, MgO-NPs do not need to be photoactivated to trigger its inherent antimicrobial activity. Multiple strategies have been have been suggested to justify the antibacterial property of Mg-ONPs, a few of which include generation of lipid peroxidation, ROS generation, enzyme inhibition, alkaline effects, and electrostatic interactions. The production of ROS results in lipid peroxidation in the bacterial cell envelope, which in turn facilitates the entry of the NPs through the plasma membrane. Consequently, the cytoplasmic pH drops, which raises the membrane potential causing the intracellular cytoplasmic components to exude out of the cell. The strong electrostatic interaction between the MgO-NPs and the bacterial cell surface also contributes to the antibacterial activity. The alkaline effect on the surface of MgO-NPs, due to the formation of a thin layer of water molecules, damages the bacterial cell membrane when it comes in contact with MgO-NPs, ultimately resulting in cell death (Rudramurthy et al., 2016).

The size of the MgO-NPs influences its antimicrobial property as it alters the surface energy. MgO-NPs with size less than 15 nm in diameter exhibit enhanced bactericidal property as compared with its large and aggregated counterparts (Huang et al., 2005). MgO-NPs were found to show strong bactericidal effect against *Salmonella* and *E. coli* O157:H7. It was found that treatment with MgO-NP causes damage and deformation to the bacterial cell membrane, causing effluence of cytoplasmic constituents and eventually cell death (Jin and He, 2011). MgO-NPs were found to exhibit antagonistic effect against both Gram-negative and Gram-positive bacteria such as *Bacillus megaterium*, *S. aureus*, *P. aeruginosa*, and *E. coli*. In addition, they also exhibit biocidal activity towards spores and thus, may be employed as an effective disinfectant against a wide range of pathogens (Rudramurthy et al, 2016).

15.6 ANTIBIOFILM ACTIVITY OF METAL-BASED NANOPARTICLES

The formation of biofilm is an important mode of growth in pathogens related to chronic infections, affecting millions of people worldwide (Shakibaie et al., 2015). The production of exopolysaccharide (EPS) leading to the subsequent formation of biofilm is one of the major virulence determinants contributing to the development of MDR strains of pathogenic organisms. EPS matrix consisting of polysaccharides, proteins, eDNA, and cellular debris makes the pathogen inaccessible to the host defense system and antibiotics. Hence, the biofilm matrix confers an increase in antibiotic resistance in the case of biofilm-associated diseases like gingivitis, otitis media, and cystic fibrosis. They are responsible for many nosocomial as well as chronic infections associated with medical devices and surgical implants (Chen et al., 2013; Pelgrift and Friedman, 2013). The difficulty in eradication of these pathogens lies in the fact that the biofilms are highly resistant and inaccessible to the antimicrobial agents.

Nanoparticles have the ability to invade the biofilm matrix and gain access to inside the microbial cells. As depicted in Fig. 15.2, they attach to the bacterial cell wall, infiltrate inside the cell membranes, and accumulate inside cells, resulting in a continuous liberation of ions inside the cell (Ansari et al., 2013). Nanoparticles have received tremendous attention for their ability to inhibit/disrupt biofilm formation for possible clinical applications. Ag-NPs effectively prevented biofilm development in extended spectrum β-lactamases producing *Klebsiella* spp. and *E. coli* by restricting bacterial growth and production of EPS. The antibiofilm property of the Ag-NPs may be exploited towards the development of Ag-NPs based coatings for medical devices and implants. These NPs may also be used in the management of various infections caused by drug-resistant biofilm forming bacteria (Ansari et al., 2013). Among the leading nosocomial pathogens, *S. aureus*, *P. aeruginosa*, and *Proteus mirabilis* are capable of colonizing medical devices and catheters resulting in biofilm-related infections of the urinary and lower respiratory tract. Selenium nanoparticles (Se-NPs) prepared using *Bacillus* sp. MSh-1 were found to exhibit antibiofilm efficacy against biofilm producing clinically isolated nosocomial pathogens. The nanoparticles

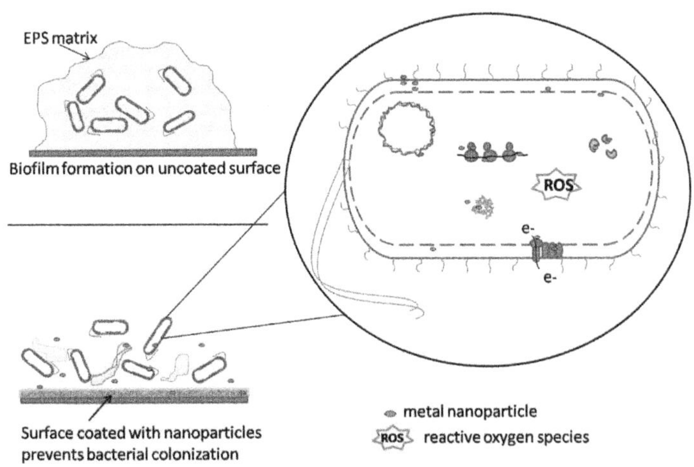

FIGURE 15.2

A diagrammatic illustration showing common mode of action of metal-based nanoparticles against biofilm forming bacteria.

significantly suppress the biofilm formation in *S. aureus*, *P. mirabilis*, and *P. aeruginosa*, by 42%, 53.4%, and 34.3%, respectively. It was also revealed that the effect of Se-NPs on the development of biofilm was considerably determined by the temperature and pH (Shakibaie et al., 2015). Selenium nanoparticles (Se-NPs) synthesized using Actinobacterium, *Streptomyces minutiscleroticus* M10A62 was able to completely inhibit biofilm formation in six *Acinetobacter* species within 48 hours of incubation. The inhibition of biofilm formation was achieved with 3.2 μg concentration of Se-NPs indicating the direct interaction of target bacteria with the nanoparticles resulted in inhibition of biofilm at low concentration (Ramya et al., 2015). Ag-NPs prevent *E. coli*, *S. aureus*, *C. albicans*, and *P. aeruginosa* from forming biofilm on plastic catheters (Pelgrift and Friedman, 2013). Ag-NPs produced by the lignin-degrading fungus *Emericella nidulans* and *A. flavus* inhibited biofilm formation in Gram-positive (*S. aureus*) and Gram-negative (*P. aeruginosa* and *E. coli*) bacteria. At a concentration of 0.5–64 $\mu g\ mL^{-1}$, the Ag-NPs significantly decreased 74%–84% biofilm formation as compared with fungal cell-free filtrate (devoid of Ag-NPs) showing 4%–6% antibiofilm activity (Barapatre et al. 2016). Zero-valent bismuth nanoparticles completely inhibited growth and formation of biofilm by *Streptococcus mutans* (Hernandez-Delgadillo et al., 2012). Selenium and tellurium nanoparticles prepared using *Stenotrophomonas maltophilia* SeITE02 and *Ochrobactrum* sp. MPV1 were found to posses antibacterial and antibiofilm activity against *P. aeruginosa* PAO1, *E. coli* JM109 and *S. aureus* ATCC-25923 (Zonaro et al., 2015).

Metal oxide nanoparticles also exhibit profound antibiofilm activity. For example, thin film composite (TFC) membranes coated with TiO_2-NP damages the cell membrane of *E. coli*, thereby preventing the adherence of bacteria onto the surface of the TFC membrane and inhibiting biofilm formation (Blecher et al., 2011). According to Hajipour et al. (2012), ZnO-NPs coated glass surfaces prevented biofilm formation by *E. coli* and *S. aureus* by generating ROS. Metal nanoparticles conjugated to photosensitizers also resulted in an increase in their antibiofilm efficiency. Photoactivated Au-NPs bound to methylene blue were found to inhibit biofilm formation in *C. albicans* (Khan, 2012). Magnetite (Fe_3O_4) nanoparticles coated with *Rosmarinus officinalis* essential oil exhibit a significant inhibitory effect on the formation of biofilm on catheters by *C. albicans* and *Candida tropicalis*. Reports suggested that the antibiofilm activity of the NPs is governed by the dimension and size of the nanoparticles. The high surface to volume ratio plays a vital role in imparting antibiofilm property to the nanoparticles. Thus, the decrease in the size of the NPs corresponds to increase in their activity. Small sized MgF_2-NPs coated glass surfaces inhibited adherence of bacteria, *S. aureus* and *E. coli*, to surfaces and subsequent biofilm formation (Pelgrift and Friedman., 2013). The colonization and growth of *Mycobacterium smegmatis* biofilm on membranes was decreased by over 98.7% when coated with Ag-NPs, with an average diameter 12.6 ± 5.7 nm at a concentration of 100 μM. The biofilm formation by *S. aureus* and *P. aeruginosa* was notably arrested by 65% and

88%, respectively, in presence of starch-stabilized nanoparticles (~20 nm in diameter), even at very low concentrations of 1–2 mM (Markowska et al., 2013).

Nanoparticles are coated and functionalized with various antimicrobial agents to enhance their antibiofilm activity. Phosphatidylcholine-coated Au nanoparticles doped with gentamicin exhibited efficient antagonistic activity against biofilm formation and established biofilm in both Gram-positive and Gram-negative bacteria (Mu et al., 2016). Bimetallic Au–Ag nanoparticles conjugate prepared using γ-proteobacterium, *Shewanella oneidensis* MR-1 remarkably arrested the biofilm production in Gram-positive bacteria (*S. aureus* and *E. faecalis*) as well as Gram-negative (*P. aeruginosa* and *E. coli*) bacteria, at 10 μM concentration (Ramasamy et al., 2016). It was found that curcumin, an antiquorum sensing agent that is known to display antibiofilm properties when fabricated with Ag-NPs exhibited profound synergistic effect against both Gram-positive (*S. aureus*) and Gram-negative (*P. aeruginosa*) bacterial pathogens. The synergistic effect of the conjugated nanoparticles resulted in disruption of 50% of the preformed biofilm, at 100 $\mu g\ mL^{-1}$ (Loo et al., 2016).

15.7 FUNCTIONALIZATION OF METAL NANOPARTICLES

Combinatorial drug therapy is expected to exhibit higher potency to combat microbial resistance as a result of the synergistic effects provided by the use of multiple drugs. The surface properties of NPs are highly flexible and can be easily manipulated using ligand engineering to impart desired properties to the NPs. Nanocomposites prepared by conjugating the nanoparticles with antibodies may be used against the antigen present on the surface of the target microbe (Fig. 15.3).

For example, Au-NPs fabricated with antibodies against protein A were used to selectivity target *S. aureus*. In another instance, the IgG in IgG–Fe_3O_4–TiO_2 magnetic nanoconjugate targets several types of bacteria, including *S. pyogenes*. Understanding the interactions of nanoparticles with the target pathogen and its biofilm mode plays a vital role in designing NP-based antimicrobial agents. Moreover, the use of NPs loaded antimicrobial agent aids in the controlled and sustainable release of the drug molecule, contributing to marked improvement in their biocompatibility and efficiency in mammalian cells (Gupta et al., 2016). According to recent reports, multiple antimicrobial agents may be fabricated within the same nanoparticle, thus making it multifunctional. The incorporation of multiple drugs in a nanoparticle results in higher antimicrobial efficacy enhanced potency and most likely overcome the existing problem of multidrug resistance in pathogenic microbes (Pelgrift and Friedman., 2013) (Table 15.4).

Among all metal nanoparticles, Au-NPs have been extensively studied and considered as a highly efficient drug delivery system due to their facile synthesis process, ease in functionalization, low toxicity, and high biocompatibility. The

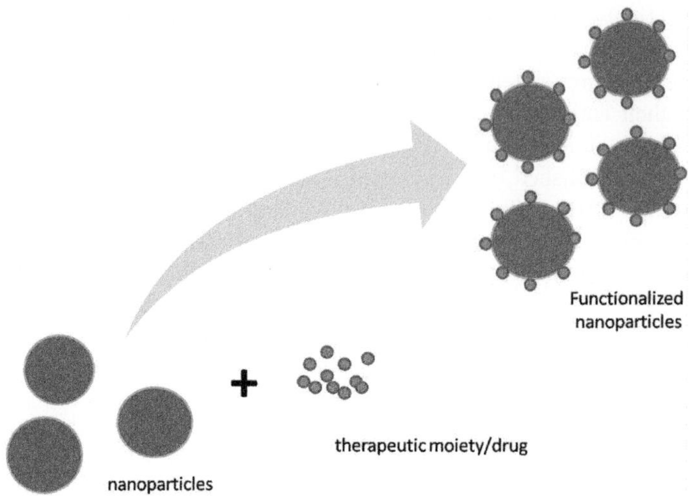

FIGURE 15.3

Schematic representation showing functionalization of metal-based nanoparticles for therapeutic application.

nontoxic and nonimmunogenic Au-NPs serves as an ideal carrier for drug delivery. Nanocomposites of Au-NPs doped with antibiotics have been employed for the selective photothermal killing of pathogenic microorganisms. Gold NPs themselves are considered inert and do not affect bacterial growth, however, conjugating antibiotics with Au-NPs significantly increases their antibacterial activity. Gold NPs synthesized using nonpathogenic fungus, *T. viride*, may be conjugated with vancomycin to form Vancomycin-bound biogenic gold nanoparticles (VBG-NPs). The VBG-NPs exhibited remarkable antimicrobial efficacy against drug-resistant *E. coli* and *S. aureus*. The antibacterial activity is proposed to be as a result of the nonspecific binding of VBG-NPs to the transpeptidase of Vancomycin-resistant *S. aureus* strain (Fayaz et al., 2011). Au-NPs have also been shown to efficiently deliver and increase the concentration of drug at the targeted site. Hence, they are used as drug carriers to selectively and specifically target cells, thereby increasing the efficiency of the drug in treating acute infections, without causing adverse effects (Shahverdi et al., 2007). Au-NPs when coated with Cefaclor, a second-generation β-lactam antibiotic, and ampicillin results in a profound increase in antibacterial activity against Gram-positive (*S. aureus*) and Gram-negative (*E. coli*) bacteria (Yin et al., 2015).

Nanoparticles may be effectively used in conjugation with antibiotics to enhance their antagonistic efficacy against various pathogens. Brown et al. (2012) determined that Au-NPs and Ag-NPs functionalized with ampicillin act as an effective broad-spectrum bactericidal agent against both Gram-positive and Gram-negative bacteria. The conjugation of ampicillin onto the surfaces of

Table 15.4 A Summary of Various Nanoparticles Conjugated With Antibiotics and Their Role as Antibacterial Agents

Nanoparticles	Drug Conjugate	Target	Activity	Reference
Ag-NPs	Ciprofloxacin, gentamycin, rifampicin, vancomycin	*Staphylococcus epidermidis Staphylococcus haemolyticus*	Antibacterial	Roshmi et al. (2015)
Ag-NPs Au-NPs	Ampicillin	*E. coli*, *V. cholerae*, *Enterobacter aerogenes*, MRSA, *Pseudomonas aeruginosa*	Antibacterial	Brown et al. (2012)
Au-NPs	Vancomycin	*E. coli*	Antibacterial	Fayaz et al. (2011)
Au-NPs	Gentamicin	*E. coli* K12	Antibacterial	Burygin et al. (2009)
Ag-NPs	Penicillin G, amoxicillin, erythromycin, clindamycin, vancomycin	*E. coli*, *Staphylococcus aureus*	Antibacterial	Shahverdi et al. (2007)
Au-NPs	Streptomycin, gentamycin, neomycin	*S. aureus*, *Micrococcus luteus*, *E. coli*, *P. aeruginosa*	Antibacterial	Grace and Pandian (2007)
Au-NPs	Ampicillin, streptomycin, kanamycin	*E. coli* DH5α, *M. luteus*, *S. aureus*	Antibacterial	Saha et al. (2007)
Au-NPs	Kanamycin	*Y. pestis* CO92, *S. epidermidis*, *P. aeruginosa* PA01, *E. aerogenes*, *S. bovis*	Antibacterial	Payne et al. (2016)
Ag-NP	Imipenem, gentamycin, vancomycin, ciprofloxacin	*P. aeruginosa*, *M. luteusS. aureus*, *E. coli*, *A. baumanii*, *E. faecalis*, *Bacillus* spp., *K. pneumoniae*	Antibacterial	Naqvi et al. (2012)

Ag-NPs and Au-NPs results in significant increase in their intrinsic antibacterial activity, which resulted in the antibacterial activity against MRSA, *P. aeruginosa*, *E. coli*, *V. cholera*, and *E. aerogenes*. Gold nanoparticles synthesized using *Bacillus* sp. when functionalized with antibiotics (gentamycin, ciprofloxacin, rifamoicin, and vancomycin) inhibited the growth of *S. epidermidis* and *S. haemolyticus* (Roshmi et al., 2015). Gold nanoparticles possess a large surface to volume ratio, thereby providing a large surface area for a number of drug molecules to get attached onto its surface. When these antimicrobial agents come in a close proximity with a nanogold core, it is believed to affect the growth of the microorganisms more effectively (Burygin et al., 2009).

Gold nanoparticles act as an efficient vehicle for drug delivery systems. The in vitro antibacterial activities of Au nanoparticles capped with aminoglycosidic antibiotics showed profound antibacterial efficiency against both Gram-positive and Gram-negative bacteria, for example, *Micrococcus luteus*, *P. aeruginosa*, *S. aureus*, and *E. coli* (Grace and Pandian, 2007). Au-NPs when conjugated with kanamycin, streptomycin, and ampicillin enhance its bactericidal efficacy. The antibiotic-capped Au-NPs showed greater bactericidal activity against *E. coli* DH5α, *M. luteus*, and *S. aureus* (Saha et al. 2007). Au-NPs conjugated with an antibiotic, kanamycin, exhibited dose-dependent broad-spectrum antibacterial activity against both Gram-positive and Gram-negative bacteria, including bacteria resistant to kanamycin. It was indicated that the increased efficacy might have resulted due to the damage of the bacterial cell envelope, consequently resulting in the leakage of intracellular components and cell death (Payne et al., 2016).

Fungi have been found to be relatively more efficient to bacteria in the synthesis of nanoparticles. Extracellularly synthesized silver nanoparticles (Ag-NPs) using culture filtrate of *A. flavus* have been successfully conjugated with conventional antibiotics. Shahverdi et al. (2007) reported the synthesis of Ag-NPs using the culture supernatant of *K. pneumoniae*. The antibacterial efficacy of amoxicillin, penicillin G, vancomycin, erythromycin, and clindamycin against *S. aureus* and *E. coli* were significantly enhanced in the presence of Ag-NPs. The highest effect was found in case of vancomycin, amoxicillin, and penicillin G against *S. aureus*. The synergistic effect of antibiotics in conjugation with biologically synthesized Ag-NPs significantly increased the susceptibility of the test pathogens. The combined effect of Ag-NPs and antibiotics, imipenem, gentamycin, and ciprofloxacin was notably exhibited against *S. aureus*, *E. coli*, *Bacillus* spp., *K. pneumonia*, *P. aeruginosa*, and *M. luteus*. Antibiotic–AgNPs conjugates were prepared in which an Ag-NP core was surrounded by antibiotic molecules. Thus, the antibiotic concentration was considerably increased at the target sites, leading to the destruction of the bacteria cell. Moreover, if a bacterium develops resistance to one of them, the other bactericidal agent would kill the bacteria (Naqvi et al., 2012).

15.8 NANOPARTICLES IN ANTIMICROBIAL PHOTODYNAMIC THERAPY

APDT has been suggested as a promising nonantibiotic approach to control and treat numerous infectious diseases caused by both Gram-positive and Gram-negative bacteria. APDT involves the use of a photosensitizer (PS), usually a nontoxic dye in combination with a harmless visible light. The irradiation of the PS generates ROS that selectively destroy microbial cells and biofilm. ROS such as hydroxyl radicals ($HO^{\bullet}$), superoxide (O_2^{-}) and hydrogen peroxide (H_2O_2) cause the oxidation of biomolecules, that is, proteins, lipids, and DNA inside the cell. It is also responsible for the damage to the cell wall and increases in membrane permeability, which results in leakage of cytoplasmic components. Since the microbial inactivation is a multitarget process, the possibility of development of resistant strains is minimal. Owing to the differences in the cell wall structures, the Gram-negative bacteria are found to be less susceptible to APDT as compared with the Gram-positive ones. Microorganisms such as yeasts, fungi, and viruses were also found to be susceptible to APDT (Yin et al., 2015).

Delivery of drug to the target site is essential to provide maximum bioavailability and therapeutic effect. Nanoparticles are used as a drug carrier to enhance the solubility of hydrophobic drugs, decrease side effects, minimize enzymatic degradation, prolong their circulation, and increase their bioavailability. It was established that the antimicrobial activity was greatly enhanced when PS was covalently bound to nanoparticles. At present, inorganic metal oxide (ZnO, TiO_2, and MgO) nanoparticles have gained much interest as antimicrobial agents due to their high biocompatibility and stability. A significant bactericidal activity of ZnO, Ag, and TiO_2 in the presence of UV light against *P. aeruginosa* and MRSA has been documented. The use of nanoparticles aids in the attachment and uptake of PS by the microbial cells and they are more easily exposed to the lethal ROS. APDT holds promising applications in self-sterilizing urinary catheters, antibacterial surgical, and dental implants. APDT may also be successfully employed in the management of periodontitis, root canal disinfection, gingivitis, oral candidosis, burn wounds, and acne (Perni et al., 2011; Yin et al., 2015).

Exploitation of metals and metal oxide nanoparticles, particularly those that produce ROS when exposed to ultraviolet (UV) light, such as zinc oxide (ZnO) and titanium dioxide (TiO_2), are finding increasing application in antimicrobial therapy. These nanoparticles act as the PS itself or interact with the PS either by conjugation or the nanoparticles. Titanium dioxide nanoparticles (TiO_2-NPs) are the most common PS and their ability to act as a photocatalytic antimicrobial agent is well established. The high reactivity of nanobased TiO_2 is extensively exploited for their bactericidal activity in the development of filters and coatings of glasses, ceramics, alumina, and polymers. As a photosensitizer, it possesses photocatalytic properties that allow the destruction of pathogenic microbes. When irradiated with UV light of wavelength less than 385 nm, TiO_2 results in the photocatalytic generation of active free hydroxyl radicals (OH^-), which catalyze the killing of both Gram-positive and Gram-negative bacteria. It also brings about photocatalytic oxidation of biofilm components on TiO_2-coated surfaces and prevents biofilm formation. Suspensions containing TiO_2 were found possess bactericidal effect on *E. coli*, *P. aeruginosa*, *Acinetobacter*, planktonic cells, and spores of *Bacillus cereus*. It has been suggested that in an aqueous medium, nanostructured TiO_2 irradiated with UV light may be employed as an efficient way to eliminate virus such as hepatitis B, influenza A/H1N1, influenza A/H3N2, polio virus type 1, SARS, and coronavirus. TiO_2 can successfully kill fungi, including *A. niger*, *C. albicans*, and *Cryptosporidium parvum* (Ravishankar and Jamuna, 2011; Yin et al., 2015).

Gold nanoparticles are generally considered to be biologically inert but they can be engineered to impart chemical or photothermal functionality. Au-NPs in conjugation with PS have been employed to attain greater bactericidal effect through APDT. The biologically inert Au-NPs absorb the UV-light and generate a significant photothermal effect, which damages the microbial cell. Gold nanoparticles have been shown to possess intrinsic antibacterial activity against *E. coli*, VRE, MRSA, and *P. aeruginosa*. Au-nanoparticles enhance the antimicrobial properties of PS such as toluidine blue O and methylene blue. According to Narband et al. (2008) the hydrophilic photosensitizer, toluidine blue O when fabricated on the surface of Au nanoparticles results in photodynamic inactivation and hyperthermia against methicillin-resistant *S. aureus*. The enhancement in the lethal properties might be the result of higher light capturing ability of the PS when adsorbed on the surface of the Au nanoparticles. Au nanorods coated with PS destroy MRSA cells by synergistic antimicrobial and photothermal effect. *S. aureus* can be inhibited using laser activated Au-NPs fabricated with antibodies. Au-nanoparticles were used against *C. albicans* biofilms using methylene blue-induced APDT (Yin et al., 2015).

15.9 NANOPARTICLES FOR ANTIMICROBIAL DRUG DELIVERY

A wide range of antimicrobial drugs are available and prescribed to treat infectious diseases caused by various fungi, bacteria, and viruses. However, the inefficient delivery of these conventional drugs to the target site results in inadequate therapeutic effects and undesirable side effects. Other limitations include difficulty in administration due to their low solubility, lack of target specificity, cytotoxic to host tissues, and rapid degradation and clearance from the body (Ranghar et al., 2014). Delivering the right dose of drug precisely and safely to the specific target at the right time is the main goal in the designing of an ideal drug delivery system (Li et al., 2011). In this regard, NPs have emerged as promising drug delivery systems. It is well reported that the metal-based nanoparticles serve as an effective antimicrobial agent against common pathogens. Owing to their small size, they are highly reactive and readily taken up by the host as well as microbial cells (Zhang et al., 2010). The surface properties of the NPs can be tuned to attain controlled release of the pharmacologically active agent at a specific site at a therapeutically optimal rate and dose (Ranghar et al., 2014). The drug molecules are generally loaded to the nanoparticles adsorption, physical encapsulation, or chemical conjugation to increase the therapeutic index and pharmacokinetics of the drugs (Zhang et al., 2010). Many advantages of nanoparticles based drug delivery have been recognized, including improvement in solubility of the drugs, increased circulation lifetime, controlled release of drugs, specificity, and selectivity of the target (Zhang et al., 2010). The use of nanoparticles as carrier allows delivery of optimum doses of a drug to the targeted site, thereby reducing the risk of resistance mechanisms and minimizing toxicity to patients (Pelgrift and Friedman, 2013) They also increase the stability of a variety of therapeutic agents such as oligonucleotides and peptides. The surfaces of Au-NPs can be readily functionalized with ligands containing phosphines, thiols, and amines, which exhibit high affinity for gold surfaces (Li et al., 2011). When administered, the drug-loaded nanoparticles can enter the host cells via endocytosis. Once inside the host cell, these nanoparticles release drugs during the course of time, that then kill the intracellular pathogen (Zhang et al., 2010). The accumulation of required dose of the drug at the site of infection eliminates the intracellular bacteria before the development of resistance. Nanoparticles may be fabricated with antibodies to specifically target antigen present on the surface of the targeted microbe. These nanoparticles based antimicrobial drug delivery systems help in distinguishing the target microbes or infectious cells from healthy cells, allowing specific targeting ability. For example, Au-NPs coated with antibodies against protein A have been successfully employed to selectivity kill *S. aureus*. Functionalized TiO_2 magnetic NPs are used to target several types of bacteria, including *S. pyogenes* (Pelgrift and Friedman, 2013). Aptamers also exhibit in vitro antimicrobial activity against beta-lactamase producing Gram-positive and Gram-negative bacteria. They also possess antiviral efficacy against reverse transcriptase of HIV and inhibits replication of Vaccinia virus (Pelgrift and Friedman, 2013). Due to the above-mentioned reasons, numerous nanoparticle-based drug delivery systems have been approved to treat a variety of diseases and many others are currently under various stages of clinical trials.

15.10 CONCLUSION

The ability to form biofilm and development of drug resistance in many pathogenic microorganisms greatly reduced the efficacy of currently available antibiotics, leading to a serious problem in public health management. Hence, the use nanoparticles serve as a novel strategy and a potential alternative to antibiotics in dealing with the serious medical scenario of the development of MDR strains.

This chapter discusses the synthesis of nanoparticles by various microorganisms, the synthesis mechanisms, and their potential applications to solve the problem of drug resistance in numerous pathogens. Over the last decade, there has been a tremendous development in the field of microorganism mediated synthesis of nanoparticles and their applications. Owing to the rich biodiversity, microorganisms as a biofactory serve as an environmentally safe and sustainable method for the synthesis of nanoparticles. Actinomycetes, yeast, algae, fungi, and bacteria have been successfully employed for the enzymatic synthesis of safe and stable metal and metal oxide nanoparticles. The ease of cultivation and manipulation of microorganisms also gives the advantage to manipulate the synthesis rate and the associated parameters. The synthesis process takes place either extracellularly or intracellularly by the aid of microbial enzymes. The biosynthesized nanoparticles naturally capped with protein/lipid layer imparts stability and physiological solubility to the nanoparticles, making them suitable for biomedical applications. Metals and metal oxide nanoparticles possess remarkable bactericidal, fungicidal, and antiviral properties, demonstrating their potential as a promising tool in antimicrobial therapy. The potency of these nanoparticles is significantly governed by physical characteristics including shape, size, and stability. Different metallic NPs, such as silver, gold titanium, copper, and magnesium, are known for their antimicrobial efficacies. Many nanoparticles exhibit bactericidal properties by generating ROS and in some cases penetrate inside the cell membrane and release metal ion. Due to their small, controllable size and property, functionalized nanoparticles have emerged as an effective carrier in drug delivery. The recent progress and the ongoing research in functionalization gave way to new avenues for commercial applications of functionalized nanoparticles in medicine and healthcare. Nanoparticles show a significant increase in biocompatibility and activity when conjugated with other materials. Nanoparticles fabricated with antibiotics, dyes and antibodies exhibit enhanced antimicrobial activity. The nanoparticles exhibiting antimicrobial property are being explored for their application in prosthetic device coatings, catheters, ointments and gels, and in dental implants. Further, the application of nanoscale antimicrobials in the management of life-threatening diseases, facilitated by their fascinating microbicidal and drug delivery capabilities, is of ongoing interest. Further, research on microbially synthesized nanoparticles possessing unique optical, physicochemical, and electronic properties is of great significance for application in medicine. Thus, nanoparticles in the field of antimicrobial chemotherapy have gained much attention as an eco-friendly agent to fight microbial resistance and prevent diseases.

REFERENCES

Abboud, Y., Saffaj, T., Chagraoui, A., El Bouari, A., Brouzi, K., Tanane, O., et al., 2014. Biosynthesis, characterization and antimicrobial activity of copper oxide nanoparticles

(CO-NPs) produced using brown alga extract (*Bifurcaria bifurcata*). Appl. Nanosci. 4, 571–576.

Ahangari, A., Salouti, M., Heidari, Z., Kazemizadeh, A.R., Safari, A.A., 2013. Development of gentamicin–gold nanospheres for antimicrobial drug delivery to *Staphylococcal* infected foci. Drug Deliv. 20, 34–39.

Allaker, R.P., Memarzadeh, K., 2014. Nanoparticles and the control of oral infections. Int. J. Antimicrob. Agents 43 (2), 95–104.

Ansari, A.M., Khan, H.M., Khan, A.A., Cameotra, W.S., Pal, R., 2013. Antibiofilm efficacy of silver nanoparticles against biofilm of extended spectrum β-lactamase isolates of *Escherichia coli* and *Klebsiella pneumoniae*. Appl. Nanosci. 2013.

Ansari, M.A., Khan, H.M., Khan, A.A., Cameotra, S.S., Saquib, Q., Musarrat, J., 2014. Interaction of Al_2O_3 nanoparticles with *Escherichia coli* and their cell envelope biomolecules. J. Appl. Microbiol. 116, 772–783.

Ansari, M.A., Khan, H.M., Alzohairy, M.A., Jalal, M., Ali, S.G., Pal, R., et al., 2015. Green synthesis of Al_2O_3 nanoparticles and their bactericidal potential against clinical isolates of multi-drug resistant *Pseudomonas aeruginosa*. World J. Microbiol. Biotechnol. 31 (1), 153–164.

Arun, G., Eyini, M., Gunasekaran, P., 2014. Green synthesis of silver nanoparticles using the mushroom fungus *Schizophyllum commune* and its biomedical applications. Biotechnol. Bioprocess Eng. 19 (6), 1083–1090.

Azizi, S., Ahmad, M.B., Namvar, F., Mohamad, R., 2014. Green biosynthesis and characterization of zinc oxide nanoparticles using brown marine macroalga *Sargassum muticum* aqueous extract. Mater. Lett. 116, 275–277.

Bala, T., Armstrong, G., Laffir, F., Thornton, R., 2011. Titania-silver and alumina–silver composite nanoparticles: novel, versatile synthesis, reaction mechanism and potential antimicrobial application. Adv. Colloid Interface Sci. 356, 395–403.

Balagurunathan, R., Radhakrishnan, M., Rajendran, R.B., Velmurugan, D., 2011. Biosynthesis of gold nanoparticles by actinomycete *Streptomyces viridogens* strain HM10. Indian J. Biochem. Biophys. 48 (5), 331–335.

Balasubramanyam, A., Sailaja, N., Mahboob, M., Rahman, M.F., Hussain, S.M., Grover, P., 2010. In vitro mutagenicity assessment of aluminium oxide nanomaterials using the Salmonella/microsome assay. Toxicol. In Vitro 24, 1871–1876.

Bansal, V., Rautaray, D., Ahmad, A., Sastry, M., 2004. Biosynthesis of zirconia nanoparticles using the fungus *Fusarium oxysporum*. J. Mater. Chem. 14 (22), 3303–3305.

Bansal, V., Rautaray, D., Bharde, A., et al., 2005. Fungus-mediated biosynthesis of silica and titania particles. J. Mater. Chem. 15 (26), 2583–2589.

Barapatre, A., Aadil, K.R., Jha, H., 2016. Synergistic antibacterial and antibiofilm activity of silver nanoparticles biosynthesized by lignin-degrading fungus. Bioresour. Bioprocess 3, 8.

Bard, A.J., Holt, K.B., 2005. Interaction of silver(I) ions with the respiratory chain of *Escherichia coli*: an electrochemical and scanning electrochemical microscopy study of the antimicrobial mechanism of micromolar Ag + . Biochemistry 44, 13214–13223.

Barwal, I., Ranjan, P., Kateriya, S., Yadav, S.C., 2011. Cellular oxido-reductive proteins of *Chlamydomonas reinhardtii* control the biosynthesis of silver nanoparticles. J. Nanobiotechnology. 9, 56.

Basavaraja, S., Balaji, S.D., Lagashetty, A., Rajasab, A.H., Venkataraman, A., 2008. Extracellular biosynthesis of silver nanoparticles using the fungus *Fusarium semitectum*. Mater. Res. Bull. 43 (5), 1164–1170.

Behera, S.S., Patra, J.K., Pramanik, K., Panda, N., Thatoi, H., 2012. Characterization and evaluation of antibacterial activities of chemically synthesized iron oxide nanoparticles. World J. Nano Sci. Eng. 2, 196–200.

Blecher, K., Nasir, A., Friedman, A., 2011. The growing role of nanotechnology in combating infectious disease. Virulence 2 (5), 395–401.

Borkow, G., 2014. Using copper to improve the well-being of the skin. Curr. Chem. Biol. 8 (2), 89–102.

Bose, S., Hochella, M.F., Gorby, Y.A., et al., 2009. Bioreduction of hematite nanoparticles by the dissimilatory iron reducing bacterium *Shewanella oneidensis* MR-1. Geochim. Cosmochim. Acta 73 (4), 962–976.

Brayner, R., Ferrari-Iliou, R., Brivois, N., Djediat, S., Benedetti, M.F., Fievet, F., 2006. Toxicological impact studies based on *Escherichia coli* bacteria in ultrafine ZnO nanoparticles colloidal medium. Nano Lett. 6, 866–870.

Brown, A.N., Smith, K., Samuels, T.A., Lu, J., Obare, S.O., Scott, M.E., 2012. Nanoparticles functionalized with ampicillin destroy multiple-antibiotic-resistant isolates of *Pseudomonas aeruginosa* and *Enterobacter aerogenes* and methicillin-resistant *Staphylococcus aureus*. Appl. Environ. Microbiol. 78 (8), 2768–2774.

Burygin, G.L., Khlebtsov, B.N., Shantrokha, A.N., Dykman, L.A., Bogatyrev, V.A., Khlebtsov, N.G., 2009. On the enhanced antibacterial activity of antibiotics mixed with gold nanoparticles. Nanoscale Res. Lett. 4 (8), 794–801.

Castro-Longoria, E., Vilchis-Nestor, A.R., Avalos-Borja, M., 2011. Biosynthesis of silver, gold and bimetallic nanoparticles using the filamentous fungus *Neurospora crassa*. Colloids Surf. B: Biointerfaces 83 (1), 42–48.

Chaloupka, K., Malam, Y., Seifalian, A.M., 2010. Nanosilver as a new generation of nanoproduct in biomedical applications. Trends Biotechnol. 28, 580–588.

Chatterjee, A.K., Sarkar, R.K., Chattopadhyay, A.P., Aich, P., Chakraborty, R., Basu, T.A., 2012. Simple robust method for synthesis of metallic copper nanoparticles of high antibacterial potency against *E. coli*. Nanotechnology 23, 85–103.

Chen, W.J., Tsai, P.J., Chen, Y.C., 2008. Functional Fe_3O_4/TiO_2 core/shell magnetic nanoparticles as photokilling agents for pathogenic bacteria. Small 4, 485–491.

Chen, M., Yu, Q., Sun, H., 2013. Novel strategies for the prevention and treatment of biofilm related infections. Int. J. Mol. Sci. 14 (9), 18488–18501.

Chitra, K., Annadurai, G., 2013. Bioengineered silver nanobowls using *Trichoderma viride* and its antibacterial activity against Gram-positive and Gram-negative bacteria. J. Nanostruct. Chem. 3, 9.

Chowdhury, S., Basu, A., Kundu, S., 2014. Green synthesis of protein capped silver nanoparticles from phytopathogenic fungus *Macrophomina phaseolina* (Tassi) Goid with antimicrobial properties against multidrug-resistant bacteria. Nanoscale Res. Lett. 9 (1), 365.

Correa-Llanten, D.N., Munoz-Ibacache, S.A., Castro, M.E., Munoz, P.A., Blamey, J.M., 2013. Gold nanoparticles synthesized by *Geobacillus* sp. strain ID17 a thermophilic bacterium isolated from Deception Island, Antarctica. Microb. Cell Fact. 12, 75.

Dibrov, P., Dzioba, J., Gosink, K.K., Hase, C.C., 2002. Chemiosmotic mechanism of antimicrobial activity of Ag + in *Vibrio cholerae*. Antimicrob. Agents Chemother. 46, 2668–2670.

Duran, N., Marcato, P.D., De Conti, R., Alves, O.L., Costa, F.T.M., Brocchi, M., 2010. Potential use of silver nanoparticles on pathogenic bacteria, their toxicity and possible mechanisms of action. J. Braz. Chem. Soc. 21, 6.

Erasmus, M., Duncan Cason, E., Marwijk, J., Botes, E., Gericke, M., Heerden, E., 2014. Gold nanoparticle synthesis using the thermophilic bacterium *Thermus scotoductus* SA-01 and the purification and characterization of its unusual gold reducing protein. Gold Bull. 47 (4), 245–253.

Esteban, T.L., Malpartida, F., Esteban, C.A., Pecharromán, C., Moya, J.S., 2009. Antibacterial and antifungal activity of a soda-lime glass containing copper nanoparticles. Nanotechnology 20 (50), 505–701.

Fayaz, A.M., Balaji, K., Girilal, M., Yadav, R., Kalaichelvan, P.T., Venketesan, R., 2010. Biogenic synthesis of silver nanoparticles and their synergistic effect with antibiotics: a study against Gram-positive and Gram-negative bacteria. Nanomedicine 6 (1), 103–109.

Fayaz, A.M., Girilal, M., Mahdy, S.A., Somsundar, S.S., Venkatesan, R., Kalaichelvan, P.T., 2011. Vancomycin bound biogenic gold nanoparticles: a different perspective for development of anti VRSA agents. Process Biochem. 46, 636–641.

Feng, Q.L., Wu, J., Chen, G.Q., Cui, F.Z., Kim, T.N., Kim, J.O., 2000. A mechanistic study of the antibacterial effect of silver ions on *Escherichia coli* and *Staphylococcus aureus*. J. Biomed. Mater. Res. 52, 662–668.

Gahlawat, G., Shikha, S., Chaddha, B.S., Chaudhuri, S.R., Mayilraj, S., Choudhury, A.R., 2016. Microbial glycolipoprotein-capped silver nanoparticles as emerging antibacterial agents against cholera. Microb. Cell Fact. 15, 1–14.

Gajbhiye, M., Kesharwani, J., Ingle, A., Gade, A., Rai, M., 2009. Fungus-mediated synthesis of silver nanoparticles and their activity against pathogenic fungi in combination with fluconazole. Nanomedicine 5 (4), 382–386.

Galib, Barve, M., Mashru, M., Jagtap, C., Patgiri, B.J., Prajapati, P.K., 2011. Therapeutic potentials of metals in ancient India: a review through Charaka Samhita. J. Ayurveda Integr. Med. 2 (2), 55–63.

Gericke, M., Pinches, A., 2006. Biological synthesis of metal nanoparticles. Hydrometallurgy 83 (1–4), 132–140.

Gil-Tomas, J., 2007. Lethal photosensitisation of *Staphylococcus aureus* using a toluidine blue O–tiopronin–gold nanoparticle conjugate. J. Mater. Chem. 17, 3739–3746.

Gogoi, S.K., Gopinath, P., Paul, A., Ramesh, A., Ghosh, S.S., Chattopadhyay, A., 2006. Green fluorescent protein-expressing *Escherichia coli* as a model system for investigating the antimicrobial activities of silver nanoparticles. Langmuir 22, 9322–9328.

Grace, N.A., Pandian, K., 2007. Antibacterial efficacy of aminoglycosidic antibiotics protected gold nanoparticles—a brief study. Colloids Surf. A Physicochem. Eng. Asp. 297, 63–70.

Gu, H., Ho, P.L., Tong, E., Wang, L., Xu, B., 2003. Presenting vancomycin on nanoparticles to enhance antimicrobial activities. Nano Lett. 3, 1261–1263.

Gupta, M., Mahajan, V.K., Mehta, K.S., Chauhan, P.S., 2014. Zinc therapy in dermatology: a review. Dermatol. Res. Pract. 2014, 709152.

Gupta, A., Landis, R.F., Rotello, V.M., 2016. Nanoparticle-based antimicrobials: surface functionality is critical. F1000 Research. 5.

Gurunathan, S., Kalishwaralal, K., Vaidyanathan, R., Venkataraman, D., Pandian, S.R., Muniyandi, J., et al., 2009. Biosynthesis, purification and characterization of silver nanoparticles using *Escherichia coli*. Colloids Surf. B: Biointerfaces 74 (1), 328–335.

Gutierrez, F.M., Olive, P.L., Banuelos, A., Orrantia, E., Nino, N., Sanchez, E.M., et al., 2010. Synthesis, characterization, and evaluation of antimicrobial and cytotoxic effect of silver and titanium nanoparticles. Nanomedicine 6, 681–688.

Hajipour, M.J., Fromm, K.M., AkbarAshkarran, A., Jimenez de Aberasturi, D., Larramendi, I.R.D., Rojo, T., 2012. Antibacterial properties of nanoparticles. Trends 30 (10), 499–511.

Hamouda, I.M., 2012. Current perspectives of nanoparticles in medical and dental biomaterials. J. Biomed. Res. 26 (3), 143–151.

He, S., Zhang, Y., Guo, Z., Gu, N., 2008. Biological synthesis of gold nanowires using extract of *Rhodopseudomonas capsulata*. Biotechnol. Prog. 24 (2), 476–480.

Hernandez-Delgadillo, R., Velasco-Arias, D., Diaz, D., Arevalo-Nino, K., Garza-Enriquez, M., De la Garza-Ramos, M.A., et al., 2012. Zerovalent bismuth nanoparticles inhibit *Streptococcus mutans* growth and formation of biofilm. Int. J. Nanomed. 7, 2109–2113.

Huang, L., Li, D.Q., Lin, Y.J., Wei, M., Evans, D.G., Duan, X., 2005. Controllable preparation of nano-MgO and investigation of its bactericidal properties. J. Inorg. Biochem. 99, 986–993.

Huh, A.J., Kwon, Y.J., 2011. "Nanoantibiotics": a new paradigm for treating infectious diseases using nanomaterials in the antibiotics resistant era. J. Control. Release 156 (2), 128–145.

Ishida, K., Cipriano, T.F., Rocha, G.M., Weissmuller, G., Gomes, F., Miranda, K., et al., 2014. Silver nanoparticle production by the fungus *Fusarium oxysporum*: nanoparticle characterisation and analysis of antifungal activity against pathogenic yeasts. Mem. Inst. Oswaldo Cruz 109 (2), 220–228.

Jain, N., Bhargava, A., Majumdar, S., Tarafdar, J.C., Panwar, J., 2011. Extracellular biosynthesis and characterization of silver nanoparticles using *Aspergillus flavus* NJP08: a mechanism perspective. Nanoscale 3 (2), 635–641.

Jayaseelan, C., Rahuman, A.A., Kirthi, A.V., Marimuthu, S., Santhosh kumar, T., Bagavan, A., et al., 2012. Novel microbial route to synthesize ZnO nanoparticles using *Aeromonas hydrophila* and their activity against pathogenic bacteria and fungi. Spectrochim. Acta A: Mol. Biomol. Spectrosc. 90, 78–84.

Jha, A.K., Prasad, K., Kulkarni, A.R., 2009. Synthesis of TiO_2 nanoparticles using microorganisms. Colloids Surf. B 71 (2), 226–229.

Jin, T., He, Y., 2011. Antibacterial activities of magnesium oxide (MgO) nanoparticles against foodborne pathogens. J. Nanopart. Res. 13, 6877–6885.

Jin, T., Sun, D., Su, Y., Zhang, H., Sue, H.J., 2009. Antimicrobial efficacy of zinc oxide quantum dots against *Listeria monocytogenes*, *Salmonella enteritidis* and *Escherichia coli* O157:H7. J. Food Sci. 74, 46–52.

Kalishwaralal, K., Deepak, V., Ram Kumar Pandian, S., Kottaisamy, M., Barathmani Kanth, S., Kartikeyan, B., et al., 2010. Biosynthesis of silver and gold nanoparticles using *Brevibacterium casei*. Colloids Surf. B: Biointerfaces 77 (2), 257–262.

Kalpana, D., Lee, Y.S., 2013. Synthesis and characterization of bactericidal silver nanoparticles using cultural filtrate of simulated microgravity grown *Klebsiella pneumoniae*. Enzyme Microb.Technol. 52 (3), 151–156.

Karthik, L., Kumar, G., Kirthi, A.V., Rahuman, A.A., Bhaskara Rao, K.V., 2014. *Streptomyces* sp. LK3 mediated synthesis of silver nanoparticles and its biomedical application. Bioprocess Biosyst. Eng. 37 (2), 261–267.

Kathiraven, T., Shanmugam, A.N., Balasubramanian, T., 2015. Green synthesis of silver nanoparticles using marine algae *Caulerpa racemosa* and their antibacterial activity against some human pathogens. Appl. Nanosci. 5 (4), 499–504.

Khan, A.U., 2012. Medicine at nanoscale: a new horizon. Int. J. Nanomed. 7, 2997–2998.

Khan, S., Alam, F., Azam, A., Khan, A.U., 2012. Gold nanoparticles enhance methylene blue-induced photodynamic therapy: a novel therapeutic approach to inhibit *Candida albicans* biofilm. Int. J. Nanomed. 7, 3245–3257.

Khashan, K.S., Sulaiman, G.M., Abdulameer, F.A., 2016. Synthesis and antibacterial activity of CuO nanoparticles suspension induced by laser ablation in liquid. Arab. J. Sci. Eng. 41, 301–310.

Kim, J.S., Kuk, E., Yu, K.N., Kim, J.H., Park, S.J., Lee, H.J., et al., 2007. Antimicrobial effects of silver nanoparticles. Nanomedicine 3, 95–101.

Korbekandi, H., Iravani, S., Abbasi, S., 2012. Optimization of biological synthesis of silver nanoparticles using *Lactobacillus casei* subsp. casei. J. Chem. Technol. Biotechnol. 87, 932–937.

Kuo, W.S., 2009. Antimicrobial gold nanorods with dual-modality photodynamic inactivation and hyperthermia. Chem. Commun. 32, 4853–4855.

Kwak, K., Kim, C., 2005. Viscosity and thermal conductivity of copper oxide nanofluid dispersed in ethylene glycol. Korea—Australia Rheol. J. 17, 35–40.

Lara, H.H., Ayala-Nunez, N.V., Ixtepan Turrent, L.D.C., Rodriguez Padilla, C., 2010. Bactericidal effect of silver nanoparticles against multidrug-resistant bacteria. World J. Microbiol. Biotechnol. 26 (4), 615–621.

Lee, C., Kim, J.Y., Lee, W.I., Nelson, K.L., Yoon, J., Sedlak, D.L., 2008. Bactericidal effect of zero-valent iron nanoparticles on *Escherichia coli*. Environ. Technol. 42, 4927–4933.

Li, B., Logan, B.E., 2004. Bacterial adhesion to glass and metal oxide surfaces. Colloids Surf. B 36, 81–90.

Li, X., Xu, H., Chen, Z.S., Chen, G., 2011. Biosynthesis of nanoparticles by microorganisms and their applications. J. Nanomater. 2011.

Li, J., Li, Q., Ma, X., Tian, B., Li, T., Yu, J., et al., 2016. Biosynthesis of gold nanoparticles by the extreme bacterium *Deinococcus radiodurans* and an evaluation of their antibacterial properties. Int. J. Nanomed. 11, 5931–5944.

Liu, Y., He, L., Mustapha, A., Li, H., Hu, Z.Q., Lin, M., 2009. Antibacterial activities of zinc oxide nanoparticles against *Escherichia coli* O157:H7. J. Appl. Microbiol. 107, 1193–1201.

Loo, C.Y., Rohanizadeh, R., Young, P.M., Traini, D., Cavaliere, R., Whitchurch, C.B., et al., 2016. Combination of silver nanoparticles and curcumin nanoparticles for enhanced anti-biofilm activities. J. Agric. Food Chem. 64 (12), 2513–2522.

Mahdavi, M., Namvar, F., Ahmad, M.B., Mohamad, R., 2013. Green biosynthesis and characterization of magnetic iron oxide (Fe_3O_4) nanoparticles using seaweed (*Sargassum muticum*) aqueous extract. Molecules 18, 5954–5964.

Manivasagan, P., Venkatesan, J., Senthilkumar, K., Sivakumar, K., Kim, S.K., 2013. Biosynthesis, antimicrobial and cytotoxic effect of silver nanoparticles using a novel *Nocardiopsis* sp. MBRC-1. Biomed. Res. Int. 2013, 287638.

Markowska, K., Grudniak, A.M., Wolska, K.I., 2013. Silver nanoparticles as an alternative strategy against bacterial biofilms. Acta Biochim. Pol. 60 (4), 523–530.

Martinez Flores, E., Negrete, J., Torres Villasenor, G., 2003. Structure and properties of Zn–Al–Cu alloy reinforced with alumina particles. Mater. Des. 24, 281–286.

McDonnell, G., Denver Russell, A., 1999. Antiseptics and disinfectants: activity, action, and resistance. Clin. Microbiol. Rev. 12 (1), 147–179.

Morones, J.R., Elechiguerra, J.L., Camacho, A., Holt, K., Kouri, J.B., Yacaman, M.J., 2005. The bactericidal effect of silver nanoparticles. Nanotechnology 16, 2346–2353.

Mu, H., Tang, J., Liu, Q., Sun, C., Wang, T., Duan, J., 2016. Potent antibacterial nanoparticles against biofilm and intracellular bacteria. Sci. Rep. 6, 18877.

Mukherjee, P., Senapati, S., Mandal, D., Ahmad, A., Khan, M.I., Kumar, R., et al., 2002. Extracellular synthesis of gold nanoparticles by the fungus *Fusarium oxysporum*. Chembio Chem. 3, 461–463.

Mukherjee, A., Mohammed Sadiq, I., Prathna, T.C., Chandrasekaran, N., 2011. Antimicrobial activity of aluminium oxide nanoparticles for potential clinical applications. Sci. Microb. Pathog. Commun. Curr. Res. Technol. Adv. 1, 245–251.

Namvar, F., Azizi, S., Ahmad, M., Shameli, K., Mohamad, R., Mahdavi, M., et al., 2015. Green synthesis and characterization of gold nanoparticles using the marine macroalgae *Sargassum muticum*. Res. Chem. Intermed. 41, 5723–5730.

Nanda, A., Saravanan, M., 2009. Biosynthesis of silver nanoparticles from *Staphylococcus aureus* and its antimicrobial activity against MRSA and MRSE. Nanomedicine 5, 452–456.

Naqvi, S.Z., Kiran, U., Ali, M.I., Jamal, A., Hameed, A., Ahmed, S., et al., 2012. Combined efficacy of biologically synthesized silver nanoparticles and different antibiotics against multidrug-resistant bacteria. Int. J. Nanomed. 8, 3187–3195.

Narband, N., Tubby, S., Parkin, I.P., Gil-Tomas, J., Ready, D., Nair, S.,P., et al., 2008. Gold nanoparticles enhance the toluidine blue-induced lethal photosensitisation of *Staphylococcus aureus*. Curr. Nanosci. 4, 409–414.

Padmavathy, N., Vijayaraghavan, R., 2008. Enhanced bioactivity of ZnO nanoparticles—an antimicrobial study. Sci. Technol. Adv. Mater. 9.

Pal, S., Tak, Y.K., Song, J.M., 2007. Does the antibacterial activity of silver nanoparticles depend on the shape of the nanoparticle? A study of the Gram-negative bacterium *Escherichia coli*. Appl. Environ. Microbiol. 73 (6), 1712–1720.

Pati, R., Mehta, R.K., Mohanty, S., Padhi, A., Sengupta, M., Vaseeharan, B., et al., 2014. Topical application of zinc oxide nanoparticles reduces bacterial skin infection in mice and exhibits antibacterial activity by inducing oxidative stress response and cell membrane disintegration in macrophages. Nanomed. Nanotech. Biol. Med. 10, 1195–1208.

Payne, J.N., Waghwani, H.K., Connor, M.G., Hamilton, W., Tockstein, S., Moolani, H., et al., 2016. Novel synthesis of kanamycin conjugated gold nanoparticles with potent antibacterial activity. Front. Microbiol. 7, 607.

Pelgrift, R.Y., Friedman, A.J., 2013. Nanotechnology as a therapeutic tool to combat microbial resistance. Adv. Drug Deliv. Rev. 65 (13-14), 1803–1815.

Perez-Gonzalez, T., Jimenez-Lopez, C., Neal, A.L., et al., 2010. Magnetite biomineralization induced by *Shewanella oneidensis*. Geochim. Cosmochim. Acta 74 (3), 967–979.

Perni, S., 2009. The antimicrobial properties of light-activated polymers containing methylene blue and gold nanoparticles. Biomaterials 30, 89–93.

Perni, S., Prokopovich, P., Pratten, J., Parkin, I.P., Wilson, M., 2011. Nanoparticles: their potential use in antibacterial photodynamic therapy. Photochem. Photobiol. Sci. 10 (5), 712–720.

Perni, S., Hakala, V., Prokopovich, P., 2014. Biogenic synthesis of antimicrobial silver nanoparticles capped with L-cysteine. Colloids Surf. A 460, 219–224.

Pissuwan, D., Cortie, C.H., Valenzuela, S.M., Cortie, M.B., 2009. Functionalised gold nanoparticles for controlling pathogenic bacteria. Trends Biotechnol. 28, 207–213.

Prabhu, Y.T., Rao, K.V., Kumari, B.S., Kumar, V.S.S., Pavani, T., 2015. Synthesis of Fe_3O_4 nanoparticles and its antibacterial application. Int. Nano Lett. 5, 85–92.

Pramanik, A., Laha, D., Bhattacharya, D., Pramanik, P., Karmakar, P., 2012. A novel study of antibacterial activity of copper iodide nanoparticle mediated by DNA and membrane damage. Colloids Surf. B 96, 50–55.

Prasad, T.N.V.K.V., Kambala, V.S.R., Naidu, R., 2013. Phyconanotechnology: synthesis of silver nanoparticles using brown marine algae *Cystophora moniliformis* and their characterization. J. Appl. Phycol. 25 (1), 177–182.

Raffi, M., Mehrwan, S., Bhatti, T.M., Akhter, J.I., Hameed, A., Yawar, W., et al., 2010. Investigations into the antibacterial behavior of copper nanoparticles against *Escherichia coli*. Ann. Microbiol. 60, 75–80.

Raheman, F., Deshmukh, S., Ingle, A., Gade, A., Rai, M., 2011. Silver nanoparticles: novel antimicrobial agent synthesized from an endophytic fungus *Pestalotia* sp. isolated from leaves of *Syzygium cumini* (L). Nano Biomed. Eng. 3, 174–178.

Rai, M., Yadav, A., Gade, A., 2009. Silver nanoparticles as a new generation of antimicrobials. Biotechnol. Adv. 27, 76–83.

Rajathi, F.A.A., Parthiban, C., Ganesh Kumar, V., Anantharaman, P., 2012. Biosynthesis of antibacterial gold nanoparticles using brown alga, *Stoechospermum marginatum* (kutzing). Spectrochim. Acta A: Mol. Biomol. Spectrosc. 99, 166–173.

Ramasamy, M., Lee, J.H., Lee, J., 2016. Potent antimicrobial and antibiofilm activities of bacteriogenically synthesized gold–silver nanoparticles against pathogenic bacteria and their physiochemical characterizations. J. Biomater. Appl. 31 (3), 366–378.

Ramya, S., Shanmugasundaram, T., Balagurunathan, R., 2015. Biomedical potential of actinobacterially synthesized selenium nanoparticles with special reference to anti-biofilm, anti-oxidant, wound healing, cytotoxic and anti-viral activities. J. Trace Elem. Med. Biol. 32, 30–39.

Ranghar, S., Sirohi, P., Verma, P., Agarwal, V., 2014. Nanoparticle-based drug delivery systems: promising approaches against infections. Braz. Arch. Biol. Technol. 57 (2), 209–222.

Ravishankar, R.V., Jamuna, B.A., 2011. Nanoparticles and their potential application as antimicrobials. In: Science Against Microbial Pathogens: Communicating Current Research and Technological Advances. Microbiol. Ser. 3 (1), 197.

Reddy, A.S., Chen, C.Y., Chen, C.C., Jean, J.S., Chen, H.R., Tseng, M.J., et al., 2010. Biological synthesis of gold and silver nanoparticles mediated by the bacteria *Bacillus subtilis*. J. Nanosci. Nanotechnol. 10 (10), 6567–6574.

Ren, G., Hu, D., Cheng, E.W.C., Vargas-Reus, M.A., Reip, P., Allaker, R.P., 2009. Characterisation of copper oxide nanoparticles for antimicrobial applications. Int. J. Antimicrob. Agents 33, 587–590.

Roshmi, T., Soumya, K.R., Jyothis, M., Radhakrishnan, E.K., 2015. Effect of biofabricated gold nanoparticle-based antibiotic conjugates on minimum inhibitory concentration of bacterial isolates of clinical origin. Gold Bull. 48, 63–71.

Rudramurthy, G.R., Swamy, M.K., Sinniah, U.R., Ghasemzadeh, A., 2016. Nanoparticles: alternatives against drug-resistant pathogenic microbes. Molecules 21 (7), E836.

Rupareli, J.P., Chatterjee, A.K., Duttagupta, S.P., Mukherji, S., 2008. Strain specificity in antimicrobial activity of silver and copper nanoparticles. Acta Biomater. 4, 707–771.

Sadiq, M., Chowdhury, B., Chandrasekaran, N., Mukherjee, A., 2009. Antimicrobial sensitivity of *Escherichia coli* to alumina nanoparticles. Nanomed. Nanotechnol. Biol. Med. 5, 282–286.

Saha, B., 2007. In vitro structural and functional evaluation of gold nanoparticles conjugated antibiotics. Nanoscale Res. Lett. 2, 614–622.

Saha, B., Bhattacharya, J., Mukherjee, A., Ghosh, A., Santra, C., Dasgupta, A., et al., 2007. In vitro structural and functional evaluation of gold nanoparticles conjugated antibiotics. Nanoscale Res. Lett. 2, 614–622.

Salari, Z., Danafar, F., Dabaghi, S., Ahmad Ataei, S., 2016. Sustainable synthesis of silver nanoparticles using macroalgae *Spirogyra varians* and analysis of their antibacterial activity. J. Saudi Chem. Soc. 20 (4), 459–464.

Samanipour, A., Dashti-Khavidaki, S., Abbasi, M.R., Abdollahi, A., 2016. Antibiotic resistance patterns of microorganisms isolated from nephrology and kidney transplant wards of a referral academic hospital. J. Res. Pharm. Pract. 5 (1), 43–51.

Sanghi, R., Verma, P., 2009. Biomimetic synthesis and characterisation of protein capped silver nanoparticles. Bioresour. Technol. 100 (1), 501–504.

Saravanan, M., Nanda, A., 2010. Extracellular synthesis of silver bionanoparticles from *Aspergillus clavatus* and its antimicrobial activity against MRSA and MRSE. Colloids Surf. B: Biointerfaces 77 (2), 214–218.

Schrofel, A., Kratosova, G., Safarik, I., Safarikova, M., Raska, I., Shor, L.M., 2014. Applications of biosynthesized metallic nanoparticles—a review. Acta. Biomater. 10 (10), 4023–4042.

Sen, K., Sinha, P., Lahiri, S., 2011. Time dependent formation of gold nanoparticles in yeast cells: a comparative study. Biochem. Eng. J. 55, 1–6.

Senapati, S., Syed, A., Moeez, S., Kumar, A., Ahmad, A., 2012. Intracellular synthesis of gold nanoparticles using alga *Tetraselmis kochinensis*. Mater. Lett. 79, 116–118.

Shahverdi, A.R., Fakhimi, A., Shahverdi, H.R., Minaian, S., 2007. Synthesis and effect of silver nanoparticles on the antibacterial activity of different antibiotics against *Staphylococcus aureus* and *Escherichia coli*. Nanomedicine 2, 168–171.

Shakibaie, M., Forootanfar, H., Golkari, Y., Mohammadi-Khorsand, T., Shakibaie, M.R., 2015. Anti-biofilm activity of biogenic selenium nanoparticles and selenium dioxide against clinical isolates of *Staphylococcus aureus*, *Pseudomonas aeruginosa*, and *Proteus mirabilis*. J. Trace Elem. Med. Biol. 29, 235–241.

Shamsuzzaman, A., Mashrai, H., Khanam Aljawfi, R.N., 2013. Biological synthesis of ZnO nanoparticles using *C. albicans* and studying their catalytic performance in the synthesis of steroidal pyrazolines. Arab. J. Chem. 2013.

Sharma, D., Kanchi, S., Bisetty, K., 2015. Biogenic synthesis of nanoparticles: a review. Arab. J. Chem. Available from: https://doi.org/10.1016/j.arabjc.2015.11.002.

Sharma, N., Pinnaka, A.K., Raje, M., Ashish, F.N.U., Bhattacharyya, M.S., Choudhury, A. R., 2012. Exploitation of marine bacteria for production of gold nanoparticles. Microb. Cell Fact. 11, 86.

Siddiqi, K.S., Husen, A., 2016. Engineered gold nanoparticles and plant adaptation potential. Nano Res. Lett. 11, 400.

Singh, G., Babele, P.K., Kumar, A., Srivastava, A., Sinha, R.P., Tyagi, M.B., 2014. Synthesis of ZnO nanoparticles using the cell extract of the *Cyanobacterium*,

Anabaena strain L31 and its conjugation with UV-B absorbing compound shinorine. J. Photochem. Photobiol. B 138, 55–62.

Singh, P., Kim, Y.J., Singh, H., Wang, C., Hwang, K.H., El-Agamy Farh, M., et al., 2015. Biosynthesis, characterization, and antimicrobial applications of silver nanoparticles. Int. J. Nanomed. 10, 2567–2577.

Sinha, S.N., Paul, D., Halder, N., Sengupta, D., Patra, S.K., 2015. Green synthesis of silver nanoparticles using fresh water green alga *Pithophora oedogonia* (Mont.) Wittrock and evaluation of their antibacterial activity. Appl. Nanosci. 5 (6), 703–709.

Sondi, I., Salopek-Sondi, B., 2004. Silver nanoparticles as an antimicrobial agent: a case study on *E. coli* as a model for Gram-negative bacteria. J. Colloid Interface Sci. 275, 177–182.

Stoimenov, P.K., 2002. Metal oxide nanoparticles as bactericidal agents. Langmuir 18, 6679–6686.

Suresh, A.K., Pelletier, D.A., Wang, W., Broich, M.L., Moon, J.W., Gu, B., et al., 2011. Biofabrication of discrete spherical gold nanoparticles using the metal-reducing bacterium *Shewanella oneidensis*. Acta Biomater. 7 (5), 2148–2152.

Syed, A., Saraswati, S., Kundu, G.C., Ahmad, A., 2013. Biological synthesis of silver nanoparticles using the fungus *Humicola* sp. and evaluation of their cytoxicity using normal and cancer cell lines. Spectrochim. Acta A: Mol. Biomol. Spectrosc. 114, 144–147.

Tak, Y.K., Pal, S., Naoghare, P.K., Rangasamy, S., Song, J.M., 2015. Shape-dependent skin penetration of silver nanoparticles: does it really matter? Sci. Rep. 5, 16908.

Thomas, R., Janardhanan, A., Varghese, R.T., Soniya, E.V., Mathew, J., Radhakrishnan, E. K., 2014. Antibacterial properties of silver nanoparticles synthesized by marine *Ochrobactrum* sp. Braz. J. Microbiol. 45 (4), 1221–1227.

Tran, N., Mir, A., Mallik, D., Sinha, A., Nayar, S., Webster, T.J., 2010. Bactericidal effect of iron oxide nanoparticles on *Staphylococcus aureus*. Int. J. Nanomed. 5, 277–284.

Velmurugan, P., Iydroose, M., Mohideen, M.H., Mohan, T.S., Cho, M., Oh, B.T., 2014. Biosynthesis of silver nanoparticles using *Bacillus subtilis* EWP-46 cell-free extract and evaluation of its antibacterial activity. Bioprocess Biosyst. Eng. 37 (8), 1527–1534.

Venkatesan, J., Manivasagan, P., Kim, S.K., Kirthi, A.V., Marimuthu, S., Rahuman, A.A., 2014. Marine algae-mediated synthesisn of gold nanoparticles using a novel *Ecklonia cava*. Bioprocess Biosyst. Eng. 37, 1591–1597.

Verma, V.C., Singh, S.K., Solanki, R., Prakash, S., 2011. Biofabrication of anisotropic gold nanotriangles using extract of endophytic *Aspergillus clavatus* as a dual functional reductant and stabilizer. Nanoscale Res. Lett. 6, 16–22.

Vidic, J., Stankic, S., Haque, F., Ciric, D., Le Goffic, R., Vidy, A., et al., 2013. Selective antibacterial effects of mixed ZnMgO nanoparticles. J. Nanopart. Res. 15, 1–10.

Vinay Gopal, J., Thenmozhi, M., Kannabiran, K., et al., 2013. Actinobacteria mediated synthesis of gold nanoparticles using *Streptomyces* sp. VITDDK3 and its antifungal activity. Mater. Lett. 93, 360–362.

Waghmare, S.R., Mulla, M.N., Marathe, S.R., Sonawane, K.D., 2015. Ecofriendly production of silver nanoparticles using *Candida utilis* and its mechanistic action against pathogenic microorganisms. 3 Biotech 5 (1), 33–38.

Weitz, I.S., Maoz, M., Panitz, D., Eichler, S., Segal, E., 2015. Combination of CuO nanoparticles and fluconazole: preparation, characterization, and antifungal activity against *Candida albicans*. J. Nanopart. Res. 2015.

Xue, B., He, D., Gao, S., Wang, D., Yokoyama, K., Wang, L., 2016. Biosynthesis of silver nanoparticles by the fungus *Arthroderma fulvum* and its antifungal activity against genera of *Candida*, *Aspergillus* and *Fusarium*. Int. J. Nanomed. 11, 1899–1906.

Yamamoto, O., 2001. Influence of particle size on the antibacterial activity of zinc oxide. Int. J. Inorg. Mater. 3, 643–646.

Yin, R., Agrawal, T., Khan, U., Gupta, G.K., Rai, V., Huang, Y.Y., et al., 2015. Antimicrobial photodynamic inactivation in nanomedicine: small light strides against bad bugs. Nanomed. (Lond). 10 (15), 2379–2404.

Zhang, L., Pornpattananangku, D., Hu, C.M., Huang, C.M., 2010. Development of nanoparticles for antimicrobial drug delivery. Curr. Med. Chem. 17 (6), 585–594.

Zhang, X., Yan, S., Tyagi, R.D., Surampalli, R.Y., 2011. Synthesis of nanoparticles by microorganisms and their application in enhancing microbiological reaction rates. Chemosphere 82 (4), 489–494.

Zharov, V.P., 2006. Photothermal nanotherapeutics and nanodiagnostics for selective killing of bacteria targeted with gold nanoparticles. Biophys. J. 90, 619–627.

Zonaro, E., Lampis, S., Turner, R.J., Qazi, S.J., Vallini, G., 2015. Biogenic selenium and tellurium nanoparticles synthesized by environmental microbial isolates efficaciously inhibit bacterial planktonic cultures and biofilms. Front. Microbiol. 6, 584.

CHAPTER

16 Biosensors for monitoring pharmaceutical nanocontaminants and drug resistant bacteria in surface water, subsurface water and wastewater effluent for reuse

A.A. Adegoke[1,2], G. Singh[1] and T.A. Stenström[1]

[1]*SARChI, Institute for Water and Wastewater Technology, Durban University of Technology, Durban, South Africa* [2]*Department of Microbiology, University of Uyo, Uyo, Nigeria*

CHAPTER OUTLINE

Nanoparticles in Pharmacotherapy. DOI: https://doi.org/10.1016/B978-0-12-816504-1.00014-4

16.1 INTRODUCTION

Pharmaceuticals and personal care products (PPCPs) are a wide range of chemicals, recognized as part of the daily consumer products across the globe and are categorized by chemists, pharmacists, as well as environmental scientists based on their usage and configurations (Daughton and Ternes, 1999). They include a large variety of drugs (atenolol (antihypertensive), acetaminophen (analgesic), norfloxacin and ciprofloxacin (antibiotics), carbamazepine (antiepileptic) and ketoprofen and diclofenac (antiinflammatory) cosmetics (moisturizers, lipsticks, hair colors), cleansers (shampoos and toothpastes), detergents, and fragrance (deodorants) products. They are released into the water environment directly or indirectly but usually in low concentration as waste after usage (Kolpin et al., 2002; Barnes et al., 2008; Focazio et al., 2008; Walraven and Laane, 2009; Loos et al., 2010; Kuroda et al., 2011). The detection and impact of PPCPs in water environment are emerging and of enhanced concern as part of the research agenda, just as PPCPs are now classified as emerging contaminants. The awareness of the risks associated with the presence of the PPCPs in water environments so far is growing; so also is the knowledge of mitigating the ecological and public health risk associated with their presence in the aquatic environment.

PPCPs themselves are emerging contaminants where many were previously unknown or unrecognized as pollutants. They were not covered by older legislation in most countries and are still not covered in several third world countries. Chemicals with new structures (PPCPs) may be represented by the original compound and derivate thereof due to secondary reactions in wastewater or receiving water bodies. With their new discovered effects or sources PPCPs can be safely classified as emerging contaminants (Daughton, 2001, 2004; Mottaleb et al., 2015).

PPCPs are released into aquatic ecosystems by both point- and nonpoint sources (Mottaleb et al., 2015; Daughton and Ternes, 1999) where domestic wastes are an important source. The use of the pharmaceuticals and the release of the partially metabolized products containing active ingredients in excretory wastes of human or animal serve as a major way for contributing the PPCPs to the environment (Williams, 2005). Halling-Sørensen et al. (1998) reported that

1:10 proportion of an administered pharmaceutical is excreted in active form. Other indirect contamination sources include landfill leachate, biosolid application, while direct disposal of PPCPs into wastewater remains a major source (Conn et al., 2006) for consideration. The contribution of biosolid used as organic fertilizers, which may partly end up in surface water by run-off, is an additional important source of some PPCPs (Adegoke et al., 2016).

The fate of these PPCPs is usually determined by the type of wastewater technologies used. In developing countries, septic systems, cesspits, soakpits, pit latrines, and other onsite wastewater treatment systems are being used mainly in rural areas. The effluents containing the PPCPs may percolate to the water table, contaminating the groundwater, or be trapped within the soil where they are absorbed by root of plants. Although PPCPs trapped within organic matter may be degraded by some microorganisms, it may also result in toxicity, mutation, and formation of undesirable end products with adverse environment effects (Lapworth et al., 2012). In municipal wastewater treatment, degradation of some PPCPs may occur while large proportions (90%) pass through either in the liquid phase or are absorbed to solid matter (Halling-Sørensen et al., 1998; Aga, 2008; Heidler et al., 2006). Besides degradation, removal rate for PPCPs depends on plant operating conditions and the wastewater technology being used, microbial community composition, and methods of disinfection (Khanal et al., 2006; Radjenovic et al., 2009; Suarez et al., 2008).

Pharmaceuticals do not always remain the way they were before their release to a multimatrix environment like wastewater. Some pharmaceutical products are manufactured as complexes, where a ligand acts to detoxify the mother compound, such that the complex can be used without much toxic effect on the tissue (Kocsis et al., 2016). The degradation process in wastewater treatment opens for decomposition, where detoxifying ligands are removed from the complex, allowing more toxic components to flow with the treated effluents, which potentially cause higher level of ecological toxicity (Aga, 2008; Huang and Sedlak, 2001; Ternes, 1998).

The wastewater treatment plants are known to be major point-sources of PPCPs and microbial contaminants into the environment with adverse downstream effects. Some wastewater treatment technologies, both on-site and central, retain high concentrations of PPCPs as for example demonstrated in septic systems (Swartz et al., 2006). This is further affirming that the reduction of PPCPs is dictated by the treatment processes involved (Conn et al., 2010a,b; Dougherty et al., 2010). Their effects call for further needs of more sensitive detection and assessments of the removal and transformation obtained with different treatment technologies. The concern about the associated risks with the probable presence of PPCPs, occasioned by weak capacity of the treatment technology, can be addressed following the detection of quantity and types of PPCPs, as well as the frequency of release in a particular location. This is also of emerging concern to health authorities and enlightened residents of such locations.

Consequently, there is need for highly sensitive technology for the detection and quantification of these contaminants, even when they occur in very small concentration. Such technology must have high specificity, sensitivity, and accurate detection capacity and be related to relevant target values. Various technologies are now being used for the detection of PPCPs in water environment. Some of them include the following or variants thereof: high performance liquid chromatography–time-of-flight mass spectrometry (HPLC–TOF-MS), gas chromatography–mass spectrometry (GC–MS), and biosensors (Rajasulochana and Preethy, 2016; Rushing et al., 2016). Recent biosensors have some overriding advantages in detecting PPCPs from environmental samples as compared with conventional LC–MS-based methods. They are cost-effective, user friendly, and possess high detection capabilities when measuring wide ranges of pollutants in complex matrices. They may be applied as portable instruments, which allows their in situ application. The evolution of biosensors with potential for determining the biological effects of pollutants, such as toxicity, cytotoxicity, genotoxicity, or endocrine disrupting (which will further be discussed in this chapter) makes them relevant for water monitoring and assessments of public health implications. This chapter will critically examine the PPCPs and microbial contaminants in water environment and the application of biosensors in monitoring their path in various water environments.

16.2 PHARMACEUTICAL INDICATORS OF WASTEWATER CONTAMINATION OF SURFACE WATER FOR DRINKING WATER PRODUCTION

The presence of pharmaceuticals in municipal drinking water may be interpreted as the result of contamination with waste from domestic, communal, industrial, or animal sources. The drinking water treatment will determine the residual concentrations as compared with the respective raw water. Caffeine has been used as a chemical wastewater indicator from domestic sources in untreated and treated drinking water (Daneshvar et al., 2012). In the latter context it can also serve as an indicator or treatment efficiency. Daneshvar et al. (2012) surveyed “PPCPs (caffeine, carbamazepine, naproxen, gemfibrozil, and trimethoprim), hormones (progesterone, estrone, and estradiol), and triazine herbicides and their metabolites (atrazine, deethylatrazine, deisopropylatrazine, simazine, and cyanazine)” from four river systems. They found that a high ratio of caffeine/carbamazepine might imply that more raw sewage might be present in surface waters than being impacted by the treated wastewater effluent. This is because caffeine (or some other related pharmaceuticals) are human pharmaceuticals. Their presence can be used to identify the human-source contamination (Seiler et al., 1999; Wu et al., 2008; Daneshvar et al., 2012).

16.2.1 PHARMACEUTICALS IN SEDIMENT

Particulate matter in surface water is known to settle and form part of the sediment. This includes free or attached nanosized matter like the PPCPs, hormones, and different metabolites. Sediment thus functions as a sink and a reservoir that can serve as a source for recontaminating and resuspension as a result of rapid changes of river water flow (Liang et al., 2013). The contaminants may also be resuspended by tidal changes affecting the outfall areas of large rivers as well as sea sediments. A cocktail of tens of PPCP compounds have been reported in sediments (Klosterhaus et al., 2013; Liang et al., 2013). Le and Munekage (2004) detected extremely high concentrations of antimicrobials in marine shrimp aquaculture pond sediments. A concentration range of 2–4 ng/L of Simvastatin was found by Langford and Thomas (2011) in sediment samples. Table 16.1 exemplifies this and other PPCP studies in aquatic biota including the sediments. Human pharmaceuticals like pharmaceutical-grade caffeine, thiamphenicol, sulfamethazine, salicylic acid, etc., were found in drinking (tap) water in China (Leung et al., 2013), showing penetration through the drinking water treatment barriers from the waste contaminated raw water (Wu et al., 2008; Daneshvar et al., 2012).

16.3 IMPACT OF SELECTED PPCPS ON THE AQUATIC ENVIRONMENT

16.3.1 HEALTH RISK

PPCPs and related contaminants in water exhibit wide ranges of long- and short-term health effects on users. The effects may be acute (short-term periods) or cumulative/chronic (gradual, slow incidence for extended period). The effects may, in the worst case, be fatal in either of the cases (Larson et al., 2014; Quinn et al., 2008; Crane et al., 2006). Though the knowledge on the long term impact of exposure to nanoconcentration of complex pharmaceutical contaminants in reused or recycled wastewater may not be fully understood, the evidence and impact are reported through accumulation in tissues (Schultz et al., 2010; Brooks et al., 2003), reproductive damage (Larson et al., 2014; Nentwig, 2007), inhibition of cell proliferation (Pomati et al., 2006), and behavioral changes (Stanley et al., 2007; Gaworecki and Klaine, 2008). Several studies have demonstrated that fish exposed to contamination with wastewater effluents containing the PPCPs can exhibit reproductive abnormalities. Contamination of the aquatic habitat with pharmaceuticals for birth control may have a tendency to reduce the population density of some aquatic animals (Sanderson et al., 2004) and further be accumulated in the food chain, exemplified by pharmaceuticals with potential accumulated long term effect due to human consumption (Quinn et al., 2008).

Bacterial exposure to sublethal concentration of antibiotics in water biota leads to the emergence of resistance as an adaptation to the antagonistic compound

Table 16.1 Some PPCPs in Various Water Environment, Reporting Nanoconcentration and Potential Effects

Category	PPCPs Name	Water Environment	Reported Concentration (Approx.) ng/L	Location of Research	Reference
Pharmaceuticals	Caffeine	Tap	25	China	Leung et al. (2013)
	Thiamphenicol		20		
	Sulfamethazine		10		
	Salicylic acid		20		
	Acetaminophen	Lake	4.2–21	Michigan, USA	Blair et al. (2013)
	Carbadox		7.2–20		
	Codeine		11		
	Gemfibrozil		9.1–42		
	Estrone	Municipal source water	1	New York, USA	Strickland and Rush (2011)
	Diazepam		4		
	Butalbital		30		
	Caffeine		20		
	Carbamazepime	River	41	Serbia	Radović et al. (2015)
	Metoprolol	Aquifer-groundwater	9	Switzerland	Morasch (2013)
	Erythromycin	Sediment (ng/g)	0.7–14	San Francisco Bay, USA	Klosterhaus et al. (2013)
			0.7–14	Pearl River Estuary, China	Liang et al. (2013)
Personal care products	Galaxolide	Treated wastewater and groundwater	42.9	Spain	Cabezaa et al. (2012)
	Octocrylene		8.42		
	Tonalide		7.5		
	Ethylhexyl methoxycinnamate		35.31		

(Taylor et al., 2011). This might further be a causal factor for high antibiotic resistant profile of some bacteria from fresh surface water environment as reported by Stenström et al. (2016). While the emergence of resistance by microbial exposure to antibiotics is known, it might be very complex in vivo and also a form of adaptation. The antagonistic effect of the antibiotic is peripherally illustrated in Fig. 16.1. Residual antibiotics, for example, released into the aquatic environment may serve as signaling molecules to the resident bacterial species, leading to the formation of N-acyl homoserine lactones (AHLs) or the diffusible signal factor cis-2-dodecenoic acid (BDSF), depending on the bacterial species (Schmid et al., 2012). This leads to the development of quorum sensing in a rather more complex process. Quorum sensing is important in the biofilm in aquatic environment leading to emergence of antibiotic resistance (Fig. 16.1) and enhanced virulence (Deng et al., 2012; Balcázar et al., 2015). In fact, biofilm has been described as the environmental reservoirs of antibiotic resistance (Balcázar et al., 2015).

Multiple antibiotic resistant bacteria have been isolated in fish, marine mammals, and seabirds (Rose et al., 2009). Broad spectrum antibiotic resistance has also been reported in marine bacteria with prior exposure to sewage (Blackburn et al., 2010) and there is evidence to suggest that the antibiotic resistant bacteria present in seabirds are of human origin (Bonnedahl et al., 2009). The presence of antibiotic resistance genes in marine ecosystems may be implied selection process occasioned by exposure to available residual pharmaceuticals (Moore et al., 2010).

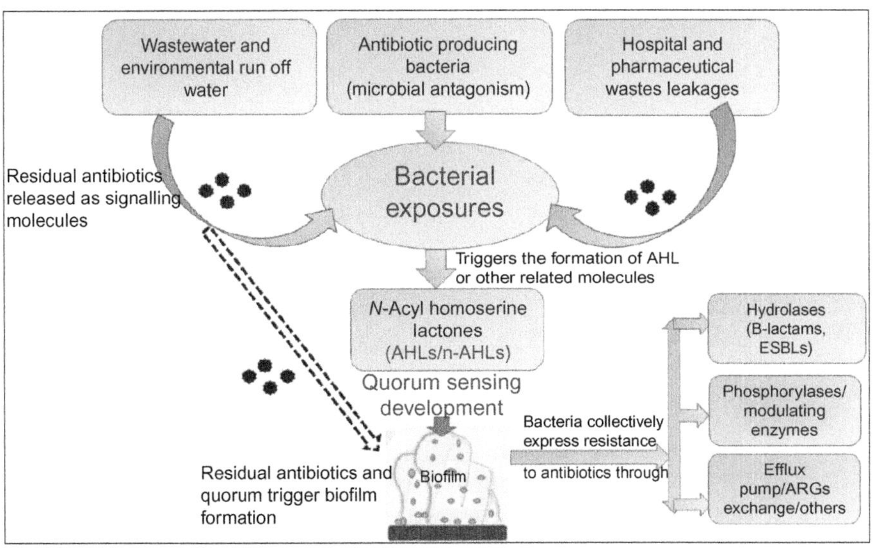

FIGURE 16.1

Sketched potential pathway for the emergence of antibiotic resistant via exposure to antibiotics in aquatic environment.

16.3.2 ENVIRONMENTAL HAZARD

Most of the pharmaceutical contaminants exhibit diverse effects that affect their survival and the entire ecological balance (Bossus et al., 2013; Di-Poi et al., 2013). PPCPs bioaccumulate in aquatic life, and may result in toxicity that potentially can be fatal. Ortiz de García et al. (2014) reported that 65.4% of all the PPCPs studied showed high toxicity or were harmful to aquatic organisms. Toxicity to the aquatic organisms was also reported by Méndez et al. (2013). Toxic effect of the PPCPs extends between various organisms in their aquatic niches, causing ecological imbalance and unpleasant shifts in the food web. Analgesics, for example, bring about interference in the appetite of marine organisms, which in turn affect their populations (Solé et al., 2010). A study (Fig. 16.2) showed that the concentration of oxytetracycline in aquatic shrimp muscles reached the peak within 8 days (Bermúdez-Almada and Espinosa-Plascencia, 2012). A further exemplification is that when a residual PPCPs exceeds the total tolerable bioaccumulated concentration, it results in oxidative stress and necrosis in mussels (Schmidt et al., 2011).

While pharmaceuticals are present in marine ecosystems as well as in surface water, groundwater, and drinking water (Kim and Carlson, 2007; Yang et al., 2011; Barnes et al., 2008; Ye et al., 2007) with attending risk, their ecological effects vary with location and the types of pharmaceuticals. Mixture of several pharmaceuticals may exhibit additive effects, synergistic effects, or antagonistic effects, which may kill or impair large range of exposed organisms, depending on the effect (DeLorenzo and Fleming, 2008; Schnell et al., 2009; Kidd et al., 2014).

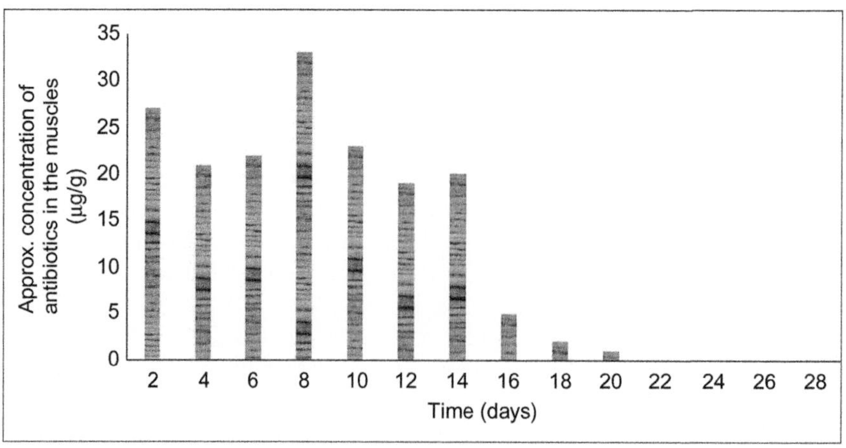

FIGURE 16.2

Muscle oxytetracycline levels/time profiles of shrimps.

Adapted from Bermúdez-Almada and Espinosa-Plascencia, 2012.

16.4 BIOSENSORS AS MONITORING TOOLS IN AQUATIC ENVIRONMENTS

In general terms, biosensors are analytical devices integrating materials originating from biological sources or biomimics within a physicochemical transducer or transducing microsystem (González-Martínez et al., 1999). Unlike bioassays or bioanalytical systems, the sensors do not necessarily require several subsidiary steps of adding reagents (Patel, 2002), but some bioassays can be further developed to biosensors by automating their process (Rogers, 2006). The overriding inherent advantages in biosensors over conventional methods or batch systems include this comparatively cost-effective automation as well as high detection capacity and user friendly automated (robotic) analytic procedures (Onnerfjord et al., 1998). Biosensors integrate immobilization techniques and materials to bring about improved stability of their biomolecular components (Malhotra et al., 2006).

The applicability of biosensors in detecting both pharmaceutical and microbial contaminants hinges on the advantages of low sample and reagent volumes, guaranteeing high detection capacity even when the PPCPs are in low nanoconcentration (Samendra et al., 2014). Their biological integration makes it possible to utilize antibodies, molecular receptors, or DNA as biorecognition elements. They can also function through interference or consideration of enzyme–analyte action. Biosensors, therefore, are classified based on the factors aiding their recognition of the analytes into:

a. immunochemical biosensors,
b. enzymatic biosensors,
c. nonenzymatic receptor biosensors,
d. whole-cell biosensors, and
e. DNA biosensors.

Biosensors can also be classified based on the type of signal transducer they contain, for example, a physical transducer can be based on electrochemical, spectroscopic, thermal, piezoelectric, and/or surface wave technologies (Malhotra et al., 2006). The applicability of different biosensors based on transduction or biorecognition element has been reported (Rodriguez-Mozaz, et al., 2004; Marco et al., 1996). The contribution of other analytical methods like GC–MS, GC–MS/MS, LC–MS, and LC–MS/MS in detection of PPCPs in environmental samples is additionally of crucial importance. Besides the recent use of biosensors, it is worth noting that LC–MS/MS remains the most sensitive methodology to determine PPCPs in the nanogram per liter range in water and wastewater. Biosensors have more diverse application as they are integrated with antibodies (more rampantly) as well as enzymes, bacteria, receptors, and DNA as recognition elements, and the use of optical and electrochemical transducing elements.

16.4.1 CARBON NANOTUBES AS BIOSENSORS

Carbon nanotubes (CNnTs) have been recognized as biosensors based on their ease of aligning with other molecules and their well appraised and recognized electrical transmission properties (Ajayan, 1999). Most atoms in CNnTs are surface based, making the impact of molecular modifications reflective of electrical properties (Yang et al., 2007). They are known for their high surface area to volume ratio with array of binding sites for a wide range of nanoparticles and molecules. Due to this CNnTs have high sensitivity, an overriding advantage of biosensors (An et al., 2004; Ali et al., 2007). The aforementioned attributes made the CNnT the choice used by Chopra et al. (2003) and Guiseppi-Elie et al. (2002), which these authors regarded as novel biosensors.

In biosensor application, CNnTs are being harnessed in addressing the problem associated with microbial contamination, especially antibiotic resistant microbial contaminants in water sources for drinking. Due to high surface active site to volume ratio, as well as high sorption efficiency, CNnTs were originally used in wastewater treatment (Ong et al., 2010). This has drawn extensive attention to the use of the CNnTs as building blocks in the design of electrochemical biosensors (Prieto-Simón et al., 2015). Several pertinent factors that favor the use of the CNnTs in treatment have also been harnessed in using them as the functional unit in electrochemical biosensors. Their surface acidity favors the adsorption of polar compounds (Rao et al., 2007) and this might be a favorable factor for detection. It also has conditional but effective adherence of nonpolar compounds (Yang et al., 2006; Wang et al., 2007). Prieto-Simón et al. (2015) optimized two potential single-walled CNnT architectures as immunosensors for the bacteriophage MS2, a virus associated with sewage-impacted water. The research outcome, which was positive for the two detection approaches used, confirmed the ease of usage associated with biosensors.

As it was applied to viruses, it can also be applied for bacterial detection. Yamada et al. (2014) integrated single-walled carbon nanotubes (SWCNTs) and immobilized antibodies into a disposable bio-nano combinatorial junction sensor to detect *Escherichia coli* K-12. Upadhyayula (2008) earlier had used SWCNT aggregates as fluorescence sensors to detect *E. coli* pKV-11 and found the SWCNTs to be excellent for bacterial adsorption and for fluorescent detection capability. As a response to the challenge of rapid and sensitive detection of pathogens, the advent of Fourier transform infrared spectroscopy (FTIS) as a technique to detect *E. coli* in just 6 hours was reportedly developed (Davis et al., 2010). There is now the possibility to identify these bacterial contaminants in water or other samples before their generation time of 1200 seconds. This served as the beginning of applying biosensors to a wide range of other microbial contaminants especially in wastewater for reuse.

Further in the detection of bacteria, several other more related but specific and timely technologies (like CNnTs) use other recognition systems for their detection, for example, a system that uses cell specific molecular recognition systems.

Olsen et al. (2003) showed improvement in *Salmonella* identification while other researchers (Kerr et al., 2001; Gehring et al., 2006) utilized specific antibody-based detection systems to identify *E. coli* O157:H7 strains. An improvement in detection involving the use of antibodies enhances specificity as better target recognition agent (Lee et al., 2008), with a higher degree of specificity, better stability, wide array of binding processes (Yao et al., 2010; Hernandez and Ozalp, 2012). Combined possibilities to detect a wide range of water-based pathogens as contaminants, parallel to detecting PPCPs in the same water environment thus exist.

16.5 APPLICATION OF BIOSENSOR IN THE DETECTION OF PPCPS AS WATER CONTAMINANTS

Different types of biosensors have been developed for environmental applications. Table 16.2 gives a brief overview of some that have been applied and their working principles.

The biosensor river analyzer (RIANA) is an extraordinarily sensitive optical immunosensor with capacity to detect multiple organic targets promptly and simultaneously (Tschmelak et al., 2005, 2006). The system has been used extensively to analyze various steroidal hormones as well as other organic pollutants in

Table 16.2 List of Few Biosensors With Environmental Application and Their Principles

Biosensor Type	Principle	Specific Application	Reference
Acetylcholinesterase inhibition-based biosensors	Electrochemistry	Pesticidal contamination and impact	Pundir and Chauhan (2012)
Genetically encoded or probe-based, or fluorescence-tagged biosensor, for example, GFP-based bacterial biosensor	Fluorescence	Bioaccumulation	Priyadarshi et al. (2012)
Nanomaterials-based biosensors	Wide-range: Electrochemical, optical, visual, or fluorescence	Detection of various contaminants: pharmaceuticals, microbes, using an appropriate aptamer	Da Costa Silva et al. (2013), Guo (2013), Hutter and Maysinger (2013), Senveli and Tigli (2013), Valentini et al. (2013), Govindhan et al. (2014), Sang et al. (2015)
Microbial fuel cell-based biosensors	Optical	Monitor ecotoxicity with PPCPs, heavy metals and other related contaminants; BOD.	Sun et al. (2015)

situ without prior sample concentration, which is necessary to apply in other systems (e.g., HPLC). The detection limits of subnanograms per liter exhibited by this system (RIANA) make it an excellent candidate among onsite detection technologies (Rodriguez-Mozaz et al., 2004; Tschmelak et al., 2004; Hua et al., 2005; Tschmelak et al., 2006). In bacterial analysis of river water, *E. coli* bioreporters are more sensitive than HPLC as they can detect 70 parts per trillion (ppt) (Lee et al., 2013).

The removal of some PPCPs from water during treatment for subsequent reuse has been determined using RIANA (Rodriguez-Mozaz et al., 2006, 2005). PPCPs like estrone and bisphenol A have been specifically detected in this process (Rodriguez-Mozaz et al., 2006, 2005). The PPCPs may also be quantified in river water or groundwater using an antibody-based biosensor (Bromage et al., 2007). Some other biosensors with ability to detect residual bisphenol A, for example, immunosensors have also been applied (Soh et al., 2003; Marchesini et al., 2005). The detection of sulfamethazine in animal urine using an optical immunosensor by Akkoyun et al. (2000) showed that urine can be one origin of the pharmaceuticals to wastewater and that it is possible to monitor the pathways of PPCPs in water cycle using immunosensors. These biosensors can easily be adapted and applied to the detection of PPCPs in water samples and have undergone further development during recent years (Prieto-Simón et al., 2015).

The same concern extends toward removing microbial contaminants from water as for PPCPs. Choffnes et al. (2012) expressed concern that this should further extend to microbial toxic products as contaminants in the water environment. Toxins in water by cyanobacteria pose a serious health threat to aquatic lives, as well as to animals and humans using the water.

16.5.1 MOLECULARLY IMPRINTED POLYMERS BIOANALYTICAL SENSORS

Biosensors based on the molecularly imprinted polymers (MIPs) (Cheong et al., 2012; Li et al., 2012; Lulinski, 2013) are gaining increased recognition. As described by Tse Sum Bui and Haupt (2010) "MIPs are tailor-made synthetic materials possessing specific cavities designed for a target molecule. They recognize their target analyte with affinities and selectivity's comparable to those of antibody–antigen, enzyme–substrate and ligand–receptor interactions." Caffeine, for example, has been analyzed using MIP-based biosensors (Zougagh et al., 2005).

The efficiency of the MIP sensors is based on the interaction between the template molecule and the complementary functional monomer group (Whitcombe and Vulfson, 2001). The minimum sample required for some MIP-based biosensor is 2.5 mL, the assay time 2 min, while the detection limits can sometimes be less than half of permitted concentration for some compounds in water.

One major challenge associated with the use of MIPs for biosensors is water compatibility with the equipment, which necessitates the need for making the surface of the system hydrophilic and requires postmodification of the MIPs (Pérez-Moral and Mayes, 2007). Appreciable improvement has therefore been made in this area as well as in size, binding sites' control, and overall system's processes (Pan et al., 2011).

16.6 BIOSENSOR FOR DETECTION OF PATHOGENS IN WATER

Several biosensors can be optimized for their use in detecting microbial water contaminants or properties, which may be drug resistant or susceptible microbial contaminants. Table 16.3 exemplifies organisms that have been detected through the use of biosensors and the specific applications used.

As stated in the introduction, one major advantage of biosensors is the incorporation of biological receptor compounds with physicochemical transducers. This allows the detection of a broad spectrum of analytes, especially microbial pathogens in water samples (Shruthi et al., 2014). Out of the four groups of biosensors, that is, optical, mass, electrochemical, and thermal sensors, only the optical transducers are designed to effectively detect bacteria, which suggests future cost effectiveness, promptness, and portability.

Advances in sensing techniques have brought together several integrative tools like piezoelectrical, calorimetrical, enthalpimetrical, DNA microarray, surface plasmon resonance (SPR), impedance spectroscopy, scanning probe microscopy (SPM), atomic force microscopy (AFM), quartz crystal microbalance (QCM), and surface enhanced Raman spectroscopy (SERS), all potentially applicable for microbial detection. Their principles have been combined to make appropriate improved biosensors with their attributes. The most commonly used for detection of microbial pathogens are electrochemical and optical techniques. In these cases, the transducers may be potentiometric, amperometric, optical, magnetic, or colorimetric devices. The amperometric detection of microorganisms basically deals

Table 16.3 Some Applications Used as Biosensors for Detecting the Microbial Contaminants in Water

Microbial Contaminant	Application	Reference
Escherichia coli (tolC strains)	Mitomycin C	Norman et al. (2005)
E. coli (tolC strains)	Formaldehyde	
E. coli (tolC strains)	Cerulinin	Bechor et al. (2002)
Pseudomonas sp.	p-Nitrophenol	Zhang et al. (2012)
Moraxella sp.		Mulchandani et al. (2005)
Recombinant luminescent bacteria	Organotin compounds	Durand et al. (2003)
Spirulina subsalsa	Chlorophenols	Tonnina et al. (2002)
Chlorella vulgaris	Isoproturon	Vedrine et al. (2003)
Baker's yeast	Benzalkonium chloride	Campanella et al. (1997)

with enzyme-catalyzed electroredox reaction in which silver ions are used as electrodes (Newman and Setford, 2006; Koyun et al., 2008). As advantageous as the electrochemical biosensors (amperometric sensors) appear to be, there can be challenges of poor selectivity with potential for oxygen/hydrogen peroxide detection especially in wastewater, except when mediators are utilized (Grieshaber et al., 2008; Pohanka and Skládal, 2008).

The opportunity of an online monitor with biosensor to detect pathogens rapidly and precisely is being investigated in water application (Mikol et al., 2007). This is exemplified by the use of β-D-glucuronidase (GUD) activity, which is a characteristic of most strains of *E. coli*. The presence of *E. coli* in water is interpreted as evidence of recent fecal contamination and the majority of guidelines and water quality standards are based on it (Bain et al., 2014). GUD-based assays for the detection of *E. coli* have no limitation regarding the chlorine-affected viable but nonculturable (VBNC) strains (Fiksdal and Tryland, 2008; Servais et al., 2009). *E. coli* in water can also be detected using 4-methylumbelliferyl-β-D-glucuronide (MUG), which liberates fluorescent 4-methylumbelliferone (4-MU) on hydrolysis. Extrapolating the signal of fluorescence makes it possible to quantify the *E. coli* count (Servais et al., 2009). An online water quality monitoring and control system for swimming pools was investigated by Duffy et al. (2009) to detect *E. coli*. The successful research outcomes led to recommendations for the application in drinking water monitoring.

Online monitoring with biosensors is of future interest for in situ detection of bacteria, and may then form a base for epidemiological tracking of pathogens. A factor to consider concerns the sensitivity of the sensor systems and that this remains a function of the transducer's properties. Additional factors to consider for the future include improvements in flow cell design, transducer sensitivity, miniaturization, as well as antibody design and production. These factors relate to optimization and calibration for effective and correct results.

16.7 GRAPHENE-BASED SENSORS AND DEVICES IN DETECTING PHARMACEUTICALS AND MICROORGANISMS IN WASTEWATER EFFLUENT FOR REUSE

Fig. 16.3 gives a brief summary of some nanomaterial enabled biosensors for the detection of pharmaceutical nanocontaminants and bacterial detection. Graphene is a monolayer 2D-allotrope of carbon atoms with unique nanostructure as well as extraordinary properties (Craciun et al., 2011; Kang et al., 2010) that make applications utilizable in many areas including as biosensors. Its exciting applications in nanoelectronics, nanocomposites, optoelectronic devices, electrochemical supercapacitor devices, fabricated field-effect transistors, among others make it a reference point for more novel applications (Kuila et al., 2012;

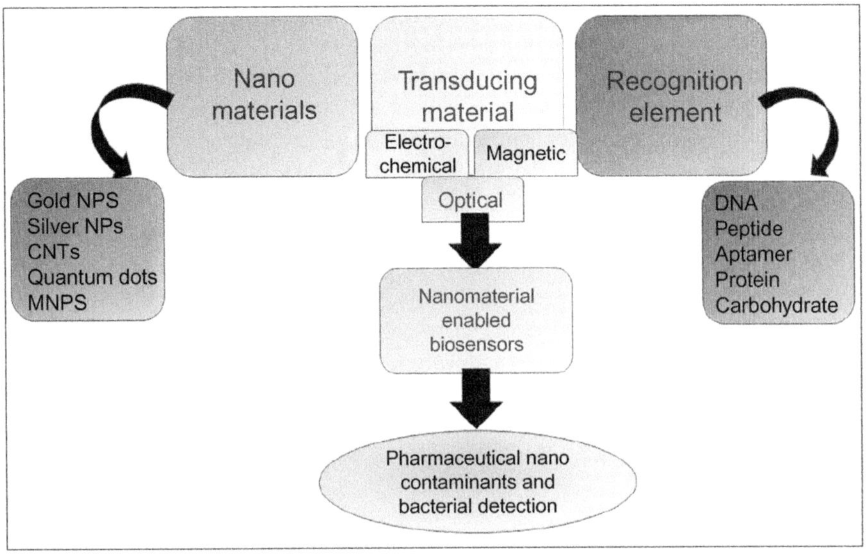

FIGURE 16.3

Nanomaterial enabled biosensors for the detection of pharmaceutical nanocontaminants and bacterial detection.

Craciun et al., 2011; Kang et al., 2010). Some of these areas of application are also directly or indirectly beneficial to the environment. More importantly is their role in sensors and biosensors (Kuila et al., 2012). Specifically, graphene has been used in optoelectronic devices, supercapacitors, and other high performance sensors. This is due to unique characteristics like peculiar optical properties, high electrical conductivity, magnificent carrier mobility, large surface-to-volume ratio, appreciable carrier density, and remarkably good mechanical strength (Zhu et al., 2010). Graphene is better than carbon nanotubes as a biosensor because of its lower cost, larger surface-area-to-volume ratio, and greater homogeneous surface for proper functionalization.

To detect wide ranges of pharmaceutical contaminants with known functional groups, graphene-based sensors (GBS) have wide ranges of chemical modification and biomolecular binding schemes to apply (Soldano et al., 2010). The use of GBS in diagnostics (Deshpande and LeRoy, 2012; Liu et al., 2012; Kuila et al., 2011) can be harnessed in monitoring wastewater effluent for reuse, to detect pharmaceuticals or PPCPs contaminants. It may also be applicable for other over-the-counter (OTC) drugs, glutamate, cholesterol, protein biomarkers, and organic components (Kang et al., 2010). The graphene-based electrodes have excellent antifouling ability, which is a relevant advantageous attribute in wastewater analysis (Zheng et al., 2012). Graphene-based nanoelectronic devices, used to make DNA sensors, may be applied to detect specific nucleotides of microorganisms or specific genes of interest (e.g., antibiotic resistant genes, ARGs) in wastewater effluent. It can also be used to make pH or gas sensors that can detect both pharmaceuticals and some waterborne bacteria like *E. coli*. The attributes of oxides of graphene to also kill bacterial pathogens besides detecting them via GBS would be an added advantage of graphene (Nanda et al., 2016).

One additional credence encouraging the use of graphene in monitoring contaminants in water is its antibacterial effect against both Gram-positive and Gram-negative bacteria without affecting normal multicellular cell culture (Fiorillo et al., 2015). The impact of GBS will surely open a new era in environmental tracking of PPCPs. The sensitivity of the modified graphene-based electrochemical detection assay was reportedly higher than the existing chromatography techniques (GC or HPLC) (Sharma et al., 2012). Meanwhile, graphene oxide-based sensors for water have been developed for the detection of heavy-metal ions. These sensors work based on fluorescence quenching of fluorophore-labeled materials, such as single-stranded DNA (ssDNA) when they bind with graphene oxide (Chang et al., 2014).

16.8 GOLD NANOPARTICLES AS SENSORS FOR DETECTING MICROBIAL AND PHARMACEUTICAL CONTAMINANTS IN WATER

Gold nanoparticles (AuNPs) based sensors are also candidates for the detection of microbial and pharmaceutical contaminants in water. The modification of their surfaces is part of addressing the functionality to make room for the attachment of the biological or PPCPs' molecules to the nanoparticles. AuNPs and gold nanorods as magnetic nanoparticles (MNPs) have been applied to detect contaminants in water samples (Leonard et al., 2003; Wang and Yu, 2013). Their potential to timely and effectively detect toxins, heavy metals, as well as inorganic and organic pollutants from pharmaceutical origin make them good candidates for different applications as biosensors (Wang and Yu, 2013). In a study to detect waterborne *E. coli* reported by Inbaraj and Chen (2016), "a thiolated single-stranded DNA probe specific to the *E. coli* O157:H7 eaeA gene was self-assembled onto the QCM sensor." The following steps included inducing hybridization through exposure of this single-stranded DNA probe to the complementary target DNA and amplification using asymmetric PCR with biotin-labeled primers. This resulted in a change in mass with a concomitant change in QCM frequency for detection of *E. coli* O157:H7. The detection limit obtained was 2.67×10^2 colony forming units (CFU)/mL (Mao et al., 2006). Based on a similar approach, a circulating-flow piezoelectric biosensor (PEB) was developed for detection of *E. coli* O157:H7 using AuNPs-conjugated thiolated probe as mass enhancer and sequence verifier (Chen et al., 2008). An *E. coli* O157:H7 eaeA gene-specific thiolated probe conjugated to PEB was exposed to *E. coli* gene fragment amplified by PCR and the resultant mass change was measured as frequency shift of

PEB, with the detection limit being 1.2×10^2 CFU/mL in the linear working range of 10^2-10^6 CFU/mL (Chen et al., 2008).

The main role of AuNPs in electrochemical sensors, which is assessed for environmental monitoring, is to facilitate the detection process (Rassaei et al., 2011). The AuNPs are used to increase the surface area of the electrodes, thus creating better affinity for the analytes (Rassaei et al., 2011). Specificity is enhanced by pairing the AuNPs with antibodies, DNA sequences, or other macromolecules. This will improve detection of both microbial and pharmaceutical contaminants in water (Gooding et al., 2009; Zhang et al., 2010). In summary, the reliability of AuNPs-sensing technology in detecting wide arrays of toxins and pollutants in both laboratory and in situ makes AuNPs acceptable for monitoring hazard components in environmental matrices, including water (Wang et al., 2010).

An AuNP-based sensor coupled with specific DNA sequence complementary to the target can be used to detect or monitor the presence of some specific pathogens in water samples (Sattarahmady et al., 2015). Sattarahmady et al. (2015) applied this to *Brucella* species in another sample matrix but the method and approach can be optimized for wastewater effluent for reuse, surface water, or groundwater applications, with the potential advantage of its shorter assay time than PCR or other microbial detection methods.

Application of nanoscale materials like AuNPs as functional units in biosensors' application is of advantage due to their portability, disposability, and high sensitivity as multiplexed sensors. The specificity of AuNPs can be improved by coupling aptamers, which can be multiple for multiplex detection (Wu et al., 2014). This will broaden detection responses. When immune-sensing electrochemical methods are involved, the use of recognized NPs like AuNPs on the electrode surface is imperative. The NPs can be immobilized on the electrode by adsorption, entrapment, cross-linking, and covalent coupling (Xu et al., 2014; Palma et al., 2006; Charles et al., 2004; Giacomelli et al., 2000). A self-assembled monolayer (SAM) on a gold electrode surface serves as the most prominently applied way of refinement. To maximize the use of AuNPs Zhang et al. (2005) coupled a self-sampling-and-flow biosensor by interjecting a nitrocellulose strip with double-sided microporous gold electrodes, which was made up of gold layer on a nylon substrate.

Toxins from *Clostridium botulinum* and *enterotoxigenic Escherichia coli* (ETEC) are seen as potential contaminating agents in water, with defined health consequences (Wang et al., 2010). There are specific AuNP-based molecular probes as well as lateral flow strip assay to detect toxin contamination of water. Currently strips containing a combination of chromatography technology and conventional immunoassays as well as AuNPs are being used for toxin-specific detection with colorimetric visualization.

Besides the biological contaminants, pharmaceuticals in water can also be detected with AuNPs-based biosensors. It should be noted that several compounds belonging to the PPCPs are aromatic ones, widely used in the pharmaceutical, plastic, dyes, rubber, and pesticide industries. Their potential impacts at nanoconcentration (Khadikar et al., 2002; Bhatkhande et al., 2003) have led to the development of AuNP-based molecular probing techniques with high and specific detection efficiencies. Currently a further development occurs where AuNPs coupled with single or multiple walled carbon nanotubes aims toward improved detection sensitivity. It has been applied by Liu and Lin (2006) to detect nitrobenzene in water samples. Further studies in the use of AuNPs-based biosensors are considering the use of isotopes or alloy of Au while a wide range of tools and platforms are being explored for more effective, efficient, timely, and cost-effective next-generation biosensors based on AuNPs.

The detection specificity is enhanced by use of various nanomaterial supports like AuNPs (already discussed in 14.8), MNPs, CNnTs, gold nanorods (AuNr), silver nanoparticles (AgNPs), silica nanoparticles (SiNPs), etc. in which antibodies take unique roles as recognition elements conjugated or coupled together. This is of special importance in detection of pharmaceuticals in drinking water, as the contaminants are usually in nanoquantities (Schaider et al., 2014), below the detection limits of most drinking water treatment technologies.

16.9 OTHER FUNCTIONALIZED NANOPARTICLES AS BIOSENSORS FOR RAPID MONITORING OF DRINKING

16.9.1 WATER QUALITY

Other functionalized nanoparticles can be used to detect pharmaceutical nanocontaminants as well as waterborne pathogens in drinking water. Table 16.4 exemplifies such functionalized nanoparticles that have been utilized to detect selected bacteria associated with water pollution. Though biosensors may not have overbearing advantages in the area of source tracking (Lazcka et al., 2007), it is however the fastest approach to give results based on correct and prompt detection, portability, and automation. Singh et al. (2017) also highlighted some aptamer-based nanoconjugates (aptasensor) for the detection of waterborne bacterial pathogens.

16.10 DEVELOPING THE GUIDELINES ON THE USE OF BIOSENSORS IN THE AQUATIC ENVIRONMENT OR DRINKING WATER SOURCE

With the emergence of different variants of biosensors and the increasing range of applicabilities, there is also a growing need to ensure standard validation for the existing variants. Onsite use of biosensors should be properly guided with

Table 16.4 Some Reported Nanomaterial-Based Sensors for Detection of Some Waterborne Bacterial Species

Potential Water-Borne Bacteria	Nanomaterial Conjugate	Unique Recognition	[a]Detection Limit or Range by Technology (CFU/mL) (^Cell)	Reference
Salmonella sp.	AuNPs	Ab–PPE–AuNPs	1.E + 02	Kalele et al. (2006)
	MNPs	Ab–MNPs	1.E + 02	Maurer et al. (2012)
	CNTs	MAb–CNTs	1.6E + 04	Dungchai et al. (2008)
	GO–AuNPs	GO–AuNPs–Ab	3	Ma et al. (2014)
Pseudomonas aeruginosa	AuNr	Ab–AuNr	Negligible detection due to cell inactivation	Grossman et al. (2004)
E. coli	AgNPs	Ab–AgNPs	5.E + 00 or range above'	Lin et al. (2008)
	[b]SiNPs	Ab–SiNPs	≤4.E + 02	Li and Church (2014)
	[b]AuNPs	Ab–AuNPs	≤1.E + 07	Nath et al. (2008)
	[b]AuNPs	Au-Synthesized thiolated probe	≤1.E + 06	Chen et al. (2008)
Staphylococcus aureus	AuNPs	Ab–AuNPs	1^	Chang et al. (2013)
			9	Yuan et al. (2014)
	AgNPs	AgNPs-probe	1	Abbaspour et al. (2015)
Listeria monocytogenes	MNPs	Ab–MIONPs	≤5.6E + 06	Jain et al. (2012)
S. aureus	Quantum dot	Luminescence labels coupled with aptamers	8	Xu et al. (2015)
E. coli O157:H7			6	
S. typhimurium (multiplex)			4	
L. monocytogenes			20	

[a]*The reported detection is not necessarily water, but the general detection as quantified by the technology reported by authors.*
[b]E. coli *O157:H7;* Ab, *antibody;* AuNPs, *gold nanoparticles;* MNPs, *magnetic nanoparticles;* CNTs, *carbon nanotubes;* AuNr, *gold nanorods;* AgNPs, *silver nanoparticles;* SiNPs, *silica nanoparticles;* IgG, *immunoglobulin;* PPE, *paraphenylene ethynylene;* MAb, *monoclonal antibody;* GO–AuNPs, *graphene oxide and gold nanoparticles; matrix dependent, it may be better in water.*

recourse to safety of the functional subunit or support. This is of particular relevance where nanoparticle support is utilized along with unique recognition elements. Onsite use of toxic nanoparticles as nanomaterial support in biosensor monitoring of drinking water sources and systems should be discouraged due to enhanced exposure risk of the consuming human population.

The regulations related to pharmaceutical contaminants in drinking water have not gained the attention they deserve in many countries, but the awareness is emerging. As the knowledge grows, more active ingredients or compounds are added to the already known list of PPCPs (USEPA 2009). This is exemplified by the addition of seven hormones in 2011 (USEPA 2011) and several others over time. Addition of more compounds necessitates the optimization or development of new sets of biosensors' aptamers and/or functional units, for example, nanomaterial support. Regulation in this area should prioritize the safety of any novel material that is introduced into the drinking water arsenal. This also encompasses the wastewater effluents for reuse, especially in irrigation, bearing in mind the implication for health (Buttke, 2011).

Health authorities are encouraged to be more proactive in monitoring the PPCPs and other recently classified toxic compounds, while the appropriately regulated use of biosensors would help the monitoring process. It is advisable to have an intermittent pilot monitoring scheme like the one organized by the New York Environmental Protection Department (NYCDEP) on pharmaceuticals in source water in 2010 (NYCDEP 2010).

Contributions for regulation of the contaminants as well as monitoring tools by independent groups are praiseworthy, but more is required. The existing provisional safety levels for pharmaceuticals in drinking water called drinking-water equivalent levels (DWELs) are products of these independent researchers, and depend on several unique factors (Bruce et al., 2010; Schwab et al., 2005). The effect of the pharmaceutical contaminants might be regionally dependent, but this monitoring can be brought together under centrally guided standards for overall human safety (WHO, 2012).

16.11 CONCLUSION

Pharmaceutical nanocontaminants may arise as byproducts from therapeutic or prophylactic use of pharmaceutical products and PPCPs, released into wastewater. These nanocontaminants and some antibiotic resistant microorganisms are not easily detected during routine and microbial contaminants checks in these waters. They are therefore usually conveyed through the treatment chains into the drinking water or food chains. These pharmaceutical nanocontaminants result in human health hazards and ecological toxicity. Various biosensors including immunochemical biosensors, enzymatic biosensors, nonenzymatic receptor biosensors, whole-cell biosensors, and DNA biosensors have been reported in effective detection of contaminants. The use of carbon nanotubes, MIPs, GBS, gold nanoparticles, as well as functionalized nanoparticles (nanomaterial-based sensors) have proven to be much more effective tools for detecting pharmaceutical nanocontaminants in water matrices. Reported efficiencies of these nanomaterial-based sensors have been very high in detecting and removing antibiotic resistant pathogens from water matrices compared with conventional tools. Antibiotic resistant pathogens listed as critical threats to public health by the WHO have been reportedly tracked by nanomaterial-based sensors in which unique recognition points were built into appropriate nanomaterial conjugates. These affirm these tools as prominent next generational tools for future detection and removal of pharmaceutical contaminants and pathogens.

REFERENCES

Abbaspour, A., Norouz-Sarvestani, F., Noori, A., Soltani, N., 2015. Aptamer-conjugated silver nanoparticles for electrochemical dual-aptamer-based sandwich detection of *Staphylococcus aureus*. Biosens. Bioelectron. 68, 149–155. Available from: https://doi.org/10.1016/j.bios.2014.12.040.

Adegoke, A.A., Awolusi, O.O., Stenström, T.A., 2016. Organic fertilizers: public health intricacies. In: Larramendy M., (Ed.), Organic Fertilizers—From Basic Concepts to Applied Outcomes. InTech. https://doi.org/10.5772/64195. Available from: <http://www.intechopen.com/books/organic-fertilizers-from-basic-concepts-to-applied-outcomes/organic-fertilizers-public-health-intricacies>.

Aga, D.S., 2008. Fate of Pharmaceuticals in the Environment and in Water Treatment Systems. CRC Press, Boca Raton.

Ajayan, P.M., 1999. Nanotubes from carbon. Chem. Rev. 99, 1787–1799.

Akkoyun, A., Kohen, V.F., Bilitewski, U., 2000. Detection of sulphamethazine with an optical biosensor and anti-idiotypic antibodies. Sens. Actuators B: Chem. 70, 12–18.

Ali, S.R., Ma, Y., Parajuli, R.R., Balogun, Y., Lai, W.Y.-C., He, H., 2007. A nonoxidative sensor based on a self-doped polyaniline/carbon nanotube composite for sensitive and selective detection of the neurotransmitter dopamine. Anal. Chem. 79, 2583–2587.

An, K.H., Jeong, S.Y., Hwang, H.R., Lee, Y.H., 2004. Enhanced sensitivity of a gas sensor incorporating single-walled carbon nanotube–polypyrrole nanocomposites. Adv. Mater. 16, 1005–1009.

Bain, R., Cronk, R., Wright, J., Yang, H., Slaymaker, T., Bartram, J., 2014. Fecal contamination of drinking-water in low- and middle-income countries: a systematic review and meta-analysis. PLoS Med. 11 (5), e1001644. Available from: https://doi.org/10.1371/journal.pmed.1001644.

Balcázar, J.L., Subirats, J., Borrego, C.M., 2015. The role of biofilms as environmental reservoirs of antibiotic resistance. Front. Microbiol. 2015, . Available from: https://doi.org/10.3389/fmicb.2015.01216. Available from:.

Barnes, K.K., Kolpin, D.W., Furlong, E.T., Zaugg, S.D., Meyer, M.T., Barber, L.B., 2008. A national reconnaissance of pharmaceuticals and other organic wastewater contaminants in the United States—I. Groundwater. Sci. Total Environ. 402, 192–200.

Bechor, O., Smulski, D.R., Van Dyk, T.K., LaRossa, R.A., Belkin, S., 2002. Recombinant microorganisms as environmental biosensors: pollutants detection by *Escherichia coli* bearing fabA: lux fusions. J. Biotechnol. 94, 125–132.

Bermúdez-Almada M.C., Espinosa-Plascencia, A., 2012. The Use of Antibiotics in Shrimp Farming, Health and Environment in Aquaculture. In: Dr. Edmir Carvalho (Ed.), ISBN: 978-953-51-0497-1, InTech, Available from: http://www.intechopen.com/books/health-and-environment-in-aquaculture/the-use-of-antibiotics-in-shrimpfarming.

Bhatkhande, D.S., Pangarkar, V.G., Beenackers, A.A.C.M., 2003. Photocatalytic degradation of nitrobenzene using TiO_2: chemical effects. Water Res. 37, 1223–1230.

Blackburn, J.K., Mitchell, M.A., Blackburn, M.-C.H., Curtis, A., Thompson, B.A., 2010. Evidence of antibiotic resistance in free-swimming, top-level marine predatory fishes. J. Zoo Wildlife Med. 41, 7–16. Available from: https://doi.org/10.1638/2007-0061.1.

Blair, B.D., Crago, J.P., Hedman, C.J., Klaper, R.D., 2013. Pharmaceuticals and personal care products found in the Great Lakes above concentrations of environmental concern. Chemosphere 93, 2116–2123.

Bonnedahl, J., Drobni, M., Gauthier-Clerc, M., Hernandez, J., Granholm, S., Kayser, Y., et al., 2009. Dissemination of *Escherichia coli* with CTX-M type ESBL between humans and yellow-legged gulls in the south of France. PLoS ONE 2009 (4), e5958. Available from: https://doi.org/10.1371/journal.pone.0005958.

Bossus, M.C., Guler, Y.Z., Short, S.J., Morrison, E.R., Ford, A.T., 2013. Behavioural and transcriptional changes in the amphipod *Echinogammarus marinus* exposed to two antidepressants, fluoxetine and sertraline. Aquat. Toxicol. 151, 46–56.

Bromage, E.R., Vadas, G.G., Harvey, E., Unger, M.A., Kaattari, S.L., 2007. Validation of an antibody-based biosensor for rapid quantification of 2,4,6-trinitrotoluene (TNT) contamination in ground water and river water. Environ. Sci. Technol. 41 (20), 7067–7072.

Brooks, B.W., Turner, P.K., Stanley, J.K., Weston, J.J., Glidewell, E.A., Foran, C.M., et al., 2003. Waterborne and sediment toxicity of fluoxetine to select organisms. Chemosphere 52, 135–142.

Bruce, G.M., Pleus, R.C., Snyder, S.A., 2010. Toxicological relevance of pharmaceuticals in drinking water. Environ. Sci. Technol. 44, 5619–5626.

Buttke, D.E., 2011. Toxicology, environmental health, and the "One Health" concept. J. Med. Toxicol. 7 (4), 329–332. Available from: https://doi.org/10.1007/s13181-011-0172-4.

Cabezaa, Y., Candelaa, L., Ronenb, D., Teijon, G., 2012. Monitoring the occurrence of emerging contaminants in treated wastewater and groundwater between 2008 and 2010. The Baix Llobregat (Barcelona, Spain). J. Hazard. Mater. 239, 32–39.

Campanella, L., Favero, G., Mastrofini, D., Tomassetti, M., 1997. Further developments in toxicity cell biosensors. Sens. Actuators B 44, 279–285.

Chang, Y.C., Yang, C.Y., Sun, R.L., Cheng, Y.F., Kao, W.C., Yang, P.C., 2013. Rapid single cell detection of *Staphylococcus aureus* by aptamer-conjugated gold. Sci. Rep. 3, 1–7.

Chang, J., Zhou, G., Christensen, E.R., Heideman, R., Chen, J., 2014. Graphene-based sensors for detection of heavy metals in water: a review. Anal. Bioanal. Chem. 406, 3957–3975.

Charles, P.T., Goldman, E.R., Rangasammy, J.G., Schauer, C.L., Chen, M.S., Taitt, C.R., 2004. Fabrication and characterization of 3D hydrogel microarrays to measure antigenicity and antibody functionality for biosensor application. Biosens. Bioelectron. 20, 753–764.

Chen, S.H., Wu, V.C., Chuang, Y.C., 2008. Using oligonucleotide-functionalized Au nanoparticles to rapidly detect foodborne pathogens on a piezoelectric biosensor. J. Microbiol. Methods 73, 7–17.

Cheong, W.J., Yang, S.H., Ali, F., 2012. Molecular imprinted polymers for separation science: review of reviews. J. Sep. Sci. 36, 609–628.

Choffnes, E.R., Relman, D.A., Olsen, L.A., Hutton, R., Mack, A., 2012. Improving Food Safety Through a One Health Approach: Workshop Summary. The National Academic Press, Washington, DC, pp. 115–130.

Chopra, S., McGuire, K., Gothard, N., Roa, A.M., Pham, A., 2003. Selective gas detection using a carbon nanotube sensor. Appl. Phys. Lett. 83, 2280–2282.

Conn, K.E., Barber, L.B., Brown, G.K., Siegrist, R.L., 2006. Occurrence and fate of organic contaminants during onsite wastewater treatment. Environ. Sci. Technol. 4023, 7358–7366.

Conn, K.E., Lowe, K.S., Drewes, J.E., Hoppe-Jones, C., Tucholke, M.B., 2010a. Occurrence of pharmaceuticals and consumer product chemicals in raw wastewater and septic tank effluent from single-family homes. Environ. Eng. Sci. 27 (4), 347–356.

Conn, K.E., Siegrist, R.L., Barber, L.B., Meyer, M.T., 2010b. Fate of trace organic compounds during vadose zone soil treatment in an onsite wastewater system. Environ. Toxicol. Chem. 29 (2), 285–293.

Craciun, M., Russo, S., Yamamoto, M., Tarucha, S., 2011. Tuneable electronic properties in graphene. Nano Today 6, 42–60. Available from: https://doi.org/10.1016/j.nantod.2010.12.001.

Crane, M., Watts, C., Boucard, T., 2006. Chronic aquatic environmental risks from exposure to human pharmaceuticals. Sci. Total Environ. 367, 23–41.

Da Costa Silva, L.M., dos Santos, S.V.P., Salgado, A.M., Pereira, K.S., 2013. Biosensors for Contaminants Monitoring in Food and Environment for Human and Environmental Health. In Tech. doi:10.5772/55617.

Daneshvar, A., Aboulfadl, K., Viglino, L., Broséus, R., Sauvé, S., Madoux-Humery, A.S., et al., 2012. Evaluating pharmaceuticals and caffeine as indicators of fecal contamination in drinking water sources of the Greater Montreal region. Chemosphere 88 (1), 131–139. Available from: https://doi.org/10.1016/j.chemosphere.2012.03.016.

Daughton, G.C., 2001. Emerging pollutants, and communicating the science of environmental chemistry and mass spectrometry: pharmaceuticals in the environment. J. Am. Soc. Mass Spectrom. 12 (10), 1067–1076.

Daughton, C.G., 2004. Non-regulated water contaminants: emerging research. Environ. Impact Assess. Rev. 24, 711–732.

Daughton, C.G., Ternes, T.A., 1999. Pharmaceuticals and personal care products in the environment: agents of subtle change? Environ. Health Perspect. 107 (6), 907–938.

Davis, R., Irudayaraj, J., Reuhs, B.L., Mauer, L.J., 2010. Detection of *E. coli* O157:H7 from ground beef using Fourier Transform Infrared (FT-IR) spectroscopy and chemometrics. J. Food Sci. 75, 340–346.

DeLorenzo, M.E., Fleming, J., 2008. Individual and mixture effects of selected pharmaceuticals and personal care products on the marine phytoplankton species *Dunaliella tertiolecta*. Arch. Environ. Contam. Toxicol. 54, 203–210.

Deng, Y., Schmid, N., Wang, C., Wang, J., Pessi, G., Wu, D., et al., 2012. Cis-2-dodecenoic acid receptor RpfR links quorum-sensing signal perception with regulation of virulence through cyclic dimeric guanosine monophosphate turnover. Proc. Natl. Acad. Sci. USA 108, 15479–15484. Available from: https://doi.org/10.1073/pnas.120503710.

Deshpande, A., LeRoy, B., 2012. Scanning probe microscopy of graphene. Physica E44, 743–759. Available from: https://doi.org/10.1016/j.physe.2011.11.024.

Di-Poi, C., Darmaillacq, A.-S., Dickel, L., Boulouard, M., Bellanger, C., 2013. Effects of perinatal exposure to waterborne fluoxetine on memory processing in the cuttlefish Sepia officinalis. Aquat. Toxicol. 132, 84–91. Available from: https://doi.org/10.1016/j.aquatox.2013.02.004.

Dougherty, J.A., Swarenzki, P.W., Dinicola, R.S., Reinhard, M., 2010. Occurrence of herbicides andpharmaceutical and personal care products in surface water and groundwater around Liberty Bay, Puget Sound, Washington. J. Environ. Qual. 39, 1173–1180.

Duffy, P., Woods, G., O'Hogain, S., Walsh, J., Caplier, C., 2009. On-line realtime water quality monitoring and control for swimming pools. Conference Papers School of Manufacturing and Design Engineering, at ARROW@DIT, 2009-09-01. Available from: <http://arrow.dit.ie/cgi/viewcontent.cgi?article = 1016&context = engschmanconn> (accessed 12.01.17).

Dungchai, W., Siangproh, W., Chaicumpa, W., Tongtawe, P., Chailapakul, O., 2008. *Salmonella typhi* determination using voltammetric amplification of nanoparticles: a highly sensitive strategy for metalloimmunoassay based on a copper-enhanced gold label. Talanta 77, 727e32.

Durand, M.J., Thouand, G., Dancheva-Ivanova, T., Vachon, P., DuBow, M., 2003. Specific detection of organotin compounds with a recombinant luminescent bacteria. Chemosphere 52, 103–111.

Fiksdal, L., Tryland, I., 2008. Application of rapid enzyme assay techniques for monitoring of microbial water quality. Curr. Opin. Biotechnol. 19, 289–294.

Fiorillo, M., Verre, A.F., Iliut, M., Peiris-Pagés, M., Ozsvari, B., Gandara, R., et al., 2015. Graphene oxide selectively targets cancer stem cells, across multiple tumor types: implications for non-toxic cancer treatment, via "differentiation-based nano-therapy". Oncotarget 6 (6), 3553–3562.

Focazio, M.J., Kolpin, D.W., Barnes, K.K., Furlong, E.T., Meyer, M.T., Zaugg, S.D., et al., 2008. A national reconnaissance for pharmaceuticals and other organic wastewater contaminants in the United States—II. Untreated drinking water sources. Sci. Total. Environ. 402 (2–3), 201–216.

Gaworecki, K.M., Klaine, S.J., 2008. Behavioral and biochemical responses of hybrid striped bass during and after fluoxetine exposure. Aquat. Toxicol. 88, 207–213.

Gehring, A.G., Albin, D.M., Bhunia, A.K., Reed, S.A., Tu, S.I., Uknalis, J., 2006. Antibody microarray detection of *Escherichia coli* O157:H7: quantification, assay limitations, and capture efficiency. Anal. Chem. 78, 6601–6607.

Giacomelli, C.E., Vermeer, A.W.P., Norde, W., 2000. Adsorption of immunoglobulin G on core–shell latex particles precoated with chaps. J. Colloid Interf. Sci. 231, 283–288.

González-Martínez, M.A., Puchades, R., Maquieira, A., Ferrer, I., Marco, M.P., Barceló, D., 1999. Reversible immunosensor for the automatic determination of atrazine. Selection of performance of three polyclonal antisera. Anal. Chim. Acta 1999 (386), 201–210.

Gooding, J.J., Shein, J., Lai, L.M.H., 2009. Using nanoparticle aggregation to give an ultrasensitive amperometric metal ion sensor. Electrochem. Commun. 11, 2015–2018.

Govindhan, M., Adhikaria, B.-R., Chen, A., 2014. Nanomaterials-based electrochemical detection of chemical contaminants. RSC Adv. 4, 63741–63760. Available from: https://doi.org/10.1039/C4RA10399H.

Grieshaber, D., Mac Kenzie, R., Voeroes, J., Reimhult, E., 2008. Electrochemical biosensors—sensor principles and architectures. Sensors 8 (3), 1400–1458.

Grossman, H.L., Myers, W.R., Vreeland, V.J., Bruehl, R., Alper, M.D., Bertozzi, C.R., et al., 2004. Detection of bacteria in suspension by using a superconducting quantum interference device. Proc. Natl. Acad. Sci. USA 101, 129–134.

Guiseppi-Elie, A., Lei, C., Baughman, R.H., 2002. Direct electron transfer of glucose oxidase on carbon nanotubes. Nanotechnology 13, 559–564.

Guo, X., 2013. Single-molecule electrical biosensors based on single-walled carbon nanotubes. Adv. Mater. 25, 3397–3408. Available from: https://doi.org/10.1002/adma.201301219.

Halling-Sørensen, B., Nors-Nielsen, S., Lanzky, P.F., Ingerslev, F., Holten-Lützhøft, H.C., Jørgensen, S.E., 1998. Occurrence, fate and effects of pharmaceutical substances in the environment—a review. Chemosphere 36 (2), 357–393.

Heidler, J., Sapkota, A., Halden, R.U., 2006. Partitioning, persistence, and accumulation in digested sludge of the topical antiseptic triclocarban during wastewater treatment. Environ. Sci. Technol. 40 (11), 3634–3639.

Hernandez, F.J., Ozalp, V.C., 2012. Graphene and other nanomaterial-based electrochemical aptasensors. Biosensors 2, 1–14.

Hua, P., Hole, J.P., Wilkinson, J.S., Proll, G., Tschmelak, J., Gauglitz, G., et al., 2005. Integrated optical fluorescence multisensor for water pollution. Opt. Express 13, 1124–1130.

Huang, C.H., Sedlak, D.L., 2001. Analysis of estrogenic hormones in municipal wastewater effluent and surface water using enzyme-linked immunosorbent assay and gas chromatography/tandem mass spectrometry. Environ. Toxicol. Chem. 20 (1), 133–139.

Hutter, E., Maysinger, D., 2013. Gold-nanoparticle-based biosensors for detection of enzyme activity. Trends Pharmacol. Sci. 34, 497–507. Available from: https://doi.org/10.1016/j.tips.2013.07.002.

Inbaraj, B.S., Chen, B.H., 2016. Nanomaterial based sensors for detection of foodborne bacterial pathogens and toxinx as well as pork adulteration in meat products. J. Food Drug Anal. 24, 15–28.

Jain, S., Singh, S.R., Horn, D.W., Davis, V.A., Ram, M.J., Pillai, S.R., 2012. Development of an antibody functionalized carbon nanotube biosensor for foodborne bacterial pathogens. J. Biosens. Bioelectron. S11 (002), 1–7.

Kalele, S.A., Kundu, A.A., Gosavi, S.W., Deobagkar, D.N., Deobagkar, D.D., Kulkarni, S. K., 2006. Rapid detection of *Escherichia coli* by using antibody-conjugated silver nanoshells. Small 2, 335–338.

Kang, X., Wang, J., Wu, H., Liu, J., Aksay, I., Lin, Y., 2010. A graphene-based electrochemical sensor for sensitive detection of paracetamol. Talanta 81, 754–759. Available from: https://doi.org/10.1016/j.talanta.2010.01.009.

Kerr, P., Chart, H., Finlay, D., Pollock, D.A., MacKie, D.P., Ball, H.J., 2001. Development of a monoclonal sandwich ELISA for the detection of animal and human *Escherichia coli* O157 strains. J. Appl. Microbiol. 90, 543–549.

Khadikar, P.V., Karmarkar, S., Singh, S., Shrivastava, A., 2002. Use of the PI index in predicting toxicity of nitrobenzene derivatives. Bioorg. Med. Chem. 10, 3163–3170.

Khanal, S.K., Xie, B., Thompson, M.L., Sung, S., Ong, S.-K., van Leeuwen, J., 2006. Fate, transport, and biodegradation of natural estrogens in the environment and engineered systems. Environ. Sci. Technol. 40 (21), 6537–6546.

Kidd, K.A., Paterson, M.J., Rennie, M.D., Podemski, C.L., Findlay, D.L., Blanchfield, P.J., et al., 2014. Direct and indirect responses of a freshwater food web to a potent synthetic oestrogen. Philos. Trans. Royal Soc. B 369, 20130578. Available from: https://doi.org/10.1098/rstb.2013.0578.

Kim, S.C., Carlson, K., 2007. Temporal and spatial trends in the occurrence of human and veterinary antibiotics in aqueous and river sediment matrices. Environ. Sci. Technol. 41 (1), 50–57.

Klosterhaus, S.L., Grace, R., Hamilton, M.C., Yee, D., 2013. Method validation and reconnaissance of pharmaceuticals, personal care products, and alkylphenols in surface waters, sediments, and mussels in an urban estuary. Environ. Int. 54, 92–99. Available from: https://doi.org/10.1016/j.envint.2013.01.009.

Kocsis, B., Domokos, J., Szabo, D., 2016. Chemical structure and pharmacokinetics of novel quinolone agents represented by avarofloxacin, delafloxacin, finafloxacin, zabofloxacin and nemonoxacin. Ann. Clin. Microbiol. Antimicrob. 15, 34. Available from: https://doi.org/10.1186/s12941-016-0150-4.

Kolpin, D.W., Furlong, T., Meyer, M.T., Thurman, E.M., Zaugg, S.D., Barber, L.B., et al., 2002. Pharmaceuticals, hormones, and other organic wastewater contaminants in U.S. streams, 1999–2000: a national reconnaissance. Environ. Sci. Technol. 36 (6), 1202–1211.

Koyun, A., Ahlatcıoğlu, E., İpek, Y.K., Biosensors and Their Principles, Kahraman, C., 2008. Fuzzy Multi-Criteria Decision Making: Theory and Applications With Recent Developments, vol. 16. Springer. Available from: <http://www.sciepub.com/reference/85688> (accessed 15.12.16).

Kuila, T., Bose, S., Khanra, P., Mishra, A.K., Kim, N.H., Lee, J.H., 2011. Recent advances in graphene-based biosensors. Biosens. Bioelectron. 26, 4637–4648.

Kuila, T., Bose, S., Mishra, A., Khanra, P., Kim, N., Lee, J., 2012. Chemical functionalization of grapheme and its applications. Prog. Mater. Sci. 57, 1061–1105. Available from: https://doi.org/10.1016/j.pmatsci.2012.03.002.

Kuroda, K., Murakami, M., Oguma, K., Muramatsu, Y., Takada, H., Takizawa, S., 2011. Assessment of groundwater pollution in Tokyo using PPCPs as sewage markers. Environ. Sci. Technol. 46, 1455–1464.

Langford, K., Thomas, K.V., 2011. Input of selected human pharmaceutical metabolites into the Norwegian aquatic environment. J. Environ. Monitor 13, 416–421. Available from: https://doi.org/10.1039/c0em00342e.

Lapworth, D.J., Baran, N., Stuart, M.E., Ward, R.S., 2012. Emerging organic contaminants in groundwater: a review of sources, fate, and occurrence. Environ. Pollut. 163, 287–303.

Larson, J.K., Carvan, M., Hutz, R.J., 2014. Engineered Nanomaterials: An Emerging Class of Novel Endocrine Disruptors. Biol. Reprod. 91 (1). Available from: https://doi.org/10.1095/biolreprod.113.116244.

Lazcka, O., Del Campo, F., Muñoz, F., 2007. Pathogen detection: A perspective of traditional methods and biosensors. Biosens. Bioelectron. 22 (7), 1205–1217. Available from: https://doi.org/10.1016/j.bios.2006.06.036.

Le, T.X., Munekage, Y., 2004. Residues of selected antibiotics in water and mud from shrimp ponds in mangrove areas in Viet Nam. Mar. Pollut. Bull. 49, 922–929. Available from: https://doi.org/10.1016/j.marpolbul.2004.06.016.

Lee, J.O., So, H.M., Jeon, E.K., Chang, H., Won, K., Kim, Y.H., 2008. Aptamers as molecular recognition elements for electrical nanobiosensors. Anal. Bioanal. Chem. 390, 1023–1032.

Lee, S., Liu, X., Takeda, S., Choi, K., 2013. Genotoxic potentials and related mechanisms of bisphenol A and other bisphenol compounds: a comparison study employing chicken DT40 cells. Chemosphere 93 (2), 434–440.

Leonard, A.S., Yermolaieva, O., Hruska-Hageman, A., Askwith, C.C., Price, M.P., Wemmie, J.A., et al., 2003. cAMP-dependent protein kinase phosphorylation of the acid-sensing ion channel-1 regulates its binding to the protein interacting with C-kinase-1. Proc. Natl. Acad. Sci. U S A. 100, 2029–2034.

Leung, H.W., Jin, L., Wei, S., Tsui, M.M.P., Zhou, B., Jiao, L., et al., 2013. Pharmaceuticals in tap water: human health risk assessment and proposed monitoring framework in China. Environ. Health Perspect. 121 (7), 839–846.

Li, Y.S., Church, J.S., 2014. Raman spectroscopy in the analysis of food and pharmaceutical nanomaterials. J. Food Drug Anal. 22, 29–48.

Li, S., Ge, Y., Piletsky, S.A., Lunec, J., 2012. Molecularly Imprinted Sensors: Overview and Applications. Elsevier, UK.

Liang, X., Chen, B., Nie, X., Shi, Z., Huang, X., Li, X., 2013. The distribution and partitioning of common antibiotics in water and sediment of the Pearl River Estuary, South China. Chemosphere 92, 1410–1416. Available from: https://doi.org/10.1016/j.chemosphere.2013.03.044.

Lin, Y.H., Chen, S.H., Chuang, Y.C., Lu, Y.C., Shen, T.Y., Chang, C.A., et al., 2008. Disposable amperometric immunosensing strips fabricated by Au nanoparticles-modified screen-printed carbon electrodes for the detection of foodborne pathogen *Escherichia coli* O157:H7. Biosens. Bioelectron. 23, 1832–1837.

Liu, G., Lin, Y., 2006. Biosensor based on self-assembling acetylcholinesterase on carbon nanotubes for flow injection/amperometric detection of organophosphate pesticides and nerve agents. Anal. Chem. 78, 835–843.

Liu, Y., Dong, X., Chen, P., 2012. Biological and chemical sensors based on graphene materials. Chem. Soc. Rev. 41, 2283–2307.

Loos, R., Locoro, G., Comero, S., Contini, S., Schwesig, D., Werres, F., et al., 2010. Pan-European survey on the occurrence of selected polar organic persistent pollutants in ground water. Water Res. 44, 4115–4126.

Lulinski, P., 2013. Molecularly imprinted polymers as the future drug delivery devices. Acta Polish. J. Pharm. Res. 70, 601–609.

Ma, X., Jiang, Y., Jia, F., Yu, Y., Chen, J., Wang, Z., 2014. An aptamer based electrochemical biosensor for the detection of Salmonella. J. Microbiol. Methods 98, 94–98. Available from: https://doi.org/10.1016/j.mimet.2014.01.003.

Malhotra, B.D., Chaubey, A., Singh, S.P., 2006. Prospects of conducting polymers in biosensors. Anal. Chim. Acta 578, 59.

Mao, X., Yang, L., Su, X., Yi, Y., 2006. A nanoparticle amplification based quartz crystal microbalance DNA sensor for detection of *Escherichia coli* O157:H7. Biosens. Bioelectron. 1, 1178–1185.

Marchesini, G.R., Meulenberg, E., Haasnoot, W., Irth, H., 2005. Biosensor immunoassays for the detection of bisphenol A. Anal. Chim. Acta 528, 37.

Marco, M.-P., Barcelo, D., Meas, D., 1996. Environmental applications of analytical biosensors. Sci. Technol. 7, 1547–1562.

Maurer, E.I., Comfort, K.K., Hussain, S.M., Schlager, J.J., Mukhopadhyay, S.M., 2012. Novel platform development using an assembly of carbon nanotube, nanogold and immobilized RNA capture element towards rapid, selective sensing of bacteria. Sensors (Basel) 12, 8135–8144.

Méndez, N., Lacorte, S., Barata, C., 2013. Effects of the pharmaceutical fluoxetine in spiked-sediments on feeding activity and growth of the polychaete *Capitella teleta*. Mar. Environ. Res. 89, 76–82. Available from: https://doi.org/10.1016/j.marenvres.2013.05.004.

Mikol, Y.B., Richardson, W.R., Vander-Schalie, W.H., Shedd, T.R., Widder, M.W., 2007. An online real-time biomonitor for contaminant surveillance in water supplies. J. AWWA 99, 107–115.

Moore, J.E., Rao, J.R., Moore, P.J., Millar, B.C., Goldsmith, C.E., Loughrey, A., et al., 2010. Determination of total antibiotic resistance in waterborne bacteria in rivers and streams in Northern Ireland: can antibiotic-resistant bacteria be an indicator of ecological change? Aquat. Ecol. 44, 349–358. Available from: https://doi.org/10.1007/s10452-009-9294-z.

Morasch, B., 2013. Occurrence and dynamics of micropollutants in a karst aquifer. Environ. Pollut. 173, 133–137.

Mottaleb, M.A., Meziani, M.J., Matin, M.A., Arafat, M.M., Wahab, M.A., 2015. Emerging micro-pollutants pharmaceuticals and personal care products (PPCPs) contamination concerns in aquatic organisms—LC/MS and GC/MS analysis. In: ACS Symposium Series, Emerging Micro-Pollutants in the Environment: Occurrence, Fate, and Distribution, vol. 1198. pp. 43–74 (Chapter 3). doi:10.1021/bk-2015- 1198.ch003.

Mulchandani, P., Hangarter, C.M., Lei, Y., Chen, W., Mulchandani, A., 2005. Amperometric microbial biosensor for p-nitrophenol using *Moraxella* sp. modified carbon paste electrode. Biosens. Bioelectron. 21, 523–527.

Nanda, S.S., Yi, D.K., Kim, K., 2016. Study of antibacterial mechanism of graphene oxide using Raman spectroscopy. Sci. Rep. 6 (28443). Available from: https://doi.org/10.1038/srep28443.

Nath, S., Kaittanis, C., Tinkham, A., Perez, J.M., 2008. Dextran-coated gold nanoparticles for the assessment of antimicrobial susceptibility. Anal. Chem. 80, 1033–1038.

Nentwig, G., 2007. Effects of pharmaceuticals on aquatic invertebrates. Part II: The antidepressant drug fluoxetine. Arch. Environ. Contam. Toxicol. 52, 163–167.

Newman, J.D., Setford, S.J., 2006. Enzymatic biosensors. Mol. Biotechnol. 32 (3), 249–268.

Norman, A., Hansen, L.H., Sørensen, S.J., 2005. Construction of a ColD cda promoter-based SOS-green fluorescent protein whole-cell biosensor with higher sensitivity toward genotoxic compounds than constructs based on recA, umuDC, or sulA promoters. Appl. Environ. Microbiol. 71, 2338–2346.

Olsen, E.V., Pathirana, S.T., Samoylov, A.M., Barbaree, J.M., Chin, B.A., Neely, W.C., et al., 2003. Specific and selective biosensor for Salmonella and its detection in the environment. J. Microbiol. Meth. 53, 273–285.

Ong, Y., Thai, A., Abdul, L., Zein, S.H., Tan, S.H., 2010. A review on carbon nanotubes in an environmental protection and green engineering perspective. Braz. J. Chem. Eng. 27 (2), 227–242.

Onnerfjord, P., Eremin, S.A., Emneus, J., Marko-Varga, G., 1998. Fluorescence polarisation for immunoreagent characterisation. J. Chromatogr. A 800, 219.

Ortiz de García, S.A., Pinto Pinto, G., García-Encina, P.A., Irusta-Mata, R., 2014. Ecotoxicology ecotoxicity and environmental risk assessment of pharmaceuticals and personal care products in aquatic environments and wastewater treatment plants. Ecotoxicology 23, 1517–1533.

Palma, R., Borghs, G., Declerck, P., Goddeeris, B., 2006. Comparison of random and oriented immobilization of antibody fragments on mixed self-assembled monolayers. J. Immunol. Methods 312, 167–181.

Pan, G., Zhang, Y., Ma, Y., Li, C., Zhang, H., 2011. Efficient one-pot synthesis of water compatible molecularly imprinted polymer microspheres by facile RAFT precipitation polymerization. Angew. Chem. Int. Ed. Engl. 50, 11731–11734.

Patel, P.D., 2002. (Bio)sensors for measurement of analytes implicated in food safety: a review. TrAC Trends Anal. Chem. 21, 96–115.
Pérez-Moral, N., Mayes, A.G., 2007. Molecularly imprinted multi-layer core–shell nanoparticles—a surface grafting approach. Macromol. Rap. Commun. 28, 2170–2175.
Pohanka, M., Skládal, P., 2008. Electrochemical biosensors—principles and applications. J. Appl. Biomed. 6 (2), 57–64.
Pomati, F., Castiglioni, S., Zuccato, E., Fanelli, R., Vigetti, D., Rossetti, C., et al., 2006. Effects of a complex mixture of therapeutic drugs at environmental levels on human embryonic cells. Environ. Sci. Technol. 40, 2442–2447.
Prieto-Simón, B., Bandaru, N.M., Saint, C., Voelcker, N.H., 2015. Tailored carbon nanotube immunosensors for the detection of microbial contamination. Biosens. Bioelectron. 67, 642–648.
Priyadarshi, H., Alam, A., Gireesh-Babu, P., Das, R., Kishore, P., Kumar, S., et al., 2012. A GFP-based bacterial biosensor with chromosomally integrated sensing cassette for quantitative detection of Hg (II) in environment. J. Environ. Sci. (China). 24, 963–968.
Pundir, C.S., Chauhan, N., 2012. Acetylcholinesterase inhibition-based biosensors for pesticide determination: a review. Anal Biochem. 429 (1), 19–31. Available from: https://doi.org/10.1016/j.ab.2012.06.025.
Quinn, B., Gagne, F., Blaise, C., 2008. An investigation into the acute and chronic toxicity of eleven pharmaceuticals (and their solvents) found in wastewater effluent on the cnidarian, Hydra attenuate. Sci. Total Environ. 389, 306–314.
Radjenovic, J., Petrovic, M., Barceló, D., 2009. Fate and distribution of pharmaceuticals in wastewater and sewage sludge of the conventional activated sludge (CAS) and advanced membrane bioreactor (MBR) treatment. Water Res. 43 (3), 831–841.
Radović, T., Grujić, S., Petković, A., Dimkić, M., Laušević, M., 2015. Determination of pharmaceuticals and pesticides in river sediments and corresponding surface and ground water in the Danube River and tributaries in Serbia. Environ. Monit. Assess. 187, 4092.
Rajasulochana, P., Preethy, V., 2016. Comparison on efficiency of various techniques in treatment of waste and sewage water—a comprehensive review. Resour. Effic. Technol. 2 (4), 175–184.
Rao, A.D., Dash, Sujata, Babu, S.V., Jain, I., 2007. Numerical modeling of cyclone impact on the ocean e a case study of the Orissa Supper Cyclone. J. Coastal. Res. 23 (5), 1245e–1250e.
Rassaei, L., Amiri, M., Cirtiu, C.M., Sillanpaa, M., Marken, F., Sillanpaa, M., 2011. Nanoparticles in electrochemical sensors for environmental monitoring. TRAC Trends Anal. Chem. 30, 1704–1715.
Rodriguez-Mozaz, S., Lopez de Alda, M., Barceľo, D., 2005. Analysis of bisphenol A in natural waters by means of an optical immunosensor. Water Res. 39 (20), 5071–5079.
Rodriguez-Mozaz, S., Lopez de Alda, M., Barceľo, D., 2006. Biosensors as useful tools for environmental analysis and monitoring. Anal. Bioanal. Chem. 386, 1025–1041.
Rodriguez-Mozaz, S., Marco, M.-P., Lopez de Alda, M., Barceľo, D., 2004. Biosensors for environmental monitoring of endocrine disruptors: a review article. Anal. Bioanal. Chem. 378, 588–598.
Rogers, K.R., 2006. Recent advances in biosensor techniques for environmental monitoring. Anal. Chim. Acta 568, 222–231.
Rose, J.M., Gast, R.J., Bogomolni, A., Ellis, J.C., Lentell, B.J., Touhey, K., et al., 2009. Occurrence and patterns of antibiotic resistance in vertebrates off the Northeastern United States coast. FEMS Microbiol. Ecol. 67, 421–431. Available from: https://doi.org/10.1111/j.1574-6941.2009.00648.x.

Rushing, B., Wooten, A., Shawky, M., Selim, M.I., 2016. Comparison of LC–MS and GC–MS for the analysis of pharmaceuticals and personal care products in surface water and treated wastewaters. Spectroscopy 14, 7–14.

Samendra, P.S., Masaaki, K., Charles, P.G., Ian, L.P., 2014. Rapid detection technologies for monitoring microorganisms in water. Biosens. J. 3, 109. Available from: https://doi.org/10.4172/2090-4967.1000109.

Sanderson, J.T., Hordijk, J., Denison, M.S., Springsteel, M.F., Nantz, M.H., van den Berg, M., 2004. Induction and inhibition of aromatase (CYP19) activity by natural and synthetic flavonoid compounds in H295R human adrenocortical carcinoma cells. Toxicol. Sci. 8, 270–279.

Sang, S., Wang, Y., Feng, Q., Wei, Y., Ji, J., Zhang, W., 2015. Progress of new label-free techniques for biosensors: a review. Crit. Rev. Biotechnol 15, 1–7. Available from: https://doi.org/10.3109/07388551.2014.991270.

Sattarahmady, N., Tondroc, G.H., Gholchinc, M., Heli, H., 2015. Gold nanoparticles biosensor of *Brucella* spp. genomic DNA: visual and spectrophotometric detections. Biochem. Eng. J. 97, 1–7.

Schaider, L.A., Rudel, R.A., Ackerman, J.M., Dunagan, S.M., Brody, J.G., 2014. Pharmaceuticals, perfluorosurfactants, and other organic wastewater compounds in public drinking water wells in a shallow sand and gravel aquifer. Sci. Total Environ. 468–469, 384–393. Available from: https://doi.org/10.1016/j.scitotenv.2013.08.067.

Schmid, N., Pessi, G., Deng, Y., Aguilar, C., Carlier, A.L., Grunau, A., et al., 2012. The AHL- and BDSF-dependent quorum sensing systems control specific and overlapping sets of genes in Burkholderia cenocepacia H111. PLoS ONE 7 (11), e49966. Available from: https://doi.org/10.1371/journal.pone.0049966.

Schmidt, W., O'Rourke, K., Hernan, R., Quinn, B., 2011. Effects of the pharmaceuticals gemfibrozil and diclofenac on the marine mussel (*Mytilus* spp.) and their comparison with standardized toxicity tests. Mar. Pollut. Bull. 62, 1389–1395. Available from: https://doi.org/10.1016/j.ma-rpolbul.2011.04.043.

Schnell, S., Bols, N.C., Barata, C., Porte, C., 2009. Single and combined toxicity of pharmaceuticals and personal care products (PPCPs) on the rainbow trout liver cell line RTL-W1. Aquat. Toxicol. 93, 244–252. Available from: https://doi.org/10.1016/j.aquatox.2009.05.007.

Schultz, M.M., Furlong, E.T., Kolpin, D.W., Werner, S.L., Schoenfuss, H.L., Barber, L.B., et al., 2010. Antidepressant pharmaceuticals in two U.S. effluent-impacted streams: Occurrence and fate in water and sediment, and selective uptake in fish neural tissue. Environ. Sci. Technol 44, 1918–1925.

Schwab, B.W., Hayes, E.P., Fiori, J.M., Mastrocco, F.J., Roden, N.M., Cragin, D., et al., 2005. Human Pharmaceuticals in US surface waters: a human health risk assessment. Regul. Toxicol. Pharmacol. 42, 296–312.

Seiler, R.L., Zaugg, S.D., Thomas, J.M., Howcroft, D.L., 1999. Caffeine and pharmaceuticals as indicators of waste water contamination in wells. USGS 37 (3), 405–410.

Senveli, S.U., Tigli, O., 2013. Biosensors in the small scale: methods and technology trends. IET Nanobiotechnol. 7, 7–21. Available from: https://doi.org/10.1049/iet-nbt.2012.0005.

Servais, P., Prats, J., Passerat, J., Garcia-Armisen, T., 2009. Abundance of culturable versus viable *E. coli* in freshwater. Can. J. Microbiol. 55, 905–909.

Sharma, P., Bhalla, V., Dravid, V., Shekhawat, G., Jinsong-Wu, E., Prasad, S., et al., 2012. Enhancing electrochemical detection on graphene oxide–CNT nanostructured electrodes using magneto-nanobioprobes. Sci. Rep. 2, 877. Available from: https://doi.org/10.1038/srep00877.

Shruthi, G.S., Amitha, C.V., Blessy, B.M., 2014. Biosensors: a modern day achievement. J. Inst. Technol. 2 (1), 26–39.

Singh, G., Manohar, M., Adegoke, A.A., Stenström, T.A., Shanker, R., 2017. Novel aptamer-linked nanoconjugate approach for detection of waterborne bacterial pathogens: an update. J. Nanopart. Res. 19, 4. Available from: https://doi.org/10.1007/s11051-016-3688-3.

Soh, N., Watanabe, T., Asano, Y., Imato, T., 2003. Indirect competitive immunoassay for bisphenol A based on a surface plasmon resonance sensor. Sensor. Mater. 15, 423–438.

Soldano, C., Mahmood, A., Dujardin, E., 2010. Production, properties and potential of grapheme. Carbon 48, 2127–2150.

Solé, M., Baena, M., Arnau, S., Carrasson, M., Maynou, F., Cartes, J.E., 2010. Muscular cholinesterase activities and lipid peroxidation levels as biomarkers in several Mediterranean marine fish species and their relationship with ecological variables. Environ. Inter. 36, 202–211.

Stanley, J.K., Ramirez, A.J., Chambliss, C.K., Brooks, B.W., 2007. Enantiospecific sublethal effects of the antidepressant fluoxetine to a model aquatic vertebrate and invertebrate. Chemosphere 69, 9–16.

Stenström, T.A., Okoh, A.I., Adegoke, A.A., 2016. Antibiogram of environmental isolates of *Acinetobacter calcoaceticus* from Nkonkobe Municipality, South Africa. Fresenius Environ. Bull. 25, 3059–3065.

Strickland, C.H., Rush P., 2011. Occurrence of pharmaceutical and personal care products (PPCPs) in source water of the New York City water supply. National environmental protection (NEP). Available from: <http://www.nyc.gov/html/dep/pdf/quality/nyc_dep_2010_ppcpreport.pdf> (accessed 15.01.17).

Suarez, S., Carballa, M., Omil, F., Lema, J.M., 2008. How are pharmaceutical and personal care products (PPCPs) removed from urban wastewaters? Rev. Environ. Sci. Biotechnol. 7 (2), 125–138.

Sun, J.Z., Peter, K.G., Si, R.W., Zhai, D.D., Liao, Z.H., Sun, D.Z., et al., 2015. Microbial fuel cell-based biosensors for environmental monitoring: a review. Water Sci. Technol. 71, 801–809.

Swartz, C.H., Reddy, S., Benotti, M.J., Yin, H., Barber, L.B., Brownawell, B.J., et al., 2006. Steroid estrogens, nonylphenol ethoxylate metabolites, and other wastewater contaminants in groundwater affected by a residential septic system on Cape Cod, MA. Environ. Sci. Technol. 40 (16), 4894–4902.

Taylor, N.G., Verner-Jeffreys, D.W., Baker-Austin, C., 2011. Aquatic systems: maintaining, mixing and mobilising antimicrobial resistance? Trends Ecol. Evol. 26, 278–284. Available from: https://doi.org/10.1016/j.tree.2011.03.004.

Ternes, T.A., 1998. Occurrence of drugs in German sewage treatment plants and rivers. Water Res. 32 (11), 3245–3260.

Tonnina, D., Campanella, L., Sammartino, M.P., Visco, G., 2002. Integral toxicity test of sea waters by an algal biosensor. Anal. Chim. (Rome) 92, 477–484.

Tschmelak, J., Proll, G., Gauglitz, G., 2004. Sub-nanogram per litre detection of the emerging contaminant progesterone with a fully automated immunosensor based on evanescent field techniques. Anal. Chim. Acta 519 (2), 143–146.

Tschmelak, J., Proll, G., Riedt, J., Kaiser, J., Kraemmer, P., Barzaga, L., et al., 2005. Automated water analyser computer supported system (AWACSS): part II. Intelligent, remote-controlled, cost-effective, on-line, water-monitoring measurement system. Biosens. Bioelectron. 20, 1509–1519.

Tschmelak, J., Kumpf, M., Kappel, N., Proll, G., Gauglitz, G., Talanta, G., 2006. Total internal reflectance fluorescence (TIRF) biosensor for environmental monitoring of testosterone with commercially available immunochemistry: antibody characterization, assay development and real sample measurements. Talanta 69 (2), 343–350. Available from: https://doi.org/10.1016/j.talanta.2005.09.048.

Tse Sum Bui, B., Haupt, K., 2010. Molecularly imprinted polymers: synthetic receptors in bioanalysis. Anal. Bioanal. Chem. 398, 2481–2492.

Upadhyayula, V.K.K., Deng, S., Mitchell, M.C., Smith, G.B., Nair V.K., et al., 2008. Adsorption kinetics of *Escherichia coli* and *Staphylococcus aureus* on single-walled carbon nanotube aggregates. Water Sci. Technol. 58 (1), 179–184. https://doi.org/10.2166/wst.2008.634.

Valentini, F., Galache, F.L., Tamburri, E., Palleschi, G., 2013. Single walled carbon nanotubes/polypyrrole–GO_x composite films to modify gold microelectrodes for glucose biosensors: study of the extended linearity. Biosens. Bioelectron. 43, 75–78. Available from: https://doi.org/10.1016/j.bios.2012.11.019.

Vedrine, C., Leclerc, J.C., Durrieu, C., Tran-Minh, C., 2003. Optical whole-cell biosensor using *Chlorella vulgaris* designed for monitoring herbicides. Biosens. Bioelectron. 18, 457–463.

Walraven, N., Laane, R.W.P.M., 2009. Assessing the discharge of pharmaceuticals along the Dutch coast of the North Sea. Rev. Environ. Contam. Toxicol. 199, 1–18.

Wang, C., Yu, C., 2013. Detection of chemical pollutants in water using gold nanoparticles as sensors: a review. Rev. Anal Chem. 32 (1), 1–14.

Wang, X.H., Liu, T., Xu, N., Zhang, Y., Wang, S., 2007. Enzyme-linked immunosorbent assay and colloidal gold immunoassay for ochratoxin A: investigation of analytical conditions and sample matrix on assay performance. Anal. Bioanal. Chem. 289, 903–911.

Wang, L., Ma, W., Xu, L., Chen, W., Zhu, Y., Xu, C., et al., 2010. Nanoparticle-based environmental sensors. Mater. Sci. Eng. R. R70, 265–274.

Whitcombe, M.J., Vulfson, E.N., 2001. Imprinting polymer. Adv. Mater. 13 (7), 467–478.

WHO. 2012. Animal waste, water quality and human health. In: Dufour, A., Bartram, J., Bos, R., Gannon, V. (Eds.), IWA Publishing, London, UK, ISBN: 9781780401232.

Williams, R.T. (Ed.), 2005. Human Pharmaceuticals: Assessing the Impacts on Aquatic Ecosystems. SETAC Press, Pensacola, FL, pp. 149–181.

Wu, J., Yue, J., Hu, R., Yang, Z., Zhang, L., 2008. Use of caffeine and human pharmaceutical compounds to identify sewage contamination. WASET 44, 438–442.

Wu, W., Li, J., Pan, D., Li, J., Song, S., Rong, M., et al., 2014. Gold nanoparticle-based enzyme-linked antibodyaptamer sandwich assay for detection of *Salmonella typhimurium*. ACS Appl. Mater. Interfaces 6 (19), 16974–16981. Available from: https://doi.org/10.1021/am5045828.

Xu, Z., Yin, H., Tian, Z., Zhou, Y., Ai, S., 2014. Electrochemical immunoassays for the detection of the activity of DNA methyltransferase by using the rolling circle amplification technique. Microchim. Acta 181, 471–477.

Xu, L., Wang, R., Callaway, Z., Li, Y., 2015. A fluorescent aptasensor coupled with nanobeads-based immunomagnetic separator for simultaneous detection of four foodborne pathogenic bacteria. ASABE 58, 891–906. Available from: https://doi.org/10.13031/trans.58.11089.

Yamada, K., Kim, C.-T., Kim, J.-H., Chung, J.-H., Lee, H.G., Jun, S., 2014. Single walled carbon nanotube-based junction biosensor for detection of *Escherichia coli*. PLoS ONE 9 (9), e105767. Available from: https://doi.org/10.1371/journal.pone.0105767.

Yang, K., Zhu, L., Xing, B., 2006. Adsorption of polycyclic aromatic hydrocarbons by carbon nanomaterials. Environ. Sci. Technol. 40 (6), 1855.

Yang, J.F., Ying, G.G., Zhao, J.L., Tao, R., Su, H.C., Liu, Y.S., 2011. Spatial and seasonal distribution of selected antibiotics in surface waters of the Pearl Bivers, China. J. Environ. Sci. Health B 46 (3), 272–280. Available from: https://doi.org/10.12691/jit-2-1-5.

Yang, W., Thordarson, P., Gooding, J.J., Ringer, S.P., Braet, F., 2007. Carbon nanotubes for biological and biomedical applications. Nanotechnology 18, 1–12.

Yao, C., Zhu, T., Qi, Y., Zhao, Y., Xia, H., Fu, W., 2010. Development of a quartz crystal microbalance biosensor with aptamers as bio-recognition element. Sensors 10, 5859–5871.

Ye, Z.Q., Weinberg, H.S., Meyer, M.T., 2007. Trace analysis of trimethoprim and sulfonamide, macrolide, quinolone, and tetracycline antibiotics in chlorinated drinking water using liquid chromatography electrospray tandem mass spectrometry. Anal. Chem. 79 (3), 1135–1144.

Yuan, J., Wu, S., Duan, N., Ma, X., Yu, X., Chen, J., et al., 2014. A sensitive gold nanoparticle-based colorimetric aptasensor for *Staphylococcus aureus*. Talanta 127, 163–168. Available from: https://doi.org/10.1016/j.talanta.2014.04.013.

Zhang, S.X., Wang, N., Yu, H.J., Niu, Y.M., Sun, C.Q., 2005. Bioelectrochem. 67 (1), 15–22.

Zhang, J., Lei, J., Pan, R., Xue, Y., Ju, H., 2010. Highly sensitive electrocatalytic biosensing of hypoxanthine based on functionalization of graphene sheets with water-soluble conducting graft copolymer. Biosens. Bioelectron. 26, 371–376. Available from: https://doi.org/10.1016/j.bios.2010.07.127.

Zhang, S., Sun, W., Xu, L.I., Zheng, X., Chu, X., Tian, J., et al., 2012. Identification of the para-nitrophenol catabolic pathway, and characterization of three enzymes involved in the hydroquinone pathway, in *Pseudomonas* sp. 1–7. BMC Microbiol. 12, 27.

Zheng, D., Vashist, S.K., Al-Rubeaan, K., Luong, J.H.T., Sheu, F.S., 2012. Mediatorless amperometric glucose biosensing using 3-aminopropyltriethoxysilanefunctionalized graphene. Talanta 99, 22–28.

Zhu, Y., Murali, S., Cai, W., Li, X., Suk, J.W., Potts, J.R., et al., 2010. Graphene and graphene oxide: Synthesis, properties, and applications. Adv. Mater. 22 (35), 3906–3924.

Zougagh, M., Ríos, A., Valcárcel, M., 2005. Automatic selective determination of caffeine in coffee and tea samples by using a supported liquid membrane-modified piezoelectric flow sensor with molecularly imprinted polymer. Anal. Chim. Acta 539, 117.

FURTHER READING

Adegoke, A.A., Faleye, A.C., Singh, G., Stenström, T.A., 2017. Antibiotic resistant superbugs: assessment of the interrelationship of occurrence in clinical settings and environmental niches. Molecules 22, 29.

Backhaus, T., 2014. Medicines, shaken and stirred: a critical review on the ecotoxicology of pharmaceutical mixtures. Phil. Trans. R Soc. B 369, 2013058. Available from: https://doi.org/10.1098/rstb.2013.0585.

Bo, S., Ciccone, G., Durazzo, M., Gambino, R., Massarenti, P., Baldi, I., et al., 2007. Efficacy of antioxidant treatment in reducing resistin serum levels: a randomized study. PLoS Clin. Trials 2007 (2), 1629–1637.

Carter, R.M., Lubrano, G.J., Guilbault, G.G., 1994. Immunobio-sensors for toxins and bacteria. Life Chem. Rep. 11, 271–277.

Charelier, R.C., Gengenbach, T.R., Griesser, H.J., Brigham-Burke, M., O'Shannessy, D.J., 1995. A general method to recondition and reuse BIAcore sensor chips fouled with covalently immobilized protein/peptide. Anal. Biochem. 229, 112–118.

Choffnes, E.R., Relman, D.A., Olsen, L.A., Hutton, R., Rapporteurs, A.M., 2011. Improving food safety through a one health approach: workshop summary. Forum on Microbial Threats; Board on Global Health; Institute of Medicine. Available from: <http://www.Onehealthinitiative.com/publications/IOM%20One%20Health%20Workshop%20Summaryt.pdf> (accesses 15.10.16).

Delehanty, J.B., Ligler, F.S., 2002. A microarray immunoassay for simultaneous detection of proteins and bacteria. Anal. Chem. 74, 5681–5687.

Hansen, L.H., Sorensen, S.J., 2000. Detection and quantification of tetracyclines by whole cell biosensors. FEMS Microbiol. Lett. 190, 273–278.

Liu, B.H., Tsao, Z.J., Wang, J.J., Yu, F.Y., 2008. Development of a monoclonal antibody against ochratoxin A and its application in enzyme-linked immunosorbent assay and gold nanoparticle immunochromatographic strip. Anal. Chem. 80, 7029–7035.

McDaniel, L.D., Young, E., Delaney, J., Ruhnau, F., Ritchie, K.B., Paul, J.H., 2010. High frequency of horizontal gene transfer in the oceans. Science 330, 50. Available from: https://doi.org/10.1126/science.1192243.

Norman, R.S., Stone, J.W., Gole, A., Murphy, C.J., Sabo-Attwood, T.L., 2008. Targeted photothermal lysis of the pathogenic bacteria, *Pseudomonas aeruginosa*, with gold nanorods. Nano Lett. 8, 302–306.

Palleschi, G., Compagnone, D., Moscone, D., 1997. Electrochemical biosensors: potential and application in the food industry. In: Biotechnology in the Food Chain VTT Symposium, 177, pp. 141–160 (Copyright 2003 ACS).

Pinckney, J.L., Hagenbuch, I.M., Long, R.A., Lovell, C.R., 2013. Sublethal effects of the antibiotic tylosin on estuarine benthic microalgal communities. Mar. Pollut. Bull. 68, 8–12. Available from: https://doi.org/10.1016/j.marpolbul.2013.01.006.

Setford, S.J., Van, Es, R.M., Blankwater, Y.J., Kroger, S., 1999. Receptor binding protein amperometric affinity sensor for rapid b-lactam quantification in milk. Anal. Chim. Acta 398, 13–22.

Sheng, J.-W., He, M., Shi, H.-C., Qian, Y., 2006. A comprehensive immunoassay for the detection of microcystins in waters based on polyclonal antibodies. Anal. Chim. Acta 572, 309–315.

Shim, W.B.O., Kim, K.Y., Chung, D.H., 2009. Development and validation of a gold nanoparticle immunochromatographic assay (ICG) for the detection of zearalenone. J. Agric. Food Chem. 57, 4035–4041.

Suedee, R., Intakong, W., Dickert, F.L., 2006. Molecularly imprinted polymer-modified electrode for on-line conductometric monitoring of haloacetic acids in chlorinated water. Anal. Chim. Acta 569 (1-2), 66–75.

Sui, Q., Cao, X., Lu, S., Zhao, W., Qiu, Z., Yu, G., 2015. Occurrence, sources and fate of pharmaceuticals and personal care products in the groundwater: a review. Emerg. Contam. 1 (1), 14–24.

Tran Minh, C., Pandey, P.C., 1992. Biosensors and toxins detection. Bull. Electrochem. 8, 199.

Uzawa, H., Ohga, K., Shinozaki, Y., Ohsawa, I., Nagatsuka, T., Seto, Y., et al., 2008. A novel sugar-probe biosensor for the deadly plant proteinous toxin, ricin. Biosens. Bioelectron. 24, 923–927.

CHAPTER

17

Metal oxide nanoparticles in removing residual pharmaceutical products and pathogens from water and wastewater

A.A. Adegoke[1,2] **and T.A. Stenström**[1]

[1]*SARChI, Institute for Water and Wastewater Technology, Durban University of Technology, Durban, South Africa* [2]*Department of Microbiology, University of Uyo, Uyo, Nigeria*

CHAPTER OUTLINE

Nanoparticles in Pharmacotherapy. DOI: https://doi.org/10.1016/B978-0-12-816504-1.00016-8

17.1 INTRODUCTION

Wastewater, surface water, and drinking water have been reported to contain both chemical and microbial contaminants known for adverse health effects on human beings and the environment (Giger et al., 2003; Gobel et al., 2004; Forslund et al., 2010; Farahat and Linderholm, 2013). Pharmaceuticals like antibiotics are released to the environment through the excreta via the sewer systems in a partially digested active form or as derivatives and are finally released into both surface water and ground water in varying amounts, which are often also significant in minute quantities. Manures, wastewater sludge, and excreta from humans harbor emerging contaminants that will reach the environment (Kim and Carson, 2005; Kim et al., 2009; Burkhardt et al., 2005; Kay et al., 2005; Zhang et al., 2008). Zhang et al. (2008) reported on the impact of residual antibiotics on the soil and water environment from farm animals, where the concentration depends on the magnitude of the farmyard and the herds raised.

Fig. 17.1 summarizes the pathways of pharmaceutical contaminants into the aquatic environment and depicts the sources as well as the routes of entry for the contaminants into the water cycle. The relative impact from different sources, their routes of entry, as well as the efficiency and applicability of different

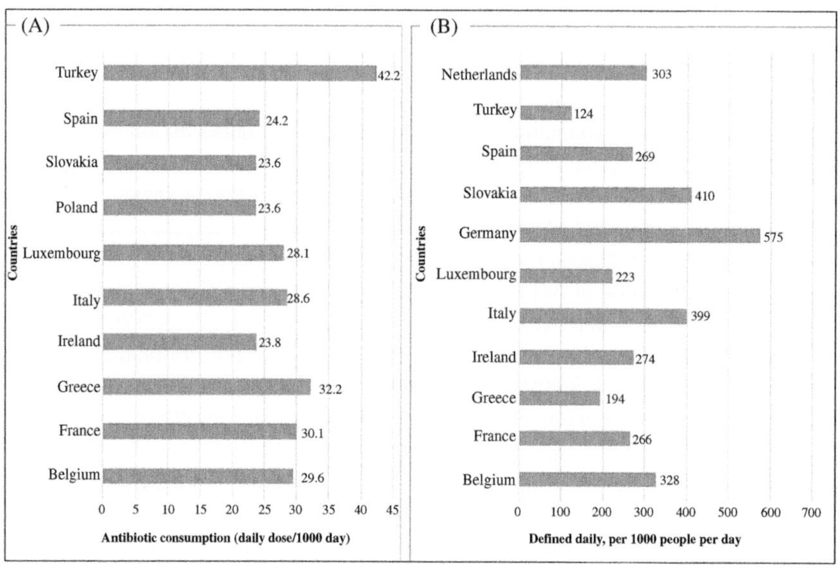

FIGURE 17.1

(A) Antibiotic consumption (daily dose/1000 day^{-1}); (B) antihypertensive drugs consumption 2013 (or nearest years) in selected countries.

Adapted from OECD Health Statistics 2015, https://doi.org/10.1787/health-data-en. https://doi.org/10.1787/888933281342.

treatment techniques are interlinked and important for the assessment and relevance in assuring water safety.

The main sources of the pharmaceutical contaminants like antibiotics, analgesics and antiinflammatory drugs, beta-blockers, lipid regulators, phenol, and neuroactive compounds are the hospitals where they are used in large quantities for a broad range of therapies, prophylaxis, or care management (Verlicchi et al., 2010). Globally, large quantities of pharmaceuticals are administered and their consumption has escalated in recent years. In South Africa, Health24.com reported that South Africa's antibiotic consumption has increased by 60% over the last 10 years. Fig. 17.2 illustrates the high consumption of both antibiotics (A) and antihypertensive drugs in some selected (mainly industrialized) countries. This further reflects the contributory effect of the release of the pharmaceutical contaminants into the water environment via the sewer system. Neither hospitals' onsite sanitation systems nor most municipal (centralized) WWTPs are designed to or have treatment steps that will remove the pharmaceutical contaminants to a larger extent (Verlicchi et al., 2015). Only 10% of most emerging contaminants are removed by basic sewer system without modifications according to Chang et al. (2010). The concentration of these emerging contaminants however is reduced with various conventional treatment and with time. This differs between different compounds and the cumulative effects on cocktails of these on the environment and in the development of multiresistant bacteria is poorly studied. The detected residual antibiotic concentration in surface water in Kwazulu-Natal, South Africa, for example, as reported by Matongo et al. (2015) includes

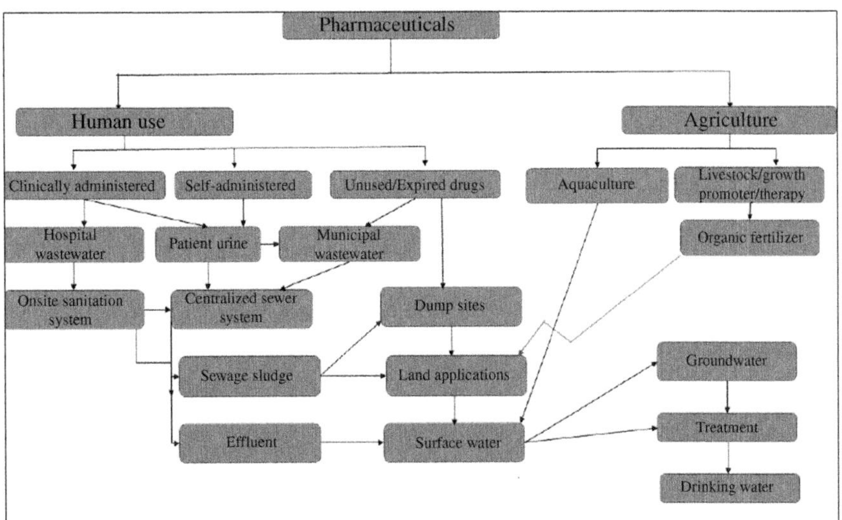

FIGURE 17.2

Pathway of pharmaceutical contaminations in the aquatic environment.

13.6 $\mu g\ L^{-1}$ for erythromycin; 6.0–13.6 $\mu g\ L^{-1}$ for sulfamethoxazole; 0.8–13.6 $\mu g\ L^{-1}$ for trimethoprim, and 4.6–13.6 $\mu g\ L^{-1}$ for sulfamethazine. This has the potential to induce resistance to antibiotics and encourages succession of the aquatic microbiomes by antimicrobial resistant microorganisms (Cantón and Ruiz-Garbajosa, 2011). Some pharmaceutical and personal care contaminants found in surface water are exemplified in Table 17.1. Most of these contaminants are reported in nanoquantities, showing the need for appropriate technology to remove them to avoid their potential ecological effects (Table 17.2).

Due to environmental side effects associated with the release of pharmaceutical contaminants into the aquatic environment (Table 17.2), there is a need for effective methods that can reduce their concentration or lead to a complete removal. Their impact affects the aquatic organisms and extends to humans through the food chain.

Table 17.1 Some Pharmaceutical Contaminants and Personal Care Products: Detection and Their Concentrations in Surface Water Illustrated With Examples From China, United States, and United Kingdom

Contaminants	Reported (Max) Concentration ($ng\ L^{-1}$)	Country	References
Atorvastatin	7.3	United States	Vanderford and Snyder (2006)
Caffeine	7110	United States	Young et al. (2008)
Dextropropoxyphene	682	United Kingdom	Roberts and Thomas (2006)
Dehydronifedipine	30	United States	Kolpin et al. (2002)
Mefenamic acid	366	United Kingdom	Roberts and Thomas (2006)
Nonylphenol	372	China	Duong et al. (2010)
Diazepam	2.6	United States	Vanderford and Snyder (2006)
Dibutyl phthalate[a]	40,680	China	Chi (2009)
Tamoxifen	42	United States	Ashton et al. (2004)
Synthetic musks	34.6	China	Hu et al. (2011a, b)
Ibuprofen	2796	United States	Wu et al. (2009)
	2370	United Kingdom	Roberts and Thomas (2006)
Indomethacin	26	United States	Wu et al. (2009)

[a] *Dibutyl phthalate has pharmaceutical grades.*

Table 17.2 Effects of Some Pharmaceutical Nanocontaminants on Aquatic Organisms to Justify Needs for Nanotechnology for Total Removal

Class	Compound(s)	Reported Effects	Affecting Aquatic Plant (P), Irrigated Plant (IP), or Animal (A)	References
Antibiotics	Enrofloxacin	Reducing growth	IP (*Daucus carota* Carrot), *Lactuca sativa* (lettuce)	Boxall et al. (2006), Jin et al. (2009)
	Ofloxacin	Neurotoxicity (encephalopathy, seizures, orofacial dyskinesias)	A (Fish)	Chen et al. (2015), Grill and Maganti (2011)
	Tetracycline	Cranial nerve toxicity and neuromuscular blockage	A (Fish)	Berrada et al. (2008), Grill and Magantz (2011)
	Sulfamethoxazole, Lomefloxacin, chlortetracycline	Affected wet mass, frond number, chlorophylls a and b, and carotenoid content	P (*Lemna gibba*)	Brain et al. (2004)
Antiinflammatory	Ibuprofen	Inhibition of growth of aquatic plants	P (several aquatic plants)	Pomati et al. (2004)
Analgesic	Carbamazepine	Inhibition of basal EROD activity in fish liver	A (Fish)	Laville et al. (2004), Nentwig et al. (2004)
		Interfere and inhibit metamorphosis	A (*Chironomus riparius*)	
Beta blocker	Propanolol	Weak EROD inducer in cultures of rainbow trout hepatocytes	A (Fish)	Laville et al. (2004)
Cardiac glycoside	Digoxin	Inhibit regeneration in some aquatic organisms	A (Hydra)	Pascoe et al. (2003)
Hormones	Estrogen	Increased feminization of exposed fish and other aquatic organisms	A (Fish, toad/frog)	Orn et al. (2003)

17.1.1 EMERGENCE OF ANTIBIOTIC RESISTANCE AS AN EFFECT OF RESIDUAL ANTIBIOTICS AS CONTAMINANTS IN AQUATIC ENVIRONMENT

Exposure to antibiotics has been identified as a major presumed factor predicating the emergence and spread of antibiotic resistance (Chang et al., 2010). There have been clear links between the frequency of antibiotic resistance and the use of antimicrobials. This occurs both in hospitals, generally in society and in animal husbandry, for example, in the cattle industry (Shryock and Richwine, 2010).

Li et al. (2010) observed that bacterial isolates in downstream parts of rivers affected by wastewater outlets exhibit higher multiple antibiotic resistance index (MARI) and harbor more ARGs than those upstream of the outlets. This can be explained with the probable duration of exposure since the bacteria probably remain in the water containing the RAbs throughout the treatment stage in WWTPs till the discharge point. This explanation may relate also to the exposure time required for the selection of ARGs (Garcillan-Barcia et al., 2011; Partridge, 2011). Exposure to antibiotic contaminants in the environment is drawing an equality in the trend of resistance between the clinical isolates and the environmental isolates (Faria et al., 2009; Xi et al., 2009; Figueira et al., 2012; Stenström et al., 2016). It is therefore pertinent to track and reduce the levels and concentration of the contaminants by introducing reliable technologies to remove them from aquatic biomes.

As emphasized earlier, most municipal WWTPs are not designed to remove persistent pharmaceuticals, but the more advanced and efficient treatment steps in developed countries as compared with developing ones will to a higher extent reduce the concentration of pharmaceutical contaminants received in the influents and thereby also reduce the impact on the receiving environment from the treated effluent. The long-term and synergistic environmental impact is yet to be documented in more detail. The compounds in Table 17.2 exemplify various effects. Additionally painkillers like diclofenac as well as hormones may have a fatal impact on fish, crustaceans, and algae even at very minute concentration emanating from different sources as exemplified in Fig. 17.1 (Klavarioti et al., 2009; Tambosi et al., 2010; Balcıoğlu and Otker, 2003; Fatta-Kassinos et al., 2011).

17.2 CONVENTIONAL TREATMENT FOR REMOVING PHARMACEUTICAL CONTAMINANTS

Conventional wastewater treatment methods were not designed to remove pharmaceutical contaminants but essentially configured to remove pathogens and nutrients that encourage the regrowth of pathogens after the discharge into the receiving water bodies (Dalrymple et al., 2007; Nikolaou et al., 2007).

Table 17.3 Exemplified Advanced Treatment Technologies to Improve Removal of Pharmaceutical Contaminants From Wastewater

Applications	Technology	References
Chemical oxidation	Coagulation, ozone and ozone/ hydrogen peroxide	Ternes et al. (2003), Zwiener and Frimmel (2003)
Biological treatment	Activated sludge process	Hartig et al. (2001), Nghiem et al. (2005)
Membrane bioreactor	Sequencing batch biological reactor (SBR) process	Hartig et al. (2001), Nghiem et al. (2005)
Reverse osmosis	Membrane filtration	Hartig et al. (2001), Nghiem et al. (2005)
Adsorption	Activated carbon	Hartig et al. (2001), Ternes et al. (2003)

It is imperative to quickly relate to the developments and modifications done in various WWTPs to substantiate improvement in the removal efficiency before addressing the further improvements, and preferably total removal, that nanotechnology may achieve.

As the awareness of the detrimental effects (e.g., Table 17.2) of pharmaceutical contaminants increases, the necessities to incorporate additional advanced treatment technologies has been recognized as exemplified in Table 17.3.

With improvement in technologies higher removal efficiencies are seen (Suhua et al., 2010), like an efficiency of 25% in sulfonamides removal in improved conventional treatment plants (Watkinson et al., 2007). Meanwhile, nonionic iodinated X-ray contrast media (ICM) that could remove higher than 80% compared with chemical oxidation through ionic (ozone) pathway could only yield improvement of $\leq 14\%$ (Ternes et al., 2003). Yet, this leaves some milestones of the pharmaceutical contaminants to be removed. Advanced oxidation/reduction processes (AO/RPs) have potential to eliminate tetracycline for instance from water, and this is advantageous in that it leaves antimicrobial radicals ($OH^{\bullet}$) behind that are usually effective against the contaminant antibiotic resistant waterborne organisms.

Limitations of the treatment methods in removing antibiotics and other pharmaceutical products from wastewater/water include the following:

1. High energy and material intensive processes involving ozone/hydrogen peroxide (Larsen et al., 2004). They are applicable only to clean water without substantial organic matter.
2. Some of these old technologies (activated sludge system) encourage an increase in solid retention time, which enhances the emergence of resistance to antibiotics as the bacteria in the wastewater/water get exposed to residual antibiotics for longer periods of time.

3. The physical treatment processes require the disposal of wastes such as membrane retentate and spend activated carbon generated during the treatment.
4. Membrane bioreactor for integration of biodegradation and membrane filtration can bring about low sludge settling rate, membrane fouling and higher cost for the same volume treated.
5. Dosed powdered activated carbon to activated sludge treatment can lead to large volumes of adsorbent waste and regeneration costs.
6. Activated carbon adsorption is less capable to remove polar organic compounds because of hydrophobic interactions (Snyder et al., 2003).

None of the technologies provide for total removal for pharmaceutical contaminants. To remove pharmaceutical contaminants with very low molecular weight, the most reliable method is nanotechnology. Nanotechnology is promising, with a large potential to utilize high reactivity due to large surface to volume ratio (Tiwari et al., 2008; Amin et al., 2014) to completely remove pharmaceuticals. Improvements in the field of nanotechnology have led to the development of adsorbents that broaden the range of utilizable and attractable materials, while problems associated with toxicity can easily be resolved by careful selection and design of appropriate nanomaterials with low toxicity as well as secondary clearance of nanomaterials for water purification. The metal oxide nanoparticles (MONs) may come as the augmentation of the existing advanced treatment systems. The following parts of this chapter will focus on the basic application of MONs in removing pharmaceutical contaminants from the wastewater/water for reuse.

17.3 PHARMACEUTICAL NANOTECHNOLOGY

Wastewater and water treatment applications to further reduce or get rid of pharmaceutical contaminants can include various methods like photocatalysis, nanofiltration (NF), adsorption, and electrochemical oxidation. These may involve the use of titanium dioxide and zinc oxide (TiO_2 and ZnO), ceramic membranes, nanowire membranes, polymer membranes, carbon nanotubes, submicron nanopowder, metal (oxides), magnetic nanoparticles, or nanostructured boron doped diamond (Schierz and Zanker, 2009; Chen et al., 2009). Oxidation through the MONs can also be utilized for eradicating the antibiotic resistant bacteria. This section will specifically address removal of pharmaceutical contaminants, personal care products, and pathogens.

MONs can be utilized to effect the removal of pharmaceutical contaminants in very minute quantities below the removal capacities of most advanced treatment processes (Amin et al., 2014). Several designs can be made to effect total removal, which may be a modification of Fig. 17.3. The treatment processes involved are with consideration of the phase of the contaminants either organic or

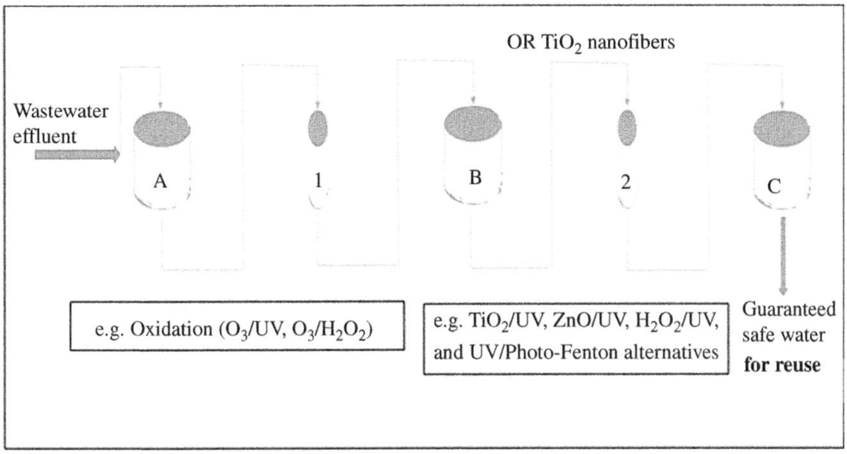

FIGURE 17.3

Schematics of possible postconventional treatment of wastewater (filtration series) using Advanced Stage A with its filter 1 and nanofiltration using metal oxide treated-adsorption stage B with its filter possibly at 2. Additional stage C may be for final assessment.

inorganic. It is also dependent on charge repulsion, size exclusion, hydrophobicity, hydrogen bonding capacity, and dipole moment (Steinle-Darling and Reinhard, 2008). This becomes more important because certain contaminants (e.g., sulphamethoxazole) may bind to organic matter, which determines their fate. The organic matters may compete with the contaminants for the adsorptive sites. This in turn reduces the removal efficiency as it affects the retention of the micro or nanopollutants.

17.4 MEMBRANE FILTRATION (NANOFILTRATION) SUPPLEMENTED WITH MONS FOR REMOVING PHARMACEUTICAL CONTAMINANTS

It is generally believed that the retention of micropollutants in membrane-based systems is affected by the availability of NOM. Processes that can incorporate NF with adsorption processes may theoretically achieve close to 100% removal efficiency. Adsorption of the nanocontaminants can be facilitated with treated MONs on the filters to be used (Anjum et al., 2016). Potential nanomembranes that can be used for effective removal of pharmaceutical nanocontaminants along with MONs (e.g., TiO_2) include carbon nanotube membranes, electrospun nanofibers, and hybrid nanomembranes, as these have been used for a wide range of chemical pollutants and some non-pharmaceutical grade nanocontaminants like dyes, heavy metals, foulants, etc. with efficiency of about 99% (Anjum et al., 2016; Felizardo et al., 2016).

The metal oxide usually serves as separator when it is conjugated with the aptamer with affinity to the antibiotic(s) to be separated. An example of these includes the use of iron oxide nanoparticles in magnetic separation hand-in-hand with gold shell along with specific aptamer to remove selected antibiotics (Nikolaus and Strehlitz, 2014). Graphite oxides, for example, are also employed as quencher depending on the aptamer of targets to detect antibiotics or particular pharmaceutical contaminant (Zhang et al., 2015). It is possible to detect multiple chemical contaminants together in one analysis using this method or in its modified form, especially in wastewater environment where there are possibilities of varying categories of contaminants at the same time. Zhang et al. (2015) reported the prior development of a microfluidic device made from cheap paper with the advantage of detecting multiple chemical contaminants in food samples for removal. This can be applied to wastewater and water. The applications can be for biosensing as well as removal of the contaminants in question (Xu et al., 2015).

A conventional WWTP treating effluent for reuse may be connected with a typical model of nanofiltration (e.g., Fig. 17.3 or with modification) that allows at least two layers of advanced treatment in which one of them incorporates a magnetic iron oxide nanoparticles (MIONs) filter and an appropriate aptamer for prompt adherence. The other method can be any of those listed in Table 17.3. Fig. 17.3 is a typical setup reflecting at least one stage (A with filter 1 or with modifications) with an advanced treatment for wastewater effluents before the stage involving electrospun nanofibers. Hybrid nanomembranes are recommended, for instance for safe reuse, especially when there is stage C where evaluation can be made before release for irrigation, recreation, pets, house sanitation water, etc. Hu et al. (2011a, b) achieved a high removal capacity (88%–95%) of cocaine and diclofenac, using a single-stranded DNA/RNA aptamer column. A combined and enhanced effect would be achievable for pharmaceutical contaminants, even if they are available in trace quantities, together with the removal of a wide range of heavy metals in the effluent, to produce a more reliable output for reuse or for the environment. These type of applications have however not reached the full-scale applications.

Nanofilter membranes of varying molecular weight have proven to be efficient to remove various pharmaceuticals like antibiotics, hormones, contraceptives, and analgesics, in water (Radjenović et al., 2008; Koyuncu et al., 2008). High and timely adsorption of tetracyclines (close to 70% of chlorotetracycline) was reported in ½ hour. The rejection of tetracycline, which was reportedly (Koyuncu et al., 2008) due to high molecular weight and the rejection of sulphonamides, was occasioned by the presence of NOM. This justified the need for a setup that posited a previous advanced removal stage before the nanofiltration stage as illustrated in Fig. 17.3. An advanced stage should preferably be designed to both remove the high molecular weight pharmaceutical contaminants as well as the existing NOM as advocated for by Liu et al. (2014). They suggested a combination of oxidation along with nanofiltration to effectively remove antibiotics like norfloxacin, ofloxacin, roxithromycin, and azithromycin. The ozonation process should precede the concentrated nanofiltration, whereby it serves as a complementary step for effective reduction or total removal of the pharmaceuticals from wastewater effluent for reuse.

17.5 METAL OXIDE NANOPARTICLES AS MORE DEPENDENT TREATMENT OPTION AGAINST ANTIBIOTICS AND ANTIBIOTIC RESISTANT WATERBORNE BACTERIA

MONs are an emerging treatment option for removal of pharmaceuticals as well as antibiotic resistant bacteria in water. The application is with membranes, for selective permeability and resistance against fouling (Li et al., 2009a, 2009b). Both polymeric and inorganic membranes are manufactured by either assembling nanoparticles into porous membranes or in a blending process (Kim et al., 2003; Li et al., 2009a, 2009b). An example is TiO_2 with improved selective permeability, antibacterial activity, as well as with additional benefits (Li et al., 2009a, 2009b).

17.5.1 TIO_2 NANOFIBERS: PHOTOCATALYTIC OXIDATION OF TETRACYCLINE AND INACTIVATION OF ANTIBIOTIC RESISTANT PATHOGENS IN WASTEWATER AND SURFACE WATER

TiO_2 is a metal oxide reported for its ability to effectively bring about oxidation of antibiotics in water and with an antimicrobial potential (Allahverdiyev et al., 2011). TiO_2 possesses high photocatalytic activity and is a useful photocatalysts for water treatment. TiO_2 nanoparticles are proven photocatalysts for removal of wide range of organic pollutants from pharmaceuticals and personal care products (Kabra et al., 2004). Chitose et al. (2003) reported that TiO_2 facilitated the effective removal of total organic carbon (TOC) from wastewater. TiO_2 has additional competitive advantages over other MONs due to safety (nontoxic), high energy efficiency, cost effectiveness, appreciable chemical and biological stability, and minimum waste production (Wang et al., 2011). TiO_2 can kill various types of bacteria and parasites, as well as inactivate viral particles (Wei et al., 1994; Brady-Estévez et al., 2008; Zan et al., 2007; Allahverdiyev et al., 2013). Antibiotics and other pharmaceutical products in water are degraded, though natural photodegradation of UV from the sunlight reduces the concentration of wide range of antibiotics in surface water (Ding et al., 2013; Zhang et al., 2016).

Photocatalytic oxidation of tetracycline using TiO_2 has been reported in several studies over a number of years (Addamo et al., 2005; Reyes et al., 2006; Palominos et al., 2009; Wang et al., 2011). The earlier studies (Wang et al., 2011; Reyes et al., 2006) showed the importance of catalysts and pH in their experiments where pH optimization and time is necessary [a range of concentration of the TET in the water has been assessed using response surface methodology (RSM)]. Degradation of tetracyclines (TET) was investigated by Reyes et al. (2006) and Bautitz and Nogueira (2007) using UV/TiO_2 processes. Reports by Addamo et al. (2005) and Reyes et al. (2006) showed that the removal kinetics for TET using UV/TiO_2 followed first order kinetics of photolysis. The degradation processes in this case were monitored by determining tetracycline disappearance. It should also be noted that photocatalytic disinfection utilizing TiO_2 can effectively inactivate a wide range of waterborne pathogens including viruses and viral indicators such as phage MS2 (Watts et al., 1995; Ryu et al., 2008), *Bacteroides fragilis* bacteriophage (Sjogren and Sierka, 1994), Poliovirus I (Watts et al., 1995), Helminthic eggs (Orta et al., 2004), *Cryptosporidium parvum* (Ryu et al., 2008), and *Giardia intestinalis* (Sokmen et al., 2008). Maness et al. (1999) also reported lethal effects (killing index of 0.98) of 25 nm sized TiO_2-NPs on *Escherichia coli* at concentration of 100 ppm.

The utilization of TiO_2-NPs for antibacterial activity research in medicine and pharmacy is favored due to their very low toxicity in comparison with Ag or Cu-NPs. This antibiotic activity will further be discussed with respect to other MONs in Section 16.5.2.

As it can be seen in Table 17.4 for some other MONs besides TiO_2, decomposition of contaminants is facilitated by doping into TiO_2 due to enhanced OH^- production and other factors (Murakami et al., 2009; Liu et al., 2011). The doping of Si into TiO_2 nanoparticles improves the decomposition of some organic contaminants of pharmaceutical relevance. This is exemplified by the advantage of factors like increase in surface area and crystallinity brought about by doping process (Iwamoto et al., 2005). Aware and Jadhav (2016) reported that there was an increase in the absorption edge wavelength and a decrease in the band gap energy when TiO_2 was doped with other metals. The researchers observed that the photocatalytic activity of the doped nanoparticles showed higher activity than undoped titania (Aware and Jadhav, 2016). These researchers envisaged that this higher photocatalytic activity by the doped TiO_2 was occasioned by several factors in favor of the doped TiO_2 including, but may not be limited to the small grain size, high crystallinity, high specific surface area, and decrease in the band gap energy.

Degradation of recalcitrant compounds in water, originating from the personal care products, is easier with doped TiO_2 than ordinary (Wu et al., 2006). Methylene blue in the visible light range was degraded by TiO_2 nanocrystals doped with noble metal (Wu et al., 2006). The degradation or photocatalytic activity can be improved to its optimum performance by doping with transition metal ions like Fe^{2+}, Mn^{2+}, Cu^{2+}, Zn^{2+} (Kubacka et al., 2009). Fe (III)-doped TiO_2 nanoparticles gave better removal efficiency of phenol than commercially available TiO_2 nanoparticles (Nahar et al., 2006). Zn^{2+} has best doping with TiO_2 due to very close ionic radii of 0.74 Å and 0.75 Å for Zn^{2+} and Ti^{4+}, respectively (Rauf et al., 2011). This makes it possible to interchange the two metals and especially replacing Ti^{4+} with Zn^{2+} without affecting the crystal structure. As will be

Table 17.4 Some Selected Single and Mixed/Doped MONs and Their Antibacterial Activity on Waterborne Bacteria

Metal Oxide Nanoparticles		Synthesis/Doping Method(s)	Applicable Water-Borne Microbial Strain Tested	References
Single	ZnO	Biological method, hydrothermal synthesis, sol–gel technique, simple thermal sublimation, self-combustion, polymerized complex method, vapor–liquid–solid technique, double-jet precipitation, and solution synthesis	*Escherichia coli*, *Pseudomonas aeruginosa*, pathogenic aquatic fungi	Raliya and Tarafdar (2014), Xu et al. (2008), Yahya et al. (2010), Jeng and Swanson (2006)
	MgO	Biological method, hydroxide precipitation–thermal decomposition	*E. coli* and *Bacillus* species	Camtakan et al. (2012), Raliya and Tarafdar (2014), Guo et al. (2015)
	TiO_2	Biological method, hydrolysis–peptization, via sol–gel, hydrothermal method	*E. coli*, *P. aeruginosa*, *Enterococcus hirae*, *Bacteroides fragilis*, viruses such as phage MS2, *B. fragilis bacteriophage*, *Poliovirus I*, *Cryptosporidium parvum* and *Giardia intestinalis.*	Sjogren and Sierka (1994), Watts et al. (1995), Mahshid et al. (2007), Ryu et al. (2008), Sokmen et al. (2008), Allahverdiyev et al. (2011), Vijayalakshmi and Rajendran (2012)
	Ag_2O	Electrochemical methods	*Bacillus* species, *E. coli*, *Proteus vulgaris*, *P. putida*, *Salmonella typhi*, *Entamoeba histolytica*, and *C. parvum*	Allahverdiyev et al. (2011)
Mixed/ doped	Zn/Fe oxide	Sol–gel	*E. coli; P. aeruginosa; B. species.*	Gordon et al. (2011), Xu et al. (2008), Yahya et al. (2010), Jeng and Swanson (2006)
	Zn/Mg oxide	Combustion	*E. coli*	Guo et al. (2015), Vidic et al. (2013)
	ZnO/Au	Photoreduction	*E. coli*, *P. aeruginosa*	Xu et al. (2008), Yahya et al. (2010), Jeng and Swanson (2006)
	Mn- and Fe-doped ZnO	Coprecipitation	*E. coli*, *S. typhi*, *P. aeruginosa*	Sharma et al. (2016)
	Fe_3O_4/Ag	Catalytic template based	*Staphylococcus aureus* (fecal-based)	Xiong et al. (2013)
	Ag–TiO_2-doped SiO_2	Sol–gel	*E. coli*	Mukhopadhyay et al. (2010)

seen for several other doped MONs in Table 17.4, Yang et al. (2010) and Zhu et al. (2012) reported the use of the sol-hydrothermal technique to synthesize Zn-doped TiO_2 powder while Mahata et al. (2011) was able to utilize controlled hydrolysis of titanium(IV) butoxide with varying improvement in activity. Several other techniques were utilized including electrospinning (Khajavi and Abbasipour, 2012), assembly process based on ligand exchange reaction (Xu et al., 2005), microemulsion (Sanchez-Dominguez et al., 2015), and sol−gel method (Seabra et al., 2011; Chauhan et al., 2012). The most versatile and widely reported technique for the synthesis of metal-doped-TiO_2 (Table 17.4) is the sol−gel method. Following its use to synthesize unique mesoporous materials (Yang et al., 2015), Zn-doped-TiO_2 continues to gain recognition in potential application in bioremediation.

17.5.2 METAL OXIDE NANOPARTICLES AS TOOLS TO REMOVE THE ANTIBIOTIC RESISTANT AND CHLORINE RESISTANT WATERBORNE ORGANISMS

MONs are acknowledged for their high antimicrobial activity against a wide range of bacterial species. Many of them have shown tremendous antimicrobial effects on drug-resistant bacteria and parasites (Allahverdiyev et al., 2011). Effective MONs include titanium oxide (TiO_2), which was discussed in 16.5.1, iron oxide (Fe_3O_4), copper oxide (CuO), and zinc oxide (ZnO). Most MONs exhibit bactericidal properties through reactive oxygen species (ROS) generation although some are effective due to their physical structure and metal ion release. The mechanism for antimicrobial activity of MONs may involve a number of possibilities that may include cascade of reactions.

1. The release of the metal ion will lead to protein overexpression, metal ion binding to intracellular macromolecules, and gene toxicity.
2. There will be formation ROS, which will lead to phagocytic oxidative outburst, lipid peroxidation, mitochondrion dysfunction, necrosis, and apoptosis.
3. Their binding to the bacterial membrane or cell wall may initiate cell content leakages and death.

Mixed metal oxides nanoparticles are also synthesized for high antimicrobial effectiveness and low toxicity. The mixed or doped MONs may also be more stable and also exhibit broader range of activity, which would be advantageous in eradicating the extreme drug resistant waterborne pathogen. Table 17.4 depicts some selected single and mixed/doped MONs and their antibacterial activity.

MONs, especially TiO_2 and Ag_2O nanoparticles, are recognized for their activity against antibiotic resistant pathogens (Allahverdiyev et al., 2011). These might also be novel compounds in removing such difficult-to-treat chlorine resistant pathogens from wastewater and surface water catchment. So, the property of

MONs can be effectively utilized to remove both the pharmaceutical chemicals and pathogens simultaneously as a novel solution strategy. Allahverdiyev et al. (2011) highlighted "the unique properties of nanoparticles, their mechanism of action as antibacterial agents and recent studies in which the effects of visible and UV-light induced TiO_2 and Ag_2O nanoparticles on drug-resistant bacteria."

Clinical applications of MONs revealed high effectiveness of several MONs on antimicrobial resistant pathogens (Gokulakrishnan et al., 2012). In one instance, Ravikumar et al. (2011) reported that Fe_3O_4, ZrO_2 and MgO showed antibacterial activities on ophthalmic pathogens. Gokulakrishnan et al. (2012) demonstrated that 10 μg MgO nanoparticles showed maximum antibacterial activity against *Streptococcus pneumoniae*. The mechanisms behind the activity involve the binding of these MONs with the thiol (−SH) groups of protein thereby destroying the cell wall (Hosseinkhani et al., 2011). The potential of MgO to inhibit the production of β-lactamase enzyme and to block the efflux pump pathway makes it a potent viable treatment option for aquatic antibiotic resistant pathogens. This is because the production of β-lactamase enzyme and the efflux pump mechanisms are vital to the emergence of antibiotic resistance (Jayaraman, 2009). Excellent antibacterial activity of TiO_2 and CdO already reported on *E. coli* known for both stated mechanisms of antibiotic resistance makes these MONs recommendable for eradication of similar waterborne (Zhang and Chen, 2009; Kim et al., 2007; Rezaei-Zarchi et al., 2010).

17.5.2.1 Zinc oxide nanoparticles

The Zinc oxide nanoparticles (ZnO-NPs) form Zn^{2+} in the aquatic environment. Though the ZnO-NPs are stable based on pH of the reacting medium, they have zero charge at pH 8 and 9. Their inhibitory mechanism include the binding of Zn^{2+} ions to the sulfhydryl (−SH) groups of enzymes, leading to cell damage and death of the sensitive types of aquatic microorganisms. Based on a number of factors including size and concentration, Zn^{2+} ions possess effective mechanism of forming ROS that interferes with the normal functions of the microbial organelles (Wilson et al., 2007; Liu et al., 2014). It produces very effective antibacterial activity even when they are available in very small quantity (Zhang et al., 2007). High antibacterial activities of ZnO-NPs have been reported among a wide range of waterborne and environmental bacteria such as *Salmonella typhi*, *E. coli*, and *Pseudomonas aeruginosa* (Narayanan et al., 2012; Ramani et al., 2012; Ishtaiwa, 2013). Ishtaiwa (2013) reported that 100% of both waterborne *E. coli* and *P. aeruginosa* are killed in the presence of ZnO/UV. A report (Colon et al., 2006) confirmed the nontoxicity of ZnO-NPs to human cells. This attribute coupled with their activity against antibiotic resistant makes ZnO-NPs a very important solution to the setback with respect to the required selective toxicity (Seil et al., 2009).

In vitro studies (Banoee et al., 2010) revealed that ZnO-NPs exhibit synergistic antibacterial activity with ciprofloxacin, in which the particle size between 20 nm and 45 nm enhanced the effectiveness of ciprofloxacin killing both

Staphylococcus aureus and *E. coli* under both environmental and body temperature. Their safety for use has been approved by the U.S. Food and Drug Administration (FDA, 2011), with need to pay attention to their size, concentration, synthesis method, shape, and life functions of the bacteria for safe application (Adams et al., 2006). The limitation against Zn effectiveness as antibacterial in water is in its insolubility of the p-type dopant (Sevinc, Hanley, 2010) but this can be circumvented by stabilizing the nanoparticles.

17.5.2.2 Iron oxide nanoparticles

Iron oxide nanoparticles (Fe_2O_3-NPs) interact and reduce a wide range of pathogens. Behera et al. (2012) reported that the activity of the iron oxide nanoparticles against two Gram negative bacteria, *Shigella flexneri* and *P. aeruginosa*. The activity of Fe_2O_3-NPs can be compared with that of neomycin (30 μg per disc). The nanoparticle was reported by Chen et al. (2011) to be highly effective against organisms like *E. coli*. Fe_2O_3-NPs have high affinity to adhere to the microbial cell wall due to high isoelectric point (7). So, Fe_2O_3-NPs penetrate the organisms and kill them through oxidative stress generated by ROS (Mohapatra and Anand, 2010; Tran et al., 2010) as discussed earlier regarding TiO_2-nanoparticles.

Appreciable inhibitory activity against waterborne microorganisms of public health importance by Fe_2O_3-NPs with particle size of 6 nm was reported (García et al., 2011). The Fe_2O_3-NPs was, for instance, reported to have inhibited *Vibrio fischeri* (Chen et al., 2011). Fe_2O_3-NPs also has high magnetic property. So, this exemplifies the reasons for the description of the MION known for large surface-to-volume ratio and magnetic property as veritable tools for wastewater treatment (Guo et al., 2013; Qu et al., 2103).

MONs like MIONs that possess large surface-to-volume ratio and magnetic attributes may be utilized with reactive filter materials for wastewater treatment nanotechnology (Tang and Lo, 2013; Guo et al., 2013). Functionalized MONs also possess appreciable physicochemical attributes like surface charge, molecular interaction and size, with heterogeneous wastes within the wastewater (Qu et al., 2013). In removing pharmaceutical wastes and microbial contaminants from wastewater, MION has the benefits of fast separation of aptamer-bound MION, bringing significant reduction of sludge volume.

17.5.2.3 Copper oxide nanoparticles

Copper oxide nanoparticles (CuO-NPs) have shown a strong antimicrobial activity against a wide range of microbial pathogens, some of which can be found in water (Abboud et al., 2014). CuO-NPs can be synthesized from algae. When synthesized, they exhibit very much better antimicrobial activity than a typical algal extract. In Abboud et al. (2014) study the CuO-NPs showed diameter of the inhibition zone of 14 mm and 16 mm against *Enterobacter aerogenes* and *S. aureus* respectively, unlike algal extract, bearing MON, without any zone in this case. Saad et al. (2015) has also demonstrated the potential of CuO-NPs to inactivate the protozoan parasites, *Entamoeba histolytica* and *C. parvum* cysts, which gives the MON overriding advantage over several other applications, removing the chlorine resistant parasite from water, reducing the risk of amoebiasis and cryptosporidiosis among the exposed group. CuO-NPs at varying concentration and particle sizes have been optimized with the high potential to inhibit growth of *P. aeruginosa* and *Proteus* species (Ren et al., 2009). Zarrindokht and Chehrazi (2011) and Abboud et al. (2014) reported separately that Gram negative bacteria are more susceptible to the CuO-NPs than the Gram positive. A report by Heinlaan et al. (2008) showed that 200 ppm of CuO-NPs with the particle size of 30 nM has immense removal activity within ½ hour. The report showed further that the MON showed its inhibitory activity even at very low concentration >0.5 ppm. Activity of CuO-NPs cut across a wide range of microbial contaminants (Heinlaan et al., 2008). They share an equivalent order of activity as silver oxide nanoparticles (Ren et al., 2009), showing that a wide range of MONs are potential solutions to the emerging trend of antimicrobial resistance.

17.5.2.4 Chitosan–metal oxide nanocomposites: applications in removing antibacterial resistant in water

To further drive home the effective application of MON in the removal of potential antibacterial resistant pathogens from the water, it is imperative to consider the widely reported potentials of chitosan–silver oxide nanocomposite (CSON) (Khawaja et al., 2018). Chitosan, alone, is recognized for broad spectrum antimicrobial activity by destruction of the bacterial cell wall integrity (Goy et al., 2016). Khawaja et al. (2018) further reported very wide zones of clearings by the CSON films showing a distinct inhibitory effect against *P. aeruginosa*, *S. aureus*, *E. coli*, and *Bacillus subtilis*. The inhibitory activity against *P. aeruginosa* and *E. coli* gives a hopeful optimism for the potential application in removing the same from the aquatic environment with slight optimization.

Polymeric methods are usually used to synthesize chitosan–ZnO nanoparticle (CS–ZnONPs) composite beads, which have been reported to have promising water treatment applications (Dehaghi et al., 2014).

17.6 TOXICITY ISSUES AND GUIDELINES ON THE USE OF NANOPARTICLES FOR WATER TREATMENT

The in vivo and in vitro toxicity of CuO, SiO_2, ZnO, and Co_3O_4 nanoparticles were examined by Cho et al. (2013) who concluded that CuO was the most toxic nanoparticle. An FDA assessment has been done where the conclusion was that ZnO and TiO_2 nanoparticles had very low toxicity with a judgment that ZnO was safe enough to be used following a careful optimization (FDA, 2011). Currently, ZnO is one of the five zinc compounds that are listed as a generally recognized as safe (GRAS) material by the U.S. Food and Drug Administration (21CFR182.8991) (FDA, 2011). TiO_2 nanoparticles were grafted with a dense polymer brush and it was demonstrated that these nanoparticle–polymer conjugates were harmless to fibroblast cells. It is thought that nanoparticle–polymer conjugates can be further used in antimicrobial application since they are not toxic and they have significant inhibitory effects on bacteria owing to unique properties of MONs.

The use of the nanoparticles should be guided by approved standards. The United States Environmental Protection Agency (USEPA) research on nanomaterials made reference to the guidance manual for the testing of manufactured nanomaterials compiled under the OECD sponsorship program as well as cooperative agreement of UNEP, ILO, FAO, WHO, UNIDO, UNITAR, and OECD (IOMC, 2010). The guidelines of the Inter-Organization Programme for the Sound Management of Chemicals (IOMC, 2010) contain the reports, identification, risk, analysis, and preliminary guidance on the use of nanoparticles. This should be perused and be adhered to very strictly to ensure safe usage of the nanoparticles.

17.7 CONCLUSIONS

Pharmaceutical contaminants in aquatic environment as well as microbial pathogens are of immense concern to public health. They constitute a broad range of toxicity to the environment with direct and indirect effects on humans. The technology being used by the conventional WWTP has not been effective to remove a substantial amount of the contaminants. The use of MON in form of MIONs as effective adsorbents for example or MON conjugated with specific aptamer to form column or any related modification has been veritably promising for removing a wide range of pharmaceutical contaminants and antibiotic resistant pathogens from wastewater and water for drinking water production and irrigation.

The use of MONs–aptamer column may be designed on treating the effluents from WWTPs, after the effluents have been subjected to advanced pretreatment stage(s). Various reports have shown complete removal with these or related models. The use of doped MONs has also been positively appraised to possess higher effectiveness to degrade and remove a wide range of pharmaceutical contaminants and personal care products from the wastewater and water. Meanwhile, it is imperative to understand the reactivity of nanoparticles in the environment, their impacts, risks, etc. as they guide us on the choice of nanoparticles to be made as well as the need to strictly adhere to the existing guidelines on their use.

REFERENCES

Abboud, Y., Saffaj, T., Chagraoui, A., El Bouari, A., Brouzi, K., Tanane, O., et al., 2014. Biosynthesis, characterization and antimicrobial activity of copper oxide nanoparticles (CONPs) produced using brown alga extract (*Bifurcaria bifurcata*). Appl. Nanosci. 4, 571. Available from: https://doi.org/10.1007/s13204-013-0233-x.

Addamo, M., Augugliaro, V., Di Paola, A., Gracía-López, E., Loddo, V., Marcí, G., et al., 2005. Removal of drugs in aqueous systems by photoassisted degradation. J. Appl. Electrochem. 35 (7–8), 765–774.

Adams, L.K., Lyon, D.Y., Alvarez, P.J.J., 2006. Comparativeeco-toxicity of nanoscale TiO2, SiO2, and ZnO water suspensions. Water Research 40 (19), 3527–3532.

Allahverdiyev, A.M., Abamor, E.S., Bagirova, M., Rafailovich, M., 2011. Antimicrobial effects of TiO_2 and Ag_2O nanoparticles against drug-resistant bacteria and leishmania parasites. Fut. Microbiol. 6 (8), 933–940. Available from: https://doi.org/10.2217/fmb.11.78.

Allahverdiyev, A.M., Abamor, E.S., Bagirova, M., et al., 2013. Investigation of antileishmanial activities of TiO_2@Ag nanoparticles on biological properties of *L. tropica* and *L. infantum* parasites, in vitro. Expl. Parasitol. 135 (1), 55–63.

Amin, M.T., Alazba, A.A., Manzoor, U., 2014. A review of removal of pollutants from water/wastewater using different types of nanomaterials. Adv. Mater. Sci. Eng. Available from: https://doi.org/10.1155/2014/825910Article ID 825910.

Anjum, M., Miandad, R., Waqas, M., Gehany, F., Barakat, M.A., 2016. Remediation of wastewater using various nano-materials. Arab. J. Chem. Available from: https://doi.org/10.1016/j.arabjc.2016.10.004.

Ashton, D., Hilton, M., Thomas, K.V., 2004. Investigating the environmental transport of human pharmaceuticals to streams in the United Kingdom. Sci. Total Environ. 333, 167–184. Available from: https://doi.org/10.1016/j.scitotenv.2004.04.062.

Aware, D.V., Jadhav, S.S., 2016. Synthesis, characterization and photocatalytic applications of Zn-doped TiO_2 nanoparticles by sol–gel method. Appl. Nanosci. 6, 965. Available from: https://doi.org/10.1007/s13204-015-0513-8.

Balcıoğlu, I.A., Otker, M., 2003. Treatment of pharmaceutical wastewater containing antibiotics by O3 and O3/H2O2 processes. Chemosphere 50 (1), 85–95. Available from: https://doi.org/10.1016/s0045-6535(02)00534-9.

Banoee, M., Seif, S., Nazari, Z.E., Jafari-Fesharaki, P., Shahverdi, H.R., Moballegh, A., et al., 2010. ZnO nanoparticles enhanced antibacterial activity of ciprofloxacin against *Staphylococcus aureus* and *Escherichia coli*. J. Biomed. Mater. Res. B: Appl. Biomater. 93 (2), 557–561. Available from: https://doi.org/10.1002/jbm.b.31615.

Bautitz, I.R., Nogueira, R.F.P., 2007. Degradation of tetracycline by photo-Fenton process—solar irradiation and matrix effects. J. Photochem. Photobiol. A. 187, 33–39.

Behera, S.S., Patra, J.K., Pramanik, K., Panda, N., Thatoi, H., 2012. Characterization and evaluation of antibacterial activities of chemically synthesized iron oxide nanoparticles. World J. Nano Sci. Eng. 2, 196–200. Available from: https://doi.org/10.4236/wjnse.2012.24026.

Berrada, H., Borrull, F., Font, G., Marcé, R.M., 2008. Determination of macrolide antibiotics in meat and fish using pressurized liquid extraction and liquid chromatography-mass spectrometry. J. Chromatogr A 1208 (1-2), 83–89. Available from: https://doi.org/10.1016/j.chroma.2008.08.107.

Boxall, A.B.A., Johnson, P., Smith, E.J., Sinclair, C.J., Stutt, E., Levy, L., 2006. Uptake of veterinary medicines from soils into plants. J. Agric. Food Chem. 163 (3), 2288–2297.

Bouwman, G.M., Reus, J.A.W.A., 1994. Persistence of Medicines in Manure, 163. Centre for Agriculture and Environment, CLM, Belfeld, Netherlands, p. 26, 1994.

Brady-Estévez, A.S., Kang, S., Elimelech, M., 2008. A single-walled-carbon-nanotube filter for removal of viral and bacterial pathogens. Small 4 (4), 481–484.

Brain, R.A., Johnson, D.J., Richards, S.M., Sanderson, H., Sibley, P.K., Solomon, K.R., 2004. Effects of 25 pharmaceutical compounds to Lemna gibba using a seven-day static-renewal test. Environ Toxicol Chem 23 (2), 371–382.

Burkhardt, M., Stamm, C., Waul, C., Singer, H., Muller, S., 2005. Surface runoff and transport of sulfonamide antibiotics and tracers on manured grassland. Journal of Environmental Quality 34, 1363–1371.

Camtakan, Z., Erenturk, S., Yusan, S., 2012. Magnesium oxide nanoparticles: preparation, characterization, and uranium sorption properties. Environ. Prog. Sustain. Energy 31, 536–543. Available from: https://doi.org/10.1002/ep.10575.

Cantón, R., Ruiz-Garbajosa, P., 2011. Co-resistance: an opportunity for the bacteria and resistance genes. Curr. Opin. Pharmacol. 11, 477–485. Available from: https://doi.org/10.1016/j.coph.2011.07.007.

Chang, X., Meyer, M.T., Liu, X., Zhao, Q., Chen, H., Chen, J., et al., 2010. Determination of antibiotics in sewage from hospitals, nursery and slaughter house, wastewater treatment plant and source water in Chongqing region of Three Gorge Reservoir in China. Environ. Pollut. 158, 1444–1450.

Chauhan, R., Kumar, A., Chaudhary, R., 2012. Structural and optical characterization of Zn doped TiO_2 nanoparticles prepared by sol–gel method. J. Sol–Gel Sci. 585–591. Available from: https://doi.org/10.1007/s10971-011-2664-2668.

Chen, G.C., Shan, X.Q., Wang, Y.S., Wen, B., Pei, Z.G., Xie, Y.N., et al., 2009. Adsorption of 2,4,6-trichlorophenol by multiwalled carbon nanotubes as affected by Cu (II). Water Res. 43 (9), 2409–2418.

Chen, J., Xiu, Z., Lowry, G.V., Alvarez, P.J.J., 2011. Effect of natural organic matter on toxicity and reactivity of nano-scale zero-valent iron. Water Res 45, 1995–2001.

Chen, M., Yi, Q., Hong, J., Zhang, L., Lin, K., Yuan, D., 2015. Simultaneous Determination of 32 Antibiotics and 12 Pesticides in Sediment Using Ultrasonic-assisted Extraction and High Performance Liquid Chromatography-tandem Mass Spectrometry. Analytical Methods .

Chi, J., 2009. Phthalate acid esters in *Potamogeton crispus* L. from Haihe River, China. Chemosphere 77, 48–52.

Chitose, N., Ueta, S., Seino, S., Yamamoto, T.A., 2003. Radiolysis of aqueous phenol solutions with nanoparticles. 1. Phenol degradation and TOC removal in solutions containing TiO_2 induced by UV, γ-ray and electron beams. Chemosphere 50 (8), 1007–1013.

Cho, W.-S., Duffin, R., Bradley, M., Megson, I.L., MacNee, W., Lee, J.K., et al., 2013. Predictive value of in vitro assays depends on the mechanism of toxicity of metal oxide nanoparticles. Part. Fibre Toxiol. 10. Available from: https://doi.org/10.1186/1743-8977-10-55.

Colon, G., Ward, B.C., Webster, T.J., 2006. Increased osteoblast and decreased *S. epidermidis* functions on nanophase ZnO and TiO_2. J. Biomed. Mater. Res. 78 (3), 595–604. Available from: https://doi.org/10.1002/jbm.a.30789.

Dalrymple, O.K., Yeh, D.H., Trotz, M.A., 2007. Removing pharmaceuticals and endocrine-disrupting compounds from wastewater by photocatalysis: a review. J. Chem. Tech. Biotech. 82 (2), 121–134.

Dehaghi, S.M., Rahmanifar, B., Moradi, A.M., Azar, P.A., 2014. Removal of permethrin pesticide from water by chitosan–zinc oxide nanoparticles composite as an adsorbent. J. Saudi. Chem. Soc. 18 (4), 348–355.

Ding, S.L., Wang, X.K., Jiang, W.Q., 2013. Photodegradation of the antimicrobial triclocarban in aqueous systems under ultraviolet radiation. Environ. Sci. Pollut. Res. 20, 3195. Available from: https://doi.org/10.1007/s11356-012-1239-8.

Duong, C.N., Ra, J.S., Cho, J., Kim, S.D., Choi, H.K., Park, J.H., et al., 2010. Estrogenic chemicals and estrogenicity in river waters of South Korea and seven Asian countries. Chemosphere 78, 286–293.

Farahat, E., Linderholm, H.W., 2013. Effects of treated wastewater irrigation on sizestructure, biochemical products and mineral content of native medicinal shrubs. Ecol. Eng. 60, 235–241.

Faria, C., Vaz-Moreira, I., Serapicos, E., Nunes, O.C., Manaia, C.M., 2009. Antibiotic resistance in coagulase negative staphylococci isolated from wastewater and drinking water. Sci. Total Environ. 407, 3876–3882.

Fatta-Kassinos, D., Meric, S., Nikolaou, A., 2011. Pharmaceutical residues in environmental waters and wastewater: current state of knowledge and future research. Anal. Bioanal. Chem. 399, 251.

FDA. Part 182—substances generally recognized as safe.Food and drug administration, Washington DC, USA. Avail-able at: http://ecfr.gpoaccess.gov/cgi/t/text/textidx?c0ecfr&sid0786bafc6f6343634fbf79fcdca7061e1&rgn0div5&-view0text&node021:3.0.1.1.13&idno021#21:3.0.1.1.13.9. Accessed 28 March 2016.

Felizardo, E., Chauque, C., Ngila, J.C., Dlamini, L.N., Greyling C.J., 2016. TiO_2 nanoparticles uniquely supported on functionalized electrospun polyacrylonitrile nanofibers for photodegradation of organic pollutants from dye effluents. In: 17th WaterNet/WARFSA/GWP-SA Symposium Programme, p. 4.

Figueira, V., Serra, E.A., Vaz-Moreira, I., Brandão, T.R.S., Manaia, C.M., 2012. Comparison of ubiquitous antibiotic-resistant Enterobacteriaceae populations isolated from wastewaters, surface waters and drinking waters. J. Water. Health 10 (1), 1–10.

Forslund, A., Ensink, J.H.J., Battilani, A., Kljujev, I., Gola, S., Raicevic, V., 2010. Faecal contamination and hygiene aspect associated with the use of treated wastewater and canal water for irrigation of potatoes (*Solanum tuberosum*). Agric. Water Manage. 98, 440–450.

Garcia-Armisen, T., Vercammen, K., Passerat, J., Triest, D., Servais, P., Cornelis, P., 2011. Antimicrobial resistance of heterotrophic bacteria in sewage-contaminated rivers. Water Res. 45, 788–796.

García, A., Espinosa, R., Delgado, L., Casals, E., González, E., Puntes, V., et al., 2011. Acute toxicity of cerium oxide, titanium oxide and iron oxide nanoparticles using standardized tests. Desalination 269 (1–3), 2136–2141.

Garcillan-Barcia, M.P., Alvarado, A., de la Cruz, F., 2011. Identification of bacterial plasmids based on mobility and plasmid population biology. FEMS Microbiol. Rev. 35, 936–956.

Giger, W., Alder, A.C., Golet, E.M., Kohler, H.P.E., McArdell, C.S., Molnar, E., Siegrist, H., et al., 2003. Occurrence and fate of antibiotics as trace contaminants in wastewaters sewage sludge and surface waters. Chimia 57, 485–491.

Gobel, A., McArdell, C.S., Suter, M.J.F., Giger, W., 2004. Trace determination of macrolide and sulfonamide antimicrobials a human sulfonamide metabolite and trimetoprim in wastewater using liquid chromatography coupled to electrospray tandem mass spectrometry. An. Chem. 76, 4756–4764.

Gokulakrishnan, R., Ravikumar, S., Raj, J.A., 2012. In vitroantibacterial potential of metal oxide nanoparticles againstantibiotic resistant bacterial pathogens. Asian Pacific Journal of Tropical Disease 2 (5), 411–413.

Gordon, T., Perlstein, B., Houbara, O., Felner, I., Banin, E., Margel, S., 2011. Synthesis and characterization of zinc/iron oxide composite nanoparticles and their antibacterial properties. Colloid Surf. A. 374, 1–8. Available from: https://doi.org/10.1016/j.colsurfa.2010.10.015.

Goy, R.C., Morais, S.T.B., Assis, O.B.G., 2016. Evaluation of the antimicrobial activity of chitosan and its quaternized derivative on *E. coli* and *S. aureus* growth. Revista Brasileira de Farmacognosia 26 (1), 122–127. 2016.

Grill, M.F., Maganti, R.K., 2011. Neurotoxic effects associated with antibiotic use: management considerations. Br J. Clin. Pharmacol 72 (3), 381–393.

Guo, B.L., Han, P., Guo, L.C., Cao, Y.Q., Li, A.D., Kong, J.Z., et al., 2015. The antibacterial activity of Ta-doped ZnO nanoparticles. Nanoscale Res. Lett. 10 (1), 1047.

Guo, H., Ma, F., Feng, X., Lou, X., Tade, M.O., 2013. Application of iron oxide based nanomaterials (NMs) in magnetic assisted chemical separation (MACS) processes for water/wastewater treatment. Adv. Mater. Res. 610–613, 1242–1251.

Hartig, C., Ernst, M., Jekel, M., 2001. Membrane filtration of two sulphonamides in tertiary effluents and subsequent adsorption on activated carbon. Water Res. 35 (16), 3998–4003.

Heinlaan, M., Ivask, A., Blinova, I., Dubourguier, H.-C., Kahru, A., 2008. Toxicity of nanosized and bulk ZnO, CuO and TiO_2 to bacteria *Vibrio fischeri* and crustaceans *Daphnia magna* and *Thamnocephalus platyurus*. Chemosphere 71 (7), 1308–1316.

Hosseinkhani, P., Zand, A.M., Imani, S., Rezayi, M., RezaeiZarchi, S., 2011. Determining the antibacterial effect of ZnO nanoparticle against the pathogenic bacterium, Shigella dysenteriae (type 1). Int. J. Nano. Dim. 1 (4), 279–285.

Hu, X., Mu, L., Zhou, Q., Wen, J., Pawliszyn, J., 2011a. ssDNA aptamer-based column for simultaneous removal of nanogram per liter level of illicit and analgesic pharmaceuticals in drinking water. Environ. Sci. Technol. 45 (11), 4890–4895.

Hu, Z.J., Shi, T.L., Cai, Y.Q., 2011b. Concentrations, distribution, and bioaccumulation of synthetic musks in the Haihe River of China. Chemosphere 84 (11), 1630–1635.

IOMC. IOMC Inter-Organization Programme For The Sound Management Of Chemicals. A cooperative agreement among UNEP, ILO, FAO, WHO, UNIDO, UNITAR and OECD 2010.

Ishtawa, 2013. Understanding mode of action of nanoparticles in water disinfection: ZnO in Bacteria Killing vs. complete photodegradation. Masters Degree thesis submitted to Faculty of Graduate Studies. AnNajah National University, Nablus, Palestine.

Iwamoto, S., Iwamoto, S., Inoue, M., Yoshida, H., Tanaka, T., Kagawa, K., 2005. XANES and XPS study of silica-modified titanias prepared by the glycothermal method. Chem. Mater. 17 (3), 650–655.

Jayaraman, R., 2009. Antibiotic resistance: an overview of mechanisms and a paradigm shift. Curr. Sci 96 (11), 1475–1484.

Jeng, H.A., Swanson, J., 2006. Toxicity of MONs in mammalian cells. J. Environ. Sci. Health A 41 (12), 2699–2711. Available from: https://doi.org/10.1080/10934520600966177.

Jin, C.X., Chen, Q.Y., Sun, R.L., Zhou, Q.X., Liu, J.J., 2009. Eco-toxic effects of sulfadiazine sodium, sulfamonomethoxine sodium and enrofloxacin on wheat, Chinese cabbage and tomato. Ecotoxicol 18, 878–885.

Kabra, K., Chaudhary, R., Sawhney, R.L., 2004. Treatment of hazardous organic and inorganic compounds through aqueousphase photocatalysis: a review. Ind. Eng. Chem. Res. 43 (24), 7683–7696.

Kay, P., Blackwell, P.A., Boxall, A.B.A., 2005. Transport of veterinary antibiotics in overland flow following the application of slurry to arable land. Chemosphere 59, 951–959.

Khajavi, R., Abbasipour, M., 2012. Electrospinning as a versatile method for fabricating coreshell, hollow and porous nanofibers. Sci. Iran 19, 2029–2034. Available from: https://doi.org/10.1016/j.scient.2012.10.037.

Khawaja, H., Zahira, E., Asghar, M.A., Asgha, M.A., 2018. Graphene oxide, chitosan and silver nanocomposite as a highly effective antibacterial agent against pathogenic strains. Colloids and Surfaces A: Physicochemical and Engineering Aspects 555, 246–255.

Kim, S.H., Kwak, S.-Y., Sohn, B.-H., Park, T.H., 2003. Design of TiO_2 nanoparticle self-assembled aromatic polyamide thinfilm-composite (TFC) membrane as an approach to solve biofouling problem. J. Membr. Sci. 211 (1), 157–165.

Kim, S.V., Carlson, K., 2005. LC-MS2 for quantifying trace amounts of pharmaceuti- cals compounds in soil and sediment matrices. Trends in Analytical Chemistry 24 (7), 635–644.

Kim, H., Ryu, J.H., Beuchat, L.R., 2007. Effectiveness of disinfectants in killing Enterobacter sakazakii in suspension, dried on the surface of stainless steel, and in a biofilm. Appl. Environ. Microbiol. 73, 1256–1265.

Kim, J., Luo, F., Jiang, X., 2009. Factors impacting the regrowth of E. coli O157:H7 in dairy manure compost. Journal of Food Protection 72, 1576–1584.

Klavarioti, M., Mantzavinos, D., Kassinos, D., 2009. Removal of residual pharmaceuticals from aqueous systems by advanced oxidation processes. Environ. Inter. 35, 402–417.

Kolpin, D.W., Furlong, E.T., Meyer, M.T., Thurman, E.M., Zaugg, S.D., Barber, L.B., 2002. Pharmaceuticals, hormones, and other organic wastewater contaminants in US streams, 1999–2000: a national reconnaissance. Environ. Sci. Technol. 36, 1202–1211.

Koyuncu, I., Arikan, O.A., Wiesner, M.R., Rice, C., 2008. Removal of hormones and antibiotics by nanofiltration membranes. J. Membr. Sci. 309, 94–101.

Kubacka, A., Colón, G., Fernández-García, M., 2009. Cationic (V, Mo, Nb, W) doping of TiO_2-anatase: a real alternative for visible light-driven photocatalysts. Catal. Today 143, 286–292.

Larsen, T.A., Lienert, J., Joss, A., Siegrist, H., 2004. How to Avoid Pharmaceuticals in the Aquatic Environment. J. Biotechnol. 113 (1–3), 295–304.

Laville, N., Aït-Aïssa, S., Gomez, E., Casellas, C., Porcher, J.M., 2004. Effects of human pharmaceuticals on cytotoxicity, EROD activity and ROS production in fish hepatocytes. Toxicology 196, 41–55.

Li, J.-F., Xu, Z.-L., Yang, H., Yu, L.-Y., Liu, M., 2009a. Effect of TiO_2 nanoparticles on the surface morphology and performance of microporous PES membrane. Appl. Surf. Sci. 255 (9), 4725–4732.

Li, J.-H., Xu, Y.-Y., Zhu, L.-P., Wang, J.-H., Du, C.-H., 2009b. Fabrication and characterization of a novel TiO_2 nanoparticle self-assembly membrane with improved fouling resistance. J. Membr. Sci. 326 (2), 659–666.

Li, D., Yu, T., Zhang, Y., Yang, M., Li, Z., 2010. Antibiotic resistance characteristics of environmental bacteria from an oxytetracycline production wastewater treatment plant and the receiving river. Appl. Environ. Microbiol. 76 (11), 3444–3451.

Liu, S., Yu, J., Jaroniec, M., 2011. Anatase TiO_2 with dominant high-energy 0 0 1 facets: synthesis, properties, and applications. Chem. Mater. 23 (18), 4085–4093.

Liu, P., Zhang, H., Feng, Y., Yang, F., Zhang, J., 2014. Removal of trace antibiotics from advanced oxidation processes. Chem. Eng. J. 240, 211–220.

Mahata, S., Mahato, S.S., Nandi, M.M., Mondal, B., 2011. Synthesis of TiO_2 nanoparticles by hydrolysis and peptization of titanium isopropoxide solution. AIP Conf. Proc. 1461, 225–228. Available from: https://doi.org/10.1063/1.4736892.

Mahshid, S., Askari, M., Ghamsari, M.S., 2007. Synthesis of TiO_2 nanoparticles by hydrolysis and peptization of titanium isopropoxide solution. J. Mater. Proc. Technol. 189, 296–300.

Maness, P.C., Smolinski, S., Blake, D.M., Huang, Z., Wolfrum, E., Jacoby, W.A., 1999. Bactericidal activity of photocatalytic TiO_2: toward an understanding of its killing mechanism. Appl. Environ. Microbiol. 65, 4094–4098.

Matongo, S., Birungi, G., Moodley, B., Ndungu, P., 2015. Occurrence of selected pharmaceuticals in water and sediment of Umgeni River, KwaZulu-Natal, South Africa. Environ. Sci. Pollut. Res. 22 (13), 1–11.

Mohapatra, M., Anand, S., 2010. Synthesis and application of nano-structured iron oxide/hydroxides—a review. Int. J. Eng. Technol. 2 (8), 127–146.

Mukhopadhyay, A., Basak, S., Das, J.K., Medda, S.K., Chattopadhyay, K., De, G., 2010. Ag–TiO_2 nanoparticle codoped SiO_2 films on ZrO_2 barrier-coated glass substrates with antibacterial activity in ambient condition. ACS Appl. Mater. Interfaces 2, 2540–2546. Available from: https://doi.org/10.1021/am100363d.

Murakami, N., Kurihara, Y., Tsubota, T., Ohno, T., 2009. Shape controlled anatase titanium(IV) oxide particles prepared by hydrothermal treatment of peroxo titanic acid in the presence of polyvinyl alcohol. J. Phys. Chem. C 113 (8), 3062–3069.

Nahar, M.S., Hasegawa, K., Kagaya, S., 2006. Photocatalytic degradation of phenol by visible light-responsive iron-doped TiO_2 and spontaneous sedimentation of the TiO_2 particles. Chemosphere 65 (11), 1976–1982.

Narayanan, P.M., Wilson, W.S., Abraham, A.T., Sevana, M., 2012. Synthesis, Characterization, and Antimicrobial Activity of Zinc Oxide Nanoparticles Against Human Pathogens. BioNanoSci 2, 329–335. Available from: https://doi.org/10.1007/s12668-012-0061-6.

Nentwig, G., Oetken, M., Oehlmann, J., 2004. Effects of pharmaceuticals on aquatic invertebrates — the example of carbamazepine and clofibric acid. In: Kümmerer, K. (Ed.), Pharmaceuticals in the Environment. Sources, Fate, Effects and Risks, 2nd edition Springer-Verlag, Berlin, Heidelberg.

Nghiem, L.D., Schäfer, A.I., Elimelech, M., 2005. Pharmaceutical retention mechanisms by nanofiltration membranes. Environ. Sci. Technol. 39 (19), 7698–7705. Available from: https://doi.org/10.1021/es0507665.

Nikolaus, N., Strehlitz, B., 2014. DNA-Aptamers Binding Aminoglycoside Antibiotics. Sensors (Basel) 14 (2), 3737–3755.

Nikolaou, A., Meric, S., Fatta, D., 2007. Occurrence patterns of pharmaceuticals in water and wastewater environments. Anal. Bioanal. Chem 387, 1225. Available from: https://doi.org/10.1007/s00216-006-1035-8.

Orn, S., Holbech, H., Madsen, T., Norrgren, L., Petersen, G., 2003. Gonad development and vitellogenin production in zebrafish exposed to ethinylestradiol and methyltestosterone. Aquat. Toxicol. 65 (4), 397–411.

Orta, M.T., Martinez, J., Monje, I., Rojas, M.N., 2004. Destruction of helminto (*Ascaris suum*) eggs by ozone. Ozone-Sci. Eng. 26, 359–366.

Palominos, R.A., Mondaca, M.A., Giraldo, A., Penuela, G., Perez-Moya, M., Mansilla, H. D., 2009. Photocatalytic oxidation of the antibiotic tetracycline on TiO_2 and ZnO suspensions. Catal. Today 144, 100–105.

Partridge, S.R., 2011. Analysis of antibiotic resistance regions in gram-negative bacteria. FEMS Microbiol. Rev. 35, 820–855.

Pascoe, D., Karntanut, W., Müller, C.T., 2003. Do pharmaceuticals affect freshwater invertebrates? A study with the cnidarian Hydra vulgaris. Chemosphere 51, 521–528.

Pomati, F., Netting, A.G., Calamari, D., Neilan, B.A., 2004. Effects of erythromycin, tetracycline and ibuprofen on the growth of Synechocystis sp. and Lemna minor. Aquat. Toxicol. 67, 387–396.

Qu, X., Alvarez, P.J.J., Li, Q., 2013. Applications of nanotechnology in water and wastewater treatment. Water Res. 47, 3931–3946.

Radjenović, J., Petrović, M., Ventura, F., Barceló, D., 2008. Rejection of pharmaceuticals in nanofiltration and reverse osmosis membrane drinking water treatment. Water Res. 42 (14), 3601–3610.

Raliya, R., Tarafdar, J.C., 2014. Biosynthesis and characterization of zinc, magnesium and titanium nanoparticles: an ecofriendly approach. Int. Nao. Lett. 4, 93.

Ramani, M., Ponnusamy, S., Muthami Zhchelvan, C., et al., 2012. From Zinc Oxide nanoparticles to microflowers: A study of growth Kinetics and biocidal activity. Mterials science and Engineering 32 (8), 2381–2389.

Rauf, M.A., Meetani, M.A., Hisaindee, S., 2011. An overview on the photocatalytic degradation of azo dyes in the presence of TiO_2 doped with selective transition metals. Desalination 276, 13–27. Available from: https://doi.org/10.1016/j.desal.2011.03.071.

Ravikumar, S., Gokulakrishnan, R., Selvanathan, K., Selvam, S., 2011. Antibacterial activity of metal oxide nanoparticles against ophthalmic pathogens. Int. J. Pharm. Res. Dev. 3 (5), 122–127.

Rezaei-Zarchi, S., Aisha, J., Madiha, J.G., Safieh, S., Fatemeh, B.F., Abdolmajid, B.M., 2010. Comparative study of antimicrobial activities of TiO_2 and CdO nanoparticles against the pathogenic strain of *Escherichia coli*. Iran. J. Pathol. 5 (2), 83–89.

Reyes, C., Fernandez, J., Freer, J., Mondaca, M.A., Zaror, C., Malato, S., et al., 2006. Degradation and inactivation of tetracycline by TiO_2 photocatalysis. J. Photochem. Photobiol. A: Chem. 184, 141–146.

Ren, G., Hu, D., Cheng, E.W.C., Vargas-Reusc, M.A., Reipd, P., Allaker, R.P., 2009. Characterisation of copper oxide nanoparticles for antimicrobial applications. International Journal of Antimicrobial Agents 33 (2009), 587–590.

Roberts, P.H., Thomas, K.V., 2006. The occurrence of selected pharmaceuticals in wastewater effluent and surface waters of the lower tyne catchment. Sci. Total Environ. 356, 143–153.

Ryu, H., Gerrity, D., Crittenden, J.C., Abbaszadegan, M., 2008. Photocatalytic inactivation of *Cryptosporidium parvum* with TiO_2 and low-pressure ultraviolet irradiation. Water Res. 42 (6–7), 1523–1530.

Saad, A.H., Solaman, M.I., Azzam, A.M., Mostafa, A.B., 2015. Antiparasitic activity of silver and copper oxide nanoparticles against *Entamoeba hystolytica* and *Cryptospiridium parvum* cysts. J. Egypt Soc. Parasitol. (JESP) 45 (3), 593–602.

Sanchez-Dominguez, M., Morales-Mendoza, G., Rodriguez-Vargas, M.J., Ibarra-Malo, C. C., Rodriguez-Rodriguez, A.A., Vela-Gonzalez, A.V., et al., 2015. Synthesis of Zn-doped TiO_2 nanoparticles by the novel oil-in-water (O/W) microemulsion method and their use for the photocatalytic degradation of phenol. J. Environ. Chem. Eng. Available from: https://doi.org/10.1016/j.jece.2015.03.010.

Schierz, S.N., Zanker, H., 2009. Aqueous suspensions of carbon nanotubes: surface oxidation, colloidal stability and uranium sorption. Environ. Pollut. 157 (4), 1088–1094.

Seabra, M.P., Salvado, I.M.M., Labrincha, J.A., 2011. Pure and (zinc or iron) doped titania powders prepared by sol–gel and used as photocatalyst. Ceram. Int. 37, 3317–3322. Available from: https://doi.org/10.1016/j.ceramint.2011.04.127.

Seil, J.T., Taylor, E.N., Webster, T.J., 2009. Reduced activity of *Staphylococcus epidermidis* in the presence of sonicated piezoelectric zinc oxide nanoparticles. In: 2009 IEEE 35th Annual Northeast Bioengineering Conference, Boston, MA, USA, 3–5 April 2009 (IEEE, 2009). pp. 1–2. doi:10.1109/NEBC.2009.4967674.

Sevinc, B.A., Hanley, L., 2010. Antibacterial activity of dental composites containing zinc oxide nanoparticles. J. Biomed. Mater. Res. B 94 (1), 22–31. Available from: https://doi.org/10.1002/jbm.b.31620.

Sharma, N., Jandaik, S., Kumar, S., Chitkara, M., Sandhu, I.S., 2016. Synthesis, characterisation and antimicrobial activity of manganese- and iron-doped zinc oxide nanoparticles. J. Exp. Nanosci. 11, 54–71. Available from: https://doi.org/10.1080/17458080.2015.1025302.

Shryock, T.R., Richwine, A., 2010. The interface between veterinary and human antibiotic use. Ann. N.Y. Acad. Sci. 1213, 92–105.

Sjogren, J.C., Sierka, R.A., 1994. Inactivation of phage MS2 by iron-aided titanium dioxide photocatalysis. Appl. Environ. Microbiol 60 (1), 344–347.

Snyder, S.A., Westerhoff, P., Yoon, Y., Sedlak, D.L., 2003. Pharmaceuticals, Personal Care Products, and Endocrine Disruptors in Water: Implications for the Water Industry. Environ. Eng. Sci. 20 (5), 449–469.

Sökmen, M., Degerli, S., Aslan, A., 2008. Photocatalytic disinfection of Giardia intestinalis and Acanthamoeba castellani cysts in water. Exp. Parasitol. 119 (1), 44–48.

Stenström, T.A., Okoh, A.I., Adegoke, A.A., 2016. Antibiogram of environmental isolates of *Acinetobacter calcoaceticus* from Nkonkobe Municipality, South Africa. Frezen Environ. Bull. 25 (8), 3059–3065.

Steinle-Darling, E., Reinhard, M., 2008. Nanofiltration for Trace Organic Contaminant Removal: Structure, Solution, and Membrane Fouling Effects on the Rejection of Perfluorochemicals. Environ. Sci. Technol. 42 (14), 5292–5297. Available from: https://doi.org/10.1021/es703207s.

Su-Hua, W., Bing-zhi, D., Yu, H., 2010. Adsorption of bisphenol A by polysulphone membrane. Desalination 253 (1–3), 22–29.

Tambosi, J.L., Yamanaka, L.Y., José, H.J., Moreira, R.D.F.P.M., Schröder, H.F., 2010. Recent research data on the removal of pharmaceuticals from sewage treatment plants (STP). Quím. Nova 33, 411.

Tang, S.C.N., Lo, I.M.C., 2013. Magnetic nanoparticles: essential factors for sustainable environmental applications. Water Res. 47, 2613–2632.

Ternes, T.A., Stüber, J., Herrmann, N., McDowell, D., Ried, A., Kampmann, M., et al., 2003. Ozonation: a tool for removal of pharmaceuticals, contrast media and musk fragrances from wastewater? Water Res. 37 (8), 1976–1982.

Tiwari, D.K., Behari, J., Sen, P., 2008. Application of nanoparticles in waste water slaughter house, wastewater treatment plant and source water in Chongqing region of Three Gorge Reservoir in China. Environ. Pollut. 158, 1444–1450.

Tran, N., Mir, A., Mallik, D., Sinha, A., Nayar, S., Webster, T.J., 2010. Bactericidal effect of iron oxide nanoparticles on *Staphylococcus aureus*. Int. J. Nanomed. 5 (1), 277–283.

Vanderford, B.J., Snyder, S.A., 2006. Analysis of pharmaceuticals in water by isotope dilution liquid chromatography/tandem mass spectrometry. Environ. Sci. Technol. 40, 7312–7320.

Verlicchi, P., Galletti, A., Petrovic, M., Barceló, D., 2010. Hospital effluents as a source of emerging pollutants: an overview of micropollutants and sustainable treatment options. J. Hydrol. 389, 416–428. Available from: https://doi.org/10.1016/j.jhydrol.2010.06.005.

Verlicchi, P., Al Aukidy, M., Zambello, E., 2015. What have we learned from worldwide experiences on the management and treatment of hospital effluent?—an overview and a discussion on perspectives. Sci. Total Environ. 514, 467–491. Available from: https://doi.org/10.1016/j.scitotenv.2015.02.020.

Vidic, J., Stankic, S., Haque, F., Ciric, D., Le Goffic, R., Vidy, A., et al., 2013. Selective antibacterial effects of mixed ZnMgO nanoparticles. J. Nanopart. Res. 15, 1595. Available from: https://doi.org/10.1007/s11051-013-1595-4.

Vijayalakshmi, R., Rajendran, V., 2012. Synthesis and characterization of nano-TiO_2 via different methods. Arch. Appl. Sci. Res. 4 (2), 1183–1190.

Wang, P., Yap, P.-S., Lim, T.T., 2011. C–N–S tridoped TiO_2 for photocatalytic degradation of tetracycline under visible-light irradiation. Appl. Catal. Gen. 399, 252–261.

Watts, R.J., Kong, S., Orr, M.P., Miller, G.C., Henry, B.E., 1995. Photocatalytic inactivation of coliform bacteria and viruses in secondary wastewater effluent. Water Res. 29 (1), 95–100.

Watkinson, A.J., Murby, E.J., Costanzo, S.D., 2007. Removal of antibiotics in conventional and advanced wastewater treatment: Implications for environmental discharge and wastewater recycling. Water Res. 41 (18), 4164–4176.

Wei, C., Lin, W.-Y., Zalnal, Z., et al., 1994. Bactericidal activity of TiO_2 photocatalyst in aqueous media: toward a solar-assisted water disinfection system. Environ. Sci. Technol. 28 (5), 934–938.

Wilson, M.R., Foucaud, L., Barlow, P.G., Hutchison, G.R., Sales, J., Simpson, R.J., et al., 2007. Nanoparticleinteractions with zinc and iron: implications for toxicology and inflammation. Toxicol ApplPharmacol 225, 80–89 [PubMed: 17900645].

Wu, L., Yu, J.C., Fu, X., 2006. Characterization and photocatalytic mechanism of nanosized CdS coupled TiO_2 nanocrystals under visible light irradiation. J. Mol. Catal. A: Chem. 244 (1–2), 25–32.

Wu, C.X., Witter, J.D., Spongberg, A.L., Czajkowski, K.P., 2009. Occurrence of selected pharmaceuticals in an agricultural landscape, western Lake Erie basin. Water Res. 43, 3407–3416.

Xiong, R., Lu, C., Wang, Y., Zhou, Z., Zhang, X., 2013. Nanofibrillated cellulose as the support and reductant for the facile synthesis of Fe_3O_4/Ag nanocomposites with catalytic and antibacterial activity. J. Mater. Chem. A 1, 14910–14918.

Xi, C., Zhang, Y., Marrs, C.F., Ye, W., Simon, C., Foxman, B., et al., 2009. Prevalence of antibiotic resistance in drinking water treatment and distribution systems. Appl Environ Microbiol 75 (17), 5714–5718. Available from: https://doi.org/10.1128/AEM.00382-09.

Xu, J., Lu, M., Guo, X., Li, H., 2005. Zinc ions surface-doped titanium dioxide nanotubes and its photocatalysis activity for degradation of methyl orange in water. J. Mol. Catal. A: Chem. 226, 123–127. Available from: https://doi.org/10.1016/j.molcata.2004.09.051.

Xu, Z., Hwang, J.-Y., Li, B., Huang, X., Wang, H., 2008. The characterization of various ZnO nanostructures using field-emission SEM. JOM 60 (4), 29–32. Available from: https://doi.org/10.1007/s11837-008-0044-9.

Xu, Y.Y., Han, T., Li, X.Q., Sun, L.H., Zhang, Y.J., Zhang, Y.S., 2015. Colorimetric detection of kanamycin based on analyte-protected silver nanoparticles and aptamer-selective sensing mechanism. Anal. Chim. Acta. 891, 298–303. Available from: https://doi.org/10.1016/j.aca.2015.08.013.

Yang, L., Zhang, Y., Ruan, W., Zhao, B., Xu, W., Lombardi, J.R., 2010. Improved surface-enhanced Raman scattering properties of TiO_2 nanoparticles by Zn dopant. J. Raman Spectrosc. 41, 721–726. Available from: https://doi.org/10.1002/jrs.2511.

Yang, Y.-J., Kelkar, A.V., Zhu, X., Bai, G., Ng, H.T., Corti, D.S., et al., 2015. Effect of sodium dodecylsulfate monomers and micelles on the stability of aqueous dispersions of titanium dioxide pigment nanoparticles against agglomeration and sedimentation. J. Colloid Interface Sci. 450, 434–445. Available from: https://doi.org/10.1016/j.jcis.2015.02.051.

Yahya, N., Daud, H., Tajuddin, N.A., Daud, H.M., Shafie, A., Puspitasari, P., 2010. Application of ZnO nanoparticles EM wave detector prepared by sol–gel and self-combustion techniques. J. Nano Res. 11, 25–34.

Young, T.A., Heidler, J., Matos-Perez, C.R., Sapkota, A., Toler, T., Gibson, K.E., et al., 2008. Ab initio and in situ comparison of caffeine, triclosan, and triclocarban as indicators of sewage-derived microbes in surface waters. Environ. Sci. Technol. 42, 3335–3340.

Zan, L., Fa, W., Peng, T., Gong, Z.-K., 2007. Photocatalysis effect of nanometer TiO_2 and TiO_2-coated ceramic plate on hepatitis B virus. J. Photochem. Photobiol. B: Biol. 86 (2), 165–169.

Zarrindokht, E.K., Chehrazi, P., 2011. Antibacterial activity of ZnO nanoparticle on Gram-positive and Gram-negative bacteria. Afr. J. Microbiol. Res. 5, 1368–1377.

Zhang, L.L., Jiang, Y.H., Ding, Y.L., Povey, M., York, D., 2007. Investigation into the antibacterial behaviour of suspensions of ZnO nanoparticles (ZnO nanofluids). J. Nanopart Res 9, 479–489.

Zhang, H., Luo, Y., Zhou, Q.X., 2008. Research advancement of ecotoxicity of tetracycline antibiotics. Journal of Agro-Environmental Science 27 (2), 407–413.

Zhang, H., Chen, G., 2009. Potent antibacterial activities of Ag/TiO_2 nanocomposite powders synthesized by a one-pot sol–gel method. Environ. Sci. Technol. 43 (8), 2905–2910.

Zhang, Y., Zuo, P., Ye, B.C., 2015. A low-cost and simple paper-based microfluidic device for simultaneous multiplex determination of different types of chemical contaminants in food. Biosens. Bioelectron. 68, 14–19. Available from: https://doi.org/10.1016/j.bios.2014.12.042.

Zhang, Y., Li, J., Zhou, L., et al., 2016. Aqueous photodegradation of antibiotic florfenicol: kinetics and degradation pathway studies. Environ. Sci. Pollut. Res. 23, 6982. Available from: https://doi.org/10.1007/s11356-015-5897-1.

Zhu, F., Zhang, P., Wu, X., Fu, L., Zhang, J., Xu, D., 2012. The origin ofhigher open-circuit voltage in Zn-doped TiO_2 nanoparticle-baseddye-sensitized solar cells. ChemPhysChem 13, 3731–3737. Available from: https://doi.org/10.1002/cphc.201200362. Appl Nanosci123.

Zwiener, C., Frimmel, F.H., 2003. Short-term tests with a pilot sewage plant and biofilm reactors for the biological degradation of the pharmaceutical compounds clofibric acid, ibuprofen, and diclofenac. Sci. Total Environ. 309 (1–3), 201–211.

FURTHER READING

Akmehmet-Balcıoglu, L., Otker, M., 2003. Treatment of pharmaceutical wastewater containing antibiotics by O_3 and O_3/H_2O_2 processes. Chemosphere 50, 85.

Arakha, M., Pal, S., Samantarrai, D., Panigrahi, T.K., Mallick, B.C., Pramanik, K., et al., 2013. Antimicrobial activity of iron oxide nanoparticle upon modulation of nanoparticle–bacteria interface. Sci. Rep. 5, 14813. Available from: https://doi.org/10.1038/srep14813.

Fang, J., Leufke, P.M., Kruk, R., Wang, D., Scherer, T., Hahn, H., 2010. External electric field driven 3D ordering architecture of silver(I) oxide meso-superstructures. Nano-Today 5 (3), 175–182.

He, W., Kim, H.K., Wamer, W.G., Melka, D., Callahan, J.H., Yin, J.J., 2014. Photogenerated charge carriers and reactive oxygen species in ZnO/Au hybrid nanostructures with enhanced photocatalytic and antibacterial activity. J. Am. Chem. Soc. 136, 750–757. Available from: https://doi.org/10.1021/ja410800y.

Lushniak, B.D., 2014. Antibiotic resistance: a public health crisis. Public Health Rep. 129 (4), 314–316.

Nawrocki, J., Kasprzyk-Hordern, B., 2010. The efficiency and mechanisms of catalytic ozonation. Appl. Catal. B: Environ. 99 (1–2), 27–42.

Orge, C.A., Orfão, J.J.M., Pereira, M.F.R., Duarte, A.M., de Farias, R., Neto, C.R., et al., 2011. Ozonation of model organic compounds catalysed by nanostructured cerium oxides. Appl. Catal. B: Environ. 103 (1–2), 190–199.

Wei, W., Mao, X., Ortiz, L.A., Sadoway, D.R., 2011. Oriented silver oxide nanostructures synthesized through a template-free electrochemical route. J. Mater. Chem. 21 (2), 432–438.

Xu, L.H., 2005. Zinc ions surface-doped titanium dioxide nanotubes and its photocatalysis activity for degradation of methyl orange in water. J. Mol. Catal. A: Chem. 226, 123–127. Available from: https://doi.org/10.1016/j.molcata.2004.09.051.

Index

Note: Page numbers followed by "*f*" and "*t*" refer to figures and tables, respectively.

C

N

O

Q

R

S

$240

CPI Antony Rowe
Eastbourne, UK
November 19, 2019